Fundamentals of Thermodynamics

8/e

SI Version

Claus Borgnakke

Richard E. Sonntag

University of Michigan

WILEY

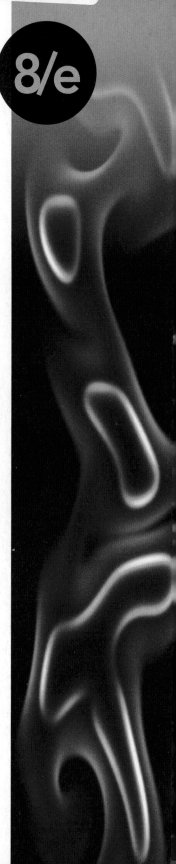

ISBN: 978-1-118-32177-5

Printed in Asia

10 9 8 7 6 5 4

Preface

In this eighth edition the basic objective of the earlier editions have been retained:

- to present a comprehensive and rigorous treatment of classical thermodynamics while retaining an engineering perspective, and in doing so
- to lay the groundwork for subsequent studies in such fields as fluid mechanics, heat transfer, and statistical thermodynamics, and also
- to prepare the student to effectively use thermodynamics in the practice of engineering.

The presentation is deliberately directed to students. New concepts and definitions are presented in the context where they are first relevant in a natural progression. The introduction has been reorganized with a very short introduction followed by the first thermodynamic properties to be defined (Chapter 1), which are those that can be readily measured: pressure, specific volume, and temperature. In Chapter 2, tables of thermodynamic properties are introduced, but only in regard to these measurable properties. Internal energy and enthalpy are introduced in connection with the energy equation and the first law, entropy with the second law, and the Helmholtz and Gibbs functions in the chapter on thermodynamic relations. Many real-world realistic examples have been included in the book to assist the student in gaining an understanding of thermodynamics, and the problems at the end of each chapter have been carefully sequenced to correlate with the subject matter, and are grouped and identified as such. The early chapters in particular contain a large number of examples, illustrations, and problems, and throughout the book, chapter-end summaries are included, followed by a set of concept/study problems that should be of benefit to the students.

This is the first edition I have prepared without the thoughtful comments from my colleague and coauthor, the late Professor Richard E. Sonntag, who substantially contributed to earlier versions of this textbook. I am grateful for the collaboration and fruitful discussions with my friend and trusted colleague, whom I have enjoyed the privilege of working with over the last three decades. Professor Sonntag consistently shared generously his vast knowledge and experience in conjunction with our mutual work on previous editions of this book and on various research projects, advising PhD students and performing general professional tasks at our department. In honor of my colleague's many contributions, Professor Sonntag still appears as a coauthor of this edition.

NEW FEATURES IN THIS EDITION

Chapter Reorganization and Revisions

The introduction and the first five chapters in the seventh edition have been completely reorganized. A much shorter introduction leads into the description of some background material from physics, thermodynamic properties, and units all of which is in the new Chapter 1. To have the tools for the analysis, the order of the presentation has been kept

from the previous editions, so the behavior of pure substances is presented in Chapter 2, with a slight expansion and separation of the different domains for solid, liquid, and gas phase behavior. Some new figures and explanations have been added to show the ideal gas region as a limit behavior for a vapor at low density.

Discussion about work and heat is now included in Chapter 3 with the energy equation to emphasize that they are transfer terms of energy explaining how energy for mass at one location can change because of energy exchange with a mass at another location. The energy equation is presented first for a control mass as a basic principle accounting for energy in a control volume as

$$\text{Change of storage} = \text{transfer in} - \text{transfer out}$$

The chapter then discusses the form of energy storage as various internal energies associated with the mass and its structure to better understand how the energy is actually stored. This also helps in understanding why internal energy and enthalpy can vary nonlinearly with temperature, leading to nonconstant specific heats. Macroscopic potential and kinetic energy then naturally add to the internal energy for the total energy. The first law of thermodynamics, which often is taken as synonymous with the energy equation, is shown as a natural consequence of the energy equation applied to a cyclic process. In this respect, the current presentation follows modern physics rather than the historical development presented in the previous editions.

After discussion about the storage of energy, the left-hand side of the energy equation, the transfer terms as work and heat transfer are discussed, so the whole presentation is shorter than that in the previous editions. This allows less time to be spent on the material used for preparation before the energy equation is applied to real systems.

All the balance equations for mass, momentum, energy, and entropy follow the same format to show the uniformity in the basic principles and make the concept something to be understood and not merely memorized. This is also the reason to use the names *energy equation* and *entropy equation* for the first and second laws of thermodynamics to stress that they are universally valid, not just used in the field of thermodynamics but apply to all situations and fields of study with no exceptions. Clearly, special cases require extensions not covered in this text, like effects of surface tension in drops or for liquid in small pores, relativity, and nuclear processes, to mention a few.

The energy equation applied to a general control volume is retained from the previous edition with the addition of a section on multiflow devices. Again, this is done to reinforce to students that the analysis is done by applying the basic principles to systems under investigation. This means that the actual mathematical form of the general laws follows the sketches and figures of the system, and the analysis is not a question about finding a suitable formula in the text.

To show the generality of the entropy equation, a small example is presented applying the energy and entropy equations to heat engines and heat pumps shown in Chapter 6. This demonstrates that the historical presentation of the second law in Chapter 5 can be completely substituted by the postulation of the entropy equation and the existence of the absolute temperature scale. Carnot cycle efficiencies and the fact that real devices have lower efficiency follow from the basic general laws. Also, the direction of heat transfer from a higher temperature domain toward a lower temperature domain is predicted by the entropy equation due to the requirement of a positive entropy generation. These are examples that show the application of the general laws for specific cases and improve the student's understanding of the material.

The rest of the chapters have been updated to improve the student's understanding of the material. The word *availability* has been substituted by *exergy* as a general concept, though it is not strictly in accordance with the original definition. The chapters concerning cycles have been expanded, with a few details for specific cycles and some extensions shown to tie the theory to industrial applications with real systems. The same is done for Chapter 13 with combustion to emphasize an understanding of the basic physics of what happens, which may not be evident in the more abstract definition of terms like *enthalpy of combustion*.

Web-Based Material

Several new documents will be available from Wiley's website for the book. The following material will be accessible for students, with additional material reserved for instructors of the course.

Notes for classical thermodynamics.   A very short set of notes covers the basic thermodynamic analysis with the general laws (continuity, energy, and entropy equations) and some of the specific laws like device equations, process equations, and so on. This is useful for students doing review of the course or for exam preparation, as it gives a comprehensive presentation in a condensed form.

Extended set of study examples.   This document includes a collection of additional examples for students to study. These examples have slightly longer and more detailed solutions than the examples printed in the book and thus are excellent for self-study. There are about eight SI unit problems for each chapter covering most of the material in the chapters.

How-to notes.   Frequently asked questions are listed for each of the set of subject areas in the book with detailed answers. These are questions that are difficult to accommodate in the book. Examples:

> How do I find a certain state for R-410A in the B-section tables?
>
> How do I make a linear interpolation?
>
> Should I use internal energy (u) or enthalpy (h) in the energy equation?
>
> When can I use the ideal gas law?

Instructor material.   The material for instructors covers typical syllabus and homework assignments for a first and a second course in thermodynamics. Additionally, examples of two standard 1-hour midterm exams and a 2-hour final exam are given for typical Thermodynamics I and Thermodynamics II classes.

FEATURES CONTINUED FROM THE SEVENTH EDITION

In-Text Concept Questions

The in-text concept questions appear in the text after major sections of material to allow student to reflect on the material just presented. These questions are intended to be quick self-tests for students or used by teachers as wrap-up checks for each of the subjects covered, and most of them emphasize the understanding of the material without being memory facts.

End-of-Chapter Engineering Applications

The last section in each chapter, called "Engineering Applications," has been revised with updated illustrations and a few more examples. These sections are intended to be motivating material, consisting mostly of informative examples of how this particular chapter material is being used in actual engineering. The vast majority of these sections do not have any material with equations or developments of theory, but they do contain figures and explanations of a few real physical systems where the chapter material is relevant for the engineering analysis and design. These sections are deliberately kept short and not all the details in the devices shown are explained, but the reader can get an idea about the applications relatively quickly.

End-of-Chapter Summaries with Main Concepts and Formulas

The end-of-chapter summaries provide a review of the main concepts covered in the chapter, with highlighted key words. To further enhance the summary, a list of skills that the student should have mastered after studying the chapter is presented. These skills are among the outcomes that can be tested with the accompanying set of study-guide problems in addition to the main set of homework problems. Main concepts and formulas are included after the summary for reference, and a collection of these will be available on Wiley's website.

Concept-Study Guide Problems

Additional concept questions are placed as problems in the first section of the end-of-chapter homework problems. These problems are similar to the in-text concept questions and serve as study guide problems for each chapter. They are a little like homework problems with numbers to provide a quick check of the chapter material. These questions are short and directed toward very specific concepts. Students can answer all of these questions to assess their level of understanding and determine if any of the subjects need to be studied further. These problems are also suitable for use with the rest of the homework problems in assignments and are included in the solution manual.

Homework Problems

The number of homework problems now exceeds 2000, with many new and modified problems. A large number of introductory problems cover all aspects of the chapter material and are listed according to the subject sections for easy selection according to the particular coverage given. They are generally ordered to be progressively more complex and involved. The later problems in many sections are related to real industrial processes and devices, and the more comprehensive problems are retained and grouped at the end as **review problems**.

Tables

The tables of the substances have been carried over from the seventh edition with **alternative refrigerant R-410A**, which is the replacement for R-22, and **carbon dioxide**, which is a natural refrigerant. Several more substances have been included in the software. The ideal gas tables have been printed on a mass basis as well as a mole basis, to reflect their use on a mass basis early in the text and a mole basis for the combustion and chemical equilibrium chapters.

Software Included

The software **CATT3** includes a number of additional substances besides those included in the printed tables in Appendix B. The current set of substances for which the software can provide the complete tables are:

Water Refrigerants:	R-11, 12, 13, 14, 21, 22, 23, 113, 114, 123, 134a, 152a, 404a, 407c, 410A, 500, 502, 507a, and C318
Cryogenics:	Ammonia, argon, ethane, ethylene, isobutane, methane, neon, nitrogen, oxygen, and propane
Ideal Gases:	air, CO_2, CO, N, N_2, NO, NO_2, H, H_2, H_2O, O, O_2, and OH

Besides the properties of the substances just mentioned, the software can provide the psychrometric chart and the compressibility and generalized charts using the Lee-Keslers equation-of-state, including an extension for increased accuracy with the acentric factor. The software can also plot a limited number of processes in the T–s and log P–log v diagrams, giving the real process curves instead of the sketches presented in the text material.

FLEXIBILITY IN COVERAGE AND SCOPE

The book attempts to cover fairly comprehensively the basic subject matter of classical thermodynamics, and I believe that it provides adequate preparation for study of the application of thermodynamics to the various professional fields as well as for study of more advanced topics in thermodynamics, such as those related to materials, surface phenomena, plasmas, and cryogenics. I also recognize that a number of colleges offer a single introductory course in thermodynamics for all departments, and I have tried to cover those topics that the various departments might wish to have included in such a course. However, since specific courses vary considerably in prerequisites, specific objectives, duration, and background of the students, the material is arranged in sections, particularly in the later chapters, so considerable flexibility exists in the amount of material that may be covered.

The book covers more material than required for a two-semester course sequence, which provides flexibility for specific choices of topic coverage. Instructors may want to visit the publisher's website at www.wiley.com/college/borgnakke for information and suggestions on possible course structure and schedules, and the additional material mentioned as Web material that will be updated to include current errata for the book.

ACKNOWLEDGMENTS

I acknowledge with appreciation the suggestions, counsel, and encouragement of many colleagues, both at the University of Michigan and elsewhere. This assistance has been very helpful to me during the writing of this edition, as it was with the earlier editions of the book. Both undergraduate and graduate students have been of particular assistance, for their perceptive questions have often caused me to rewrite or rethink a given portion of the text, or to try to develop a better way of presenting the material in order to anticipate such questions or difficulties. Finally, the encouragement and patience of my wife and family have been indispensable, and have made this time of writing pleasant and enjoyable, in spite of the pressures of the project. A special thanks to a number of colleagues at other institutions who have reviewed the earlier editions of the book and provided input to the revisions. Some of the reviewers are

Ruhul Amin, *Montana State University*
Edward E. Anderson, *Texas Tech University*
Cory Berkland, *University of Kansas*
Eugene Brown, *Virginia Polytechnic Institute and State University*
Sung Kwon Cho, *University of Pittsburgh*
Sarah Codd, *Montana State University*
Ram Devireddy, *Louisiana State University*
Fokion Egolfopoulos, *University of Southern California*
Harry Hardee, *New Mexico State University*
Hong Huang, *Wright State University*
Satish Ketkar, *Wayne State University*
Boris Khusid, *New Jersey Institute of Technology*
Joseph F. Kmec, *Purdue University*
Roy W. Knight, *Auburn University*
Daniela Mainardi, *Louisiana Tech University*
Randall Manteufel, *University of Texas, San Antonio*
Harry J. Sauer, Jr., *Missouri University of Science and Technology*
J. A. Sekhar, *University of Cincinnati*
Ahmed Soliman, *University of North Carolina, Charlotte*
Reza Toossi, *California State University, Long Beach*
Thomas Twardowski, *Widener University*
Etim U. Ubong, *Kettering University*
Yanhua Wu, *Wright State University*
Walter Yuen, *University of California at Santa Barbara*

I also wish to welcome the new editor, Linda Ratts, and thank her for encouragement and help during the production of this edition.

I hope that this book will contribute to the effective teaching of thermodynamics to students who face very significant challenges and opportunities during their professional careers. Your comments, criticism, and suggestions will also be appreciated, and you may communicate those to me at claus@umich.edu.

CLAUS BORGNAKKE
Ann Arbor, Michigan
July 2012

Contents

10 Power and Refrigeration Systems—Gaseous Working Fluids   400

11 Ideal Gas Mixtures   448

Contents of Appendix 671

Answers to Selected Problems 757

Index 765

Symbols

a	acceleration
A	area
a, A	specific Helmholtz function and total Helmholtz function
AF	air-fuel ratio
B_S	adiabatic bulk modulus
B_T	isothermal bulk modulus
c	velocity of sound
c	mass fraction
C_D	coefficient of discharge
C_p	constant-pressure specific heat
C_v	constant-volume specific heat
C_{po}	zero-pressure constant-pressure specific heat
C_{vo}	zero-pressure constant-volume specific heat
COP	coefficient of performance
CR	compression ratio
e, E	specific energy and total energy
EMF	electromotive force
F	force
FA	fuel-air ratio
g	acceleration due to gravity
g, G	specific Gibbs function and total Gibbs function
h, H	specific enthalpy and total enthalpy
HV	heating value
i	electrical current
I	irreversibility
J	proportionality factor to relate units of work to units of heat
k	specific heat ratio: C_p/C_v
K	equilibrium constant
KE	kinetic energy
L	length
m	mass
$\dot{m}$	mass flow rate
M	molecular mass
M	Mach number
n	number of moles
n	polytropic exponent
P	pressure
P_i	partial pressure of component i in a mixture
PE	potential energy

P_r	reduced pressure P/P_c
P_r	relative pressure as used in gas tables
q, Q	heat transfer per unit mass and total heat transfer
$\dot{Q}$	rate of heat transfer
Q_H, Q_L	heat transfer with high-temperature body and heat transfer with low-temperature body; sign determined from context
R	gas constant
$\overline{R}$	universal gas constant
s, S	specific entropy and total entropy
S_{gen}	entropy generation
$\dot{S}_{\text{gen}}$	rate of entropy generation
t	time
T	temperature
T_r	reduced temperature T/T_c
u, U	specific internal energy and total internal energy
v, V	specific volume and total volume
v_r	relative specific volume as used in gas tables
$\mathbf{V}$	velocity
w, W	work per unit mass and total work
$\dot{W}$	rate of work, or power
w^{rev}	reversible work between two states
x	quality
y	gas-phase mole fraction
y	extraction fraction
Z	elevation
Z	compressibility factor
Z	electrical charge

Script Letters

$\mathscr{E}$	electrical potential
$\mathscr{S}$	surface tension
$\mathscr{T}$	tension

Greek Letters

α	residual volume
α	dimensionless Helmholtz function a/RT
α_p	volume expansivity
β	coefficient of performance for a refrigerator
β'	coefficient of performance for a heat pump
β_S	adiabatic compressibility
β_T	isothermal compressibility
δ	dimensionless density ρ/ρ_c
η	efficiency
μ	chemical potential
ν	stoichiometric coefficient
ρ	density
τ	dimensionless temperature variable T_c/T
τ_0	dimensionless temperature variable $1 - T_r$
Φ	equivalence ratio
ϕ	relative humidity

ϕ, Φ		exergy or availability for a control mass
ψ		exergy, flow availability
ω		humidity ratio or specific humidity
ω		acentric factor

Subscripts

c	property at the critical point
c.v.	control volume
e	state of a substance leaving a control volume
f	formation
f	property of saturated liquid
fg	difference in property for saturated vapor and saturated liquid
g	property of saturated vapor
i	state of a substance entering a control volume
i	property of saturated solid
if	difference in property for saturated liquid and saturated solid
ig	difference in property for saturated vapor and saturated solid
r	reduced property
s	isentropic process
0	property of the surroundings
0	stagnation property

Superscripts

$\overline{}$	bar over symbol denotes property on a molal basis (over V, H, S, U, A, G, the bar denotes partial molal property)
$\circ$	property at standard-state condition
$*$	ideal gas
$*$	property at the throat of a nozzle
irr	irreversible
r	real gas part
rev	reversible

Introduction

1

The field of thermodynamics is concerned with the science of energy focusing on energy storage and energy conversion processes. We will study the effects on different substances, as we may expose a mass to heating/cooling or to volumetric compression/expansion. During such processes we are transferring energy into or out of the mass, so it changes its conditions expressed by properties like temperature, pressure, and volume. We use several processes similar to this in our daily lives; we heat water to make coffee or tea or cool it in a refrigerator to make cold water or ice cubes in a freezer. In nature, water evaporates from oceans and lakes and mixes with air where the wind can transport it, and later the water may drop out of the air as either rain (liquid water) or snow (solid water). As we study these processes in detail, we will focus on situations that are physically simple and yet typical of real-life situations in industry or nature.

By a combination of processes, we are able to illustrate more complex devices or complete systems—for instance, a simple steam power plant that is the basic system that generates the majority of our electric power. A power plant that produces electric power and hot water for district heating burns coal, as shown in Fig. 1.1. The coal is supplied by ship, and the district heating pipes are located in underground tunnels and thus are not visible. A more technical description and a better understanding are obtained from the simple schematic of the power plant, as shown in Fig. 1.2. This includes various outputs from the plant as electric power to the net, warm water for district heating, slag from burning coal, and other materials like ash and gypsum; the last output is a flow of exhaust gases out of the chimney.

Another set of processes forms a good description of a refrigerator that we use to cool food or apply it at very low temperatures to produce a flow of cold fluid for cryogenic surgery by freezing tissue for minimal bleeding. A simple schematic for such a system is shown in Fig. 1.3. The same system can also function as an air conditioner with the dual purpose of cooling a building in summer and heating it in winter; in this last mode of use, it is also called a *heat pump*. For mobile applications, we can make simple models for gasoline and diesel engines typically used for ground transportation and gas turbines in jet engines used in aircraft, where low weight and volume are of prime concern. These are just a few examples of familiar systems that the theory of thermodynamics allows us to analyze. Once we learn and understand the theory, we will be able to extend the analysis to other cases we may not be familiar with.

Beyond the description of basic processes and systems, thermodynamics is extended to cover special situations like moist atmospheric air, which is a mixture of gases, and the combustion of fuels for use in the burning of coal, oil, or natural gas, which is a chemical and energy conversion process used in nearly all power-generating devices. Many other extensions are known; these can be studied in specialty texts. Since all the processes engineers deal with have an impact on the environment, we must be acutely aware of the ways in which we can optimize the use of our natural resources and produce the minimal amount of negative consequences for our environment. For this reason, the

(Courtesy of Dong Energy A/S, Denmark.)

FIGURE 1.1 A power station in Esbjerg, Denmark.

treatment of efficiencies for processes and devices is important in a modern analysis and is required knowledge for a complete engineering consideration of system performance and operation.

Before considering the application of the theory, we will cover a few basic concepts and definitions for our analysis and review some material from physics and chemistry that we will need.

1.1 A THERMODYNAMIC SYSTEM AND THE CONTROL VOLUME

A thermodynamic system is a device or combination of devices containing a quantity of matter that is being studied. To define this more precisely, a control volume is chosen so that it contains the matter and devices inside a control surface. Everything external to the

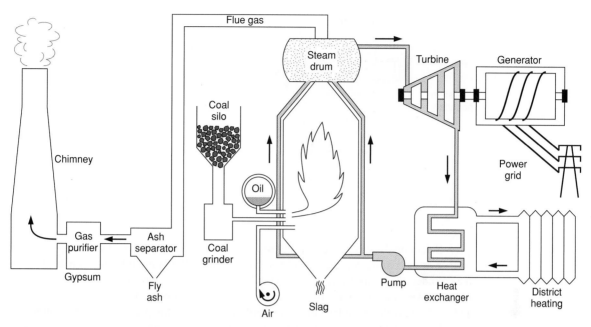

FIGURE 1.2 Schematic diagram of a steam power plant.

control volume is the surroundings, with the separation provided by the control surface. The surface may be open or closed to mass flows, and it may have flows of energy in terms of heat transfer and work across it. The boundaries may be movable or stationary. In the case of a control surface that is closed to mass flow, so that no mass can escape or enter

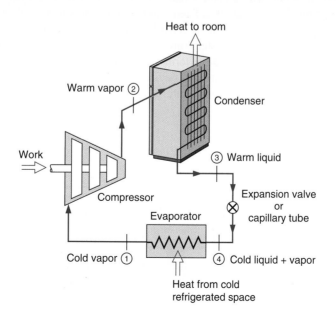

FIGURE 1.3
Schematic diagram
of a refrigerator.

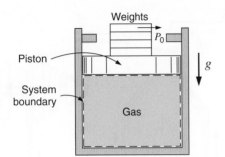

FIGURE 1.4 Example of a control mass.

the control volume, it is called a control mass containing the same amount of matter at all times.

Selecting the gas in the cylinder of Fig. 1.4 as a control volume by placing a control surface around it, we recognize this as a control mass. If a Bunsen burner is placed under the cylinder, the temperature of the gas will increase and the piston will move out. As the piston moves, the boundary of the control mass also changes. As we will see later, heat and work cross the boundary of the control mass during this process, but the matter that composes the control mass can always be identified and remains the same.

An isolated system is one that is not influenced in any way by the surroundings so that no mass, heat, or work is transferred across the boundary of the system. In a more typical case, a thermodynamic analysis must be made of a device like an air compressor which has a flow of mass into and out of it, as shown schematically in Fig. 1.5. The real system includes possibly a storage tank, as shown later in Fig. 1.20. In such an analysis, we specify a control volume that surrounds the compressor with a surface that is called the control surface, across which there may be a transfer of mass, and momentum, as well as heat and work.

Thus, the more general control surface defines a control volume, where mass may flow in or out, with a control mass as the special case of no mass flow in or out. Hence, the control mass contains a fixed mass at all times, which explains its name. The general formulation of the analysis is considered in detail in Chapter 4. The terms closed system (fixed mass) and open system (involving a flow of mass) are sometimes used to make this distinction. Here, we use the term system as a more general and loose description for a mass, device, or combination of devices that then is more precisely defined when a control volume is selected. The procedure that will be followed in presenting the first and second

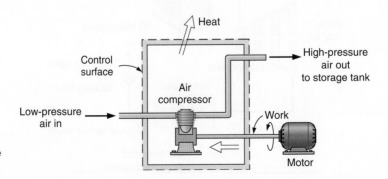

FIGURE 1.5 Example of a control volume.

laws of thermodynamics is first to present these laws for a control mass and then to extend the analysis to the more general control volume.

 ## 1.2 MACROSCOPIC VERSUS MICROSCOPIC POINTS OF VIEW

The behavior of a system may be investigated from either a microscopic or macroscopic point of view. Let us briefly describe a system from a microscopic point of view. Consider a system consisting of a cube 25 mm on a side and containing a monatomic gas at atmospheric pressure and temperature. This volume contains approximately 10^{20} atoms. To describe the position of each atom, we need to specify three coordinates; to describe the velocity of each atom, we specify three velocity components.

Thus, to describe completely the behavior of this system from a microscopic point of view, we must deal with at least 6×10^{20} equations. Even with a modern computer, this is a hopeless computational task. However, there are two approaches to this problem that reduce the number of equations and variables to a few that can be computed relatively easily. One is the statistical approach, in which, on the basis of statistical considerations and probability theory, we deal with average values for all particles under consideration. This is usually done in connection with a model of the atom under consideration. This is the approach used in the disciplines of kinetic theory and statistical mechanics.

The other approach to reducing the number of variables to a few that can be handled relatively easily involves the macroscopic point of view of classical thermodynamics. As the word *macroscopic* implies, we are concerned with the gross or average effects of many molecules. These effects can be perceived by our senses and measured by instruments. However, what we really perceive and measure is the time-averaged influence of many molecules. For example, consider the pressure a gas exerts on the walls of its container. This pressure results from the change in momentum of the molecules as they collide with the wall. From a macroscopic point of view, however, we are concerned not with the action of the individual molecules but with the time-averaged force on a given area, which can be measured by a pressure gauge. In fact, these macroscopic observations are completely independent of our assumptions regarding the nature of matter.

Although the theory and development in this book are presented from a macroscopic point of view, a few supplementary remarks regarding the significance of the microscopic perspective are included as an aid to understanding the physical processes involved. Another book in this series, *Introduction to Thermodynamics: Classical and Statistical*, by R. E. Sonntag and G. J. Van Wylen, includes thermodynamics from the microscopic and statistical point of view.

A few remarks should be made regarding the continuum approach. We are normally concerned with volumes that are very large compared to molecular dimensions and with time scales that are very large compared to intermolecular collision frequencies. For this reason, we deal with very large numbers of molecules that interact extremely often during our observation period, so we sense the system as a simple uniformly distributed mass in the volume called a continuum. This concept, of course, is only a convenient assumption that loses validity when the mean free path of the molecules approaches the order of magnitude of the dimensions of the vessel, as, for example, in high-vacuum technology. In much engineering work the assumption of a continuum is valid and convenient, consistent with the macroscopic point of view.

1.3 PROPERTIES AND STATE OF A SUBSTANCE

If we consider a given mass of water, we recognize that this water can exist in various forms. If it is a liquid initially, it may become a vapor when it is heated or a solid when it is cooled. Thus, we speak of the different phases of a substance. A phase is defined as a quantity of matter that is homogeneous throughout. When more than one phase is present, the phases are separated from each other by the phase boundaries. In each phase the substance may exist at various pressures and temperatures or, to use the thermodynamic term, in various states. The state may be identified or described by certain observable, macroscopic properties; some familiar ones are temperature, pressure, and density. In later chapters, other properties will be introduced. Each of the properties of a substance in a given state has only one definite value, and these properties always have the same value for a given state, regardless of how the substance arrived at the state. In fact, a property can be defined as any quantity that depends on the state of the system and is independent of the path (that is, the prior history) by which the system arrived at the given state. Conversely, the state is specified or described by the properties. Later we will consider the number of independent properties a substance can have, that is, the minimum number of properties that must be specified to fix the state of the substance.

Thermodynamic properties can be divided into two general classes: intensive and extensive. An intensive property is independent of the mass; the value of an extensive property varies directly with the mass. Thus, if a quantity of matter in a given state is divided into two equal parts, each part will have the same value of intensive properties as the original and half the value of the extensive properties. Pressure, temperature, and density are examples of intensive properties. Mass and total volume are examples of extensive properties. Extensive properties per unit mass, such as specific volume, are intensive properties.

Frequently we will refer not only to the properties of a substance but also to the properties of a system. When we do so, we necessarily imply that the value of the property has significance for the entire system, and this implies equilibrium. For example, if the gas that composes the system (control mass) in Fig. 1.4 is in thermal equilibrium, the temperature will be the same throughout the entire system, and we may speak of the temperature as a property of the system. We may also consider mechanical equilibrium, which is related to pressure. If a system is in mechanical equilibrium, there is no tendency for the pressure at any point to change with time as long as the system is isolated from the surroundings. There will be variation in pressure with elevation because of the influence of gravitational forces, although under equilibrium conditions there will be no tendency for the pressure at any location to change. However, in many thermodynamic problems, this variation in pressure with elevation is so small that it can be neglected. Chemical equilibrium is also important and will be considered in Chapter 14. When a system is in equilibrium regarding all possible changes of state, we say that the system is in thermodynamic equilibrium.

1.4 PROCESSES AND CYCLES

Whenever one or more of the properties of a system change, we say that a change in state has occurred. For example, when one of the weights on the piston in Fig. 1.6 is removed, the piston rises and a change in state occurs, for the pressure decreases and the specific

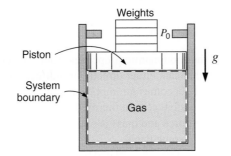

FIGURE 1.6 Example of a system that may undergo a quasi-equilibrium process.

volume increases. The path of the succession of states through which the system passes is called the process.

Let us consider the equilibrium of a system as it undergoes a change in state. The moment the weight is removed from the piston in Fig. 1.6, mechanical equilibrium does not exist; as a result, the piston is moved upward until mechanical equilibrium is restored. The question is this: Since the properties describe the state of a system only when it is in equilibrium, how can we describe the states of a system during a process if the actual process occurs only when equilibrium does not exist? One step in finding the answer to this question concerns the definition of an ideal process, which we call a *quasi-equilibrium* process. A quasi-equilibrium process is one in which the deviation from thermodynamic equilibrium is infinitesimal, and all the states the system passes through during a quasi-equilibrium process may be considered equilibrium states. Many actual processes closely approach a quasi-equilibrium process and may be so treated with essentially no error. If the weights on the piston in Fig. 1.6 are small and are taken off one by one, the process could be considered quasi-equilibrium. However, if all the weights are removed at once, the piston will rise rapidly until it hits the stops. This would be a nonequilibrium process, and the system would not be in equilibrium at any time during this change of state.

For nonequilibrium processes, we are limited to a description of the system before the process occurs and after the process is completed and equilibrium is restored. We are unable to specify each state through which the system passes or the rate at which the process occurs. However, as we will see later, we are able to describe certain overall effects that occur during the process.

Several processes are described by the fact that one property remains constant. The prefix *iso-* is used to describe such a process. An isothermal process is a constant-temperature process, an isobaric process is a constant-pressure process, and an isochoric process is a constant-volume process.

When a system in a given initial state goes through a number of different changes of state or processes and finally returns to its initial state, the system has undergone a cycle. Therefore, at the conclusion of a cycle, all the properties have the same value they had at the beginning. Steam (water) that circulates through a steam power plant undergoes a cycle.

A distinction should be made between a thermodynamic cycle, which has just been described, and a mechanical cycle. A four-stroke-cycle internal-combustion engine goes through a mechanical cycle once every two revolutions. However, the working fluid does not go through a thermodynamic cycle in the engine, since air and fuel are burned and changed to products of combustion that are exhausted to the atmosphere. In this book, the term *cycle* will refer to a thermodynamic cycle unless otherwise designated.

1.5 UNITS FOR MASS, LENGTH, TIME, AND FORCE

Since we are considering thermodynamic properties from a macroscopic perspective, we are dealing with quantities that can, either directly or indirectly, be measured and counted. Therefore, the matter of units becomes an important consideration. In the remaining sections of this chapter we will define certain thermodynamic properties and the basic units. Because the relation between force and mass is often difficult for students to understand, it is considered in this section in some detail.

Force, mass, length, and time are related by Newton's second law of motion, which states that the force acting on a body is proportional to the product of the mass and the acceleration in the direction of the force:

$$F \propto ma$$

The concept of time is well established. The basic unit of time is the second (s), which in the past was defined in terms of the solar day, the time interval for one complete revolution of the earth relative to the sun. Since this period varies with the season of the year, an average value over a one-year period is called the *mean solar day*, and the mean solar second is 1/86 400 of the mean solar day. In 1967, the General Conference of Weights and Measures (CGPM) adopted a definition of the second as the time required for a beam of cesium-133 atoms to resonate 9 192 631 770 cycles in a cesium resonator.

For periods of time less than 1 s, the prefixes *milli*, *micro*, *nano*, *pico*, or *femto*, as listed in Table 1.1, are commonly used. For longer periods of time, the units minute (min), hour (h), or day (day) are frequently used. It should be pointed out that the prefixes in Table 1.1 are used with many other units as well.

The concept of length is also well established. The basic unit of length is the meter (m), which used to be marked on a platinum–iridium bar. Currently, the CGPM has adopted a more precise definition of the meter in terms of the speed of light (which is now a fixed constant): The meter is the length of the path traveled by light in a vacuum during a time interval of 1/299 792 458 of a second.

The fundamental unit of mass is the kilogram (kg). As adopted by the first CGPM in 1889 and restated in 1901, it is the mass of a certain platinum–iridium cylinder maintained under prescribed conditions at the International Bureau of Weights and Measures. A related unit that is used frequently in thermodynamics is the mole (mol), defined as an amount of substance containing as many elementary entities as there are atoms in 0.012 kg of carbon-12. These elementary entities must be specified; they may be atoms, molecules, electrons, ions,

TABLE 1.1
Unit Prefixes

Factor	Prefix	Symbol	Factor	Prefix	Symbol
10^{15}	peta	P	10^{-3}	milli	m
10^{12}	tera	T	10^{-6}	micro	μ
10^{9}	giga	G	10^{-9}	nano	n
10^{6}	mega	M	10^{-12}	pico	p
10^{3}	kilo	k	10^{-15}	femto	f

or other particles or specific groups. For example, 1 mol of diatomic oxygen, having a molecular mass of 32 (compared to 12 for carbon), has a mass of 0.032 kg. The mole is often termed a *gram mole*, since it is an amount of substance in grams numerically equal to the molecular mass. In this book, when using the metric SI system, we will find it preferable to use the kilomole (kmol), the amount of substance in kilograms numerically equal to the molecular mass, rather than the mole.

The system of units in use presently throughout most of the world is the metric International System, commonly referred to as *SI units* (from Le Système International d'Unités). In this system, the second, meter, and kilogram are the basic units for time, length, and mass, respectively, as just defined, and the unit of force is defined directly from Newton's second law.

Therefore, a proportionality constant is unnecessary, and we may write that law as an equality:

$$F = ma \qquad (1.1)$$

The unit of force is the newton (N), which by definition is the force required to accelerate a mass of 1 kg at the rate of 1 m/s^2:

$$1\,\text{N} = 1\,\text{kg}\,\text{m/s}^2$$

It is worth noting that SI units derived from proper nouns use capital letters for symbols; others use lowercase letters. The liter, with the symbol L, is an exception.

The term *weight* is often used with respect to a body and is sometimes confused with mass. Weight is really correctly used only as a force. When we say that a body weighs so much, we mean that this is the force with which it is attracted to the earth (or some other body), that is, the product of its mass and the local gravitational acceleration. The mass of a substance remains constant with elevation, but its weight varies with elevation.

Example 1.1

What is the weight of a 1 kg mass at an altitude where the local acceleration of gravity is 9.75 m/s^2?

Solution

Weight is the force acting on the mass, which from Newton's second law is

$$F = mg = 1\,\text{kg} \times 9.75\,\text{m/s}^2 \times [1\,\text{N}\,\text{s}^2/\text{kg}\cdot\text{m}] = 9.75\,\text{N}$$

In-Text Concept Questions

   a. Make a control volume around the turbine in the steam power plant in Fig. 1.2 and list the flows of mass and energy located there.

   b. Take a control volume around your kitchen refrigerator, indicate where the components shown in Fig. 1.3 are located, and show all energy transfers.

1.6 SPECIFIC VOLUME AND DENSITY

The specific volume of a substance is defined as the volume per unit mass and is given the symbol v. The density of a substance is defined as the mass per unit volume, and it is therefore the reciprocal of the specific volume. Density is designated by the symbol ρ. Specific volume and density are intensive properties.

The specific volume of a system in a gravitational field may vary from point to point. For example, if the atmosphere is considered a system, the specific volume increases as the elevation increases. Therefore, the definition of specific volume involves the specific volume of a substance at a point in a system.

Consider a small volume δV of a system, and let the mass be designated δm. The specific volume is defined by the relation

$$v = \lim_{\delta V \to \delta V'} \frac{\delta V}{\delta m}$$

where $\delta V'$ is the smallest volume for which the mass can be considered a continuum. Volumes smaller than this will lead to the recognition that mass is not evenly distributed in space but is concentrated in particles as molecules, atoms, electrons, and so on. This is tentatively indicated in Fig. 1.7, where in the limit of a zero volume the specific volume may be infinite (the volume does not contain any mass) or very small (the volume is part of a nucleus).

Thus, in a given system, we should speak of the specific volume or density at a point in the system and recognize that this may vary with elevation. However, most of the systems that we consider are relatively small, and the change in specific volume with elevation is not significant. Therefore, we can speak of one value of specific volume or density for the entire system.

In this book, the specific volume and density will be given either on a mass or a mole basis. A bar over the symbol (lowercase) will be used to designate the property on a mole basis. Thus, $\bar{v}$ will designate molal specific volume and $\bar{\rho}$ will designate molal density. In SI units, those for specific volume are m^3/kg and m^3/mol (or $m^3/kmol$); for density, the corresponding units are kg/m^3 and mol/m^3 (or $kmol/m^3$).

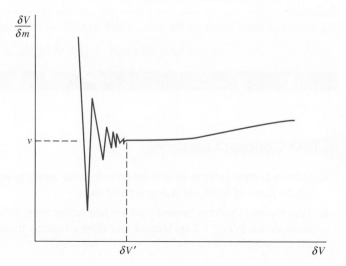

FIGURE 1.7 The continuum limit for the specific volume.

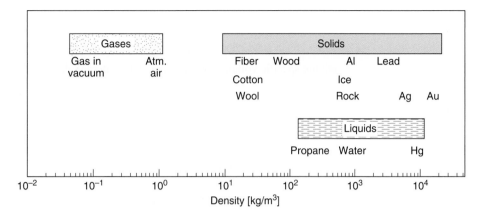

FIGURE 1.8 Density of common substances.

Although the SI unit for volume is the cubic meter, a commonly used volume unit is the liter (L), which is a special name given to a volume of 0.001 m³; that is, $1\,L = 10^{-3}\,m^3$. The general ranges of density for some common solids, liquids, and gases are shown in Fig. 1.8. Specific values for various solids, liquids, and gases in SI units are listed in Tables A.3, A.4, and A.5, respectively.

Example 1.2

A 1 m³ container, shown in Fig. 1.9, is filled with 0.12 m³ of granite, 0.15 m³ of sand, and 0.2 m³ of liquid 25 °C water; the rest of the volume, 0.53 m³, is air with a density of 1.15 kg/m³. Find the overall (average) specific volume and density.

Solution

From the definition of specific volume and density, we have

$$v = V/m \quad \text{and} \quad \rho = m/V = 1/v$$

We need to find the total mass, taking density from Tables A.3 and A.4:

$$m_{\text{granite}} = \rho V_{\text{granite}} = 2750\,\text{kg/m}^3 \times 0.12\,\text{m}^3 = 330\,\text{kg}$$

$$m_{\text{sand}} = \rho_{\text{sand}}\,V_{\text{sand}} = 1500\,\text{kg/m}^3 \times 0.15\,\text{m}^3 = 225\,\text{kg}$$

$$m_{\text{water}} = \rho_{\text{water}}\,V_{\text{water}} = 997\,\text{kg/m}^3 \times 0.2\,\text{m}^3 = 199.4\,\text{kg}$$

$$m_{\text{air}} = \rho_{\text{air}}\,V_{\text{air}} = 1.15\,\text{kg/m}^3 \times 0.53\,\text{m}^3 = 0.61\,\text{kg}$$

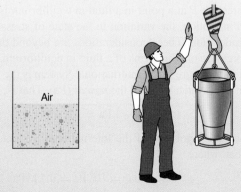

FIGURE 1.9 Sketch for Example 1.2.

Now the total mass becomes

$$m_{\text{tot}} = m_{\text{granite}} + m_{\text{sand}} + m_{\text{water}} + m_{\text{air}} = 755 \text{ kg}$$

and the specific volume and density can be calculated:

$$v = V_{\text{tot}}/m_{\text{tot}} = 1 \text{ m}^3/755 \text{ kg} = 0.001\,325 \text{ m}^3/\text{kg}$$

$$\rho = m_{\text{tot}}/V_{\text{tot}} = 755 \text{ kg}/1 \text{ m}^3 = 755 \text{ kg/m}^3$$

Remark: It is misleading to include air in the numbers for ρ and V, as the air is separate from the rest of the mass.

In-Text Concept Questions

c. Why do people float high in the water when swimming in the Dead Sea as compared with swimming in a freshwater lake?

d. The density of liquid water is $\rho = 1008 - T/2$ [kg/m^3] with T in °C. If the temperature increases, what happens to the density and specific volume?

1.7 PRESSURE

When dealing with liquids and gases, we ordinarily speak of pressure; for solids we speak of stresses. The pressure in a fluid at rest at a given point is the same in all directions, and we define pressure as the normal component of force per unit area. More specifically, if δA is a small area, $\delta A'$ is the smallest area over which we can consider the fluid a continuum, and δF_n is the component of force normal to δA, we define pressure, P, as

$$P = \lim_{\delta A \to \delta A'} \frac{\delta F_n}{\delta A}$$

where the lower limit corresponds to sizes as mentioned for the specific volume, shown in Fig. 1.7. The pressure P at a point in a fluid in equilibrium is the same in all directions. In a viscous fluid in motion, the variation in the state of stress with orientation becomes an important consideration. These considerations are beyond the scope of this book, and we will consider pressure only in terms of a fluid in equilibrium.

The unit for pressure in the International System is the force of one newton acting on a square meter area, which is called the *pascal* (Pa). That is,

$$1 \text{ Pa} = 1 \text{ N/m}^2$$

Two other units, not part of the International System, continue to be widely used. These are the bar, where

$$1 \text{ bar} = 10^5 \text{ Pa} = 0.1 \text{ MPa}$$

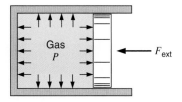

FIGURE 1.10 The balance of forces on a movable boundary relates to inside gas pressure.

and the standard atmosphere, where

$$1 \text{ atm} = 101\ 325 \text{ Pa} = 14.696 \text{ lbf/in.}^2$$

which is slightly larger than the bar. In this book, we will normally use the SI unit, the pascal, and especially the multiples of kilopascal and megapascal. The bar will be utilized often in the examples and problems, but the atmosphere will not be used, except in specifying certain reference points.

Consider a gas contained in a cylinder fitted with a movable piston, as shown in Fig. 1.10. The pressure exerted by the gas on all of its boundaries is the same, assuming that the gas is in an equilibrium state. This pressure is fixed by the external force acting on the piston, since there must be a balance of forces for the piston to remain stationary. Thus, the product of the pressure and the movable piston area must be equal to the external force. If the external force is now changed in either direction, the gas pressure inside must accordingly adjust, with appropriate movement of the piston, to establish a force balance at a new equilibrium state. As another example, if the gas in the cylinder is heated by an outside body, which tends to increase the gas pressure, the piston will move instead, such that the pressure remains equal to whatever value is required by the external force.

Example 1.3

The hydraulic piston/cylinder system shown in Fig. 1.11 has a cylinder diameter of $D = 0.1$ m with a piston and rod mass of 25 kg. The rod has a diameter of 0.01 m with an outside atmospheric pressure of 101 kPa. The inside hydraulic fluid pressure is 250 kPa. How large a force can the rod push with in the upward direction?

Solution

We will assume a static balance of forces on the piston (positive upward), so

$$F_{\text{net}} = ma = 0$$

$$= P_{\text{cyl}} A_{\text{cyl}} - P_0(A_{\text{cyl}} - A_{\text{rod}}) - F - m_p g$$

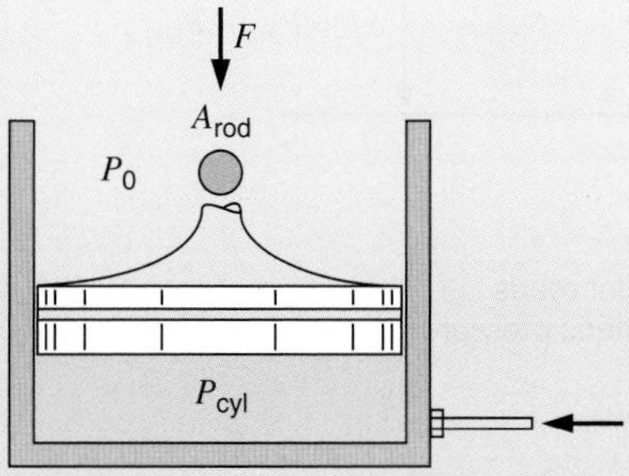

FIGURE 1.11 Sketch for Example 1.3.

Solve for F:

$$F = P_{\text{cyl}} A_{\text{cyl}} - P_0(A_{\text{cyl}} - A_{\text{rod}}) - m_p g$$

The areas are

$$A_{\text{cyl}} = \pi r^2 = \pi D^2/4 = \frac{\pi}{4} 0.1^2 \, \text{m}^2 = 0.007\,854 \, \text{m}^2$$

$$A_{\text{rod}} = \pi r^2 = \pi D^2/4 = \frac{\pi}{4} 0.01^2 \, \text{m}^2 = 0.000\,078\,54 \, \text{m}^2$$

So the force becomes

$$F = [250 \times 0.007\,854 - 101(0.007\,854 - 0.000\,078\,54)]1000 - 25 \times 9.81$$

$$= 1963.5 - 785.32 - 245.25$$

$$= 932.9 \, \text{N}$$

Note that we must convert kPa to Pa to get units of N.

In most thermodynamic investigations we are concerned with absolute pressure. Most pressure and vacuum gauges, however, read the difference between the absolute pressure and the atmospheric pressure existing at the gauge. This is referred to as *gauge pressure*. It is shown graphically in Fig. 1.12, and the following examples illustrate the principles. Pressures below atmospheric and slightly above atmospheric, and pressure differences (for example, across an orifice in a pipe), are frequently measured with a manometer, which contains water, mercury, alcohol, oil, or other fluids.

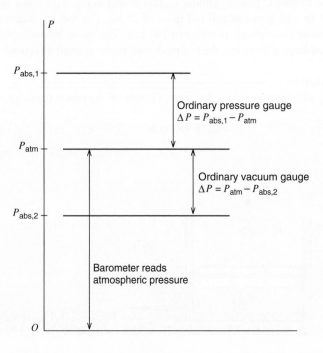

FIGURE 1.12
Illustration of terms used in pressure measurement.

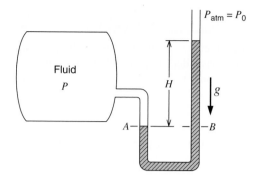

FIGURE 1.13

Example of pressure measurement using a column of fluid.

Consider the column of fluid of height H standing above point B in the manometer shown in Fig. 1.13. The force acting downward at the bottom of the column is

$$P_0 A + mg = P_0 A + \rho A g H$$

where m is the mass of the fluid column, A is its cross-sectional area, and ρ is its density. This force must be balanced by the upward force at the bottom of the column, which is $P_B A$. Therefore,

$$P_B - P_0 = \rho g H$$

Since points A and B are at the same elevation in columns of the same fluid, their pressures must be equal (the fluid being measured in the vessel has a much lower density, such that its pressure P is equal to P_A). Overall,

$$\Delta P = P - P_0 = \rho g H \qquad (1.2)$$

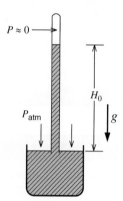

FIGURE 1.14

Barometer.

For distinguishing between absolute and gauge pressure in this book, the term *pascal* will always refer to absolute pressure. Any gauge pressure will be indicated as such.

Consider the barometer used to measure atmospheric pressure, as shown in Fig. 1.14. Since there is a near vacuum in the closed tube above the vertical column of fluid, usually mercury, the height of the fluid column gives the atmospheric pressure directly from Eq. 1.2:

$$P_{atm} = \rho g H_0 \qquad (1.3)$$

Example 1.4

A mercury barometer located in a room at 25 °C has a height of 750 mm. What is the atmospheric pressure in kPa?

Solution

The density of mercury at 25 °C is found from Table A.4 to be 13 534 kg/m³. Using Eq. 1.3,

$$P_{atm} = \rho g H_0 = 13\,534 \text{ kg/m}^3 \times 9.807 \text{ m/s}^2 \times 0.750 \text{ m}/1000$$

$$= 99.54 \text{ kPa}$$

Example 1.5

A mercury (Hg) manometer is used to measure the pressure in a vessel as shown in Fig. 1.13. The mercury has a density of 13 590 kg/m^3, and the height difference between the two columns is measured to be 24 cm. We want to determine the pressure inside the vessel.

Solution

The manometer measures the gauge pressure as a pressure difference. From Eq. 1.2,

$$\Delta P = P_{\text{gauge}} = \rho g H = 13\,590\ \text{kg/m}^3 \times 9.807\ \text{m/s}^2 \times 0.24\ \text{m}$$

$$= 31\,985\ \text{Pa} = 31.985\ \text{kPa}$$

$$= 0.316\ \text{atm}$$

To get the absolute pressure inside the vessel, we have

$$P_A = P_{\text{vessel}} = P_B = \Delta P + P_{\text{atm}}$$

We need to know the atmospheric pressure measured by a barometer (absolute pressure). Assume that this pressure is known to be 750 mm Hg. The absolute pressure in the vessel becomes

$$P_{\text{vessel}} = \Delta P + P_{\text{atm}} = 31\,985\ \text{Pa} + 13\,590\ \text{kg/m}^3 \times 0.750\ \text{m} \times 9.807\ \text{m/s}^2$$

$$= 31\,985 + 99\,954 = 131\,940\ \text{Pa} = 1.302\ \text{atm}$$

Example 1.6

What is the pressure at the bottom of the 7.5 m tall storage tank of fluid at 25 °C shown in Fig. 1.15? Assume that the fluid is gasoline with atmospheric pressure 101 kPa on the top surface. Repeat the question for the liquid refrigerant R-134a when the top surface pressure is 1 MPa.

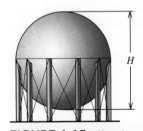

FIGURE 1.15 Sketch for Example 1.6.

Solution

The densities of the liquids are listed in Table A.4:

$$\rho_{\text{gasoline}} = 750\ \text{kg/m}^3; \quad \rho_{\text{R-134a}} = 1206\ \text{kg/m}^3$$

The pressure difference due to gravity is, from Eq. 1.2,

$$\Delta P = \rho g H$$

The total pressure is

$$P = P_{\text{top}} + \Delta P$$

For the gasoline, we get

$$\Delta P = \rho g H = 750\ \text{kg/m}^3 \times 9.807\ \text{m/s}^2 \times 7.5\ \text{m} = 55\,164\ \text{Pa}$$

Now convert all pressures to kPa:

$$P = 101 + 55.164 = 156.2\ \text{kPa}$$

For the R-134a, we get

$$\Delta P = \rho g H = 1206 \, \text{kg/m}^3 \times 9.807 \, \text{m/s}^2 \times 7.5 \, \text{m} = 88\,704 \, \text{Pa}$$

Now convert all pressures to kPa:

$$P = 1000 + 88.704 = 1089 \, \text{kPa}$$

Example 1.7

A piston/cylinder with a cross-sectional area of 0.01 m² is connected with a hydraulic line to another piston/cylinder with a cross-sectional area of 0.05 m². Assume that both chambers and the line are filled with hydraulic fluid of density 900 kg/m³ and the larger second piston/cylinder is 6 m higher up in elevation. The telescope arm and the buckets have hydraulic piston/cylinders moving them, as seen in Fig. 1.16. With an outside atmospheric pressure of 100 kPa and a net force of 25 kN on the smallest piston, what is the balancing force on the second larger piston?

FIGURE 1.16   Sketch for Example 1.7.

Solution

When the fluid is stagnant and at the same elevation, we have the same pressure throughout the fluid. The force balance on the smaller piston is then related to the pressure (we neglect the rod area) as

$$F_1 + P_0 A_1 = P_1 A_1$$

from which the fluid pressure is

$$P_1 = P_0 + F_1/A_1 = 100 \, \text{kPa} + 25 \, \text{kN}/0.01 \, \text{m}^2 = 2600 \, \text{kPa}$$

The pressure at the higher elevation in piston/cylinder 2 is, from Eq. 1.2,

$$P_2 = P_1 - \rho g H = 2600 \, \text{kPa} - 900 \, \text{kg/m}^3 \times 9.81 \, \text{m/s}^2 \times 6 \, \text{m}/(1000 \, \text{Pa/kPa})$$

$$= 2547 \, \text{kPa}$$

where the second term is divided by 1000 to convert from Pa to kPa. Then the force balance on the second piston gives

$$F_2 + P_0 A_2 = P_2 A_2$$

$$F_2 = (P_2 - P_0)A_2 = (2547 - 100) \, \text{kPa} \times 0.05 \, \text{m}^2 = 122.4 \, \text{kN}$$

If the density is variable, we should consider Eq. 1.2 in differential form as

$$dP = -\rho g\, dh$$

including the sign, so pressure drops with increasing height. Now the finite difference becomes

$$P = P_0 - \int_0^H \rho g\, dh \tag{1.4}$$

with the pressure P_0 at zero height.

In-Text Concept Questions

e. A car tire gauge indicates 195 kPa; what is the air pressure inside?

f. Can I always neglect ΔP in the fluid above location A in Fig. 1.13? What circumstances does that depend on?

g. A U-tube manometer has the left branch connected to a box with a pressure of 110 kPa and the right branch open. Which side has a higher column of fluid?

1.8 ENERGY

A macroscopic amount of mass can possess energy in the form of internal energy inherent in its internal structure, kinetic energy in its motion, and potential energy associated with external forces acting on the mass. We write the total energy as

$$E = \text{Internal} + \text{Kinetic} + \text{Potential} = U + \text{KE} + \text{PE}$$

and the specific total energy becomes

$$e = E/m = u + ke + pe = u + \tfrac{1}{2}V^2 + gz \tag{1.5}$$

where the kinetic energy is taken as the translational energy and the potential energy is written for the external force being the gravitational force assumed constant. If the mass is rotating, we should add a rotational kinetic energy ($\tfrac{1}{2}I\omega^2$) to the translational term. What is called internal energy on the *macroscale* has a similar set of energies associated with the *microscale* motion of the individual molecules. This enables us to write

$$u = u_{\text{ext molecule}} + u_{\text{translation}} + u_{\text{int molecule}} \tag{1.6}$$

as a sum of the potential energy from intermolecular forces between molecules, the molecule translational kinetic energy, and the energy associated with the molecular internal and atomic structure.

Without going into detail, we realize that there is a difference between the intermolecular forces. Thus, the first term of the energy for a configuration where the molecules are close together, as in a solid or liquid (high density), contrasts with the situation for a gas like air, where the distance between the molecules is large (low density). In the limit of a very thin gas, the molecules are so far apart that they do not sense each other, unless they collide and the first term becomes near zero. This is the limit we have when we consider a substance to be an ideal gas, as will be covered in Chapter 2.

The translational energy depends only on the mass and center of mass velocity of the molecules, whereas the last energy term depends on the detailed structure. In general, we

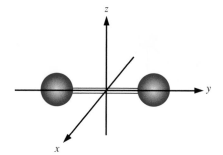

FIGURE 1.17 The coordinate system for a diatomic molecule.

can write the energy as

$$u_{\text{int molecule}} = u_{\text{potential}} + u_{\text{rotation}} + u_{\text{vibration}} + u_{\text{atoms}} \tag{1.7}$$

To illustrate the potential energy associated with the intermolecular forces, consider an oxygen molecule of two atoms, as shown in Fig. 1.17. If we want to separate the two atoms, we pull them apart with a force and thereby we do some work on the system, as explained in Chapter 3. That amount of work equals the binding (potential) energy associated with the two atoms as they are held together in the oxygen molecule.

Consider a simple monatomic gas such as helium. Each molecule consists of a helium atom. Such an atom possesses electronic energy as a result of both orbital angular momentum of the electrons about the nucleus and angular momentum of the electrons spinning on their axes. The electronic energy is commonly very small compared with the translational energies. (Atoms also possess nuclear energy, which, except in the case of nuclear reactions, is constant. We are not concerned with nuclear energy at this time.) When we consider more complex molecules, such as those composed of two or three atoms, additional factors must be considered. In addition to having electronic energy, a molecule can rotate about its center of gravity and thus have rotational energy. Furthermore, the atoms may vibrate with respect to each other and have vibrational energy. In some situations, there may be an interaction between the rotational and vibrational modes of energy.

In evaluating the energy of a molecule, we often refer to the degree of freedom, f, of these energy modes. For a monatomic molecule such as helium, $f = 3$, which represents the three directions x, y, and z in which the molecule can move. For a diatomic molecule, such as oxygen, $f = 6$. Three of these are the translation of the molecule as a whole in the x, y, and z directions, and two are for rotation. The reason that there are only two modes of rotational energy is evident from Fig. 1.17, where we take the origin of the coordinate system at the center of gravity of the molecule and the y-axis along the molecule's internuclear axis. The molecule will then have an appreciable moment of inertia about the x-axis and the z-axis but not about the y-axis. The sixth degree of freedom of the molecule is vibration, which relates to stretching of the bond joining the atoms.

For a more complex molecule such as H_2O, there are additional vibrational degrees of freedom. Fig. 1.18 shows a model of the H_2O molecule. From this diagram, it is evident that there are three vibrational degrees of freedom. It is also possible to have rotational energy about all three axes. Thus, for the H_2O molecule, there are nine degrees of freedom ($f = 9$): three translational, three rotational, and three vibrational.

Most complex molecules, such as typical polyatomic molecules, are usually three-dimensional in structure and have multiple vibrational modes, each of which contributes to

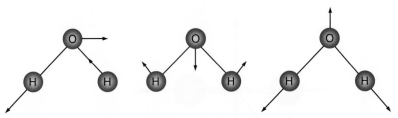

FIGURE 1.18   The three principal vibrational modes for the H_2O molecule.

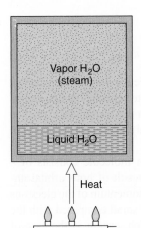

FIGURE 1.19   Heat transfer to H_2O.

the energy storage of the molecule. The more complicated the molecule is, the larger the number of degrees of freedom that exist for energy storage. The modes of energy storage and their evaluation are discussed in some detail in Appendix C, for those interested in further development of the quantitative effects from a molecular viewpoint.

This general discussion can be summarized by referring to Fig. 1.19. Let heat be transferred to H_2O. During this process, the temperature of the liquid and vapor (steam) will increase, and eventually all the liquid will become vapor. From the macroscopic point of view, we are concerned only with the energy that is transferred as heat, the change in properties such as temperature and pressure, and the total amount of energy (relative to some base) that the H_2O contains at any instant. Thus, questions about how energy is stored in the H_2O do not concern us. From a microscopic viewpoint, we are concerned about the way in which energy is stored in the molecules. We might be interested in developing a model of the molecule so that we can predict the amount of energy required to change the temperature a given amount. Although the focus in this book is on the macroscopic or classical viewpoint, it is helpful to keep in mind the microscopic or statistical perspective, as well as the relationship between the two, which helps us understand basic concepts such as energy.

1.9   EQUALITY OF TEMPERATURE

Although temperature is a familiar property, defining it exactly is difficult. We are aware of temperature first of all as a sense of hotness or coldness when we touch an object. We also learn early that when a hot body and a cold body are brought into contact, the hot body becomes cooler and the cold body becomes warmer. If these bodies remain in contact for some time, they usually appear to have the same hotness or coldness. However, we also realize that our sense of hotness or coldness is very unreliable. Sometimes very cold bodies may seem hot, and bodies of different materials that are at the same temperature appear to be at different temperatures.

Because of these difficulties in defining temperature, we define equality of temperature. Consider two blocks of copper, one hot and the other cold, each of which is in contact with a mercury-in-glass thermometer. If these two blocks of copper are brought into thermal communication, we observe that the electrical resistance of the hot block decreases with time and that of the cold block increases with time. After a period of time has elapsed, however, no further changes in resistance are observed. Similarly, when the blocks are first brought in thermal communication, the length of a side of the hot block decreases with time but the length of a side of the cold block increases with time. After a period of time, no further change in the length of either block is perceived. In addition, the mercury column of the thermometer in the hot block drops at first and that in the cold block rises, but after a period of time no further changes in height are observed. We may say, therefore, that two bodies have equality of temperature if, when they are in thermal communication, no change in any observable property occurs.

 1.10 THE ZEROTH LAW OF THERMODYNAMICS

Now consider the same two blocks of copper and another thermometer. Let one block of copper be brought into contact with the thermometer until equality of temperature is established, and then remove it. Then let the second block of copper be brought into contact with the thermometer. Suppose that no change in the mercury level of the thermometer occurs during this operation with the second block. We then can say that both blocks are in thermal equilibrium with the given thermometer.

The zeroth law of thermodynamics states that when two bodies have equality of temperature with a third body, they in turn have equality of temperature with each other. This seems obvious to us because we are so familiar with this experiment. Because the principle is not derivable from other laws, and because it precedes the first and second laws of thermodynamics in the logical presentation of thermodynamics, it is called the zeroth law of thermodynamics. This law is really the basis of temperature measurement. Every time a body has equality of temperature with the thermometer, we can say that the body has the temperature we read on the thermometer. The problem remains of how to relate temperatures that we might read on different mercury thermometers or obtain from different temperature-measuring devices, such as thermocouples and resistance thermometers. This observation suggests the need for a standard scale for temperature measurements.

 1.11 TEMPERATURE SCALES

Two scales are commonly used for measuring temperature, namely, the Fahrenheit (after Gabriel Fahrenheit, 1686–1736) and the Celsius. The Celsius scale was formerly called the centigrade scale but is now designated the Celsius scale after Anders Celsius (1701–1744), the Swedish astronomer who devised this scale.

The Fahrenheit temperature scale is used with the English Engineering System of Units and the Celsius scale with the SI unit system. Until 1954 both of these scales were based on two fixed, easily duplicated points: the ice point and the steam point. The temperature of the ice point is defined as the temperature of a mixture of ice and water that is in equilibrium with saturated air at a pressure of 1 atm. The temperature of the steam point is the temperature of water and steam, which are in equilibrium at a pressure of 1 atm. On the Fahrenheit scale these two points are assigned the numbers 32 and 212, respectively, and on the Celsius scale the points are 0 and 100, respectively. Why Fahrenheit chose these numbers is an interesting story. In searching for an easily reproducible point, Fahrenheit selected the temperature of the human body and assigned it the number 96. He assigned the number 0 to the temperature of a certain mixture of salt, ice, and salt solution. On this scale, the ice point was approximately 32. When this scale was slightly revised and fixed in terms of the ice point and steam point, the normal temperature of the human body was found to be 98.6 F.

In this book, the symbols F and °C will denote the Fahrenheit and Celsius scales, respectively (the Celsius scale symbol includes the degree symbol since the letter C alone denotes Coulomb, the unit of electrical charge in the SI system of units). The symbol T will refer to temperature on all temperature scales.

At the tenth CGPM meeting in 1954, the Celsius scale was redefined in terms of a single fixed point and the ideal-gas temperature scale. The single fixed point is the triple point of water (the state in which the solid, liquid, and vapor phases of water exist together in equilibrium). The magnitude of the degree is defined in terms of the ideal-gas temperature scale, which is discussed in Chapter 5. The essential features of this new scale are a single fixed point and a definition of the magnitude of the degree. The triple point of water is assigned the value of 0.01 °C. On this scale, the steam point is experimentally found to be 100.00 °C. Thus, there is essential agreement between the old and new temperature scales.

We have not yet considered an absolute scale of temperature. The possibility of such a scale comes from the second law of thermodynamics and is discussed in Chapter 5. On the basis of the second law of thermodynamics, a temperature scale that is independent of any thermometric substance can be defined. This absolute scale is usually referred to as the *thermodynamic scale of temperature.* However, it is difficult to use this scale directly; therefore, a more practical scale, the International Temperature Scale, which closely represents the thermodynamic scale, has been adopted.

The absolute scale related to the Celsius scale is the Kelvin scale (after William Thomson, 1824–1907, who is also known as Lord Kelvin), and is designated K (without the degree symbol). The relation between these scales is

$$K = {}^\circ C + 273.15 \tag{1.8}$$

In 1967, the CGPM defined the kelvin as 1/273.16 of the temperature at the triple point of water. The Celsius scale is now defined by this equation instead of by its earlier definition.

A number of empirically based temperature scales, to standardize temperature measurement and calibration, have been in use during the last 70 years. The most recent of these is the International Temperature Scale of 1990, or ITS-90. It is based on a number of fixed and easily reproducible points that are assigned definite numerical values of temperature, and on specified formulas relating temperature to the readings on certain temperature-measuring instruments for the purpose of interpolation between the defining fixed points. Details of the ITS-90 are not considered further in this book. This scale is a practical means for establishing measurements that conform closely to the absolute thermodynamic temperature scale.

1.12 ENGINEERING APPLICATIONS

When we deal with materials to move or trade them, we need to specify the amount; that is often done as either the total mass or volume. For substances with reasonably well-defined density we can use either measure. For instance, water, gasoline, oil, natural gas, and many food items are common examples of materials for which we use volume to express the amount. Other examples are amounts of gold, coal, and food items where we use mass as the amount. To store or transport materials, we often need to know both measures to be able to size the equipment appropriately.

FIGURE 1.20 Air compressor with tank.

(© zilli/iStockphoto)

Pressure is used in applications for process control or limit control for safety reasons. In most cases, this is the gauge pressure. For instance, a storage tank has a pressure indicator to show how close it is to being full, but it may also have a pressure-sensitive safety valve that will open and let material escape if the pressure exceeds a preset value. An air tank with a compressor on top is shown in Fig. 1.20. As a portable unit, it is used to drive air tools, such as nailers. A pressure gauge will activate a switch to start the compressor when the pressure drops below a preset value, and it will disengage the compressor when a preset high value is reached.

Tire pressure gauges, shown in Fig. 1.21, are connected to the valve stem on the tire. Some gauges have a digital readout. The tire pressure is important for the safety and durability of automobile tires. Too low a pressure causes large deflections and the tire may overheat; too high a pressure leads to excessive wear in the center.

A spring-loaded pressure relief valve is shown in Fig. 1.22. With the cap the spring can be compressed to make the valve open at a higher pressure, or the opposite. This valve is used for pneumatic systems.

FIGURE 1.21 Automotive tire pressure gauges.

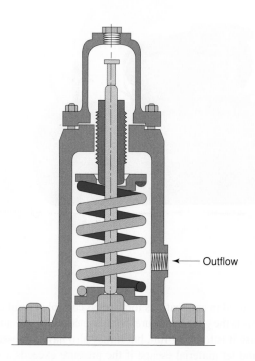

FIGURE 1.22
Schematic of a pressure relief valve.

Outflow

When a throttle plate in an intake system for an automotive engine restricts the flow (Fig. 1.23), it creates a vacuum behind it that is measured by a pressure gauge sending a signal to the computer control. The smallest absolute pressure (highest vacuum) occurs when the engine idles, and the highest pressure (smallest vacuum) occurs when the engine is at full throttle. In Fig. 1.23, the throttle is shown completely closed.

A pressure difference, ΔP, can be used to measure flow velocity indirectly, as shown schematically in Fig. 1.24 (this effect is felt when you hold your hand out of a car window, with a higher pressure on the side facing forward and a lower pressure on the other side, giving a net force on your hand). The engineering analysis of such processes is developed and presented in Chapter 7. In a speedboat, a small pipe has its end pointing forward, feeling the higher pressure due to the relative velocity between the boat and the water. The other end goes to a speedometer transmitting the pressure signal to an indicator.

An aneroid barometer, shown in Fig. 1.25, measures the absolute pressure used for weather predictions. It consists of a thin metal capsule or bellows that expands or contracts

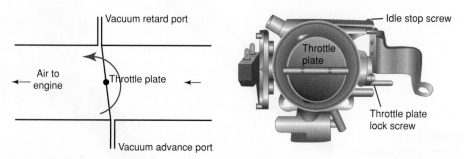

FIGURE 1.23 Automotive engine intake throttle.

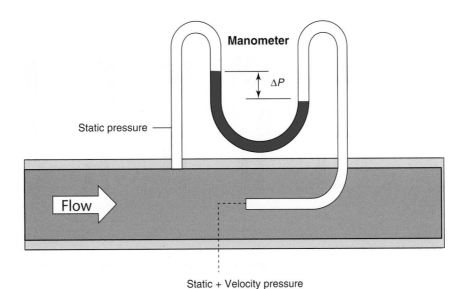

FIGURE 1.24 Schematic of flow velocity measurement.

with atmospheric pressure. Measurement is by a mechanical pointer or by a change in electrical capacitance with distance between two plates.

Numerous types of devices are used to measure temperature. Perhaps the most familiar of these is the liquid-in-glass thermometer, in which the liquid is commonly mercury. Since the density of the liquid decreases with temperature, the height of the liquid column rises accordingly. Other liquids are also used in such thermometers, depending on the range of temperature to be measured.

Two types of devices commonly used in temperature measurement are thermocouples and thermistors. Examples of thermocouples are shown in Fig. 1.26. A thermocouple consists of a pair of junctions of two dissimilar metals that creates an electrical potential (voltage) that increases with the temperature difference between the junctions. One junction is maintained at a known reference temperature (for example, in an ice bath), such

FIGURE 1.25
Aneroid barometer.

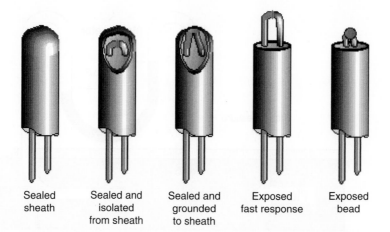

FIGURE 1.26
Thermocouples.

Sealed
sheath

Sealed and
isolated
from sheath

Sealed and
grounded
to sheath

Exposed
fast response

Exposed
bead

that the voltage measured indicates the temperature of the other junction. Different material combinations are used for different temperature ranges, and the size of the junction is kept small to have a short response time. Thermistors change their electrical resistance with temperature, so if a known current is passed through the thermistor, the voltage across it becomes proportional to the resistance. The output signal is improved if this is arranged in an electrical bridge that provides input to an instrument. The small signal from these sensors is amplified and scaled so that a meter can show the temperature or the signal can be sent to a computer or a control system. High-precision temperature measurements are made in a similar manner using a platinum resistance thermometer. A large portion of the ITS-90 (13.8033 K to 1234.93 K) is measured in such a manner. Higher temperatures are determined from visible-spectrum radiation intensity observations.

It is also possible to measure temperature indirectly by certain pressure measurements. If the vapor pressure, discussed in Chapter 2, is accurately known as a function of temperature, then this value can be used to indicate the temperature. Also, under certain conditions, a constant-volume gas thermometer, discussed in Chapter 5, can be used to determine temperature by a series of pressure measurements.

SUMMARY

We introduce a thermodynamic system as a control volume, which for a fixed mass is a control mass. Such a system can be isolated, exchanging neither mass, momentum, nor energy with its surroundings. A closed system versus an open system refers to the ability of mass exchange with the surroundings. If properties for a substance change, the state changes and a process occurs. When a substance has gone through several processes, returning to the same initial state, it has completed a cycle.

Basic units for thermodynamic and physical properties are mentioned, and most are covered in Table A.1. Thermodynamic properties such as density ρ, specific volume v, pressure P, and temperature T are introduced together with units for these properties. Properties are classified as intensive, independent of mass (like v), or extensive, proportional to mass (like V). Students should already be familiar with other concepts from physics such as force F, velocity $\mathbf{V}$, and acceleration a. Application of Newton's law of motion leads to the variation of static pressure in a column of fluid and the measurements of pressure (absolute and gauge) by barometers and manometers. The normal temperature scale and the absolute temperature scale are introduced.

You should have learned a number of skills and acquired abilities from studying this chapter that will allow you to

- Define (choose) a control volume (C.V.) around some matter; sketch the content and identify storage locations for mass; and identify mass and energy flows crossing the C.V. surface.
- Know properties P, T, v, and ρ and their units.
- Know how to look up conversion of units in Table A.1.
- Know that energy is stored as kinetic, potential, or internal (in molecules).
- Know that energy can be transferred.
- Know the difference between (v, ρ) and (V, m) intensive and extensive properties.
- Apply a force balance to a given system and relate it to pressure P.
- Know the difference between relative (gauge) and absolute pressure P.
- Understand the working of a manometer or a barometer and derive ΔP or P from height H.
- Know the difference between a relative and an absolute temperature T.
- Be familiar with magnitudes (v, ρ, P, T).

Most of these concepts will be repeated and reinforced in the following chapters, such as properties in Chapter 2, energy transfer as heat and work, and internal energy in Chapter 3, together with their applications.

KEY CONCEPTS AND FORMULAS

Control volume	everything inside a control surface
Pressure definition	$P = \dfrac{F}{A}$ (mathematical limit for small A)
Specific volume	$v = \dfrac{V}{m}$
Density	$\rho = \dfrac{m}{V}$ (Tables A.3, A.4, and A.5)
Static pressure variation	$\Delta P = \rho g H = -\int \rho g\, dh$
Absolute temperature	$T[\text{K}] = T[°\text{C}] + 273.15$
	$T[\text{R}] = T[\text{F}] + 459.67$
Units	Table A.1
Specific total energy	$e = u + \dfrac{1}{2}V^2 + gz$

Concepts from Physics

Newton's law of motion	$F = ma$
Acceleration	$a = \dfrac{d^2x}{dt^2} = \dfrac{d\mathbf{V}}{dt}$
Velocity	$\mathbf{V} = \dfrac{dx}{dt}$

CONCEPT-STUDY GUIDE PROBLEMS

1.1 Make a control volume around the whole power plant in Fig. 1.1 and list the flows of mass and energy in or out and any storage of energy. Make sure you know what is inside and what is outside your chosen control volume.

1.2 Make a control volume around the refrigerator in Fig. 1.3. Identify the mass flow of external air and show where you have significant heat transfer and where storage changes.

1.3 Separate the list P, F, V, v, ρ, T, a, m, L, t, and **V** into intensive properties, extensive properties, and nonproperties.

1.4 A tray of liquid water is placed in a freezer where it cools from $20\,°C$ to $-5\,°C$. Show the energy flow(s) and storage and explain what changes.

1.5 The overall density of fibers, rock wool insulation, foams, and cotton is fairly low. Why?

1.6 Is density a unique measure of mass distribution in a volume? Does it vary? If so, on what kind of scale (distance)?

1.7 Water in nature exists in three different phases: solid, liquid, and vapor (gas). Indicate the relative magnitude of density and the specific volume for the three phases.

1.8 What is the approximate mass of 1 L of gasoline? Of helium in a balloon at T_0, P_0?

1.9 Can you carry $1\ m^3$ of liquid water?

1.10 A heavy refrigerator has four height-adjustable feet. What feature of the feet will ensure that they do not make dents in the floor?

1.11 A swimming pool has an evenly distributed pressure at the bottom. Consider a stiff steel plate lying on the ground. Is the pressure below it just as evenly distributed?

1.12 What physically determines the variation of the atmospheric pressure with elevation?

1.13 Two divers swim at a depth of 20 m. One of them swims directly under a supertanker; the other avoids the tanker. Who feels a greater pressure?

1.14 A manometer with water shows a ΔP of $P_0/20$; what is the column height difference?

1.15 Does the pressure have to be uniform for equilibrium to exist?

1.16 A water skier does not sink too far down in the water if the speed is high enough. What makes that situation different from our static pressure calculations?

1.17 What is the lowest temperature in degrees Celsius? In kelvin?

1.18 Convert the formula for water density in In-Text Concept Problem **d** to be for T in kelvin.

1.19 A thermometer that indicates the temperature with a liquid column has a bulb with a larger volume of liquid. Why?

1.20 What is the main difference between the macroscopic kinetic energy in a motion like the blowing of wind versus the microscopic kinetic energy of individual molecules? Which one can you sense with your hand?

1.21 How can you illustrate the binding energy between the three atoms in water as they sit in a triatomic water molecule? *Hint:* imagine what must happen to create three separate atoms.

HOMEWORK PROBLEMS

Properties, Units, and Force

1.22 An apple "weighs" 60 g and has a volume of $75\ cm^3$ in a refrigerator at $8\,°C$. What is the apple's density? List three intensive and two extensive properties of the apple.

1.23 A stainless steel storage tank contains 5 kg of oxygen gas and 7 kg of nitrogen gas. How many kilomoles are in the tank?

1.24 A steel cylinder of mass 4 kg contains 4 L of water at $25\,°C$ at 100 kPa. Find the total mass and volume of the system. List two extensive and three intensive properties of the water.

1.25 The standard acceleration (at sea level and $45°$ latitude) due to gravity is $9.806\ 65\ m/s^2$. What is the force needed to hold a mass of 2 kg at rest in this gravitational field? How much mass can a force of 1 N support?

1.26 When you move up from the surface of the earth, the gravitation is reduced as $g = 9.807 - 3.32 \times 10^{-6} z$, with z being the elevation in meters. By what percentage is the weight of an airplane reduced when it cruises at 11 000 m?

1.27 A car rolls down a hill with a slope such that the gravitational "pull" in the direction of motion is one-tenth of the standard gravitational force (see Problem 1.25). If the car has a mass of 2500 kg, find the acceleration.

1.28 A 1500 kg car moving at 20 km/h is accelerated at a constant rate of 4 m/s² up to a speed of 75 km/h. What are the force and total time required?

1.29 The elevator in a hotel has a mass of 750 kg, and it carries six people with a total mass of 450 kg. How much force should the cable pull up with to have an acceleration of 1 m/s² in the upward direction?

1.30 One of the people in the previous problem weighs 80 kg standing still. How much weight does this person feel when the elevator starts moving?

1.31 A bottle of 12 kg steel has 1.75 kmol of liquid propane. It accelerates horizontally at a rate of 3 m/s². What is the needed force?

Specific Volume

1.32 A 1 m³ container is filled with 400 kg of granite stone, 200 kg of dry sand, and 0.2 m³ of liquid 25 °C water. Using properties from Tables A.3 and A.4, find the average specific volume and density of the masses when you exclude air mass and volume.

1.33 A power plant that separates carbon dioxide from the exhaust gases compresses it to a density of 110 kg/m³ and stores it in an unminable coal seam with a porous volume of 100 000 m³. Find the mass that can be stored.

1.34 A 5 m³ container is filled with 900 kg of granite (density of 2400 kg/m³). The rest of the volume is air, with density equal to 1.15 kg/m³. Find the mass of air and the overall (average) specific volume.

1.35 A tank has two rooms separated by a membrane. Room A has 1.5 kg of air and a volume of 0.5 m³; room B has 0.75 m³ of air with density 0.8 kg/m³. The membrane is broken, and the air comes to a uniform state. Find the final density of the air.

1.36 One kilogram of diatomic oxygen (O_2, molecular mass of 32) is contained in a 500 L tank. Find the specific volume on both a mass and a mole basis (v and $\bar{v}$).

Pressure

1.37 A 5000 kg elephant has a cross-sectional area of 0.02 m² on each foot. Assuming an even distribution, what is the pressure under its feet?

1.38 A valve in the cylinder shown in Fig. P1.38 has a cross-sectional area of 11 cm² with a pressure of 735 kPa inside the cylinder and 99 kPa outside. How large a force is needed to open the valve?

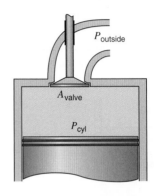

FIGURE P1.38

1.39 The hydraulic lift in an auto repair shop has a cylinder diameter of 0.2 m. To what pressure should the hydraulic fluid be pumped to lift 80 kg of piston/arms and 700 kg of a car?

1.40 A laboratory room has a vacuum of 0.1 kPa. What net force does that put on the door of size 2 m by 1 m?

1.41 A vertical hydraulic cylinder has a 125 mm diameter piston with hydraulic fluid inside the cylinder and an ambient pressure of 1 bar. Assuming standard gravity, find the piston mass that will create an inside pressure of 1500 kPa.

1.42 A piston/cylinder with a cross-sectional area of 0.01 m² has a piston mass of 100 kg resting on the stops, as shown in Fig. P1.42. With an outside atmospheric pressure of 100 kPa, what should the water pressure be to lift the piston?

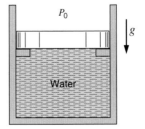

FIGURE P1.42

1.43 A large exhaust fan in a laboratory room keeps the pressure inside at 15 cm of water vacuum relative to the hallway. What is the net force on the door measuring 1.9 m by 1.1 m?

1.44 A tornado rips off a 100 m² roof with a mass of 1000 kg. What is the minimum vacuum pressure needed to do that if we neglect the anchoring forces?

1.45 A 2.5 m tall steel cylinder has a cross-sectional area of 1.5 m². At the bottom, with a height of 0.5 m, is liquid water, on top of which is a 1 m high layer of gasoline. This is shown in Fig. P1.45. The gasoline surface is exposed to atmospheric air at 101 kPa. What is the highest pressure in the water?

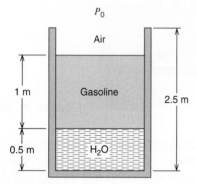

FIGURE P1.45

1.46 An underwater buoy is anchored at the seabed with a cable, and it contains a total mass of 250 kg. What should the volume be so that the cable holds it down with a force of 1200 N?

1.47 At the beach, atmospheric pressure is 1025 mbar. You dive 15 m down in the ocean, and you later climb a hill up to 250 m in elevation. Assume that the density of water is about 1000 kg/m³ and the density of air is 1.18 kg/m³. What pressure do you feel at each place?

1.48 A steel tank of cross-sectional area 3 m² and height 16 m weighs 8000 kg and is open at the top, as shown in Fig. P1.48. We want to float it in the ocean

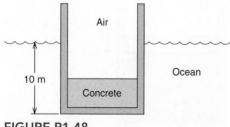

FIGURE P1.48

so that it is positioned 10 m straight down by pouring concrete into its bottom. How much concrete should we use?

1.49 Liquid water with density ρ is filled on top of a thin piston in a cylinder with cross-sectional area A and total height H, as shown in Fig. P1.49. Air is let in under the piston so that it pushes up, causing the water to spill over the edge. Derive the formula for the air pressure as a function of piston elevation from the bottom, h.

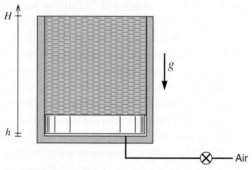

FIGURE P1.49

Manometers and Barometers

1.50 A probe is lowered 16 m into a lake. Find the absolute pressure there.

1.51 The density of atmospheric air is about 1.15 kg/m³, which we assume is constant. How large an absolute pressure will a pilot encounter when flying 3500 m above ground level, where the pressure is 101 kPa?

1.52 The standard pressure in the atmosphere with elevation (H) above sea level can be correlated as $P = P_0 (1 - H/L)^{5.26}$ with $L = 44\,300$ m. With the local sea level pressure P_0 at 101 kPa, what is the pressure at 10 000 m elevation?

1.53 A barometer to measure absolute pressure shows a mercury column height of 725 mm. The temperature is such that the density of the mercury is 13 550 kg/m³. Find the ambient pressure.

1.54 A differential pressure gauge mounted on a vessel shows 1.25 MPa, and a local barometer gives atmospheric pressure as 0.96 bar. Find the absolute pressure inside the vessel.

1.55 Blue manometer fluid of density 925 kg/m³ shows a column height difference of 3 cm vacuum with one end attached to a pipe and the other open to $P_0 = 101$ kPa. What is the absolute pressure in the pipe?

1.56 What pressure difference does a 10 m column of atmospheric air show?

1.57 A barometer measures 760 mm Hg at street level and 735 mm Hg on top of a building. How tall is the building if we assume air density of 1.15 kg/m^3?

1.58 The pressure gauge on an air tank shows 50 kPa when the diver is 10 m down in the ocean. At what depth will the gauge pressure be zero? What does that mean?

1.59 A submarine maintains an internal pressure of 101 kPa and dives 240 m down in the ocean, which has an average density of 1030 kg/m^3. What is the pressure difference between the inside and the outside of the submarine hull?

1.60 Assume that we use a pressure gauge to measure the air pressure at street level and at the roof of a tall building. If the pressure difference can be determined with an accuracy of 1 mbar (0.001 bar), what uncertainty in the height estimate does that correspond to?

1.61 The absolute pressure in a tank is 115 kPa and the local ambient absolute pressure is 97 kPa. If a U-tube with mercury (density = 13 550 kg/m^3) is attached to the tank to measure the gauge pressure, what column height difference will it show?

1.62 A U-tube manometer filled with water (density = 1000 kg/m^3) shows a height difference of 25 cm. What is the gauge pressure? If the right branch is tilted to make an angle of 30° with the horizontal, as shown in Fig. P1.62, what should the length of the column in the tilted tube be relative to the U-tube?

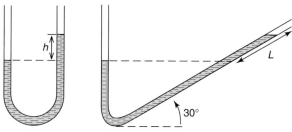

FIGURE P1.62

1.63 A pipe flowing light oil has a manometer attached, as shown in Fig. P1.63. What is the absolute pressure in the pipe flow?

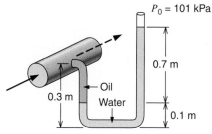

FIGURE P1.63

1.64 The difference in height between the columns of a manometer is 250 mm, with a fluid of density 900 kg/m^3. What is the pressure difference? What is the height difference if the same pressure difference is measured using mercury (density = 13 600 kg/m^3) as manometer fluid?

1.65 Two cylinders are filled with liquid water, $\rho = 1000$ kg/m^3, and connected by a line with a closed valve, as shown in Fig. P1.65. A has 100 kg and B has 500 kg of water, their cross-sectional areas are $A_A = 0.1$ m^2 and $A_B = 0.25$ m^2, and the height h is 1 m. Find the pressure on either side of the valve. The valve is opened, and water flows to an equilibrium. Find the final pressure at the valve location.

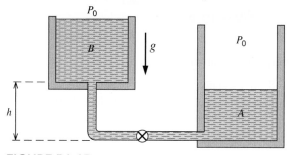

FIGURE P1.65

1.66 Two piston/cylinder arrangements, A and B, have their gas chambers connected by a pipe, as shown in Fig. P1.66. The cross-sectional areas are $A_A = 75$ cm^2 and $A_B = 25$ cm^2, with the piston mass in A being $m_A = 25$ kg. Assume an outside pressure of 100 kPa and standard gravitation. Find the mass m_B so that none of the pistons have to rest on the bottom.

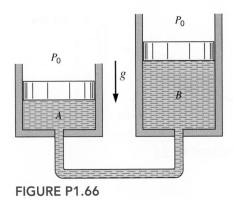

FIGURE P1.66

1.67 A piece of experimental apparatus, Fig. P1.67, is located where $g = 9.5$ m/s^2 and the temperature is 5 °C. Air flow inside the apparatus is determined by measuring the pressure drop across an orifice with a mercury manometer (see Problem 1.73 for density) showing a height difference of 200 mm. What is the pressure drop in kPa?

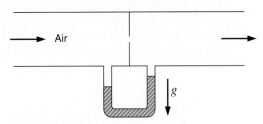

FIGURE P1.67

Energy and Temperature

1.68 An escalator brings four people, whose total mass is 300 kg, 25 m up in a building. Explain what happens with respect to energy transfer and stored energy.

1.69 A 42 kg package is lifted up to the top shelf in a storage bin that is 4 m above the ground floor. How much increase in potential energy does the package get?

1.70 A car of mass 1775 kg travels with a velocity of 100 km/h. Find the kinetic energy. How high should the car be lifted in the standard gravitational field to have a potential energy that equals the kinetic energy?

1.71 An oxygen molecule with mass $m = M \, m_o = 32 \times 1.66 \times 10^{-27}$ kg moves with a velocity of 360 m/s. What is the kinetic energy of the molecule? What temperature does that correspond to if it has to equal $(3/2) \, kT$, where k is Boltzmans constant and T is absolute temperature in kelvin?

1.72 The human comfort zone is between 18 °C and 24 °C. What is the range in kelvin? What is the maximum relative change from the low to the high temperature?

1.73 The density of mercury changes approximately linearly with temperature as $\rho_{Hg} = 13\,595 - 2.5\,T$ kg/m^3 (T in Celsius), so the same pressure difference will result in a manometer reading that is influenced by temperature. If a pressure difference of 100 kPa is measured in the summer at 35 °C and in the winter at -15 °C, what is the difference in column height between the two measurements?

1.74 A mercury thermometer measures temperature by measuring the volume expansion of a fixed mass of liquid mercury due to a change in density (see Problem 1.73). Find the relative change (%) in volume for a change in temperature from 10 °C to 20 °C.

1.75 The density of liquid water is $\rho = 1008 - T/2$ [kg/m^3] with T in °C. If the temperature increases 10 °C, how much deeper does a 1 m layer of water become?

1.76 The atmosphere becomes colder at higher elevations. As an average, the standard atmospheric absolute temperature can be expressed as $T_{atm} = 288 - 6.5 \times 10^{-3} \, z$, where z is the elevation in meters. How cold is it outside an airplane cruising at 12 000 m, expressed in kelvin and degrees Celsius?

Review Problems

1.77 Repeat Problem 1.67 if the flow inside the apparatus is liquid water ($\rho = 1000$ kg/m^3) instead of air. Find the pressure difference between the two holes flush with the bottom of the channel. You cannot neglect the two unequal water columns.

1.78 In the city water tower, water is pumped up to a level 25 m above ground in a pressurized tank with air at 150 kPa over the water surface. This is illustrated in Fig. P1.78. Assuming water density of 1000 kg/m^3 and standard gravity, find the pressure required to pump more water in at ground level.

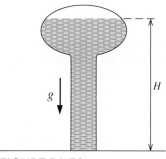

FIGURE P1.78

1.79 The main waterline into a tall building has a pressure of 600 kPa at 5 m elevation below ground level. The building is shown in Fig. P1.79. How much extra pressure does a pump need to add to ensure a waterline pressure of 150 kPa at the top floor 150 m aboveground?

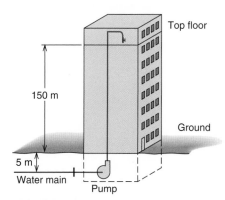

FIGURE P1.79

1.80 Two cylinders are connected by a piston, as shown in Fig. P1.80. Cylinder A is used as a hydraulic

lift and pumped up to 500 kPa. The piston mass is 25 kg, and there is standard gravity. What is the gas pressure in cylinder B?

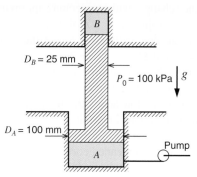

FIGURE P1.80

1.81 A 5 kg piston in a cylinder with a diameter of 100 mm is loaded with a linear spring and the outside atmospheric pressure is 100 kPa, as shown in Fig. P1.81. The spring exerts no force on the piston when it is at the bottom of the cylinder, and for the state shown, the pressure is 400 kPa with volume 0.4 L. The valve is opened to let some air in, causing the piston to rise 2 cm. Find the new pressure.

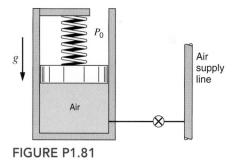

FIGURE P1.81

COMPUTER, DESIGN, AND OPEN-ENDED PROBLEMS

1.82 Plot the atmospheric pressure as a function of elevation (0–20 000 m) at a location where the ground pressure is 100 kPa at 500 m elevation. Use the variation shown in Problem 1.52.

1.83 Write a program to do the temperature correction on a mercury barometer reading (see Problem 1.57). Input the reading and temperature and output the corrected reading at 20 °C and pressure in kPa.

1.84 Make a list of different weights and scales that are used to measure mass directly or indirectly. Investigate the ranges of mass and the accuracy that can be obtained.

1.85 Thermometers are based on several principles. Expansion of a liquid with a rise in temperature is used in many applications. Electrical resistance, thermistors, and thermocouples are common in instrumentation and remote probes. Investigate a variety of thermometers and list their range, accuracy, advantages, and disadvantages.

1.86 A thermistor is used as a temperature transducer. Its resistance changes with temperature approximately as $R = R_0 \exp[\alpha(1/T - 1/T_0)]$ where it has resistance R_0 at temperature T_0. Select the constants as $R_0 = 3000 \ \Omega$ and $T_0 = 298$ K and compute α so that it has a resistance of 200 Ω at 100 °C. Write a program to convert a measured resistance, R, into information about the temperature. Find information for actual thermistors and plot the calibration curves with the formula given in this problem and the recommended correction given by the manufacturer.

1.87 Blood pressure is measured with a sphygmomanometer while the sound from the pulse is checked. Investigate how this works, list the range of pressures normally recorded as the systolic (high) and diastolic (low) pressures, and present your findings in a short report.

1.88 A micromanometer uses a fluid with density 1000 kg/m³, and it is able to measure height difference with an accuracy of ±0.5 mm. Its range is a maximum height difference of 0.5 m. Investigate to determined if any transducers are available to replace the micromanometer.

Pure Substance Behavior

In the previous chapter we considered three familiar properties of a substance: specific volume, pressure, and temperature. We now turn our attention to pure substances and consider some of the phases in which a pure substance may exist, the number of independent properties a pure substance may have, and methods of presenting thermodynamic properties.

Properties and the behavior of substances are very important for our studies of devices and thermodynamic systems. The steam power plant shown in Fig. 1.1 and other power plants using different fuels such as oil, natural gas, or nuclear energy have very similar processes, using water as the working substance. Water vapor (steam) is made by boiling water at high pressure in the steam generator followed by expansion in the turbine to a lower pressure, cooling in the condenser, and returning it to the boiler by a pump that raises the pressure, as shown in Fig. 1.2. We must know the properties of water to properly size equipment such as the burners or heat exchangers, the turbine, and the pump for the desired transfer of energy and the flow of water. As the water is transformed from liquid to vapor, we need to know the temperature for the given pressure, and we must know the density or specific volume so that the piping can be properly dimensioned for the flow. If the pipes are too small, the expansion creates excessive velocities, leading to pressure losses and increased friction, and thus demanding a larger pump and reducing the turbine's work output.

Another example is a refrigerator, shown in Fig. 1.3, where we need a substance that will boil from liquid to vapor at a low temperature, say $-20\,°C$. This process absorbs energy from the cold space, keeping it cold. Inside the black grille in the back or at the bottom of the refrigerator, the now hot substance is cooled by air flowing around the grille, so it condenses from vapor to liquid at a temperature slightly higher than room temperature. When such a system is designed, we need to know the pressures at which these processes take place and the amount of energy involved; these topics are covered in Chapters 3 and 4. We also need to know how much volume the substance occupies, that is, the specific volume, so that the piping diameters can be selected as mentioned for the steam power plant. The substance is selected so that the pressure is reasonable during these processes; it should not be too high, due to leakage and safety concerns, and not too low, as air might leak into the system.

A final example of a system in which we need to know the properties of the substance is the gas turbine and a variation thereof, namely, a jet engine. In these systems, the working substance is a gas (very similar to air) and no phase change takes place. A combustion process burns fuel and air, freeing a large amount of energy, which heats the gas so that it expands. We need to know how hot the gas gets and how large the expansion is so that we can analyze the expansion process in the turbine and the exit nozzle of the jet engine. In this device, large velocities are needed inside the turbine section and for the exit of the jet engine. This high-velocity flow pushes on the blades in the turbine to create shaft work or pushes on the compressor blades in the jet engine (giving *thrust*) to move the aircraft forward.

These are just a few examples of complete thermodynamic systems in which a substance goes through several processes involving changes in its thermodynamic state and therefore its properties. As your studies progress, many other examples will be used to illustrate the general subjects.

2.1 THE PURE SUBSTANCE

A pure substance is one that has a homogeneous and invariable chemical composition. It may exist in more than one phase, but the chemical composition is the same in all phases. Thus, liquid water, a mixture of liquid water and water vapor (steam), and a mixture of ice and liquid water are all pure substances; every phase has the same chemical composition. In contrast, a mixture of liquid air and gaseous air is not a pure substance because the composition of the liquid phase is different from that of the vapor phase.

Sometimes a mixture of gases, such as air, is considered a pure substance as long as there is no change of phase. Strictly speaking, this is not true. As we will see later, we should say that a mixture of gases such as air exhibits some of the characteristics of a pure substance as long as there is no change of phase.

In this book, the emphasis will be on simple compressible substances. This term designates substances whose surface effects, magnetic effects, and electrical effects are insignificant when dealing with the substances. But changes in volume, such as those associated with the expansion of a gas in a cylinder, are very important. Reference will be made, however, to other substances for which surface, magnetic, and electrical effects are important. We will refer to a system consisting of a simple compressible substance as a *simple compressible system*.

2.2 THE PHASE BOUNDARIES

Consider an amount of water contained in a piston/cylinder arrangement that keeps a set constant pressure, as in Fig. 2.1a, and whose temperature we can monitor. Assume that the water starts out at room conditions of P_0, T_0, in which state it is a liquid. If the water is slowly heated, the temperature increases and the volume increases slightly, but by design, the pressure stays constant. When the temperature reaches 99.6 °C, additional heat transfer results in a phase change with formation of some vapor, as indicated in Fig. 2.1b, while the

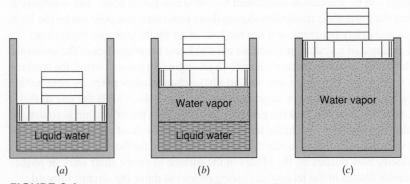

FIGURE 2.1 Constant pressure change from a liquid to a vapor.

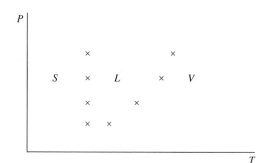

FIGURE 2.2 The separation of the phases in a *P–T* diagram.

temperature stays constant and the volume increases considerably. Further heating generates more and more vapor and a large increase in the volume until the last drop of liquid is vaporized. Subsequent heating results in an increase in both the temperature and volume of the vapor, as shown in Fig. 2.1*c*.

The term saturation temperature designates the temperature at which vaporization takes place at a given pressure; it is also commonly known as the boiling temperature. If the experiment is repeated for a different pressure, we get a different saturation temperature that may be marked as in Fig. 2.2, separating the liquid (*L*) and vapor (*V*) regions. If the experiment is done with cooling instead of heating, we will find that as the temperature is decreased, we reach a point at which ice (*S* for solid state) starts to form, with an associated increase in volume. While cooling, the system forms more ice and less liquid at a constant temperature, which is a different saturation temperature commonly called the freezing point. When all the liquid is changed to ice, further cooling will reduce the temperature and the volume is nearly constant. The freezing point is also marked in Fig. 2.2 for each set pressure, and these points separate the liquid region from the solid region. Each of the two sets of marks, if formed sufficiently close, forms a curve and both are saturation curves; the left one is called the fusion line (it is nearly straight) as the border between the solid phase and the liquid phase, whereas the right one is called the vaporization curve.

If the experiment is repeated for lower and lower pressures, it is observed that the two saturation curves meet and further reduction in the pressure results in a single saturation curve called the sublimation line separating the solid phase from the vapor phase. The point where the curves meet is called the triple point and is the only *P*, *T* combination in which all three phases (solid, liquid, and vapor) can coexist; below the triple point in *T* or *P* no liquid phase can exist. The three different saturation curves are shown in Fig. 2.3, which is called the phase diagram. This diagram shows the distinct set of saturation properties (T_{sat}, P_{sat}) for which it is possible to have two phases in equilibrium with one another. For a high pressure, 22.09 MPa for water, the vaporization curve stops at a point called the critical point.

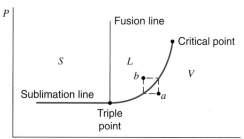

FIGURE 2.3 Sketch of a water phase diagram.

TABLE 2.1
Some Solid–Liquid–Vapor Triple-Point Data

	Temperature, °C	**Pressure, kPa**
Hydrogen (normal)	−259	7.194
Oxygen	−219	0.15
Nitrogen	−210	12.53
Carbon dioxide	−56.4	520.8
Mercury	−39	0.000 000 13
Water	0.01	0.6113
Zinc	419	5.066
Silver	961	0.01
Copper	1083	0.000 079

Above this pressure there is no boiling phenomenon, and heating of a liquid will produce a vapor without boiling in a smooth transition.

The properties at the triple point can vary quite significantly among substances, as is evident from Table 2.1. Mercury, like other metals, has a particularly low triple-point pressure, and carbon dioxide has an unusually high one; recall the use of mercury as a barometer fluid in Chapter 1, where it is useful because its vapor pressure is low. A small sample of critical-point data is shown in Table 2.2, with a more extensive set given in Table A.2 in Appendix A. The knowledge about the two endpoints of the vaporization curve does give some indication of where the liquid and vapor phases intersect, but more detailed information is needed to make a phase determination from a given pressure and temperature.

Whereas Fig. 2.3 is a sketch in linear coordinates, the real curves are plotted to scale in Fig. 2.4 for water and the scale for pressure is done logarithmically to cover the wide range. In this phase diagram, several different solid phase regions are shown; this can be the case for other substances as well. All the solids are ice but each phase region has a different crystal structure and they share a number of phase boundaries with several triple points; however, there is only one triple point where a solid–liquid–vapor equilibrium is possible.

So far we have mainly discussed the substance water, but all pure substances exhibits the same general behavior. It was mentioned earlier that the triple-point data vary significantly among substances; this is also true for the critical-point data. For example, the critical-point temperature for helium (see Table A.2) is 5.3 K, and room temperature is thus more than 50 times larger. Water has a critical temperature of 647.29 K, which is more than twice the room temperature, and most metals have an even higher critical temperature than water.

TABLE 2.2
Some Critical-Point Data

	Critical Temperature, °C	**Critical Pressure, MPa**	**Critical Volume, m³/kg**
Water	374.14	22.09	0.003 155
Carbon dioxide	31.05	7.39	0.002 143
Oxygen	−118.35	5.08	0.002 438
Hydrogen	−239.85	1.30	0.032 192

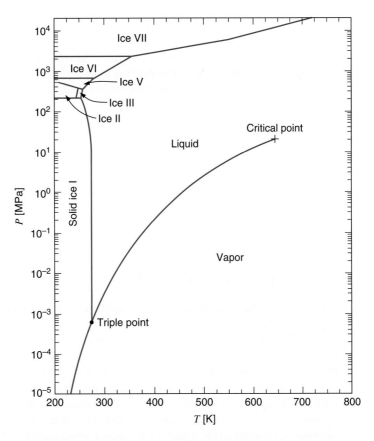

FIGURE 2.4 Water phase diagram.

The phase diagram for carbon dioxide plotted to scale is shown in Fig. 2.5, and again, the pressure axis is logarithmic to cover the large range of values. This is unusual in that the triple-point pressure is above atmospheric pressure (see also Table 2.2) and the fusion line slopes to the right, the opposite of the behavior of water. Thus, at atmospheric pressure

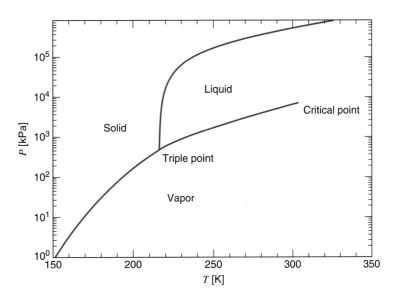

FIGURE 2.5 Carbon dioxide phase diagram.

of 100 kPa, solid carbon dioxide will make a phase transition directly to vapor without becoming liquid first, a process called sublimation. This is used to ship frozen meat in a package where solid carbon dioxide is added instead of ice, so as it changes phase the package stays dry; hence, it is also referred to as *dry ice*. Figure 2.5 shows that this phase change happens at about 200 K and thus is quite cold.

In-Text Concept Questions

a. If the pressure is smaller than the smallest P_{sat} at a given T, what is the phase?

b. An external water tap has the valve activated by a long spindle, so the closing mechanism is located well inside the wall. Why?

c. What is the lowest temperature (approximately) at which water can be liquid?

2.3   THE *P–v–T* SURFACE

Let us consider the experiment in Fig. 2.1 again but now also assume that we have measured the total water volume, which, together with mass, gives the property specific volume. We can then plot the temperature as a function of volume following the constant pressure process. Assuming that we start at room conditions and heat the liquid water, the temperature goes up and the volume expands slightly, as indicated in Fig. 2.6, starting from state *A* and going toward state *B*. As state *B* is reached, we have liquid water at 99.6 °C, which is called saturated liquid. Further heating increases the volume at constant temperature (the boiling temperature), producing more vapor and less liquid that eventually reaches state *C*, called saturated vapor after all the liquid is vaporized. Further heating will produce vapor at higher temperatures in states called superheated vapor, where the temperature is higher than the saturated temperature for the given pressure. The difference between a given T and the saturated temperature for the same pressure is called the degree of superheat.

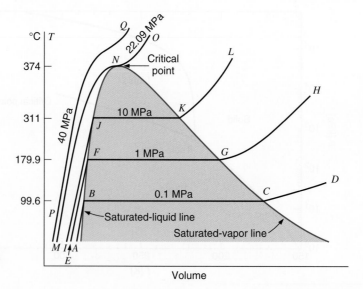

FIGURE 2.6   *T–v* diagram for water showing liquid and vapor phases (not to scale).

For higher pressures the saturation temperature is higher, like 179.9 °C at state *F* for a pressure of 1 MPa, and so forth. At the critical pressure of 22.09 MPa, heating proceeds from state *M* to state *N* to state *O* in a smooth transition from a liquid state to a vapor state without undergoing the constant temperature vaporization (boiling) process. During heating at this and higher pressures two phases are never present at the same time, and in the region where the substance changes distinctly from a liquid to a vapor state, it is called a dense fluid. The states in which the saturation temperature is reached for the liquid (*B*, *F*, *J*) are saturated liquid states forming the saturated liquid line. Similarly, the states along the other border of the two-phase region (*N*, *K*, *G*, *C*) are saturated vapor states forming the saturated vapor line.

We can now display the possible *P–v–T* combinations for typical substances as a surface in a *P-v-T* diagram shown in Figs. 2.7 and 2.8. Figure 2.7 shows a substance such

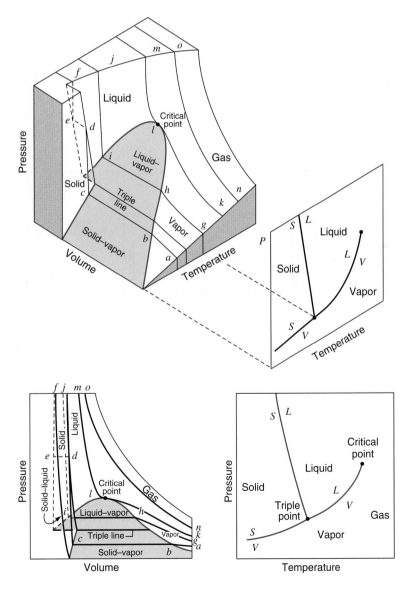

FIGURE 2.7 *P–v–T* surface for a substance that expands on freezing.

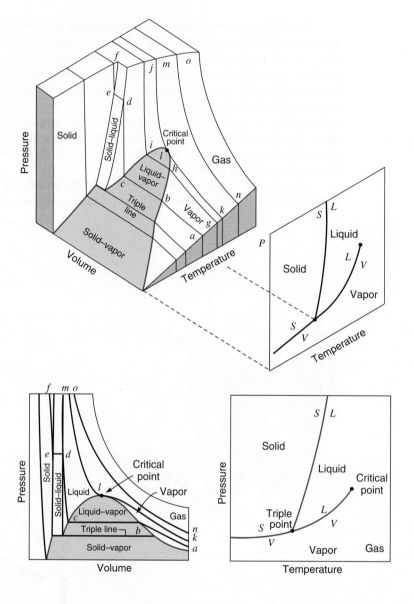

FIGURE 2.8 *P–v–T* surface for a substance that contracts on freezing.

as water that increases in volume during freezing, so the solid surface had a higher specific volume than the liquid surface. Figure 2.8 shows a surface for a substance that decreases in volume upon freezing, a condition that is more common.

As you look at the two three-dimensional surfaces, observe that the *P–T* phase diagram can be seen when looking at the surface parallel with the volume axis; the liquid–vapor surface is flat in that direction, so it collapses to the vaporization curve. The same happens with the solid–vapor surface, which is shown as the sublimation line, and the solid–liquid surface becomes the fusion line. For these three surfaces, it cannot be determined where on the surface a state is by having the (*P*, *T*) coordinates alone. The two properties are not independent; they are a pair of saturated *P* and *T*. A property like specific volume is needed to indicate where on the surface the state is for a given *T* or *P*.

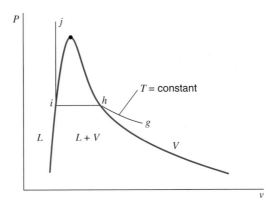

FIGURE 2.9 *P–v* diagram for water showing liquid and vapor phases.

If the surface is viewed from the top down parallel with the pressure axis, we see the *T–v* diagram, a sketch of which was shown in Fig. 2.6 without the complexities associated with the solid phase. Cutting the surface at a constant *P*, *T*, or *v* will leave a trace showing one property as a function of another with the third property constant. One such example is illustrated with the curve *g–h–i–j*, which shows *P* as a function of *v* following a constant *T* curve. This is more clearly indicated in a *P–v* diagram showing the two-phase *L–V* region of the *P–v–T* surface when viewed parallel with the *T* axis, as shown in Fig. 2.9.

Since the three-dimensional surface is fairly complicated, we will indicate processes and states in *P–v*, *T–v*, or *P–T* diagrams to get a visual impression of how the state changes during a process. Of these diagrams, the *P–v* diagram will be particularly useful when we talk about work done during a process in the following chapter.

Looking down at the *P–v–T* surface from above parallel with the pressure axis, the whole surface is visible and not overlapping. That is, for every coordinate pair (*T*, *v*) there is one and only one state on the surface, so *P* is then a unique function of *T* and *v*. This is a general principle that states that for a simple pure substance, the state is defined by two independent properties.

To understand the significance of the term *independent property*, consider the saturated-liquid and saturated-vapor states of a pure substance. These two states have the same pressure and the same temperature, but they are definitely not the same state. In a saturation state, therefore, pressure and temperature are not independent properties. Two independent properties, such as pressure and specific volume or pressure and quality, are required to specify a saturation state of a pure substance.

The reason for mentioning previously that a mixture of gases, such as air, has the same characteristics as a pure substance as long as only one phase is present concerns precisely this point. The state of air, which is a mixture of gases of definite composition, is determined by specifying two properties as long as it remains in the gaseous phase. Air then can be treated as a pure substance.

2.4 TABLES OF THERMODYNAMIC PROPERTIES

Tables of thermodynamic properties of many substances are available, and in general, all these tables have the same form. In this section we will refer to the steam tables. The steam tables are selected both because they are a vehicle for presenting thermodynamic tables and

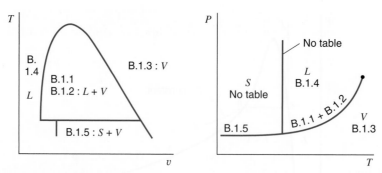

FIGURE 2.10 Listing of the steam tables.

because steam is used extensively in power plants and industrial processes. Once the steam tables are understood, other thermodynamic tables can be readily used.

Several different versions of steam tables have been published over the years. The set included in Table B.1 in Appendix B is a summary based on a complicated fit to the behavior of water. It is very similar to the *Steam Tables* by Keenan, Keyes, Hill, and Moore, published in 1969 and 1978. We will concentrate here on the three properties already discussed in Chapter 1 and in Section 2.3, namely, T, P, and v, and note that the other properties listed in the set of Tables B.1—u, h, and s—will be introduced later.

The steam tables in Appendix B consist of five separate tables, as indicated in Fig. 2.10. The superheated vapor region is given in Table B.1.3, and that of compressed liquid is given in Table B.1.4. The compressed-solid region shown in Fig. 2.4 is not presented in Appendix B. The saturated-liquid and saturated-vapor region, as seen in Figs. 2.6 and 2.9 (and as the vaporization line in Fig. 2.4), is listed according to the values of T in Table B.1.1 and according to the values of P (T and P are not independent in the two-phase regions) in Table B.1.2. Similarly, the saturated-solid and saturated-vapor region is listed according to T in Table B.1.5, but the saturated-solid and saturated-liquid region, the third phase boundary line shown in Fig. 2.4, is not listed in Appendix B.

In Table B.1.1, the first column after the temperature gives the corresponding saturation pressure in kilopascals. The next three columns give the specific volume in cubic meters per kilogram. The first of these columns gives the specific volume of the saturated liquid, v_f; the third column gives the specific volume of the saturated vapor, v_g; and the second column gives the difference between the two, v_{fg}, as defined in Section 2.5. Table B.1.2 lists the same information as Table B.1.1, but the data are listed according to pressure, as mentioned earlier.

Table B.1.5 of the steam tables gives the properties of saturated solid and saturated vapor that are in equilibrium. The first column gives the temperature, and the second column gives the corresponding saturation pressure. As would be expected, all these pressures are less than the triple-point pressure. The next three columns give the specific volume of the saturated solid, v_i, saturated vapor, v_g and the difference v_{ig}.

Appendix B also includes thermodynamic tables for several other substances; refrigerant fluids R-134a and R-410A, ammonia and carbon dioxide, and the cryogenic fluids nitrogen and methane. In each case, only two tables are given: saturated liquid vapor listed by temperature (equivalent to Table B.1.1 for water) and superheated vapor (equivalent to Table B.1.3).

Example 2.1

Determine the phase for each of the following water states using the tables in Appendix B and indicate the relative position in the P–v, T–v, and P–T diagrams.

a. 120 °C, 500 kPa
b. 120 °C, 0.5 m³/kg

Solution

a. Enter Table B.1.1 with 120 °C. The saturation pressure is 198.5 kPa, so we have a compressed liquid, point a in Fig. 2.11. That is above the saturation line for 120 °C. We could also have entered Table B.1.2 with 500 kPa and found the saturation temperature as 151.86 °C, so we would say that it is a subcooled liquid. That is to the left of the saturation line for 500 kPa, as seen in the P–T diagram.

b. Enter Table B.1.1 with 120 °C and notice that

$$v_f = 0.001\ 06 < v < v_g = 0.891\ 86\ \text{m}^3/\text{kg}$$

so the state is a two-phase mixture of liquid and vapor, point b in Fig. 2.11. The state is to the left of the saturated vapor state and to the right of the saturated liquid state, both seen in the T–v diagram.

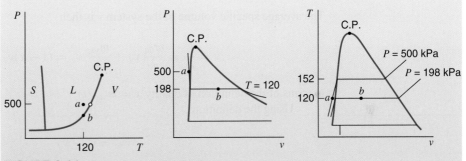

FIGURE 2.11   Diagram for Example 2.1.

2.5 THE TWO-PHASE STATES

The two-phase states have already been shown in the P–v–T diagrams and the corresponding two-dimensional projections of it in the P–T, T–v, and P–v diagrams. Each of these surfaces describes a mixture of substance in two phases, such as a combination of some liquid with some vapor, as shown in Fig. 2.1b. We assume for such mixtures that the two phases are in equilibrium at the same P and T and that each of the masses is in a state of saturated liquid, saturated solid, or saturated vapor, according to which mixture it is. We will treat the liquid–vapor mixture in detail, as this is the most common one in technical applications; the other two-phase mixtures can be treated exactly the same way.

FIGURE 2.12

T–v diagram for the two-phase liquid–vapor region showing the quality–specific volume relation.

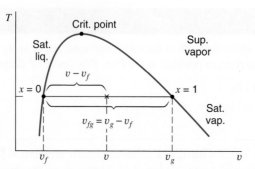

By convention, the subscript f is used to designate a property of a saturated liquid and the subscript g a property of a saturated vapor (the subscript g being used to denote saturation temperature and pressure). Thus, a saturation condition involving part liquid and part vapor, such as that shown in Fig. 2.1b, can be shown on *T–v* coordinates, as in Fig. 2.12. All of the liquid present is at state f with specific volume v_f, and all of the vapor present is at state g with specific volume v_g. The total volume is the sum of the liquid volume and the vapor volume, or

$$V = V_{\text{liq}} + V_{\text{vap}} = m_{\text{liq}} v_f + m_{\text{vap}} v_g$$

The average specific volume of the system v is then

$$v = \frac{V}{m} = \frac{m_{\text{liq}}}{m} v_f + \frac{m_{\text{vap}}}{m} v_g = (1 - x)v_f + x v_g \tag{2.1}$$

in terms of the definition of quality $x = m_{\text{vap}}/m$.

Using the definition

$$v_{fg} = v_g - v_f$$

Eq. 2.1 can also be written as

$$v = v_f + x v_{fg} \tag{2.2}$$

Now the quality x can be viewed as the fraction $(v - v_f)/v_{fg}$ of the distance between saturated liquid and saturated vapor, as indicated in Fig. 2.12.

To illustrate the use of quality, let us find the overall specific volume for a saturated mixture of water at 200 °C and a quality of 70 %. From Table B.1.1 we get the specific volume for the saturated liquid and vapor at 200 °C and then use Eq. 2.1.

$$v = (1 - x)v_f + x\,v_g = 0.3 \times 0.001\,156\ \text{m}^3/\text{kg} + 0.7 \times 0.127\,36\ \text{m}^3/\text{kg}$$

$$= 0.0895\ \text{m}^3/\text{kg}$$

There is no mass of water with that value of specific volume. It represents an average for the two masses, one with a state of $x = 0$ and the other with the state $x = 1$, both shown in Fig. 2.12 as the border points of the two-phase region.

Example 2.2

A closed vessel contains 0.1 m³ of saturated liquid and 0.9 m³ of saturated vapor R-134a in equilibrium at 30 °C. Determine the percent vapor on a mass basis.

Solution

Values of the saturation properties for R-134a are found from Table B.5.1. The mass–volume relations then give

$$V_{\text{liq}} = m_{\text{liq}} v_f, \qquad m_{\text{liq}} = \frac{0.1}{0.000\,843} = 118.6\,\text{kg}$$

$$V_{\text{vap}} = m_{\text{vap}} v_g, \qquad m_{\text{vap}} = \frac{0.9}{0.026\,71} = 33.7\,\text{kg}$$

$$m = 152.3\,\text{kg}$$

$$x = \frac{m_{\text{vap}}}{m} = \frac{33.7}{152.3} = 0.221$$

That is, the vessel contains 90 % vapor by volume but only 22.1 % vapor by mass.

2.6 THE LIQUID AND SOLID STATES

When a liquid has a pressure higher than the saturation pressure (see Fig. 2.3, state *b*) for a given temperature, the state is a compressed liquid state. If we look at the same state but compare it to the saturated liquid state at the same pressure, we notice that the temperature is lower than the saturation temperature; thus, we call it a subcooled liquid. For these liquid states we will use the term compressed liquid in the remainder of the text. Similar to the solid states, the liquid P–v–T surface for lower temperatures is fairly steep and flat, so this region also describes an incompressible substance with a specific volume that is only a weak function of T, which we can write

$$v \approx v(T) = v_f \qquad (2.3)$$

where the saturated liquid specific volume v_f at T is found in the Appendix B tables as the first part of the tables for each substance. A few other entries are found as density (1/v) for some common liquids in Table A.3.

A state with a temperature lower than the saturated temperature for a given pressure on the fusion or the sublimation line gives a solid state that could then also be called a subcooled solid. If for a given temperature the pressure is higher than the saturated sublimation pressure, we have a compressed solid unless the pressure is so high that it exceeds the saturation pressure on the fusion line. This upper limit is seen in Fig. 2.4 for water since the fusion line has a negative slope. This is not the case for most other substances, as in Fig. 2.5, where the fusion line has a positive slope. The properties of a solid are mainly a function of the temperature, as the solid is nearly incompressible, meaning that the pressure cannot change the intermolecular distances and the volume is unaffected by pressure. This is evident from the P–v–T surface for the solid, which is near vertical in Figs. 2.7 and 2.8.

This behavior can be written as an equation

$$v \approx v(T) = v_i \qquad (2.4)$$

with the saturated solid specific volume v_i seen in Table B.1.5 for water. Such a table is not printed for any other substance, but a few entries for density $(1/v)$ are found in Table A.4.

Table B.1.4 gives the properties of the compressed liquid. To demonstrate the use of this table, consider some mass of saturated-liquid water at $100\,°C$. Its properties are given in Table B.1.1, and we note that the pressure is 0.1013 MPa and the specific volume is $0.001\,044\ m^3/kg$. Suppose the pressure is increased to 10 MPa while the temperature is held constant at $100\,°C$ by the necessary transfer of heat, Q. Since water is slightly compressible, we would expect a slight decrease in specific volume during this process. Table B.1.4 gives this specific volume as $0.001\,039\ m^3/kg$. This is only a slight decrease, and only a small error would be made if one assumed that the volume of a compressed liquid is equal to the specific volume of the saturated liquid at the same temperature. In many situations this is the most convenient procedure, particularly when compressed-liquid data are not available. It is very important to note, however, that the specific volume of saturated liquid at the given pressure, 10 MPa, does not give a good approximation. This value, from Table B.1.2, at a temperature of $311.1\,°C$, is $0.001\,452\ m^3/kg$, which is in error by almost 40 %.

2.7 THE SUPERHEATED VAPOR STATES

A state with a pressure lower than the saturated pressure for a given T (see Fig. 2.3, state a) is an expanded vapor or, if compared to the saturated state at the same pressure, it has a higher temperature and thus called superheated vapor. We generally use the latter designation for these states and for states close to the saturated vapor curve. The tables in Appendix B are used to find the properties.

The superheated water vapor properties are arranged in Table B.1.3 as subsections for a given pressure listed in the heading. The properties are shown as a function of temperature along a curve like K–L in Fig. 2.6, starting with the saturation temperature for the given pressure given in parentheses after the pressure. As an example, consider a state of 500 kPa and $200\,°C$, for which we see the boiling temperature of $151.86\,°C$ in the heading, so the state is superheated $48\,°C$ and the specific volume is $0.4249\ m^3/kg$. If the pressure is higher than the critical pressure, as for the curve P–Q in Fig. 2.6, the saturation temperature is not listed. The low-temperature end of the P–Q curve is listed in Table B.1.4, as those states are compressed liquid.

A few examples of the use of the superheated vapor tables, including possible interpolations, are presented below.

Example 2.3

Determine the phase for each of the following states using the tables in Appendix B and indicate the relative position in the P–v, T–v, and P–T diagrams, as in Fig. 2.11.

a. Ammonia 30 °C, 1000 kPa
b. R-134a   200 kPa, 0.125 m^3/kg

Solution ———

a. Enter Table B.2.1 with 30 °C. The saturation pressure is 1167 kPa. As we have a lower P (see Fig. 2.13), it is a superheated vapor state. We could also have entered with 1000 kPa and found a saturation temperature of slightly less than 25 °C, so we have a state that is superheated about 5 °C.

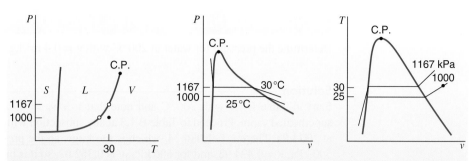

FIGURE 2.13 Diagram for Example 2.3a.

b. Enter Table B.5.2 (or B.5.1) with 200 kPa and notice that

$$v > v_g = 0.1000 \, \text{m}^3/\text{kg}$$

so from the P–v diagram in Fig. 2.14 the state is superheated vapor. We can find the state in Table B.5.2 between 40 °C and 50 °C.

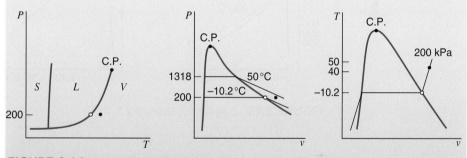

FIGURE 2.14 Diagram for Example 2.3b.

Example 2.4

A rigid vessel contains saturated ammonia vapor at 20 °C. Heat is transferred to the system until the temperature reaches 40 °C. What is the final pressure?

Solution

Since the volume does not change during this process, the specific volume also remains constant. From the ammonia tables, Table B.2.1, we have

$$v_1 = v_2 = 0.149 \, 22 \, \text{m}^3/\text{kg}$$

Since v_g at 40 °C is less than 0.149 22 m³/kg, it is evident that in the final state the ammonia is superheated vapor. By interpolating between the 800 kPa and 1000 kPa columns of Table B.2.2, we find that

$$P_2 = 945 \, \text{kPa}$$

Example 2.5

Determine the pressure for water at 200 °C with $v = 0.4$ m³/kg.

Solution

Start in Table B.1.1 with 200 °C and note that $v > v_g = 0.127\ 36$ m³/kg, so we have superheated vapor. Proceed to Table B.1.3 at any subsection with 200 °C; suppose we start at 200 kPa. There $v = 1.080\ 34$, which is too large, so the pressure must be higher. For 500 kPa, $v = 0.424\ 92$, and for 600 kPa, $v = 0.352\ 02$, so it is bracketed. This is shown in Fig. 2.15.

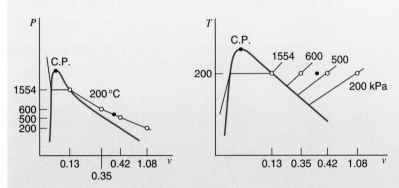

FIGURE 2.15 Diagram for Example 2.5.

A linear interpolation, Fig. 2.16, between the two pressures is done to get P at the desired v.

$$P = 500 + (600 - 500)\frac{0.4 - 0.424\ 92}{0.352\ 02 - 0.424\ 92} = 534.2 \text{ kPa}$$

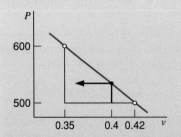

The real constant-T curve is slightly curved and not linear, but for manual interpolation we assume a linear variation.

FIGURE 2.16 Linear interpolation for Example 2.5.

d. Some tools should be cleaned in liquid water at 150 °C. How high a P is needed?

e. Water at 200 kPa has a quality of 50 %. Is the volume fraction V_g/V_{tot} <50 % or >50 %?

f. Why are most of the compressed liquid or solid regions not included in the printed tables?

g. Why is it not typical to find tables for argon, helium, neon, or air in an Appendix B table?

h. What is the percent change in volume as liquid water freezes? Mention some effects the volume change can have in nature and in our households.

2.8 THE IDEAL GAS STATES

Further away from the saturated vapor curve at a given temperature the pressure is lower and the specific volume is higher, so the forces between the molecules are weaker, resulting in a simple correlation among the properties. If we plot the constant P, T, or v curves in the two-dimensional projections of the three-dimensional surfaces, we get curves as shown in Fig. 2.17.

The constant-pressure curve in the T–v diagram and the constant–specific volume curve in the P–T diagram move toward straight lines further out in the superheated vapor region. A second observation is that the lines extend back through the origin, which gives not just a linear relation but one without an offset. This can be expressed mathematically as

$$T = Av \quad \text{for} \quad P = \text{constant} \tag{2.5}$$

$$P = BT \quad \text{for} \quad v = \text{constant} \tag{2.6}$$

The final observation is that the multiplier A increases with P and the multiplier B decreases with v following simple mathematical functions:

$$A = A_0 P \quad \text{and} \quad B = B_0/v$$

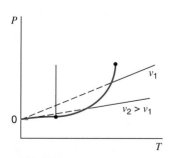
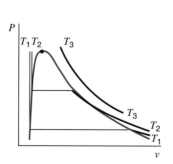
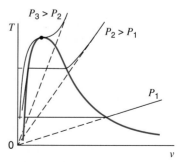

FIGURE 2.17 The isometric, isothermal, and isobaric curves.

Both of these relations are satisfied by the ideal gas equation of state

$$Pv = RT \qquad (2.7)$$

where the constant R is the ideal gas constant and T is the absolute temperature in kelvins or rankines called the ideal gas scale. We will discuss absolute temperature further in Chapter 5, showing that it equals the thermodynamic scale temperature. Comparing different gases gives a further simplification such that R scales with the molecular mass:

$$R = \frac{\overline{R}}{M} \qquad (2.8)$$

In this relation, $\overline{R}$ is the universal gas constant with the value

$$\overline{R} = 8.3145 \frac{\text{kJ}}{\text{kmol·K}}$$

Values of the gas constant for different gases are listed in Table A.5.

The behavior described by the ideal gas law in Eq. 2.7 is very different from the behavior described by similar laws for liquid or solid states, as in Eqs. 2.3 and 2.4. An ideal gas has a specific volume that is very sensitive to both P and T, varying linearly with temperature and inversely with pressure, and the sensitivity to pressure is characteristic of a highly compressible substance. If the temperature is doubled for a given pressure the volume doubles, and if the pressure is doubled for a given temperature the volume is reduced to half the value.

Multiplying Eq. 2.7 by mass gives a scaled version of the ideal gas law as

$$PV = mRT = n\overline{R}T \qquad (2.9)$$

using it on a mass basis or a mole basis, where n is the number of moles:

$$n = m/M \qquad (2.10)$$

Based on the ideal gas law given in Eq. 2.9, it is noticed that one mole of substance occupies the same volume for a given state (P, T) regardless of its molecular mass. Light, small molecules like H_2 take up the same volume as much heavier and larger molecules like R-134a for the same (P, T).

In later applications, we analyze situations with a mass flow rate ($\dot{m}$ in kg/s) entering or leaving the control volume. Having an ideal gas flow with a state (P, T), we can differentiate Eq. 2.9 with time to get

$$P\dot{V} = \dot{m}RT = \dot{n}\overline{R}T \qquad (2.11)$$

Example 2.6

What is the mass of air contained in a room 6 m × 10 m × 4 m if the pressure is 100 kPa and the temperature is 25 °C?

Solution

Assume air to be an ideal gas. By using Eq. 2.9 and the value of R from Table A.5, we have

$$m = \frac{PV}{RT} = \frac{100 \text{ kN/m}^2 \times 240 \text{ m}^3}{0.287 \text{ kN·m/kg·K} \times 298.2 \text{ K}} = 280.5 \text{ kg}$$

Example 2.7

A tank has a volume of 0.5 m³ and contains 10 kg of an ideal gas having a molecular mass of 24. The temperature is 25 °C. What is the pressure?

Solution

The gas constant is determined first:

$$R = \frac{\overline{R}}{M} = \frac{8.3145 \text{ kN·m/kmol·K}}{24 \text{ kg/kmol}}$$

$$= 0.346\,44 \text{ kN·m/kg·K}$$

We now solve for P:

$$P = \frac{mRT}{V} = \frac{10 \text{ kg} \times 0.346\,44 \text{ kN·m/kg·K} \times 298.2 \text{ K}}{0.5 \text{ m}^3}$$

$$= 2066 \text{ kPa}$$

Example 2.8

A gas bell is submerged in liquid water, with its mass counterbalanced with rope and pulleys, as shown in Fig. 2.18. The pressure inside is measured carefully to be 105 kPa, and the temperature is 21 °C. A volume increase is measured to be 0.75 m³ over a period of 185 s. What is the volume flow rate and the mass flow rate of the flow into the bell, assuming it is carbon dioxide gas?

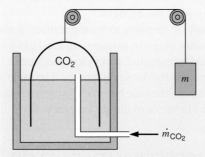

FIGURE 2.18   Sketch for Example 2.8.

Solution

The volume flow rate is

$$\dot{V} = \frac{dV}{dt} = \frac{\Delta V}{\Delta t} = \frac{0.75}{185} = 0.004\,054 \text{ m}^3/\text{s}$$

and the mass flow rate is $\dot{m} = \rho \dot{V} = \dot{V}/v$. At close to room conditions, the carbon dioxide is an ideal gas, so $PV = mRT$ or $v = RT/P$, and from Table A.5 we have the ideal gas constant $R = 0.1889$ kJ/kg·K. The mass flow rate becomes

$$\dot{m} = \frac{P\dot{V}}{RT} = \frac{105 \times 0.004\,054}{0.1889 \times (273.15 + 21)} \frac{\text{kPa·m}^3/\text{s}}{\text{kJ/kg}} = 0.007\,66 \text{ kg/s}$$

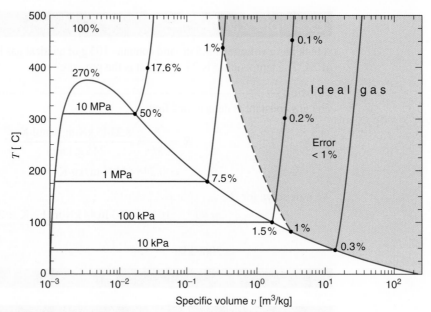

FIGURE 2.19 Temperature-specific volume diagram for water.

Because of its simplicity, the ideal gas equation of state is very convenient to use in thermodynamic calculations. However, two questions are now appropriate. The ideal gas equation of state is a good approximation at low density. But what constitutes low density? In other words, over what range of density will the ideal gas equation of state hold with accuracy? The second question is, how much does an actual gas at a given pressure and temperature deviate from ideal gas behavior?

One specific example in response to these questions is shown in Fig. 2.19, a T–v diagram for water that indicates the error in assuming ideal gas for saturated vapor and for superheated vapor. As would be expected, at very low pressure or high temperature the error is small, but it becomes severe as the density increases. The same general trend would occur in referring to Figs. 2.7 or 2.8. As the state becomes further removed from the saturation region (i.e., high T or low P), the behavior of the gas becomes closer to that of the ideal gas model.

2.9 THE COMPRESSIBILITY FACTOR

A more quantitative study of the question of the ideal gas approximation can be conducted by introducing the compressibility factor Z, defined as

$$Z = \frac{Pv}{RT}$$

or

$$Pv = ZRT \tag{2.12}$$

Note that for an ideal gas $Z = 1$, and the deviation of Z from unity is a measure of the deviation of the actual relation from the ideal gas equation of state.

Figure 2.20 shows a skeleton compressibility chart for nitrogen. From this chart we make three observations. The first is that at all temperatures, $Z \rightarrow 1$ as $P \rightarrow 0$. That is,

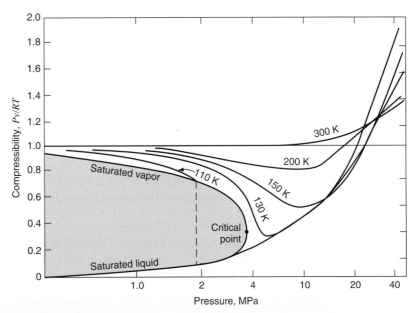

FIGURE 2.20 Compressibility of nitrogen.

as the pressure approaches zero, the *P–v–T* behavior closely approaches that predicted by the ideal gas equation of state. Second, at temperatures of 300 K and above (that is, room temperature and above), the compressibility factor is near unity up to a pressure of about 10 MPa. This means that the ideal gas equation of state can be used for nitrogen (and, as it happens, air) over this range with considerable accuracy.

Third, at lower temperatures or at very high pressures, the compressibility factor deviates significantly from the ideal gas value. Moderate-density forces of attraction tend to pull molecules together, resulting in a value of $Z < 1$, whereas very-high-density forces of repulsion tend to have the opposite effect.

If we examine compressibility diagrams for other pure substances, we find that the diagrams are all similar in the characteristics described above for nitrogen, at least in a qualitative sense. Quantitatively the diagrams are all different, since the critical temperatures and pressures of different substances vary over wide ranges, as indicated by the values listed in Table A.2. Is there a way we can put all of these substances on a common basis? To do so, we "reduce" the properties with respect to the values at the critical point. The reduced properties are defined as

$$\text{reduced pressure} = P_r = \frac{P}{P_c}, \qquad P_c = \text{critical pressure}$$

$$\text{reduced temperature} = T_r = \frac{T}{T_c}, \qquad T_c = \text{critical temperature} \qquad (2.13)$$

These equations state that the reduced property for a given state is the value of this property in this state divided by the value of this same property at the critical point.

If lines of constant T_r are plotted on a Z versus P_r diagram, a plot such as that in Fig. D.1 is obtained. The striking fact is that when such Z versus P_r diagrams are prepared for a number of substances, all of them nearly coincide, especially when the substances have simple, essentially spherical molecules. Correlations for substances with more complicated molecules are reasonably close, except near or at saturation or at high density. Thus, Fig. D.1

is actually a generalized diagram for simple molecules, which means that it represents the average behavior for a number of simple substances. When such a diagram is used for a particular substance, the results will generally be somewhat in error. However, if P–v–T information is required for a substance in a region where no experimental measurements have been made, this generalized compressibility diagram will give reasonably accurate results. We need to know only the critical pressure and critical temperature to use this basic generalized chart.

In our study of thermodynamics, we will use Fig. D.1 primarily to help us decide whether, in a given circumstance, it is reasonable to assume ideal gas behavior as a model. For example, we note from the chart that if the pressure is very low (that is, $<<P_c$), the ideal gas model can be assumed with good accuracy, regardless of the temperature. Furthermore, at high temperatures (that is, greater than about twice T_c), the ideal gas model can be assumed with good accuracy up to pressures as high as four or five times P_c. When the temperature is less than about twice the critical temperature and the pressure is not extremely low, we are in a region, commonly termed the *superheated vapor* region, in which the deviation from ideal gas behavior may be considerable. In this region it is preferable to use tables of thermodynamic properties or charts for a particular substance, as discussed in Section 2.4.

Example 2.9

Is it reasonable to assume ideal gas behavior at each of the given states?

a. Nitrogen at 20 °C, 1.0 MPa

b. Carbon dioxide at 20 °C, 1.0 MPa

c. Ammonia at 20 °C, 1.0 MPa

Solution
In each case, it is first necessary to check phase boundary and critical state data.

a. For nitrogen, the critical properties are, from Table A.2, 126.2 K, 3.39 MPa. Since the given temperature, 293.2 K, is more than twice T_c and the reduced pressure is less than 0.3, ideal gas behavior is a very good assumption.

b. For carbon dioxide, the critical properties are 304.1 K, 7.38 MPa. Therefore, the reduced properties are 0.96 and 0.136. From Fig. D.1, carbon dioxide is a gas (although $T < T_c$) with a Z of about 0.95, so the ideal gas model is accurate to within about 5 % in this case.

c. The ammonia tables, Table B.2, give the most accurate information. From Table B.2.1 at 20 °C, $P_g = 858$ kPa. Since the given pressure of 1 MPa is greater than P_g, this state is a compressed liquid, not a gas.

Example 2.10

Determine the specific volume for R-134a at 100 °C, 3.0 MPa for the following models:

a. The R-134a tables, Table B.5

b. Ideal gas

c. The generalized chart, Fig. D.1

Solution

a. From Table B.5.2 at $100\,°C$, 3 MPa

$$v = 0.006\,65\,\text{m}^3/\text{kg (most accurate value)}$$

b. Assuming ideal gas, we have

$$R = \frac{\overline{R}}{M} = \frac{8.3145}{102.03} = 0.081\,49\,\frac{\text{kJ}}{\text{kg·K}}$$

$$v = \frac{RT}{P} = \frac{0.081\,49 \times 373.2}{3000} = 0.010\,14\,\text{m}^3/\text{kg}$$

which is more than 50 % too large.

c. Using the generalized chart, Fig. D.1, we obtain

$$T_r = \frac{373.2}{374.2} = 1.0, \qquad P_r = \frac{3}{4.06} = 0.74, \qquad Z = 0.67$$

$$v = Z \times \frac{RT}{P} = 0.67 \times 0.010\,14 = 0.006\,79\,\text{m}^3/\text{kg}$$

which is only 2 % too large.

Example 2.11

Propane in a steel bottle of volume 0.1 m³ has a quality of 10 % at a temperature of 15 °C. Use the generalized compressibility chart to estimate the total propane mass and to find the pressure.

Solution

To use Fig. D.1, we need the reduced pressure and temperature. From Table A.2 for propane, $P_c = 4250$ kPa and $T_c = 369.8$ K. The reduced temperature is, from Eq. 2.13,

$$T_r = \frac{T}{T_c} = \frac{273.15 + 15}{369.8} = 0.7792 = 0.78$$

From Fig. D.1, shown in Fig. 2.21, we can read for the saturated states:

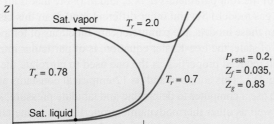

FIGURE 2.21 Diagram for Example 2.11.

For the two-phase state, the pressure is the saturated pressure:

$$P = P_{r\,sat} \times P_c = 0.2 \times 4250\,\text{kPa} = 850\,\text{kPa}$$

The overall compressibility factor becomes, as Eq. 2.1 for v,

$$Z = (1-x)Z_f + xZ_g = 0.9 \times 0.035 + 0.1 \times 0.83 = 0.1145$$

The gas constant from Table A.5 is $R = 0.1886$ kJ/kg·K, so the gas law is Eq. 2.12:

$$PV = mZRT$$

$$m = \frac{PV}{ZRT} = \frac{850 \times 0.1}{0.1145 \times 0.1886 \times 288.15}\,\frac{\text{kPa·m}^3}{\text{kJ/kg}} = 13.66\,\text{kg}$$

In-Text Concept Questions

i. How accurate is it to assume that methane is an ideal gas at room conditions?

j. I want to determine a state of some substance, and I know that $P = 200$ kPa; is it helpful to write $PV = mRT$ to find the second property?

k. A bottle at 298 K should have liquid propane; how high a pressure is needed? (Use Fig. D.1.)

2.10 EQUATIONS OF STATE

Instead of the ideal gas model to represent gas behavior, or even the generalized compressibility chart, which is approximate, it is desirable to have an equation of state that accurately represents the P–v–T behavior for a particular gas over the entire superheated vapor region. Such an equation is necessarily more complicated and consequently more difficult to use. Many such equations have been proposed and used to correlate the observed behavior of gases. As an example, consider the class of relatively simple equation known as *cubic equations of state*

$$P = \frac{RT}{v-b} - \frac{a}{v^2 + cbv + db^2} \tag{2.14}$$

in terms of the four parameters a, b, c, and d. (Note that if all four are zero, this reduces to the ideal gas model.) Several other different models in this class are given in Appendix D. In some of these models, the parameters are functions of temperature. A more complicated equation of state, the Lee-Kesler equation, is of particular interest, since this equation, expressed in reduced properties, is the one used to correlate the generalized compressibility chart, Fig. D.1. This equation and its 12 empirical constants are also given in Appendix D. When we use a computer to determine and tabulate pressure, specific volume, and temperature, as well as other thermodynamic properties, as in the tables in Appendix B, modern equations are much more complicated, often containing 40 or more empirical constants. This subject is discussed in detail in Chapter 12.

2.11 COMPUTERIZED TABLES

Most of the tables in the Appendix are supplied in a computer program on the disk accompanying this book. The main program operates with a visual interface in the Windows environment on a PC-type computer and is generally self-explanatory.

The main program covers the full set of tables for water, refrigerants, and cryogenic fluids, as in Tables B.1 to B.7, including the compressed liquid region, which is printed only for water. For these substances, a small graph with the P–v diagram shows the region around the critical point down toward the triple line covering the compressed liquid, two-phase liquid–vapor, dense fluid, and superheated vapor regions. As a state is selected and the properties are computed, a thin crosshair set of lines indicates the state in the diagram so that this can be seen with a visual impression of the state's location.

Ideal gases corresponding to Table A.7 for air and Tables A.8 or A.9 for other ideal gases are covered. You can select the substance and the units to work in for all the table sections, providing a wider choice than the printed tables. Metric units (SI) or standard English units for the properties can be used, as well as a mass basis (kg or lbm) or a mole basis, satisfying the need for the most common applications.

The generalized chart, Fig. D.1, with the compressibility factor, is included to allow a more accurate value of Z to be obtained than can be read from the graph. This is particularly useful for a two-phase mixture where the saturated liquid and saturated vapor values are needed. Besides the compressibility factor, this part of the program includes correction terms beyond ideal gas approximations for changes in the other thermodynamic properties.

The only mixture application that is included with the program is moist air.

Example 2.12

Find the states in Examples 2.1 and 2.3 with the computer-aided thermodynamics tables (CATT) and list the missing property of P–v–T and x if applicable.

Solution

Water states from Example 2.1: Click Water, click Calculator, and then select Case 1 (T, P). Input $(T, P) = (120, 0.5)$. The result is as shown in Fig. 2.22.

$\Rightarrow$ Compressed liquid $\qquad v = 0.0106\,\text{m}^3/\text{kg}$ (as in Table B.1.4)

Click Calculator and then select Case 2 (T, v). Input $(T, v) = (120, 0.5)$:

$\Rightarrow$ Two-phase $\qquad x = 0.5601, P = 198.5\,\text{kPa}$

Ammonia state from Example 2.3: Click Cryogenics; check that it is ammonia. Otherwise, select Ammonia, click Calculator, and then select Case 1 (T, P). Input $(T, P) = (30, 1)$:

$\Rightarrow$ Superheated vapor $\qquad v = 0.1321\,\text{m}^3/\text{kg}$ (as in Table B.2.2)

R-134a state from Example 2.3: Click Refrigerants; check that it is R-134a. Otherwise, select R-134a (Alt-R), click Calculator, and then select Case 5 (P, v). Input $(P, v) = (0.2, 0.125)$:

$\Rightarrow$ Superheated vapor $\qquad T = 44.0\,^\circ\text{C}$

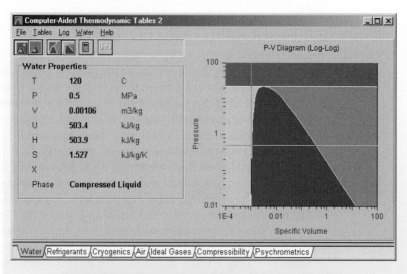

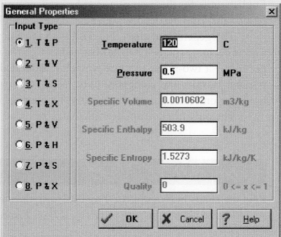

FIGURE 2.22 CATT result for Example 2.1.

In-Text Concept Questions

l. A bottle at 298 K should have liquid propane; how high a pressure is needed? (Use the software.)

2.12 ENGINEERING APPLICATIONS

Information about the phase boundaries is important for storage of substances in a two-phase state like a bottle of gas. The pressure in the container is the saturation pressure for the prevailing temperature, so an estimate of the maximum temperature the system will be

(a) Stainless steel tanks

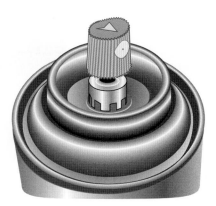
(b) Top of aerosol can

FIGURE 2.23
Storage tanks.

subject to gives the maximum pressure for which the container must be dimensioned (Figs. 2.23 and 2.24).

In a refrigerator a compressor pushes the refrigerant through the system, and this determines the highest fluid pressure. The harder the compressor is driven, the higher the pressure becomes. When the refrigerant condenses, the temperature is determined by the saturation temperature for that pressure, so the system must be designed to hold the temperature and pressure within a desirable range (Fig. 2.25).

The effect of expansion-contraction of matter with temperature is important in many different situations. Two of those are shown in Fig. 2.26; the railroad tracks have small gaps to allow for expansion, which leads to the familiar clunk-clunk sound from the train wheels when they roll over the gap. A bridge may have a finger joint that provides a continuous support surface for automobile tires so that they do not bump, as the train does.

FIGURE 2.24
A tanker to transport liquefied natural gas (LNG), which is mainly methane.

(NATALIA KOLESNIKOVA/AFP//Getty Images, Inc.)

FIGURE 2.25
Household refrigerator
components.

(*a*) Compressor

(*b*) Condenser

(© Victor Maffe/iStockphoto)

(© David R. Frazier Photolibrary, Inc. /Alamy)

FIGURE 2.26
Thermal expansion
joints.

(*a*) Railroad tracks

(*b*) Bridge expansion joint

When air expands at constant pressure, it occupies a larger volume; thus, the density is smaller. This is how a hot air balloon can lift a gondola and people with a total mass equal to the difference in air mass between the hot air inside the balloon and the surrounding colder air; this effect is called *buoyancy* (Fig. 2.27).

FIGURE 2.27
Hot air balloon.

(© ElementalImaging/iStockphoto)

SUMMARY

Thermodynamic properties of a pure substance and the phase boundaries for solid, liquid, and vapor states are discussed. Phase equilibrium for vaporization (boiling liquid to vapor), or the opposite, condensation (vapor to liquid); sublimation (solid to vapor) or the opposite, solidification (vapor to solid); and melting (solid to liquid) or the opposite, solidifying (liquid to solid), should be recognized. The three-dimensional P–v–T surface and the two-dimensional representations in the (P, T), (T, v), and (P, v) diagrams, and the vaporization, sublimation, and fusion lines, are related to the printed tables in Appendix B. Properties from printed and computer tables covering a number of substances are introduced, including two-phase mixtures, for which we use the mass fraction of vapor (quality). The ideal gas law approximates the limiting behavior for low density. An extension of the ideal gas law is shown with the compressibility factor Z, and other, more complicated equations of state are mentioned.

You should have learned a number of skills and acquired abilities from studying this chapter that will allow you to

- Know phases and the nomenclature used for states and interphases.
- Identify a phase given a state (T, P).
- Locate states relative to the critical point and know Tables 2.2 and A.2.
- Recognize phase diagrams and interphase locations.
- Locate states in the Appendix B tables with any entry: (T, P), (T, v), or (P, v).
- Recognize how the tables show parts of the (T, P), (T, v), or (P, v) diagrams.
- Find properties in the two-phase regions; use quality x.
- Locate states using any combination of (T, P, v, x) including linear interpolation.
- Know when you have a liquid or solid and the properties in Tables A.3 and A.4.
- Know when a vapor is an ideal gas (or how to find out).
- Know the ideal gas law and Table A.5.
- Know the compressibility factor Z and the compressibility chart, Fig. D.1.
- Know the existence of more general equations of state.
- Know how to get properties from the computer program.

KEY CONCEPTS AND FORMULAS

Phases	Solid, liquid, and vapor (gas)
Phase equilibrium	T_{sat}, P_{sat}, v_f, v_g, v_i
Multiphase boundaries	Vaporization, sublimation, and fusion lines:
	general (Fig. 2.3), water (Fig. 2.4), and CO_2 (Fig. 2.5)
	Triple point: Table 2.1
	Critical point: Table 2.2, Table A.2
Equilibrium state	Two independent properties (#1, #2)
Quality	$x = m_{vap}/m$ (vapor mass fraction)
	$1 - x = m_{liq}/m$ (liquid mass fraction)
Average specific volume	$v = (1 - x)v_f + xv_g$ (only two-phase mixture)
Equilibrium surface	P–v–T Tables or equation of state
Ideal gas law	$Pv = RT$ $PV = mRT = n\overline{R}T$

Universal gas constant	$\overline{R} = 8.3145 \text{ kJ/kmol·K}$
Gas constant	$R = \overline{R}/M$ kJ/kg·K, Table A.5 or M from Table A.2
Compressibility factor Z	$Pv = ZRT$ Chart for Z in Fig. D.1
Reduced properties	$P_r = \dfrac{P}{P_c}$ $T_r = \dfrac{T}{T_c}$ Entry to compressibility chart
Equations of state	Cubic, pressure explicit: Appendix D, Table D.1
	Lee Kesler: Appendix D, Table D.2, and Fig. D.1

CONCEPT-STUDY GUIDE PROBLEMS

2.1 Are the pressures in the tables absolute or gauge pressures?

2.2 What is the minimum pressure for liquid carbon dioxide?

2.3 When you skate on ice, a thin liquid film forms under the skate; why?

2.4 At higher elevations, as in mountains, air pressure is lower; how does that affect the cooking of food?

2.5 Water at room temperature and room pressure has $v \approx 1 \times 10^n \text{ m}^3/\text{kg}$; what is n?

2.6 Can a vapor exist below the triple point temperature?

2.7 In Example 2.1b, is there any mass at the indicated specific volume? Explain.

2.8 Sketch two constant-pressure curves (500 kPa and 30 000 kPa) in a T–v diagram and indicate on the curves where in the water tables the properties are found.

2.9 If I have 1 L of R-410A at 1 MPa, 20 °C, what is the mass?

2.10 Locate the state of ammonia at 200 kPa, -10 °C. Indicate in both the P–v and T–v diagrams the location of the nearest states listed in Table B.2.

2.11 Why are most compressed liquid or solid regions not included in the printed tables?

2.12 How does a constant-v process for an ideal gas appear in a P–T diagram?

2.13 If $v = RT/P$ for an ideal gas, what is the similar equation for a liquid?

2.14 To solve for v given (P, T) in Eq. 2.14, what is the mathematical problem?

2.15 As the pressure of a gas becomes larger, Z becomes larger than 1. What does that imply?

HOMEWORK PROBLEMS

Phase Diagrams, Triple and Critical Points

2.16 Carbon dioxide at 280 K can be in three different phases: vapor, liquid, and solid. Indicate the pressure range for each phase.

2.17 Modern extraction techniques can be based on dissolving material in supercritical fluids such as carbon dioxide. How high are the pressure and density of carbon dioxide when the pressure and temperature are around the critical point? Repeat for ethyl alcohol.

2.18 The ice cap at the North Pole may be 1000 m thick, with a density of 920 kg/m^3. Find the pressure at the bottom and the corresponding melting temperature.

2.19 Find the lowest temperature at which it is possible to have water in the liquid phase. At what pressure must the liquid exist?

2.20 Water at 27 °C can exist in different phases, depending on the pressure. Give the approximate pressure range in kPa for water in each of the three phases: vapor, liquid, and solid.

2.21 Find the lowest temperature in kelvin for which metal can exist as a liquid if the metal is (a) mercury or (b) zinc.

2.22 A substance is at 2 MPa and 17 °C in a rigid tank. Using only the critical properties, can the phase of the mass be determined if the substance is oxygen, water, or propane?

2.23 Give the phase for the following states:
  a. CO_2 at $T = 40\,°C$ and $P = 0.5$ MPa
  b. Air at $T = 20\,°C$ and $P = 200$ kPa
  c. NH_3 at $T = 170\,°C$ and $P = 600$ kPa

General Tables

2.24 Determine the phase of water at
  a. $T = 260\,°C, P = 5$ MPa
  b. $T = -2\,°C, P = 100$ kPa

2.25 Determine the phase of the substance at the given state using Appendix B tables.
  a. Water: $100\,°C$, 500 kPa
  b. Ammonia: $-10\,°C$, 150 kPa
  c. R-410A: $0\,°C$, 350 kPa

2.26 Give the missing property of P–v–T and x for water at
  a. 10 MPa, 0.001 04 m³/kg
  b. 1 MPa, $190\,°C$
  c. $400\,°C$, 0.19 m³/kg
  d. 10 kPa, $10\,°C$

2.27 For water at 200 kPa with a quality of 10 %, find the volume fraction of vapor.

2.28 Determine whether refrigerant R-410A in each of the following states is a compressed liquid, a superheated vapor, or a mixture of saturated liquid and vapor.
  a. $50\,°C$, 0.05 m³/kg
  b. 1.0 MPa, $20\,°C$
  c. 0.1 MPa, 0.1 m³/kg
  d. $-20\,°C$, 200 kPa

2.29 Show the states in Problem 2.28 in a sketch of the P–v diagram.

2.30 How great is the change in the liquid specific volume for water at $20\,°C$ as you move up from state i toward state j in Fig. 2.14, reaching 15 000 kPa?

2.31 Fill out the following table for substance ammonia:

	P[kPa]	T[°C]	v[m³/kg]	x
a.		20	0.1185	
b.		−20		0.5

2.32 Place the two states a–b listed in Problem 2.31 as labeled dots in a sketch of the P–v and T–v diagrams.

2.33 Give the missing property of P, T, v, and x for R-410A at
  a. $T = -20\,°C, P = 450$ kPa
  b. $P = 300$ kPa, $v = 0.092$ m³/kg

2.34 Determine the specific volume for R-410A at these states:
  a. $-10\,°C$, 500 kPa
  b. $20\,°C$, 1500 kPa
  c. $20\,°C$, quality 15 %

2.35 Give the missing property of P, T, v, and x for CH_4 at
  a. $T = 155$ K, $v = 0.04$ m³/kg
  b. $T = 350$ K, $v = 0.25$ m³/kg

2.36 Give the specific volume of carbon dioxide at $-20\,°C$ for 2000 kPa and for 1400 kPa.

2.37 Calculate the following specific volumes:
  a. Carbon dioxide: $10\,°C$, 80 % quality
  b. Water: 2 MPa, 45 % quality
  c. Nitrogen: 120 K, 60 % quality

2.38 You want a pot of water to boil at $105\,°C$. How heavy a lid should you put on the 15 cm diameter pot when $P_{atm} = 101$ kPa?

2.39 Water at 600 kPa with a quality of 25 % has its pressure raised 50 kPa in a constant-volume process. What is the new quality and temperature?

2.40 A sealed rigid vessel has volume of 1 m³ and contains 2 kg of water at $100\,°C$. The vessel is now heated. If a safety pressure valve is installed, at what pressure should the valve be set to have a maximum temperature of $200\,°C$?

2.41 Saturated water vapor at 200 kPa is in a constant-pressure piston/cylinder assembly. At this state the piston is 0.1 m from the cylinder bottom. How much is this distance, and what is the temperature if the water is cooled to occupy half of the original volume?

2.42 Saturated liquid water at $60\,°C$ is put under pressure to decrease the volume by 1 % while keeping the temperature constant. To what pressure should it be compressed?

2.43 In your refrigerator, the working substance evaporates from liquid to vapor at $-20\,°C$ inside a pipe around the cold section. Outside (on the back or below) is a black grille, inside of which the working substance condenses from vapor to liquid at $+45\,°C$. For each location, find the pressure and the change in specific volume (v) if the substance is ammonia.

2.44 Repeat the previous problem with the substances
  a. R-134a
  b. R-410A

2.45 A glass jar is filled with saturated water at 500 kPa of quality 15 %, and a tight lid is put on. Now it is cooled to −10 °C. What is the mass fraction of solid at this temperature?

2.46 Two tanks are connected as shown in Fig. P2.46, both containing water. Tank A is at 200 kPa, $v = 0.5$ m³/kg, $V_A = 1$ m³, and tank B contains 3.5 kg at 0.5 MPa and 400 °C. The valve is now opened and the two tanks come to a uniform state. Find the final specific volume.

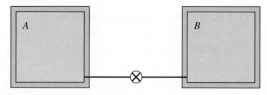

FIGURE P2.46

2.47 Saturated vapor R-410A at 60 °C changes volume at constant temperature. Find the new pressure, and quality if saturated, if the volume doubles. Repeat the problem for the case where the volume is reduced to half of the original volume.

2.48 A steel tank contains 6 kg of propane (liquid + vapor) at 20 °C with a volume of 0.015 m³. The tank is now slowly heated. Will the liquid level inside eventually rise to the top or drop to the bottom of the tank? What if the initial mass is 1 kg instead of 6 kg?

2.49 Saturated water vapor at 60 °C has its pressure decreased to increase the volume by 10 % while keeping the temperature constant. To what pressure should it be expanded?

2.50 Ammonia at 20 °C with a quality of 50 % and a total mass of 2 kg is in a rigid tank with an outlet valve at the top. How much vapor mass can be removed through the valve until the liquid is gone, assuming that the temperature stays constant?

2.51 A sealed, rigid vessel of 2 m³ contains a saturated mixture of liquid and vapor R-134a at 20 °C. If it is heated to 50 °C, the liquid phase disappears. Find the pressure at 50 °C and the initial mass of the liquid.

2.52 A storage tank holds methane at 120 K, with a quality of 25 %, and it warms up by 5 °C per hour due to a failure in the refrigeration system. How much time will it take before the methane becomes single phase, and what is the pressure then?

2.53 A 400 m³ storage tank is being constructed to hold liquified natural gas (LNG), which may be assumed to be essentially pure methane. If the tank is to contain 98 % liquid and 2 % vapor, by volume, at 100 kPa, what mass of LNG (kg) will the tank hold? What is the quality in the tank?

2.54 A piston/cylinder arrangement is loaded with a linear spring and the outside atmosphere. It contains water at 5 MPa, 400 °C, with the volume being 0.1 m³, as shown in Fig. P2.54. If the piston is at the bottom, the spring exerts a force such that $P_{\text{lift}} = 200$ kPa. The system now cools until the pressure reaches 1200 kPa. Find the mass of water and the final state (T_2, v_2) and plot the P–v diagram for the process.

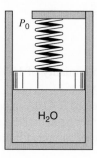

P_0

H_2O

FIGURE P2.54

2.55 A pressure cooker (closed tank) contains water at 100 °C, with the liquid volume being 1/20th of the vapor volume. It is heated until the pressure reaches 2.0 MPa. Find the final temperature. Has the final state more or less vapor than the initial state?

2.56 A pressure cooker has the lid screwed on tight. A small opening with $A = 5$ mm² is covered with a petcock that can be lifted to let steam escape. How much mass should the petcock have to allow boiling at 120 °C with an outside atmosphere at 101.3 kPa?

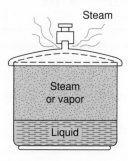

Steam

Steam or vapor

Liquid

FIGURE P2.56

Ideal Gas

2.57 What is the relative (%) change in P if we double the absolute temperature of an ideal gas, keeping the mass and volume constant? Repeat if we double V, keeping m and T constant.

2.58 A 1 m³ tank is filled with a gas at room temperature (20 °C) and pressure (450 kPa). How much mass is there if the gas is (a) air, (b) neon, or (c) propane?

2.59 Calculate the ideal gas constant for argon and hydrogen based on Table A.2 and verify the value with Table A.5.

2.60 A pneumatic cylinder (a piston/cylinder with air) must close a door with a force of 500 N. The cylinder's cross-sectional area is 5 cm². With $V = 50$ cm³, $T = 20$ °C, what is the air pressure and its mass?

2.61 Is it reasonable to assume that at the given states the substance behaves as an ideal gas?
a. Oxygen at 30 °C, 3 MPa
b. Methane at 30 °C, 3 MPa
c. Water at 30 °C, 3 MPa
d. R-134a at 30 °C, 3 MPa
e. R-134a at 30 °C, 100 kPa

2.62 Helium in a steel tank is at 250 kPa, 300 K with a volume of 0.1 m³. It is used to fill a balloon. When the pressure drops to 125 kPa, the flow of helium stops by itself. If all the helium is still at 300 K, how big a balloon is produced?

2.63 A spherical helium balloon 10 m in diameter is at ambient T and P, 20 °C and 100 kPa. How much helium does it contain? It can lift a total mass that equals the mass of displaced atmospheric air. How much mass of the balloon fabric and cage can then be lifted?

2.64 A glass is cleaned in hot water at 45 °C and placed on the table, bottom up. The room air at 20 °C that was trapped in the glass is heated up to 40 °C and some of it leaks out, so the net resulting pressure inside is 2 kPa above the ambient pressure of 101 kPa. Now the glass and the air inside cool down to room temperature. What is the pressure inside the glass?

2.65 Air in an automobile tire is initially at -10 °C and 230 kPa. After the automobile is driven awhile, the temperature rises to 10 °C. Find the new pressure. You must make one assumption on your own.

FIGURE P2.65

2.66 A rigid tank of 1 m³ contains nitrogen gas at 600 kPa, 400 K. By mistake, someone lets 0.5 kg flow out. If the final temperature is 375 K, what is the final pressure?

2.67 Assume we have three states of saturated vapor R-134a at $+40$ °C, 0 °C, and -40 °C. Calculate the specific volume at the set of temperatures and corresponding saturated pressure, assuming ideal gas behavior. Find the percent relative error $= 100(v - v_g)/v_g$ with v_g from the saturated R-134a table.

2.68 Do Problem 2.67 for R-410A.

2.69 A 1 m³ rigid tank has propane at 125 kPa, 300 K and connected by a valve to another tank of 0.5 m³ with propane at 250 kPa, 400 K. The valve is opened, and the two tanks come to a uniform state at 325 K. What is the final pressure?

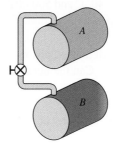

FIGURE P2.69

2.70 A 1 m³ rigid tank with air at 1 MPa and 400 K is connected to an air line as shown in Fig. P2.70. The valve is opened and air flows into the tank until the pressure reaches 5 MPa, at which point the valve is closed and the temperature inside is 450 K.
a. What is the mass of air in the tank before and after the process?
b. The tank eventually cools to room temperature, 300 K. What is the pressure inside the tank then?

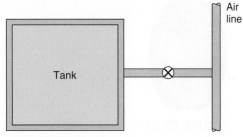

FIGURE P2.70

2.71 A cylindrical gas tank 1 m long, with an inside diameter of 30 cm, is evacuated and then filled with carbon dioxide gas at 20 °C. To what pressure should it be charged if there is 2 kg of carbon dioxide?

Compressibility Factor

2.72 Find the compressibility factor (Z) for saturated vapor ammonia at 100 kPa and at 2000 kPa.

2.73 Find the compressibility factor for nitrogen at
a. 2000 kPa, 120 K
b. 2000 kPa, 300 K
c. 120 K, $v = 0.005$ m^3/kg

2.74 Find the compressibility for carbon dioxide at 40 °C and 10 MPa using Fig. D.1.

2.75 What is the percent error in specific volume if the ideal gas model is used to represent the behavior of superheated ammonia at 40 °C and 500 kPa? What if the generalized compressibility chart, Fig. D.1, is used instead?

2.76 A cylinder fitted with a frictionless piston contains butane at 25 °C, 500 kPa. Can the butane reasonably be assumed to behave as an ideal gas at this state?

2.77 Estimate the saturation pressure of chlorine at 300 K.

2.78 A bottle with a volume of 0.1 m^3 contains butane with a quality of 75 % and a temperature of 300 K. Estimate the total butane mass in the bottle using the generalized compressibility chart.

2.79 Find the volume of 2 kg of ethylene at 270 K, 2500 kPa using Z from Fig. D.1.

2.80 For $T_r = 0.7$, what is the ratio v_g/v_f using Fig. D.1 compared to Table D.3?

2.81 Refrigerant R-32 is at −10 °C with a quality of 15 %. Find the pressure and specific volume.

2.82 To plan a commercial refrigeration system using R-123, we would like to know how much more volume saturated vapor R-123 occupies per kg at −30 °C compared to the saturated liquid state.

2.83 A new refrigerant, R-125, is stored as a liquid at −20 °C with a small amount of vapor. For 1.5 kg of R-125, find the pressure and volume.

Equations of State

For these problems, see Appendix D for the equation of state (EOS) and Chapter 12.

2.84 Determine the pressure of nitrogen at 160 K, $v = 0.002\,91$ m^3/kg using ideal gas, the van der Waals EOS, and the nitrogen table.

2.85 Determine the pressure of nitrogen at 160 K, $v = 0.002\,91$ m^3/kg using the Redlich-Kwong EOS and the nitrogen table.

2.86 Determine the pressure of nitrogen at 160 K, $v = 0.002\,91$ m^3/kg using the Soave EOS and the nitrogen table.

2.87 Carbon dioxide at 60 °C is pumped at a very high pressure, 10 MPa, into an oil well to reduce the viscosity of oil for better flow. Find its specific volume from the carbon dioxide table, ideal gas, and van der Waals EOS by iteration.

2.88 Solve the previous problem using the Redlich-Kwong EOS. Notice that this becomes a trial-and-error process.

2.89 Solve Problem 2.87 using the Soave EOS. Notice that this becomes a trial-and-error process.

2.90 A tank contains 8.35 kg of methane in 0.1 m^3 at 250 K. Find the pressure using ideal gas, the van der Waals EOS, and the methane table.

2.91 Do the previous problem using the Redlich-Kwong EOS.

2.92 Do Problem 2.90 using the Soave EOS.

Review Problems

2.93 Determine the quality (if saturated) or temperature (if superheated) of the following substances at the given two states:
a. Water at
   1: 120 °C, 1 m^3/kg; 2: 10 MPa, 0.01 m^3/kg
b. Nitrogen at
   1: 1 MPa, 0.03 m^3/kg; 2: 100 K, 0.03 m^3/kg

2.94 Give the phase and the missing properties of P, T, v, and x for

a. R-410A at $10\,°C$ with $v = 0.01$ m³/kg

b. Water at $T = 350\,°C$ with $v = 0.2$ m³/kg

c. R-410A at $5\,°C$ and $P = 1000$ kPa

d. R-134a at 294 kPa and $v = 0.05$ m³/kg

2.95 Give the phase and the missing properties of P, T, v, and x. These may be a little more difficult to determine if the appendix tables are used instead of the software.

a. R-410A, $T = 20\,°C$, $v = 0.03$ m³/kg

b. H₂O, $v = 0.2$ m³/kg, $x = 0.5$

c. H₂O, $T = 60\,°C$, $v = 0.001\,016$ m³/kg

d. NH₃, $T = -10\,°C$, $P = 60$ kPa

e. R-134a, $v = 0.005$ m³/kg, $x = 0.5$

2.96 The refrigerant R-410A in a piston/cylinder arrangement is initially at $15\,°C$ with $x = 1$. It is then expanded in a process so that $P = Cv^{-1}$ to a pressure of 200 kPa. Find the final temperature and specific volume.

2.97 Consider two tanks, A and B, connected by a valve, as shown in Fig. P2.97. Each has a volume of 200 L, and tank A has R-410A at $25\,°C$, 10 % liquid and 90 % vapor by volume, while tank B is evacuated. The valve is now opened, and saturated vapor flows from A to B until the pressure in B has reached that in A, at which point the valve is closed. This process occurs slowly such that all temperatures stay at $25\,°C$ throughout the process. How much has the quality changed in tank A during the process?

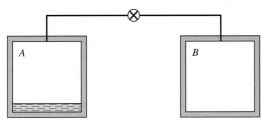

FIGURE P2.97

2.98 A tank contains 2 kg of nitrogen at 100 K with a quality of 50 %. Through a volume flowmeter and valve, 0.5 kg is now removed while the temperature remains constant. Find the final state inside the tank and the volume of nitrogen removed if the valve/meter is located at

a. the top of the tank

b. the bottom of the tank

2.99 A spring-loaded piston/cylinder assembly contains water at $500\,°C$ and 3 MPa. The setup is such that pressure is proportional to volume, $P = CV$. It

is now cooled until the water becomes saturated vapor. Sketch the P–v diagram and find the final pressure.

2.100 For a certain experiment, R-410A vapor is contained in a sealed glass tube at $20\,°C$. We want to know the pressure at this condition, but there is no means of measuring it, since the tube is sealed. However, if the tube is cooled to $-10\,°C$, small droplets of liquid are observed on the glass walls. What is the initial pressure?

2.101 A cylinder/piston arrangement contains water at $105\,°C$, 85 % quality, with a volume of 1 L. The system is heated, causing the piston to rise and encounter a linear spring, as shown in Fig. P2.101. At this point the volume is 1.5 L, the piston diameter is 150 mm, and the spring constant is 100 N/mm. The heating continues, so the piston compresses the spring. What is the cylinder temperature when the pressure reaches 200 kPa?

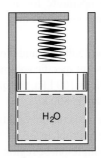

FIGURE P2.101

2.102 Determine the mass of methane gas stored in a 2 m³ tank at $-30\,°C$, 1.5 MPa. Estimate the percent error in the mass determination if the ideal gas model is used.

2.103 A cylinder containing ammonia is fitted with a piston restrained by an external force that is proportional to the cylinder volume squared. Initial conditions are $10\,°C$, 90 % quality, and a volume of 5 L. A valve on the cylinder is opened and additional ammonia flows into the cylinder until the mass inside has doubled. If at this point the pressure is 1.2 MPa, what is the final temperature?

2.104 What is the percent error in pressure if the ideal gas model is used to represent the behavior of superheated vapor R-410A at $60\,°C$, $0.034\,70$ m³/kg? What if the generalized compressibility chart, Fig. D.1, is used instead? (Note that iterations are needed.)

Linear Interpolation

2.105 Find the pressure and temperature for saturated vapor R-410A with $v = 0.1$ m³/kg.

2.106 Use a linear interpolation to estimate T_{sat} at 900 kPa for nitrogen. Sketch by hand the curve $P_{sat}(T)$ by using a few table entries around 900 kPa from Table B.6.1. Is your linear interpolation above or below the actual curve?

2.107 Use a double linear interpolation to find the pressure for superheated R-134a at 13 °C with $v = 0.3$ m³/kg.

2.108 Find the specific volume for carbon dioxide at 0 °C and 625 kPa.

Computer Tables

2.109 Use the computer software to find the properties for ammonia at the two states listed in Problem 2.31.

2.110 Find the value of the saturated temperature for nitrogen by linear interpolation in Table B.6.1 for a pressure of 900 kPa. Compare this to the value given by the computer software.

2.111 Use the computer software to sketch the variation of pressure with temperature in Problem 2.40. Extend the curve slightly into the single-phase region.

COMPUTER, DESIGN, AND OPEN-ENDED PROBLEMS

2.112 Make a spreadsheet that will tabulate and plot saturated pressure versus temperature for ammonia starting at $T = -40$ °C and ending at the critical point in steps of 10 °C.

2.113 Make a spreadsheet that will tabulate and plot values of P and T along a constant specific volume line for water. The starting state is 100 kPa, the quality is 50 %, and the ending state is 800 kPa.

2.114 Use the computer software to sketch the variation of pressure with temperature in Problem 2.52. Extend the curve a little into the single-phase region.

2.115 In Problem 2.96, follow the path of the process for the R-410A for any state between the initial and final states inside the cylinder.

2.116 As the atmospheric temperature and pressure vary with elevation, the air density varies and the pressure is therefore not linear in elevation, as it is in a liquid. Develop an expression for the pressure variation containing an integral over an expression containing T. *Hint*: Start with Eq. 1.2 in differential form and use the ideal gas law, assuming we know the temperature as a function of elevation $T(h)$.

2.117 Saturated pressure as a function of temperature follows the correlation developed by Wagner as

$$\ln P_r = [w_1 \tau + w_2 \tau^{1.5} + w_3 \tau^3 + w_4 \tau^6]/T_r$$

where the reduced pressure and temperature are $P_r = P/P_c$ and $T_r = T/T_c$. The temperature variable is $\tau = 1 - T_r$. The parameters are found for R-134a as

	w_1	w_2	w_3	w_4
R-134a	−7.598 84	1.488 86	−3.798 73	1.813 79

Compare this correlation to the table in Appendix B.

2.118 Find the constants in the curve fit for the saturation pressure using Wagner's correlation, as shown in the previous problem for water and methane. Find other correlations in the literature, compare them to the tables, and give the maximum deviation.

2.119 The specific volume of saturated liquid can be approximated by the Rackett equation as

$$v_f = \frac{\overline{R} T_c}{M P_c} Z_c^n; \quad n = 1 + (1 - T_r)^{2/7}$$

with the reduced temperature, $T_r = T/T_c$, and the compressibility factor, $Z_c = P_c v_c / R T_c$. Using values from Table A.2 with the critical constants, compare the formula to the tables for substances where the saturated specific volume is available.

First Law of Thermodynamics and Energy Equation

Having completed our review of basic definitions and concepts, we are ready to discuss the *first law of thermodynamics* and the *energy equation*. These are alternative expressions for the same fundamental physical law. Later we will see the actual difference in the expression of the first law and the energy equation and recognize that they are consistent with one another. Our procedure will be to state the energy equation for a system (control mass) undergoing a process with a change of state of the system with time. We then look at the same law expressed for a complete cycle and recognize the first law of thermodynamics, which is the historically first formulation of the law.

After the energy equation is formulated, we will use it to relate the change of state inside a control volume to the amount of energy that is transferred in a process as work or heat transfer. As a car engine transfers some work to the car, the car's speed increases, and we can relate the kinetic energy increase to the work; or, if a stove provides a certain amount of heat transfer to a pot with water, we can relate the water temperature increase to the heat transfer. More complicated processes can also occur, such as the expansion of very hot gases in a piston cylinder, as in a car engine, in which work is given out and at the same time, heat is transferred to the colder walls. In other applications, we can also see a change in state without any work or heat transfer, such as a falling object that changes kinetic energy at the same time it is changing elevation. For all cases, the energy equation relates the various forms of energy of the control mass to the transfers of energy by heat or work.

3.1 THE ENERGY EQUATION

In Chapter 1 we discussed the energy associated with a substance and its thermodynamic state, which was called the internal energy U and included some additional energy forms, such as kinetic and potential energies. The combination is the *total energy E*, which we wrote as

$$E = me = U + \text{KE} + \text{PE} = m(u + ke + pe) \qquad (3.1)$$

showing that all terms scale with total mass, so *u*, *ke*, and *pe* are specific energies.

Before proceeding with development of the energy equation with the analysis and examples, let us look at the various terms of the total energy. The total energy is written with the kinetic energy and the potential energy associated with the gravitational field as

$$E = mu + \frac{1}{2}m\mathbf{V}^2 + mgZ \tag{3.2}$$

and in a process it is possible to see changes in any of the energy forms. A ball rolling up a hill will slow down as it gains height, thus lowering the kinetic energy and increasing the potential energy during the process, which is a simple energy conversion process. The kinetic and potential energies are associated with the physical state and location of the mass and generally are labeled mechanical energy to distinguish them from the internal energy, which is characteristic of the thermodynamic state of the mass and thus is labeled thermal energy.

For a control volume with constant mass, a control mass, we express the *conservation of energy* as a basic physical principle in a mathematical equation. This principle states that you cannot create or destroy energy within the limits of classical physics. This limitation means that quantum mechanical effects, which would change the energy associated with a change in mass, are ignored, as well as relativity, so we assume that any velocity is significantly smaller than the speed of light. From this we deduce that if the control mass has a change in energy, the change must be due to an energy transfer into or out of the mass. Such energy transfers are not related to any mass transfer (we look at a control mass), and they can only occur as work or heat transfers. Writing this as an instantaneous rate process, we get

$$\frac{dE_{cv}}{dt} = \dot{E}_{cv} = \dot{Q} - \dot{W} = +\text{in} - \text{out} \tag{3.3}$$

where the sign convention follows the historical development counting heat transfer as positive in and work positive out of the control volume, as illustrated in Fig. 3.1. Notice that the sign convention is a choice, and in more complicated systems you may decide differently; the important concept to understand is that Eq. 3.3 and Fig. 3.1 belong together, so if an arrow in the figure changes direction, the corresponding sign in the equation switches. This equation gives the rate of change of the stored total energy as equal to the rate at which energy is added minus the rate at which energy is removed. Net changes in storage are explained by the transfers on the right-hand side of the equation, and there can be no other explanation. Notice that the transfers must come from or go to the surroundings of the control volume and thus affect the storage in the surroundings in the opposite direction compared to the control volume. A process can move energy from one place to another, but it cannot change the total energy.

In many cases, we are interested in finite changes from the beginning to the end of a process and not in focusing on the instantaneous rate at which the process takes place. For these cases, we integrate the energy equation, Eq. 3.3, with time from the beginning of the process t_1 to the end of the process t_2 by multiplying by dt to get

$$dE_{cv} = dU + d(\text{KE}) + d(\text{PE}) = \delta Q - \delta W \tag{3.4}$$

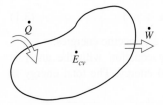

FIGURE 3.1 Sign convention for energy terms.

and integrating

$$\int dE_{cv} = E(t_2) - E(t_1) = E_2 - E_1$$

Now the right-hand side terms are integrated as

$$\int \left[\dot{Q} - \dot{W} \right] dt = \int_{\text{path}} \delta Q - \int_{\text{path}} \delta W = {}_1Q_2 - {}_1W_2$$

Here the integration depends not only on the starting and ending states but also on the process path in between; thus, δQ is used instead of dQ to indicate an inexact differential. Subscripts are placed to indicate the difference, so E_1 refers to the total energy for the control volume at state 1 and thus to only a function of the state. However, ${}_1Q_2$ indicates the cumulative (integrated) heat transfer during the process, which is a function not only of states 1 and 2 but also of the path the process followed; the same applies to the work term ${}_1W_2$. Section 3.4 discusses the integration of the work and heat transfer terms in detail to further explain this process. The energy equation for finite changes become

$$E_2 - E_1 = {}_1Q_2 - {}_1W_2 \tag{3.5}$$

accompanied by

$$E_2 - E_1 = U_2 - U_1 + \frac{1}{2}m(\mathbf{V}_2^2 - \mathbf{V}_1^2) + mg(Z_2 - Z_1)$$

In general, we will refer to both equations Eq. 3.3 and Eq. 3.5 as the energy equation, depending on the analysis: whether we want the rate form or the form with the finite changes. This is similar to expressing a salary for work as pay per hour or a finite amount over a specified time period, such as monthly or yearly. Both versions of the energy equation can be shown as

$$\text{Change of storage} = +\text{in} - \text{out}$$

which is a basic balance equation accounting for the changes, such as those in a bank account. If you make a deposit, the balance goes up (an "in" term); if you make a withdrawal, the balance goes down (an "out" term). Similar equations are presented in subsequent chapters for other quantities, such as mass, momentum, and entropy.

To illustrate the connection between the sketch of the real system and the energy equation, look at Fig. 3.2. For this control volume, the energy equation on a rate form, Eq. 3.3 is

$$\dot{E}_{cv} = \dot{E}_A + \dot{E}_B + \dot{E}_C = \dot{Q}_A + \dot{Q}_C - \dot{W}_B \tag{3.6}$$

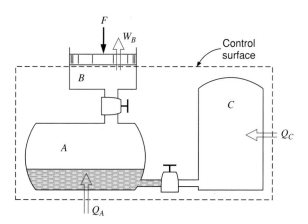

FIGURE 3.2 A control volume with several different subsystems.

and the conservation of mass becomes

$$\dot{m}_{cv} = \dot{m}_A + \dot{m}_B + \dot{m}_C = 0 \tag{3.7}$$

Each of the three energy storage terms is then written out as in Eq. 3.2 to include the different kinds of energy that could be in containers A, B, and C. The form for the finite changes corresponding to Eq. 3.5 now has a nontrivial expression for the conservation of mass together with the energy equation as

$$m_2 - m_1 = (m_{2A} + m_{2B} + m_{2C}) - (m_{1A} + m_{1B} + m_{1C}) = 0 \tag{3.8}$$

$$(E_{2A} + E_{2B} + E_{2C}) - (E_{1A} + E_{1B} + E_{1C}) = {}_1Q_{2A} + {}_1Q_{2C} - {}_1W_{2B} \tag{3.9}$$

Total mass is not changed; however, the distribution between the A, B, and C subdomains may have changed during the process, so if one has a mass increase, the others have a matching decrease in mass. The same applies to the energy, with the added effect that total energy is changed by the heat and work transferred across the control volume boundary.

Example 3.1

A tank containing a fluid is stirred by a paddle wheel. The work input to the paddle wheel is 5090 kJ. The heat transfer from the tank is 1500 kJ. Consider the tank and the fluid inside a control surface and determine the change in internal energy of this control mass.

The energy equation is (Eq. 3.5)

$$U_2 - U_1 + \frac{1}{2}m(\mathbf{V}_2^2 - \mathbf{V}_1^2) + mg(Z_2 - Z_1) = {}_1Q_2 - {}_1W_2$$

Since there is no change in kinetic and potential energy, this reduces to

$$U_2 - U_1 = {}_1Q_2 - {}_1W_2$$

$$U_2 - U_1 = -1500 - (-5090) = 3590 \text{ kJ}$$

3.2 THE FIRST LAW OF THERMODYNAMICS

Consider a control mass where the substance inside goes through a cycle. This can be the water in the steam power plant in Fig. 1.2 or a substance in a piston/cylinder, as in Fig. 1.6, going through several processes that are repeated. As the substance returns to its original state, there is no net change in the control volume's total energy and the rate of change is thus zero. The net sum of the right-hand-side terms gives the energy equation as

$$0 = \oint \delta Q - \oint \delta W \tag{3.10}$$

The symbol $\oint \delta Q$, which is called the cyclic integral of the heat transfer, represents the net heat transfer in the cycle and $\oint \delta W$ is the cyclic integral of the work representing the net

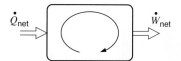

FIGURE 3.3 A cyclic machine.

work given out during the cycle. Rewriting the equation as

$$\oint \delta Q = \oint \delta W \tag{3.11}$$

gives the statement of the first law of thermodynamics. Whereas this was shown as a consequence of the energy equation, historically this was postulated first and the energy equation was derived from it. The above equations could also be written with rates and shown as in Fig. 3.3, where the integrals imply the summation over all the boundaries of the control volume as

$$\text{Cycle:} \quad \dot{Q}_{\text{net in}} = \dot{W}_{\text{net out}} \tag{3.12}$$

This equation was originally stated for heat engines, where the purpose is to get some work output with a heat input from a source that explains the traditional sign convention for heat and work. Modern applications include heat pumps and refrigerators, where the work is the driving input and the heat transfer is net out. One can therefore characterize all cycles as energy conversion devices; the energy is conserved, but it comes out in a form different from that of the input. Further discussion of such cycles is provided in Chapter 5, and the details of the cycles are presented in Chapters 9 and 10.

Before we can apply the energy equation or the first law of thermodynamics, we need to elaborate on the work and heat transfer terms as well as the internal energy.

3.3 THE DEFINITION OF WORK

The classical definition of work is mechanical work defined as a force F acting through a displacement x, so incrementally

$$\delta W = F\, dx$$

and the finite work becomes

$$_1W_2 = \int_1^2 F\, dx \tag{3.13}$$

To evaluate the work, it is necessary to know the force F as a function of x. In this section, we show examples with physical arrangements that lead to simple evaluations of the force, so the integration is straightforward. Real systems can be very complex, and some mathematical examples will be shown without a mechanical explanation.

Work is energy in transfer and thus crosses the control volume boundary. In addition to mechanical work by a single point force, it can be a rotating shaft, as in a car's transmission system; electrical work, as from a battery or a power outlet; or chemical work, to mention a few other possibilities. Look at Fig. 3.4, with a simple system of a battery, a motor, and a pulley. Depending upon the choice of control volume, the work crossing the surface, as in

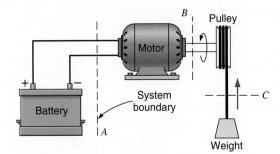

FIGURE 3.4 Example of work crossing the boundary of a system.

sections A, B, or C, can be electrical through the wires, mechanical by a rotating shaft out of the motor, or a force from the rope on the pulley.

The potential energy expressed in Eq. 3.2 comes from the energy exchanged with the gravitational field as a mass changes elevation. Consider the weight in Fig. 3.4, initially at rest and held at some height measured from a reference level. If the pulley now slowly turns, raising the weight, we have a force and a displacement expressed as

$$F = ma = mg$$

$$\delta W = -F\,dZ = -d\,\mathrm{PE}$$

with the negative sign as work goes in to raise the weight. Then we get

$$d\,\mathrm{PE} = F\,dZ = mg\,dZ$$

and integration gives

$$\int_{\mathrm{PE}_1}^{\mathrm{PE}_2} d\,\mathrm{PE} = m\int_{Z_1}^{Z_2} g\,dZ$$

Assuming that g does not vary with Z (which is a very reasonable assumption for moderate changes in elevation), we obtain

$$\mathrm{PE}_2 - \mathrm{PE}_1 = mg(Z_2 - Z_1) \tag{3.14}$$

When the potential energy is included in the total energy, as in Eq. 3.2, the gravitational force is not included in work calculated from Eq. 3.13. The other energy term in the energy equation is the kinetic energy of the control mass, which is generated from a force applied to the mass. Consider the horizontal motion of a mass initially at rest to which we apply a force F in the x direction. Assume that we have no heat transfer and no change in internal energy. The energy equation, Eq. 3.4, will then become

$$\delta W = -F\,dx = -d\,\mathrm{KE}$$

But

$$F = ma = m\frac{d\mathbf{V}}{dt} = m\frac{dx}{dt}\frac{d\mathbf{V}}{dx} = m\mathbf{V}\frac{d\mathbf{V}}{dx}$$

Then

$$d\,\mathrm{KE} = F\,dx = m\mathbf{V}\,d\mathbf{V}$$

Integrating, we obtain

$$\int_{KE=0}^{KE} d\,KE = \int_{\mathbf{V}=0}^{\mathbf{V}} m\,\mathbf{V}\,d\mathbf{V}$$

$$KE = \frac{1}{2}m\mathbf{V}^2$$

(3.15)

Units of Work

Our definition of work involves the product of a unit force (one newton) acting through a unit distance (one meter). This unit for work in SI units is called the joule (J).

$$1\,J = 1\,N\,m$$

Power is the time rate of doing work and is designated by the symbol $\dot{W}$:

$$\dot{W} \equiv \frac{\delta W}{dt}$$

The unit for power is a rate of work of one joule per second, which is a watt (W):

$$1\,W = 1\,J/s$$

A commonly used unit for power is the horsepower (hp), where

$$1\,hp = 0.7355\,kW$$

Note that the work crossing the boundary of the system in Fig. 3.4 is that associated with a rotating shaft. To derive the expression for power, we use the differential work

$$\delta W = F\,dx = Fr\,d\theta = T\,d\theta$$

that is, force acting through a distance dx or a torque ($T = Fr$) acting through an angle of rotation, as shown in Fig. 3.5. Now the power becomes

$$\dot{W} = \frac{\delta W}{dt} = F\frac{dx}{dt} = F\mathbf{V} = Fr\frac{d\theta}{dt} = T\omega$$

(3.16)

that is, force times rate of displacement (velocity) or torque times angular velocity.

It is often convenient to speak of the work per unit mass of the system, often termed *specific work*. This quantity is designated w and is defined as

$$w \equiv \frac{W}{m}$$

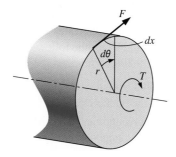

FIGURE 3.5 Force acting at radius r gives a torque $T = F_r$.

Example 3.2

A car of mass 1100 kg drives with a velocity such that it has a kinetic energy of 400 kJ (see Fig. 3.6). Find the velocity. If the car is raised with a crane, how high should it be lifted in the standard gravitational field to have a potential energy that equals the kinetic energy?

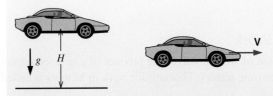

FIGURE 3.6 Sketch for Example 3.2.

Solution

The standard kinetic energy of the mass is

$$\mathrm{KE} = \frac{1}{2}m\mathbf{V}^2 = 400 \, \mathrm{kJ}$$

From this we can solve for the velocity:

$$\mathbf{V} = \sqrt{\frac{2 \, \mathrm{KE}}{m}} = \sqrt{\frac{2 \times 400 \, \mathrm{kJ}}{1100 \, \mathrm{kg}}}$$

$$= \sqrt{\frac{800 \times 1000 \, \mathrm{N \cdot m}}{1100 \, \mathrm{kg}}} = \sqrt{\frac{8000 \, \mathrm{kg \cdot m \, s^{-2} \, m}}{11 \, \mathrm{kg}}} = 27 \, \mathrm{m/s}$$

Standard potential energy is

$$\mathrm{PE} = mgH$$

so when this is equal to the kinetic energy, we get

$$H = \frac{\mathrm{KE}}{mg} = \frac{400\,000 \, \mathrm{N \cdot m}}{1100 \, \mathrm{kg} \times 9.807 \, \mathrm{m \, s^{-2}}} = 37.1 \, \mathrm{m}$$

Notice the necessity of converting the kJ to J in both calculations.

Example 3.3

Consider a stone having a mass of 10 kg and a bucket containing 100 kg of liquid water. Initially the stone is 10.2 m above the water, and the stone and the water are at the same temperature, state 1. The stone then falls into the water.

Determine ΔU, $\Delta\mathrm{KE}$, $\Delta\mathrm{PE}$, Q, and W for the following changes of state, assuming standard gravitational acceleration of 9.806 65 m/s^2.

a. The stone is about to enter the water, state 2.

b. The stone has just come to rest in the bucket, state 3.

c. Heat has been transferred to the surroundings in such an amount that the stone and water are at the same temperature, T_1, state 4.

Analysis and Solution

The energy equation for any of the steps is

$$\Delta U + \Delta KE + \Delta PE = Q - W$$

and each term can be identified for each of the changes of state.

a. The stone has fallen from Z_1 to Z_2, and we assume no heat transfer as it falls. The water has not changed state; thus

$$\Delta U = 0, \qquad {}_1Q_2 = 0, \qquad {}_1W_2 = 0$$

and the first law reduces to

$$\Delta KE + \Delta PE = 0$$

$$\Delta KE = -\Delta PE = -mg(Z_2 - Z_1)$$

$$= -10\,\text{kg} \times 9.806\,65\,\text{m/s}^2 \times (-10.2\,\text{m})$$

$$= 1000\,\text{J} = 1\,\text{kJ}$$

That is, for the process from state 1 to state 2,

$$\Delta KE = 1\,\text{kJ} \quad \text{and} \quad \Delta PE = -1\,\text{kJ}$$

b. For the process from state 2 to state 3 with zero kinetic energy, we have

$$\Delta PE = 0, \qquad {}_2Q_3 = 0, \qquad {}_2W_3 = 0$$

Then

$$\Delta U + \Delta KE = 0$$

$$\Delta U = -\Delta KE = 1\,\text{kJ}$$

c. In the final state, there is no kinetic or potential energy, and the internal energy is the same as in state 1.

$$\Delta U = -1\,\text{kJ}, \qquad \Delta KE = 0, \qquad \Delta PE = 0, \qquad {}_3W_4 = 0$$

$${}_3Q_4 = \Delta U = -1\,\text{kJ}$$

In-Text Concept Questions

a. In a complete cycle, what is the net change in energy and in volume?

b. Explain in words what happens with the energy terms for the stone in Example 3.3. What would happen if the object was a bouncing ball falling to a hard surface?

c. Make a list of at least five systems that store energy, explaining which form of energy is involved.

d. A constant mass goes through a process in which 100 J of heat transfer comes in and 100 J of work leaves. Does the mass change state?

e. The electric company charges the customers per kW·hour. What is that in SI units?

f. Torque, energy, and work have the same units (Nm). Explain the difference.

3.4 WORK DONE AT THE MOVING BOUNDARY OF A SIMPLE COMPRESSIBLE SYSTEM

We have already noted that there are a variety of ways in which work can be done on or by a system. These include work done by a rotating shaft, electrical work, and work done by the movement of the system boundary, such as the work done in moving the piston in a cylinder. In this section, we will consider in some detail the work done at the moving boundary of a simple compressible system during a quasi-equilibrium process.

Consider as a system the gas contained in a cylinder and piston, as in Fig. 3.7. Remove one of the small weights from the piston, which will cause the piston to move upward a distance dL. We can consider this quasi-equilibrium process and calculate the amount of work W done by the system during this process. The total force on the piston is PA, where P is the pressure of the gas and A is the area of the piston. Therefore, the work δW is

$$\delta W = P A \, dL$$

But $A \, dL = dV$, the change in volume of the gas. Therefore,

$$\delta W = P \, dV \qquad (3.17)$$

The work done at the moving boundary during a given quasi-equilibrium process can be found by integrating Eq. 3.17. However, this integration can be performed only if we know the relationship between P and V during this process. This relationship may be expressed as an equation or it may be shown as a graph.

Let us consider a graphical solution first. We use as an example a compression process such as occurs during the compression of air in a cylinder, Fig. 3.8. At the beginning of the process the piston is at position 1, and the pressure is relatively low. This state is represented on a pressure–volume diagram (usually referred to as a *P–V diagram*). At the conclusion of the process the piston is in position 2, and the corresponding state of the gas is shown at point 2 on the *P–V* diagram. Let us assume that this compression was a quasi-equilibrium process and that during the process the system passed through the states shown by the line connecting states 1 and 2 on the *P–V* diagram. The assumption of a quasi-equilibrium process is essential here because each point on line 1–2 represents a definite state, and these states correspond to the actual state of the system only if the deviation from equilibrium is

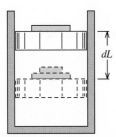

FIGURE 3.7

Example of work done at the moving boundary of a system in a quasi-equilibrium process.

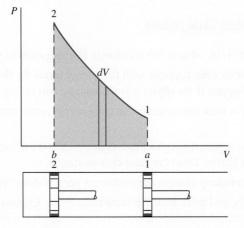

FIGURE 3.8 Use of a *P–V* diagram to show work done at the moving boundary of a system in a quasi-equilibrium process.

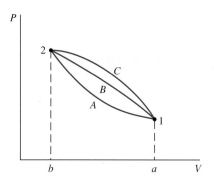

FIGURE 3.9 Various quasi-equilibrium processes between two given states, indicating that work is a path function.

infinitesimal. The work done on the air during this compression process can be found by integrating Eq. 3.17:

$$_1W_2 = \int_1^2 \delta W = \int_1^2 P\, dV \tag{3.18}$$

The symbol $_1W_2$ is to be interpreted as the work done during the process from state 1 to state 2. It is clear from the P–V diagram that the work done during this process,

$$\int_1^2 P\, dV$$

is represented by the area under curve 1–2, area a–1–2–b–a. In this example, the volume decreased, and area a–1–2–b–a represents work done on the system. If the process had proceeded from state 2 to state 1 along the same path, the same area would represent work done by the system.

Further consideration of a P–V diagram, such as Fig. 3.9, leads to another important conclusion. It is possible to go from state 1 to state 2 along many different quasi-equilibrium paths, such as A, B, or C. Since the area under each curve represents the work for each process, the amount of work done during each process not only is a function of the end states of the process but also depends on the path followed in going from one state to another. For this reason, work is called a *path function* or, in mathematical parlance, δW is an inexact differential.

This concept leads to a brief consideration of point and path functions or, to use other terms, *exact* and *inexact differentials*. Thermodynamic properties are *point functions*, a name that comes from the fact that for a given point on a diagram (such as Fig. 3.9) or surface (such as Fig. 2.7) the state is fixed, and thus there is a definite value for each property corresponding to this point. The differentials of point functions are exact differentials, and the integration is simply

$$\int_1^2 dV = V_2 - V_1$$

Thus, we can speak of the volume in state 2 and the volume in state 1, and the change in volume depends only on the initial and final states.

Work, however, is a path function, for, as has been indicated, the work done in a quasi-equilibrium process between two given states depends on the path followed. The differentials of path functions are inexact differentials, and the symbol δ will be used in this book to designate inexact differentials (in contrast to d for exact differentials). Thus, for work, we write

$$\int_1^2 \delta W = {}_1W_2$$

It would be more precise to use the notation ${}_1W_{2A}$, which would indicate the work done during the change from state 1 to state 2 along path A. However, the notation ${}_1W_2$ implies that the process between states 1 and 2 has been specified. Note that we never speak about the work in the system in state 1 or state 2, and thus we never write $W_2 - W_1$.

So far, we have discussed boundary movement work in a quasi-equilibrium process. We should also realize that there may very well be boundary movement work in a nonequilibrium process. Then the total force exerted on the piston by the gas inside the cylinder, PA, does not equal the external force, F_{ext}, and the work is not given by Eq. 3.17. The work can, however, be evaluated in terms of F_{ext} or, dividing by area, an equivalent external pressure, P_{ext}. The work done at the moving boundary in this case is

$$\delta W = F_{\text{ext}}\, dL = P_{\text{ext}}\, dV \qquad (3.19)$$

Evaluation of Eq. 3.19 in any particular instance requires a knowledge of how the external force or pressure changes during the process. For this reason, the integral in Eq. 3.18 is often called the indicated work.

Example 3.4

Consider a slightly different piston/cylinder arrangement, as shown in Fig. 3.10. In this example, the piston is loaded with a mass m_p, the outside atmosphere P_0, a linear spring, and a single point force F_1. The piston traps the gas inside with a pressure P. A force balance on the piston in the direction of motion yields

$$m_p a \cong 0 = \sum F_\uparrow - \sum F_\downarrow$$

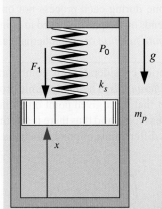

FIGURE 3.10 Sketch of the physical system for Example 3.4.

with a zero acceleration in a quasi-equilibrium process. The forces, when the spring is in contact with the piston, are

$$\sum F_\uparrow = PA, \qquad \sum F_\downarrow = m_p g + P_0 A + k_s(x - x_0) + F_1$$

with the linear spring constant, k_s. The piston position for a relaxed spring is x_0, which depends on how the spring is installed. The force balance then gives the gas pressure by division with area A as

$$P = P_0 + [m_p g + F_1 + k_s(x - x_0)]/A$$

To illustrate the process in a P–V diagram, the distance x is converted to volume by division and multiplication with A:

$$P = P_0 + \frac{m_p g}{A} + \frac{F_1}{A} + \frac{k_s}{A^2}(V - V_0) = C_1 + C_2 V$$

This relation gives the pressure as a linear function of the volume, with the line having a slope of $C_2 = k_s/A^2$. Possible values of P and V are as shown in Fig. 3.11 for an expansion. Regardless of what substance is inside, any process must proceed along the line in the P–V diagram. The work term in a quasi-equilibrium process then follows as

$$_1W_2 = \int_1^2 P\,dV = \text{area under the process curve}$$

$$_1W_2 = \frac{1}{2}(P_1 + P_2)(V_2 - V_1)$$

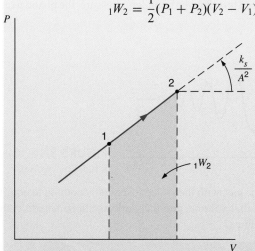

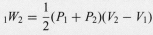

FIGURE 3.11 The process curve showing possible P–V combinations for Example 3.4.

For a contraction instead of an expansion, the process would proceed in the opposite direction from the initial point 1 along a line of the same slope shown in Fig. 3.11.

Example 3.5

Consider the system shown in Fig. 3.12, in which the piston of mass m_p is initially held in place by a pin. The gas inside the cylinder is initially at pressure P_1 and volume V_1. When the pin is released, the external force per unit area acting on the system (gas) boundary is comprised of two parts:

$$P_{\text{ext}} = F_{\text{ext}}/A = P_0 + m_p g/A$$

Calculate the work done by the system when the piston has come to rest.

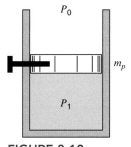

FIGURE 3.12
Example of a nonequilibrium process.

After the piston is released, the system is exposed to the boundary pressure equal to P_{ext}, which dictates the pressure inside the system, as discussed in Section 1.7 in connection with Fig. 1.10. We further note that neither of the two components of this external force will change with a boundary movement, since the cylinder is vertical (gravitational force) and the top is open to the ambient surroundings (movement upward merely pushes the air out of the way). If the initial pressure P_1 is greater than that resisting the boundary, the piston will move upward at a finite rate, that is, in a nonequilibrium process, with the cylinder pressure eventually coming to equilibrium at the value P_{ext}. If we were able to trace the average cylinder pressure as a function of time, it would typically behave as shown in Fig. 3.13. However, the work done by the system during this process is done against the force resisting the boundary movement and is therefore given by Eq. 3.19. Also, since the external force is constant during this process, the result is

$$_1W_2 = \int_1^2 P_{ext}\, dV = P_{ext}(V_2 - V_1)$$

where V_2 is greater than V_1, and the work done by the system is positive. If the initial pressure had been less than the boundary pressure, the piston would have moved downward,

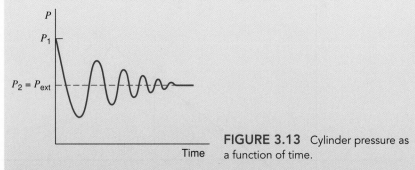

FIGURE 3.13 Cylinder pressure as a function of time.

compressing the gas, with the system eventually coming to equilibrium at P_{ext}, at a volume less than the initial volume, and the work would be negative, that is, done on the system by its surroundings.

The work term can be examined by measuring the pressure and volume during a process from which we can evaluate the integral in Eq. 3.14. Using curve fitting and numerical methods, we can estimate the work term as the area below the process curve in the P–V diagram. However, it is also useful if the whole process curve can be approximated with an analytical function. In that case, the integration can be done mathematically, knowing the values of the parameters in the function. For this purpose, a simple mathematical form of the curve called a polytropic process has been used, with just two parameters, an exponent and a constant, as

$$PV^n = \text{constant} \tag{3.20}$$

The polytropic exponent n is indicative of the type of process, and it can vary from minus to plus infinity. Several simple processes fall into this class of functions. For instance, for $n = 0$ we have a constant pressure process, and for the limits of $n \rightarrow \pm\infty$ we have a constant volume process. For this process equation, we calculate the integral in Eq. 3.14 as

$$PV^n = \text{constant} = P_1 V_1^n = P_2 V_2^n$$

$$P = \frac{\text{constant}}{V^n} = \frac{P_1 V_1^n}{V^n} = \frac{P_2 V_2^n}{V^n}$$

$$\int_1^2 P\, dV = \text{constant} \int_1^2 \frac{dV}{V^n} = \text{constant} \left(\frac{V^{-n+1}}{-n+1} \right) \Bigg|_1^2$$

$$\int_1^2 P\, dV = \frac{\text{constant}}{1-n} \left(V_2^{1-n} - V_1^{1-n} \right) = \frac{P_2 V_2^n V_2^{1-n} - P_1 V_1^n V_1^{1-n}}{1-n}$$

$$= \frac{P_2 V_2 - P_1 V_1}{1-n} \tag{3.21}$$

Note that the resulting equation, Eq. 3.21, is valid for any exponent n except $n = 1$. Where $n = 1$,

$$PV = \text{constant} = P_1 V_1 = P_2 V_2$$

and

$$\int_1^2 P\, dV = P_1 V_1 \int_1^2 \frac{dV}{V} = P_1 V_1 \ln \frac{V_2}{V_1} \tag{3.22}$$

Note that in Eqs. 3.21 and 3.22, we did not say that the work is equal to the expressions given in these equations. These expressions give us the value of a certain integral, that is, a mathematical result. Whether or not that integral equals the work in a particular process depends on the result of a thermodynamic analysis of that process. It is important to keep the mathematical result separate from the thermodynamic analysis, for there are many situations in which work is not given by Eq. 3.18.

The polytropic process as described demonstrates one special functional relationship between P and V during a process. There are many other possible relations, some of which will be examined in the problems at the end of this chapter.

Example 3.6

Consider as a system the gas in the cylinder shown in Fig. 3.14; the cylinder is fitted with a piston on which a number of small weights are placed. The initial pressure is 200 kPa, and the initial volume of the gas is 0.04 m^3.

a. Let a Bunsen burner be placed under the cylinder, and let the volume of the gas increase to 0.1 m^3 while the pressure remains constant. Calculate the work done by the system during this process.

$$_1 W_2 = \int_1^2 P\, dV$$

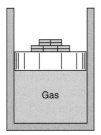

Gas

FIGURE 3.14
Sketch for
Example 3.6.

Since the pressure is constant, we conclude from Eq. 3.18 and Eq. 3.21 with $n = 0$ that

$$_1W_2 = P\int_1^2 dV = P(V_2 - V_1)$$

$$_1W_2 = 200\,\text{kPa} \times (0.1 - 0.04)\text{m}^3 = 12.0\,\text{kJ}$$

b. Consider the same system and initial conditions, but at the same time that the Bunsen burner is under the cylinder and the piston is rising. Remove weights from the piston at such a rate that, during the process, the temperature of the gas remains constant.

   If we assume that the ideal-gas model is valid, then, from Eq. 2.9,

$$PV = mRT$$

We note that this is a polytropic process with exponent $n = 1$. From our analysis, we conclude that the work is given by Eq. 3.18 and that the integral in this equation is given by Eq. 3.22. Therefore,

$$_1W_2 = \int_1^2 P\,dV = P_1V_1 \ln \frac{V_2}{V_1}$$

$$= 200\,\text{kPa} \times 0.04\,\text{m}^3 \times \ln \frac{0.10}{0.04} = 7.33\,\text{kJ}$$

c. Consider the same system, but during the heat transfer remove the weights at such a rate that the expression $PV^{1.3} = \text{constant}$ describes the relation between pressure and volume during the process. Again, the final volume is $0.1\,\text{m}^3$. Calculate the work.

   This is a polytropic process in which $n = 1.3$. Analyzing the process, we conclude again that the work is given by Eq. 3.18 and that the integral is given by Eq. 3.21. Therefore,

$$P_2 = 200\left(\frac{0.04}{0.10}\right)^{1.3} = 60.77\,\text{kPa}$$

$$_1W_2 = \int_1^2 P\,dV = \frac{P_2V_2 - P_1V_1}{1 - 1.3} = \frac{60.77 \times 0.1 - 200 \times 0.04}{1 - 1.3}\,\text{kPa·m}^3$$

$$= 6.41\,\text{kJ}$$

d. Consider the system and the initial state given in the first three examples, but let the piston be held by a pin so that the volume remains constant. In addition, let heat be transferred from the system until the pressure drops to 100 kPa. Calculate the work.

   Since $\delta W = P\,dV$ for a quasi-equilibrium process, the work is zero, because there is no change in volume. This can also be viewed as a limit of a polytropic process for $n \rightarrow \infty$, and thus Eq. 3.21 gives zero work.

   The process for each of the four examples is shown on the P–V diagram of Fig. 3.15. Process 1–2a is a constant-pressure process, and area 1–2a–f–e–1 represents the work. Similarly, line 1–2b represents the process in which $PV = \text{constant}$, line 1–2c the process in which $PV^{1.3} = \text{constant}$, and line 1–2d the constant-volume process. The student should compare the relative areas under each curve with the numerical results obtained for the amounts of work done.

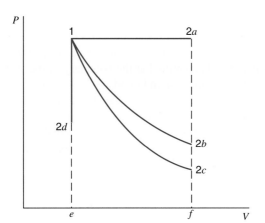

FIGURE 3.15 *P–V* diagram showing work done in the various processes of Example 3.6.

In-Text Concept Questions

g. What is roughly the relative magnitude of the work in process 1–2*c* versus process 1–2*a* shown in Fig. 3.15?

h. Helium gas expands from 125 kPa, 350 K and from 0.25 m³ to 100 kPa in a polytropic process with $n = 1.667$. Is the work positive, negative, or zero?

i. An ideal gas goes through an expansion process in which the volume doubles. Which process will lead to the larger work output: an isothermal process or a polytropic process with $n = 1.25$?

3.5 DEFINITION OF HEAT

The thermodynamic definition of heat is somewhat different from the everyday understanding of the word. It is essential to understand clearly the definition of heat given here, because it plays a part in many thermodynamic problems.

If a block of hot copper is placed in a beaker of cold water, we know from experience that the block of copper cools down and the water warms up until the copper and water reach the same temperature. What causes this decrease in the temperature of the copper and the increase in the temperature of the water? We say that it is the result of the transfer of energy from the copper block to the water. It is from such a transfer of energy that we arrive at a definition of heat.

Heat is defined as the form of energy that is transferred across the boundary of a system at a given temperature to another system (or the surroundings) at a lower temperature by virtue of the temperature difference between the two systems. That is, heat is transferred from the system at the higher temperature to the system at the lower temperature, and the heat transfer occurs solely because of the temperature difference between the two systems. Another aspect of this definition of heat is that a body never contains heat. Rather, heat can be identified only as it crosses the boundary. Thus, heat is a transient phenomenon. If we consider the hot block of copper as one system and the cold water in the beaker as another system, we recognize that originally neither system contains any heat (they do contain energy, of course). When the copper block is placed in the water and the two are in thermal communication, heat is transferred from the copper to the water until equilibrium of temperature is established. At this point we no longer have heat transfer, because there is no temperature difference. Neither system contains heat at the conclusion of the process. It also follows that heat is identified at the boundary of the system, for heat is defined as energy transferred across the system boundary.

Heat, like work, is a form of energy transfer to or from a system. Therefore, the units for heat, and for any other form of energy as well, are the same as the units for work, or at least are directly proportional to them. In the International System, the unit for heat (energy) is the joule. The calorie is defined as the amount of heat required to raise 1 g of water from 14.5 °C to 15.5 °C.

Heat transferred *to* a system is considered positive, and heat transferred *from* a system is considered negative. Thus, positive heat represents energy transferred to a system, and negative heat represents energy transferred from a system. The symbol Q represents heat. A process in which there is no heat transfer ($Q = 0$) is called an adiabatic process.

From a mathematical perspective, heat, like work, is a path function and is recognized as an inexact differential. That is, the amount of heat transferred when a system undergoes a change from state 1 to state 2 depends on the path that the system follows during the change of state. Since heat is an inexact differential, the differential is written as δQ. On integrating, we write

$$\int_1^2 \delta Q = {}_1Q_2$$

In words, ${}_1Q_2$ is the heat transferred during the given process between states 1 and 2.

It is also convenient to speak of the heat transfer per unit mass of the system, q, often termed specific heat transfer, which is defined as

$$q \equiv \frac{Q}{m}$$

3.6 HEAT TRANSFER MODES

Heat transfer is the transport of energy due to a temperature difference between different amounts of matter. We know that an ice cube taken out of the freezer will melt when it is placed in a warmer environment such as a glass of liquid water or on a plate with room air around it. From the discussion about energy in Section 1.8, we realize that molecules of matter have translational (kinetic), rotational, and vibrational energy. Energy in these modes can be transmitted to the nearby molecules by interactions (collisions) or by exchange of molecules such that energy is emitted by molecules that have more on average (higher temperature) to those that have less on average (lower temperature). This energy exchange between molecules is heat transfer by conduction, and it increases with the temperature difference and the ability of the substance to make the transfer. This is expressed in Fourier's law of conduction,

$$\dot{Q} = -kA\frac{dT}{dx} \tag{3.23}$$

giving the rate of heat transfer as proportional to the conductivity, k, the total area, A, and the temperature gradient. The minus sign indicates the direction of the heat transfer from a higher-temperature to a lower-temperature region. Often the gradient is evaluated as a temperature difference divided by a distance when an estimate has to be made if a mathematical or numerical solution is not available.

Values of conductivity, k, are on the order of 100 W/m·K for metals, 1 to 10 for nonmetallic solids as glass, ice, and rock, 0.1 to 10 for liquids, around 0.1 for insulation materials, and 0.1 down to less than 0.01 for gases.

A different mode of heat transfer takes place when a medium is flowing, called convective heat transfer. In this mode, the bulk motion of a substance moves matter with a certain energy level over or near a surface with a different temperature. Now the heat transfer by conduction is dominated by the manner in which the bulk motion brings the two substances in contact or close proximity. Examples are the wind blowing over a building or flow through heat exchangers, which can be air flowing over/through a radiator with water flowing inside the radiator piping. The overall heat transfer is typically correlated with Newton's law of cooling as

$$\dot{Q} = Ah\,\Delta T \qquad (3.24)$$

where the transfer properties are lumped into the heat transfer coefficient, h, which then becomes a function of the media properties, the flow and geometry. A more detailed study of fluid mechanics and heat transfer aspects of the overall process is necessary to evaluate the heat transfer coefficient for a given situation.

Typical values for the convection coefficient (all in W/m^2·K) are:

Natural convection	$h = 5$–25, gas	$h = 50$–1000, liquid
Forced convection	$h = 25$–250, gas	$h = 50$–20 000, liquid
Boiling phase change	$h = 2500$–100 000	

The final mode of heat transfer is radiation, which transmits energy as electromagnetic waves in space. The transfer can happen in empty space and does not require any matter, but the emission (generation) of the radiation and the absorption do require a substance to be present. Surface emission is usually written as a fraction, emissivity ε, of a perfect black body emission as

$$\dot{Q} = \varepsilon\sigma A T_s^4 \qquad (3.25)$$

with the surface temperature, T_s, and the Stefan-Boltzmann constant, σ. Typical values of emissivity range from 0.92 for nonmetallic surfaces to 0.6 to 0.9 for nonpolished metallic surfaces to less than 0.1 for highly polished metallic surfaces. Radiation is distributed over a range of wavelengths and it is emitted and absorbed differently for different surfaces, but such a description is beyond the scope of this book.

Example 3.7

Consider the constant transfer of energy from a warm room at 20 °C inside a house to the colder ambient temperature of −10 °C through a single-pane window, as shown in Fig. 3.16.

The temperature variation with distance from the outside glass surface is shown by an outside convection heat transfer layer, but no such layer is inside the room (as a simplification). The glass pane has a thickness of 5 mm (0.005 m) with a conductivity of 1.4 W/m·K and a total surface area of 0.5 m^2. The outside wind is blowing, so the convective heat transfer coefficient is 100 W/m^2·K. With an outer glass surface temperature of 12.1 °C, we would like to know the rate of heat transfer in the glass and the convective layer.

For the conduction through the glass, we have

$$\dot{Q} = -kA\frac{dT}{dx} = -kA\frac{\Delta T}{\Delta x} = -1.4\frac{\text{W}}{\text{m}\cdot\text{K}} \times 0.5\ \text{m}^2\frac{20 - 12.1}{0.005}\frac{\text{K}}{\text{m}} = -1106\ \text{W}$$

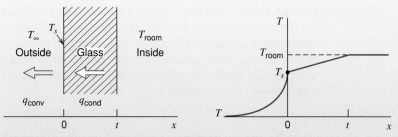

FIGURE 3.16 Conduction and convection heat transfer through a window pane.

and the negative sign shows that energy is leaving the room. For the outside convection layer, we have

$$\dot{Q} = hA\ \Delta T = 100\frac{\text{W}}{\text{m}^2\cdot\text{K}} \times 0.5\ \text{m}^2[12.1 - (-10)]\ \text{K} = 1105\ \text{W}$$

with a direction from the higher to the lower temperature, that is, toward the outside.

3.7 INTERNAL ENERGY—A THERMODYNAMIC PROPERTY

Internal energy is an extensive property because it depends on the mass of the system. Kinetic and potential energies are also extensive properties.

The symbol U designates the internal energy of a given mass of a substance. Following the convention used with other extensive properties, the symbol u designates the internal energy per unit mass. We could speak of u as the specific internal energy, as we do with specific volume. However, because the context will usually make it clear whether u or U is referred to, we will use the term *internal energy* to refer to both internal energy per unit mass and the total internal energy.

In Chapter 2 we noted that in the absence of motion, gravity, surface effects, electricity, or other effects, the state of a pure substance is specified by two independent properties. It is very significant that, with these restrictions, the internal energy may be one of the independent properties of a pure substance. This means, for example, that if we specify the pressure and internal energy (with reference to an arbitrary base) of superheated steam, the temperature is also specified.

Thus, in tables of thermodynamic properties such as the steam tables, the value of internal energy can be tabulated along with other thermodynamic properties. Tables 1 and 2 of the steam tables (Tables B.1.1 and B.1.2) list the internal energy for saturated states.

Included are the internal energy of saturated liquid u_f, the internal energy of saturated vapor u_g, and the difference between the internal energy of saturated liquid and saturated vapor u_{fg}. The values are given in relation to an arbitrarily assumed reference state, which, for water in the steam tables, is taken as zero for saturated liquid at the triple-point temperature, $0.01\,^\circ$C. All values of internal energy in the steam tables are then calculated relative to this reference (note that the reference state cancels out when finding a difference in u between any two states). Values for internal energy are found in the steam tables in the same manner as for specific volume. In the liquid–vapor saturation region,

$$U = U_{\text{liq}} + U_{\text{vap}}$$

or

$$mu = m_{\text{liq}}u_f + m_{\text{vap}}u_g$$

Dividing by m and introducing the quality x gives

$$u = (1 - x)u_f + xu_g$$

$$u = u_f + xu_{fg}$$

As an example, the specific internal energy of saturated steam having a pressure of 0.6 MPa and a quality of 95 % can be calculated as

$$u = u_f + xu_{fg} = 669.9 + 0.95(1897.5) = 2472.5 \text{ kJ/kg}$$

Values for u in the superheated vapor region are tabulated in Table B.1.3, for compressed liquid in Table B.1.4, and for solid–vapor in Table B.1.5.

Example 3.8

Determine the missing property (P, T, or x) and v for water at each of the following states:

a. $T = 300\,^\circ$C, $u = 2780$ kJ/kg
b. $P = 2000$ kPa, $u = 2000$ kJ/kg

For each case, the two properties given are independent properties and therefore fix the state. For each, we must first determine the phase by comparison of the given information with phase boundary values.

a. At $300\,^\circ$C, from Table B.1.1, $u_g = 2563.0$ kJ/kg. The given $u > u_g$, so the state is in the superheated vapor region at some P less than P_g, which is 8581 kPa. Searching through Table B.1.3 at $300\,^\circ$C, we find that the value $u = 2780$ is between given values of u at 1600 kPa (2781.0) and 1800 kPa (2776.8). Interpolating linearly, we obtain

$$P = 1648 \text{ kPa}$$

Note that quality is undefined in the superheated vapor region. At this pressure, by linear interpolation, we have $v = 0.1542$ m^3/kg.

b. At $P = 2000$ kPa, from Table B.1.2, the given u of 2000 kJ/kg is greater than u_f (906.4) but less than u_g (2600.3). Therefore, this state is in the two-phase region with $T = T_g = 212.4\,^\circ$C, and

$$u = 2000 = 906.4 + x\,1693.8, \qquad x = 0.6456$$

Then,

$$v = 0.001\,177 + 0.6456 \times 0.098\,45 = 0.064\,74 \text{ m}^3/\text{kg}.$$

In-Text Concept Questions

j. Water is heated from 100 kPa, 20 °C to 1000 kPa, 200 °C. In one case, pressure is raised at $T = C$; then T is raised at $P = C$. In a second case, the opposite order is used. Does that make a difference for $_1Q_2$ and $_1W_2$?

k. A rigid insulated tank A contains water at 400 kPa, 800 °C. A pipe and valve connect this to another rigid insulated tank B of equal volume having saturated water vapor at 100 kPa. The valve is opened and stays open while the water in the two tanks comes to a uniform final state. Which two properties determine the final state?

3.8 PROBLEM ANALYSIS AND SOLUTION TECHNIQUE

At this point in our study of thermodynamics, we have progressed sufficiently far (that is, we have accumulated sufficient tools with which to work) that it is worthwhile to develop a somewhat formal technique or procedure for analyzing and solving thermodynamic problems. For the time being, it may not seem entirely necessary to use such a rigorous procedure for many of our problems, but we should keep in mind that as we acquire more analytical tools, the problems that we are capable of dealing with will become much more complicated. Thus, it is appropriate that we begin to practice this technique now in anticipation of these future problems.

The following steps show a systematic formulation of thermodynamics problems so that it can be understood by others and it ensures that no shortcuts are taken, thus eliminating many errors that otherwise occur due to oversight of basic assumptions that may not apply.

1. Make a sketch of the physical system with components and illustrate all mass flows, heat flows, and work rates. Include an indication of forces like external pressures and single-point forces.

2. Define (i.e., choose) a control mass or control volume by placing a control surface that contains the substance/device you want to analyze. Indicate the presence of all the transfer terms into and out of the control volume and label different parts of the system if they do not have the same thermodynamic state.

3. Write the general laws for each of the chosen control volumes (for now we just use the energy equation, but later we will use several laws). If a transfer term leaves one control volume and enters another, you should have one term in each equation with the opposite sign.

4. Write down the auxiliary or particular laws for whatever is inside each of the control volumes. The constitution of a substance is either written down or referenced to a table. The equation for a given process is normally easily written down; it is given by

the way the system or device is constructed and often is an approximation to reality. That is, we make a simplified mathematical model of the real-world behavior.

5. Finish the formulation by combining all the equations (don't use numbers yet); then check which quantities are known and which are unknown. It is important to specify all the states by determining which two independent properties determine any given state. This task is most easily done by illustrating all the processes and states in a $(P–v)$, $(T–v)$, or similar diagram. These diagrams are also helpful in table lookup and locating a state.

As we write the energy equation

$$U_2 - U_1 + \frac{1}{2}m(V_2^2 - V_1^2) + mg(Z_2 - Z_1) = {}_1Q_2 - {}_1W_2 \qquad (3.26)$$

we must also consider the various terms of the storage. If the mass does not move significantly, either with a high speed or in elevation, then assume that the changes in kinetic energy and/or potential energy are small.

It is not always necessary to write out all these steps, and in the majority of the examples throughout this book we will not do so. However, when faced with a new and unfamiliar problem, the student should always at least think through this set of questions to develop the ability to solve more challenging problems. In solving the problem in the following example, we will use this technique in detail.

Example 3.9

A vessel having a volume of 5 m³ contains 0.05 m³ of saturated liquid water and 4.95 m³ of saturated water vapor at 0.1 MPa. Heat is transferred until the vessel is filled with saturated vapor. Determine the heat transfer for this process.

Control mass:	All the water inside the vessel.
Sketch:	Fig. 3.17.
Initial state:	Pressure, volume of liquid, volume of vapor; therefore, state 1 is fixed.
Final state:	Somewhere along the saturated-vapor curve; the water was heated, so $P_2 > P_1$.
Process:	Constant volume and mass; therefore, constant specific volume.
Diagram:	Fig. 3.18.
Model:	Steam tables.

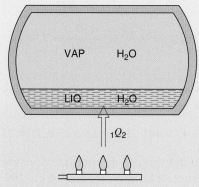

VAP   H₂O

LIQ   H₂O

$_1Q_2$

FIGURE 3.17 Sketch for Example 3.9.

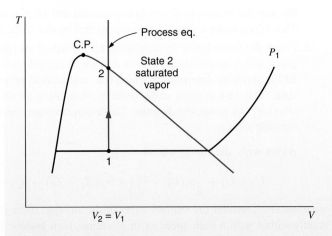

FIGURE 3.18 Diagram for Example 3.9.

Analysis

From the energy equation we have

$$U_2 - U_1 + m\frac{\mathbf{V}_2^2 - \mathbf{V}_1^2}{2} + mg(Z_2 - Z_1) = {}_1Q_2 - {}_1W_2$$

By examining the control surface for various work modes, we conclude that the work for this process is zero. Furthermore, the system is not moving, so there is no change in kinetic energy. There is a small change in the center of mass of the system, but we will assume that the corresponding change in potential energy (in kilojoules) is negligible. Therefore,

$$ {}_1Q_2 = U_2 - U_1 $$

Solution

The heat transfer will be found from the energy equation. State 1 is known, so U_1 can be calculated. The specific volume at state 2 is also known (from state 1 and the process). Since state 2 is saturated vapor, state 2 is fixed, as is seen in Fig. 3.18. Therefore, U_2 can also be found.

The solution proceeds as follows:

$$ m_{1\,\text{liq}} = \frac{V_{\text{liq}}}{v_f} = \frac{0.05}{0.001\,043} = 47.94\,\text{kg} $$

$$ m_{1\,\text{vap}} = \frac{V_{\text{vap}}}{v_g} = \frac{4.95}{1.6940} = 2.92\,\text{kg} $$

Then

$$ U_1 = m_{1\,\text{liq}}u_{1\,\text{liq}} + m_{1\,\text{vap}}u_{1\,\text{vap}} $$

$$ = 47.94(417.36) + 2.92(2506.1) = 27\,326\,\text{kJ} $$

To determine u_2 we need to know two thermodynamic properties, since this determines the final state. The properties we know are the quality, $x = 100\,\%$, and v_2, the final specific volume, which can readily be determined.

$$m = m_{1\,\text{liq}} + m_{1\,\text{vap}} = 47.94 + 2.92 = 50.86\,\text{kg}$$

$$v_2 = \frac{V}{m} = \frac{5.0}{50.86} = 0.098\,31\,\text{m}^3/\text{kg}$$

In Table B.1.2 we find, by interpolation, that at a pressure of 2.03 MPa, $v_g = 0.098\,31$ m^3/kg. The final pressure of the steam is therefore 2.03 MPa. Then

$$u_2 = 2600.5\,\text{kJ/kg}$$

$$U_2 = mu_2 = 50.86(2600.5) = 132\,261\,\text{kJ}$$

$${}_1Q_2 = U_2 - U_1 = 132\,261 - 27\,326 = 104\,935\,\text{kJ}$$

Example 3.10

The piston/cylinder setup of Example 3.4 contains 0.5 kg of ammonia at $-20\,^\circ$C with a quality of 25 %. The ammonia is now heated to $+20\,^\circ$C, at which state the volume is observed to be 1.41 times larger. Find the final pressure, the work the ammonia produced, and the heat transfer.

Analysis

The forces acting on the piston, the gravitation constant, the external atmosphere at constant pressure, and the linear spring give a linear relation between P and $v(V)$.

Process: $P = C_1 + C_2 v$; see Example 3.5, Fig. 3.12

State 1: (T_1, x_1) from Table B.2.1

$$P_1 = P_{\text{sat}} = 190.2\,\text{kPa}$$

$$v_1 = v_f + x_1 v_{fg} = 0.001\,504 + 0.25 \times 0.621\,84 = 0.156\,96\,\text{m}^3/\text{kg}$$

$$u_1 = u_f + x_1 u_{fg} = 88.76 + 0.25 \times 1210.7 = 391.44\,\text{kJ/kg}$$

State 2: $(T_2, v_2 = 1.41\,v_1 = 1.41 \times 0.156\,96 = 0.2213\,\text{m}^3/\text{kg})$

Table B.2.2 state very close to $P_2 = 600$ kPa, $u_2 \simeq 1347.9$ kJ/kg

The work term can now be integrated, knowing P versus v, and can be seen as the area in the P–v diagram, shown in Fig. 3.19.

$${}_1W_2 = \int_1^2 P\,dV = \int_1^2 Pm\,dv = \text{area} = m\frac{1}{2}(P_1 + P_2)(v_2 - v_1)$$

$$= 0.5\,\text{kg} \times \frac{1}{2}(190.2 + 600)\,\text{kPa} \times (0.2213 - 0.156\,96)\,\text{m}^3/\text{kg}$$

$$= 12.71\,\text{kJ}$$

$$_1Q_2 = m(u_2 - u_1) + {}_1W_2$$
$$= 0.5\,\text{kg}\,(1347.9 - 391.44)\frac{\text{kJ}}{\text{kg}} + 12.71\,\text{kJ}$$
$$= 490.94\,\text{kJ}$$

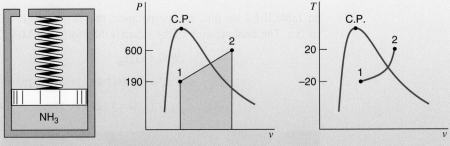

FIGURE 3.19 Diagrams for Example 3.10.

Example 3.11

The piston/cylinder setup shown in Fig. 3.20 contains 0.1 kg of water at 1000 kPa, 500 °C. The water is now cooled with a constant force on the piston until it reaches half of the initial volume. After this, it cools to 25 °C while the piston is against the stops. Find the final water pressure and the work and heat transfer in the overall process, and show the process in a P–v diagram.

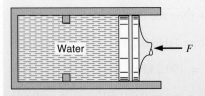

FIGURE 3.20 Sketch for Example 3.11.

Analysis

We recognize that this is a two-step process, one of constant P and one of constant V. This behavior is dictated by the construction of the device.

$\quad$ *State 1:* $\quad (P, T)$ $\quad$ From Table B.1.3; $v_1 = 0.354\,11\ \text{m}^3/\text{kg}$, $u_1 = 3124.34\ \text{kJ/kg}$

$\quad$ *Process 1–1a:* $\quad P = \text{constant} = F/A$

$\qquad$ 1a–2: $\quad v = \text{constant} = v_{1a} = v_2 = v_1/2$

$\quad$ *State 2:* $\quad (T, v_2 = v_1/2 = 0.177\,06\ \text{m}^3/\text{kg})$

$$X_2 = (v_2 - v_f)/v_{fg} = \frac{0.17706 - 0.001\,003}{43.3583}$$

$$= 0.004\,060\,5$$

$$u_2 = u_f + x_2 u_{fg} = 104.86 + 0.004\,060\,5 \times 2304.9$$
$$= 114.219\,\text{kJ/kg}$$

From Table B.1.1, $v_2 < v_g$, so the state is two phase and $P_2 = P_{\text{sat}} = 3.169$ kPa.

$$_1W_2 = \int_1^2 P\,dV = m \int_1^2 P\,dv = m P_1(v_{1a} - v_1) + 0$$

$$= 0.1\,\text{kg} \times 1000\,\text{kPa} \times (0.177\,06 - 0.345\,11)\,\text{m}^3/\text{kg} = -17.7\,\text{kJ}$$

Note that the work done from $1a$ to 2 is zero (no change in volume), as shown in Fig. 3.21.

$$_1Q_2 = m\,(u_2 - u_1) + {_1W_2}$$

$$= 0.1\,\text{kg} \times (114.219 - 3124.34)\,\text{kJ/kg} - 17.7\,\text{kJ}$$

$$= -318.71\,\text{kJ}$$

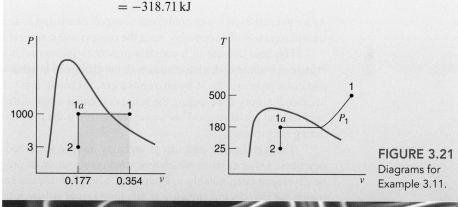

FIGURE 3.21 Diagrams for Example 3.11.

3.9 THE THERMODYNAMIC PROPERTY ENTHALPY

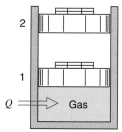

FIGURE 3.22 The constant-pressure quasi-equilibrium process.

In analyzing specific types of processes, we frequently encounter certain combinations of thermodynamic properties, which are therefore also properties of the substance undergoing the change of state. To demonstrate one such situation, let us consider a control mass undergoing a quasi-equilibrium constant-pressure process, as shown in Fig. 3.22. Assume that there are no changes in kinetic or potential energy and that the only work done during the process is that associated with the boundary movement. Taking the gas as our control mass and applying the energy equation, Eq. 3.5, we have, in terms of Q,

$$U_2 - U_1 = {_1Q_2} - {_1W_2}$$

The work done can be calculated from the relation

$$_1W_2 = \int_1^2 P\,dV$$

Since the pressure is constant,

$$_1W_2 = P \int_1^2 dV = P(V_2 - V_1)$$

Therefore,

$$_1Q_2 = U_2 - U_1 + P_2V_2 - P_1V_1$$
$$= (U_2 + P_2V_2) - (U_1 + P_1V_1)$$

We find that, in this very restricted case, the heat transfer during the process is given in terms of the change in the quantity $U + PV$ between the initial and final states. Because all these quantities are thermodynamic properties, that is, functions only of the state of the system, their combination must also have these same characteristics. Therefore, we find it convenient to define a new extensive property, the enthalpy,

$$H \equiv U + PV \qquad (3.27)$$

or, per unit mass,

$$h \equiv u + Pv \qquad (3.28)$$

As for internal energy, we could speak of specific enthalpy, h, and total enthalpy, H. However, we will refer to both as enthalpy, since the context will make it clear which is being discussed.

The heat transfer in a constant-pressure, quasi-equilibrium process is equal to the change in enthalpy, which includes both the change in internal energy and the work for this particular process. This is by no means a general result. It is valid for this special case only because the work done during the process is equal to the difference in the PV product for the final and initial states. This would not be true if the pressure had not remained constant during the process.

The significance and use of enthalpy are not restricted to the special process just described. Other cases in which this same combination of properties $u + Pv$ appears will be developed later, notably in Chapter 4, where we discuss control volume analyses. Our reason for introducing enthalpy at this time is that although the tables in Appendix B list values for internal energy, many other tables and charts of thermodynamic properties give values for enthalpy but not for internal energy. Therefore, it is necessary to calculate internal energy at a state using the tabulated values and Eq. 3.28:

$$u = h - Pv$$

Students often become confused about the validity of this calculation when analyzing system processes that do not occur at constant pressure, for which enthalpy has no physical significance. We must keep in mind that enthalpy, being a property, is a state or point function, and its use in calculating internal energy at the same state is not related to, or dependent on, any process that may be taking place.

Tabular values of internal energy and enthalpy, such as those included in Tables B.1 through B.7, are all relative to some arbitrarily selected base. In the steam tables, the internal energy of saturated liquid at $0.01\,°C$ is the reference state and is given a value of zero. For refrigerants, such as R-134a, R-410A, and ammonia, the reference state is arbitrarily taken as saturated liquid at $-40\,°C$. The enthalpy in this *reference state* is assigned the value of zero. Cryogenic fluids, such as nitrogen, have other arbitrary reference states chosen for enthalpy values listed in their tables. Because each of these reference states is arbitrarily selected, it is always possible to have negative values for enthalpy, as for saturated-solid water in Table B.1.5. When enthalpy and internal energy are given values relative to the same reference state, as they are in essentially all thermodynamic tables, the difference between internal energy and enthalpy at the reference state is equal to Pv. Since the specific volume of the liquid is very small, this product is negligible as far as the significant figures of the tables are concerned, but the principle should be kept in mind, for in certain cases it is significant.

The enthalpy of a substance in a saturation state and with a given quality is found in the same way as the specific volume and internal energy. The enthalpy of saturated liquid has the symbol h_f, saturated vapor h_g, and the increase in enthalpy during vaporization h_{fg}. For a saturation state, the enthalpy can be calculated by one of the following relations:

$$h = (1 - x)h_f + xh_g$$
$$h = h_f + xh_{fg}$$

The enthalpy of compressed liquid water may be found from Table B.1.4. For substances for which compressed-liquid tables are not available, the enthalpy is taken as that of saturated liquid at the same temperature.

Example 3.12

A cylinder fitted with a piston has a volume of 0.1 m^3 and contains 0.5 kg of steam at 0.4 MPa. Heat is transferred to the steam until the temperature is 300 °C, while the pressure remains constant.

Determine the heat transfer and the work for this process.

Control mass: Water inside cylinder.

Process: Constant pressure, $P_2 = P_1$

Initial state: P_1, V_1, m; therefore, v_1 is known, state 1 is fixed (at P_1, v_1, check steam tables—two-phase region).

Final state: P_2, T_2; therefore, state 2 is fixed (superheated).

Diagram: Fig. 3.23.

Model: Steam tables.

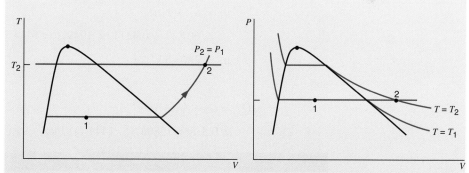

FIGURE 3.23   The constant-pressure quasi-equilibrium process.

Analysis

There is no change in kinetic energy or potential energy. Work is done by movement at the boundary. Assume the process to be quasi-equilibrium. Since the pressure is constant, we have

$$_1W_2 = \int_1^2 P\,dV = P\int_1^2 dV = P(V_2 - V_1) = m(P_2 v_2 - P_1 v_1)$$

Therefore, the energy equation is, in terms of Q,

$$_1Q_2 = m(u_2 - u_1) + {}_1W_2$$
$$= m(u_2 - u_1) + m(P_2v_2 - P_1v_1) = m(h_2 - h_1)$$

Solution

There is a choice of procedures to follow. State 1 is known, so v_1 and h_1 (or u_1) can be found. State 2 is also known, so v_2 and h_2 (or u_2) can be found. Using the first law of thermodynamics and the work equation, we can calculate the heat transfer and work. Using the enthalpies, we have

$$v_1 = \frac{V_1}{m} = \frac{0.1\ m^3}{0.5\ \text{kg}} = 0.2 = (0.001\ 084 + x_1 0.4614)\frac{m^3}{\text{kg}}$$

$$x_1 = \frac{0.1989}{0.4614} = 0.4311$$

$$h_1 = h_f + x_1 h_{fg}$$
$$= 604.74 + 0.4311 \times 2133.8 = 1524.7\ \text{kJ/kg}$$

$$h_2 = 3066.8\ \text{kJ/kg}$$

$$_1Q_2 = 0.5\ \text{kg} \times (3066.8 - 1524.7)\ \text{kJ/kg} = 771.1\ \text{kJ}$$

$$_1W_2 = mP(v_2 - v_1) = 0.5 \times 400(0.6548 - 0.2) = 91.0\ \text{kJ}$$

Therefore,

$$U_2 - U_1 = {}_1Q_2 - {}_1W_2 = 771.1 - 91.0 = 680.1\ \text{kJ}$$

The heat transfer could also have been found from u_1 and u_2:

$$u_1 = u_f + x_1 u_{fg}$$
$$= 604.31 + 0.4311 \times 1949.3 = 1444.7\ \text{kJ/kg}$$

$$u_2 = 2804.8\ \text{kJ/kg}$$

and

$$_1Q_2 = U_2 - U_1 + {}_1W_2$$
$$= 0.5\ \text{kg} \times (2804.8 - 1444.7)\ \text{kJ/kg} + 91.0 = 771.1\ \text{kJ}$$

3.10 THE CONSTANT-VOLUME AND CONSTANT-PRESSURE SPECIFIC HEATS

In this section we will consider a homogeneous phase of a substance of constant composition. This phase may be a solid, a liquid, or a gas, but no change of phase will occur. We will then define a variable termed the *specific heat*, the amount of heat required per unit mass to raise the temperature by one degree. Since it would be of interest to examine the relation between the specific heat and other thermodynamic variables, we note first that the heat transfer is given by Eq. 3.4. Neglecting changes in kinetic and potential energies, and assuming a simple compressible substance and a quasi-equilibrium process, for which the work in Eq. 3.4 is given by Eq. 3.16, we have

$$\delta Q = dU + \delta W = dU + P\,dV$$

We find that this expression can be evaluated for two separate special cases:

1. Constant volume, for which the work term ($P\,dV$) is zero, so that the specific heat (at constant volume) is

$$C_v = \frac{1}{m}\left(\frac{\delta Q}{\delta T}\right)_v = \frac{1}{m}\left(\frac{\partial U}{\partial T}\right)_v = \left(\frac{\partial u}{\partial T}\right)_v \qquad (3.29)$$

2. Constant pressure, for which the work term can be integrated and the resulting PV terms at the initial and final states can be associated with the internal energy terms, as in Section 3.9, thereby leading to the conclusion that the heat transfer can be expressed in terms of the enthalpy change. The corresponding specific heat (at constant pressure) is

$$C_p = \frac{1}{m}\left(\frac{\delta Q}{\delta T}\right)_p = \frac{1}{m}\left(\frac{\partial H}{\partial T}\right)_p = \left(\frac{\partial h}{\partial T}\right)_p \qquad (3.30)$$

Note that in each of these special cases, the resulting expression, Eq. 3.29 or 3.30, contains only thermodynamic properties, from which we conclude that the constant-volume and constant-pressure specific heats must themselves be thermodynamic properties. This means that, although we began this discussion by considering the amount of heat transfer required to cause a unit temperature change and then proceeded through a very specific development leading to Eq. 3.29 (or 3.30), the result ultimately expresses a relation among a set of thermodynamic properties and therefore constitutes a definition that is independent of the particular process leading to it (in the same sense that the definition of enthalpy in the previous section is independent of the process used to illustrate one situation in which the property is useful in a thermodynamic analysis). As an example, consider the two identical fluid masses shown in Fig. 3.24. In the first system 100 kJ of heat is transferred to it, and in the second system 100 kJ of work is done on it. Thus, the change of internal energy is the same for each, and therefore the final state and the final temperature are the same in each. In accordance with Eq. 3.29, therefore, exactly the same value for the average constant-volume specific heat would be found for this substance for the two processes, even though the two processes are very different as far as heat transfer is concerned.

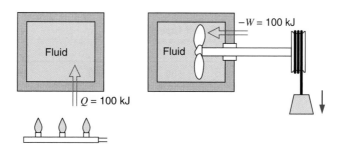

FIGURE 3.24 Sketch showing two ways in which a given ΔU may be achieved.

Solids and Liquids

As a special case, consider either a solid or a liquid. Since both of these phases are nearly incompressible,

$$dh = du + d(Pv) \approx du + v\,dP \tag{3.31}$$

Also, for both of these phases, the specific volume is very small, such that in many cases

$$dh \approx du \approx C\,dT \tag{3.32}$$

where C is either the constant-volume or the constant-pressure specific heat, as the two would be nearly the same. In many processes involving a solid or a liquid, we might further assume that the specific heat in Eq. 3.32 is constant (unless the process occurs at low temperature or over a wide range of temperatures). Equation 3.32 can then be integrated to

$$h_2 - h_1 \simeq u_2 - u_1 \simeq C(T_2 - T_1) \tag{3.33}$$

Specific heats for various solids and liquids are listed in Tables A.3 and A.4.

In other processes for which it is not possible to assume constant specific heat, there may be a known relation for C as a function of temperature. Equation 3.32 could then also be integrated.

 3.11 THE INTERNAL ENERGY, ENTHALPY, AND SPECIFIC HEAT OF IDEAL GASES

In general, for any substance the internal energy u depends on the two independent properties specifying the state. For a low-density gas, however, u depends primarily on T and much less on the second property, P or v. For example, consider several values for superheated vapor steam from Table B.1.3, shown in Table 3.1. From these values, it is evident that u depends strongly on T but not much on P. Also, we note that the dependence of u on P is less at low pressure and is much less at high temperature; that is, as the density decreases, so does dependence of u on P (or v). It is therefore reasonable to extrapolate this behavior to very low density and to assume that as gas density becomes so low that the ideal-gas model is appropriate, internal energy does not depend on pressure at all but is a function only of temperature. That is, for an ideal gas,

$$Pv = RT \quad \text{and} \quad u = f(T)\,\text{only} \tag{3.34}$$

TABLE 3.1

Internal Energy for Superheated Vapor Steam

	P, kPa			
T, °C	10	100	500	1000
200	2661.3	2658.1	2642.9	2621.9
700	3479.6	3479.2	3477.5	3475.4
1200	4467.9	4467.7	4466.8	4465.6

The relation between the internal energy u and the temperature can be established by using the definition of constant-volume specific heat given by Eq. 3.29:

$$C_v = \left(\frac{\partial u}{\partial T}\right)_v$$

Because the internal energy of an ideal gas is not a function of specific volume, for an ideal gas we can write

$$C_{v0} = \frac{du}{dT}$$

$$du = C_{v0}\, dT \tag{3.35}$$

where the subscript 0 denotes the specific heat of an ideal gas. For a given mass m,

$$dU = mC_{v0}\, dT \tag{3.36}$$

From the definition of enthalpy and the equation of state of an ideal gas, it follows that

$$h = u + Pv = u + RT \tag{3.37}$$

Since R is a constant and u is a function of temperature only, it follows that the enthalpy, h, of an ideal gas is also a function of temperature only. That is,

$$h = f(T) \tag{3.38}$$

The relation between enthalpy and temperature is found from the constant-pressure specific heat as defined by Eq. 3.30:

$$C_p = \left(\frac{\partial h}{\partial T}\right)_p$$

Since the enthalpy of an ideal gas is a function of the temperature only and is independent of the pressure, it follows that

$$C_{p0} = \frac{dh}{dT}$$

$$dh = C_{p0}\, dT \tag{3.39}$$

For a given mass m,

$$dH = mC_{p0}\, dT \tag{3.40}$$

The consequences of Eqs. 3.35 and 3.39 are demonstrated in Fig. 3.25, which shows two lines of constant temperature. Since internal energy and enthalpy are functions of temperature only, these lines of constant temperature are also lines of constant internal energy and constant enthalpy. From state 1 the high temperature can be reached by a variety of paths, and in each case the final state is different. However, regardless of the path, the change in internal energy is the same, as is the change in enthalpy, for lines of constant temperature are also lines of constant u and constant h.

Because the internal energy and enthalpy of an ideal gas are functions of temperature only, it also follows that the constant-volume and constant-pressure specific heats are also functions of temperature only. That is,

$$C_{v0} = f(T), \quad C_{p0} = f(T) \tag{3.41}$$

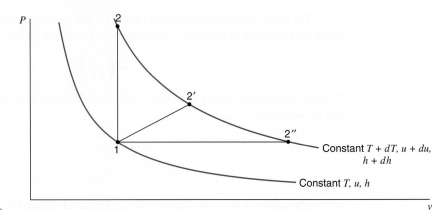

FIGURE 3.25 P–v diagram for an ideal gas.

Because all gases approach ideal-gas behavior as the pressure approaches zero, the ideal-gas specific heat for a given substance is often called the *zero-pressure specific heat*, and the zero-pressure, constant-pressure specific heat is given the symbol C_{p0}. The zero-pressure, constant-volume specific heat is given the symbol C_{v0}. Figure 3.26 shows C_{p0} as a function of temperature for a number of substances. These values are determined by the techniques of statistical thermodynamics and will not be discussed here. A brief summary presentation of this subject is given in Appendix C. It is noted there that the principal factor causing specific heat to vary with temperature is molecular vibration. More complex molecules have multiple vibrational modes and therefore show greater temperature dependency, as is seen

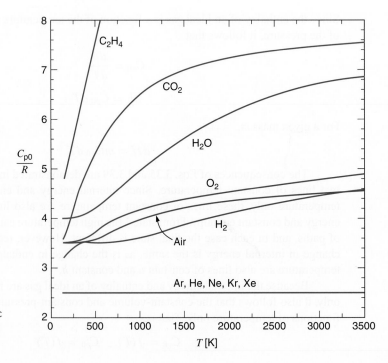

FIGURE 3.26 Specific heat for some gases as a function of temperature.

in Fig. 3.26. This is an important consideration when deciding whether or not to account for specific heat variation with temperature in any particular application.

A very important relation between the constant-pressure and constant-volume specific heats of an ideal gas may be developed from the definition of enthalpy:

$$h = u + Pv = u + RT$$

Differentiating and substituting Eqs. 3.35 and 3.39, we have

$$dh = du + R \, dT$$

$$C_{p0} \, dT = C_{v0} \, dT + R \, dT$$

Therefore,

$$C_{p0} - C_{v0} = R \tag{3.42}$$

On a mole basis, this equation is written

$$\overline{C}_{p0} - \overline{C}_{v0} = \overline{R} \tag{3.43}$$

This tells us that the difference between the constant-pressure and constant-volume specific heats of an ideal gas is always constant, though both are functions of temperature. Thus, we need examine only the temperature dependency of one, and the other is given by Eq. 3.42.

Let us consider the specific heat C_{p0}. There are three possibilities to examine. The situation is simplest if we assume constant specific heat, that is, no temperature dependence. Then it is possible to integrate Eq. 3.39 directly to

$$h_2 - h_1 = C_{p0}(T_2 - T_1) \tag{3.44}$$

We note from Fig. 3.26 the circumstances under which this will be an accurate model. It should be added, however, that it may be a reasonable approximation under other conditions, especially if an average specific heat in the particular temperature range is used in Eq. 3.44. Values of specific heat at room temperature and gas constants for various gases are given in Table A.5.

The second possibility for the specific heat is to use an analytical equation for C_{p0} as a function of temperature. Because the results of specific-heat calculations from statistical thermodynamics do not lend themselves to convenient mathematical forms, these results have been approximated empirically. The equations for C_{p0} as a function of temperature are listed in Table A.6 for a number of gases.

The third possibility is to integrate the results of the calculations of statistical thermodynamics from an arbitrary reference temperature to any other temperature T and to define a function

$$h_T = \int_{T_0}^{T} C_{p0} \, dT$$

This function can then be tabulated in a single-entry (temperature) table. Then, between any two states 1 and 2,

$$h_2 - h_1 = \int_{T_0}^{T_2} C_{p0} \, dT - \int_{T_0}^{T_1} C_{p0} \, dT = h_{T_2} - h_{T_1} \tag{3.45}$$

and it is seen that the reference temperature cancels out. This function h_T (and a similar function $u_T = h_T - RT$) is listed for air in Table A.7 and for other gases in Table A.8.

To summarize the three possibilities, we note that using the ideal-gas tables, Tables A.7 and A.8, gives us the most accurate answer, but that the equations in Table A.6 would give a close empirical approximation. Constant specific heat would be less accurate, except for monatomic gases and gases below room temperature. It should be remembered that all these results are part of the ideal-gas model, which in many of our problems is not a valid assumption for the behavior of the substance.

Example 3.13

Calculate the change of enthalpy as 1 kg of oxygen is heated from 300 to 1500 K. Assume ideal-gas behavior.

Solution

For an ideal gas, the enthalpy change is given by Eq. 3.39. However, we also need to make an assumption about the dependence of specific heat on temperature. Let us solve this problem in several ways and compare the answers.

Our most accurate answer for the ideal-gas enthalpy change for oxygen between 300 and 1500 K would be from the ideal-gas tables, Table A.8. This result is, using Eq. 3.45,

$$h_2 - h_1 = 1540.2 - 273.2 = 1267.0 \text{ kJ/kg}$$

The empirical equation from Table A.6 should give a good approximation to this result. Integrating Eq. 3.39, we have

$$h_2 - h_1 = \int_{T_1}^{T_2} C_{p0} \, dT = \int_{\theta_1}^{\theta_2} C_{p0}(\theta) \times 1000 \, d\theta$$

$$= 1000 \left[0.88\theta - \frac{0.0001}{2}\theta^2 + \frac{0.54}{3}\theta^3 - \frac{0.33}{4}\theta^4 \right]_{\theta_1=0.3}^{\theta_2=1.5}$$

$$= 1241.5 \text{ kJ/kg}$$

which is lower than the first result by 2.0%.

If we assume constant specific heat, we must be concerned about what value we are going to use. If we use the value at 300 K from Table A.5, we find, from Eq. 3.44, that

$$h_2 - h_1 = C_{p0}(T_2 - T_1) = 0.922 \times 1200 = 1106.4 \text{ kJ/kg}$$

which is low by 12.7%. However, suppose we assume that the specific heat is constant at its value at 900 K, the average temperature. Substituting 900 K into the equation for specific heat from Table A.6, we have

$$C_{p0} = 0.88 - 0.0001(0.9) + 0.54(0.9)^2 - 0.33(0.9)^3$$

$$= 1.0767 \text{ kJ/kg·K}$$

Substituting this value into Eq. 3.44 gives the result

$$h_2 - h_1 = 1.0767 \times 1200 = 1292.1 \text{ kJ/kg}$$

which is high by about 2.0%, a much closer result than the one using the room temperature specific heat. It should be kept in mind that part of the model involving ideal gas with constant specific heat also involves a choice of what value is to be used.

Example 3.14

A cylinder fitted with a piston has an initial volume of 0.1 m³ and contains nitrogen at 150 kPa, 25 °C. The piston is moved, compressing the nitrogen until the pressure is 1 MPa and the temperature is 150 °C. During this compression process heat is transferred from the nitrogen, and the work done on the nitrogen is 20 kJ. Determine the amount of this heat transfer.

Control mass: Nitrogen.
Process: Work input known.
Initial state: P_1, T_1, V_1; state 1 fixed.
Final state: P_2, T_2; state 2 fixed.
Model: Ideal gas, constant specific heat with value at 300 K, Table A.5.

Analysis
From the energy equation we have

$$_1Q_2 = m(u_2 - u_1) + {_1}W_2$$

Solution
The mass of nitrogen is found from the equation of state with the value of R from Table A.5:

$$m = \frac{PV}{RT} = \frac{150\,\text{kPa} \times 0.1\,\text{m}^3}{0.2968\frac{\text{kJ}}{\text{kg·K}} \times 298.15\,\text{K}} = 0.1695\,\text{kg}$$

Assuming constant specific heat as given in Table A.5, we have

$$_1Q_2 = mC_{v0}(T_2 - T_1) + {_1}W_2$$
$$= 0.1695\,\text{kg} \times 0.745\frac{\text{kJ}}{\text{kg·K}} \times (150 - 25)\,\text{K} - 20.0\,\text{kJ}$$
$$= 15.8 - 20.0 = -4.2\,\text{kJ}$$

It would, of course, be somewhat more accurate to use Table A.8 than to assume constant specific heat (room temperature value), but often the slight increase in accuracy does not warrant the added difficulties of manually interpolating the tables.

In-Text Concept Questions

l. To determine v or u for some liquid or solid, is it more important that I know P or T?

m. To determine v or u for an ideal gas, is it more important that I know P or T?

n. I heat 1 kg of a substance at constant pressure (200 kPa) until the temperature increases by one degree. How much heat is needed if the substance is water at 10 °C, steel at 25 °C, air at 325 K, or ice at −10 °C?

Example 3.15

A 25 kg cast-iron wood-burning stove, shown in Fig. 3.27, contains 5 kg of soft pine wood and 1 kg of air. All the masses are at room temperature, 20 °C, and pressure, 101 kPa. The wood now burns and heats all the mass uniformly, releasing 1500 W. Neglect any air flow and changes in mass and heat losses. Find the rate of change of the temperature (dT/dt) and estimate the time it will take to reach a temperature of 75 °C.

FIGURE 3.27 Sketch for Example 3.15.

Solution

C.V.: The iron, wood, and air.
This is a control mass.

Energy equation rate form:
$$\dot{E} = \dot{Q} - \dot{W}$$

We have no changes in kinetic or potential energy and no change in mass, so

$$U = m_{air}u_{air} + m_{wood}u_{wood} + m_{iron}u_{iron}$$

$$\dot{E} = \dot{U} = m_{air}\dot{u}_{air} + m_{wood}\dot{u}_{wood} + m_{iron}\dot{u}_{iron}$$

$$= (m_{air}C_{V\,air} + m_{wood}C_{wood} + m_{iron}C_{iron})\frac{dT}{dt}$$

Now the energy equation has zero work, an energy release of $\dot{Q}$, and becomes

$$(m_{air}C_{V\,air} + m_{wood}C_{wood} + m_{iron}C_{iron})\frac{dT}{dt} = \dot{Q} - 0$$

$$\frac{dT}{dt} = \frac{\dot{Q}}{(m_{air}C_{V\,air} + m_{wood}C_{wood} + m_{iron}C_{iron})}$$

$$= \frac{1500}{1 \times 0.717 + 5 \times 1.38 + 25 \times 0.42}\frac{W}{kg\,(kJ/kg)} = 0.0828\ \text{K/s}$$

Assuming the rate of temperature rise is constant, we can find the elapsed time as

$$\Delta T = \int \frac{dT}{dt}dt = \frac{dT}{dt}\Delta t$$

$$\Rightarrow \Delta t = \frac{\Delta T}{\frac{dT}{dt}} = \frac{75 - 20}{0.0828} = 664\ \text{s} = 11\ \text{min}$$

3.12 GENERAL SYSTEMS THAT INVOLVE WORK

In the preceding discussion about work, we focused on work from a single point force or a distributed force over an area as pressure. There are other types of forces and displacements that differ by the nature of the force and the displacement. We will mention a few of the more typical situations that commonly arises and write the work term as

$$_1W_2 = \int_1^2 F_{gen}\, dx_{gen} \tag{3.46}$$

TABLE 3.2
Generalized Work Term

System	Force	Unit	Displacement	Unit
Simple force	F	N	dx	m
Pressure	P	Pa	dV	m^3
Spring	$\mathcal{T} = k_s(x - x_0)$	N	dx	m
Stretched wire	F	N	$dx = x_0\, de$	m
Surface tension	$\mathcal{S} = AEe$	N/m	dA	m^2
Electrical	$\mathcal{E}$	volt	dZ^*	Coulon

*Notice the time derivative $dZ/dt = i$ (current in amps).

In this expression, we have a generalized force and a generalized displacement. For each case, we must know the expression for both and also know how the force changes during the process. Simple examples are listed in Table 3.2, and the resulting expressions for the work term can be worked out once the function $F_{gen}(x_{gen})$ is known. For many of these systems, the sign notation is such that the force is positive when work goes into the system, so with our sign definition we would have the general form

$$\delta W = P\, dV - \mathcal{T}\, dL - \mathcal{S}\, dA - \mathcal{E}\, dZ + \cdots \tag{3.47}$$

where other terms are possible. The rate of work from this form represents power as

$$\dot{W} = \frac{dW}{dt} = P\dot{V} - \mathcal{T}\mathbf{V} - \mathcal{S}\dot{A} - \mathcal{E}\dot{Z} + \cdots \tag{3.48}$$

It should also be noted that many other forms of work can be identified in processes that are not quasi-equilibrium processes. For example, there is the work done by shearing forces in the friction in a viscous fluid or the work done by a rotating shaft that crosses the system boundary.

The identification of work is an important aspect of many thermodynamic problems. We have already noted that work can be identified only at the boundaries of the system. For example, consider Fig. 3.28, which shows a gas separated from the vacuum by a membrane. Let the membrane rupture and the gas fill the entire volume. Neglecting any work associated with the rupturing of the membrane, we can ask whether work is done in the process. If we take as our system the gas and the vacuum space, we readily conclude that no work is done because no work can be identified at the system boundary. If we take the gas as a system, we do have a change of volume, and we might be tempted to calculate the work from the integral

$$\int_1^2 P\, dV$$

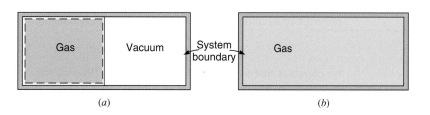

FIGURE 3.28
Example of a process involving a change of volume for which the work is zero.

(a)   (b)

However, this is not a quasi-equilibrium process, and therefore the work cannot be calculated from this relation. Because there is no resistance at the system boundary as the volume increases, we conclude that for this system no work is done in this process of filling the vacuum.

Example 3.16

During the charging of a storage battery, the current i is 20 A and the voltage $\mathscr{E}$ is 12.8 V. The rate of heat transfer from the battery is 10 W. At what rate is the internal energy increasing?

Solution

Since changes in kinetic and potential energy are insignificant, the first law of thermodynamics can be written as a rate equation in the form of Eq. 5.31:

$$\frac{dU}{dt} = \dot{Q} - \dot{W}$$

$$\dot{W} = \mathscr{E}i = -12.8 \times 20 = -256\,\text{W} = -256\,\text{J/s}$$

Therefore,

$$\frac{dU}{dt} = \dot{Q} - \dot{W} = -10 - (-256) = 246\,\text{J/s}$$

3.13 CONSERVATION OF MASS

In the previous sections, we considered the first law of thermodynamics for a control mass undergoing a change of state. A control mass is defined as a fixed quantity of mass. The question now is whether the mass of such a system changes when its energy changes. If it does, our definition of a control mass as a fixed quantity of mass is no longer valid when the energy changes.

We know from relativistic considerations that mass and energy are related by the well-known equation

$$E = mc^2 \qquad (3.49)$$

where c = velocity of light and E = energy. We conclude from this equation that the mass of a control mass does change when its energy changes. Let us calculate the magnitude of this change of mass for a typical problem and determine whether this change in mass is significant.

Consider a rigid vessel that contains a 1 kg stoichiometric mixture of a hydrocarbon fuel (such as gasoline) and air. From our knowledge of combustion, we know that after combustion takes place, it will be necessary to transfer about 2900 kJ from the system to restore it to its initial temperature. From the energy equation

$$U_2 - U_1 = {}_1Q_2 - {}_1W_2$$

we conclude that since ${}_1W_2 = 0$ and ${}_1Q_2 = -2900$ kJ, the internal energy of this system decreases by 2900 kJ during the heat transfer process. Let us now calculate the decrease in mass during this process using Eq. 3.49.

The velocity of light, c, is 2.9979×10^8 m/s. Therefore,

$$2900 \text{ kJ} = 2\,900\,000 \text{ J} = m(\text{kg}) \times (2.9979 \times 10^8 \text{ m/s})^2$$

and so

$$m = 3.23 \times 10^{-11} \text{ kg}$$

Thus, when the energy of the control mass decreases by 2900 kJ, the decrease in mass is 3.23×10^{-11} kg.

A change in mass of this magnitude cannot be detected by even our most accurate instrumentation. Certainly, a fractional change in mass of this magnitude is beyond the accuracy required in essentially all engineering calculations. Therefore, if we use the laws of conservation of mass and conservation of energy as separate laws, we will not introduce significant error into most thermodynamic problems and our definition of a control mass as having a fixed mass can be used even though the energy changes.

Example 3.17

Consider 1 kg of water on a table at room conditions $20\,^\circ$C, 100 kPa. We want to examine the energy changes for each of three processes: accelerate it from rest to 10 m/s, raise it 10 m, and heat it $10\,^\circ$C.

For this control mass, the energy changes become

$$\Delta \text{KE} = \frac{1}{2} m \left(\mathbf{V}_2^2 - \mathbf{V}_1^2\right) = \frac{1}{2} \times 1 \text{ kg} \times (10^2 - 0)\, \text{m}^2/\text{s}^2 = 50 \text{ kg}\cdot\text{m}^2/\text{s}^2 = 50 \text{ J}$$

$$\Delta \text{PE} = mg(Z_2 - Z_1) = 1 \text{ kg} \times 9.81 \text{ m/s}^2 \times (10 - 0)\,\text{m} = 98.1 \text{ J}$$

$$\Delta U = m(u_2 - u_1) = mC_v(T_2 - T_1) = 1 \text{ kg} \times 4.18 \frac{\text{kJ}}{\text{kg}\cdot\text{K}} \times 10 \text{ K} = 41.8 \text{ kJ}$$

Notice how much smaller the kinetic and potential energy changes are compared to the change in the internal energy due to the raised temperature. For the kinetic and potential energies to be significant, say 10 % of ΔU, the velocity must be much higher, such as 100 m/s, and the height difference much greater, such as 500 m. In most engineering applications such levels are very uncommon, so the kinetic and potential energies are often neglected.

All the previous discussion and examples dealt with a single mass at a uniform state. This is not always the case, so let us look at a situation where we have a container separated into two compartments by a valve, as shown in Fig. 3.29. Assume that the two masses have different starting states and that, after opening the valve, the masses mix to form a single, uniform final state.

FIGURE 3.29 Two connected tanks with different initial states.

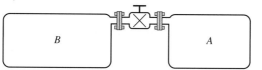

For a control mass that is the combination of A and B, we get the conservation of mass written out in the continuity equation

$$\text{Conservation of mass:} \quad m_2 = m_1 = m_{1A} + m_{1B} \tag{3.50}$$

$$\text{Energy Eq:} \quad m_2 e_2 - (m_{1A}e_{1A} + m_{1B}e_{1B}) = {}_1Q_2 - {}_1W_2 \tag{3.51}$$

The general form of the mass conservation equation is called the *continuity equation*, a common name for it in fluid mechanics, and it is covered in the following chapter.

As the system does not change elevation and we look at state 2 after any motion has ended, we have zero kinetic energy in all states and no changes in potential energy, so the left-hand side in the energy equation becomes

$$m_2 e_2 - (m_{1A}e_{1A} + m_{1B}e_{1B})$$
$$= m_2(u_2 + 0 + gZ_1) - m_{1A}(u_{1A} + 0 + gZ_1) - m_{1B}(u_{1B} + 0 + gZ_1)$$
$$= m_2 u_2 - m_{1A}(u_{1A} + u_{1B}) + [m_2 - (m_{1A} + m_{1B})]gZ_1$$
$$= m_2 u_2 - m_{1A}(u_{1A} + u_{1B})$$

Notice how the factor for the potential energy is zero from the conservation of mass and the left-hand side terms simplify to contain internal energy only. If we take the energy equation and divide by the total mass, we get

$$u_2 - (m_{1A}u_{1A} + m_{1B}u_{1B})/m_2 = ({}_1Q_2 - {}_1W_2)/m_2$$

or

$$u_2 = \frac{m_{1A}}{m_2}u_{1A} + \frac{m_{1B}}{m_2}u_{1B} + ({}_1Q_2 - {}_1W_2)/m_2 \tag{3.52}$$

For an insulated (${}_1Q_2 = 0$) and rigid ($V = C$, so ${}_1W_2 = 0$) container the last terms vanish and the final specific internal energy is the mass-weighted average of the initial values. The mass weighing factors are dimensionless and they correspond to the relative contribution to the total mass from each part, so they sum to 1, which is seen by dividing the continuity equation by the total mass.

For the stated process, $V = C$, the second property that determines the final state is the specific volume as

$$v_2 = V_2/m_2 = \frac{m_{1A}}{m_2}v_{1A} + \frac{m_{1B}}{m_2}v_{1B} \tag{3.53}$$

which also is a mass-weighted average of the initial values.

3.14 ENGINEERING APPLICATIONS

Energy Storage and Conversion

Energy can be stored in a number of different forms by various physical implementations, which have different characteristics with respect to storage efficiency, rate of energy transfer, and size (Figs. 3.30–3.33). These systems can also include a possible energy conversion that consists of a change of one form of energy to another form of energy. The storage is usually temporary, lasting for periods ranging from a fraction of a second to days or years, and can be for very small or large amounts of energy. Also, it is basically a shift of the energy transfer from a time when it is unwanted and thus inexpensive to a time when it is wanted and then often expensive. It is also very important to consider the maximum rate

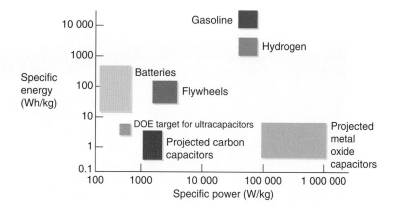

FIGURE 3.30
Specific energy versus specific power.

of energy transfer in the charging or discharging process, as size and possible losses are sensitive to that rate.

Notice from Fig. 3.30 that it is difficult to have high power and high energy storage in the same device. It is also difficult to store energy more compactly than in gasoline.

Mechanical Systems

A flywheel stores energy and momentum in its angular motion $\frac{1}{2}I\omega^2$. It is used to dampen out fluctuations arising from single (or few) cylinder engines that otherwise would give an uneven rotational speed. The storage is for only a very short time. A modern flywheel is used to dampen fluctuations in intermittent power supplies like a wind turbine. It can store more energy than the flywheel shown in Fig. 3.31. A bank of several flywheels can provide substantial power for 5 to 10 minutes.

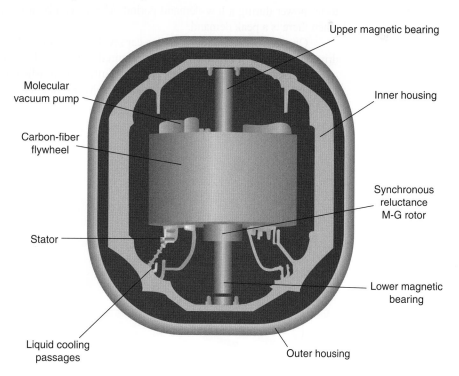

FIGURE 3.31
Modern flywheel.

FIGURE 3.32 The world's largest artificial hydro-storage facility in Ludington, Michigan, pumps water 100 m above Lake Michigan when excess power is available. It can deliver 1800 MW when needed from reversible pumps/turbines.

A fraction of the kinetic energy in air can be captured and converted into electrical power by wind turbines, or the power can be used directly to drive a water pump or other equipment.

When excess power is available, it can be used to pump water up to a reservoir at a higher elevation (see Fig. 3.32) and later can be allowed to run out through a turbine, providing a variable time shift in the power going to the electrical grid.

Air can be compressed into large tanks or volumes (as in an abandoned salt mine) using power during a low-demand period. The air can be used later in power production when there is a peak demand.

One form of hybrid engine for a car involves coupling a hydraulic pump/motor to the drive shaft. When a braking action is required, the drive shaft pumps hydraulic fluid into a high-pressure tank that has nitrogen as a buffer. Then, when acceleration is needed, the high-pressure fluid runs backward through the hydraulic motor, adding power to the drive

FIGURE 3.33
Examples of different types of batteries.

(© sciencephotos/Alamy Limited)

shaft in the process. This combination is highly beneficial for city driving, such as for a bus that stops and starts many times, whereas there is virtually no gain for a truck driving long distances on the highway at nearly constant speed.

Thermal Systems

Water can be heated by solar influx, or by some other source, to provide heat at a time when this source is not available. Similarly, water can be chilled or frozen at night to be used the next day for air-conditioning purposes. A cool-pack is placed in the freezer so that the next day it can be used in a lunch box to keep it cool. This is a gel with a high specific heat or a substance that undergoes a phase change.

Electrical Systems

Some batteries can only be discharged once, but others can be reused and go through many cycles of charging-discharging. A chemical process frees electrons on one of two poles that are separated by an electrolyte. The type of pole and the electrolyte give the name to the battery, such as a zinc-carbon battery (typical AA battery) or a lead-acid battery (typical automobile battery). Newer types of batteries like a Ni-hydride or a lithium-ion battery are more expensive but have higher energy storage, and they can provide higher bursts of power (Fig. 3.33).

Chemical Systems

Various chemical reactions can be made to operate under conditions such that energy can be stored at one time and recovered at another time. Small heat packs can be broken to mix some chemicals that react and release energy in the form of heat; in other cases, they can be glow-sticks that provide light. A fuel cell is also an energy conversion device that converts a flow of hydrogen and oxygen into a flow of water plus heat and electricity. High-temperature fuel cells can use natural gas or methanol as the fuel; in this case, carbon dioxide is also a product.

The latest technology for a solar-driven power plant consists of a large number of adjustable mirrors tracking the sun so that the sunlight is focused on the top of a tower. The light heats a flow of molten salt that flows to storage tanks and the power plant. At times when the sunlight is absent the storage tanks provides the energy buffer to keep the power plant running, thus increasing the utilization of the plant. Earlier versions of such technology used water or other substances to capture the energy, but the higher specific heat of salt provides an economical buffer system.

When work needs to be transferred from one body to another, a moving part is required, which can be a piston/cylinder combination. Examples are shown in Fig. 3.34. If the substance that generates the motion is a gas, it is a pneumatic system, and if the substance is a liquid, it is a hydraulic system. The gas or vapor is typically used when the motion has

FIGURE 3.34 Basic hydraulic or pneumatic cylinders.

(a) Hydraulic cylinder

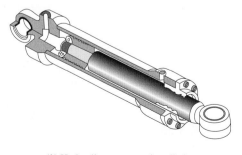

(b) Hydraulic or pneumatic cylinder

FIGURE 3.35
Heavy-duty equipment
using hydraulic cylinders.

(*a*) Forklift

(*b*) Construction frontloader

to be fast or the volume change large and the pressures moderate. For high-pressure (large-force) displacements a hydraulic cylinder is used (examples include a bulldozer, forklift, frontloader, and backhoe. Also, see Example 1.7). Two of these large pieces of equipment are shown in Fig. 3.35.

We also consider cases where the substance inside the piston/cylinder undergoes a combustion process, as in gasoline and diesel engines. A schematic of an engine cylinder and a photo of a modern V6 automotive engine are shown in Fig. 3.36. This subject is discussed in detail in Chapter 10.

Many other transfers of work involve rotating shafts, such as the transmission and drive shaft in a car or a chain and rotating gears in a bicycle or motorcycle. For transmission of power over long distances, the most convenient and efficient form is electricity. A transmission tower and line are shown in Fig. 3.37.

Heat transfer occurs between domains at different temperatures, as in a building with different inside and outside temperatures. The double set of window panes shown in Fig. 3.38 is used to reduce the rate of heat transfer through the window. In situations where an increased rate of heat transfer is desirable, fins are often used to increase the surface area for heat transfer to occur. Examples are shown in Fig. 3.39.

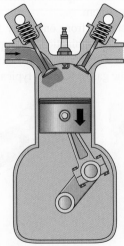

FIGURE 3.36
Schematic and photo of
an automotive engine.

(*a*) Schematic of engine cylinder

(*b*) V6 automotive engine

FIGURE 3.37
Electrical power transmission tower and line.

(© Sergey Peterman/iStockphoto)

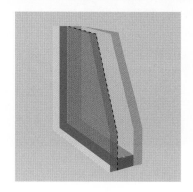

FIGURE 3.38
Thermopane window.

The last example of a finned heat exchanger is a heat pipe or a thermosyphon used for an enhanced cooling capacity of a central processing unit (CPU) in a computer (Fig. 3.40). The small aluminum block with the copper piping attaches to the top of the CPU unit. Inside the copper tubing is a liquid that boils at a temperature of about 60 °C. The vapor rises to the top, where the copper piping is connected to the fins, and a fan blows air through the fins, thus cooling and condensing the vapor. The liquid falls with gravity or is transported by a wick back to the volume on top of the CPU unit. The heat pipe allows the boiling heat transfer with the high transfer coefficient to act on the small area of the CPU. The less effective vapor to air heat transfer takes place further away, with more room for a larger area. Similar heat pipes are used in solar heat collectors and in the support pillars for the Alaskan oil pipeline, where they keep the permafrost ground frozen while the pipeline is warm.

FIGURE 3.39
Examples of fin-enhanced heat transfer.

(© Baloncici/iStockphoto)

(*a*) Motorcycle engine cylinder

(© C. Borgnakke)

(*b*) Inside of a baseboard heater

(Martin Leigh/Getty Images, Inc.)

(*c*) Air cooled heavy-equipment oil coolers

(© C. Borgnakke.)

FIGURE 3.40 A thermosyphon with a fan on the back for CPU cooling.

When heat transfer calculations are done in practice, it is convenient to use a common form for all modes of heat transfer:

$$\dot{Q} = C_q A\, \Delta T = \Delta T / R_t \tag{3.54}$$

The heat transfer scales with the cross-sectional area perpendicular to the direction of $\dot{Q}$, and the rest of the information is in the constant C_q. With a rewrite, this equation is also used to define the thermal resistance, $R_t = 1/C_q A$, so for a high resistance the heat transfer is small for a given temperature difference, ΔT. This form corresponds to conduction, Eq. 3.23 with $dT/dx \approx \Delta T/\Delta x$, so $C_q = k/\Delta x$, and to convection, Eq. 3.24 $C_q = h$. Finally, the radiation expression in Eq. 3.25 can be factored out to show a temperature difference, and then the factor C_q depends on the temperature in a nonlinear manner.

An application that involves heat transfer and the unsteady form of the energy equation is the following, where we would like to know how fast a given mass can adjust to an external temperature. Assume we have a mass, m, with a uniform temperature T_0 that we lower into a water bath with temperature T_∞, and the heat transfer between the mass and water has a heat transfer coefficient C_q with surface area A.

The energy equation for the mass becomes

$$\frac{dE_{cv}}{dt} = \frac{dU_{cv}}{dt} = mC_v\frac{dT}{dt} = \dot{Q} = -C_q A(T - T_\infty)$$

where kinetic and potential energy are neglected and there is no work involved. It is also assumed that the change in internal energy can be expressed with a specific heat, so this expression does not apply to a phase change process. This is a first-order differential equation in T, so select a transformation $\theta = T - T_\infty$ to get

$$mC_v\frac{d\theta}{dt} = -C_q A\theta \quad \text{or} \quad \theta^{-1}d\theta = -\frac{C_q A}{mC_v}\, dt$$

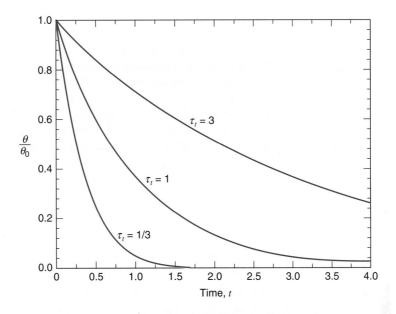

FIGURE 3.41 The exponential time decay of dimensionless temperature θ/θ_0.

Integrating the equation from $t = 0$ where $T = T_0$ ($\theta = \theta_0$), we get

$$\ln\left(\frac{\theta}{\theta_0}\right) = -\frac{C_q A}{m C_v}t \quad \text{or} \quad \theta = \theta_0 \exp\left(-\frac{t}{\tau}\right)$$

with the thermal time constant

$$\tau = \frac{m C_v}{C_q A} = m\, C_v R_t \tag{3.55}$$

Expressing the solution in temperature

$$T - T_\infty = (T_0 - T_\infty)\exp\left(-\frac{t}{\tau}\right) \tag{3.56}$$

shows an exponential time decay of the temperature difference between the mass and its surroundings with a time scale of τ (Fig. 3.41). If the mass is the tip of a thermocouple, we obtain a fast response for a small thermal time constant (small mC_v, high C_qA). However, if the mass is a house (given mC_v), we want a large time constant so that we lower the effective C_qA by having insulation.

SUMMARY

Conservation of energy is expressed as an equation for change of the total energy written for a control mass, and then the first law of thermodynamics is shown as a logical consequence of the energy equation. The energy equation is shown in a rate form to cover transient processes and then is also integrated with time for finite changes. The concept of work is introduced, and its relation to kinetic and potential energy is shown since they are part of the total energy. Work is a function of the process path as well as the beginning state and end state. The displacement work is equal to the area below the process curve drawn in a P–V diagram in an equilibrium process. A number of ordinary processes can be expressed as polytropic processes having a particular simple mathematical form for the P–V relation. Work involving the action of surface tension, single-point forces, or electrical systems should be recognized and treated separately. Any nonequilibrium processes (say, dynamic forces, which are important due to accelerations) should be identified so that only equilibrium force or pressure is used to evaluate the work term. Heat transfer is energy transferred due to a temperature difference, and the conduction, convection, and radiation modes are discussed.

Internal energy and enthalpy are introduced as substance properties with specific heats (heat capacity) as derivatives of these properties with temperature. Property variations for limited cases are presented for incompressible states of a substance such as liquids and solids and for a highly compressible state such as an ideal gas. The specific heat for solids and liquids changes little with temperature, whereas the specific heat for a gas can change substantially with temperature.

You should have learned a number of skills and acquired abilities from studying this chapter that will allow you to

- Recognize the components of total energy stored in a control mass.
- Write the energy equation for a single, uniform control mass.
- Recognize force and displacement in a system.
- Understand power as the rate of work (force $\times$ velocity, torque $\times$ angular velocity).
- Know that work is a function of the end states and the path followed in a process.
- Know that work is the area under the process curve in a P–V diagram.
- Calculate the work term knowing the P–V or F–x relationship.
- Evaluate the work involved in a polytropic process between two states.
- Distinguish between an equilibrium process and a nonequilibrium process.
- Recognize the three modes of heat transfer: conduction, convection, and radiation.
- Be familiar with Fourier's law of conduction and its use in simple applications.
- Know the simple models for convection and radiation heat transfer.
- Find the properties u and h for a given state in the tables in Appendix B.
- Locate a state in the tables with an entry such as (P, h).
- Find changes in u and h for liquid or solid states using Tables A.3 and A.4.
- Find changes in u and h for ideal-gas states using Table A.5.
- Find changes in u and h for ideal-gas states using Tables A.7 and A.8.
- Recognize that forms for C_p in Table A.6 are approximations to curves in Fig. 3.26 and that more accurate tabulations are in Tables A.7 and A.8.
- Formulate the conservation of mass and energy for a more complex control mass where there are different masses with different states.
- Know the difference between the general laws as the conservation of mass (continuity equation), conservation of energy (first law), and a specific law that describes a device behavior or process.

KEY CONCEPTS AND FORMULAS

Total energy	$E = U + \mathrm{KE} + \mathrm{PE} = mu + \dfrac{1}{2}m\mathbf{V}^2 + mgZ$
Kinetic energy	$\mathrm{KE} = \dfrac{1}{2}m\mathbf{V}^2$
Potential energy	$\mathrm{PE} = mgZ$
Specific energy	$e = u + \dfrac{1}{2}\mathbf{V}^2 + gZ$

Enthalpy	$h \equiv u + Pv$
Two-phase mass average	$u = u_f + x u_{fg} = (1 - x)u_f + x u_g$
	$h = h_f + x h_{fg} = (1 - x)h_f + x h_g$
Specific heat, heat capacity	$C_v = \left(\dfrac{\partial u}{\partial T}\right)_v$; $C_p = \left(\dfrac{\partial h}{\partial T}\right)_p$
Solids and liquids	Incompressible, so $v = $ constant $\cong v_f$ (or v_i) and v small
	$C = C_v = C_p$ (Tables A.3 and A.4)
	$u_2 - u_1 = C(T_2 - T_1)$
	$h_2 - h_1 = u_2 - u_1 + v(P_2 - P_1)$ (Often the second
	term is small.)
	$h = h_f + v_f(P - P_{sat})$; $u \cong u_f$ (saturated at same T)
Ideal gas	$h = u + Pv = u + RT$ (only functions of T)
	$C_v = \dfrac{du}{dT}$; $C_p = \dfrac{dh}{dT} = C_v + R$
	$u_2 - u_1 = \displaystyle\int C_v\,dT \cong C_v(T_2 - T_1)$
	$h_2 - h_1 = \displaystyle\int C_p\,dT \cong C_p(T_2 - T_1)$
	Left-hand side from Table A.7 or A.8, middle from Table A.6, and right-hand side from Table A.6 at T_{avg} or from Table A.5 at $25\,^\circ$C
Energy equation rate form	$\dot{E} = \dot{Q} - \dot{W}$ (rate = +in − out)
Energy equation integrated	$E_2 - E_1 = {}_1Q_2 - {}_1W_2$ (change = +in − out)
	$m(e_2 - e_1) = m(u_2 - u_1) + \dfrac{1}{2}m(\mathbf{V}_2^2 - \mathbf{V}_1^2) + mg(Z_2 - Z_1)$
Multiple masses, states	$E = m_A e_A + m_B e_B + m_C e_C + \cdots$
Work	Energy in transfer: mechanical, electrical, and chemical
Heat	Energy in transfer caused by ΔT
Displacement work	$W = \displaystyle\int_1^2 F\,dx = \int_1^2 P\,dV = \int_1^2 \mathscr{S}\,dA = \int_1^2 T\,d\theta$
Specific work	$w = W/m$ (work per unit mass)
Power, rate of work	$\dot{W} = F\mathbf{V} = P\dot{V} = T\omega$ ($\dot{V}$ displacement rate)
	Velocity $\mathbf{V} = r\omega$, torque $T = Fr$, angular velocity $= \omega$
Polytropic process	$PV^n = $ constant or $Pv^n = $ constant
Polytropic process work	${}_1W_2 = \dfrac{1}{1 - n}(P_2 V_2 - P_1 V_1)$ (if $n \neq 1$)
	${}_1W_2 = P_1 V_1 \ln \dfrac{V_2}{V_1}$ (if $n = 1$)
Conduction heat transfer	$\dot{Q} = -kA\dfrac{dT}{dx} \simeq kA\dfrac{\Delta T}{L}$
Conductivity	k (W/m·K)
Convection heat transfer	$\dot{Q} = hA\,\Delta T$

Convection coefficient	$h \, (\mathrm{W/m^2 \cdot K})$
Radiation heat transfer	$\dot{Q} = \varepsilon \sigma A (T_s^4 - T_{\mathrm{amb}}^4) \quad (\sigma = 5.67 \times 10^{-8} \, \mathrm{W/m^2 \cdot K^4})$
(net to ambient)	
Rate integration	$_1Q_2 = \int \dot{Q} \, dt \approx \dot{Q}_{\mathrm{avg}} \, \Delta t$

CONCEPT-STUDY GUIDE PROBLEMS

3.1 What is 1 cal in SI units and what is the name given to 1 N·m?

3.2 A car engine is rated at 110 kW. What is the power in hp?

3.3 Why do we write ΔE or $E_2 - E_1$, whereas we write $_1Q_2$ and $_1W_2$?

3.4 If a process in a control mass increases energy $E_2 - E_1 > 0$, can you say anything about the sign for $_1Q_2$ and $_1W_2$?

3.5 In Fig. P3.5, CV A is the mass inside a piston/cylinder, and CV B is the mass plus the piston outside, which is the standard atmosphere. Write the energy equation and work term for the two CVs, assuming we have a nonzero Q between state 1 and state 2.

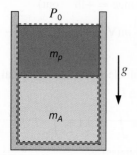

FIGURE P3.5

3.6 A 500 W electric space heater with a small fan inside heats air by blowing it over a hot electrical wire. For each control volume—(a) wire only, (b) all the room air, and (c) total room air plus the heater—specify the storage, work, and heat transfer terms as +500 W or −500 W or 0 (neglect any $\dot{Q}$ through the room walls or windows).

3.7 Two engines provide the same amount of work to lift a hoist. One engine provides a force of 3 F in a cable and the other provided 1 F. What can you say about the motion of the point where the force acts in the two engines?

3.8 Two hydraulic piston/cylinders are connected through a hydraulic line, so they have roughly the same pressure. If they have diameters of D_1 and $D_2 = 2D_1$, respectively, what can you say about the piston forces F_1 and F_2?

3.9 Assume a physical setup as in Fig. P3.5. We now heat the cylinder. What happens to P, T, and v (up, down, or constant)? What transfers do we have for Q and W (positive, negative, or zero)?

3.10 A drag force on an object moving through a medium (like a car through air or a submarine through water) is $F_d = 0.225 \, A \, \rho \mathbf{V}^2$. Verify that the unit becomes N.

3.11 Figure P3.11 shows three physical situations. Show the possible process in a P–v diagram.

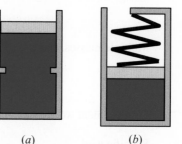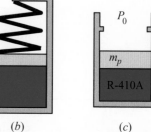

(a) (b) (c)

FIGURE P3.11

3.12 For the indicated physical setup in (a), (b), and (c) in Fig. P3.11, write a process equation and the expression for work.

3.13 Assume the physical situation in Fig. P3.11b; what is the work term a, b, c, or d?

a. $_1w_2 = P_1(v_2 - v_1)$

b. $_1w_2 = v_1(P_2 - P_1)$

c. $_1w_2 = \frac{1}{2}(P_1 + P_2)(v_2 - v_1)$

d. $_1w_2 = \frac{1}{2}(P_1 - P_2)(v_2 + v_1)$

3.14 Figure P3.14 shows three physical situations; show the possible process in a *P*–*v* diagram.

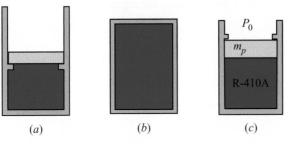

(*a*)   (*b*)   (*c*)

FIGURE P3.14

3.15 What can you say about the beginning state of the R-410A in Fig. P3.11*c* versus that in Fig. P3.14*c* for the same piston/cylinder?

3.16 A piece of steel has a conductivity of $k = 15$ W/*mK* and a brick has $k = 1$ W/*mK*. How thick a steel wall will provide the same insulation as a 10 cm thick brick?

3.17 A thermopane window (see Fig. 3.38) traps some gas between the two glass panes. Why is this beneficial?

3.18 On a chilly, 10 °C fall day a house, with an indoor temperature of 20 °C, loses 6 kW by heat transfer. What transfer happens on a warm summer day with an indoor temperature of 30 °C, assuming everything else is the same?

3.19 Verify that a surface tension $\mathscr{S}$ with units N/m also can be called a *surface energy* with units J/m². The latter is useful for consideration of a liquid drop or liquid in small pores (capillary).

3.20 Liquid water is heated, so it becomes superheated vapor. Should *u* or *h* be used in the energy equation? Explain.

3.21 Liquid water is heated, so it becomes superheated vapor. Can specific heat be used to find the heat transfer? Explain.

3.22 Look at the R-410A value for u_f at -50 °C. Can the energy really be negative? Explain.

3.23 A rigid tank with pressurized air is used (a) to increase the volume of a linear spring-loaded piston/cylinder (cylindrical geometry) arrangement and (b) to blow up a spherical balloon. Assume that in both cases, $P = A + BV$ with the same *A* and *B*. What is the expression for the work term in each situation?

3.24 An ideal gas in a piston/cylinder is heated with 2 kJ during an isothermal process. How much work is involved?

3.25 An ideal gas in a piston/cylinder is heated with 2 kJ during an isobaric process. Is the work positive, negative, or zero?

3.26 You heat a gas 10 K at $P = C$. Which one in Table A.5 requires most energy? Why?

3.27 You mix 20 °C water with 50 °C water in an open container. What do you need to know to determine the final temperature?

HOMEWORK PROBLEMS

Kinetic and Potential Energy

3.28 A piston motion moves a 25 kg hammerhead vertically down 1 m from rest to a velocity of 50 m/s in a stamping machine. What is the change in total energy of the hammerhead?

3.29 A 1200 kg car accelerates from 30 to 50 km/h in 5 s. How much work input does that require? If it continues to accelerate from 50 to 70 km/h in 5 s, is that the same?

3.30 The rolling resistance of a car depends on its weight as $F = 0.006\ m_{car}g$. How far will a 1200 kg car roll if the gear is put in neutral when it drives at 90 km/h on a level road without air resistance?

3.31 A piston of mass 2 kg is lowered 0.5 m in the standard gravitational field. Find the required force and the work involved in the process.

3.32 A hydraulic hoist raises a 1750 kg car 1.8 m in an auto repair shop. The hydraulic pump has a constant pressure of 800 kPa on its piston. What is the increase in potential energy of the car and how much volume should the pump displace to deliver that amount of work?

3.33 Airplane takeoff from an aircraft carrier is assisted by a steam-driven piston/cylinder with an average pressure of 1250 kPa. A 17 500 kg airplane should accelerate from zero to 30 m/s, with 30 % of the energy coming from the steam piston. Find the needed piston displacement volume.

3.34 Solve Problem 3.33, but assume that the steam pressure in the cylinder starts at 1000 kPa, dropping linearly with volume to reach 100 kPa at the end of the process.

3.35 A steel ball weighing 5 kg rolls horizontally at a rate of 10 m/s. If it rolls up an incline, how high up will it be when it comes to rest, assuming standard gravitation?

Force Displacement Work

3.36 A hydraulic cylinder of area 0.005 m² must push a 1000 kg arm and shovel 0.5 m straight up. What pressure is needed and how much work is done?

3.37 Two hydraulic piston/cylinders are connected with a line. The master cylinder has an area of 5 cm², creating a pressure of 1000 kPa. The slave cylinder has an area of 3 cm². If 25 J is the work input to the master cylinder, what is the force and displacement of each piston and the work output of the slave cylinder piston?

3.38 The air drag force on a car is $0.225\,A\,\rho\mathbf{V}^2$. Assume air at 290 K, 100 kPa, and a car frontal area of 4 m² driving at 90 km/h. How much energy is used to overcome the air drag driving for 30 min?

3.39 Two hydraulic cylinders maintain a pressure of 800 kPa. One has a cross-sectional area of 0.01 m², the other 0.03 m². To deliver work of 1 kJ to the piston, how large a displacement V and piston motion H are needed for each cylinder? Neglect P_{atm}.

3.40 A linear spring, $F = k_s(x - x_0)$ with spring constant $k_s = 500$ N/m, is stretched until it is 100 mm longer. Find the required force and the work input.

Boundary Work

3.41 The R-410A in Problem 3.14c is at 1000 kPa, 50 °C with a mass of 0.1 kg. It is cooled so that the volume is reduced to half the initial volume. The piston mass and gravitation are such that a pressure of 400 kPa will float the piston. Find the work in the process.

3.42 A 400 L tank, A (see Fig. P3.42), contains argon gas at 200 kPa and 30 °C. Cylinder B, having a frictionless piston of such mass that a pressure of 150 kPa will float it, is initially empty. The valve is opened,

and argon flows into B and eventually reaches a uniform state of 150 kPa and 30 °C throughout. What is the work done by the argon?

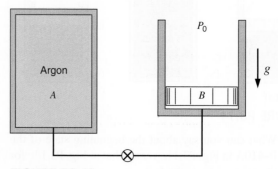

FIGURE P3.42

3.43 A piston/cylinder assembly contains 2 kg of liquid water at 20 °C and 300 kPa, as shown in Fig. P3.43. There is a linear spring mounted on the piston such that when the water is heated, the pressure reaches 3 MPa with a volume of 0.1 m³.
a. Find the final temperature.
b. Plot the process in a P–v diagram.
c. Find the work in the process.

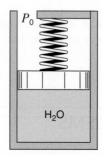

FIGURE P3.43

3.44 Air in a spring-loaded piston/cylinder setup has a pressure that is linear with volume, $P = A + BV$. With an initial state of $P = 150$ kPa, $V = 1$ L and a final state of 800 kPa, $V = 2$ L, it is similar to the setup in Problem 3.43. Find the work done by the air.

3.45 Heat transfer to a 1.5 kg block of ice at -10 °C melts it to liquid at 10 °C in a kitchen. How much work does the water gives out?

3.46 A cylinder fitted with a frictionless piston contains 5 kg of superheated R-134a vapor at 1000 kPa and

140 °C. The setup is cooled at constant pressure until the R-134a reaches a quality of 25 %. Calculate the work done in the process.

3.47 A nitrogen gas goes through a polytropic process with $n = 1.3$ in a piston/cylinder. It starts out at 600 K, 600 kPa, and ends at 800 K. Is the work positive, negative, or zero?

3.48 Helium gas expands from 135 kPa, 350 K and 0.25 m³ to 100 kPa in a polytropic process with $n = 1.667$. How much work does it give out?

3.49 Air goes through a polytropic process from 125 kPa and 325 K to 300 kPa and 500 K. Find the polytropic exponent n and the specific work in the process.

3.50 A balloon behaves so that the pressure is $P = C_2 V^{1/3}$ and $C_2 = 100$ kPa/m. The balloon is blown up with air from a starting volume of 1 m³ to a volume of 4 m³. Find the final mass of air, assuming it is at 25 °C, and the work done by the air.

Heat Transfer

3.51 The brake shoe and steel drum of a car continuously absorb 75 W as the car slows down. Assume a total outside surface area of 0.1 m² with a convective heat transfer coefficient of 10 W/m²·K to the air at 20 °C. How hot does the outside brake and drum surface become when steady conditions are reached?

3.52 A water heater is covered with insulation boards over a total surface area of 3 m². The inside board surface is at 80 °C, the outside surface is at 15 °C, and the board material has a conductivity of 0.08 W/m·K. How thick should the board be to limit the heat transfer loss to 200 W?

3.53 A 2 m² window has a surface temperature of 15 °C, and the outside wind is blowing air at 7 °C across it with a convection heat transfer coefficient of $h = 125$ W/m²·K. What is the total heat transfer loss?

3.54 A steel pot, with conductivity of 50 W/m·K and a 5 mm thick bottom, is filled with 15 °C liquid water. The pot has a diameter of 20 cm and is now placed on an electric stove that delivers 500 W as heat transfer. Find the temperature on the outer pot bottom surface, assuming the inner surface is at 15 °C.

3.55 A log of burning wood in the fireplace has a surface temperature of 480 °C. Assume that the emissivity is 1 (a perfect black body) and find the radiant emission of energy per unit surface area.

3.56 A wall surface on a house is 30 °C with an emissivity of $\varepsilon = 0.7$. The surrounding ambient air is at 15 °C with an average emissivity of 0.9. Find the rate of radiation energy from each of those surfaces per unit area.

3.57 A radiant heat lamp is a rod, 0.5 m long and 0.5 cm in diameter, through which 250 W of electric energy is deposited. Assume that the surface has an emissivity of 0.9 and neglect incoming radiation. What will the rod surface temperature be?

Properties (u, h) from General Tables

3.58 Determine the phase of the following substances and find the values of the unknown quantities.
  a. Nitrogen: $P = 2000$ kPa, 120 K, $v = ?$, $Z = ?$
  b. Nitrogen: 120 K, $v = 0.0050$ m³/kg, $Z = ?$
  c. Air: $T = 100$ °C, $v = 0.500$ m³/kg, $P = ?$
  d. R-410A: $T = 25$ °C, $v = 0.01$ m³/kg, $P = ?$, $h = ?$

3.59 Find the phase and the missing properties of P, T, v, u, and x for water at
  a. 500 kPa, 100 °C
  b. 5000 kPa, $u = 800$ kJ/kg
  c. 5000 kPa, $v = 0.06$ m³/kg
  d. −6 °C, $v = 1$ m³/kg

3.60 Indicate the location of the four states in Problem 3.59 as points in both the P–v and T–v diagrams.

3.61 Find the missing properties of P, v, u, and x and the phase of ammonia, NH_3.
  a. $T = 65$ °C, $P = 600$ kPa
  b. $T = 20$ °C, $P = 100$ kPa
  c. $T = 50$ °C, $v = 0.1185$ m³/kg

3.62 Find the missing property of P, T, v, u, h, and x and indicate the states in a P–v and a T–v diagram for
  a. Water at 5000 kPa, $u = 750$ kJ/kg
  b. R-134a at 20 °C, $u = 300$ kJ/kg
  c. Nitrogen at 250 K, 200 kPa

3.63 Determine the phase and the missing properties.
  a. H_2O 20 °C, $v = 0.001\ 000$ m³/kg $P = ?$, $u = ?$
  b. R-410A 400 kPa, $v = 0.075$ m³/kg $T = ?$, $u = ?$
  c. NH_3 10 °C, $v = 0.1$ m³/kg $P = ?$, $u = ?$
  d. N_2 101.3 kPa, $u = 50$ kJ/kg $T = ?$, $v = ?$

3.64 Determine the phase of the following substances and find the values of the unknown quantities.
  a. R-410A: $T = -20$ °C, $u = 220$ kJ/kg, $P = ?$, $x = ?$
  b. Ammonia: $T = -10$ °C, $v = 0.35$ m³/kg, $P = ?$, $u = ?$
  c. Water: $P = 400$ kPa, $h = 2800$ kJ/kg, $T = ?$, $v = ?$

3.65 Find the missing property of P, T, v, u, h, and x and indicate the states in a P–v and a T–v diagram for
  a. R-410A at 500 kPa, $h = 300$ kJ/kg
  b. R-410A at 10 °C, $u = 250$ kJ/kg
  c. R-134a at 40 °C, $h = 400$ kJ/kg

3.66 Saturated liquid water at 20 °C is compressed to a higher pressure with constant temperature. Find the changes in u and h from the initial state when the final pressure is
  a. 500 kPa
  b. 2000 kPa

Problem Analysis

3.67 Consider Problem 3.81. Take the whole room as a C.V. and write both conservation of mass and conservation of energy equations. Write equations for the process (two are needed) and use them in the conservation equations. Now specify the four properties that determine the initial state (two) and the final state (two); do you have them all? Count unknowns and match them with the equations to determine those.

3.68 Consider a steel bottle as a CV. It contains carbon dioxide at −20 °C, quality 20 %. It has a safety valve that opens at 6 MPa. The bottle is now accidentally heated until the safety valve opens. Write the process equation that is valid until the valve opens and plot the P–v diagram for the process.

3.69 A piston/cylinder contains water with quality 75 % at 200 kPa. Slow expansion is performed while there is heat transfer and the water is at constant pressure. The process stops when the volume has doubled. How do you determine the final state and the heat transfer?

3.70 Two rigid insulated tanks are connected with a pipe and valve. One tank has 0.5 kg air at 200 kPa, 300 K and the other has 0.75 kg air at 100 kPa, 400 K. The valve is opened and the air comes to a single uniform state without any heat transfer. How do you determine the final temperature and pressure?

3.71 Look at Problem 3.138 and plot the P–v diagram for the process. Only T_2 is given; how do you determine the second property of the final state? What do you need to check, and does it influence the work term?

Simple Processes

3.72 A 125 L rigid tank contains nitrogen (N_2) at 900 K and 3 MPa. The tank is now cooled to 100 K. What are the work and heat transfer for the process?

3.73 A constant-pressure piston/cylinder assembly contains 0.2 kg of water as saturated vapor at 400 kPa. It is now cooled so that the water occupies half of the original volume. Find the work and heat transfer done in the process.

3.74 Saturated vapor R-410A at 0 °C in a rigid tank is cooled to −10 °C. Find the specific heat transfer.

3.75 A rigid tank contains 0.5 kg of R-134a at 40 °C, 500 kPa. The tank is placed in a refrigerator that brings it to −20 °C. Find the process heat transfer and show the process in a P–v diagram.

3.76 A piston/cylinder contains air at 600 kPa, 290 K and a volume of 0.01 m³. A constant-pressure process gives 54 kJ of work out. Find the final volume, the temperature of the air, and the heat transfer.

3.77 A piston/cylinder contains 1.5 kg of water at 200 kPa, 150 °C. It is now heated by a process in which pressure is linearly related to volume to a state of 600 kPa, 350 °C. Find the final volume, the heat transfer, and the work in the process.

3.78 A piston/cylinder device contains 50 kg water at 200 kPa with a volume of 0.1 m³. Stops in the cylinder are placed to restrict the enclosed volume to a maximum of 0.5 m³. The water is now heated until the piston reaches the stops. Find the necessary heat transfer.

3.79 Ammonia (0.5 kg) in a piston/cylinder at 200 kPa, −10 °C is heated by a process in which pressure varies linearly with volume to a state of 100 °C, 300 kPa. Find the work and heat transfer for the ammonia in the process.

3.80 A piston/cylinder contains 1 kg water at 20 °C with volume 0.1 m³. By mistake someone locks the piston, preventing it from moving while we heat the water to saturated vapor. Find the final temperature and the amount of heat transfer in the process.

3.81 A water-filled reactor with a volume of 1 m³ is at 20 MPa and 360 °C and is placed inside a containment room, as shown in Fig. P3.81. The room is well insulated and initially evacuated. Due to a failure, the reactor ruptures and the water fills the containment room. Find the minimum room volume so that the final pressure does not exceed 200 kPa.

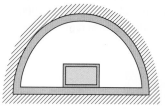

FIGURE P3.81

3.82 A 25 kg mass moves at 25 m/s. Now a brake system brings the mass to a complete stop with a constant deceleration over a period of 5 s. Assume the mass is at constant P and T. The brake energy is absorbed by 0.5 kg of water initially at 20 °C and 100 kPa. Find the energy the brake removes from the mass and the temperature increase of the water, assuming its pressure is constant.

3.83 A piston/cylinder arrangement with a linear spring as shown in Fig. P3.83 contains R-134a at 15 °C, $x = 0.4$ and a volume of 0.02 m^3. It is heated to 60 °C, at which point the specific volume is 0.030 02 m^3/kg. Find the final pressure, the work, and the heat transfer in the process.

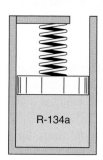

FIGURE P3.83

3.84 Assume the same setup as in Problem 3.81, but the room has a volume of 100 m^3. Show that the final state is two phase and find the final pressure by trial and error.

3.85 A piston/cylinder contains carbon dioxide at -20 °C and quality 75 %. It is compressed in a process wherein pressure is linear in volume to a state of 3 MPa and 20 °C. Find specific heat transfer.

3.86 A rigid steel tank of mass 2.1 kg contains 0.5 kg R-410A at 0 °C with a specific volume of 0.01 m^3/kg.

The whole system is now heated to a room temperature of 25 °C.
a. Find the volume of the tank.
b. Find the final P.
c. Find the process heat transfer.

3.87 The piston/cylinder in Fig. P3.87 contains 0.1 kg water at 500 °C, 1000 kPa. The piston has a stop at half of the original volume. The water now cools to a room temperature of 25 °C.
a. Sketch the possible water states in a P–v diagram.
b. Find the final pressure and volume.
c. Find the heat transfer and work in the process.

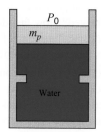

FIGURE P3.87

3.88 A spring-loaded piston/cylinder assembly contains 1 kg of water at 500 °C, 3 MPa. The setup is such that the pressure is proportional to the volume: $P = CV$. It is now cooled until the water becomes saturated vapor. Sketch the P–v diagram, and find the final state, the work and heat transfer in the process.

3.89 Superheated refrigerant R-134a at 20 °C and 0.5 MPa is cooled in a piston/cylinder arrangement at constant temperature to a final two-phase state with quality of 50 %. The refrigerant mass is 5 kg, and during this process 500 kJ of heat is removed. Find the initial and final volumes and the necessary work.

3.90 Two kilograms of nitrogen at 100 K, $x = 0.5$ is heated in a constant-pressure process to 300 K in a piston/cylinder arrangement. Find the initial and final volumes and the total heat transfer required.

Specific Heats: Solids and Liquids

3.91 A computer CPU chip consists of 50 g silicon, 20 g copper, and 50 g polyvinyl chloride (plastic). It now heats from 15 °C to 70 °C as the computer is turned on. How much energy did the heating require?

3.92 A copper block of volume 1 L is heat treated at 500 °C and now cooled in a 200 L oil bath initially at 20 °C, as shown in Fig. P3.92. Assuming no heat transfer with the surroundings, what is the final temperature?

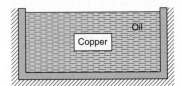

FIGURE P3.92

3.93 I have 2 kg of liquid water at 20 °C, 100 kPa. I now add 20 kJ of energy at constant pressure. How hot does the water get if it is heated? How fast does it move if it is pushed by a constant horizontal force? How high does it go if it is raised straight up?

3.94 A house is being designed to use a thick concrete floor mass as thermal storage material for solar energy heating. The concrete is 30 cm thick, and the area exposed to the sun during the daytime is 4 × 6 m. It is expected that this mass will undergo an average temperature rise of about 3 °C during the day. How much energy will be available for heating during the nighttime hours?

3.95 A car with mass 1275 kg is driven at 60 km/h when the brakes are applied quickly to decrease its speed to 20 km/h. Assume that the brake pads have a 0.5 kg mass with a specific heat of 1.1 kJ/kg·K and that the brake disks/drums are 4.0 kg of steel. Further assume that both masses are heated uniformly. Find the temperature increase in the brake assembly.

3.96 A piston/cylinder (0.5 kg steel altogether) maintaining a constant pressure has 0.2 kg R-134a as saturated vapor at 150 kPa. It is heated to 40 °C, and the steel is at the same temperature as the R-134a at any time. Find the work and heat transfer for the process.

3.97 A 15 kg steel tank initially at −10 °C is filled with 100 kg of milk (assumed to have the same properties as water) at 30 °C. The milk and the steel come to a uniform temperature of +5 °C in a storage room. How much heat transfer is needed for this process?

3.98 An engine, shown in Fig. P3.98, consists of a 100 kg cast iron block with a 20 kg aluminum head,

20 kg of steel parts, 5 kg of engine oil, and 6 kg of glycerine (antifreeze). All initial temperatures are 5 °C, and as the engine starts we want to know how hot it becomes if it absorbs a net of 7000 kJ before it reaches a steady uniform temperature.

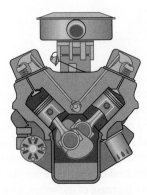

Automobile engine   **FIGURE P3.98**

Properties (u, h, C_v, C_p), Ideal Gas

3.99 An ideal gas is heated from 500 to 1500 K. Find the change in enthalpy using constant specific heat from Table A.5 (room temperature value) and discuss the accuracy of the result if the gas is
a. Argon
b. Oxygen
c. Carbon dioxide

3.100 Use the ideal-gas air Table A.7 to evaluate the specific heat C_p at 300 K as a slope of the curve $h(T)$ by $\Delta h/\Delta T$. How much larger is it at 1000 K and at 1500 K?

3.101 Estimate the constant specific heats for R-134a from Table B.5.2 at 100 kPa and 125 °C. Compare this to the specific heats in Table A.5 and explain the difference.

3.102 We want to find the change in u for carbon dioxide between 600 K and 1200 K.
a. Find it from a constant C_{v0} from Table A.5.
b. Find it from a C_{v0} evaluated from the equation in Table A.6 at the average T.
c. Find it from the values of u listed in Table A.8.

3.103 Nitrogen at 300 K, 3 MPa is heated to 500 K. Find the change in enthalpy using (a) Table B.6, (b) Table A.8, and (c) Table A.5.

3.104 Repeat Problem 3.102 for oxygen gas.

3.105 For a special application, we need to evaluate the change in enthalpy for carbon dioxide from 30 °C to 1500 °C at 100 kPa. Do this using the constant specific heat value from Table A.5 and repeat using Table A.8. Which table is more accurate?

3.106 The temperature of water at 400 kPa is raised from 150 °C to 1200 °C. Evaluate the change in specific internal energy using (a) the steam tables, (b) the ideal gas Table A.8, and (c) the specific heat Table A.5.

3.107 Water at 20 °C and 100 kPa is brought to 100 kPa and 1500 °C. Find the change in the specific internal energy, using the water tables and ideal-gas tables.

Specific Heats Ideal Gas

3.108 Air is heated from 300 to 350 K at constant volume. Find $_1q_2$. What is $_1q_2$ if the temperature rises from 1300 K to 1350 K?

3.109 Air (3 kg) is in a piston/cylinder similar to Fig. P3.5 at 27 °C, 300 kPa. It is now heated to 600 K. Plot the process path in a P–v diagram and find the work and heat transfer in the process.

3.110 A closed rigid container is filled with 1.5 kg water at 100 kPa, 55 °C; 1 kg of stainless steel, and 0.5 kg of polyvinyl chloride, both at 20 °C, and 0.1 kg air at 400 K, 100 kPa. It is now left alone, with no external heat transfer, and no water vaporizes. Find the final temperature and air pressure.

3.111 A 10 m high cylinder, with a cross-sectional area of 0.1 m², has a massless piston at the bottom with water at 20 °C on top of it, as shown in Fig. P3.111. Air at 300 K, with a volume of 0.3 m³, under the piston is heated so that the piston moves up, spilling the water out over the side. Find the total heat transfer to the air when all the water has been pushed out.

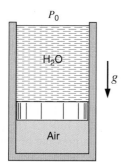

FIGURE P3.111

3.112 A cylinder with a piston restrained by a linear spring contains 2 kg of carbon dioxide at 500 kPa and 400 °C. It is cooled to 40 °C, at which point the pressure is 250 kPa. Calculate the heat transfer for the process.

3.113 A constant pressure container is filled with 1 kg of stainless steel and 0.5 kg of PVC (polyvinyl chloride) both at 25 °C and 0.25 kg of hot air at 500 K, 100 kPa. The container is now left alone with no external heat transfer.
a. Find the final temperature.
b. Find the process work.

3.114 A constant-pressure piston/cylinder contains 0.5 kg air at 300 K, 400 kPa. Assume the piston/cylinder has a total mass of 1 kg steel and is at the same temperature as the air at any time. The system is now heated to 1600 K by heat transfer.
a. Find the heat transfer using constant specific heats for air.
b. Find the heat transfer *not* using constant specific heats for air.

3.115 An insulated cylinder is divided into two parts of 1 m³ each by an initially locked piston, as shown in Fig. P3.115. Side A has air at 200 kPa, 300 K, and side B has air at 1.0 MPa, 1000 K. The piston is now unlocked so that it is free to move, and it conducts heat so that the air comes to a uniform temperature $T_A = T_B$. Find the mass in both A and B and the final T and P.

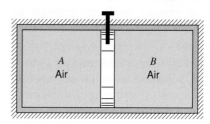

FIGURE P3.115

Polytropic Process

3.116 Air in a piston/cylinder is at 1800 K, 7 MPa and expands in a polytropic process with $n = 1.5$ to a volume eight times larger. Find the specific work and specific heat transfer in the process and draw the P–v diagram. Use constant specific heats to solve the problem.

3.117 Helium gas expands from 135 kPa, 350 K and 0.25 m³ to 100 kPa in a polytropic process with $n = 1.667$. How much heat transfer is involved?

3.118 A gasoline engine has a piston/cylinder with 0.1 kg air at 4 MPa, 1527 °C after combustion, and this is expanded in a polytropic process with $n = 1.5$ to a volume 10 times larger. Find the expansion work and heat transfer using the specific heat value in Table A.5.

3.119 Solve the previous problem using Table A.7.

3.120 Find the specific heat transfer in Problem 3.49.

3.121 A piston/cylinder has nitrogen gas at 750 K and 1500 kPa, as shown in Fig. P3.121. Now it is expanded in a polytropic process with $n = 1.2$ to $P = 750$ kPa. Find the final temperature, the specific work, and the specific heat transfer in the process.

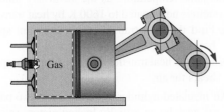

FIGURE P3.121

3.122 A piston/cylinder assembly has 1 kg of propane gas at 700 kPa and 40 °C. The piston cross-sectional area is 0.5 m², and the total external force restraining the piston is directly proportional to the cylinder volume squared. Heat is transferred to the propane until its temperature reaches 700 °C. Determine the final pressure inside the cylinder, the work done by the propane, and the heat transfer during the process.

3.123 A piston/cylinder arrangement of initial volume 0.025 m³ contains saturated water vapor at 180 °C. The steam now expands in a polytropic process with exponent $n = 1$ to a final pressure of 200 kPa while it does work against the piston. Determine the heat transfer for this process.

3.124 A piston/cylinder contains pure oxygen at ambient conditions 20 °C, 100 kPa. The piston is moved to a volume that is seven times smaller than the initial volume in a polytropic process with exponent $n = 1.25$. Use the constant specific heat to find the final pressure and temperature, the specific work, and the specific heat transfer.

3.125 A piston/cylinder assembly in a car contains 0.2 L of air at 90 kPa and 20 °C, as shown in Fig. P3.125. The air is compressed in a quasi-equilibrium polytropic process with polytropic exponent $n = 1.25$

to a final volume nine times smaller. Determine the final pressure and temperature, and the heat transfer for the process.

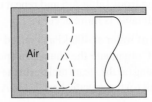

FIGURE P3.125

3.126 An air pistol contains compressed air in a small cylinder, as shown in Fig. P3.126. Assume that the volume is 1 cm³, the pressure is 1 MPa, and the temperature is 27 °C when armed. A bullet, with $m = 15$ g, acts as a piston initially held by a pin (trigger); when released, the air expands in an isothermal process (T = constant). If the air pressure is 0.1 MPa in the cylinder as the bullet leaves the gun, find

a. the final volume and the mass of air

b. the work done by the air and the work done on the atmosphere

c. the work done to the bullet and the bullet exit velocity

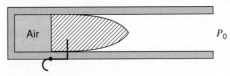

FIGURE P3.126

3.127 Air goes through a polytropic process with $n = 1.3$ in a piston/cylinder setup. It starts at 200 kPa, 300 K and ends with a pressure of 2200 kPa. Find the expansion ratio v_2/v_1, the specific work, and the specific heat transfer.

3.128 Nitrogen gas goes through a polytropic process with $n = 1.3$ in a piston/cylinder arrangement. It starts out at 600 K, 600 kPa and ends at 800 K. Find the final pressure, the process specific work and specific heat transfer.

Multistep Processes: All Substances

3.129 A piston/cylinder shown in Fig. P3.129 contains 0.5 m³ of R-410A at 2 MPa, 150 °C. The piston mass and atmosphere give a pressure of 450 kPa that will float the piston. The whole setup cools in a freezer maintained at −20 °C. Find the heat transfer and show the P–v diagram for the process when $T_2 = -20$ °C.

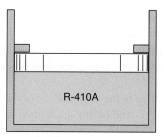

FIGURE P3.129

3.130 A cylinder containing 1 kg of ammonia has an externally loaded piston. Initially, the ammonia is at 2 MPa and 180 °C. It is now cooled to saturated vapor at 40 °C and then further cooled to 20 °C, at which point the quality is 50 %. Find the total work and the heat transfer for the process, assuming a piecewise linear variation of P versus V.

3.131 A helium gas is heated at constant volume from 100 kPa, 300 K to 600 K. A following process expands the gas at constant pressure to three times the initial volume. What is the specific work and the specific heat transfer in the combined process?

3.132 Water in a piston/cylinder (Fig. P3.132) is at 101 kPa, 25 °C, and mass 0.5 kg. The piston rests on some stops, and the pressure should be 1000 kPa to float the piston. We now heat the water, so the piston just reaches the end of the cylinder. Find the total heat transfer.

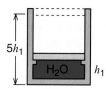

FIGURE P3.132

3.133 A setup like the one in Fig. P3.129 has the R-410A initially at 1000 kPa, 50 °C of mass 0.1 kg. The balancing equilibrium pressure is 400 kPa, and it is now cooled so that the volume is reduced to half of the starting volume. Find the heat transfer for the process.

3.134 A piston/cylinder assembly contains 1 kg of liquid water at 20 °C and 300 kPa. Initially the piston floats, as shown in Fig. P3.134, with a maximum enclosed volume of 0.002 m³ if the piston touches the stops. Now heat is added so that a final pressure of 600 kPa is reached. Find the final volume and the heat transfer in the process.

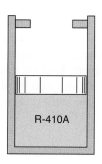

FIGURE P3.134

3.135 The piston/cylinder in Fig. P3.135 contains 0.1 kg R-410A at 600 kPa, 60 °C. It is now cooled, so the volume is reduced to half of the initial volume. The piston has upper stops mounted, and the piston mass and gravitation are such that a floating pressure is 400 kPa.
a. Find the final temperature.
b. How much work is involved?
c. What is the heat transfer in the process?
d. Show the process path in a P–v diagram.

FIGURE P3.135

3.136 A piston cylinder contains air at 1000 kPa, 800 K with a volume of 0.05 m³. The piston is pressed against the upper stops (see Fig. P3.14c) and it will float at a pressure of 700 kPa. Now the air is cooled to 400 K. What is the process work and heat transfer?

3.137 The piston/cylinder arrangement in Fig. P3.137 contains 10 g ammonia at 20 °C with a volume of 1 L. There are some stops, so if the piston is at the stops, the volume is 1.4 L. The ammonia is now heated to 200 °C. The piston and cylinder are made of 0.5 kg aluminum. Assume that the mass has the same temperature as the ammonia at any time. Find the final volume and the total heat transfer and plot the P–V diagram for the process.

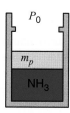

FIGURE P3.137

3.138 A piston/cylinder has 0.5 kg of air at 2000 kPa, 1000 K, as shown in Fig. P3.138. The cylinder has stops, so $V_{min} = 0.03\ m^3$. The air now cools to 400 K by heat transfer to the ambient. Find the final volume and pressure of the air (does it hit the stops?) and the work and heat transfer in the process.

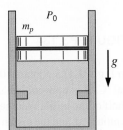

FIGURE P3.138

Energy Equation Rate Form

3.139 A 100 hp car engine has a drive shaft rotating at 2000 RPM. How much torque is on the shaft for 25% of full power?

3.140 A crane uses 2 kW to raise a 100 kg box to a height of 20 m. How much time does it take?

3.141 An escalator raises a 100 kg bucket to a height of 10 m in 1 min. Determine the rate of work in the process.

3.142 A pot of water is boiling on a stove supplying 275 W to the water. What is the rate of mass vaporization (kg/s), assuming a constant pressure process?

3.143 The heaters in a spacecraft suddenly fail. Heat is lost by radiation at the rate of 100 kJ/h, and the electric instruments generate 75 kJ/h. Initially, the air is at 100 kPa and 25 °C with a volume of 10 m^3. How long will it take to reach an air temperature of −20 °C?

3.144 As fresh-poured concrete hardens, the chemical transformation releases energy at a rate of 2 W/kg. Assume that the center of a poured layer does not have any heat loss and that it has an average specific heat of 0.9 kJ/kg·K. Find the temperature rise during 1 h of the hardening (curing) process.

3.145 A 1.2 kg pot of water at 20 °C is put on a stove supplying 250 W to the water. How long will it take to come to a boil (100 °C)?

3.146 The rate of heat transfer to the surroundings from a person at rest is about 400 kJ/h. Suppose that the ventilation system fails in an auditorium containing 100 people. Assume the energy goes into the air of volume 1500 m^3 initially at 300 K and

101 kPa. Find the rate (degrees per minute) of the air temperature change.

3.147 A 500 W heater is used to melt 2 kg of solid ice at −5 °C to liquid at +5 °C at a constant pressure of 150 kPa.
a. Find the change in the total volume of the water.
b. Find the energy the heater must provide to the water.
c. Find the time the process will take, assuming uniform T in the water.

3.148 A drag force on a car, with frontal area $A = 2\ m^2$, driving at 80 km/h in air at 20 °C, is $F_d = 0.225\ A\ \rho_{air}\mathbf{V}^2$. How much power is needed, and what is the traction force?

3.149 A 3 kg mass of nitrogen gas at 2000 K, $V = C$, cools with 500 W. What is dT/dt?

General Work

3.150 Electric power is volts times amperes ($P = Vi$). When a car battery at 12 V is charged with 6 amps for 3 h, how much energy is delivered?

3.151 A copper wire of diameter 2 mm is 10 m long and stretched out between two posts. The normal stress (pressure), $\sigma = E(L-L_0)/L_0$, depends on the length, L, versus the unstretched length, L_0, and Young's modulus, $E = 1.1 \times 10^6$ kPa. The force is $F = A\sigma$ and is measured to be 110 N. How much longer is the wire, and how much work was put in?

3.152 A film of ethanol at 20 °C has a surface tension of 22.3 mN/m and is maintained on a wire frame, as shown in Fig. P3.152. Consider the film with two surfaces as a control mass and find the work done when the wire is moved 10 mm to make the film 20 mm × 40 mm.

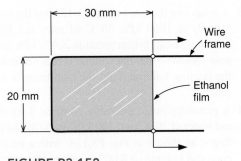

FIGURE P3.152

3.153 A battery is well insulated while being charged by 12.3 V at a current of 6 A. Take the battery as a control mass and find the instantaneous rate of work and the total work done over 4 h.

3.154 A sheet of rubber is stretched out over a ring of radius 0.25 m. I pour liquid water at 20 °C on it, as in Fig. P3.154, so that the rubber forms a half-sphere (cup). Neglect the rubber mass and find the surface tension near the ring.

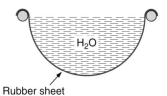

Rubber sheet

FIGURE P3.154

3.155 Assume that we fill a spherical balloon from a bottle of helium gas. The helium gas provides work $\int P\,dV$ that stretches the balloon material $\int \mathcal{S}\,dA$ and pushes back the atmosphere $\int P_0\,dV$. Write the incremental balance for $dW_{\text{helium}} = dW_{\text{stretch}} + dW_{\text{atm}}$ to establish the connection between the helium pressure, the surface tension $\mathcal{S}$, and P_0 as a function of the radius.

3.156 Assume a balloon material with a constant surface tension of $\mathcal{S} = 2$ N/m. What is the work required to stretch a spherical balloon up to a radius of $r = 0.5$ m? Neglect any effect from atmospheric pressure.

3.157 A soap bubble has a surface tension of $\mathcal{S} = 3 \times 10^{-4}$ N/cm as it sits flat on a rigid ring of diameter 5 cm. You now blow on the film to create a half-sphere surface of diameter 5 cm. How much work was done?

3.158 A 0.5 m long steel rod with a 1 cm diameter is stretched in a tensile test. What is the work required to obtain a relative strain of 0.1 %? The modulus of elasticity of steel is 2×10^8 kPa.

More Complex Devices

3.159 A piston/cylinder has a water volume separated in $V_A = 0.2$ m³ and $V_B = 0.3$ m³ by a stiff membrane (Fig. P3.159). The initial state in A is 1000 kPa, $x = 0.75$, and in B it is 1600 kPa and 250 °C. Now the membrane ruptures and the water comes to a uniform state at 200 °C. What is the final pressure? Find the work and the heat transfer in the process.

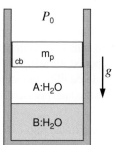

FIGURE P3.159

3.160 The cylinder volume below the constant loaded piston has two compartments, A and B, filled with water, as shown in Fig. P3.160. A has 0.5 kg at 200 kPa and 150 °C and B has 400 kPa with a quality of 50 % and a volume of 0.1 m³. The valve is opened and heat is transferred so that the water comes to a uniform state with a total volume of 1.006 m³. Find the total mass of water and the total initial volume. Find the work and the heat transfer in the process.

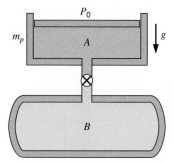

FIGURE P3.160

3.161 Two rigid tanks are filled with water (Fig. P3.161). Tank A is 0.2 m³ at 100 kPa, 150 °C and tank B is 0.3 m³ at saturated vapor of 300 kPa. The tanks are connected by a pipe with a closed valve. We open the valve and let all the water come to a single uniform state while we transfer enough heat to have a final pressure of 300 kPa. Give the two property values that determine the final state and find the heat transfer.

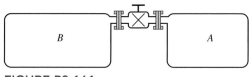

FIGURE P3.161

3.162 A tank has a volume of 1 m³ with oxygen at 15 °C, 300 kPa. Another tank contains 4 kg oxygen at 60 °C, 500 kPa. The two tanks are connected by a pipe and valve that is opened, allowing the whole system to come to a single equilibrium state with the ambient at 20 °C. Find the final pressure and the heat transfer.

3.163 A rigid insulated tank is separated into two rooms by a stiff plate. Room A, of 0.5 m³, contains air at 250 kPa and 300 K and room B, of 1 m³, has air at 500 kPa and 1000 K. The plate is removed and the air comes to a uniform state without any heat transfer. Find the final pressure and temperature.

3.164 A rigid tank A of volume 0.6 m³ contains 3 kg of water at 120 °C, and rigid tank B has a volume of 0.4 m³ with water at 600 kPa, 200 °C. The tanks are connected to a piston/cylinder initially empty with closed valves, as shown in Fig. P3.164. The pressure in the cylinder should be 800 kPa to float the piston. Now the valves are slowly opened and heat is transferred so that the water reaches a uniform state at 250 °C with the valves open. Find the final volume and pressure, and the work and heat transfer in the process.

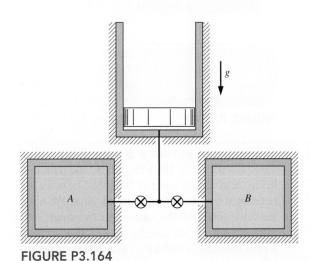

FIGURE P3.164

Review Problems

3.165 Ten kilograms of water in a piston/cylinder setup with constant pressure is at 450 °C and occupies a volume of 0.633 m³. The system is now cooled to 20 °C. Show the $P–v$ diagram, and find the work and heat transfer for the process.

3.166 A piston/cylinder setup, as shown in Fig. P3.166, contains 1 kg of water at 20 °C with a volume of 0.1 m³. Initially, the piston rests on some stops with the top surface open to the atmosphere, P_0, and a mass such that a water pressure of 400 kPa will lift it. To what temperature should the water be heated to lift the piston? If it is heated to saturated vapor, find the final temperature, volume, and work, $_1W_2$.

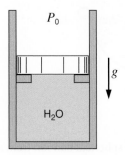

FIGURE P3.166

3.167 Two kilograms of water is contained in a piston/cylinder (Fig. P3.167) with a massless piston loaded with a linear spring and the outside atmosphere. Initially the spring force is zero and $P_1 = P_0 = 100$ kPa with a volume of 0.2 m³. If the piston just hits the upper stops, the volume is 0.8 m³ and $T = 600$ °C. Heat is now added until the pressure reaches 1.2 MPa. Show the $P–V$ diagram, and find the work and heat transfer for the process.

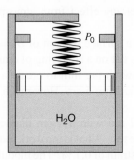

FIGURE P3.167

3.168 Ammonia (NH_3) is contained in a sealed rigid tank at 0 °C, $x = 50\%$ and is then heated to 100 °C. Find the final state P_2, u_2 and the specific work and heat transfer.

3.169 A piston/cylinder system contains 50 L of air at 300 °C, 100 kPa, with the piston initially on a set of stops. A total external constant force acts on the piston, so the balancing pressure inside should be 200 kPa. The cylinder is made of 2 kg of steel initially at 1300 °C. The system is insulated so that heat transfer occurs only between the steel cylinder and the air. The system comes to equilibrium. Find the final temperature and the work done by the air in the process and plot the process P–V diagram.

3.170 A piston/cylinder setup contains 1 kg of ammonia at $20\,^{\circ}$C with a volume of $0.1\ m^3$, as shown in Fig. P3.170. Initially the piston rests on some stops with the top surface open to the atmosphere, P_0, so that a pressure of 1400 kPa is required to lift it. To what temperature should the ammonia be heated to lift the piston? If it is heated to saturated vapor, find the final temperature, volume, and heat transfer, $_1Q_2$.

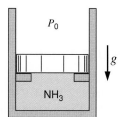

FIGURE P3.170

3.171 A piston held by a pin in an insulated cylinder, shown in Fig. P3.171, contains 2 kg of water at $100\,^{\circ}$C, with a quality of 98 %. The piston has a mass of 102 kg, with cross-sectional area of $100\ cm^2$, and the ambient pressure is 100 kPa. The pin is released, which allows the piston to move. Determine the final state of the water, assuming the process to be adiabatic.

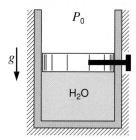

FIGURE P3.171

3.172 A vertical cylinder (Fig. P3.172) has a 61.18 kg piston locked with a pin, trapping 10 L of R-410A at $10\,^{\circ}$C with 90 % quality inside. Atmospheric pressure is 100 kPa, and the cylinder cross-sectional area is $0.006\ m^2$. The pin is removed, allowing the piston to move and come to rest with a final temperature of $10\,^{\circ}$C for the R-410A. Find the final pressure, the work done, and the heat transfer for the R-410A.

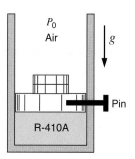

FIGURE P3.172

3.173 A spring-loaded piston/cylinder arrangement contains R-134a at $20\,^{\circ}$C, 24 % quality with a volume of 50 L. The setup is heated and thus expands, moving the piston. It is noted that when the last drop of liquid disappears, the temperature is $40\,^{\circ}$C. The heating is stopped when $T = 130\,^{\circ}$C. Verify that the final pressure is about 1200 kPa by iteration and find the work done in the process.

3.174 Water in a piston/cylinder, similar to Fig. P3.170, is at $100\,^{\circ}$C, $x = 0.5$ with mass 1 kg, and the piston rests on the stops. The equilibrium pressure that will float the piston is 300 kPa. The water is heated to $300\,^{\circ}$C by an electrical heater. At what temperature would all the liquid be gone? Find the final (P, v), the work, and the heat transfer in the process.

3.175 A piston/cylinder arrangement has a linear spring and the outside atmosphere acting on the piston shown in Fig. P3.175. It contains water at 3 MPa and $400\,^{\circ}$C with a volume of $0.1\ m^3$. If the piston is at the bottom, the spring exerts a force such that a pressure of 200 kPa inside is required to balance the forces. The system now cools until the pressure reaches 1 MPa. Find the heat transfer for the process.

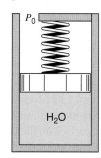

FIGURE P3.175

3.176 A $1\ m^3$ tank containing air at $25\,^{\circ}$C and 500 kPa is connected through a valve to another tank containing 4 kg of air at $60\,^{\circ}$C and 200 kPa (Fig. P3.176). Now the valve is opened and the entire system reaches thermal equilibrium with the surroundings at 20 °C. Assume constant specific heat at 25 °C and determine the final pressure and the heat transfer.

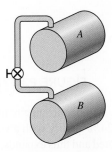

FIGURE P3.176

3.177 Consider the piston/cylinder arrangement shown in Fig. P3.177. A frictionless piston is free to move between two sets of stops. When the piston rests on the lower stops, the enclosed volume is 400 L. When the piston reaches the upper stops, the volume is 600 L. The cylinder initially contains water at 100 kPa, with 20 % quality. It is heated until the water eventually exists as saturated vapor. The mass of the piston requires 300 kPa pressure to move it against the outside ambient pressure. Determine the final pressure in the cylinder, the heat transfer, and the work for the overall process.

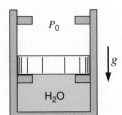

FIGURE P3.177

3.178 Ammonia (2 kg) in a piston/cylinder is at 100 kPa, −20 °C and is now heated in a polytropic process with $n = 1.3$ to a pressure of 200 kPa. Do not use the ideal gas approximation and find T_2, the work, and the heat transfer in the process.

3.179 A small, flexible bag contains 0.1 kg of ammonia at −10 °C and 300 kPa. The bag material is such that the pressure inside varies linearly with the volume. The bag is left in the sun with an incident radiation of 75 W, losing energy with an average 25 W to the ambient ground and air. After a while, the bag is heated to 30 °C, at which time the pressure is 1000 kPa. Find the work and heat transfer in the process and the elapsed time.

3.180 A piston/cylinder setup, shown in Fig. P3.180, contains R-410A at −20 °C, $x = 20$ %. The volume is 0.2 m³. It is known that $V_{stop} = 0.4$ m³, and if the piston sits at the bottom, the spring force balances the other loads on the piston. The system is now heated to 20 °C. Find the mass of the fluid and show the P–v diagram. Find the work and heat transfer.

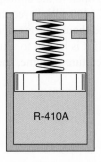

R-410A

FIGURE P3.180

3.181 A spherical balloon contains 2 kg of R-410A at 0 °C with a quality of 30 %. This system is heated until the pressure in the balloon reaches 1 MPa. For this process, it can be assumed that the pressure in the balloon is directly proportional to the balloon diameter. How does pressure vary with volume, and what is the heat transfer for the process?

3.182 A piston/cylinder arrangement B is connected to a 1 m³ tank A by a line and valve, shown in Fig. P3.182. Initially both contain water, with A at 100 kPa, saturated vapor and B at 400 °C, 300 kPa, 1 m³. The valve is now opened, and the water in both A and B comes to a uniform state.
a. Find the initial mass in A and B.
b. If the process results in $T_2 = 200$ °C, find the heat transfer and the work.

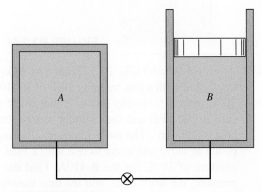

FIGURE P3.182

COMPUTER, DESIGN, AND OPEN-ENDED PROBLEMS

3.183 Reconsider the process in Problem 3.130, in which three states were specified. Solve the problem by fitting a single smooth curve (P versus v) through the three points. Map out the path followed (including temperature and quality) during the process.

3.184 Ammonia vapor is compressed inside a cylinder by an external force acting on the piston. The ammonia is initially at $30\,°C$, $500\,kPa$, and the final pressure is $1400\,kPa$. The following data have been measured for the process:

Pressure, kPa	500	653	802	945	1100	1248	1400
Volume, L	1.25	1.08	0.96	0.84	0.72	0.60	0.50

Determine the work done by the ammonia by summing the area below the P–V process curve. As you plot it, P is the height and the change in volume is the base of a number of rectangles.

3.185 Examine the sensitivity of the final pressure to the containment room volume in Problem 3.81. Solve for the volume for a range of final pressures, 100–250 kPa, and sketch the pressure versus volume curve.

3.186 Track the process described in Problem 3.90 so that you can sketch the amount of heat transfer added and the work given out as a function of the volume.

3.187 Write a program to solve Problem 3.95 for a range of initial velocities. Let the car mass and final velocity be input variables.

3.188 For one of the substances in Table A.6, compare the enthalpy change between any two temperatures, T_1 and T_2, as calculated by integrating the specific heat equation; by assuming constant specific heat at the average temperature; and by assuming constant specific heat at temperature T_1.

3.189 Write a program for Problem 3.125, where the initial state, the volume ratio, and the polytropic exponent are input variables. To simplify the formulation, use constant specific heat.

3.190 Examine a process whereby air at 300 K, 100 kPa is compressed in a piston/cylinder arrangement to 600 kPa. Assume the process is polytropic with exponents in the 1.2–1.6 range. Find the work and heat transfer per unit mass of air. Discuss the different cases and how they may be accomplished by insulating the cylinder or by providing heating or cooling.

3.191 A cylindrical tank of height 2 m with a cross-sectional area of $0.5\,m^2$ contains hot water at $80\,°C$, 125 kPa. It is in a room with temperature $T_0 = 20\,°C$, so it slowly loses energy to the room air proportional to the temperature difference as

$$\dot{Q}_{loss} = CA(T - T_0)$$

with the tank surface area, A, and C is a constant. For different values of the constant C, estimate the time it takes to bring the water to $50\,°C$. Make enough simplifying assumptions so that you can solve the problem mathematically, that is, find a formula for $T(t)$.

4 Energy Equation for a Control Volume

In the preceding chapter we developed the energy analysis for a control mass going through a process. Many applications in thermodynamics do not readily lend themselves to a control mass approach but are conveniently handled by the more general control volume technique, as discussed in Chapter 1. This chapter is concerned with development of the control volume forms of the conservation of mass and energy in situations where flows of substance are present.

4.1 CONSERVATION OF MASS AND THE CONTROL VOLUME

The control volume was introduced in Chapter 1 and serves to define the part of space that includes the volume of interest for a study or analysis. The surface of this control volume is the *control surface* that completely surrounds the volumes. Mass as well as heat and work can cross the control surface, and the mass together with its properties can change with time. Figure 4.1 shows a schematic diagram of a control volume that includes heat transfer, shaft work, moving boundary work, accumulation of mass within the control volume, and several mass flows. It is important to identify and label each flow of mass and energy and the parts of the control volume that can store (accumulate) mass.

Let us consider the conservation of mass law as it relates to the control volume. The physical law concerning mass, recalling Section 3.13, says that we cannot create or destroy mass. We will express this law in a mathematical statement about the mass in the control volume. To do this, we must consider all the mass flows into and out of the control volume and the net increase of mass within the control volume. As a somewhat simpler control volume, we consider a tank with a cylinder and piston and two pipes attached, as shown in Fig. 4.2. The rate of change of mass inside the control volume can be different from zero if we add or take a flow of mass out as

$$\text{Rate of change} = +\text{in} - \text{out}$$

With several possible flows, this is written as

$$\frac{dm_{C.V.}}{dt} = \sum \dot{m}_i - \sum \dot{m}_e \qquad (4.1)$$

which states that if the mass inside the control volume changes with time, it is because we add some mass or take some mass out. There are no other means by which the mass inside the control volume could change. Equation 4.1 expressing the conservation of mass is commonly termed the continuity equation. While this form of the equation is sufficient

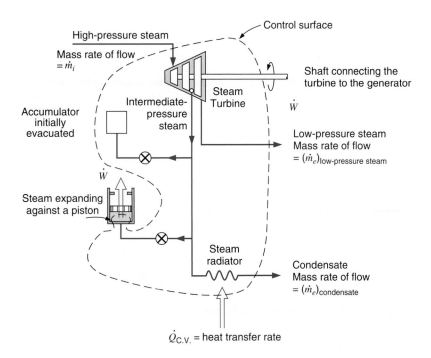

FIGURE 4.1
Schematic diagram of a control volume showing mass and energy transfers and accumulation.

for the majority of applications in thermodynamics, it is frequently rewritten in terms of the local fluid properties in the study of fluid mechanics and heat transfer. In this book, we are mainly concerned with the overall mass balance and thus consider Eq. 4.1 as the general expression for the continuity equation.

Since Eq. 4.1 is written for the total mass (lumped form) inside the control volume, we may have to consider several contributions to the mass as

$$m_{C.V.} = \int \rho \, dV = \int (1/v)dV = m_A + m_B + m_C + \cdots$$

Such a summation is needed when the control volume has several accumulation units with different states of the mass.

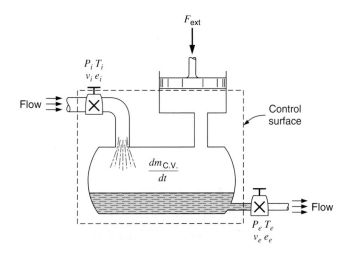

FIGURE 4.2
Schematic diagram of a control volume for the analysis of the continuity equation.

FIGURE 4.3 The flow across a control volume surface with a flow cross-sectional area of A. Average velocity is shown to the left of the valve, and a distributed flow across the area is shown to the right of the valve.

Let us now consider the mass flow rates across the control volume surface in a little more detail. For simplicity we assume that the fluid is flowing in a pipe or duct, as illustrated in Fig. 4.3. We wish to relate the total flow rate that appears in Eq. 4.1 to the local properties of the fluid state. The flow across the control volume surface can be indicated with an average velocity shown to the left of the valve or with a distributed velocity over the cross section, as shown to the right of the valve.

The volume flow rate is

$$\dot{V} = \mathbf{V}A = \int \mathbf{V}_{\text{local}}\, dA \qquad (4.2)$$

so the mass flow rate becomes

$$\dot{m} = \rho_{\text{avg}}\dot{V} = \dot{V}/v = \int (\mathbf{V}_{\text{local}}/v)dA = \mathbf{V}A/v \qquad (4.3)$$

where often the average velocity is used. It should be noted that this result, Eq. 4.3, has been developed for a stationary control surface, and we tacitly assumed that the flow was normal to the surface. This expression for the mass flow rate applies to any of the various flow streams entering or leaving the control volume, subject to the assumptions mentioned.

Example 4.1

Air is flowing in a 0.2 m diameter pipe at a uniform velocity of 0.1 m/s. The temperature is 25 °C and the pressure is 150 kPa. Determine the mass flow rate.

Solution
From Eq. 4.3 the mass flow rate is

$$\dot{m} = \mathbf{V}A/v$$

For air, using R from Table A.5, we have

$$v = \frac{RT}{P} = \frac{0.287\,\text{kJ/kg·K} \times 298.2\,\text{K}}{150\,\text{kPa}} = 0.5705\,\text{m}^3/\text{kg}$$

The cross-sectional area is

$$A = \frac{\pi}{4}(0.2)^2 = 0.0314\,\text{m}^2$$

Therefore,

$$\dot{m} = \mathbf{V}A/v = 0.1\,\text{m/s} \times 0.0314\,\text{m}^2/(0.5705\,\text{m}^3/\text{kg}) = 0.0055\,\text{kg/s}$$

In-Text Concept Question

a. A mass flow rate into a control volume requires a normal velocity component. Why?

4.2 THE ENERGY EQUATION FOR A CONTROL VOLUME

We have already considered the energy equation for a control mass, which consists of a fixed quantity of mass, and noted, in Eq. 3.5, that it may be written as

$$E_2 - E_1 = {}_1Q_2 - {}_1W_2$$

We have also noted that this may be written as an instantaneous rate equation as Eq. 3.3.

$$\frac{dE_{\text{C.M.}}}{dt} = \dot{Q} - \dot{W} \tag{4.4}$$

To write the energy equation as a rate equation for a control volume, we proceed in a manner analogous to that used in developing a rate equation for the law of conservation of mass. For this purpose, a control volume is shown in Fig. 4.4 that involves the rate of heat transfer, rates of work, and mass flows. The fundamental physical law states that we cannot create or destroy energy such that any rate of change of energy must be caused by transfer rates of energy into or out of the control volume. We have already included rates of heat transfer and work in Eq. 4.4, so the additional explanations we need are associated with the mass flow rates.

The fluid flowing across the control surface enters or leaves with an amount of energy per unit mass as

$$e = u + \frac{1}{2}\mathbf{V}^2 + gZ$$

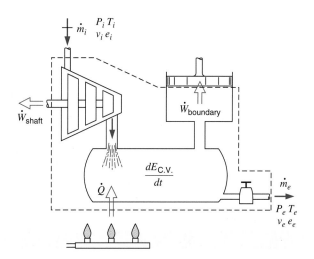

FIGURE 4.4
Schematic diagram illustrating terms in the energy equation for a general control volume.

relating to the state and position of the fluid. Whenever a fluid mass enters a control volume at state i or exits at state e, there is boundary movement work associated with that process.

To explain this in more detail, consider an amount of mass flowing into the control volume. As this mass flows in there is a pressure at its back surface, so as this mass moves into the control volume it is being pushed by the mass behind it, which is the surroundings. The net effect is that after the mass has entered the control volume, the surroundings have pushed it in against the local pressure with a velocity giving it a rate of work in the process. Similarly, a fluid exiting the control volume at state e must push the surrounding fluid ahead of it, doing work on it, which is work leaving the control volume. The velocity and the area correspond to a certain volume per unit time entering the control volume, enabling us to relate that to the mass flow rate and the specific volume at the state of the mass going in. Now we are able to express the rate of flow work as

$$\dot{W}_{\text{flow}} = F\mathbf{V} = \int P\mathbf{V}\,dA = P\dot{V} = Pv\dot{m} \tag{4.5}$$

For the flow that leaves the control volume, work is being done by the control volume, $P_e v_e \dot{m}_e$, and for the mass that enters, the surroundings do the rate of work, $P_i v_i \dot{m}_i$. The flow work per unit mass is then Pv, and the total energy associated with the flow of mass is

$$e + Pv = u + Pv + \frac{1}{2}\mathbf{V}^2 + gZ = h + \frac{1}{2}\mathbf{V}^2 + gZ \tag{4.6}$$

In this equation, we have used the definition of the thermodynamic property enthalpy, and it is the appearance of the combination $(u + Pv)$ for the energy in connection with a mass flow that is the primary reason for the definition of the property enthalpy. Its introduction earlier in conjunction with the constant-pressure process was done to facilitate use of the tables of thermodynamic properties at that time.

Example 4.2

Assume we are standing next to the local city's main water line. The liquid water inside flows at a pressure of 600 kPa with a temperature of about 10 °C. We want to add a smaller amount, 1 kg, of liquid to the line through a side pipe and valve mounted on the main line. How much work will be involved in this process?

If the 1 kg of liquid water is in a bucket and we open the valve to the water main in an attempt to pour it down into the pipe opening, we realize that the water flows the other way. The water flows from a higher to a lower pressure, that is, from inside the main line to the atmosphere (from 600 kPa to 101 kPa).

We must take the 1 kg of liquid water, put it into a piston/cylinder (like a handheld pump), and attach the cylinder to the water pipe. Now we can press on the piston until the water pressure inside is 600 kPa and then open the valve to the main line and slowly squeeze the 1 kg of water in. The work done at the piston surface to the water is

$$W = \int P\,dV = P_{\text{water}}\,mv = 600 \text{ kPa} \times 1 \text{ kg} \times 0.001 \text{ m}^3/\text{kg} = 0.6 \text{ kJ}$$

and this is the necessary flow work for adding the 1 kg of liquid.

The extension of the energy equation from Eq. 4.4 becomes

$$\frac{dE_{\text{C.V.}}}{dt} = \dot{Q}_{\text{C.V.}} - \dot{W}_{\text{C.V.}} + \dot{m}_i e_i - \dot{m}_e e_e + \dot{W}_{\text{flow in}} - \dot{W}_{\text{flow out}}$$

and the substitution of Eq. 4.5 gives

$$\frac{dE_{\text{C.V.}}}{dt} = \dot{Q}_{\text{C.V.}} - \dot{W}_{\text{C.V.}} + \dot{m}_i(e_i + P_i v_i) - \dot{m}_e(e_e + P_e v_e)$$

$$= \dot{Q}_{\text{C.V.}} - \dot{W}_{\text{C.V.}} + m_i\left(h_i + \frac{1}{2}\mathbf{V}_i^2 + gZ_i\right) - \dot{m}_e\left(h_e + \frac{1}{2}\mathbf{V}_e^2 + gZ_e\right)$$

In this form of the energy equation the rate of work term is the sum of all shaft work terms and boundary work terms and any other types of work given out by the control volume; however, the flow work is now listed separately and included with the mass flow rate terms.

For the general control volume we may have several entering or leaving mass flow rates, so a summation over those terms is often needed. The final form of the energy equation then becomes

$$\frac{dE_{\text{C.V.}}}{dt} = \dot{Q}_{\text{C.V.}} - \dot{W}_{\text{C.V.}} + \sum \dot{m}_i\left(h_i + \frac{1}{2}\mathbf{V}_i^2 + gZ_i\right) - \sum \dot{m}_e\left(h_e + \frac{1}{2}\mathbf{V}_e^2 + gZ_e\right) \quad (4.7)$$

stating that the rate of change of energy inside the control volume is due to a net rate of heat transfer, a net rate of work (measured positive out), and the summation of energy fluxes due to mass flows into and out of the control volume. As with the conservation of mass, this equation can be written for the total control volume and can therefore be put in the lumped or integral form where

$$E_{\text{C.V.}} = \int \rho e\, dV = me = m_A e_A + m_B e_B + m_C e_C + \cdots$$

As the kinetic and potential energy terms per unit mass appear together with the enthalpy in all the flow terms, a shorter notation is often used:

$$h_{\text{tot}} \equiv h + \frac{1}{2}\mathbf{V}^2 + gZ$$

$$h_{\text{stag}} \equiv h + \frac{1}{2}\mathbf{V}^2$$

defining the total enthalpy and the stagnation enthalpy (used in fluid mechanics). The shorter equation then becomes

$$\frac{dE_{\text{C.V.}}}{dt} = \dot{Q}_{\text{C.V.}} - \dot{W}_{\text{C.V.}} + \sum \dot{m}_i h_{\text{tot},i} - \sum \dot{m}_e h_{\text{tot},e} \quad (4.8)$$

giving the general energy equation on a rate form. All applications of the energy equation start with the form in Eq. 4.8, and for special cases this will result in a slightly simpler form, as shown in the subsequent sections.

4.3 THE STEADY-STATE PROCESS

Our first application of the control volume equations will be to develop a suitable analytical model for the long-term steady operation of devices such as turbines, compressors, nozzles, boilers, and condensers—a very large class of problems of interest in thermodynamic analysis. This model will not include the short-term transient startup or shutdown of such devices, but only the steady operating period of time.

Let us consider a certain set of assumptions (beyond those leading to Eqs. 4. 1 and 4.7) that lead to a reasonable model for this type of process, which we refer to as the steady-state process.

1. The control volume does not move relative to the coordinate frame.
2. The state of the mass at each point in the control volume does not vary with time.
3. As for the mass that flows across the control surface, the mass flux and the state of this mass at each discrete area of flow on the control surface do not vary with time. The rates at which heat and work cross the control surface remain constant.

As an example of a steady-state process, consider a centrifugal air compressor that operates with a constant mass rate of flow into and out of the compressor, constant properties at each point across the inlet and exit ducts, a constant rate of heat transfer to the surroundings, and a constant power input. At each point in the compressor the properties are constant with time, even though the properties of a given elemental mass of air vary as it flows through the compressor. Often, such a process is referred to as a *steady-flow process*, since we are concerned primarily with the properties of the fluid entering and leaving the control volume. However, in the analysis of certain heat transfer problems in which the same assumptions apply, we are primarily interested in the spatial distribution of properties, particularly temperature, and such a process is referred to as a *steady-state process*. Since this is an introductory book, we will use the term *steady-state process* for both. The student should realize that the terms *steady-state process* and *steady-flow process* are both used extensively in the literature.

Let us now consider the significance of each of these assumptions for the steady-state process.

1. The assumption that the control volume does not move relative to the coordinate frame means that all velocities measured relative to the coordinate frame are also velocities relative to the control surface, and there is no work associated with the acceleration of the control volume.
2. The assumption that the state of the mass at each point in the control volume does not vary with time requires that

$$\frac{dm_{C.V.}}{dt} = 0 \quad \text{and} \quad \frac{dE_{C.V.}}{dt} = 0$$

Therefore, we conclude that for the steady-state process we can write, from Eqs. 4.1 and 4.7,

Continuity equation: $$\sum \dot{m}_i = \sum \dot{m}_e \qquad (4.9)$$

Energy equation:

$$\dot{Q}_{C.V.} + \sum \dot{m}_i \left(h_i + \frac{\mathbf{V}_i^2}{2} + g Z_i \right) = \sum \dot{m}_e \left(h_e + \frac{\mathbf{V}_e^2}{2} + g Z_e \right) + \dot{W}_{C.V.} \qquad (4.10)$$

3. The assumption that the various mass flows, states, and rates at which heat and work cross the control surface remain constant requires that every quantity in Eqs. 4.9 and 4.10 be steady with time. This means that application of Eqs. 4.9 and 4.10 to the operation of some device is independent of time.

Many of the applications of the steady-state model are such that there is only one flow stream entering and one leaving the control volume. For this type of process, we can write

Continuity equation: $\quad\quad\quad\quad\quad \dot{m}_i = \dot{m}_e = \dot{m}$ $\quad\quad\quad\quad\quad$ (4.11)

Energy equation: $\dot{Q}_{C.V.} + \dot{m}\left(h_i + \dfrac{\mathbf{V}_i^2}{2} + gZ_i\right) = \dot{m}\left(h_e + \dfrac{\mathbf{V}_e^2}{2} + gZ_e\right) + \dot{W}_{C.V.}$ $\quad$ (4.12)

Rearranging this equation, we have

$$q + h_i + \frac{\mathbf{V}_i^2}{2} + gZ_i = h_e + \frac{\mathbf{V}_e^2}{2} + gZ_e + w \quad\quad\quad (4.13)$$

where, by definition,

$$q = \frac{\dot{Q}_{C.V.}}{\dot{m}} \quad \text{and} \quad w = \frac{\dot{W}_{C.V.}}{\dot{m}} \quad\quad\quad (4.14)$$

Note that the units for q and w are kJ/kg. From their definition, q and w can be thought of as the heat transfer and work (other than flow work) per unit mass flowing into and out of the control volume for this particular steady-state process.

The symbols q and w are also used for the heat transfer and work per unit mass of a control mass. However, since it is always evident from the context whether it is a control mass (fixed mass) or a control volume (involving a flow of mass) with which we are concerned, the significance of the symbols q and w will also be readily evident in each situation.

The steady-state process is often used in the analysis of reciprocating machines, such as reciprocating compressors or engines. In this case the rate of flow, which may actually be pulsating, is considered to be the average rate of flow for an integral number of cycles. A similar assumption is made regarding the properties of the fluid flowing across the control surface and the heat transfer and work crossing the control surface. It is also assumed that for an integral number of cycles the reciprocating device undergoes, the energy and mass within the control volume do not change.

A number of examples are given in the next section to illustrate the analysis of steady-state processes.

In-Text Concept Questions

b. Can a steady-state device have boundary work?

c. What can you say about changes in $\dot{m}$ and $\dot{V}$ through a steady flow device?

d. In a multiple-device flow system, I want to determine a state property. Where should I look for information—upstream or downstream?

4.4 EXAMPLES OF STEADY-STATE PROCESSES

In this section, we consider a number of examples of steady-state processes in which there is one fluid stream entering and one leaving the control volume, such that the energy equation can be written in the form of Eq. 4.13. Some may instead utilize control volumes that include more than one fluid stream, such that it is necessary to write the energy equation in the more general form of Eq. 4.10, presented in Section 4.5. A listing of many simple flow devices

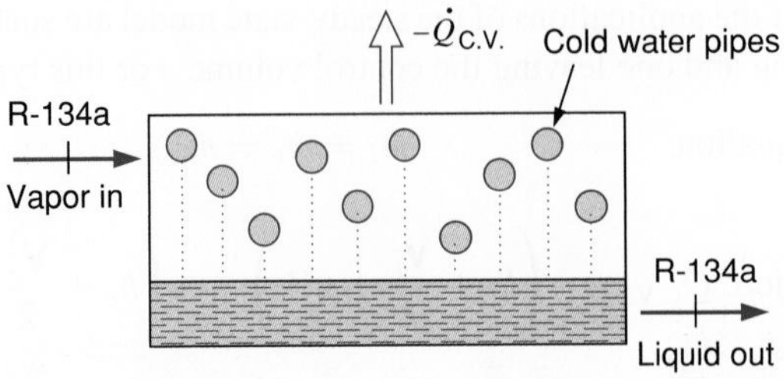

FIGURE 4.5 A refrigeration system condenser.

is given in Table 4.1 at the end of this chapter covering a few more than presented in the following sections.

Heat Exchanger

A steady-state heat exchanger is a simple fluid flow through a pipe or system of pipes, where heat is transferred to or from the fluid. The fluid may be heated or cooled, and may or may not boil, changing from liquid to vapor, or condense, changing from vapor to liquid. One such example is the condenser in an R-134a refrigeration system, as shown in Fig. 4.5. Superheated vapor enters the condenser and liquid exits. The process tends to occur at constant pressure, since a fluid flowing in a pipe usually undergoes only a small pressure drop because of fluid friction at the walls. The pressure drop may or may not be taken into account in a particular analysis. There is no means for doing any work (shaft work, electrical work, etc.), and changes in kinetic and potential energies are commonly negligibly small. (One exception may be a boiler tube in which liquid enters and vapor exits at a much larger specific volume. In such a case, it may be necessary to check the exit velocity using Eq. 4.3.) The heat transfer in most heat exchangers is then found from Eq. 4.13 as the change in enthalpy of the fluid. In the condenser shown in Fig. 4.5, the heat transfer out of the condenser then goes to whatever is receiving it, perhaps a stream of air or of cooling water. It is often simpler to write the first law around the entire heat exchanger, including both flow streams, in which case there is little or no heat transfer with the surroundings. Such a situation is the subject of the following example.

Example 4.3

Consider a water-cooled condenser in a large refrigeration system in which R-134a is the refrigerant fluid. The refrigerant enters the condenser at 1.0 MPa and 60 °C, at the rate of 0.2 kg/s, and exits as a liquid at 0.95 MPa and 35 °C. Cooling water enters the condenser at 10 °C and exits at 20 °C. Determine the rate at which cooling water flows through the condenser.

Control volume:	Condenser.
Sketch:	Fig. 4.6
Inlet states:	R-134a—fixed; water—fixed.
Exit states:	R-134a—fixed; water—fixed.
Process:	Steady-state.
Model:	R-134a tables; steam tables.

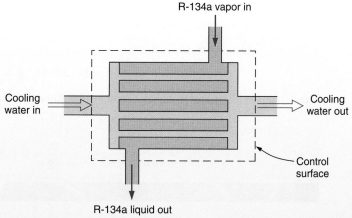

FIGURE 4.6
Schematic diagram of an R-134a condenser.

Analysis

With this control volume we have two fluid streams, the R-134a and the water, entering and leaving the control volume. It is reasonable to assume that both kinetic and potential energy changes are negligible. We note that the work is zero, and we make the other reasonable assumption that there is no heat transfer across the control surface. Therefore, the first law, Eq. 4.10, reduces to

$$\sum \dot{m}_i h_i = \sum \dot{m}_e h_e$$

Using the subscripts r for refrigerant and w for water, we write

$$\dot{m}_r (h_i)_r + \dot{m}_w (h_i)_w = \dot{m}_r (h_e)_r + \dot{m}_w (h_e)_w$$

Solution

From the R-134a and steam tables, we have

$$(h_i)_r = 441.89 \text{ kJ/kg}, \qquad (h_i)_w = 42.00 \text{ kJ/kg}$$

$$(h_e)_r = 249.10 \text{ kJ/kg}, \qquad (h_e)_w = 83.95 \text{ kJ/kg}$$

Solving the above equation for $\dot{m}_w$, the rate of flow of water, we obtain

$$\dot{m}_w = \dot{m}_r \frac{(h_i - h_e)_r}{(h_e - h_i)_w} = 0.2 \text{ kg/s} \frac{(441.89 - 249.10) \text{ kJ/kg}}{(83.95 - 42.00) \text{ kJ/kg}} = 0.919 \text{ kg/s}$$

This problem can also be solved by considering two separate control volumes, one having the flow of R-134a across its control surface and the other having the flow of water across its control surface. Further, there is heat transfer from one control volume to the other.

The heat transfer for the control volume involving R-134a is calculated first. In this case the steady-state energy equation, Eq. 4.10, reduces to

$$\dot{Q}_{C.V.} = \dot{m}_r(h_e - h_i)_r$$
$$= 0.2 \, \text{kg/s} \times (249.10 - 441.89) \, \text{kJ/kg} = -38.558 \, \text{kW}$$

This is also the heat transfer to the other control volume, for which $\dot{Q}_{C.V.} = +38.558 \, \text{kW}$.

$$\dot{Q}_{C.V.} = \dot{m}_w(h_e - h_i)_w$$
$$\dot{m}_w = \frac{38.558 \, \text{kW}}{(83.95 - 42.00) \, \text{kJ/kg}} = 0.919 \, \text{kg/s}$$

Nozzle

A nozzle is a steady-state device whose purpose is to create a high-velocity fluid stream at the expense of the fluid's pressure. It is contoured in an appropriate manner to expand a flowing fluid smoothly to a lower pressure, thereby increasing its velocity. There is no means to do any work—there are no moving parts. There is little or no change in potential energy and usually little or no heat transfer. Nozzles that are exposed to high temperatures may be cooled like the exit nozzle in a rocket or have enough heat conducted away from them as a diesel injector nozzle or a nozzle for injecting natural gas into a furnace. These situations are rather complex and require a more detailed heat transfer analysis. In addition, the kinetic energy of the fluid at the nozzle inlet is usually small and would be neglected if its value is not known.

Example 4.4

Steam at 0.6 MPa and 200 °C enters an insulated nozzle with a velocity of 50 m/s. It leaves at a pressure of 0.15 MPa and a velocity of 600 m/s. Determine the final temperature if the steam is superheated in the final state and the quality if it is saturated.

Control volume:	Nozzle.
Inlet state:	Fixed (see Fig. 4.7).
Exit state:	P_e known.
Process:	Steady-state.
Model:	Steam tables.

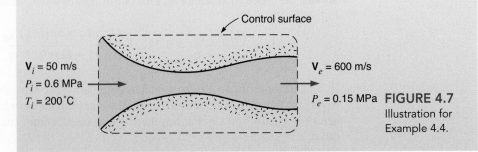

FIGURE 4.7
Illustration for Example 4.4.

Analysis

We have

$$\dot{Q}_{\text{C.V.}} = 0 \quad \text{(nozzle insulated)}$$

$$\dot{W}_{\text{C.V.}} = 0$$

$$\text{PE}_i \approx \text{PE}_e$$

The energy equation (Eq. 4.13) yields

$$h_i + \frac{\mathbf{V}_i^2}{2} = h_e + \frac{\mathbf{V}_e^2}{2}$$

Solution

Solving for h_e, we obtain

$$h_e = 2850.1 + \left[\frac{(50)^2}{2 \times 1000} - \frac{(600)^2}{2 \times 1000} \right] \frac{\text{m}^2/\text{s}^2}{\text{J/kJ}} = 2671.4 \,\text{kJ/kg}$$

The two properties of the fluid leaving that we now know are pressure and enthalpy, and therefore the state of this fluid is determined. Since h_e is less than h_g at 0.15 MPa, the quality is calculated.

$$h = h_f + x h_{fg}$$

$$2671.4 = 467.1 + x_e 2226.5$$

$$x_e = 0.99$$

Diffuser

A steady-state diffuser is a device constructed to decelerate a high-velocity fluid in a manner that results in an increase in pressure of the fluid. In essence, it is the exact opposite of a nozzle, and it may be thought of as a fluid flowing in the opposite direction through a nozzle, with the opposite effects. The assumptions are similar to those for a nozzle, with a large kinetic energy at the diffuser inlet and a small, but usually not negligible, kinetic energy at the exit being the only terms besides the enthalpies remaining in the energy equation, Eq. 4.13.

Throttle

A throttling process occurs when a fluid flowing in a line suddenly encounters a restriction in the flow passage. This may be a plate with a small hole in it, as shown in Fig. 4.8, it may

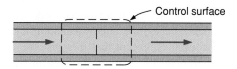

FIGURE 4.8 The throttling process.

be a partially closed valve protruding into the flow passage, or it may be a change to a tube of much smaller diameter, called a capillary tube, which is normally found on a refrigerator. The result of this restriction is an abrupt pressure drop in the fluid, as it is forced to find its way through a suddenly smaller passageway. This process is drastically different from the smoothly contoured nozzle expansion and area change, which results in a significant velocity increase. There is typically some increase in velocity in a throttle, but both inlet and exit kinetic energies are usually small enough to be neglected. There is no means for doing work and little or no change in potential energy. Usually, there is neither time nor opportunity for appreciable heat transfer, such that the only terms left in the energy equation, Eq. 4.13, are the inlet and exit enthalpies. We conclude that a steady-state throttling process is approximately a pressure drop at constant enthalpy, and we will assume this to be the case unless otherwise noted.

Frequently, a throttling process involves a change in the phase of the fluid. A typical example is the flow through the expansion valve of a vapor-compression refrigeration system, which is shown in Example 4.8.

Turbine

A turbine is a rotary steady-state machine whose purpose is to produce shaft work (power, on a rate basis) at the expense of the pressure of the working fluid. Two general classes of turbines are steam (or other working fluid) turbines, in which the steam exiting the turbine passes to a condenser, where it is condensed to liquid, and gas turbines, in which the gas usually exhausts to the atmosphere from the turbine. In either type, the turbine exit pressure is fixed by the environment into which the working fluid exhausts, and the turbine inlet pressure has been reached by previously pumping or compressing the working fluid in another process. Inside the turbine, there are two distinct processes. In the first, the working fluid passes through a set of nozzles, or the equivalent—fixed blade passages contoured to expand the fluid to a lower pressure and to a high velocity. In the second process inside the turbine, this high-velocity fluid stream is directed onto a set of moving (rotating) blades, in which the velocity is reduced before being discharged from the passage. This directed velocity decrease produces a torque on the rotating shaft, resulting in shaft work output. The low-velocity, low-pressure fluid then exhausts from the turbine.

The energy equation for this process is either Eq. 4.10 or 4.13. Usually, changes in potential energy are negligible, as is the inlet kinetic energy. It was demonstrated in Example 3.17 that for modest velocities and elevation differences the kinetic and potential energies are quite small compared to the changes in internal energy for even smaller temperature differences. Since the enthalpy is closely related to the internal energy, its change for smaller temperature differences is thus also larger than the kinetic and potential energy changes. Often, the exit kinetic energy is neglected, and any heat rejection from the turbine is undesirable and is commonly small. We therefore normally assume that a turbine process is adiabatic, and the work output in this case reduces to the decrease in enthalpy from the inlet to exit states. A turbine is analyzed in Example 4.7 as part of a power plant.

The preceding discussion concerned the turbine, which is a rotary work-producing device. There are other nonrotary devices that produce work, which can be called *expanders* as a general name. In such devices, the energy equation analysis and assumptions are generally the same as for turbines, except that in a piston/cylinder-type expander, there would in most cases be a larger heat loss or rejection during the process.

Compressor and Pump

The purpose of a steady-state compressor (gas) or pump (liquid) is the same: to increase the pressure of a fluid by putting in shaft work (power, on a rate basis). There are two fundamentally different classes of compressors. The most common is a rotary-type compressor (either axial flow or radial/centrifugal flow), in which the internal processes are essentially the opposite of the two processes occurring inside a turbine. The working fluid enters the compressor at low pressure, moving into a set of rotating blades, from which it exits at high velocity, a result of the shaft work input to the fluid. The fluid then passes through a diffuser section, in which it is decelerated in a manner that results in a pressure increase. The fluid then exits the compressor at high pressure.

The energy equation for the compressor is either Eq. 4.10 or 4.13. Usually, changes in potential energy are negligible, as is the inlet kinetic energy. Often the exit kinetic energy is neglected as well. Heat rejection from the working fluid during compression would be desirable, but it is usually small in a rotary compressor, which is a high-volume flow-rate machine, and there is not sufficient time to transfer much heat from the working fluid. We therefore normally assume that a rotary compressor process is adiabatic, and the work input in this case reduces to the change in enthalpy from the inlet to exit states.

In a piston/cylinder-type compressor, the cylinder usually contains fins to promote heat rejection during compression (or the cylinder may be water-jacketed in a large compressor for even greater cooling rates). In this type of compressor, the heat transfer from the working fluid is significant and is not neglected in the energy equation. As a general rule, in any example or problem in this book, we will assume that a compressor is adiabatic unless otherwise noted.

Example 4.5

The compressor in a plant (see Fig. 4.9) receives carbon dioxide at 100 kPa, 280 K, with a low velocity. At the compressor discharge, the carbon dioxide exits at 1100 kPa, 500 K, with a velocity of 25 m/s, and then flows into a constant-pressure aftercooler (heat exchanger), where it is cooled down to 350 K. The power input to the compressor is 50 kW. Determine the heat transfer rate in the aftercooler.

Solution
C.V. compressor, steady state, single inlet and exit flow.

Energy Eq. 4.13: $\quad q + h_1 + \frac{1}{2}\mathbf{V}_1^2 = h_2 + \frac{1}{2}\mathbf{V}_2^2 + w$

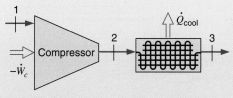

Compressor section　　Cooler section　　**FIGURE 4.9** Sketch for Example 4.5.

In this solution, let us assume that the carbon dioxide behaves as an ideal gas with variable specific heat (Appendix A.8). It would be more accurate to use Table B.3 to find the enthalpies, but the difference is fairly small in this case.

We also assume that $q \cong 0$ and $\mathbf{V}_1 \cong 0$, so, getting h from Table A. 8,

$$-w = h_2 - h_1 + \frac{1}{2}\mathbf{V}_2^2 = 401.52 - 198 + \frac{(25)^2}{2 \times 1000} = 203.5 + 0.3 = 203.8 \text{ kJ/kg}$$

Remember here to convert kinetic energy in J/kg to kJ/kg by division by 1000.

$$\dot{m} = \frac{\dot{W}_c}{w} = \frac{-50}{-203.8} \frac{\text{kW}}{\text{kJ/kg}} = 0.245 \text{ kg/s}$$

C.V. aftercooler, steady state, single inlet and exit flow, and no work.

Energy Eq. 4.13:

$$q + h_2 + \frac{1}{2}\mathbf{V}_2^2 = h_3 + \frac{1}{2}\mathbf{V}_3^2$$

Here we assume no significant change in kinetic energy (notice how unimportant it was), and again we look for h in Table A.8:

$$q = h_3 - h_2 = 257.9 - 401.5 = -143.6 \text{ kJ/kg}$$

$$\dot{Q}_{\text{cool}} = -\dot{Q}_{\text{C.V.}} = -\dot{m}q = 0.245 \text{ kg/s} \times 143.6 \text{ kJ/kg} = 35.2 \text{ kW}$$

Example 4.6

A small liquid water pump is located 15 m down in a well (see Fig. 4.10), taking water in at $10\,^{\circ}\text{C}$, 90 kPa at a rate of 1.5 kg/s. The exit line is a pipe of diameter 0.04 m that goes up to a receiver tank maintaining a gauge pressure of 400 kPa. Assume that the process is adiabatic, with the same inlet and exit velocities, and the water stays at $10\,^{\circ}\text{C}$. Find the required pump work.

C.V. pump + pipe. Steady state, one inlet, one exit flow. Assume same velocity in and out and no heat transfer.

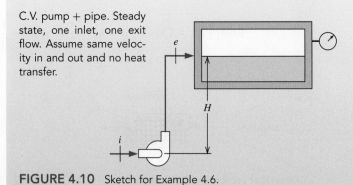

FIGURE 4.10 Sketch for Example 4.6.

Solution

Continuity equation: $\quad \dot{m}_{in} = \dot{m}_{ex} = \dot{m}$

Energy Equation. 4.12: $\quad \dot{m}\left(h_{in} + \frac{1}{2}\mathbf{V}_{in}^2 + gZ_{in}\right) = \dot{m}\left(h_{ex} + \frac{1}{2}\mathbf{V}_{ex}^2 + gZ_{ex}\right) + \dot{W}$

States: $\quad h_{ex} = h_{in} + (P_{ex} - P_{in})v \quad$ (v is constant and u is constant.)

From the energy equation

$$\dot{W} = \dot{m}(h_{in} + gZ_{in} - h_{ex} - gZ_{ex}) = \dot{m}[g(Z_{in} - Z_{ex}) - (P_{ex} - P_{in})v]$$

$$= 1.5\frac{kg}{s} \times \left[9.807\,\frac{m}{s^2} \times \frac{-15 - 0}{1000}\,\frac{m}{J/kJ} - (400 + 101.3 - 90)\,kPa \times 0.001\,001\,\frac{m^3}{kg}\right]$$

$$= 1.5 \times (-0.147 - 0.412) = -0.84\,kW$$

That is, the pump requires a power input of 840 W.

Complete Cycles: Power Plant and Refrigerator

The following examples illustrate the incorporation of several of the devices and machines already discussed in this section into a complete thermodynamic system, which is built for a specific purpose.

Example 4.7

Consider the simple steam power plant, as shown in Fig. 4.11. The following data are for such a power plant where the states are numbered and there is specific pump work as 4 kJ/kg.

State	Pressure	Temperature or Quality
1	2.0 MPa	300 °C
2	1.9 MPa	290 °C
3	15 kPa	90 %
4	14 kPa	45 °C

Determine the following quantities per kilogram flowing through the unit:

a. Heat transfer in the line between the boiler and turbine.

b. Turbine work.

c. Heat transfer in the condenser.

d. Heat transfer in the boiler.

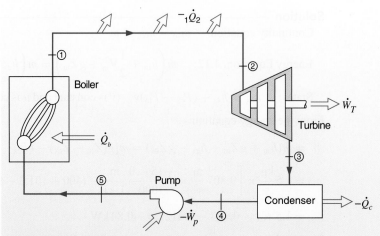

FIGURE 4.11 Simple steam power plant.

Since there are several control volumes to be considered in the solution to this problem, let us consolidate our solution procedure somewhat in this example. Using the notation of Fig. 4.11, we have:

All processes: Steady-state.
Model: Steam tables.

From the steam tables:

$$h_1 = 3023.5 \text{ kJ/kg}$$

$$h_2 = 3002.5 \text{ kJ/kg}$$

$$h_3 = 225.9 + 0.9(2373.1) = 2361.7 \text{ kJ/kg}$$

$$h_4 = 188.4 \text{ kJ/kg}$$

All analyses: No changes in kinetic or potential energy will be considered in the solution. In each case, the energy equation is given by Eq. 4.13.

Now, we proceed to answer the specific questions raised in the problem statement.

a. For the control volume for the pipeline between the boiler and the turbine, the energy equation and solution are

$$_1q_2 + h_1 = h_2$$

$$_1q_2 = h_2 - h_1 = 3002.5 - 3023.5 = -21.0 \text{ kJ/kg}$$

b. A turbine is essentially an adiabatic machine. Therefore, it is reasonable to neglect heat transfer in the energy equation, so that

$$h_2 = h_3 + {}_2w_3$$

$${}_2w_3 = 3002.5 - 2361.7 = 640.8 \text{ kJ/kg}$$

c. There is no work for the control volume enclosing the condenser. Therefore, the energy equation and solution are

$${}_3q_4 + h_3 = h_4$$

$${}_3q_4 = 188.4 - 2361.7 = -2173.3 \text{ kJ/kg}$$

d. If we consider a control volume enclosing the boiler, the work is equal to zero, so the energy equation becomes

$${}_5q_1 + h_5 = h_1$$

A solution requires a value for h_5, which can be found by taking a control volume around the pump:

$$h_4 = h_5 + {}_4w_5$$

$$h_5 = 188.4 - (-4) = 192.4 \text{ kJ/kg}$$

Therefore, for the boiler,

$${}_5q_1 + h_5 = h_1$$

$${}_5q_1 = 3023.5 - 192.4 = 2831.1 \text{ kJ/kg}$$

Example 4.8

The refrigerator shown in Fig. 4.12 uses R-134a as the working fluid. The mass flow rate through each component is 0.1 kg/s, and the power input to the compressor is 5.0 kW. The following state data are known, using the state notation of Fig. 4.12:

$$P_1 = 100 \text{ kPa}, \qquad T_1 = -20\,^\circ\text{C}$$
$$P_2 = 800 \text{ kPa}, \qquad T_2 = 50\,^\circ\text{C}$$
$$T_3 = 30\,^\circ\text{C}, \qquad x_3 = 0.0$$
$$T_4 = -25\,^\circ\text{C}$$

Determine the following:

a. The quality at the evaporator inlet.

b. The rate of heat transfer to the evaporator.

c. The rate of heat transfer from the compressor.

All processes: Steady-state.
Model: R-134a tables.
All analyses: No changes in kinetic or potential energy. The energy equation in each case is given by Eq. 4.10.

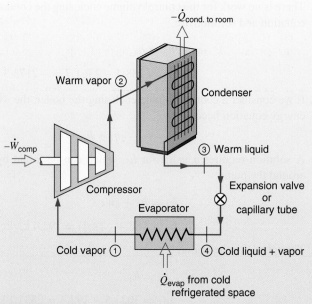

FIGURE 4.12 Refrigerator.

Solution

a. For a control volume enclosing the throttle, the energy equation gives

$$h_4 = h_3 = 241.8 \, \text{kJ/kg}$$

$$h_4 = 241.8 = h_{f4} + x_4 h_{fg4} = 167.4 + x_4 \times 215.6$$

$$x_4 = 0.345$$

b. For a control volume enclosing the evaporator, the energy equation gives

$$\dot{Q}_{\text{evap}} = \dot{m}(h_1 - h_4)$$

$$= 0.1(387.2 - 241.8) = 14.54 \, \text{kW}$$

c. And for the compressor, the energy equation gives

$$\dot{Q}_{\text{comp}} = \dot{m}(h_2 - h_1) + \dot{W}_{\text{comp}}$$

$$= 0.1(435.1 - 387.2) - 5.0 = -0.21 \, \text{kW}$$

In-Text Concept Questions

e. How does a nozzle or sprayhead generate kinetic energy?

f. What is the difference between a nozzle flow and a throttle process?

g. If you throttle a saturated liquid, what happens to the fluid state? What happens if this is done to an ideal gas?

h. A turbine at the bottom of a dam has a flow of liquid water through it. How does that produce power? Which terms in the energy equation are important if the C.V. is the turbine only? If the C.V. is the turbine plus the upstream flow up to the top of the lake, which terms in the energy equation are then important?

i. If you compress air, the temperature goes up. Why? When the hot air, at high P, flows in long pipes, it eventually cools to ambient T. How does that change the flow?

j. A mixing chamber has all flows at the same P, neglecting losses. A heat exchanger has separate flows exchanging energy, but they do not mix. Why have both kinds?

4.5 MULTIPLE FLOW DEVICES

In the previous section, we considered a number of devices and complete cycles that use a single flow through each component. Some applications have flows that separate or combine in one of the system devices. For example, *a faucet* in the kitchen or bathroom typically combines a warm and a cold flow of liquid water to produce an outlet flow at a desired temperature. In a *natural gas furnace* a small nozzle provides a gaseous fuel flow that is mixed with an air flow to produce a combustible mixture. A final example is a *flash evaporator* in a geothermal power plant where high-pressure hot liquid is throttled to a lower pressure (similar to the throttle/value in the refrigerator cycle). The resulting exit flow of two-phase fluid is then separated in a chamber to a flow of saturated vapor and a flow of saturated liquid.

For these and similar situations, the continuity and energy equations do not simplify as much as in the previous examples, so we will show the analysis for such a physical setup. Consider the mixing chamber in Fig. 4.13 with two inlet flows and a single exit flow operating in steady-state mode with no shaft; no work is involved, and we neglect kinetic and potential energies. The continuity and energy equations for this case become

$$\text{Continuity Eq. 4.9:} \qquad 0 = \dot{m}_1 + \dot{m}_2 - \dot{m}_3$$

$$\text{Energy Eq. 4.10:} \qquad 0 = \dot{m}_1 h_1 + \dot{m}_2 h_2 - \dot{m}_3 h_3 + \dot{Q}$$

We can scale the equations with the total mass flow rate of the exit flow by dividing the equations with $\dot{m}_3$ so that continuity equation gives

$$1 = \frac{\dot{m}_1}{\dot{m}_3} + \frac{\dot{m}_2}{\dot{m}_3} \qquad (4.15)$$

and the energy equation gives

$$0 = \frac{\dot{m}_1}{\dot{m}_3} h_1 + \frac{\dot{m}_2}{\dot{m}_3} h_2 - h_3 + \dot{Q}/\dot{m}_3 \qquad (4.16)$$

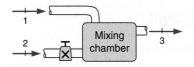

FIGURE 4.13
A mixing chamber.

In the scaled energy equation, the dimensionless mass flow ratios are factors in the flow terms and, these factors add to one according to the continuity equation. Select one as a parameter $0 < y < 1$; then we get from the continuity equation

$$y = \frac{\dot{m}_1}{\dot{m}_3}; \qquad \frac{\dot{m}_2}{\dot{m}_3} = 1 - y \qquad (4.17)$$

and the energy equation is

$$0 = yh_1 + (1 - y)h_2 - h_3 + \dot{Q}/\dot{m}_3 \qquad (4.18)$$

If the inlet states are given, it determines the enthalpy in the exit flow as

$$h_3 = yh_1 + (1 - y)h_2 + \dot{Q}/\dot{m}_3 \qquad (4.19)$$

This exit enthalpy is a mass-weighted average of the two inlet enthalpies determined by a single mass flow ratio and the possible heat transfer. If there is no heat transfer, the exit enthalpy h_3 can vary between the two inlet enthalpies h_2 and h_1 since the ratio y is between 0 and 1. This is exactly what is done when a kitchen faucet is switched between cold and hot water with the same total flow rate. Other combinations of known and unknown parameters exist, so the energy equation can determine one parameter and the continuity equation can determine the two flow rate ratios in terms of a single parameter y.

Example 4.9

We have a flow of 3 kg/s superheated steam at 300 kPa, 300 °C that we want to desuperheat by mixing it with liquid water at 300 kPa, 90 °C so that the output is a flow of saturated steam at 300 kPa. Assume the mixing chamber is insulated and find the flow rate of liquid water needed for the process.

Control volume: Mixing chamber, similar to Fig. 4.13.
Process: Steady-state adiabatic mixing.
Inlet, exit states: States 1, 2, and 3 all known.
Model: Steam tables, so

$$h_1 = 3069.28 \text{ kJ/kg}; \qquad h_2 = 376.9 \text{ kJ/kg}; \qquad h_3 = 2725.3 \text{ kJ/kg}$$

Analysis
For this situation the unknown is $\dot{m}_2$, so the continuity equation gives the output flow rate as

$$\dot{m}_3 = \dot{m}_1 + \dot{m}_2$$

which we substitute into the energy equation as

$$0 = \dot{m}_1 h_1 + \dot{m}_2 h_2 - (\dot{m}_1 - \dot{m}_2)h_3$$

Solution

The only unknown is the second mass flow rate, so use the energy equation to give

$$\dot{m}_2 = \dot{m}_1(h_3 - h_1)/(h_2 - h_3)$$

$$= 3\,\text{kg/s}\,\frac{2725.3 - 3069.28}{376.9 - 2725.3} = 0.439\,\text{kg/s}$$

4.6 THE TRANSIENT PROCESS

In the previous sections, we considered the steady-state process and several examples of its application in single-flow situations, and we extended the analysis to multiple flows. There are a number of processes of interest that do not fall in this category, and they can be characterized as those where the states and conditions change with time and thus involve an unsteady flow. This is, for example, the filling or emptying of a closed tank with a liquid or gas where the stored mass and its state in the control volume change with time. Think about a flat tire you fill with air; the mass of air and its pressure increase as the process proceeds, and the process stops when a desired pressure is reached. This type of process is called a *transient process* to distinguish it from the steady-state process. In general, the word *transient* means that something changes with time and it does not necessarily have a mass flow involved. To analyze such situations, we need some simplifying assumptions for the mathematical analysis as follows:

1. The control volume remains constant relative to the coordinate frame.
2. The state of the mass within the control volume may change with time, but at any instant of time the state is uniform throughout the entire control volume (or over several identifiable regions that make up the entire control volume).
3. The state of the mass crossing each of the areas of flow on the control surface is constant with time, although the mass flow rates may vary with time.

Let us examine the consequence of these assumptions and derive an expression for the energy equation that applies to this process. The assumption that the control volume remains stationary relative to the coordinate frame has already been discussed in Section 4.3. The remaining assumptions lead to the following simplification for the continuity equation and the energy equation.

The overall process occurs over time t, and during this time the instantaneous expression for the mass inside the control volume is given by the continuity equation in Eq. 4.1. To get the accumulated change, we integrate each term in the equation with time to get

$$\int_0^t \left(\frac{dm_{\text{C.V.}}}{dt} \right) dt = (m_2 - m_1)_{\text{C.V.}}$$

The total mass leaving the control volume during time t is

$$\int_0^t \left(\sum \dot{m}_e \right) dt = \sum m_e$$

and the total mass entering the control volume during time t is

$$\int_0^t \left(\sum \dot{m}_i \right) dt = \sum m_i$$

Therefore, for this period of time t, we can write the *continuity equation* for the transient process as

$$(m_2 - m_1)_{\text{C.V.}} = \sum m_i - \sum m_e \tag{4.20}$$

The energy equation for the changes over a finite time was presented in Eq. 3.5 for the control mass to which we have to add the flow terms. Let us integrate the energy equation in Eq. 4.8 by integration of each term as

$$\int_0^t \frac{dE_{\text{C.V.}}}{dt} dt = E_2 - E_1 = m_2 e_2 - m_1 e_1$$

$$= m_2 \left(u_2 + \frac{1}{2} \mathbf{V}_2^2 + gZ_2 \right) - m_1 \left(u_1 + \frac{1}{2} \mathbf{V}_1^2 + gZ_1 \right)$$

$$\int_0^t \dot{Q}_{\text{C.V.}} \, dt = Q_{\text{C.V.}}$$

$$\int_0^t \dot{W}_{\text{C.V.}} \, dt = W_{\text{C.V.}}$$

For the flow terms, the third assumption allows a simple integration as

$$\int_0^t \left[\sum \dot{m}_i h_{\text{tot} i} \right] dt = \sum m_i h_{\text{tot} i} = \sum m_i \left(h_i + \frac{1}{2} \mathbf{V}_i^2 + gZ_i \right)$$

$$\int_0^t \left[\sum \dot{m}_e h_{\text{tot} e} \right] dt = \sum m_e h_{\text{tot} e} = \sum m_e \left(h_e + \frac{1}{2} \mathbf{V}_e^2 + gZ_e \right)$$

For the period of time t, the transient process energy equation can now be written as

$$E_2 - E_1 = Q_{\text{C.V.}} - W_{\text{C.V.}} + \sum m_i \left(h_i + \frac{1}{2} \mathbf{V}_i^2 + gZ_i \right)$$

$$- \sum m_e \left(h_e + \frac{1}{2} \mathbf{V}_e^2 + gZ_e \right) \tag{4.21}$$

Notice how this energy equation is similar to the one for a control mass, Eq. 3.5, extended with the flow terms. Now the right-hand side explains all the possibilities for transferring energy across the control volume boundary as transfer by heat, work, or mass flows during a certain period of time. The left-hand-side storage change contains the internal energies (u_2, u_1), whereas the flow terms on the right-hand side contain enthalpies. If the state of the flow crossing the control volume boundary varies with time, an average for the exit or inlet flow properties should be used, which may not be simple to estimate.

As an example of the type of problem for which these assumptions are valid and Eq. 4.21 is appropriate, let us consider the classic problem of flow into an evacuated vessel. This is the subject of Example 4.10.

Example 4.10

Steam at a pressure of 1.4 MPa and a temperature of 300 °C is flowing in a pipe (Fig. 4.14). Connected to this pipe through a valve is an evacuated tank. The valve is opened and the tank fills with steam until the pressure is 1.4 MPa, and then the valve is closed. The process takes place adiabatically, and kinetic energies and potential energies are negligible. Determine the final temperature of the steam.

$$
\begin{aligned}
\textit{Control volume}&: \quad \text{Tank, as shown in Fig. 4.14.} \\
\textit{Initial state (in tank)}&: \quad \text{Evacuated, mass } m_1 = 0. \\
\textit{Final state}&: \quad P_2 \text{ known.} \\
\textit{Inlet state}&: \quad P_i, T_i \text{ (in line) known.} \\
\textit{Process}&: \quad \text{Transient, single flow in.} \\
\textit{Model}&: \quad \text{Steam tables.}
\end{aligned}
$$

Analysis

From the energy equation, Eq. 4.21, we have

$$
Q_{\text{C.V.}} + \sum m_i \left(h_i + \frac{\mathbf{V}_i^2}{2} + gZ_i \right)
$$

$$
= \sum m_e \left(h_e + \frac{\mathbf{V}_e^2}{2} + gZ_e \right)
$$

$$
+ \left[m_2 \left(u_2 + \frac{\mathbf{V}_2^2}{2} + gZ_2 \right) - m_1 \left(u_1 + \frac{\mathbf{V}_1^2}{2} + gZ_1 \right) \right]_{\text{C.V.}} + W_{\text{C.V.}}
$$

We note that $Q_{\text{C.V.}} = 0$, $W_{\text{C.V.}} = 0$, $m_e = 0$, and $(m_1)_{\text{C.V.}} = 0$. We further assume that changes in kinetic and potential energy are negligible. Therefore, the statement of the first law for this process reduces to

$$
m_i h_i = m_2 u_2
$$

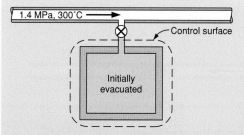

FIGURE 4.14 Flow into an evacuated vessel—control volume analysis.

From the continuity equation for this process, Eq. 4.20, we conclude that

$$
m_2 = m_i
$$

Therefore, combining the continuity equation with the energy equation, we have

$$h_i = u_2$$

That is, the final internal energy of the steam in the tank is equal to the enthalpy of the steam entering the tank.

Solution

From the steam tables, we obtain

$$h_i = u_2 = 3040.4 \text{ kJ/kg}$$

Since the final pressure is given as 1.4 MPa, we know two properties at the final state and therefore the final state is determined. The temperature corresponding to a pressure of 1.4 MPa and an internal energy of 3040.4 kJ/kg is found to be 452 °C.

This problem can also be solved by considering the steam that enters the tank and the evacuated space as a control mass, as indicated in Fig. 4.15.

The process is adiabatic, but we must examine the boundaries for work. If we visualize a piston between the steam that is included in the control mass and the steam that flows behind, we readily recognize that the boundaries move and that the steam in the pipe does work on the steam that comprises the control mass. The amount of this work is

$$-W = P_1 V_1 = m P_1 v_1$$

Writing the energy equation for the control mass, Eq. 3.5, and noting that kinetic and potential energies can be neglected, we have

$$_1Q_2 = U_2 - U_1 + {}_1W_2$$
$$0 = U_2 - U_1 - P_1 V_1$$
$$0 = mu_2 - mu_1 - m P_1 v_1 = mu_2 - mh_1$$

Therefore,

$$u_2 = h_1$$

which is the same conclusion that was reached using a control volume analysis.

The two other examples that follow illustrate further the transient process.

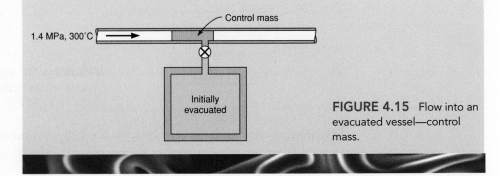

1.4 MPa, 300°C

Control mass

Initially evacuated

FIGURE 4.15 Flow into an evacuated vessel—control mass.

Example 4.11

A tank of 2 m³ volume contains saturated ammonia at a temperature of 40 °C. Initially, the tank contains 50 % liquid and 50 % vapor by volume. Vapor is withdrawn from the top of the tank until the temperature is 10 °C. Assuming that only vapor (i.e., no liquid) leaves and that the process is adiabatic, calculate the mass of ammonia that is withdrawn.

Control volume:	Tank.
Initial state:	T_1, V_{liq}, V_{vap}; state fixed.
Final state:	T_2.
Exit state:	Saturated vapor (temperature changing).
Process:	Transient.
Model:	Ammonia tables.

Analysis

In the energy equation, Eq. 4.21, we note that $Q_{\text{C.V.}} = 0$, $W_{\text{C.V.}} = 0$, and $m_i = 0$, and we assume that changes in kinetic and potential energy are negligible. However, the enthalpy of saturated vapor varies with temperature, and therefore we cannot simply assume that the enthalpy of the vapor leaving the tank remains constant. However, we note that at 40 °C, $h_g = 1470.2$ kJ/kg and at 10 °C, $h_g = 1452.0$ kJ/kg. Since the change in h_g during this process is small, we may accurately assume that h_e is the average of the two values given above. Therefore,

$$(h_e)_{\text{av}} = 1461.1 \text{ kJ/kg}$$

and the energy equation reduces to

$$m_2 u_2 - m_1 u_1 = -m_e h_e$$

and the continuity equation (from Eq. 4.20) becomes

$$(m_2 - m_1)_{\text{C.V.}} = -m_e$$

Combining these two equations, we have

$$m_2(h_e - u_2) = m_1 h_e - m_1 u_1$$

Solution

The following values are from the ammonia tables:

$$v_{f1} = 0.001\,725 \text{ m}^3/\text{kg}, \qquad v_{g1} = 0.083\,13 \text{ m}^3/\text{kg}$$

$$v_{f2} = 0.001\,60, \qquad v_{fg2} = 0.203\,81$$

$$u_{f1} = 368.7 \text{ kJ/kg}, \qquad u_{g1} = 1341.0 \text{ kJ/kg}$$

$$u_{f2} = 226.0, \qquad u_{fg2} = 1099.7$$

Calculating first the initial mass, m_1, in the tank, we find that the mass of the liquid initially present, m_{f1}, is

$$m_{f1} = \frac{V_f}{v_{f1}} = \frac{1.0}{0.001\,725} = 579.7 \text{ kg}$$

Similarly, the initial mass of vapor, m_{g1}, is

$$m_{g1} = \frac{V_g}{v_{g1}} = \frac{1.0}{0.083\ 13} = 12.0\ \text{kg}$$

$$m_1 = m_{f1} + m_{g1} = 579.7 + 12.0 = 591.7\ \text{kg}$$

$$m_1 h_e = 591.7 \times 1461.1 = 864\ 533\ \text{kJ}$$

$$m_1 u_1 = (mu)_{f1} + (mu)_{g1} = 579.7 \times 368.7 + 12.0 \times 1341.0$$
$$= 229\ 827\ \text{kJ}$$

Substituting these into the energy equation, we obtain

$$m_2(h_e - u_2) = m_1 h_e - m_1 u_1 = 864\ 533 - 229\ 827 = 634\ 706\ \text{kJ}$$

There are two unknowns, m_2 and u_2, in this equation. However,

$$m_2 = \frac{V}{v_2} = \frac{2.0}{0.001\ 60 + x_2(0.203\ 81)}$$

and

$$u_2 = 226.0 + x_2(1099.7)$$

and thus both are functions only of x_2, the quality at the final state. Consequently,

$$\frac{2.0(1461.1 - 226.0 - 1099.7 x_2)}{0.001\ 60 + 0.203\ 81 x_2} = 634\ 706$$

Solving for x_2, we get

$$x_2 = 0.011\ 057$$

Therefore,

$$v_2 = 0.001\ 60 + 0.011\ 057 \times 0.203\ 81 = 0.003\ 853\ 5\ \text{m}^3/\text{kg}$$

$$m_2 = \frac{V}{v_2} = \frac{2}{0.003\ 853\ 5} = 519\ \text{kg}$$

and the mass of ammonia withdrawn, m_e, is

$$m_e = m_1 - m_2 = 591.7 - 519 = 72.7\ \text{kg}$$

In-Text Concept Questions

k. An initially empty cylinder is filled with air coming in at $20\,°\text{C}$, 100 kPa until it is full. Assuming no heat transfer, is the final temperature larger than, equal to, or smaller than $20\,°\text{C}$? Does the final T depend on the size of the cylinder?

(4.7) ENGINEERING APPLICATIONS

Flow Systems and Flow Devices

The majority of devices and technical applications of energy conversions and transfers involve the flow of a substance. They can be passive devices like valves and pipes, active devices like turbines and pumps that involve work, or heat exchangers that involve a heat transfer into or out of the flowing fluid. Examples of these are listed in Table 4.1 together with their purpose and common assumptions, shown after the chapter summary.

Passive Devices as Nozzles, Diffusers, and Valves or Throttles

A nozzle is a passive (no moving parts) device that increases the velocity of a fluid stream at the expense of its pressure. Its shape, smoothly contoured, depends on whether the flow is subsonic or supersonic. A diffuser, basically the opposite of a nozzle, is shown in Fig. 4.16, in connection with flushing out a fire hydrant without having a high-velocity stream of water.

A flow is normally controlled by operating a valve that has a variable opening for the flow to pass through. With a small opening it represents a large restriction to the flow leading to a high pressure drop across the valve, whereas a large opening allows the flow to pass through freely with almost no restriction. There are many different types of valves in use, several of which are shown in Fig. 4.17.

Heaters/Coolers and Heat Exchangers

Two examples of heat exchangers are shown in Fig. 4.18. The aftercooler reduces the temperature of the air coming out of a compressor before it enters the engine. The purpose of the heat exchanger in Fig. 4.18*b* is to cool a hot flow or to heat a cold flow. The inner tubes act as the interphase area between the two fluids.

Active Flow Devices and Systems

A few air compressors and fans are shown in Fig. 4.19. These devices require a work input so that the compressor can deliver air flow at a higher pressure and the fan can provide air flow with some velocity. When the substance pushed to a higher pressure is a liquid, it is done with a pump, examples of which are shown in Fig. 4.20.

FIGURE 4.16
Diffuser.

(a) Ball valve   (b) Check valve   (c) Large butterfly valve

Inlet   Outlet

FIGURE 4.17 Several types of valves.

(d) Solenoid valve   (e) Pipeline gate valve

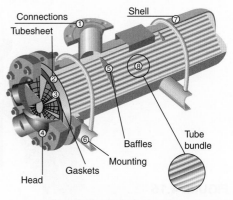

FIGURE 4.18 Heat exchangers.

(a) An aftercooler for a diesel engine   (b) A shell and tube heat exchanger

Connections

Shell

Tubesheet

Baffles

Tube bundle

Mounting

Gaskets

Head

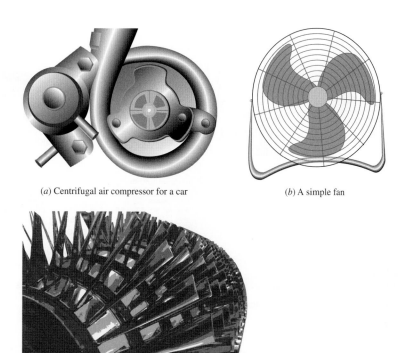

FIGURE 4.19 Air compressors and fans.

(a) Centrifugal air compressor for a car

(b) A simple fan

(c) Large axial-flow gas turbine compressor rotor

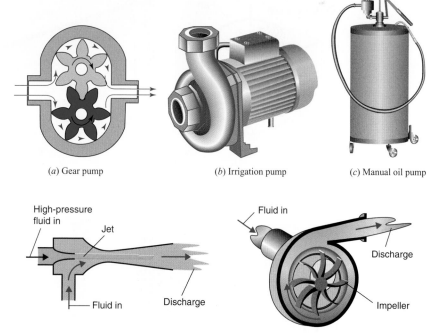

(a) Gear pump

(b) Irrigation pump

(c) Manual oil pump

High-pressure fluid in

Jet

Fluid in

Discharge

Fluid in

Discharge

Impeller

FIGURE 4.20 Liquid pumps.

(d) Jet pump and rotating pump

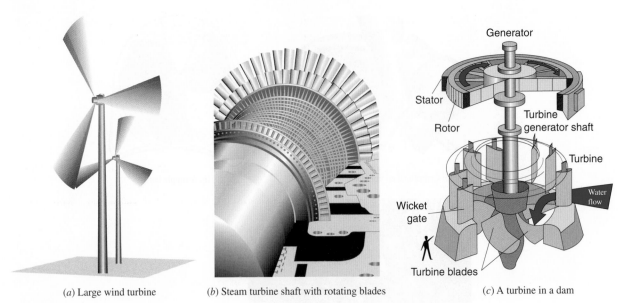

(a) Large wind turbine   (b) Steam turbine shaft with rotating blades   (c) A turbine in a dam

FIGURE 4.21 Examples of turbines.

Three different types of turbines are shown in Fig. 4.21. The steam turbine's outer stationary housing also has blades that turn the flow; these are not shown in Fig. 4.21b.

Figure 4.22 shows an air conditioner in cooling mode. It has two heat exchangers: one inside that cools the inside air and one outside that dumps energy into the outside atmosphere. This is functionally the same as what happens in a refrigerator. The same type

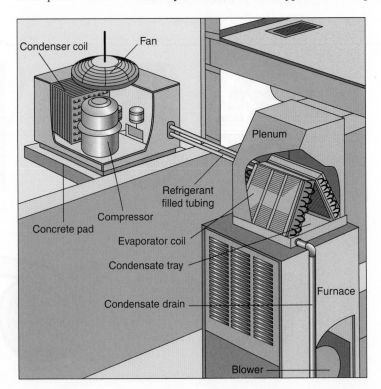

FIGURE 4.22
Household
air-conditioning system.

of system can be used as a heat pump. In heating mode, the flow is switched so that the inside heat exchanger is the hot one (condenser and heat rejecter) and the outside heat exchanger is the cold one (evaporator).

There are many types of power-producing systems. A coal-fired steam power plant was shown schematically in Figs. 1.1 and 1.2, and other types of engines were also described in Chapter 1. This subject will be developed in detail in Chapters 9 and 10.

Multiflow Devices

The text gave an example of a *mixing chamber* with two inlets and one exit flow, and Example 4.9 described a *desuperheater* often used in a power plant immediately before steam is sent to a process application or to a district heating system. In those cases, the purpose is to lower the peak temperature before distribution, which will reduce the heat transfer losses in the piping system. Exhaust systems in a building have several inlets to the ducting before reaching the exhaust fans, whereas the *heating ducts* in a building have one main inlet and many different outlets, so often the duct size is reduced along the way, as it needs to carry a smaller mass flow rate. Nearly every manufacturing plant has a *compressed air system* with a single inlet flow from the main compressor and an outlet at all of the workstations for compressed air tools and machines.

Large steam turbines can have outlets at several different pressures for various process applications, and a few outlets are used for *feedwater heaters* to boost the efficiency of the basic power cycle; see Chapter 9.

SUMMARY

Conservation of mass is expressed as a rate of change of total mass due to mass flows into or out of the control volume. The control mass energy equation is extended to include mass flows that also carry energy (internal, kinetic, and potential) and the flow work needed to push the flow into or out of the control volume against the prevailing pressure. The conservation of mass (continuity equation) and the conservation of energy (energy equation) are applied to a number of standard devices.

A steady-state device has no storage effects, with all properties constant with time, and constitutes the majority of all flow-type devices. A combination of several devices forms a complete system built for a specific purpose, such as a power plant, jet engine, or refrigerator.

A transient process with a change in mass (storage) such as filling or emptying of a container is considered based on an average description. It is also realized that the startup or shutdown of a steady-state device leads to a transient process.

You should have learned a number of skills and acquired abilities from studying this chapter that will allow you to

- Understand the physical meaning of the conservation equations. Rate = +in − out.
- Understand the concepts of mass flow rate, volume flow rate, and local velocity.
- Recognize the flow and nonflow terms in the energy equation.
- Know how the most typical devices work and if they have heat or work transfers.
- Have a sense of devices where kinetic and potential energies are important.
- Analyze steady-state single-flow devices such as nozzles, throttles, turbines, and pumps.

- Extend the application to a multiple-flow device such as a heat exchanger, mixing chamber, or turbine, given the specific setup.
- Apply the conservation equations to complete systems as a whole or to the individual devices and recognize their connections and interactions.
- Recognize and use the proper form of the equations for transient problems.
- Be able to assume a proper average value for any flow term in a transient.
- Recognize the difference between storage of energy (dE/dt) and flow ($\dot{m}h$).

A number of steady-flow devices are listed in Table 4.1 with a very short statement of each device's purpose, known facts about work and heat transfer, and a common assumption, if appropriate. This list is not complete with respect to the number of devices or with respect to the facts listed but is meant to show typical devices, some of which may be unfamiliar to many readers.

TABLE 4.1
Typical Steady-Flow Devices

Device	Purpose	Given	Assumption
Aftercooler	Cool a flow after a compressor	$w = 0$	$P = $ constant
Boiler	Bring substance to a vapor state	$w = 0$	$P = $ constant
Combustor	Burn fuel; acts like heat transfer in	$w = 0$	$P = $ constant
Compressor	Bring a substance to higher pressure	w in	$q = 0$
Condenser	Take q out to bring substance to liquid state	$w = 0$	$P = $ constant
Deaerator	Remove gases dissolved in liquids	$w = 0$	$P = $ constant
Dehumidifier	Remove water from air		$P = $ constant
Desuperheater	Add liquid water to superheated vapor steam to make it saturated vapor	$w = 0$	$P = $ constant
Diffuser	Convert KE energy to higher P	$w = 0$	$q = 0$
Economizer	Low-T, low-P heat exchanger	$w = 0$	$P = $ constant
Evaporator	Bring a substance to vapor state	$w = 0$	$P = $ constant
Expander	Similar to a turbine, but may have a q		
Fan/blower	Move a substance, typically air	w in, KE up	$P = C, q = 0$
Feedwater heater	Heat liquid water with another flow	$w = 0$	$P = $ constant
Flash evaporator	Generate vapor by expansion (throttling)	$w = 0$	$q = 0$
Heat engine	Convert part of heat into work	q in, w out	
Heat exchanger	Transfer heat from one medium to another	$w = 0$	$P = $ constant
Heat pump	Move a Q from T_{low} to T_{high}; requires a work input, refrigerator	w in	
Heater	Heat a substance	$w = 0$	$P = $ constant
Humidifier	Add water to air–water mixture	$w = 0$	$P = $ constant
Intercooler	Heat exchanger between compressor stages	$w = 0$	$P = $ constant
Mixing chamber	Mix two or more flows	$w = 0$	$q = 0$

TABLE 4.1 (*continued*)
Typical Steady-Flow Devices

Device	Purpose	Given	Assumption
Nozzle	Create KE; P drops	$w = 0$	$q = 0$
	Measure flow rate		
Pump	Same as compressor, but handles liquid	w in, P up	$q = 0$
Reactor	Allow a reaction between two or more substances	$w = 0$	$q = 0$, $P = C$
Regenerator	Usually a heat exchanger to recover energy	$w = 0$	$P = $ constant
Steam generator	Same as a boiler; heat liquid water to superheated vapor	$w = 0$	$P = $ constant
Supercharger	A compressor driven by engine shaft work to drive air into an automotive engine	w in	
Superheater	A heat exchanger that brings T up over T_{sat}	$w = 0$	$P = $ constant
Throttle	Same as a valve		
Turbine	Create shaft work from high P flow	w out	$q = 0$
Turbocharger	A compressor driven by an exhaust flow turbine to charge air into an engine	$\dot{W}_{turbine} = -\dot{W}_C$	
Valve	Control flow by restriction; P drops	$w = 0$	$q = 0$

KEY CONCEPTS AND FORMULAS

Volume flow rate	$\dot{V} = \int \mathbf{V}\, dA = A\mathbf{V}$	(using average velocity)
Mass flow rate	$\dot{m} = \int \rho \mathbf{V}\, dA = \rho A\mathbf{V} = A\mathbf{V}/v$	(using average values)
Flow work rate	$\dot{W}_{flow} = P\dot{V} = \dot{m}Pv$	
Flow direction	From higher P to lower P unless significant KE or PE exists	

Instantaneous Process

Continuity equation
$$\dot{m}_{C.V.} = \sum \dot{m}_i - \sum \dot{m}_e$$

Energy equation
$$\dot{E}_{C.V.} = \dot{Q}_{C.V.} - \dot{W}_{C.V.} + \sum \dot{m}_i h_{tot\,i} - \sum \dot{m}_e h_{tot\,e}$$

Total enthalpy
$$h_{tot} = h + \frac{1}{2}\mathbf{V}^2 + gZ = h_{stagnation} + gZ$$

Steady State

No storage: $\dot{m}_{C.V.} = 0$; $\dot{E}_{C.V.} = 0$

Continuity equation
$$\sum \dot{m}_i = \sum \dot{m}_e \quad (\text{in} = \text{out})$$

Energy equation
$$\dot{Q}_{C.V.} + \sum \dot{m}_i h_{tot\,i} = \dot{W}_{C.V.} + \sum \dot{m}_e h_{tot\,e} \quad (\text{in} = \text{out})$$

Specific heat transfer $\quad q = \dot{Q}_{C.V.}/\dot{m}$ (steady state only)

Specific work $\quad w = \dot{W}_{C.V.}/\dot{m}$ (steady state only)

Steady-state, single-flow energy equation $\quad q + h_{tot\,i} = w + h_{tot\,e}$ (in = out)

Transient Process

Continuity equation $\quad m_2 - m_1 = \sum m_i - \sum m_e$

Energy equation $\quad E_2 - E_1 = {}_1 Q_2 - {}_1 W_2 + \sum m_i h_{\text{tot } i} - \sum m_e h_{\text{tot } e}$

$$E_2 - E_1 = m_2 \left(u_2 + \frac{1}{2} \mathbf{V}_2^2 + g Z_2 \right) - m_1 \left(u_1 + \frac{1}{2} \mathbf{V}_1^2 + g Z_1 \right)$$

$$h_{\text{tot } e} = h_{\text{tot exit average}} \approx \frac{1}{2} \left(h_{\text{hot } e1} + h_{\text{tot } e2} \right)$$

CONCEPT-STUDY GUIDE PROBLEMS

4.1 A temperature difference drives a heat transfer. Does a similar concept apply to $\dot{m}$?

4.2 What effect can be felt upstream in a flow?

4.3 Which of the properties (P, v, T) can be controlled in a flow? How?

4.4 Air at 500 kPa is expanded to 100 kPa in two steady-flow cases. Case one is a nozzle and case two is a turbine; the exit state is the same for both cases. What can you say about the specific turbine work relative to the specific kinetic energy in the exit flow of the nozzle?

4.5 Pipes that carry a hot fluid like steam in a power plant, exhaust pipe for a diesel engine in a ship, etc., are often insulated. Is that done to reduce heat loss or is there another purpose?

4.6 A windmill takes out a fraction of the wind kinetic energy as power on a shaft. How do the temperature and wind velocity influence the power? Hint: write the power term as mass flow rate times specific work.

4.7 An underwater turbine extracts a fraction of the kinetic energy from the ocean current. How do the temperature and water velocity influence the power? Hint: write the power term as mass flow rate times specific work.

4.8 A liquid water turbine at the bottom of a dam takes energy out as power on a shaft. Which term(s) in the energy equation are changing and important?

4.9 You blow a balloon up with air. What kinds of work terms, including flow work, do you see in that case? Where is energy stored?

4.10 A storage tank for natural gas has a top dome that can move up or down as gas is added to or subtracted from the tank, maintaining 110 kPa, 290 K inside. A pipeline at 110 kPa, 290 K now supplies some natural gas to the tank. Does its state change during the filling process? What happens to the flow work?

HOMEWORK PROBLEMS

Continuity Equation and Flow Rates

4.11 A large brewery has a pipe of cross-sectional area 0.2 m² flowing carbon dioxide at 400 kPa, 10 °C with a volume flow rate of 0.3 m³/s. Find the velocity and the mass flow rate.

4.12 Air at 35 °C, 105 kPa flows in a 100 mm × 150 mm rectangular duct in a heating system. The mass flow rate is 0.015 kg/s. What are the velocity of the air flowing in the duct and the volume flow rate?

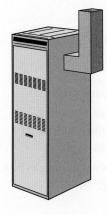

FIGURE P4.12

4.13 A pool is to be filled with 80 m³ water from a garden hose of 2 cm diameter flowing water at 2 m/s. Find the mass flow rate of water and the time it takes to fill the pool.

4.14 An empty bathtub has its drain closed and is being filled with water from the faucet at a rate of 10 kg/min. After 10 min, the drain is opened and 4 kg/min flows out; at the same time, the inlet flow is reduced to 2 kg/min. Plot the mass of the water in the bathtub versus time and determine the time from the very beginning when the tub will be empty.

4.15 A flat channel of depth 1 m has a fully developed laminar flow of air at P_0, T_0 with a velocity profile of $\mathbf{V} = 4\mathbf{V}_c x(H - x)/H^2$, where $\mathbf{V}_c$ is the velocity on the centerline and x is the distance across the channel, as shown in Fig. P4.15. Find the total mass flow rate and the average velocity both as functions of $\mathbf{V}_c$ and H.

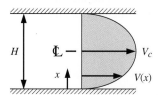

FIGURE P4.15

4.16 A boiler receives a constant flow of 5000 kg/h liquid water at 5 MPa and 20 °C, and it heats the flow such that the exit state is 400 °C with a pressure of 4.5 MPa. Determine the necessary minimum pipe flow area in both the inlet and exit pipe(s) if there should be no velocities larger than 20 m/s.

4.17 An airport ventilation system takes 2.5 m³/s air at 100 kPa, 17 °C into a furnace, heats it to 52 °C, and delivers the flow to a duct with cross-sectional area 0.4 m² at 110 kPa. Find the mass flow rate and the velocity in the duct.

Single-Flow, Single-Device Processes

Nozzles, Diffusers

4.18 Liquid water at 15 °C flows out of a nozzle straight up 15 m. What is nozzle $\mathbf{V}_{\text{exit}}$?

4.19 A nozzle receives an ideal gas flow with a velocity of 25 m/s, and the exit at 100 kPa, 300 K velocity is 250 m/s. Determine the inlet temperature if the gas is argon, helium, or nitrogen.

4.20 A diffuser receives 0.1 kg/s steam at 500 kPa, 350 °C. The exit is at 800 kPa, 400 °C with negligible kinetic energy and the flow is adiabatic. Find the diffuser inlet velocity and the inlet area.

4.21 In a jet engine a flow of air at 1000 K, 200 kPa, and 30 m/s enters a nozzle, as shown in Fig. P4.21, where the air exits at 850 K, 90 kPa. What is the exit velocity, assuming no heat loss?

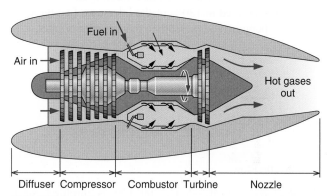

FIGURE P4.21

4.22 In a jet engine a flow of air at 1000 K, 200 kPa, and 40 m/s enters a nozzle, where the air exits at 500 m/s, 90 kPa. What is the exit temperature, assuming no heat loss?

4.23 The wind is blowing horizontally at 30 m/s in a storm at P_0, 20 °C toward a wall, where it comes to a stop (stagnation) and leaves with negligible velocity similar to a diffuser with a very large exit area. Find the stagnation temperature from the energy equation.

4.24 A diffuser, shown in Fig. P4.24, has air entering at 100 kPa and 300 K with a velocity of 200 m/s. The inlet cross-sectional area of the diffuser is 100 mm². At the exit the area is 860 mm², and the exit velocity is 20 m/s. Determine the exit pressure and temperature of the air.

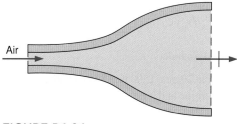

FIGURE P4.24

4.25 A meteorite hits the upper atmosphere at 3000 m/s, where the pressure is 0.1 atm and the temperature is −40 °C. How hot does the air become right in front of the meteorite, assuming no heat transfer in this adiabatic stagnation process?

4.26 The front of a jet engine acts similarly to a diffuser, receiving air at 900 km/h, −5 °C, and 50 kPa, bringing it to 80 m/s relative to the engine before entering the compressor (see Fig. P4.26). If the flow area is increased to 120 % of the inlet area, find the temperature and pressure in the compressor inlet.

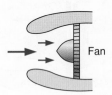

Fan

FIGURE P4.26

Throttle Flow

4.27 R-410A at −5 °C, 700 kPa is throttled, so it becomes cold at −30 °C. What is exit P?

4.28 Saturated liquid R-134a at 25 °C is throttled to 300 kPa in a refrigerator. What is the exit temperature? Find the percent increase in the volume flow rate.

4.29 Carbon dioxide used as a natural refrigerant flows out of a cooler at 10 MPa, 40 °C, after which it is throttled to 2.5 MPa. Find the state (T, x) for the exit flow.

4.30 Liquid water at 180 °C, 2000 kPa is throttled into a flash evaporator chamber having a pressure of 500 kPa. Neglect any change in the kinetic energy. What is the fraction of liquid and vapor in the chamber?

4.31 Methane at 1 MPa, 300 K is throttled through a valve to 100 kPa. Assume no change in the kinetic energy. What is the exit temperature?

4.32 R-134a is throttled in a line flowing at 25 °C, 750 kPa with negligible kinetic energy to a pressure of 165 kPa. Find the exit temperature and the ratio of the exit pipe diameter to that of the inlet pipe (D_{ex}/D_{in}) so that the velocity stays constant.

Turbines, Expanders

4.33 Air at 20 m/s, 1500 K, 875 kPa with 5 kg/s flows into a turbine and it flows out at 25 m/s, 850 K, 105 kPa. Find the power output using constant specific heats.

4.34 Solve the previous problem using Table A.7.

4.35 A wind turbine with a rotor diameter of 20 m takes 40 % of the kinetic energy out as shaft work on a day with a temperature of 20 °C and a wind speed of 35 km/h. What power is produced?

4.36 A liquid water turbine receives 2 kg/s water at 2000 kPa, 20 °C with a velocity of 30 m/s. The exit is at 100 kPa, 20 °C, and very low velocity. Find the specific work and the power produced.

4.37 A small, high-speed turbine operating on compressed air produces a power output of 100 W. The inlet state is 400 kPa, 50 °C, and the exit state is 150 kPa, −30 °C. Assuming the velocities to be low and the process to be adiabatic, find the required mass flow rate of air through the turbine.

4.38 Hoover Dam across the Colorado River dams up Lake Mead 200 m higher than the river downstream (see Fig. P4.38). The electric generators driven by water-powered turbines deliver 1300 MW of power. If the water is 17.5 °C, find the minimum amount of water running through the turbines.

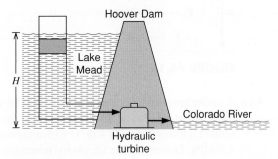

FIGURE P4.38

4.39 A small turbine, shown in Fig. P4.39, is operated at part load by throttling a 0.25 kg/s steam supply at 1.4 MPa and 250 °C down to 1.1 MPa before it enters the turbine, and the exhaust is at 10 kPa. If the turbine produces 110 kW, find the exhaust temperature (and quality if saturated).

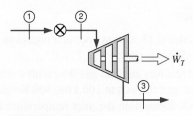

FIGURE P4.39

4.40 A small expander (a turbine with heat transfer) has 0.05 kg/s helium entering at 1000 kPa, 550 K and leaving at 250 kPa, 300 K. The power output on the shaft measures 55 kW. Find the rate of heat transfer, neglecting kinetic energies.

Compressors, Fans

4.41 A compressor in a commercial refrigerator receives R-410A at $-25\,°C$ and $x = 1$. The exit is at 1000 kPa and $40\,°C$. Neglect kinetic energies and find the specific work.

4.42 A compressor brings nitrogen from 100 kPa, 290 K to 2000 kPa. The process has a specific work input of 450 kJ/kg and the exit temperature is 500 K. Find the specific heat transfer using constant specific heats.

4.43 A refrigerator uses the natural refrigerant carbon dioxide where the compressor brings 0.02 kg/s from 1 MPa, $-20\,°C$ to 6 MPa using 2 kW of power. Find the compressor exit temperature.

4.44 A factory generates compressed air from 100 kPa, $17\,°C$ by compression to 1000 kPa, 600 K, after which it cools to 300 K in a constant pressure cooler (see Fig. P4.44). Find the specific compressor work and the specific heat transfer in the cooler.

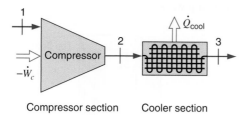

Compressor section   Cooler section

FIGURE P4.44

4.45 A compressor brings R-134a from 150 kPa, $-10\,°C$ to 1200 kPa, $50\,°C$. It is water cooled, with heat loss estimated as 30 kW, and the shaft work input is measured to be 150 kW. What is the mass flow rate through the compressor?

4.46 The compressor of a large gas turbine receives air from the ambient surroundings at 95 kPa, $20\,°C$ with low velocity. At the compressor discharge, air exits at 1.52 MPa, $430\,°C$ with a velocity of 90 m/s. The power input to the compressor is 5000 kW. Determine the mass flow rate of air through the unit.

4.47 A compressor in an industrial air conditioner compresses ammonia from a state of saturated vapor at 200 kPa to a pressure of 1000 kPa. At the exit, the temperature is measured to be $100\,°C$ and the mass flow rate is 0.5 kg/s. What is the required motor size (kW) for this compressor?

4.48 An exhaust fan in a building should be able to move 3 kg/s atmospheric pressure air at $25\,°C$ through a 0.5 m diameter vent hole. How high a velocity must it generate, and how much power is required to do that?

4.49 An air flow is brought from $20\,°C$, 100 kPa to 1000 kPa, $330\,°C$ by an adiabatic compressor driven by a 50 kW motor. What are the mass flow rate and the exit volume flow rate of air?

Heaters, Coolers

4.50 The air conditioner in a house or a car has a cooler that brings atmospheric air from $30\,°C$ to $10\,°C$, both states at 101 kPa. For a flow rate of 0.75 kg/s, find the rate of heat transfer.

4.51 A boiler section boils 3 kg/s saturated liquid water at 2500 kPa to saturated vapor in a reversible constant-pressure process. Find the specific heat transfer in the process.

4.52 A condenser (cooler) receives 0.05 kg/s of R-410A at 2000 kPa, $80\,°C$ and cools it to $10\,°C$. Assume the exit properties are as for saturated liquid with the same T. What cooling capacity (kW) must the condenser have?

4.53 Find the heat transfer in Problem 4.16.

4.54 A chiller cools liquid water for air-conditioning purposes. Assume that 2.5 kg/s water at $20\,°C$, 100 kPa is cooled to $5\,°C$ in a chiller. How much heat transfer (kW) is needed?

4.55 Saturated liquid nitrogen at 600 kPa enters a boiler at a rate of 0.008 kg/s and exits as saturated vapor (see Fig. P4.55). It then flows into a superheater also at 600 kPa, where it exits at 600 kPa, 280 K. Find the rate of heat transfer in the boiler and the superheater.

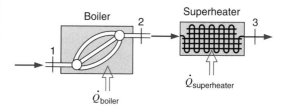

FIGURE P4.55

4.56 In a steam generator, compressed liquid water at 10 MPa, 30 °C enters a 30 mm diameter tube at a rate of 3 L/s. Steam at 9 MPa, 400 °C exits the tube. Find the rate of heat transfer to the water.

4.57 Liquid nitrogen at 90 K, 400 kPa flows into a probe used in a cryogenic survey. In the return line the nitrogen is then at 160 K, 400 kPa. Find the specific heat transfer to the nitrogen. If the return line has a cross-sectional area 100 times larger than that of the inlet line, what is the ratio of the return velocity to the inlet velocity?

4.58 An evaporator has R-410A at −20 °C and quality 20 % flowing in, with the exit flow being saturated vapor at −20 °C. Knowing that there is no work, find the specific heat transfer.

4.59 Liquid glycerine flows around an engine, cooling it as it absorbs energy. The glycerine enters the engine at 60 °C and receives 19 kW of heat transfer. What is the required mass flow rate if the glycerine should come out at a maximum temperature of 85 °C?

Pumps, Pipe and Channel Flows

4.60 An irrigation pump takes water from a river at 10 °C, 100 kPa and pumps it up to an open canal, where it flows out 100 m higher at 10 °C. The pipe diameter in and out of the pump is 0.1 m, and the motor driving the unit is 5 hp. What is the flow rate, neglecting kinetic energy and losses?

4.61 A pipe flows water at 15 °C from one building to another. In the winter, the pipe loses an estimated 500 W of heat transfer. What is the minimum required mass flow rate that will ensure that the water does not freeze (i.e., reach 0 °C)?

4.62 A steam pipe for a 300 m tall building receives superheated steam at 200 kPa at ground level. At the top floor the pressure is 125 kPa, and the heat loss in the pipe is 110 kJ/kg. What should the inlet temperature be so that no water will condense inside the pipe?

4.63 Consider a water pump that receives liquid water at 25 °C, 100 kPa and delivers it to a same-diameter short pipe having a nozzle with an exit diameter of 2 cm (0.02 m) to the atmosphere at 100 kPa (see Fig. P4.63). Neglect the kinetic energy in the pipes and assume constant u for the water. Find the exit velocity and the mass flow rate if the pump draws 1 kW of power.

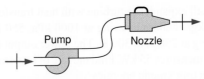

FIGURE P4.63

4.64 A small stream with water at 15 °C runs out over a cliff, creating a 50 m tall waterfall. Estimate the downstream temperature when you neglect the horizontal flow velocities upstream and downstream from the waterfall. How fast was the water dropping just before it splashed into the pool at the bottom of the waterfall?

4.65 A cutting tool uses a nozzle that generates a high-speed jet of liquid water. Assume an exit velocity of 500 m/s of 20 °C liquid water with a jet diameter of 2 mm (0.002 m). What is the mass flow rate? What size (power) pump is needed to generate this from a steady supply of 20 °C liquid water at 200 kPa?

4.66 The main water line into a tall building has a pressure of 600 kPa at 5 m below ground level, as shown in Fig. P4.66. A pump brings the pressure up so that the water can be delivered at 150 kPa at the top floor 100 m above ground level. Assume a flow rate of 10 kg/s liquid water at 10 °C and neglect any difference in kinetic energy and internal energy u. Find the pump work.

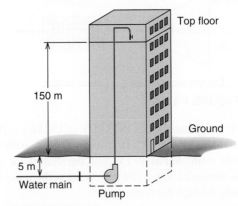

FIGURE P4.66

Multiple-Flow, Single-Device Processes

Turbines, Compressors, Expanders

4.67 An adiabatic steam turbine in a power plant receives 5 kg/s steam at 3000 kPa, 500 °C. Twenty percent of the flow is extracted at 1000 kPa, 350 °C to a feedwater heater, and the remainder flows out at 200 kPa, 200 °C (see Fig. P4.67). Find the turbine power output.

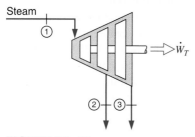

FIGURE P4.67

4.68 Cogeneration is often used where a steam supply is needed for industrial process energy. Assume that a supply of 5 kg/s steam at 0.5 MPa is needed. Rather than generating this from a pump and boiler, the setup in Fig. P4.68 is used to extract the supply from the high-pressure turbine. Find the power the turbine now cogenerates in this process.

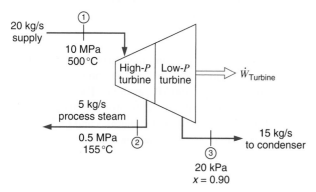

FIGURE P4.68

4.69 A steam turbine receives steam from two boilers (see Fig. P4.69). One flow is 5 kg/s at 3 MPa, 700 °C and the other flow is 10 kg/s at 800 kPa, 500 °C. The exit state is 10 kPa, with a quality of 96 %. Find the total power out of the adiabatic turbine.

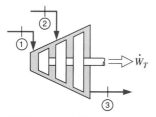

FIGURE P4.69

4.70 Two steady flows of air enter a control volume, as shown in Fig. P4.70. One is a 0.025 kg/s flow at 350 kPa, 150 °C, state 1, and the other enters at 450 kPa, 15 °C, state 2. A single flow exits at 100 kPa, −40 °C, state 3. The control volume ejects 1 kW heat to the surroundings and produces 4 kW of power output. Neglect kinetic energies and determine the mass flow rate at state 2.

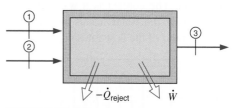

FIGURE P4.70

4.71 A large, steady expansion engine has two low-velocity flows of water entering. High-pressure steam enters at point 1 with 2.0 kg/s at 2 MPa, 500 °C, and 0.5 kg/s of cooling water at 120 kPa, 30 °C centers at point 2. A single flow exits at point 3, with 150 kPa and 80 % quality, through a 0.15 m diameter exhaust pipe. There is a heat loss of 300 kW. Find the exhaust velocity and the power output of the engine.

Heat Exchangers

4.72 A condenser (heat exchanger) brings 1 kg/s water flow at 10 kPa quality 95 % to saturated liquid at 10 kPa, as shown in Fig. P4.72. The cooling is done by lake water at 20 °C that returns to the lake at 30 °C. For an insulated condenser, find the flow rate of cooling water.

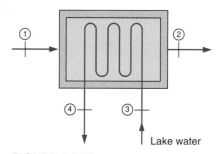

FIGURE P4.72

4.73 Air at 800 K flows with 3 kg/s into a heat exchanger and out at 100 °C. How much (kg/s) water coming in at 100 kPa, 20 °C can the air heat to the boiling point?

4.74 A dual-fluid heat exchanger has 5 kg/s water entering at 40 °C, 150 kPa and leaving at 10 °C, 150 kPa. The other fluid is glycol, entering at −10 °C, 160 kPa and leaving at 10 °C, 160 kPa. Find the required mass flow rate of glycol and the rate of internal heat transfer.

4.75 A heat exchanger, shown in Fig. P4.75, is used to cool an air flow from 800 to 360 K, with both states at 1 MPa. The coolant is a water flow at 15 °C, 0.1 MPa. If the water leaves as saturated vapor, find the ratio of the flow rates $\dot{m}_{water}/\dot{m}_{air}$.

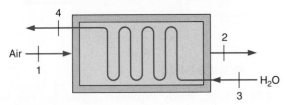

FIGURE P4.75

4.76 A superheater brings 2.5 kg/s of saturated water vapor at 2 MPa to 450 °C. The energy is provided by hot air at 1200 K flowing outside the steam tube in the opposite direction as the water, a setup known as a *counterflowing heat exchanger* (similar to Fig. P4.75). Find the smallest possible mass flow rate of the air to ensure that its exit temperature is 20 °C larger than the incoming water temperature.

4.77 In a co-flowing (same-direction) heat exchanger, 1 kg/s air at 500 K flows into one channel and 2 kg/s air flows into the neighboring channel at 300 K. If it is infinitely long, what is the exit temperature? Sketch the variation of T in the two flows.

4.78 An automotive radiator has glycerine at 95 °C enter and return at 55 °C, as shown in Fig. P4.78.

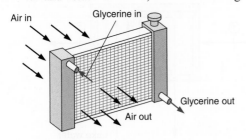

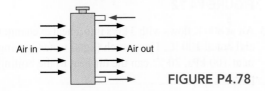

FIGURE P4.78

Air flows in at 20 °C and leaves at 25 °C. If the radiator should transfer 25 kW, what is the mass flow rate of the glycerine and what is the volume flow rate of air in at 100 kPa?

4.79 A cooler in an air conditioner brings 0.5 kg/s of air at 35 °C to 5 °C, both at 101 kPa. It then mixes the output with a flow of 0.25 kg/s air at 20 °C and 101 kPa, sending the combined flow into a duct. Find the total heat transfer in the cooler and the temperature in the duct flow.

4.80 A copper wire has been heat treated to 1100 K and is now pulled into a cooling chamber that has 1.5 kg/s air coming in at 20 °C; the air leaves the other end at 60 °C. If the wire moves 0.25 kg/s copper, how hot is the copper as it comes out?

4.81 A co-flowing (same-direction) heat exchanger has one line with 0.25 kg/s oxygen at 17 °C, 200 kPa entering and the other line has 0.6 kg/s nitrogen at 150 kPa, 500 K entering. The heat exchanger is very long, so the two flows exit at the same temperature. Use constant heat capacities and find the exit temperature.

Mixing Processes

4.82 Two air flows are combined to a single flow. One flow is 1 m³/s at 20 °C and the other is 2 m³/s at 200 °C, both at 100 kPa, as in Fig. P4.82. They mix without any heat transfer to produce an exit flow at 100 kPa. Neglect kinetic energies and find the exit temperature and volume flow rate.

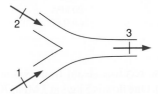

FIGURE P4.82

4.83 A flow of water at 2000 kPa, 20 °C is mixed with a flow of 2 kg/s water at 2000 kPa, 180 °C. What should the flow rate of the first flow be to produce an exit state of 200 kPa and 100 °C?

4.84 An open feedwater heater in a power plant heats 4 kg/s water at 45 °C, 100 kPa by mixing it with steam from the turbine at 100 kPa, 150 °C, as in Fig. P4.84. Assume the exit flow is saturated liquid at the given pressure and find the mass flow rate from the turbine.

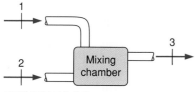

FIGURE P4.84

4.85 A desuperheater has a flow of ammonia of 1.5 kg/s at 1000 kPa, 100 °C that is mixed with another flow of ammonia at 25 °C and quality 50 % in an adiabatic mixing chamber. Find the flow rate of the second flow so that the outgoing ammonia is saturated vapor at 1000 kPa.

4.86 A mixing chamber with heat transfer, as shown in Fig. P4.86, receives 2 kg/s of R-410A, at 1 MPa, 40 °C in one line and 1 kg/s of R-410A at 15 °C with a quality of 50 % in a line with a valve. The outgoing flow is at 1 MPa, 60 °C. Find the rate of heat transfer to the mixing chamber.

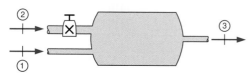

FIGURE P4.86

4.87 A geothermal supply of hot water at 500 kPa, 150 °C is fed to an insulated flash evaporator at the rate of 1.5 kg/s. A stream of saturated liquid at 200 kPa is drained from the bottom of the chamber, and a stream of saturated vapor at 200 kPa is drawn from the top and fed to a turbine. Find the mass flow rate of the two exit flows.

4.88 To keep a jet engine cool, some intake air bypasses the combustion chamber. Assume that 2 kg/s of hot air at 2000 K and 500 kPa is mixed with 1.5 kg/s air at 500 K, 500 kPa without any external heat transfer, as in Fig. P4.88. Find the exit temperature using constant specific heat from Table A.5.

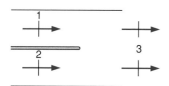

FIGURE P4.88

4.89 Solve the previous problem using values from Table A.7.

Multiple Devices, Cycle Processes

4.90 A two-stage compressor takes nitrogen in at 20 °C, 150 kPa and compresses it to 600 kPa, 450 K. Then it flows through an intercooler, where it cools to 320 K, and the second stage compresses it to 3000 kPa, 530 K. Find the specific work in each of the two compressor stages and the specific heat transfer in the intercooler.

4.91 The intercooler in the previous problem uses cold liquid water to cool the nitrogen. The nitrogen flow is 0.1 kg/s, and the liquid water inlet is 20 °C and is set up to flow in the opposite direction from the nitrogen, so the water leaves at 35 °C. Find the flow rate of the water.

4.92 The following data are for a simple steam power plant as shown in Fig. P4.92. State 6 has $x_6 = 0.92$ and velocity of 200 m/s. The rate of steam flow is 25 kg/s, with 300 kW of power input to the pump. Piping diameters are 200 mm from the steam generator to the turbine and 75 mm from the condenser to the economizer and steam generator. Determine the velocity at state 5 and the power output of the turbine.

State	1	2	3	4	5	6	7
P, kPa	6200	6100	5900	5700	5500	10	9
T, °C		45	175	500	490		40
h, kJ/kg		194	744	3426	3404		168

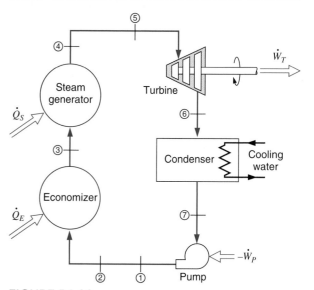

FIGURE P4.92

4.93 For the steam power plant shown in Problem 4.92, assume that the cooling water comes from a lake at 15 °C and is returned at 25 °C. Determine the rate of heat transfer in the condenser and the mass flow rate of cooling water from the lake.

4.94 For the steam power plant shown in Problem 4.92, determine the rate of heat transfer in the economizer, which is a low-temperature heat exchanger. Also find the rate of heat transfer needed in the steam generator.

4.95 A somewhat simplified flow diagram for a nuclear power plant is given in Fig. P4.95. Mass flow rates and the various states in the cycle are shown in the accompanying table.

   The cycle includes a number of heaters in which heat is transferred from steam, taken out of the turbine at some intermediate pressure, to liquid water pumped from the condenser on its way to the steam drum. The heat exchanger in the reactor supplies 157 MW, and it may be assumed that there is no heat transfer in the turbines.

State	$\dot{m}$, kg/s	P, kPa	T, °C	h, kJ/kg
1	75.6	7240	sat vap	
2	75.6	6900		2765
3	62.874	345		2517
4		310		
5		7		2279
6	75.6	7	33	138
7		415		140
8	2.772	35		2459
9	4.662	310		558
10		35	34	142
11	75.6	380	68	285
12	8.064	345		2517
13	75.6	330		
14				349
15	4.662	965	139	584
16	75.6	7930		565
17	4.662	965		2593
18	75.6	7580		688
19	1386	7240	277	1220
20	1386	7410		1221
21	1386	7310		

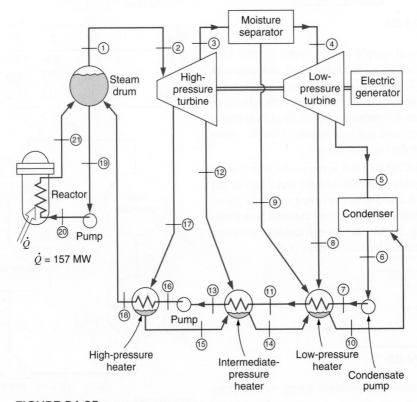

FIGURE P4.95

a. Assuming the moisture separator has no heat transfer between the two turbine sections, determine the enthalpy and quality (h_4, x_4).

b. Determine the power output of the low-pressure turbine.

c. Determine the power output of the high-pressure turbine.

d. Find the ratio of the total power output of the two turbines to the total power delivered by the reactor.

4.96 Consider the power plant described in the previous problem.

a. Determine the quality of the steam leaving the reactor.

b. What is the power to the pump that feeds water to the reactor?

4.97 An R-410A heat pump cycle shown in Fig. P4.97 has an R-410A flow rate of 0.05 kg/s with 5 kW into the compressor. The following data are given:

State	1	2	3	4	5	6
P, kPa	3100	3050	3000	420	400	390
T, °C	120	110	45		−10	−5
h, kJ/kg	377	367	134	—	280	284

Calculate the heat transfer from the compressor, the heat transfer from the R-410A in the condenser, and the heat transfer to the R-410A in the evaporator.

4.98 A modern jet engine has a temperature after combustion of about 1500 K at 3200 kPa as it enters the turbine section (see state 3, Fig. P4.98). The compressor inlet is at 80 kPa, 260 K (state 1) and the outlet (state 2) is at 3300 kPa, 780 K; the turbine outlet (state 4) into the nozzle is at 400 kPa, 900 K and the nozzle exit (state 5) is at 80 kPa, 640 K. Neglect any heat transfer and neglect kinetic energy except out of the nozzle. Find the compressor and turbine specific work terms and the nozzle exit velocity.

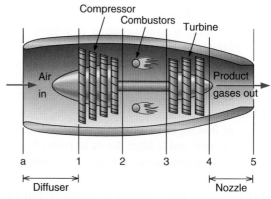

FIGURE P4.98

4.99 A proposal is made to use a geothermal supply of hot water to operate a steam turbine, as shown in Fig. P4.99. The high-pressure water at 1.5 MPa, 180 °C is throttled into a flash evaporator chamber, which forms liquid and vapor at a lower pressure

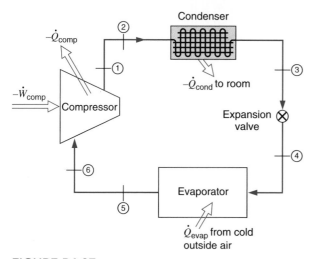

FIGURE P4.97

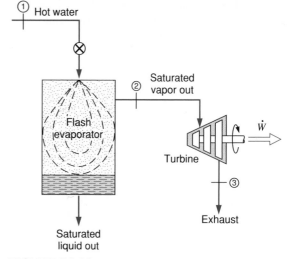

FIGURE P4.99

of 400 kPa. The liquid is discarded, while the saturated vapor feeds the turbine and exits at 10 kPa with a 90% quality. If the turbine should produce 1 MW, find the required mass flow rate of hot geothermal water in kilograms per hour.

Transient Processes

4.100 An initially empty cylinder is filled with air from 20 °C, 100 kPa until it is full. Assuming no heat transfer, is the final temperature above, equal to, or below 20 °C? Does the final T depend on the size of the cylinder?

4.101 An initially empty canister of volume 0.2 m³ is filled with carbon dioxide from a line at 800 kPa, 400 K. Assume the process runs until it stops by itself and it is adiabatic. Use constant specific heat to find the final temperature in the canister.

4.102 A tank contains 1 m³ air at 100 kPa, 300 K. A pipe of flowing air at 1000 kPa, 320 K is connected to the tank and is filled slowly to 1000 kPa. Find the heat transfer needed to reach a final temperature of 300 K.

4.103 A 1 m³ tank contains ammonia at 150 kPa and 25 °C. The tank is attached to a line flowing ammonia at 1200 kPa, 60 °C. The valve is opened, and mass flows in until the tank is half full of liquid (by volume) at 25 °C. Calculate the heat transferred from the tank during this process.

4.104 A 2.5 L tank initially is empty, and we want to fill it with 10 g of ammonia. The ammonia comes from a line with saturated vapor at 25 °C. To achieve the desired amount, we cool the tank while we fill it slowly, keeping the tank and its content at 30 °C. Find the final pressure to reach before closing the valve and the heat transfer.

4.105 An insulated 2 m³ tank is to be charged with R-134a from a line flowing the refrigerant at 3 MPa, 90 °C. The tank is initially evacuated, and the valve is closed when the pressure inside the tank reaches 3 MPa. Find the mass in the tank and its final temperature.

4.106 Helium in a steel tank is at 250 kPa, 300 K with a volume of 0.1 m³. It is used to fill a balloon. When the tank pressure drops to 150 kPa, the flow of helium stops by itself. If all the helium still is at 300 K, how big a balloon did I get? Assume the pressure in the balloon varies linearly with volume

from 100 kPa ($V = 0$) to the final 150 kPa. How much heat transfer took place?

4.107 A 25 L tank, shown in Fig. P4.107, that is initially evacuated is connected by a valve to an air supply line flowing air at 20 °C, 800 kPa. The valve is opened, and air flows into the tank until the pressure reaches 600 kPa. Determine the final temperature and mass inside the tank, assuming the process is adiabatic. Develop an expression for the relation between the line temperature and the final temperature using constant specific heats.

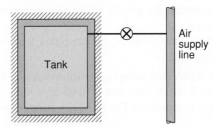

FIGURE P4.107

4.108 A nitrogen line at 300 K, 0.5 MPa, shown in Fig. P4.108, is connected to a turbine that exhausts to a closed, initially empty tank of 50 m³. The turbine operates to a tank pressure of 0.5 MPa, at which point the temperature is 250 K. Assuming the entire process is adiabatic, determine the turbine work.

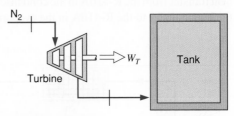

FIGURE P4.108

4.109 A 1 m³ rigid tank contains 100 kg R-410A at ambient temperature, 15 °C. A valve on top of the tank is opened, and saturated vapor is throttled to ambient pressure, 100 kPa, and flows to a collector system. During the process, the temperature inside the tank remains at 15 °C. The valve is closed when no more liquid remains inside. Calculate the heat transfer to the tank.

4.110 A 200 L tank (see Fig. P4.110) initially contains water at 100 kPa and a quality of 1 %. Heat is transferred to the water, thereby raising its pressure and temperature. At a pressure of 2 MPa, a safety valve opens and saturated vapor at 2 MPa flows out. The process continues, maintaining 2 MPa inside until the quality in the tank is 90 %, then stops. Determine the total mass of water that flowed out and the total heat transfer.

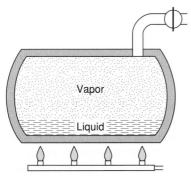

FIGURE P4.110

4.111 A 1 L can of R-134a is at room temperature, $20\,°C$, with a quality of 50 %. A leak in the top valve allows vapor to escape and heat transfer from the room takes place, so we reach a final state of $5\,°C$ with a quality of 100 %. Find the mass that escaped and the heat transfer.

4.112 A 2 m tall cylinder has a small hole in the bottom as in Fig. P4.112. It is filled with liquid water 1 m high, on top of which is a 1 m high air column at atmospheric pressure of 100 kPa. As the liquid water near the hole has a higher P than 100 kPa, it runs out. Assume a slow process with constant T. Will the flow ever stop? When?

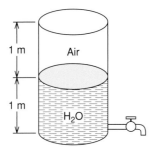

FIGURE P4.112

Review Problems

4.113 A pipe of radius R has a fully developed laminar flow of air at P_0, T_0 with a velocity profile of

$\mathbf{V} = \mathbf{V}_c[1 - (r/R)^2]$, where $\mathbf{V}_c$ is the velocity on the center-line and r is the radius, as shown in Fig. P4.113. Find the total mass flow rate and the average velocity, both as functions of $\mathbf{V}_c$ and R.

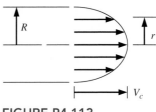

FIGURE P4.113

4.114 Steam at 3 MPa, $400\,°C$ enters a turbine with a volume flow rate of $5\ m^3/s$. An extraction of 15 % of the inlet mass flow rate exits at 600 kPa and $200\,°C$. The rest exits the turbine at 20 kPa with a quality of 90 % and a velocity of 20 m/s. Determine the volume flow rate of the extraction flow and the total turbine work.

4.115 In a glass factory, a 2 m wide sheet of glass at 1500 K comes out of the final rollers, which fix the thickness at 5 mm with a speed of 0.5 m/s (see Fig. P4.115). Cooling air in the amount of 20 kg/s comes in at $17\,°C$ from a slot 2 m wide and flows parallel with the glass. Suppose this setup is very long, so that the glass and air come to nearly the same temperature (a co-flowing heat exchanger); what is the exit temperature?

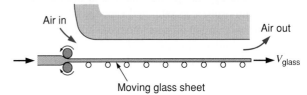

FIGURE P4.115

4.116 Assume a setup similar to that of the previous problem, but with the air flowing in the opposite direction as the glass—it comes in where the glass goes out. How much air flow at $17\,°C$ is required to cool the glass to 450 K, assuming the air must be at least 120 K cooler than the glass at any location?

4.117 A 500 L insulated tank contains air at $40\,°C$, 2 MPa. A valve on the tank is opened, and air escapes until half the original mass is gone, at which point the valve is closed. What is the pressure inside at that point?

4.118 Three air flows, all at 200 kPa, are connected to the same exit duct and mix without external heat transfer. Flow 1 has 1 kg/s at 400 K, flow 2 has 3 kg/s at 290 K, and flow 3 has 2 kg/s at 700 K. Neglect kinetic energies and find the volume flow rate in the exit flow.

4.119 Consider the power plant described in Problem 4.95.
  a. Determine the temperature of the water leaving the intermediate pressure heater, T_{13}, assuming no heat transfer to the surroundings.
  b. Determine the pump work between states 13 and 16.

4.120 A 1 m^3, 40 kg rigid steel tank contains air at 500 kPa, and both tank and air are at 20 °C. The tank is connected to a line flowing air at 2 MPa, 20 °C. The valve is opened, allowing air to flow into the tank until the pressure reaches 1.5 MPa, and is then closed. Assume the air and tank are always at the same temperature and the final temperature is 35 °C. Find the final air mass and the heat transfer.

4.121 A steam engine based on a turbine is shown in Fig. P4.121. The boiler tank has a volume of 100 L and initially contains saturated liquid with a very small amount of vapor at 100 kPa. Heat is now added by the burner. The pressure regulator, which keeps the pressure constant, does not open before the boiler pressure reaches 700 kPa. The saturated vapor enters the turbine at 700 kPa and is discharged to the atmosphere as saturated vapor at 100 kPa. The burner is turned off when no more liquid is present in the boiler. Find the total turbine work and the total heat transfer to the boiler for this process.

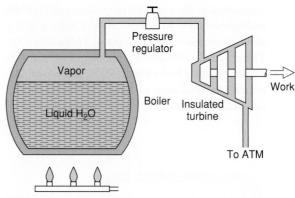

FIGURE P4.121

4.122 A 2 m^3 storage tank contains 95 % liquid and 5 % vapor by volume of liquified natural gas (LNG) at 160 K, as shown in Fig. P4.122. It may be assumed that LNG has the same properties as pure methane. Heat is transferred to the tank and saturated vapor at 160 K flows into the steady flow heater, which it leaves at 300 K. The process continues until all the liquid in the storage tank is gone. Calculate the total amount of heat transfer to the tank and the total amount of heat transferred to the heater.

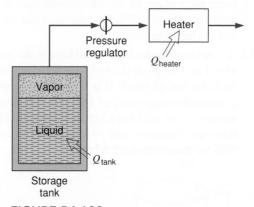

FIGURE P4.122

COMPUTER, DESIGN, AND OPEN-ENDED PROBLEMS

4.123 A 2 m^3 insulated tank contains saturated vapor steam at 4 MPa. A valve on top of the tank is opened, and saturated vapor escapes. During the process any liquid formed collects at the bottom of the tank, so only the saturated vapor exits. We want to find the mass that has escaped when the final pressure is 1 MPa. Taking an average exit enthalpy is not very accurate, so divide the process into two or three steps with piecewise average values of the exit enthalpy for a better estimate. Use, for example, (4-3), (3-2), and (2-1) MPa as the steps in which the problem is solved.

4.124 The air–water counterflowing heat exchanger given in Problem 4.75 has an air exit temperature of 360 K. Suppose the air exit temperature is listed as 300 K; then a ratio of the mass flow rates is found from the energy equation to be 5. Show that this is an impossible process by looking at air and water temperatures at several locations inside the heat exchanger. Discuss how this puts a limit on the energy that can be extracted from the air.

4.125 A co-flowing heat exchanger receives air at 800 K, 1 MPa and liquid water at 15 °C, 100 kPa, as shown in Fig. P4.125. The air line heats the water so that at the exit the air temperature is 20 °C above the water temperature. Investigate the limits for the air and water exit temperatures as a function of the ratio of the two mass flow rates. Plot the temperatures of the air and water inside the heat exchanger along the flow path.

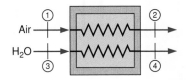

FIGURE P4.125

5 The Classical Second Law of Thermodynamics

The first law of thermodynamics states that during any cycle that a system undergoes, the cyclic integral of the heat is equal to the cyclic integral of the work. The first law, however, places no restrictions on the direction of flow of heat and work. A cycle in which a given amount of heat is transferred from the system and an equal amount of work is done on the system satisfies the first law just as well as a cycle in which the flows of heat and work are reversed. However, we know from our experience that a proposed cycle that does not violate the first law does not ensure that the cycle will actually occur. It is this kind of experimental evidence that led to the formulation of the second law of thermodynamics. Thus, a cycle will occur only if both the first and second laws of thermodynamics are satisfied.

In its broader significance, the second law acknowledges that processes proceed in a certain direction but not in the opposite direction. A hot cup of coffee cools by virtue of heat transfer to the surroundings, but heat will not flow from the cooler surroundings to the hotter cup of coffee. Gasoline is used as a car drives up a hill, but the fuel in the gasoline tank cannot be restored to its original level when the car coasts down the hill. Such familiar observations as these, and a host of others, are evidence of the validity of the second law of thermodynamics.

In this chapter we consider the second law for a system undergoing a cycle, and in the next two chapters we extend the principles to a system undergoing a change of state and then to a control volume.

5.1 HEAT ENGINES AND REFRIGERATORS

Consider the system and the surroundings previously cited in the development of the first law, as shown in Fig. 5.1. Let the gas constitute the system and, as in our discussion of the first law, let this system undergo a cycle in which work is first done on the system by the paddle wheel as the weight is lowered. Then let the cycle be completed by transferring heat to the surroundings.

We know from our experience that we cannot reverse this cycle. That is, if we transfer heat to the gas, as shown by the dotted arrow, the temperature of the gas will increase but the paddle wheel will not turn and raise the weight. With the given surroundings (the container, the paddle wheel, and the weight), this system can operate in a cycle in which the heat transfer and work are both negative, but it cannot operate in a cycle in which both the heat transfer and work are positive, even though this would not violate the first law.

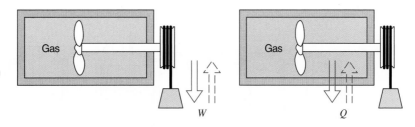

FIGURE 5.1 A system that undergoes a cycle involving work and heat.

Consider another cycle, known from our experience to be impossible to complete. Let two systems, one at a high temperature and the other at a low temperature, undergo a process in which a quantity of heat is transferred from the high-temperature system to the low-temperature system. We know that this process can take place. We also know that the reverse process, in which heat is transferred from the low-temperature system to the high-temperature system, does not occur, and that it is impossible to complete the cycle by heat transfer only. This impossibility is illustrated in Fig. 5.2.

These two examples lead us to a consideration of the heat engine and the refrigerator, which is also referred to as a heat pump. With the heat engine we can have a system that operates in a cycle and performs net positive work and net positive heat transfer. With the heat pump we can have a system that operates in a cycle and has heat transferred to it from a low-temperature body and heat transferred from it to a high-temperature body, though work is required to do this. Three simple heat engines and two simple refrigerators will be considered.

The first heat engine is shown in Fig. 5.3. It consists of a cylinder fitted with appropriate stops and a piston. Let the gas in the cylinder constitute the system. Initially, the piston rests on the lower stops, with a weight on the platform. Let the system now undergo a process in which heat is transferred from some high-temperature body to the gas, causing it to expand and raise the piston to the upper stops. At this point the weight is removed. Now let the system be restored to its initial state by transferring heat from the gas to a low-temperature body, thus completing the cycle. Since the weight was raised during the cycle, it is evident that work was done by the gas during the cycle. From the first law we conclude that the net heat transfer was positive and equal to the work done during the cycle.

Such a device is called a *heat engine*, and the substance to which and from which heat is transferred is called the *working substance* or *working fluid*. A heat engine may be defined as a device that operates in a thermodynamic cycle and does a certain amount of net positive work through the transfer of heat from a high-temperature body to a low-temperature body. Often the term *heat engine* is used in a broader sense to include all devices that produce work, either through heat transfer or through combustion, even though the device does not operate in a thermodynamic cycle. The internal combustion engine and the gas turbine are examples of such devices, and calling them *heat engines* is an acceptable use of the term. In this chapter, however, we are concerned with the more restricted form of heat engine, as just defined, one that operates on a thermodynamic cycle.

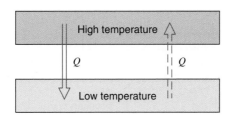

FIGURE 5.2 An example showing the impossibility of completing a cycle by transferring heat from a low-temperature body to a high-temperature body.

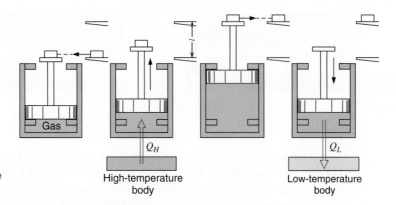

FIGURE 5.3 A simple heat engine.

A simple steam power plant is an example of a heat engine in this restricted sense. Each component in this plant may be analyzed individually as a steady-state, steady-flow process, but as a whole it may be considered a heat engine (Fig. 5.4) in which water (steam) is the working fluid. An amount of heat, Q_H, is transferred from a high-temperature body, which may be the products of combustion in a furnace, a reactor, or a secondary fluid that, in turn, has been heated in a reactor. In Fig. 5.4, the turbine is shown schematically as driving the pump. What is significant, however, is the net work that is delivered during the cycle. The quantity of heat Q_L is rejected to a low-temperature body, which is usually the cooling water in a condenser. Thus, the simple steam power plant is a heat engine in the restricted sense, for it has a working fluid, to which and from which heat is transferred, and which does a certain amount of work as it undergoes a cycle.

Thus, by means of a heat engine, we are able to have a system operate in a cycle and have both the net work and the net heat transfer positive, which we were not able to do with the system and surroundings of Fig. 5.1.

We note that in using the symbols Q_H and Q_L, we have departed from our sign connotation for heat, because for a heat engine Q_L is negative when the working fluid is considered as the system. In this chapter, it will be advantageous to use the symbol Q_H to represent the heat transfer to or from the high-temperature body and Q_L to represent the

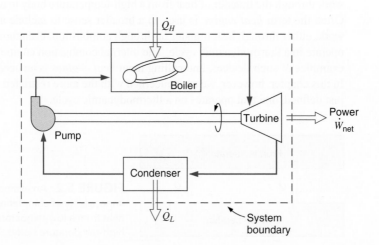

FIGURE 5.4 A heat engine involving steady-state processes.

heat transfer to or from the low-temperature body. The direction of the heat transfer will be evident from the context and indicated by arrows in the figures.

At this point, it is appropriate to introduce the concept of thermal efficiency of a heat engine. In general, we say that efficiency is the ratio of output, the energy sought, to input, the energy that costs, but the output and input must be clearly defined. At the risk of oversimplification, we may say that in a heat engine the energy sought is the work and the energy that costs money is the heat from the high-temperature source (indirectly, the cost of the fuel). Thermal efficiency is defined as

$$\eta_{\text{thermal}} = \frac{W(\text{energy sought})}{Q_H(\text{energy that costs})} = \frac{Q_H - Q_L}{Q_H} = 1 - \frac{Q_L}{Q_H} \tag{5.1}$$

Heat engines and refrigerators can also be considered in general as *energy conversion devices*, as the energy coming in is conserved but comes out in a different form. The heat engine transforms the heat input at a high temperature to a work output and a heat output at a lower temperature, whereas the refrigerator and heat pump convert the input work and heat to an output heat at an elevated temperature. The thermal efficiency is then a conversion efficiency for the process of going from the necessary input to the desired output.

Heat engines vary greatly in size and shape, from large steam engines, gas turbines, or jet engines to gasoline engines for cars and diesel engines for trucks or cars, to much smaller engines for lawn mowers or handheld devices such as chain saws or trimmers. Typical values for the thermal efficiency of real engines are about 35 %–50 % for large power plants, 30 %–35 % for gasoline engines, and 30 %–40 % for diesel engines. Smaller utility-type engines may have only about 20 % efficiency, owing to their simple carburetion and controls and to the fact that some losses scale differently with size and therefore represent a larger fraction for smaller machines.

Example 5.1

An automobile engine produces 136 hp on the output shaft with a thermal efficiency of 30 %. The fuel it burns gives 35 000 kJ/kg as energy release. Find the total rate of energy rejected to the ambient and the rate of fuel consumption in kg/s.

Solution

From the definition of a heat engine efficiency, Eq. 5.1, and the conversion of hp from Table A.1 we have

$$\dot{W} = \eta_{\text{eng}}\dot{Q}_H = 136 \text{ hp} \times 0.7355 \text{ kW/hp} = 100 \text{ kW}$$

$$\dot{Q}_H = \dot{W}/\eta_{\text{eng}} = 100/0.3 = 333 \text{ kW}$$

The energy equation for the overall engine gives

$$\dot{Q}_L = \dot{Q}_H - \dot{W} = (1 - 0.3)\dot{Q}_H = 233 \text{ kW}$$

From the energy release in the burning we have $\dot{Q}_H = \dot{m}q_H$, so

$$\dot{m} = \dot{Q}_H/q_H = \frac{333 \text{ kW}}{35\,000 \text{ kJ/kg}} = 0.0095 \text{ kg/s}$$

An actual engine shown in Fig. 5.5 rejects energy to the ambient through (a) the radiator cooled by atmospheric air, (b) heat transfer from the exhaust system, and (c) the exhaust flow of hot gases.

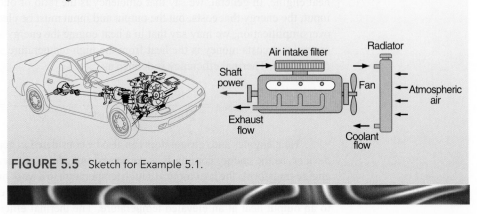

FIGURE 5.5 Sketch for Example 5.1.

The second cycle that we were not able to complete was the one indicating the impossibility of transferring heat directly from a low-temperature body to a high-temperature body. This can, of course, be done with a refrigerator or heat pump. A vapor-compression refrigerator cycle, which was introduced in Chapter 1 and shown in Fig. 1.3, is shown again in Fig. 5.6. The working fluid is the refrigerant, such as R-134a or ammonia, which goes through a thermodynamic cycle. Heat is transferred to the refrigerant in the evaporator, where its pressure and temperature are low. Work is done on the refrigerant in the compressor, and heat is transferred from it in the condenser, where its pressure and temperature are high. The pressure drops as the refrigerant flows through the throttle valve or capillary tube.

Thus, in a refrigerator or heat pump, we have a device that operates in a cycle, that requires work, and that transfers heat from a low-temperature body to a high-temperature body.

The efficiency of a refrigerator is expressed in terms of the coefficient of performance (COP), which we designate with the symbol β. For a refrigerator the objective, that is, the

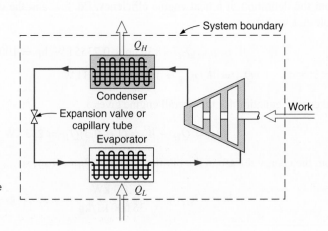

FIGURE 5.6 A simple vapor-compression refrigeration cycle.

energy sought, is Q_L, the heat transferred from the refrigerated space. The energy that costs is the work, W. Thus, the COP, $\beta,$[1] is

$$\beta = \frac{Q_L(\text{energy sought})}{W(\text{energy that costs})} = \frac{Q_L}{Q_H - Q_L} = \frac{1}{Q_H/Q_L - 1} \qquad (5.2)$$

A household refrigerator may have a COP of about 2.5, whereas that of a deep-freeze unit will be closer to 1.0. Lower cold-temperature space or higher warm-temperature space will result in lower values of COP, as will be seen in Section 5.6. For a heat pump operating over a moderate temperature range, a value of its COP can be around 4, with this value decreasing sharply as the heat pump's operating temperature range is broadened.

Example 5.2

The refrigerator in a kitchen shown in Fig. 5.7 receives electrical input power of 150 W to drive the system, and it rejects 400 W to the kitchen air. Find the rate of energy taken out of the cold space and the COP of the refrigerator.

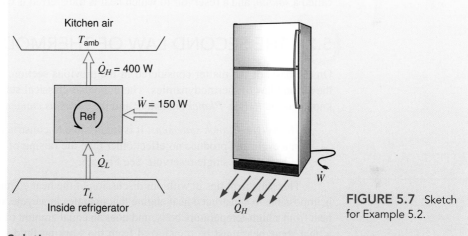

FIGURE 5.7 Sketch for Example 5.2.

Solution

C.V. refrigerator. Assume a steady state, so there is no storage of energy. The information provided is $\dot{W} = 150$ W, and the heat rejected is $\dot{Q}_H = 400$ W.

---

[1]It should be noted that a refrigeration or heat pump cycle can be used with either of two objectives. It can be used as a refrigerator, in which case the primary objective is Q_L, the heat transferred to the refrigerant from the refrigerated space. It can also be used as a heating system (in which case it is usually referred to as a *heat pump*), the objective being Q_H, the heat transferred from the refrigerant to the high-temperature body, which is the space to be heated. Q_L is transferred to the refrigerant from the ground, the atmospheric air, or well water. The COP for this case, β', is

$$\beta' = \frac{Q_H(\text{energy sought})}{W(\text{energy that costs})} = \frac{Q_H}{Q_H - Q_L} = \frac{1}{1 - Q_L/Q_H}$$

It also follows that for a given cycle,

$$\beta' - \beta = 1$$

Unless otherwise specified, the term *COP* will always refer to a refrigerator as defined by Eq. 5.2.

The energy equation gives

$$\dot{Q}_L = \dot{Q}_H - \dot{W} = 400 - 150 = 250 \text{ W}$$

This is also the rate of energy transfer into the cold space from the warmer kitchen due to heat transfer and exchange of cold air inside with warm air when you open the door.

From the definition of the COP, Eq. 5.2,

$$\beta_{\text{REFRIG}} = \frac{\dot{Q}_L}{\dot{W}} = \frac{250}{150} = 1.67$$

Before we state the second law, the concept of a *thermal reservoir* should be introduced. A thermal reservoir is a body to which and from which heat can be transferred indefinitely without change in the temperature of the reservoir. Thus, a thermal reservoir always remains at constant temperature. The ocean and the atmosphere approach this definition very closely. Frequently, it will be useful to designate a high-temperature reservoir and a low-temperature reservoir. Sometimes a reservoir from which heat is transferred is called a *source*, and a reservoir to which heat is transferred is called a *sink*.

5.2 THE SECOND LAW OF THERMODYNAMICS

On the basis of the matter considered in the previous section, we are now ready to state the second law of thermodynamics. There are two classical statements of the second law, known as the *Kelvin–Planck statement* and the *Clausius statement*.

> *The Kelvin–Planck statement*: It is impossible to construct a device that will operate in a cycle and produce no effect other than the raising of a weight and the exchange of heat with a single reservoir. See Fig. 5.8.

This statement ties in with our discussion of the heat engine. In effect, it states that it is impossible to construct a heat engine that operates in a cycle, receives a given amount of heat from a high-temperature body, and does an equal amount of work. The only alternative is that some heat must be transferred from the working fluid at a lower temperature to a low-temperature body. Thus, work can be done by the transfer of heat only if there are two temperature levels, and heat is transferred from the high-temperature body to the heat engine and also from the heat engine to the low-temperature body. This implies that it is impossible to build a heat engine that has a thermal efficiency of 100 %.

> *The Clausius statement*: It is impossible to construct a device that operates in a cycle and produces no effect other than the transfer of heat from a cooler body to a warmer body. See Fig. 5.9.

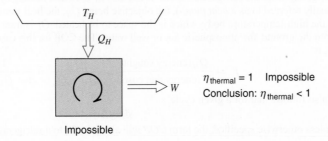

FIGURE 5.8 The Kelvin–Planck statement.

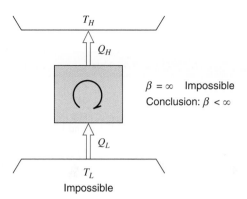

FIGURE 5.9 The Clausius statement.

This statement is related to the refrigerator or heat pump. In effect, it states that it is impossible to construct a refrigerator that operates without an input of work. This also implies that the COP is always less than infinity.

Three observations should be made about these two statements. The first observation is that both are negative statements. It is, of course, impossible to prove these negative statements. However, we can say that the second law of thermodynamics (like every other law of nature) rests on experimental evidence. Every relevant experiment that has been conducted, either directly or indirectly, verifies the second law, and no experiment has ever been conducted that contradicts the second law. The basis of the second law is therefore experimental evidence.

A second observation is that these two statements of the second law are equivalent. Two statements are equivalent if the truth of either statement implies the truth of the other or if the violation of either statement implies the violation of the other. That a violation of the Clausius statement implies a violation of the Kelvin–Planck statement may be shown. The device at the left in Fig. 5.10 is a refrigerator that requires no work and thus violates the Clausius statement. Let an amount of heat Q_L be transferred from the low-temperature reservoir to this refrigerator, and let the same amount of heat Q_L be transferred to the high-temperature reservoir. Let an amount of heat Q_H that is greater than Q_L be transferred from the high-temperature reservoir to the heat engine, and let the engine reject the amount of heat Q_L as it does an amount of work, W, that equals $Q_H - Q_L$. Because there is no net heat transfer to the low-temperature reservoir, the low-temperature reservoir, along with the heat engine and the refrigerator, can be considered together as a device that operates in a cycle

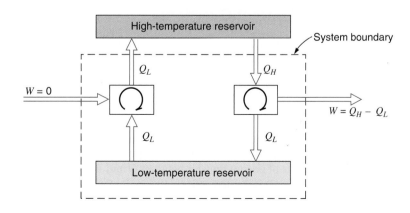

FIGURE 5.10 Demonstration of the equivalence of the two statements of the second law.

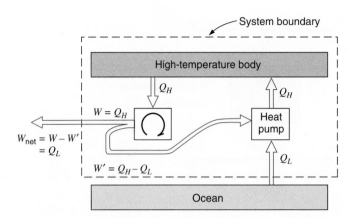

FIGURE 5.11 A perpetual-motion machine of the second kind.

and produces no effect other than the raising of a weight (work) and the exchange of heat with a single reservoir. Thus, a violation of the Clausius statement implies a violation of the Kelvin–Planck statement. The complete equivalence of these two statements is established when it is also shown that a violation of the Kelvin–Planck statement implies a violation of the Clausius statement. This is left as an exercise for the student.

The third observation is that frequently, the second law of thermodynamics has been stated as the impossibility of constructing a perpetual-motion machine of the second kind. A perpetual-motion machine of the first kind would create work from nothing or create mass or energy, thus violating the first law. A perpetual-motion machine of the second kind would extract heat from a source and then convert this heat completely into other forms of energy, thus violating the second law. A perpetual-motion machine of the third kind would have no friction, and thus would run indefinitely but produce no work.

A heat engine that violated the second law could be made into a perpetual-motion machine of the second kind by taking the following steps. Consider Fig. 5.11, which might be the power plant of a ship. An amount of heat Q_L is transferred from the ocean to a high-temperature body by means of a heat pump. The work required is W', and the heat transferred to the high-temperature body is Q_H. Let the same amount of heat be transferred to a heat engine that violates the Kelvin–Planck statement of the second law and does an amount of work, $W = Q_H$. Of this work, an amount $Q_H - Q_L$ is required to drive the heat pump, leaving the net work ($W_{\text{net}} = Q_L$) available for driving the ship. Thus, we have a perpetual-motion machine in the sense that work is done by utilizing freely available sources of energy such as the ocean or atmosphere.

In-Text Concept Questions

a. Electrical appliances (TV, stereo) use electric power as input. What happens to the power? Are those heat engines? What does the second law say about those devices?

b. Geothermal underground hot water or steam can be used to generate electric power. Does that violate the second law?

c. A windmill produces power on a shaft taking kinetic energy out of the wind. Is it a heat engine? Is it a perpetual-motion machine? Explain.

d. Heat engines and heat pumps (refrigerators) are energy conversion devices altering amounts of energy transfer between Q and W. Which conversion direction ($Q \rightarrow W$ or $W \rightarrow Q$) is limited and which is unlimited according to the second law?

5.3 THE REVERSIBLE PROCESS

The question that can now logically be posed is this: If it is impossible to have a heat engine of 100 % efficiency, what is the maximum efficiency one can have? The first step in the answer to this question is to define an ideal process, which is called a reversible process.

A reversible process for a system is defined as a process that, once having taken place, can be reversed and in so doing leave no change in either system or surroundings.

Let us illustrate the significance of this definition for a gas contained in a cylinder that is fitted with a piston. Consider first Fig. 5.12, in which a gas, which we define as the system, is restrained at high pressure by a piston that is secured by a pin. When the pin is removed, the piston is raised and forced abruptly against the stops. Some work is done by the system, since the piston has been raised a certain amount. Suppose we wish to restore the system to its initial state. One way of doing this would be to exert a force on the piston and thus compress the gas until the pin can be reinserted in the piston. Since the pressure on the face of the piston is greater on the return stroke than on the initial stroke, the work done on the gas in this reverse process is greater than the work done by the gas in the initial process. An amount of heat must be transferred from the gas during the reverse stroke so that the system has the same internal energy as it had originally. Thus, the system is restored to its initial state, but the surroundings have changed by virtue of the fact that work was required to force the piston down and heat was transferred to the surroundings. The initial process, therefore, is an irreversible one because it could not be reversed without leaving a change in the surroundings.

In Fig. 5.13, let the gas in the cylinder comprise the system, and let the piston be loaded with a number of weights. Let the weights be slid off horizontally, one at a time, allowing the gas to expand and do work in raising the weights that remain on the piston. As the size of the weights is made smaller and their number is increased, we approach a process that can be reversed, for at each level of the piston during the reverse process there will be a small weight that is exactly at the level of the platform and thus can be placed on the platform without requiring work. In the limit, therefore, as the weights become very small, the reverse process can be accomplished in such a manner that both the system and its surroundings are in exactly the same state they were in initially. Such a process is a reversible process.

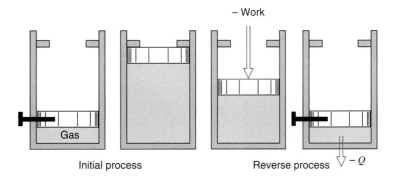

FIGURE 5.12 An example of an irreversible process.

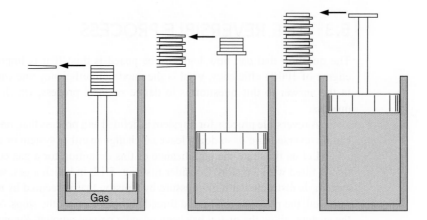

FIGURE 5.13 An example of a process that approaches reversibility.

 5.4 FACTORS THAT RENDER PROCESSES IRREVERSIBLE

There are many factors that make processes irreversible. Four of those factors—friction, unrestrained expansion, heat transfer through a finite temperature difference, and mixing of two different substances—are considered in this section.

Friction

It is readily evident that friction makes a process irreversible, but a brief illustration may amplify the point. Let a block and an inclined plane make up a system, as in Fig. 5.14, and let the block be pulled up the inclined plane by weights that are lowered. A certain amount of work is needed to do this. Some of this work is required to overcome the friction between the block and the plane, and some is required to increase the potential energy of the block. The block can be restored to its initial position by removing some of the weights and thus allowing the block to slide back down the plane. Some heat transfer from the system to the surroundings will no doubt be required to restore the block to its initial temperature. Since the surroundings are not restored to their initial state at the conclusion of the reverse process, we conclude that friction has rendered the process irreversible. Another type of frictional effect is that associated with the flow of viscous fluids in pipes and passages and in the movement of bodies through viscous fluids.

FIGURE 5.14
Demonstration of the fact that friction makes processes irreversible.

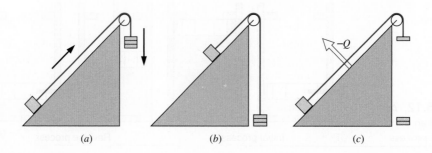

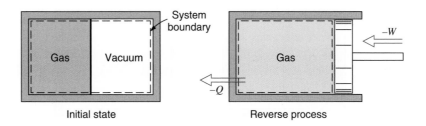

FIGURE 5.15
Demonstration of the fact that unrestrained expansion makes processes irreversible.

Unrestrained Expansion

The classic example of an unrestrained expansion, as shown in Fig. 5.15, is a gas separated from a vacuum by a membrane. Consider what happens when the membrane breaks and the gas fills the entire vessel. It can be shown that this is an irreversible process by considering what would be necessary to restore the system to its original state. The gas would have to be compressed and heat transferred from the gas until its initial state is reached. Since the work and heat transfer involve a change in the surroundings, the surroundings are not restored to their initial state, indicating that the unrestrained expansion was an irreversible process. The process described in Fig. 5.12 is also an example of an unrestrained expansion, and so is the flow through a restriction like a valve or throttle.

In the reversible expansion of a gas, there must be only an infinitesimal difference between the force exerted by the gas and the restraining force, so that the rate at which the boundary moves will be infinitesimal. In accordance with our previous definition, this is a quasi-equilibrium process. However, actual systems have a finite difference in forces, which causes a finite rate of movement of the boundary, and thus the processes are irreversible in some degree.

Heat Transfer Through a Finite Temperature Difference

Consider as a system a high-temperature body and a low-temperature body, and let heat be transferred from the high-temperature body to the low-temperature body. The only way in which the system can be restored to its initial state is to provide refrigeration, which requires work from the surroundings, and some heat transfer to the surroundings will also be necessary. Because of the heat transfer and the work, the surroundings are not restored to their original state, indicating that the process is irreversible.

An interesting question is now posed. Heat is defined as energy that is transferred through a temperature difference. We have just shown that heat transfer through a temperature difference is an irreversible process. Therefore, how can we have a reversible heat-transfer process? A heat-transfer process approaches a reversible process as the temperature difference between the two bodies approaches zero. Therefore, we define a reversible heat-transfer process as one in which heat is transferred through an infinitesimal temperature difference. We realize, of course, that to transfer a finite amount of heat through an infinitesimal temperature difference would require an infinite amount of time or an infinite area. Therefore, all actual heat transfers are through a finite temperature difference and hence are irreversible, and the greater the temperature difference, the greater the irreversibility. We will find, however, that the concept of reversible heat transfer is very useful in describing ideal processes.

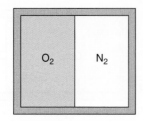
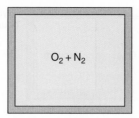

FIGURE 5.16
Demonstration of the fact that the mixing of two different substances is an irreversible process.

Mixing

Figure 5.16 illustrates the process of mixing two different gases separated by a membrane. When the membrane is broken, a homogeneous mixture of oxygen and nitrogen fills the entire volume, This process will be considered in some detail in Chapter 11. We can say here that this may be considered a special case of an unrestrained expansion, for each gas undergoes an unrestrained expansion as it fills the entire volume. An air separation plant requires an input of work to re-create masses of pure oxygen and pure nitrogen.

Mixing the same substance at two different states is also an irreversible process. Consider the mixing of hot and cold water to produce lukewarm water. The process can be reversed, but it requires a work input to a heat pump that will heat one part of the water and cool the other part.

Other Factors

A number of other factors make processes irreversible, but they will not be considered in detail here. Hysteresis effects and the i^2R loss encountered in electrical circuits are both factors that make processes irreversible. Ordinary combustion is also an irreversible process.

It is frequently advantageous to distinguish between internal and external irreversibility. Figure 5.17 shows two identical systems to which heat is transferred. Assuming each system to be a pure substance, the temperature remains constant during the heat-transfer process. In one system the heat is transferred from a reservoir at a temperature $T + dT$, and in the other the reservoir is at a much higher temperature, $T + \Delta T$, than the system. The first is a reversible heat-transfer process, and the second is an irreversible heat-transfer process. However, as far as the system itself is concerned, it passes through exactly the same states in both processes, which we assume are reversible. Thus, we can say for

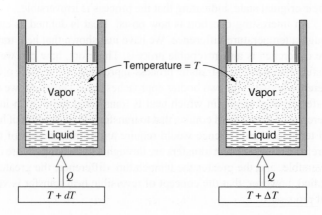

FIGURE 5.17
Illustration of the difference between an internally and an externally reversible process.

the second system that the process is internally reversible but externally irreversible because the irreversibility occurs outside the system.

We should also note the general interrelation of reversibility, equilibrium, and time. In a reversible process, the deviation from equilibrium is infinitesimal, and therefore it occurs at an infinitesimal rate. Since it is desirable that actual processes proceed at a finite rate, the deviation from equilibrium must be finite, and therefore the actual process is irreversible in some degree. The greater the deviation from equilibrium, the greater the irreversibility and the more rapidly the process will occur. It should also be noted that the quasi-equilibrium process, which was described in Chapter 1, is a reversible process, and hereafter, the term *reversible process* will be used.

In-Text Concept Questions

e. Ice cubes in a glass of liquid water will eventually melt and all the water will approach room temperature. Is this a reversible process? Why?

f. Does a process become more or less reversible with respect to heat transfer if it is fast rather than slow? *Hint:* Recall from Chapter 3 that $\dot{Q} = CA\,\Delta T$.

g. If you generated hydrogen from, say, solar power, which of these would be more efficient: (1) transport it and then burn it in an engine or (2) convert the solar power to electricity and transport that? What else would you need to know in order to give a definite answer?

5.5 THE CARNOT CYCLE

Having defined the reversible process and considered some factors that make processes irreversible, let us again pose the question raised in Section 5.3. If the efficiency of all heat engines is less than 100 %, what is the most efficient cycle we can have? Let us answer this question for a heat engine that receives heat from a high-temperature reservoir and rejects heat to a low-temperature reservoir. Since we are dealing with reservoirs, we recognize that both the high temperature and the low temperature of the reservoirs are constant and remain constant regardless of the amount of heat transferred.

Let us assume that this heat engine, which operates between the given high-temperature and low-temperature reservoirs, does so in a cycle in which every process is reversible. If every process is reversible, the cycle is also reversible; and if the cycle is reversed, the heat engine becomes a refrigerator. In the next section, we will show that this is the most efficient cycle that can operate between two constant-temperature reservoirs. It is called the Carnot cycle and is named after a French engineer, Nicolas Leonard Sadi Carnot (1796–1832), who expressed the foundations of the second law of thermodynamics in 1824.

We now turn our attention to the Carnot cycle. Figure 5.18 shows a power plant that is similar in many respects to a simple steam power plant and, we assume, operates on the Carnot cycle. Consider the working fluid to be a pure substance, such as steam. Heat is transferred from the high-temperature reservoir to the water (steam) in the boiler. For this process to be a reversible heat transfer, the temperature of the water (steam) must be only infinitesimally lower than the temperature of the reservoir. This result also implies, since the temperature of the reservoir remains constant, that the temperature of the water must remain constant. Therefore, the first process in the Carnot cycle is a reversible isothermal process in which heat is transferred from the high-temperature reservoir to the working fluid. A change of phase from liquid to vapor at constant pressure is, of course, an isothermal process for a pure substance.

The next process occurs in the turbine without heat transfer and is therefore adiabatic. Since all processes in the Carnot cycle are reversible, this must be a reversible adiabatic process, during which the temperature of the working fluid decreases from the temperature of the high-temperature reservoir to the temperature of the low-temperature reservoir.

In the next process, heat is rejected from the working fluid to the low-temperature reservoir. This must be a reversible isothermal process in which the temperature of the working fluid is infinitesimally higher than that of the low-temperature reservoir. During this isothermal process, some of the steam is condensed.

The final process, which completes the cycle, is a reversible adiabatic process in which the temperature of the working fluid increases from the low temperature to the high temperature. If this were to be done with water (steam) as the working fluid, a mixture of liquid and vapor would have to be taken from the condenser and compressed. (This would be very inconvenient in practice, and therefore in all power plants the working fluid is completely condensed in the condenser. The pump handles only the liquid phase.)

Since the Carnot heat engine cycle is reversible, every process could be reversed, in which case it would become a refrigerator. The refrigerator is shown by the dotted arrows and text in parentheses in Fig. 5.18. The temperature of the working fluid in the evaporator would be infinitesimally lower than the temperature of the low-temperature reservoir, and in the condenser it would be infinitesimally higher than that of the high-temperature reservoir.

It should be emphasized that the Carnot cycle can, in principle, be executed in many different ways. Many different working substances can be used, such as a gas or a substance with a phase change, as described in Chapter 1. There are also various possible arrangements of machinery. For example, a Carnot cycle can be devised that takes place entirely within a cylinder, using a gas as a working substance, as shown in Fig. 5.19.

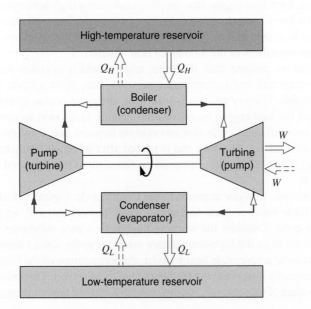

FIGURE 5.18

Example of a heat engine that operates on a Carnot cycle.

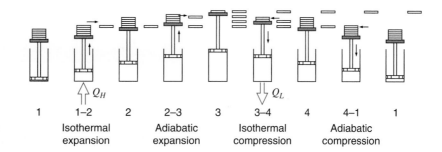

FIGURE 5.19

Example of a gaseous system operating on a Carnot cycle.

The important point to be made here is that the Carnot cycle, regardless of what the working substance may be, always has the same four basic processes. These processes are:

1. A reversible isothermal process in which heat is transferred to or from the high-temperature reservoir.
2. A reversible adiabatic process in which the temperature of the working fluid decreases from the high temperature to the low temperature.
3. A reversible isothermal process in which heat is transferred to or from the low-temperature reservoir.
4. A reversible adiabatic process in which the temperature of the working fluid increases from the low temperature to the high temperature.

5.6 TWO PROPOSITIONS REGARDING THE EFFICIENCY OF A CARNOT CYCLE

There are two important propositions regarding the efficiency of a Carnot cycle.

First Proposition

It is impossible to construct an engine that operates between two given reservoirs and is more efficient than a reversible engine operating between the same two reservoirs.

$$\text{Proposition I:} \quad \eta_{\text{any}} \leq \eta_{\text{rev}}$$

The proof of this statement is provided by a thought experiment. An initial assumption is made, and it is then shown that this assumption leads to impossible conclusions. The only possible conclusion is that the initial assumption was incorrect.

Let us assume that there is an irreversible engine operating between two given reservoirs that has a greater efficiency than a reversible engine operating between the same two reservoirs. Let the heat transfer to the irreversible engine be Q_H, the heat rejected be Q_L', and the work be W_{IE} (which equals $Q_H - Q_L'$), as shown in Fig. 5.20. Let the reversible engine operate as a refrigerator (this is possible since it is reversible). Finally, let the heat transfer with the low-temperature reservoir be Q_L, the heat transfer with the high-temperature reservoir be Q_H, and the work required be W_{RE} (which equals $Q_H - Q_L$).

Since the initial assumption was that the irreversible engine is more efficient, it follows (because Q_H is the same for both engines) that $Q_L' < Q_L$ and $W_{\text{IE}} > W_{\text{RE}}$. Now the irreversible engine can drive the reversible engine and still deliver the net work W_{net}, which equals

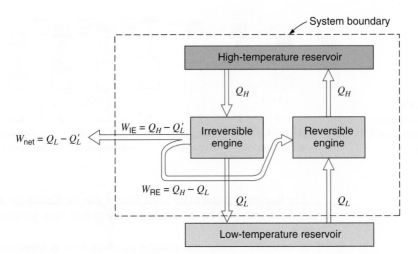

FIGURE 5.20
Demonstration of the fact that the Carnot cycle is the most efficient cycle operating between two fixed-temperature reservoirs.

$W_{\text{IE}} - W_{\text{RE}} = Q_L - Q_L'$. If we consider the two engines and the high-temperature reservoir as a system, as indicated in Fig. 5.20, we have a system that operates in a cycle, exchanges heat with a single reservoir, and does a certain amount of work. However, this would constitute a violation of the second law, and we conclude that our initial assumption (that an irreversible engine is more efficient than a reversible engine) is incorrect. Therefore, we cannot have an irreversible engine that is more efficient than a reversible engine operating between the same two reservoirs.

Second Proposition

All engines that operate on the Carnot cycle between two given constant-temperature reservoirs have the same efficiency.

$$\text{Proposition II:} \quad \eta_{\text{rev 1}} = \eta_{\text{rev 2}}$$

The proof of this proposition is similar to the proof just outlined, which assumes that there is one Carnot cycle that is more efficient than another Carnot cycle operating between the same temperature reservoirs. Let the Carnot cycle with the higher efficiency replace the irreversible cycle of the previous argument, and let the Carnot cycle with the lower efficiency operate as the refrigerator. The proof proceeds with the same line of reasoning as in the first proposition. The details are left as an exercise for the student.

5.7 THE THERMODYNAMIC TEMPERATURE SCALE

In discussing temperature in Chapter 1, we pointed out that the zeroth law of thermodynamics provides a basis for temperature measurement, but that a temperature scale must be defined in terms of a particular thermometer substance and device. A temperature scale that is independent of any particular substance, which might be called an absolute temperature scale, would be most desirable. In the preceding paragraph, we noted that the efficiency of a Carnot cycle is independent of the working substance and depends only on the reservoir temperatures. This fact provides the basis for such an absolute temperature scale called the *thermodynamic scale*. Since the efficiency of a Carnot cycle is a function only of the temperature, it follows that

$$\eta_{thermal} = 1 - \frac{Q_L}{Q_H} = 1 - \psi(T_L, T_H) \tag{5.3}$$

where ψ designates a functional relation.

There are many functional relations that could be chosen to satisfy the relation given in Eq. 5.3. For simplicity, the thermodynamic scale is defined as

$$\frac{Q_H}{Q_L} = \frac{T_H}{T_L} \tag{5.4}$$

Substituting this definition into Eq. 5.3 results in the following relation between the thermal efficiency of a Carnot cycle and the absolute temperatures of the two reservoirs:

$$\eta_{thermal} = 1 - \frac{Q_L}{Q_H} = 1 - \frac{T_L}{T_H} \tag{5.5}$$

It should be noted, however, that the definition of Eq. 5.4 is not complete since it does not specify the magnitude of the degree of temperature or a fixed reference point value. In the following section, we will discuss in greater detail the ideal-gas absolute temperature introduced in Section 2.8 and show that this scale satisfies the relation defined by Eq. 5.4.

5.8 THE IDEAL-GAS TEMPERATURE SCALE

In this section we reconsider in greater detail the ideal-gas temperature scale introduced in Section 2.8. This scale is based on the observation that as the pressure of a real gas approaches zero, its equation of state approaches that of an ideal gas:

$$Pv = RT$$

It will be shown that the ideal-gas temperature scale satisfies the definition of thermodynamic temperature given in the preceding section by Eq. 5.4. But first, let us consider how an ideal gas might be used to measure temperature in a constant-volume gas thermometer, shown schematically in Fig. 5.21.

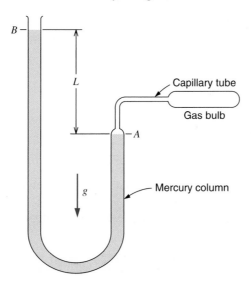

FIGURE 5.21
Schematic diagram of a constant-volume gas thermometer.

Let the gas bulb be placed in the location where the temperature is to be measured, and let the mercury column be adjusted so that the level of mercury stands at the reference mark A. Thus, the volume of the gas remains constant. Assume that the gas in the capillary tube is at the same temperature as the gas in the bulb. Then the pressure of the gas, which is indicated by the height L of the mercury column, is a measure of the temperature.

Let the pressure that is associated with the temperature of the triple point of water (273.16 K) also be measured, and let us designate this pressure $P_{t.p.}$. Then, from the definition of an ideal gas, any other temperature T could be determined from a pressure measurement P by the relation

$$T = 273.16 \left(\frac{P}{P_{t.p.}} \right)$$

Example 5.3

In a certain constant-volume ideal-gas thermometer, the measured pressure at the ice point (see Section 1.11) of water, 0 °C, is 110.9 kPa and at the steam point, 100 °C, it is 151.5 kPa. Extrapolating, at what Celsius temperature does the pressure go to zero (i.e., zero absolute temperature)?

Analysis

From the ideal-gas equation of state $PV = mRT$ at constant mass and volume, pressure is directly proportional to temperature, as shown in Fig. 5.22.

$$P = CT, \text{ where } T \text{ is the absolute ideal-gas temperature}$$

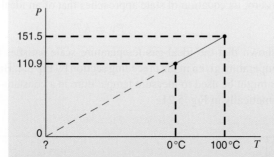

FIGURE 5.22 Plot for Example 5.3.

Solution

Slope $\dfrac{\Delta P}{\Delta T} = \dfrac{151.5 - 110.9}{100 - 0} = 0.406 \text{ kPa/°C}$

Extrapolating from the 0 °C point to $P = 0$,

$$T = 0 - \frac{110.9 \text{ kPa}}{0.406 \text{ kPa/°C}} = -273.15 \text{ °C}$$

establishing the relation between absolute ideal-gas Kelvin and Celsius temperature scales.

(*Note:* Compatible with the subsequent present-day definitions of the Kelvin scale and the Celsius scale in Section 1.11.)

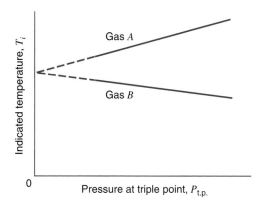

FIGURE 5.23 Sketch showing how the ideal-gas temperature is determined.

From a practical point of view, we have the problem that no gas behaves exactly like an ideal gas. However, we do know that as the pressure approaches zero, the behavior of all gases approaches that of an ideal gas. Suppose then that a series of measurements is made with varying amounts of gas in the gas bulb. This means that the pressure measured at the triple point, and also the pressure at any other temperature, will vary. If the indicated temperature T_i (obtained by assuming that the gas is ideal) is plotted against the pressure of gas with the bulb at the triple point of water, a curve like the one shown in Fig. 5.23 is obtained. When this curve is extrapolated to zero pressure, the correct ideal-gas temperature is obtained. Different curves might result from different gases, but they would all indicate the same temperature at zero pressure.

We have outlined only the general features and principles for measuring temperature on the ideal-gas scale of temperatures. Precision work in this field is difficult and laborious, and there are only a few laboratories in the world where such work is carried on. The International Temperature Scale, which was mentioned in Chapter 1, closely approximates the thermodynamic temperature scale and is much easier to work with in actual temperature measurement.

We now demonstrate that the ideal-gas temperature scale discussed earlier is, in fact, identical to the thermodynamic temperature scale, which was defined in the discussion of the Carnot cycle and the second law. Our objective can be achieved by using an ideal gas as the working fluid for a Carnot-cycle heat engine and analyzing the four processes that make up the cycle. The four state points, 1, 2, 3, and 4, and the four processes are as shown in Fig. 5.24. For convenience, let us consider a unit mass of gas inside the cylinder. Now for each of the four processes, the reversible work done at the moving boundary is given by Eq. 3.16:

$$\delta w = P\,dv$$

Similarly, for each process the gas behavior is, from the ideal-gas relation, Eq. 2.7,

$$Pv = RT$$

and the internal energy change, from Eq. 3.33, is

$$du = C_{v0}\,dT$$

Assuming no changes in kinetic or potential energies, the energy equation is, from Eq. 3.4 at unit mass,

$$\delta q = du + \delta w$$

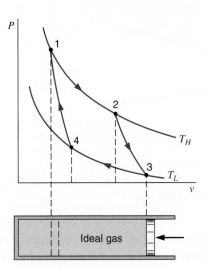

FIGURE 5.24 The ideal-gas Carnot cycle.

Substituting the three previous expressions into this equation, we have for each of the four processes

$$\delta q = C_{v0}\, dT + \frac{RT}{v}\, dv \qquad (5.6)$$

The shape of the two isothermal processes shown in Fig. 5.23 is known, since Pv is constant in each case. The isothermal heat addition process 1–2 is an expansion at T_H, such that v_2 is larger than v_1. Similarly, the isothermal heat rejection process 3–4 is a compression at a lower temperature, T_L, and v_4 is smaller than v_3. The adiabatic expansion process 2–3 is an expansion from T_H to T_L, with an increase in specific volume, while the adiabatic compression process 4–1 is a compression from T_L to T_H, with a decrease in specific volume. The area below each process line represents the work for that process, as given by Eq. 3.17.

We now proceed to integrate Eq. 5.6 for each of the four processes that make up the Carnot cycle. For the isothermal heat addition process 1–2, we have

$$q_H = {}_1q_2 = 0 + RT_H\, \ln\frac{v_2}{v_1} \qquad (5.7)$$

For the adiabatic expansion process 2–3, we divide by T to get,

$$0 = \int_{T_H}^{T_L} \frac{C_{v0}}{T}\, dT + R\, \ln\frac{v_3}{v_2} \qquad (5.8)$$

For the isothermal heat rejection process 3–4,

$$q_L = -{}_3q_4 = -0 - RT_L\, \ln\frac{v_4}{v_3}$$
$$= +RT_L\, \ln\frac{v_3}{v_4} \qquad (5.9)$$

and for the adiabatic compression process 4–1, we divide by T to get,

$$0 = \int_{T_L}^{T_H} \frac{C_{v0}}{T}\, dT + R\, \ln\frac{v_1}{v_4} \qquad (5.10)$$

From Eqs. 5.8 and 5.10, we get

$$\int_{T_L}^{T_H} \frac{C_{v0}}{T} \, dT = R \, \ln \frac{v_3}{v_2} = -R \, \ln \frac{v_1}{v_4}$$

Therefore,

$$\frac{v_3}{v_2} = \frac{v_4}{v_1}, \qquad \text{or} \qquad \frac{v_3}{v_4} = \frac{v_2}{v_1} \qquad (5.11)$$

Thus, from Eqs. 5.7 and 5.9 and substituting Eq. 5.11, we find that

$$\frac{q_H}{q_L} = \frac{RT_H \, \ln \dfrac{v_2}{v_1}}{RT_L \, \ln \dfrac{v_3}{v_4}} = \frac{T_H}{T_L}$$

which is Eq. 5.4, the definition of the thermodynamic temperature scale in connection with the second law.

5.9 IDEAL VERSUS REAL MACHINES

Following the definition of the thermodynamic temperature scale by Eq. 5.4, it was noted that the thermal efficiency of a Carnot cycle heat engine is given by Eq. 5.5. It also follows that a Carnot cycle operating as a refrigerator or heat pump will have a COP expressed as

$$\beta = \frac{Q_L}{Q_H - Q_L} \underset{\text{Carnot}}{=} \frac{T_L}{T_H - T_L} \qquad (5.12)$$

$$\beta' = \frac{Q_H}{Q_H - Q_L} \underset{\text{Carnot}}{=} \frac{T_H}{T_H - T_L} \qquad (5.13)$$

For all three "efficiencies" in Eqs. 5.5, 5.12, and 5.13, the first equality sign is the definition with the use of the energy equation and thus is always true. The second equality sign is valid only if the cycle is reversible, that is, a Carnot cycle. Any real heat engine, refrigerator, or heat pump will be less efficient, such that

$$\eta_{\text{real thermal}} = 1 - \frac{Q_L}{Q_H} \leq 1 - \frac{T_L}{T_H}$$

$$\beta_{\text{real}} = \frac{Q_L}{Q_H - Q_L} \leq \frac{T_L}{T_H - T_L}$$

$$\beta'_{\text{real}} = \frac{Q_H}{Q_H - Q_L} \leq \frac{T_H}{T_H - T_L}$$

A final point needs to be made about the significance of absolute zero temperature in connection with the second law and the thermodynamic temperature scale. Consider a Carnot-cycle heat engine that receives a given amount of heat from a given high-temperature reservoir. As the temperature at which heat is rejected from the cycle is lowered, the net work output increases and the amount of heat rejected decreases. In the limit, the heat rejected is zero, and the temperature of the reservoir corresponding to this limit is absolute zero.

Similarly, for a Carnot-cycle refrigerator, the amount of work required to produce a given amount of refrigeration increases as the temperature of the refrigerated space decreases. Absolute zero represents the limiting temperature that can be achieved, and the amount of work required to produce a finite amount of refrigeration approaches infinity as the temperature at which refrigeration is provided approaches zero.

Example 5.4

Let us consider the heat engine, shown schematically in Fig. 5.25, that receives a heat-transfer rate of 1 MW at a high temperature of 550 °C and rejects energy to the ambient surroundings at 300 K. Work is produced at a rate of 450 kW. We would like to know how much energy is discarded to the ambient surroundings and the engine efficiency and compare both of these to a Carnot heat engine operating between the same two reservoirs.

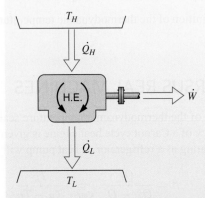

FIGURE 5.25 A heat engine operating between two constant-temperature energy reservoirs for Example 5.4.

Solution

If we take the heat engine as a control volume, the energy equation gives

$$\dot{Q}_L = \dot{Q}_H - \dot{W} = 1000 - 450 = 550 \text{ kW}$$

and from the definition of efficiency

$$\eta_{\text{thermal}} = \dot{W}/\dot{Q}_H = 450/1000 = 0.45$$

For the Carnot heat engine, the efficiency is given by the temperature of the reservoirs:

$$\eta_{\text{Carnot}} = 1 - \frac{T_L}{T_H} = 1 - \frac{300}{550 + 273} = 0.635$$

The rates of work and heat rejection become

$$\dot{W} = \eta_{\text{Carnot}} \dot{Q}_H = 0.635 \times 1000 = 635 \text{ kW}$$
$$\dot{Q}_L = \dot{Q}_H - \dot{W} = 1000 - 635 = 365 \text{ kW}$$

The actual heat engine thus has a lower efficiency than the Carnot (ideal) heat engine, with a value of 45 % typical for a modern steam power plant. This also implies that the actual engine rejects a larger amount of energy to the ambient surroundings (55 %) compared with the Carnot heat engine (36 %).

Example 5.5

As one mode of operation of an air conditioner is the cooling of a room on a hot day, it works as a refrigerator, shown in Fig. 5.26. A total of 4 kW should be removed from a room at 24 °C to the outside atmosphere at 35 °C. We would like to estimate the magnitude of the required work. To do this we will not analyze the processes inside the refrigerator, which is deferred to Chapter 9, but we can give a lower limit for the rate of work, assuming it is a Carnot-cycle refrigerator.

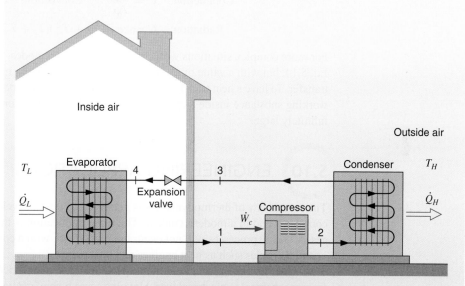

An air conditioner in cooling mode

FIGURE 5.26 An air conditioner in cooling mode where T_L is the room.

Solution

The COP is

$$\beta = \frac{\dot{Q}_L}{\dot{W}} = \frac{\dot{Q}_L}{\dot{Q}_H - \dot{Q}_L} = \frac{T_L}{T_H - T_L} = \frac{273 + 24}{35 - 24} = 27$$

so the rate of work or power input will be

$$\dot{W} = \dot{Q}_L/\beta = 4/27 = 0.15 \, \text{kW}$$

Since the power was estimated assuming a Carnot refrigerator, it is the smallest amount possible. Recall also the expressions for heat-transfer rates in Chapter 3. If the refrigerator should push 4.15 kW out to the atmosphere at 35 °C, the high-temperature side of it should be at a higher temperature, maybe 45 °C, to have a reasonably small-sized heat exchanger. As it cools the room, a flow of air of less than, say, 18 °C would be needed. Redoing the COP with a high of 45 °C and a low of 18 °C gives 10.8, which is more realistic. A real refrigerator would operate with a COP on the order of 5 or less.

In the previous discussion and examples, we considered the constant-temperature energy reservoirs and used those temperatures to calculate the Carnot-cycle efficiency. However, if we recall the expressions for the rate of heat transfer by conduction, convection, or radiation in Chapter 3, they can all be shown as

$$\dot{Q} = C \, \Delta T \qquad (5.14)$$

The constant C depends on the mode of heat transfer as

Conduction: $C = \dfrac{kA}{\Delta x}$ Convection: $C = hA$

Radiation: $C = \varepsilon \sigma A (T_s^2 + T_\infty^2)(T_s + T_\infty)$

For more complex situations with combined layers and modes, we also recover the form in Eq. 5.14, but with a value of C that depends on the geometry, materials, and modes of heat transfer. To have a heat transfer, we therefore must have a temperature difference so that the working substance inside a cycle cannot attain the reservoir temperature unless the area is infinitely large.

5.10 ENGINEERING APPLICATIONS

The second law of thermodynamics is presented as it was developed, with some additional comments and in a modern context. The main implication is the limits it imposes on processes: Some processes will not occur but others will, with a constraint on the operation of complete cycles such as heat engines and heat pumps.

Nearly all energy conversion processes that generate work (typically converted further from mechanical to electrical work) involve some type of cyclic heat engine. These include the engine in a car, a turbine in a power plant, or a windmill. The source of energy can be a storage reservoir (fossil fuels that can burn, such as gasoline or natural gas) or a more temporary form, for example, the wind kinetic energy that ultimately is driven by heat input from the sun.

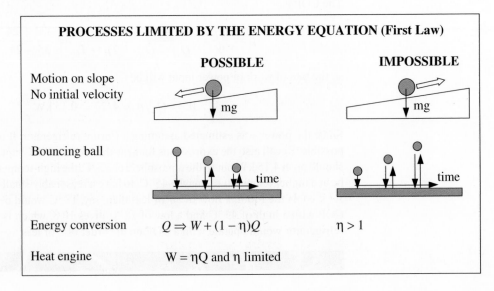

Machines that violate the energy equation, say generate energy from nothing, are called perpetual-motion machines of the first kind. Such machines have been "demonstrated" and investors asked to put money into their development, but most of them had some kind of energy input not easily observed (such as a small, compressed air line or a hidden fuel supply). Recent examples are cold fusion and electrical phase imbalance; these can be measured only by knowledgeable engineers. Today it is recognized that these processes are impossible.

Machines that violate the second law but obey the energy equation are called perpetual-motion machines of the second kind. These are a little more subtle to analyze, and for the unknowledgeable person they often look as if they should work. There are many examples of these and they are even proposed today, often hidden by a variety of complicated processes that obscure the overall process.

PROCESSES LIMITED BY THE SECOND LAW

	POSSIBLE	IMPOSSIBLE
Heat transfer No work term	$\dot{Q}$ (at T_{hot}) $\Rightarrow$ $\dot{Q}$ (at T_{cold})	$\dot{Q}$ (at T_{cold}) $\Rightarrow$ $\dot{Q}$ (at T_{hot})
Flow, $\dot{m}$ No KE, PE	$P_{high} \Rightarrow P_{low}$	$P_{low} \Rightarrow P_{high}$
Energy conversion	$W \Rightarrow Q$ (100%)	$Q \Rightarrow W$ (100%)
Energy conversion	$Q \Rightarrow W + (1 - \eta)Q$ $W = \eta Q$ and η limited	$\eta > \eta_{rev.\ heat\ eng}$
Chemical reaction like combustion	Fuel + air $\Rightarrow$ products	Products $\Rightarrow$ fuel + air
Heat exchange, mixing	hot cold $\Rightarrow$ warm	warm $\Rightarrow$ hot cold
Mixing	O_2 N_2 $\Rightarrow$ air	air $\Rightarrow$ O_2 N_2

Actual Heat Engines and Heat Pumps

The necessary heat transfer in many of these systems typically takes place in dual-fluid heat exchangers where the working substance receives or rejects heat. These heat engines typically have an external combustion of fuel, as in coal, oil, or natural gas-fired power plants, or they receive heat from a nuclear reactor or some other source. There are only a few types of movable engines with external combustion, notably a Stirling engine (see Chapter 10) that uses a light gas as a working substance. Heat pumps and refrigerators all have heat transfer external to the working substance with work input that is electrical, as in the standard household refrigerator, but it can also be shaft work from a belt, as in a car air-conditioning system. The heat transfer requires a temperature difference (recall Eq. 5.14) such that the rates become

$$\dot{Q}_H = C_H \, \Delta T_H \quad \text{and} \quad \dot{Q}_L = C_L \, \Delta T_L$$

where the C's depend on the details of the heat transfer and interface area. That is, for a heat engine, the working substance goes through a cycle that has

$$T_{\text{high}} = T_H - \Delta T_H \qquad \text{and} \qquad T_{\text{low}} = T_L + \Delta T_L$$

so the operating range that determines the cycle efficiency becomes

$$\Delta T_{\text{HE}} = T_{\text{high}} - T_{\text{low}} = T_H - T_L - (\Delta T_H + \Delta T_L) \qquad (5.15)$$

For a heat pump, the working substance must be warmer than the reservoir to which it moves $\dot{Q}_H$, and it must be colder than the reservoir from which it takes $\dot{Q}_L$, so we get

$$T_{\text{high}} = T_H + \Delta T_H \qquad \text{and} \qquad T_{\text{low}} = T_L - \Delta T_L$$

giving an operating range for the working substance as

$$\Delta T_{\text{HP}} = T_{\text{high}} - T_{\text{low}} = T_H - T_L + (\Delta T_H + \Delta T_L) \qquad (5.16)$$

This effect is illustrated in Fig 5.27 for both the heat engine and the heat pump. Notice that in both cases, the effect of the finite temperature difference due to the heat transfer is to decrease the performance. The heat engine's maximum possible efficiency is lower due to the lower T_{high} and the higher T_{low}, and the heat pump's (also the refrigerator's) COP is lower due to the higher T_{high} and the lower T_{low}.

For heat engines with an energy conversion process in the working substance such as combustion, there is no heat transfer to or from an external energy reservoir. These are typically engines that move and thus cannot have large pieces of equipment, as volume and mass are undesirable, as in car and truck engines, gas turbines, and jet engines. When the working substance becomes hot, it has a heat transfer loss to its surroundings that lowers the pressure (given the volume) and thus decreases the ability to do work on any moving boundary. These processes are more difficult to analyze and require extensive knowledge to predict any net effect like efficiency, so in later chapters we will use some simple models to describe these cycles.

A final comment about heat engines and heat pumps is that there are no practical examples of these that run in a Carnot cycle. All the cyclic devices operate in slightly different cycles determined by the behavior of the physical arrangements, as shown in Chapters 9 and 10.

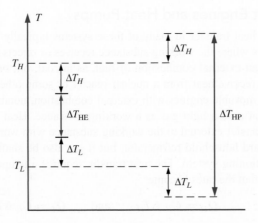

FIGURE 5.27
Temperature span for heat engines and heat pumps.

Some Historical Developments in Thermodynamics

Progress in understanding the physical sciences led to the basic development of the second law of thermodynamics before the first law. A wide variety of people with different backgrounds did work in this area, Carnot and Kelvin, among others, shown in Table 5.1, that, combined with developments in mathematics and physics, helped foster the Industrial Revolution. Much of this work took place in the second half of the 1800s followed by applications continuing into the early 1900s such as steam turbines, gasoline and diesel engines, and modern refrigerators. All of these inventions and developments, summarized in the following table, had a profound effect on our society.

TABLE 5.1

Important Historical Events in Thermodynamics

Year	Person	Event
1660	Robert Boyle	$P = C/V$ at constant T (first gas law attempt)
1687	Isaac Newton	Newton's laws, gravitation, law of motion
1712	Thomas Newcomen and Thomas Savery	First practical steam engine using the piston/cylinder
1714	Gabriel Fahrenheit	First mercury thermometer
1738	Daniel Bernoulli	Forces in hydraulics, Bernoulli's equation (Ch. 7)
1742	Anders Celsius	Proposes Celsius scale
1765	James Watt	Steam engine that includes a separate condenser (Ch. 9)
1787	Jacques A. Charles	Ideal-gas relation between V and T
1824	Sadi Carnot	Concept of heat engine, hints at second law
1827	George Ohm	Ohm's law formulated
1839	William Grove	First fuel cell (Ch. 13)
1842	Julius Robert Mayer	Conservation of energy
1843	James P. Joule	Experimentally measured equivalency of work and heat
1848	William Thomson	Lord Kelvin proposes absolute temperature scale based on the work done by Carnot and Charles
1850	Rudolf Clausius and, later, William Rankine	First law of energy conservation; thermodynamics is a new science
1865	Rudolf Clausius	Shows that entropy (Ch. 6) increases in a closed system (second law)
1877	Nikolaus Otto	Develops the Otto cycle engine (Ch. 10)
1878	J. Willard Gibbs	Heterogeneous equilibria, phase rule
1882	Joseph Fourier	Mathematical theory of heat transfer
1882		Electrical generating plant in New York (Ch. 9)
1893	Rudolf Diesel	Develops the compression-ignition engine (Ch. 10)
1896	Henry Ford	First Ford machine (quadricycle) built in Michigan
1927	General Electric Co.	First refrigerator made available to consumers (Ch. 9)

SUMMARY The classical presentation of the second law of thermodynamics starts with the concepts of heat engines and refrigerators. A heat engine produces work from heat transfer obtained from a thermal reservoir, and its operation is limited by the Kelvin–Planck statement. Refrigerators are functionally the same as heat pumps, and they drive energy by heat transfer from a colder environment to a hotter environment, something that will not happen by itself. The Clausius statement says in effect that the refrigerator or heat pump does need work input to accomplish the task. To approach the limit of these cyclic devices, the idea of reversible processes is discussed and further explained by the opposite, namely, irreversible processes and impossible machines. A perpetual-motion machine of the first kind violates the first law of thermodynamics (energy equation), and a perpetual-motion machine of the second kind violates the second law of thermodynamics.

The limitations for the performance of heat engines (thermal efficiency) and heat pumps or refrigerators (coefficient of performance or COP) are expressed by the corresponding Carnot-cycle device. Two propositions about the Carnot-cycle device are another way of expressing the second law of thermodynamics instead of the statements of Kelvin–Planck or Clausius. These propositions lead to the establishment of the thermodynamic absolute temperature, done by Lord Kelvin, and the Carnot-cycle efficiency. We show this temperature to be the same as the ideal-gas temperature introduced in Chapter 2.

You should have learned a number of skills and acquired abilities from studying this chapter that will allow you to

- Understand the concepts of heat engines, heat pumps, and refrigerators.
- Have an idea about reversible processes.
- Know a number of irreversible processes and recognize them.
- Know what a Carnot cycle is.
- Understand the definition of thermal efficiency of a heat engine.
- Understand the definition of the coefficient of performance (COP) of a heat pump.
- Know the difference between absolute and relative temperature.
- Know the limits of thermal efficiency as dictated by the thermal reservoirs and the Carnot-cycle device.
- Have an idea about the thermal efficiency of real heat engines.
- Know the limits of COP as dictated by the thermal reservoirs and the Carnot-cycle device.
- Have an idea about the COP of real refrigerators.

KEY CONCEPTS AND FORMULAS

(All W, Q can also be rates $\dot{W}$, $\dot{Q}$)

Heat engine

$$W_{\text{HE}} = Q_H - Q_L; \qquad \eta_{\text{HE}} = \frac{W_{\text{HE}}}{Q_H} = 1 - \frac{Q_L}{Q_H}$$

Heat pump

$$W_{\text{HP}} = Q_H - Q_L; \qquad \beta_{\text{HP}} = \frac{Q_H}{W_{\text{HP}}} = \frac{Q_H}{Q_H - Q_L}$$

Refrigerator

$$W_{\text{REF}} = Q_H - Q_L; \qquad \beta_{\text{REF}} = \frac{Q_L}{W_{\text{REF}}} = \frac{Q_L}{Q_H - Q_L}$$

Factors that make processes irreversible

Friction, unrestrained expansion ($W = 0$), Q over ΔT, mixing, current through a resistor, combustion, or valve flow (throttle)

Carnot cycle

1–2 Isothermal heat addition Q_H in at T_H

2–3 Adiabatic expansion process T goes down

3–4 Isothermal heat rejection Q_L out at T_L

4–1 Adiabatic compression process T goes up

Proposition I	$\eta_{\text{any}} \leq \eta_{\text{reversible}}$	Same T_H, T_L
Proposition II	$\eta_{\text{Carnot 1}} = \eta_{\text{Carnot 2}}$	Same T_H, T_L

Absolute temperature $\quad\quad \dfrac{T_L}{T_H} = \dfrac{Q_L}{Q_H}$

Real heat engine $\quad\quad \eta_{\text{HE}} = \dfrac{W_{\text{HE}}}{Q_H} \leq \eta_{\text{Carnot HE}} = 1 - \dfrac{T_L}{T_H}$

Real heat pump $\quad\quad \beta_{\text{HP}} = \dfrac{Q_H}{W_{\text{HP}}} \leq \beta_{\text{Carnot HP}} = \dfrac{T_H}{T_H - T_L}$

Real refrigerator $\quad\quad \beta_{\text{REF}} = \dfrac{Q_L}{W_{\text{REF}}} \leq \beta_{\text{Carnot REF}} = \dfrac{T_L}{T_H - T_L}$

Heat-transfer rates $\quad\quad \dot{Q} = C\,\Delta T$

CONCEPT-STUDY GUIDE PROBLEMS

5.1 Two heat engines operate between the same two energy reservoirs, and both receive the same Q_H. One engine is reversible and the other is not. What can you say about the two Q_L's?

5.2 Compare two domestic heat pumps (*A* and *B*) running with the same work input. If *A* is better than *B*, which one provides more heat?

5.3 Suppose we forget the model for heat transfer, $\dot{Q} = C A\,\Delta T$; can we draw some information about the direction of Q from the second law?

5.4 A combination of two heat engines is shown in Fig. P5.4. Find the overall thermal efficiency as a function of the two individual efficiencies.

5.5 Compare two heat engines receiving the same Q, one at 1200 K and the other at 1800 K, both of which reject heat at 500 K. Which one is better?

5.6 A car engine takes atmospheric air in at 20 °C, no fuel, and exhausts the air at −20 °C, producing work in the process. What do the first and second laws say about that?

5.7 A combination of two refrigerator cycles is shown in Fig. P5.7. Find the overall COP as a function of COP_1 and COP_2.

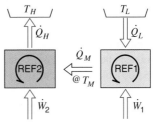

FIGURE P5.7

5.8 After you have driven a car on a trip and it is back home, the car's engine has cooled down and thus is back to the state in which it started. What happened to all the energy released in the burning of gasoline? What happened to all the work the engine gave out?

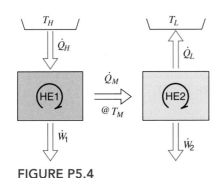

FIGURE P5.4

5.9 Does a reversible heat engine burning coal (which in practice cannot be done reversibly) have impacts on our world other than depletion of the coal reserve?

5.10 If the efficiency of a power plant goes up as the low temperature drops, why do all power plants not reject energy at, say, $-40\,°C$?

5.11 If the efficiency of a power plant goes up as the low temperature drops, why not let the heat rejection go to a refrigerator at, say, $-10\,°C$ instead of ambient $20\,°C$?

5.12 A coal-fired power plant operates with a high temperature of $600\,°C$, whereas a jet engine has about 1400 K. Does this mean that we should replace all power plants with jet engines?

5.13 Heat transfer requires a temperature difference (see Chapter 3) to push the $\dot{Q}$. What does that imply for a real heat engine? A refrigerator?

5.14 Hot combustion gases (air) at 1500 K are used as the heat source in a heat engine where the gas is cooled to 750 K and the ambient is at 300 K. This is not a constant-temperature source. How does that affect the efficiency?

HOMEWORK PROBLEMS

Heat Engines and Refrigerators

5.15 A window-mounted air conditioner removes 3.5 kJ from the inside of a home using 1.75 kJ work input. How much energy is released outside and what is its coefficient of performance?

5.16 A lawnmower tractor engine produces 12 hp using 25 kW of heat transfer from burning fuel. Find the thermal efficiency and the rate of heat transfer rejected to the ambient.

5.17 Calculate the thermal efficiency of the steam power plant cycle described in Example 4.7.

5.18 A room is heated with a 2000 W electric heater. How much power can be saved if a heat pump with a COP of 2.5 is used instead?

5.19 Calculate the COP of the R-134a refrigerator described in Example 4.8.

5.20 A large coal fired power plant has an efficiency of 45 % and produces net 1500 MW of electricity. Coal releases 25 000 kJ/kg as it burns so how much coal is used per hour?

5.21 A window air conditioner (Fig. P5.21) discards 1.7 kW to the ambient with a power input of 500 W. Find the rate of cooling and the COP.

5.22 An industrial machine is being cooled by 0.4 kg/s water at $15\,°C$ that is chilled from $30\,°C$ by a refrigeration unit with a COP of 3. Find the rate of cooling required and the power input to the unit.

5.23 Calculate the COP of the R-410A heat pump cycle described in Problem 4.97.

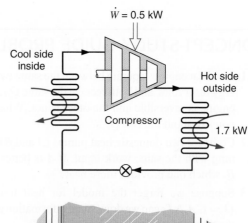

FIGURE P5.21

5.24 A farmer runs a heat pump with a 2 kW motor. It should keep a chicken hatchery at $30\,°C$, which loses energy at a rate of 10 kW to the colder ambient T_{amb}. What is the minimum COP that will be acceptable for the pump?

5.25 A sports car engine delivers 100 hp to the driveshaft with a thermal efficiency of 25 %. The fuel has a heating value of 40 000 kJ/kg. Find the rate of fuel consumption and the combined power rejected through the radiator and exhaust.

5.26 In a Rankine cycle, 0.9 MW is taken out in the condenser, 0.63 MW is taken out from the turbine, and the pump work is 0.03 MW. Find the plant's thermal efficiency. If everything could be reversed, find the COP as a refrigerator.

5.27 An experimental power plant outputs 130 MW of electrical power. It uses a supply of 1200 MW from a geothermal source and rejects energy to the atmosphere. Find the power to the air and how much air should be flowed to the cooling tower (kg/s) if its temperature cannot be increased more than 12 °C.

5.28 A water cooler for drinking water should cool 25 L/h water from 18 °C to 10 °C while the water reservoirs also gains 60 W from heat transfer. Assume that a small refrigeration unit with a COP of 2.5 does the cooling. Find the total rate of cooling required and the power input to the unit.

5.29 A large stationary diesel engine produces 5 MW with a thermal efficiency of 40 %. The exhaust gas, which we assume is air, flows out at 800 K and the temperature of the intake air is 290 K. How large a mass flow rate is that, assuming this is the only way we reject heat? Can the exhaust flow energy be used?

5.30 For each of the cases below, determine if the heat engine satisfies the first law (energy equation) and if it violates the second law.
 a. $\dot{Q}_H = 6\,\text{kW}$, $\dot{Q}_L = 4\,\text{kW}$, $\dot{W} = 2\,\text{kW}$
 b. $\dot{Q}_H = 6\,\text{kW}$, $\dot{Q}_L = 0\,\text{kW}$, $\dot{W} = 6\,\text{kW}$
 c. $\dot{Q}_H = 6\,\text{kW}$, $\dot{Q}_L = 2\,\text{kW}$, $\dot{W} = 5\,\text{kW}$
 d. $\dot{Q}_H = 6\,\text{kW}$, $\dot{Q}_L = 6\,\text{kW}$, $\dot{W} = 0\,\text{kW}$

5.31 For each of the cases in Problem 5.30, determine if a heat pump satisfies the first law (energy equation) and if it violates the second law.

5.32 Calculate the amount of work input a refrigerator needs to make ice cubes out of a tray of 0.25 kg liquid water at 10 °C. Assume that the refrigerator has $\beta = 3.5$ and a motor-compressor of 500 W. How much time does it take if this is the only cooling load?

Second Law and Processes

5.33 Prove that a cyclic device that violates the Kelvin–Planck statement of the second law also violates the Clausius statement of the second law.

5.34 Discuss the factors that would make the heat pump described in Problem 4.97 an irreversible cycle.

5.35 Assume a cyclic machine that exchanges 6 kW with a 250 °C reservoir and has
 a. $\dot{Q}_L = 0\,\text{kW}$, $\dot{W} = 6\,\text{kW}$
 b. $\dot{Q}_L = 6\,\text{kW}$, $\dot{W} = 0\,\text{kW}$
 and $\dot{Q}_L$ is exchanged with a 30 °C ambient. What can you say about the processes in the two cases, a and b, if the machine is a heat engine? Repeat the question for the case of a heat pump.

5.36 Consider a heat engine and heat pump connected as shown in Fig. P5.36. Assume that $T_{H1} = T_{H2} > T_{amb}$ and determine for each of the three cases if the setup satisfies the first law and/or violates the second law.

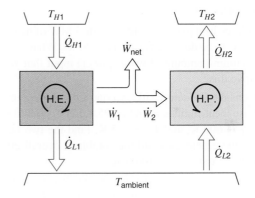

	$\dot{Q}_{H1}$	$\dot{Q}_{L1}$	$\dot{W}_1$	$\dot{Q}_{H2}$	$\dot{Q}_{L2}$	$\dot{W}_2$
a	6	4	2	3	2	1
b	6	4	2	5	4	1
c	3	2	1	4	3	1

FIGURE P5.36

5.37 The simple refrigeration cycle is shown in Problem 5.21 and in Fig. 5.6. Mention a few of the processes that are expected to be irreversible.

Carnot Cycle and Absolute Temperature

5.38 Calculate the thermal efficiency of a Carnot cycle heat engine operating between reservoirs at

300 °C and 45 °C. Compare the result to that of Example 4.7.

5.39 An ideal (Carnot) heat engine has an efficiency of 40 %. If the high temperature is raised 15 %, what is the new efficiency keeping the same low temperature?

5.40 In a few places where the air is very cold in the winter, such as −25 °C, it is possible to find a temperature of 10 °C below ground. What efficiency will a heat engine have when operating between these two thermal reservoirs?

5.41 Consider the combination of a heat engine and a heat pump, as in Problem 5.36, with a low temperature of 400 K. What should the high temperature be so that the heat engine is reversible? For that temperature, what is the COP for a reversible heat pump?

5.42 Find the power output and the low T heat rejection rate for a Carnot cycle heat engine that receives 6 kW at 250 °C and rejects heat at 30 °C, as in Problem 5.35.

5.43 A large heat pump should upgrade 4 MW of heat at 65 °C to be delivered as heat at 120 °C. What is the minimum amount of work (power) input that will drive this?

5.44 Consider the setup with two stacked (temperature-wise) heat engines, as in Fig. P5.4. Let $T_H = 850$ K, $T_M = 600$ K, and $T_L = 375$ K. Find the two heat engine efficiencies and the combined overall efficiency, assuming Carnot cycles.

5.45 Assume the refrigerator in your kitchen runs in a Carnot cycle. Estimate the maximum COP.

5.46 A car engine burns 5 kg fuel (equivalent to addition of Q_H) at 1500 K and rejects energy to the radiator and the exhaust at an average temperature of 750 K. If the fuel provides 40 000 kJ/kg, what is the maximum amount of work the engine can provide?

5.47 An air conditioner provides 1 kg/s of air at 15 °C cooled by outside atmospheric air at 35 °C. Estimate the amount of power needed to operate the air conditioner. Clearly state all assumptions made.

5.48 A refrigerator should remove 400 kJ from some food. Assume the refrigerator works in a Carnot cycle between −10 °C and 42 °C with a motor-compressor of 400 W. How much time does it take if this is the only cooling load?

5.49 Calculate the amount of work input a freezer needs to make ice cubes out of a tray of 0.25 kg liquid water at 10 °C. Assume the freezer works in a Carnot cycle between −8 °C and 35 °C with a motor-compressor of 600 W. How much time does it take if this is the only cooling load?

5.50 A heat pump is used to heat a house during the winter. The house is to be maintained at 20 °C at all times. When the ambient temperature outside drops to −10 °C, the rate at which heat is lost from the house is estimated to be 25 kW. What is the minimum electrical power required to drive the heat pump?

FIGURE P5.50

5.51 Thermal storage is made with a rock (granite) bed of 2 m³ that is heated to 400 K using solar energy. A heat engine receives a Q_H from the bed and rejects heat to the ambient at 290 K. The rock bed therefore cools down, and as it reaches 290 K the process stops. Find the energy the rock bed can give out. What is the heat engine efficiency at the beginning of the process and what is it at the end of the process?

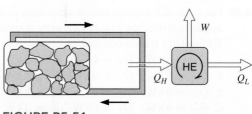

FIGURE P5.51

5.52 A proposal is to build a 1000 MW electric power plant with steam as the working fluid. The condensers are to be cooled with river water (see Fig. P5.52). The maximum steam temperature is 550 °C, and the pressure in the condensers will be 10 kPa. Estimate the temperature rise of the river downstream from the power plant.

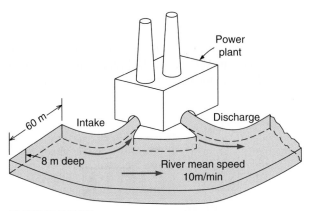

FIGURE P5.52

5.53 A certain solar-energy collector produces a maximum temperature of 100 °C. The energy is used in a cycle heat engine that operates in a 20 °C environment. What is the maximum thermal efficiency? If the collector is redesigned to focus the incoming light, what should the maximum temperature be to produce a 25 % improvement in engine efficiency?

5.54 A constant temperature of −125 °C must be maintained in a cryogenic experiment, although it gains 120 W due to heat transfer. What is the smallest motor you would need for a heat pump absorbing heat from the container and rejecting heat to the room at 20 °C?

5.55 Helium has the lowest normal boiling point of any of the elements at 4.2 K. At this temperature the enthalpy of evaporation is 83.3 kJ/kmol. A Carnot refrigeration cycle is analyzed for the production of 1 kmol of liquid helium at 4.2 K from saturated vapor at the same temperature. What is the work input to the refrigerator and the COP for the cycle with an ambient at 300 K?

5.56 R-134a fills a 0.1 m^3 capsule at 20 °C, 200 kPa. It is placed in a deep freezer, where it is cooled to −10 °C. The deep freezer sits in a room with ambient temperature of 20 °C and has an inside temperature of −10 °C. Find the amount of energy the freezer must remove from the R-134a and the extra amount of work input to the freezer to perform the process.

5.57 A heat engine has a solar collector receiving 0.2 kW/m^2 inside which a transfer medium is heated to 450 K. The collected energy powers a heat engine that rejects heat at 40 °C. If the heat engine should deliver 2.5 kW, what is the minimum size (area) of the solar collector?

5.58 Sixty kilograms per hour of water runs through a heat exchanger, entering as saturated liquid at 200 kPa and leaving as saturated vapor. The heat is supplied by a heat pump operating from a low-temperature reservoir at 16 °C with a COP of half that of the similar Carnot unit. Find the rate of work into the heat pump.

5.59 A power plant with a thermal efficiency of 40 % is located on a river similar to the arrangement in Fig. P5.52. With a total river mass flow rate of 1 × 10^5 kg/s at 15 °C, find the maximum power production allowed if the river water should not be heated more than 1 degree.

5.60 The management at a large factory cannot decide which of two fuels to purchase. The selected fuel will be used in a heat engine operating between the fuel-burning temperature and a low-exhaust temperature. Fuel A burns at 2200 K and exhausts at 450 K, delivering 30 000 kJ/kg, and costs $1.50/kg. Fuel B burns at 1200 K and exhausts at 350 K, delivering 40 000 kJ/kg, and costs $1.30/kg. Which fuel would you buy and why?

Actual Cycles

5.61 A salesperson selling refrigerators and deep freezers will guarantee a minimum COP of 4.5 year round. How would you evaluate that performance? Are they all the same?

5.62 A cyclic machine, shown in Fig. P5.62, receives 325 kJ from a 1000 K energy reservoir. It rejects 125 kJ to a 400 K energy reservoir, and the cycle produces 200 kJ of work as output. Is this cycle reversible, irreversible, or impossible?

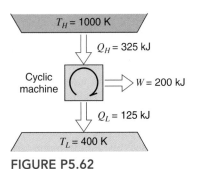

FIGURE P5.62

5.63 Consider the previous problem and assume the temperatures and heat input are as given. If the actual machine has an efficiency that is half that of the corresponding Carnot cycle, find the work out and the rejected heat transfer.

5.64 An inventor has developed a refrigeration unit that maintains the cold space at $-10\,°C$ while operating in a $25\,°C$ room. A COP of 8.5 is claimed. How do you evaluate this?

5.65 In a remote location, you run a heat engine to provide the power to run a refrigerator. The input to the heat engine is 700 K and the low T is 400 K; it has an actual efficiency equal to half of that of the corresponding Carnot unit. The refrigerator has $T_L = -10\,°C$ and $T_H = 35\,°C$, with a COP that is one-third that of the corresponding Carnot unit. Assume a cooling capacity of 2 kW is needed and find the rate of heat input to the heat engine.

5.66 A car engine with a thermal efficiency of 33 % drives the air-conditioner unit (a refrigerator) as well as powering the car and other auxiliary equipment. On a hot ($35\,°C$) summer day, the air conditioner takes outside air in and cools it to $5\,°C$, sending it into a duct using 2 kW of power input. It is assumed to be half as good as a Carnot refrigeration unit. Find the extra rate of fuel (kW) being burned just to drive the air conditioner unit and its COP. Find the flow rate of cold air the air-conditioner unit can provide.

Finite ΔT Heat Transfer

5.67 A refrigerator maintaining a $5\,°C$ inside temperature is located in a $30\,°C$ room. It must have a high temperature ΔT above room temperature and a low temperature ΔT below the refrigerated space in the cycle to actually transfer the heat. For a ΔT of $0°$, $5°$, and $10\,°C$, respectively, calculate the COP, assuming a Carnot cycle.

5.68 The ocean near Hawaii is $20\,°C$ near the surface and $5\,°C$ at some depth. A power plant based on this temperature difference is being planned. How large an efficiency could it have? If the two heat transfer terms (Q_H and Q_L) both require a 2-degree difference to operate, what is the maximum efficiency?

5.69 A house is cooled by a heat pump driven by an electric motor using the inside as the low-temperature reservoir. The house gains energy in direct proportion to the temperature difference as $\dot{Q}_{gain} = K(T_H - T_L)$. Determine the minimum electric power to drive the heat pump as a function of the two temperatures.

FIGURE P5.69

5.70 An air conditioner in a very hot region uses a power input of 2.5 kW to cool a $15\,°C$ space with the high temperature in the cycle at $40\,°C$. The Q_H is pushed to the ambient air at $30\,°C$ in a heat exchanger where the transfer coefficient is 50 W/m^{2}K. Find the required minimum heat transfer area.

5.71 Consider a room at $20\,°C$ that is cooled by an air conditioner with a COP of 3.2 using a power input of 2 kW, and the outside temperature is $35\,°C$. What is the constant in the heat transfer Eq. 5.14 for the heat transfer from the outside into the room?

5.72 A car engine operates with a thermal efficiency of 35 %. Assume the air conditioner has a COP of $\beta = 3$ working as a refrigerator cooling the inside using engine shaft work to drive it. How much extra fuel energy should be spent to remove 1 kJ from the inside?

5.73 Arctic explorers are unsure if they can use a 5 kW motor-driven heat pump to stay warm. It should keep their shelter at $15\,°C$. The shelter loses energy at a rate of 0.5 kW per degree difference to the colder ambient. The heat pump has a COP that is 50 % that of a Carnot heat pump. If the ambient temperature can fall to $-25\,°C$ at night, would you recommend this heat pump to the explorers?

5.74 Using the given heat pump in the previous problem, how warm could it make the shelter in the arctic night?

5.75 A window air conditioner cools a room at $T_L = 20\,°C$ with a maximum of 1.2 kW power input. The room gains 0.6 kW per degree temperature difference to the ambient, and the refrigeration COP is $\beta = 0.6\,\beta_{Carnot}$. Find the maximum outside temperature, T_H, for which the air conditioner provides sufficient cooling.

5.76 A heat pump has a COP that is 50 % of the theoretical maximum. It maintains a house at 20 °C, which leaks energy of 0.6 kW per degree temperature difference to the ambient. For a maximum of 1.5 kW power input, find the minimum outside temperature for which the heat pump is a sufficient heat source.

5.77 The room in Problem 5.75 has a combined thermal mass of 2000 kg wood, 250 kg steel, and 500 kg plaster board, $C_p = 1$kJ/kg · K. Estimate how quickly the room heats up if the air conditioner is turned off on a day when it is 35 °C outside.

5.78 On a cold (-10 °C) winter day, a heat pump provides 20 kW to heat a house maintained at 20 °C, and it has a COP_{HP} of 3.8. How much power does the heat pump require? The next day, a storm brings the outside temperature to -15 °C, assuming the same COP and the same house heat transfer coefficient for the heat loss to the outside air. How much power does the heat pump require then?

5.79 In the previous problem, it was assumed that the COP will be the same when the outside temperature drops. Given the temperatures and the actual COP at the -10 °C winter day, give an estimate for a more realistic COP for the outside -15 °C case.

Ideal Gas Carnot Cycles

5.80 Hydrogen gas is used in a Carnot cycle having an efficiency of 60 % with a low temperature of 300 K. During heat rejection, the pressure changes from 90 kPa to 120 kPa. Find the high- and low-temperature heat transfers and the net cycle work per unit mass of hydrogen.

5.81 Carbon dioxide is used in an ideal gas refrigeration cycle, the reverse of Fig. 5.24. Heat absorption is at 250 K and heat rejection is at 325 K where the pressure changes from 1200 kPa to 2400 kPa. Find the refrigeration COP and the specific heat transfer at the low temperature.

5.82 Air in a piston/cylinder goes through a Carnot cycle with the P–v diagram shown in Fig. 5.24. The high and low temperatures are 600 K and 300 K, respectively. The heat added at the high temperature is 250 kJ/kg, and the lowest pressure in the cycle is

75 kPa. Find the specific volume and pressure after heat rejection and the net work per unit mass.

Review Problems

5.83 A 4 L jug of milk at 25 °C is placed in your refrigerator, where it is cooled down to 5 °C. The high temperature in the Carnot refrigeration cycle is 45 °C, the low temperature is -5 °C, and the properties of milk are the same as those of liquid water. Find the amount of energy that must be removed from the milk and the additional work needed to drive the refrigerator.

5.84 Consider a combination of a gas turbine power plant and a steam power plant, as shown in Fig. P5.4. The gas turbine operates at higher temperatures (thus called a *topping cycle*) than the steam power plant (thus called a *bottom cycle*). Assume both cycles have a thermal efficiency of 32 %. What is the efficiency of the overall combination, assuming Q_L in the gas turbine equals Q_H to the steam power plant?

5.85 We wish to produce refrigeration at -30 °C. A reservoir, shown in Fig. P5.85, is available at 200 °C and the ambient temperature is 30 °C. Thus, work can be done by a cyclic heat engine operating between the 200 °C reservoir and the ambient. This work is used to drive the refrigerator. Determine the ratio of the heat transferred from the 200 °C reservoir to the heat transferred from the -30 °C reservoir, assuming all processes are reversible.

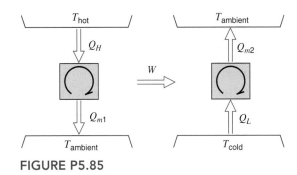

FIGURE P5.85

5.86 Redo the previous problem, assuming the actual devices both have a performance that is 60 % of the theoretical maximum.

5.87 A house should be heated by a heat pump, $\beta' = 2.2$, and maintained at 20 °C at all times. It is estimated that it loses 0.8 kW for each degree that the ambient is lower than the inside. Assume an outside temperature of -10 °C and find the needed power to drive the heat pump.

5.88 Give an estimate for the COP in the previous problem and the power needed to drive the heat pump when the outside temperature drops to -15 °C.

5.89 An air conditioner with a power input of 1.2 kW is working as a refrigerator ($\beta = 3$) or as a heat pump ($\beta' = 4$). It maintains an office at 20 °C year round that exchanges 0.5 kW per degree temperature difference with the atmosphere. Find the maximum and minimum outside temperatures for which this unit is sufficient.

5.90 An air conditioner on a hot summer day removes 8 kW of energy from a house at 21 °C and pushes energy to the outside, which is at 31 °C. The house has a mass of 15 000 kg with an average specific heat of 0.95 kJ/kg·K. In order to do this, the cold side of the air conditioner is at 5 °C and the hot side is at 40 °C. The air conditioner (refrigerator) has a COP that is 60 % that of a corresponding Carnot refrigerator. Find the actual COP of the air conditioner and the power required to run it.

5.91 The air conditioner in the previous problem is turned off. How quickly does the house heat up in degrees per second (°C/s)?

5.92 Air in a rigid 1 m³ box is at 300 K, 200 kPa. It is heated to 600 K by heat transfer from a reversible heat pump that receives energy from the ambient at 300 K besides the work input. Use constant specific heat at 300 K. Since the COP changes, write $dQ = m_{air}\, C_v\, dT$ and find dW. Integrate dW with the temperature to find the required heat pump work.

5.93 A Carnot heat engine, shown in Fig. P5.93, receives energy from a reservoir at T_{res} through a heat exchanger, where the heat transferred is proportional to the temperature difference as $\dot{Q}_H = K(T_{res} - T_H)$. It rejects heat at a given low temperature T_L. To design the heat engine for maximum work output, show that the high temperature, T_H, in the cycle should be selected as $T_H = \sqrt{T_{res} T_L}$

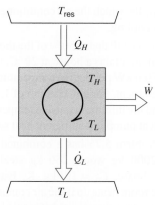

FIGURE P5.93

5.94 A combination of a heat engine driving a heat pump (see Fig. P5.94) takes waste energy at 50 °C as a source Q_{w1} to the heat engine, rejecting heat at 30 °C. The remainder, Q_{w2}, goes into the heat pump that delivers a Q_H at 150 °C. If the total waste energy is 5 MW, find the rate of energy delivered at the high temperature.

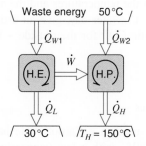

FIGURE P5.94

5.95 Consider the rock bed thermal storage in Problem 5.51. Use the specific heat so that you can write dQ_H in terms of dT_{rock} and find the expression for dW out of the heat engine. Integrate this expression over temperature and find the total heat engine work output.

5.96 Consider a Carnot cycle heat engine operating in outer space. Heat can be rejected from this engine only by thermal radiation, which is proportional to the radiator area and the fourth power of absolute temperature, $\dot{Q}_{rad} \sim KAT^4$. Show that for a given engine work output and given T_H, the radiator area will be minimum when the ratio $T_L/T_H = 3/4$.

5.97 A Carnot heat engine operating between a high T_H and low T_L energy reservoirs has an efficiency given by the temperatures. Compare this to two combined heat engines, one operating between T_H and an intermediate temperature T_M giving out work W_A and the other operating between T_M and T_L giving out work W_B. The combination must have the same efficiency as the single heat engine, so the heat transfer ratio $Q_H/Q_L = \psi(T_H, T_L) = [Q_H/Q_M] [Q_M/Q_L]$. The last two heat transfer ratios can be expressed by the same function $\psi()$ also involving the temperature T_M. Use this to show a condition that the function $\psi()$ must satisfy.

5.98 A 10 m^3 tank of air at 500 kPa, 600 K acts as the high-temperature reservoir for a Carnot heat engine that rejects heat at 300 K. A temperature difference of 25 °C between the air tank and the Carnot cycle high temperature is needed to transfer the heat. The heat engine runs until the air temperature has dropped to 400 K and then stops. Assume constant specific heat for air and find how much work is given out by the heat engine.

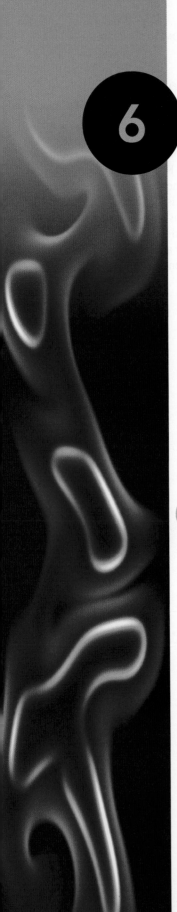

6 Entropy for a Control Mass

Up to this point in our consideration of the second law of thermodynamics, we have dealt only with thermodynamic cycles. Although this is a very important and useful approach, we are often concerned with processes rather than cycles. Thus, we might be interested in the second-law analysis of processes we encounter daily, such as the combustion process in an automobile engine, the cooling of a cup of coffee, or the chemical processes that take place in our bodies. It would also be beneficial to be able to deal with the second law quantitatively as well as qualitatively.

In our consideration of the first law, we initially stated the law in terms of a cycle, but we then defined a property, the internal energy, that enabled us to use the first law quantitatively for processes. Similarly, we have stated the second law for a cycle, and we now find that the second law leads to a property, entropy, that enables us to treat the second law quantitatively for processes. Energy and entropy are both abstract concepts that help to describe certain observations. As we noted in Chapter 1, thermodynamics can be described as the science of energy and entropy. The significance of this statement will become increasingly evident.

6.1 THE INEQUALITY OF CLAUSIUS

The first step in our consideration of the property we call entropy is to establish the inequality of Clausius, which is

$$\oint \frac{\delta Q}{T} \leq 0$$

The inequality of Clausius is a corollary or a consequence of the second law of thermodynamics. It will be demonstrated to be valid for all possible cycles, including both reversible and irreversible heat engines and refrigerators. Since any reversible cycle can be represented by a series of Carnot cycles, in this analysis we need consider only a Carnot cycle that leads to the inequality of Clausius.

Consider first a reversible (Carnot) heat engine cycle operating between reservoirs at temperatures T_H and T_L, as shown in Fig. 6.1. For this cycle, the cyclic integral of the heat transfer, $\oint \delta Q$, is greater than zero.

$$\oint \delta Q = Q_H - Q_L > 0$$

Since T_H and T_L are constant, from the definition of the absolute temperature scale and from the fact that this is a reversible cycle, it follows that

$$\oint \frac{\delta Q}{T} = \frac{Q_H}{T_H} - \frac{Q_L}{T_L} = 0$$

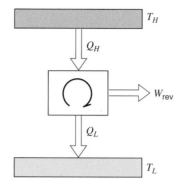

FIGURE 6.1

Reversible heat engine cycle for demonstration of the inequality of Clausius.

If $\oint \delta Q$, the cyclic integral of δQ, approaches zero (by making T_H approach T_L) and the cycle remains reversible, the cyclic integral of $\delta Q/T$ remains zero. Thus, we conclude that for all reversible heat engine cycles

$$\oint \delta Q \geq 0$$

and

$$\oint \frac{\delta Q}{T} = 0$$

Now consider an irreversible cyclic heat engine operating between the same T_H and T_L as the reversible engine of Fig. 6.1 and receiving the same quantity of heat Q_H. Comparing the irreversible cycle with the reversible one, we conclude from the second law that

$$W_{irr} < W_{rev}$$

Since $Q_H - Q_L = W$ for both the reversible and irreversible cycles, we conclude that

$$Q_H - Q_{L\ irr} < Q_H - Q_{L\ rev}$$

and therefore

$$Q_{L\ irr} > Q_{L\ rev}$$

Consequently, for the irreversible cyclic engine,

$$\oint \delta Q = Q_H - Q_{L\ irr} > 0$$

$$\oint \frac{\delta Q}{T} = \frac{Q_H}{T_H} - \frac{Q_{L\ irr}}{T_L} < 0$$

Suppose that we cause the engine to become more and more irreversible but keep Q_H, T_H, and T_L fixed. The cyclic integral of δQ then approaches zero, and that for $\delta Q/T$ becomes a progressively larger negative value. In the limit, as the work output goes to zero,

$$\oint \delta Q = 0$$

$$\oint \frac{\delta Q}{T} < 0$$

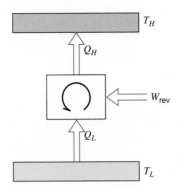

FIGURE 6.2
Reversible refrigeration
cycle for demonstration
of the inequality of
Clausius.

Thus, we conclude that for all irreversible heat engine cycles

$$\oint \delta Q \geq 0$$

$$\oint \frac{\delta Q}{T} < 0$$

To complete the demonstration of the inequality of Clausius, we must perform similar analyses for both reversible and irreversible refrigeration cycles. For the reversible refrigeration cycle shown in Fig. 6.2,

$$\oint \delta Q = -Q_H + Q_L < 0$$

and

$$\oint \frac{\delta Q}{T} = -\frac{Q_H}{T_H} + \frac{Q_L}{T_L} = 0$$

As the cyclic integral of δQ approaches zero reversibly (T_H approaches T_L), the cyclic integral of $\delta Q / T$ remains at zero. In the limit,

$$\oint \delta Q = 0$$

$$\oint \frac{\delta Q}{T} = 0$$

Thus, for all reversible refrigeration cycles,

$$\oint \delta Q \leq 0$$

$$\oint \frac{\delta Q}{T} = 0$$

Finally, let an irreversible cyclic refrigerator operate between temperatures T_H and T_L and receive the same amount of heat Q_L as the reversible refrigerator of Fig. 6.2. From the second law, we conclude that the work input required will be greater for the irreversible refrigerator, or

$$W_{\text{irr}} > W_{\text{rev}}$$

Since $Q_H - Q_L = W$ for each cycle, it follows that

$$Q_{H\text{ irr}} - Q_L > Q_{H\text{ rev}} - Q_L$$

and therefore,

$$Q_{H\text{ irr}} > Q_{H\text{ rev}}$$

That is, the heat rejected by the irreversible refrigerator to the high-temperature reservoir is greater than the heat rejected by the reversible refrigerator. Therefore, for the irreversible refrigerator,

$$\oint \delta Q = -Q_{H\text{ irr}} + Q_L < 0$$

$$\oint \frac{\delta Q}{T} = -\frac{Q_{H\text{ irr}}}{T_H} + \frac{Q_L}{T_L} < 0$$

As we make this machine progressively more irreversible but keep Q_L, T_H, and T_L constant, the cyclic integrals of δQ and $\delta Q/T$ both become larger in the negative direction. Consequently, a limiting case as the cyclic integral of δQ approaches zero does not exist for the irreversible refrigerator.

Thus, for all irreversible refrigeration cycles,

$$\oint \delta Q < 0$$

$$\oint \frac{\delta Q}{T} < 0$$

Summarizing, we note that, in regard to the sign of $\oint \delta Q$, we have considered all possible reversible cycles (i.e., $\oint \delta Q \gtrless 0$), and for each of these reversible cycles

$$\oint \frac{\delta Q}{T} = 0$$

We have also considered all possible irreversible cycles for the sign of $\oint \delta Q$ (i.e., $\oint \delta Q \gtrless 0$), and for all these irreversible cycles

$$\oint \frac{\delta Q}{T} < 0$$

Thus, for all cycles we can write

$$\oint \frac{\delta Q}{T} \le 0 \tag{6.1}$$

where the equality holds for reversible cycles and the inequality for irreversible cycles. This relation, Eq. 6.1, is known as the *inequality of Clausius*.

The significance of the inequality of Clausius may be illustrated by considering the simple steam power plant cycle shown in Fig. 6.3. This cycle is slightly different from the usual cycle for steam power plants in that the pump handles a mixture of liquid and vapor in such proportions that saturated liquid leaves the pump and enters the boiler. Suppose that someone reports that the pressure and quality at various points in the cycle are as given in Fig. 6.3. Does this cycle satisfy the inequality of Clausius?

Heat is transferred in two places, the boiler and the condenser. Therefore,

$$\oint \frac{\delta Q}{T} = \int \left(\frac{\delta Q}{T}\right)_{\text{boiler}} + \int \left(\frac{\delta Q}{T}\right)_{\text{condenser}}$$

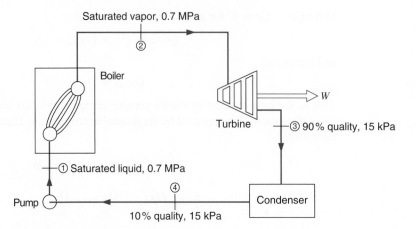

FIGURE 6.3 A simple steam power plant that demonstrates the inequality of Clausius.

Since the temperature remains constant in both the boiler and condenser, this may be integrated as follows:

$$\oint \frac{\delta Q}{T} = \frac{1}{T_1} \int_1^2 \delta Q + \frac{1}{T_3} \int_3^4 \delta Q = \frac{{}_1 Q_2}{T_1} + \frac{{}_3 Q_4}{T_3}$$

Let us consider a 1 kg mass as the working fluid. We have then

$${}_1 q_2 = h_2 - h_1 = 2066.3 \text{ kJ/kg}, \qquad T_1 = 164.97 \,^{\circ}\text{C}$$

$${}_3 q_4 = h_4 - h_3 = 463.4 - 2361.8 = -1898.4 \text{ kJ/kg}, \qquad T_3 = 53.97 \,^{\circ}\text{C}$$

Therefore,

$$\oint \frac{\delta Q}{T} = \frac{2066.3}{164.97 + 273.15} - \frac{1898.4}{53.97 + 273.15} = -1.087 \text{ kJ/kg·K}$$

Thus, this cycle satisfies the inequality of Clausius, which is equivalent to saying that it does not violate the second law of thermodynamics.

In-Text Concept Questions

a. Does Clausius say anything about the sign for $\oint \delta Q$?

b. Does the statement of Clausius require a constant T for the heat transfer as in a Carnot cycle?

6.2 ENTROPY—A PROPERTY OF A SYSTEM

By applying Eq. 6.1 and Fig. 6.4, we can demonstrate that the second law of thermodynamics leads to a property of a system that we call *entropy*. Let a system (control mass) undergo a reversible process from state 1 to state 2 along a path A, and let the cycle be completed along path B, which is also reversible.

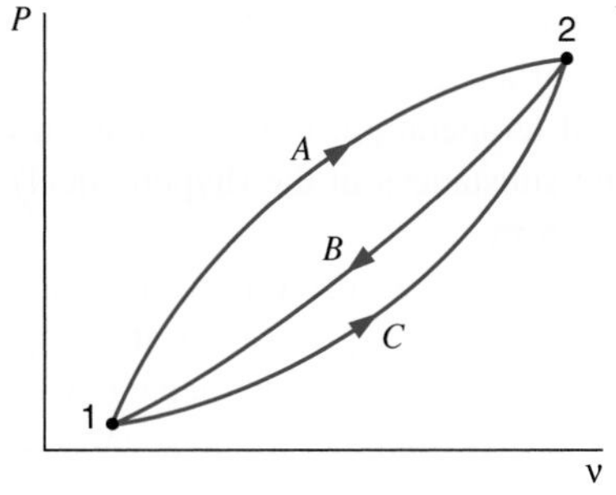

FIGURE 6.4 Two reversible cycles demonstrating that entropy is a property of a substance.

Because this is a reversible cycle, we can write

$$\oint \frac{\delta Q}{T} = 0 = \int_1^2 \left(\frac{\delta Q}{T}\right)_A + \int_2^1 \left(\frac{\delta Q}{T}\right)_B$$

Now consider another reversible cycle, which proceeds first along path C and is then completed along path B. For this cycle, we can write

$$\oint \frac{\delta Q}{T} = 0 = \int_1^2 \left(\frac{\delta Q}{T}\right)_C + \int_2^1 \left(\frac{\delta Q}{T}\right)_B$$

Subtracting the second equation from the first, we have

$$\int_1^2 \left(\frac{\delta Q}{T}\right)_A = \int_1^2 \left(\frac{\delta Q}{T}\right)_C$$

Since $\oint \delta Q/T$ is the same for all reversible paths between states 1 and 2, we conclude that this quantity is independent of the path and is a function of the end states only; it is therefore a property. This property is called entropy and is designated S. It follows that entropy may be defined as a property of a substance in accordance with the relation

$$dS \equiv \left(\frac{\delta Q}{T}\right)_{rev} \tag{6.2}$$

Entropy is an extensive property, and the entropy per unit mass is designated s. It is important to note that entropy is defined here in terms of a reversible process.

The change in the entropy of a system as it undergoes a change of state may be found by integrating Eq. 6.2. Thus,

$$S_2 - S_1 = \int_1^2 \left(\frac{\delta Q}{T}\right)_{rev} \tag{6.3}$$

To perform this integration, we must know the relation between T and Q, and illustrations will be given subsequently. The important point is that since entropy is a property, the change in the entropy of a substance in going from one state to another is the same for all processes, both reversible and irreversible, between these two states. Equation 6.3 enables us to find the change in entropy only along a reversible path. However, once the change has been evaluated, this value is the magnitude of the entropy change for all processes between these two states.

Equation 6.3 enables us to calculate changes of entropy, but it tells us nothing about absolute values of entropy. From the third law of thermodynamics, which is based on observations of low-temperature chemical reactions, it is concluded that the entropy of all pure substances (in the appropriate structural form) can be assigned the absolute value of zero at the absolute zero of temperature. It also follows from the subject of statistical thermodynamics that all pure substances in the (hypothetical) ideal-gas state at absolute zero temperature have zero entropy.

However, when there is no change of composition, as would occur in a chemical reaction, for example, it is quite adequate to give values of entropy relative to some arbitrarily selected reference state, such as was done earlier when tabulating values of internal energy and enthalpy. In each case, whatever reference value is chosen, it will cancel out when the change of property is calculated between any two states. This is the procedure followed with the thermodynamic tables to be discussed in the following section.

A word should be added here regarding the role of T as an integrating factor. We noted in Chapter 3 that Q is a path function, and therefore δQ in an inexact differential. However, since $(\delta Q/T)_{\text{rev}}$ is a thermodynamic property, it is an exact differential. From a mathematical perspective, we note that an inexact differential may be converted to an exact differential by the introduction of an integrating factor. Therefore, $1/T$ serves as the integrating factor in converting the inexact differential δQ to the exact differential $\delta Q/T$ for a reversible process.

6.3 THE ENTROPY OF A PURE SUBSTANCE

Entropy is an extensive property of a system. Values of specific entropy (entropy per unit mass) are given in tables of thermodynamic properties in the same manner as specific volume and specific enthalpy. The units of specific entropy in the steam tables, refrigerant tables, and ammonia tables are kJ/kg·K, and the values are given relative to an arbitrary reference state. In the steam tables, the entropy of saturated liquid at 0.01 °C is given the value of zero. For many refrigerants, the entropy of saturated liquid at −40 °C is assigned the value of zero.

In general, we use the term *entropy* to refer to both total entropy and entropy per unit mass, since the context or appropriate symbol will clearly indicate the precise meaning of the term.

In the saturation region, the entropy may be calculated using the quality. The relations are similar to those for specific volume, internal energy and enthalpy.

$$s = (1 - x)s_f + xs_g$$

$$s = s_f + xs_{fg}$$

The entropy of a compressed liquid is tabulated in the same manner as the other properties. These properties are primarily a function of the temperature and are not greatly different from those for saturated liquid at the same temperature. Table 4 of the steam tables, which is summarized in Table B.1.4, give the entropy of compressed liquid water in the same manner as for other properties.

The thermodynamic properties of a substance are often shown on a temperature–entropy diagram and on an enthalpy–entropy diagram, which is also called a *Mollier diagram*, after Richard Mollier (1863–1935) of Germany. Figures 6.5 and 6.6 show the essential elements of temperature–entropy and enthalpy–entropy diagrams for steam. The general features of such diagrams are the same for all pure substances. A more complete temperature–entropy diagram for steam is shown in Fig. E.1 in Appendix E.

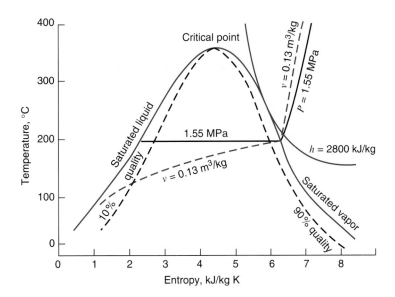

FIGURE 6.5
Temperature–entropy
diagram for steam.

These diagrams are valuable both because they present thermodynamic data and because they enable us to visualize the changes of state that occur in various processes. As our study progresses, the student should acquire facility in visualizing thermodynamic processes on these diagrams. The temperature–entropy diagram is particularly useful for this purpose.

For most substances, the difference in the entropy of a compressed liquid and a saturated liquid at the same temperature is so small that a process in which liquid is heated at constant pressure nearly coincides with the saturated-liquid line until the saturation temperature is reached (Fig. 6.7). Thus, if water at 10 MPa is heated from 0 °C to the saturation temperature, it would be shown by line *ABD*, which coincides with the saturated-liquid line.

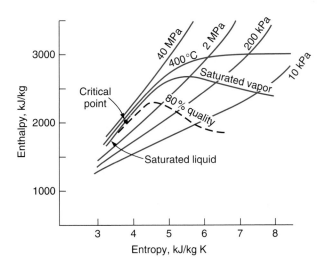

FIGURE 6.6
Enthalpy–entropy
diagram for steam.

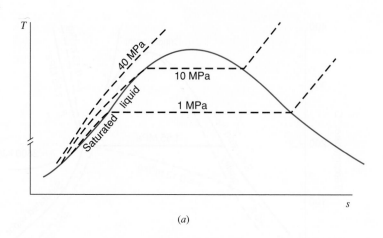

(a)

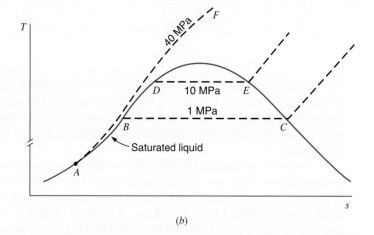

FIGURE 6.7
Temperature–entropy
diagram showing
properties of a
compressed liquid,
water.

(b)

6.4 ENTROPY CHANGE IN REVERSIBLE PROCESSES

Having established that entropy is a thermodynamic property of a system, we now consider its significance in various processes. In this section, we will limit ourselves to systems that undergo reversible processes and consider the Carnot cycle, reversible heat-transfer processes, and reversible adiabatic processes.

Let the working fluid of a heat engine operating on the Carnot cycle make up the system. The first process is the isothermal transfer of heat to the working fluid from the high-temperature reservoir. For this process, we can write

$$S_2 - S_1 = \int_1^2 \left(\frac{\delta Q}{T} \right)_{\text{rev}}$$

Since this is a reversible process in which the temperature of the working fluid remains constant, the equation can be integrated to give

$$S_2 - S_1 = \frac{1}{T_H} \int_1^2 \delta Q = \frac{{}_1 Q_2}{T_H}$$

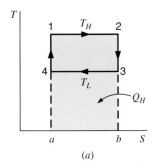 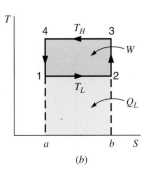

FIGURE 6.8 The Carnot cycle on the temperature–entropy diagram.

This process is shown in Fig. 6.8*a*, and the area under line 1–2, area 1–2–*b*–*a*–1, represents the heat transferred to the working fluid during the process.

The second process of a Carnot cycle is a reversible adiabatic one. From the definition of entropy,

$$dS = \left(\frac{\delta Q}{T} \right)_{\text{rev}}$$

it is evident that the entropy remains constant in a reversible adiabatic process. A constant-entropy process is called an isentropic process. Line 2–3 represents this process, and this process is concluded at state 3 when the temperature of the working fluid reaches T_L.

The third process is the reversible isothermal process in which heat is transferred from the working fluid to the low-temperature reservoir. For this process, we can write

$$S_4 - S_3 = \int_3^4 \left(\frac{\delta Q}{T} \right)_{\text{rev}} = \frac{{}_3 Q_4}{T_L}$$

Because during this process the heat transfer is negative (in regard to the working fluid), the entropy of the working fluid decreases. Moreover, because the final process 4–1, which completes the cycle, is a reversible adiabatic process (and therefore isentropic), it is evident that the entropy decrease in process 3–4 must exactly equal the entropy increase in process 1–2. The area under line 3–4, area 3–4–*a*–*b*–3, represents the heat transferred from the working fluid to the low-temperature reservoir.

Since the net work of the cycle is equal to the net heat transfer, area 1–2–3–4–1 must represent the net work of the cycle. The efficiency of the cycle may also be expressed in terms of areas:

$$\eta_{\text{th}} = \frac{W_{\text{net}}}{Q_H} = \frac{\text{area } 1\text{–}2\text{–}3\text{–}4\text{–}1}{\text{area } 1\text{–}2\text{–}b\text{–}a\text{–}1}$$

Some statements made earlier about efficiencies may now be understood graphically. For example, increasing T_H while T_L remains constant increases the efficiency. Decreasing T_L while T_H remains constant increases the efficiency. It is also evident that the efficiency approaches 100 % as the absolute temperature at which heat is rejected approaches zero.

If the cycle is reversed, we have a refrigerator or heat pump. The Carnot cycle for a refrigerator is shown in Fig. 6.8*b*. Notice that the entropy of the working fluid increases at T_L, since heat is transferred to the working fluid at T_L. The entropy decreases at T_H because of heat transfer from the working fluid.

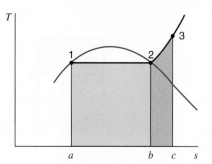

FIGURE 6.9 A temperature–entropy diagram showing areas that represent heat transfer for an internally reversible process.

Let us next consider reversible heat-transfer processes. Actually, we are concerned here with processes that are internally reversible, that is, processes that have no irreversibilities within the boundary of the system. For such processes, the heat transfer to or from a system can be shown as an area on a temperature–entropy diagram. For example, consider the change of state from saturated liquid to saturated vapor at constant pressure. This process would correspond to process 1–2 on the T–s diagram of Fig. 6.9 (note that absolute temperature is required here), and area 1–2–b–a–1 represents the heat transfer. Since this is a constant-pressure process, the heat transfer per unit mass is equal to h_{fg}. Thus,

$$s_2 - s_1 = s_{fg} = \frac{1}{m} \int_1^2 \left(\frac{\delta Q}{T} \right)_{\text{rev}} = \frac{1}{mT} \int_1^2 \delta Q = \frac{{}_1 q_2}{T} = \frac{h_{fg}}{T}$$

This relation gives a clue about how s_{fg} is calculated for tabulation in tables of thermodynamic properties. For example, consider steam at 10 MPa. From the steam tables, we have

$$h_{fg} = 1317.1 \text{ kJ/kg}$$

$$T = 311.06 + 273.15 = 584.21 \text{ K}$$

Therefore,

$$s_{fg} = \frac{h_{fg}}{T} = \frac{1317.1}{584.21} = 2.2544 \text{ kJ/kg·K}$$

This is the value listed for s_{fg} in the steam tables.

If heat is transferred to the saturated vapor at constant pressure, the steam is superheated along line 2–3. For this process, we can write

$$_2 q_3 = \frac{1}{m} \int_2^3 \delta Q = \int_2^3 T \, ds$$

Since T is not constant, this equation cannot be integrated unless we know a relation between temperature and entropy. However, we do realize that the area under line 2–3, area 2–3–c–b–2, represents $\int_2^3 T \, ds$ and therefore represents the heat transferred during this reversible process.

The important conclusion to draw here is that for processes that are internally reversible, the area underneath the process line on a temperature–entropy diagram represents the quantity of heat transferred. This is not true for irreversible processes, as will be demonstrated later.

Example 6.1

Consider a Carnot-cycle heat pump with R-134a as the working fluid. Heat is absorbed into the R-134a at $0\,°C$, during which process it changes from a two-phase state to saturated vapor. The heat is rejected from the R-134a at $60\,°C$ and ends up as saturated liquid. Find the pressure after compression, before the heat rejection process, and determine the COP for the cycle.

Solution

From the definition of the Carnot cycle, we have two constant-temperature (isothermal) processes that involve heat transfer and two adiabatic processes in which the temperature changes. The variation in s follows from Eq. 6.2:

$$ds = \delta q / T$$

and the Carnot cycle is shown in Fig. 6.8 and for this case in Fig. 6.10. We therefore have

State 4 Table B.5.1: $\quad s_4 = s_3 = s_{f@60\mathrm{deg}} = 1.2857\ \mathrm{kJ/kg \cdot K}$

State 1 Table B.5.1: $\quad s_1 = s_2 = s_{g@0\mathrm{deg}} = 1.7262\ \mathrm{kJ/kg \cdot K}$

State 2 Table B.5.2: $\quad 60\,°C,\ s_2 = s_1 = s_{g@0\mathrm{deg}}$

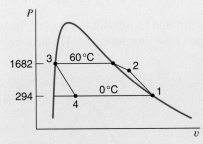 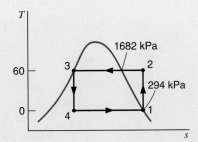

FIGURE 6.10 Diagram for Example 6.1.

Interpolate between 1400 kPa and 1600 kPa in Table B.5.2:

$$P_2 = 1400 + (1600 - 1400)\frac{1.7262 - 1.736}{1.7135 - 1.736} = 1487.1\ \mathrm{kPa}$$

From the fact that it is a Carnot cycle, the COP becomes, from Eq. 5.13,

$$\beta' = \frac{q_H}{w_{\mathrm{IN}}} = \frac{T_H}{T_H - T_L} = \frac{333.15}{60} = 5.55$$

Remark. Notice how much the pressure varies during the heat rejection process. Because this process is very difficult to accomplish in a real device, no heat pump or refrigerator is designed to attempt to approach a Carnot cycle.

Example 6.2

A cylinder/piston setup contains 1 L of saturated liquid refrigerant R-410A at 20 °C. The piston now slowly expands, maintaining constant temperature to a final pressure of 400 kPa in a reversible process. Calculate the work and heat transfer required to accomplish this process.

Solution

C.V. The refrigerant R-410A, which is a control mass, and in this case changes in kinetic and potential energies are negligible.

Continuity Eq.: $\quad m_2 = m_1 = m$

Energy Eq. 3.5: $\quad m(u_2 - u_1) = {}_1Q_2 - {}_1W_2$

Entropy Eq. 6.3: $\quad m(s_2 - s_1) = \int \delta Q/T$

Process: $\quad T = \text{constant, reversible}$

State 1 (T, P) Table B.4.1: $\quad u_1 = 87.94 \text{ kJ/kg}, \qquad s_1 = 0.3357 \text{ kJ/kg·K}$

$$m = V/v_1 = 0.001/0.000\,923 = 1.083 \text{ kg}$$

State 2 (T, P) Table B.4.2: $\quad u_2 = 276.44 \text{ kJ/kg}, \qquad s_2 = 1.2108 \text{ kJ/kg·K}$

As T is constant, we have $\int \delta Q/T = {}_1Q_2/T$, so from the entropy equation:

$${}_1Q_2 = mT(s_2 - s_1) = 1.083 \times 293.15 \times (1.2108 - 0.3357) = 277.8 \text{ kJ}$$

The work is then, from the energy equation,

$${}_1W_2 = m(u_1 - u_2) + {}_1Q_2 = 1.083 \times (87.94 - 276.44) + 277.8 = 73.7 \text{ kJ}$$

Note from Fig. 6.11 that it would be difficult to calculate the work as the area in the P–v diagram due to the shape of the process curve. The heat transfer is the area in the T–s diagram.

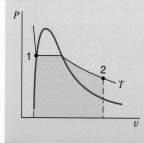
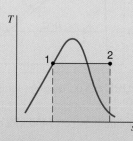

FIGURE 6.11 Diagram for Example 6.2.

In-Text Concept Questions

c. How can you change s of a substance going through a reversible process?

d. A reversible process adds heat to a substance. If T is varying, does that influence the change in s?

e. Water at 100 kPa, 150 °C receives 75 kJ/kg in a reversible process by heat transfer. Which process changes s the most: constant T, constant v, or constant P?

6.5 THE THERMODYNAMIC PROPERTY RELATION

At this point we derive two important thermodynamic relations for a simple compressible substance. These relations are

$$T\,dS = dU + P\,dV$$

$$T\,dS = dH - V\,dP$$

The first of these relations can be derived by considering a simple compressible substance in the absence of motion or gravitational effects. The first law for a change of state under these conditions can be written

$$\delta Q = dU + \delta W$$

The equations we are deriving here deal first with the changes of state in which the state of the substance can be identified at all times. Thus, we must consider a quasi-equilibrium process or, to use the term introduced in the previous chapter, a *reversible process*. For a reversible process of a simple compressible substance, we can write

$$\delta Q = T\,dS \qquad \text{and} \qquad \delta W = P\,dV$$

Substituting these relations into the energy equation, we have

$$T\,dS = dU + P\,dV \tag{6.5}$$

which is the first equation we set out to derive. Note that this equation was derived by assuming a reversible process. This equation can therefore be integrated for any reversible process, for during such a process the state of the substance can be identified at any point during the process. We also note that Eq. 6.5 deals only with properties. Suppose we have an irreversible process taking place between the given initial and final states. The properties of a substance depend only on the state, and therefore the changes in the properties during a given change of state are the same for an irreversible process as for a reversible process. Therefore, Eq. 6.5 is often applied to an irreversible process between two given states, but the integration of Eq. 6.5 is performed along a reversible path between the same two states.

Since enthalpy is defined as

$$H \equiv U + PV$$

it follows that

$$dH = dU + P\,dV + V\,dP$$

Substituting this relation into Eq. 6.5, we have

$$T\,dS = dH - V\,dP \tag{6.6}$$

which is the second relation that we set out to derive. These two expressions, Eqs. 6.5 and 6.6, are two forms of the thermodynamic property relation and are frequently called Gibbs equations.

These equations can also be written for a unit mass:

$$T\,ds = du + P\,dv \tag{6.7}$$

$$T\,ds = dh - v\,dP \tag{6.8}$$

The Gibbs equations will be used extensively in certain subsequent sections of this book.

If we consider substances of fixed composition other than a simple compressible substance, we can write "$T\,dS$" equations other than those just given for a simple compressible substance. In Eq. 3.47 we noted that for a reversible process, we can write the following expression for work:

$$\delta W = P\,dV - \mathcal{T}\,dL - \mathcal{S}\,dA - \mathcal{E}\,dZ + \cdots$$

It follows that a more general expression for the thermodynamic property relation would be

$$T\,dS = dU + P\,dV - \mathcal{T}\,dL - \mathcal{S}\,dA - \mathcal{E}\,dZ + \cdots \tag{6.9}$$

6.6 ENTROPY CHANGE OF A SOLID OR LIQUID

In Section 3.10 we considered the calculation of the internal energy and enthalpy changes with temperature for solids and liquids and found that, in general, it is possible to express both in terms of the specific heat, in the simple manner of Eq. 3.30, and in most instances, in the integrated form of Eq. 3.31. We can now use this result and the thermodynamic property relation, Eq. 6.7, to calculate the entropy change for a solid or liquid. Note that for such a phase the specific volume term in Eq. 6.7 is very small, so that substituting Eq. 3.30 yields

$$ds \simeq \frac{du}{T} \simeq \frac{C}{T}\,dT \tag{6.10}$$

Now, as was mentioned in Section 3.10, for many processes involving a solid or liquid, we may assume that the specific heat remains constant, in which case Eq. 6.10 can be integrated. The result is

$$s_2 - s_1 \simeq C \ln \frac{T_2}{T_1} \tag{6.11}$$

If the specific heat is not constant, then commonly C is known as a function of T, in which case Eq. 6.10 can also be integrated to find the entropy change. Equation 6.11 illustrates what happens in a reversible adiabatic ($dq = 0$) process, which therefore is isentropic. In this process, the approximation of constant v leads to constant temperature, which explains why pumping liquid does not change the temperature.

Example 6.3

One kilogram of liquid water is heated from 20 °C to 90 °C. Calculate the entropy change, assuming constant specific heat, and compare the result with that found when using the steam tables.

Control mass: Water.
Initial and final states: Known.
Model: Constant specific heat, value at room temperature.

Solution

For constant specific heat, from Eq. 6.11,

$$s_2 - s_1 = 4.184 \ln \left(\frac{363.2}{293.2}\right) = 0.8958 \text{ kJ/kg·K}$$

Comparing this result with that obtained by using the steam tables, we have

$$s_2 - s_1 = s_{f\,90\,°C} - s_{f\,20\,°C} = 1.1925 - 0.2966$$
$$= 0.8959 \text{ kJ/kg·K}$$

6.7 ENTROPY CHANGE OF AN IDEAL GAS

Two very useful equations for computing the entropy change of an ideal gas can be developed from Eq. 6.7 by substituting Eqs. 3.35 and 3.39:

$$T\,ds = du + P\,dv$$

For an ideal gas

$$du = C_{v0}\,dT \quad \text{and} \quad \frac{P}{T} = \frac{R}{v}$$

Therefore,

$$ds = C_{v0}\frac{dT}{T} + \frac{R\,dv}{v} \tag{6.12}$$

$$s_2 - s_1 = \int_1^2 C_{v0}\frac{dT}{T} + R\ln\frac{v_2}{v_1} \tag{6.13}$$

Similarly,

$$T\,ds = dh - v\,dP$$

For an ideal gas

$$dh = C_{p0}\,dT \quad \text{and} \quad \frac{v}{T} = \frac{R}{P}$$

Therefore,

$$ds = C_{p0}\frac{dT}{T} - R\frac{dP}{P} \tag{6.14}$$

$$s_2 - s_1 = \int_1^2 C_{p0}\frac{dT}{T} - R\ln\frac{P_2}{P_1} \tag{6.15}$$

To evaluate the integrals in Eqs. 6.13 and 6.15, we must know the temperature dependence of the specific heats. However, if we recall that their difference is always constant, as expressed by Eq. 3.42, we realize that we need to examine the temperature dependence of only one of the specific heats.

As in Section 3.11, let us consider the specific heat C_{p0}. Again, there are three possibilities to examine, the simplest of which is the assumption of constant specific heat. In this instance, it is possible to do the integral in Eq. 6.15 directly to

$$s_2 - s_1 = C_{p0}\ln\frac{T_2}{T_1} - R\ln\frac{P_2}{P_1} \tag{6.16}$$

Similarly, the integral in Eq. 6.13 for constant specific heat gives

$$s_2 - s_1 = C_{v0} \ln \frac{T_2}{T_1} + R \ln \frac{v_2}{v_1} \qquad (6.17)$$

The second possibility for the specific heat is to use an analytical equation for C_{p0} as a function of temperature, for example, one of those listed in Table A.6. The third possibility is to integrate the results of the calculations of statistical thermodynamics from reference temperature T_0 to any other temperature T and define the standard entropy

$$s_T^0 = \int_{T_0}^T \frac{C_{p0}}{T} dT \qquad (6.18)$$

This function can then be tabulated in the single-entry (temperature) ideal-gas table, as for air in Table A.7(F.5) or for other gases in Table A.8(F.6). The entropy change between any two states 1 and 2 is then given by

$$s_2 - s_1 = \left(s_{T2}^0 - s_{T1}^0\right) - R \ln \frac{P_2}{P_1} \qquad (6.19)$$

As with the energy functions discussed in Section 3.11, the ideal-gas tables, Tables A.7 and A.8, would give the most accurate results, and the equations listed in Table A.6 would give a close empirical approximation. Constant specific heat would be less accurate, except for monatomic gases and for other gases below room temperature. Again, it should be remembered that all these results are part of the ideal-gas model, which may or may not be appropriate in any particular problem.

Example 6.4

Consider Example 3.13, in which oxygen is heated from 300 to 1500 K. Assume that during this process, the pressure dropped from 200 to 150 kPa. Calculate the change in entropy per kilogram.

Solution

The most accurate answer for the entropy change, assuming ideal-gas behavior, would be found from the ideal-gas tables, Table A.8. This result is, using Eq. 6.19,

$$s_2 - s_1 = (8.0649 - 6.4168) - 0.2598 \ln \left(\frac{150}{200}\right)$$

$$= 1.7228 \text{ kJ/kg·K}$$

The empirical equation from Table A.6 should give a good approximation to this result. Integrating Eq. 6.15, we have

$$s_2 - s_1 = \int_{T_1}^{T_2} C_{p0} \frac{dT}{T} - R \ln \frac{P_2}{P_1}$$

$$s_2 - s_1 = \left[0.88 \ln \theta - 0.0001\theta + \frac{0.54}{2}\theta^2 - \frac{0.33}{3}\theta^3 \right]_{\theta_1=0.3}^{\theta_2=1.5}$$

$$- 0.2598 \ln \left(\frac{150}{200}\right)$$

$$= 1.7058 \text{ kJ/kg·K}$$

which is within 1.0 % of the previous value. For constant specific heat, using the value at 300 K from Table A.5, we have

$$s_2 - s_1 = 0.922 \ln\left(\frac{1500}{300}\right) - 0.2598 \ln\left(\frac{150}{200}\right)$$

$$= 1.5586 \text{ kJ/kg·K}$$

which is too low by 9.5 %. If, however, we assume that the specific heat is constant at its value at 900 K, the average temperature, as in Example 3.13, is

$$s_2 - s_1 = 1.0767 \ln\left(\frac{1500}{300}\right) + 0.0747 = 1.8076 \text{ kJ/kg·K}$$

which is high by 4.9 %.

Example 6.5

Calculate the change in entropy per kilogram as air is heated from 300 to 600 K while pressure drops from 400 to 300 kPa. Assume:

1. Constant specific heat.
2. Variable specific heat.

Solution

1. From Table A.5 for air at 300 K,

$$C_{p0} = 1.004 \text{ kJ/kg·K}$$

Therefore, using Eq. 6.16, we have

$$s_2 - s_1 = 1.004 \ln\left(\frac{600}{300}\right) - 0.287 \ln\left(\frac{300}{400}\right) = 0.7785 \text{ kJ/kg·K}$$

2. From Table A.7,

$$s_{T1}^0 = 6.8693 \text{ kJ/kg·K},$$
$$s_{T2}^0 = 7.5764 \text{ kJ/kg·K}$$

Using Eq. 6.19 gives

$$s_2 - s_1 = 7.5764 - 6.8693 - 0.287 \ln\left(\frac{300}{400}\right) = 0.7897 \text{ kJ/kg·K}$$

Let us now consider the case of an ideal gas undergoing an isentropic process, a situation that is analyzed frequently. We conclude that Eq. 6.15 with the left side of the equation equal to zero then expresses the relation between the pressure and temperature at the initial and final states, with the specific relation depending on the nature of the specific heat as a function of T. As was discussed following Eq. 6.15, there are three possibilities to examine. Of these, the most accurate is the third, that is, the ideal-gas Tables A.7 or A.8 and Eq. 6.19, with the integrated temperature function s_T^0 defined by Eq. 6.18. The following example illustrates the procedure.

Example 6.6

One kilogram of air is contained in a cylinder fitted with a piston at a pressure of 400 kPa and a temperature of 600 K. The air is expanded to 150 kPa in a reversible adiabatic process. Calculate the work done by the air.

Control mass: Air.
Initial state: P_1, T_1; state 1 fixed.
Final state: P_2.
Process: Reversible and adiabatic.
Model: Ideal gas and air tables, Table A.7.

Analysis

From the energy equation we have

$$0 = u_2 - u_1 + w$$

The second law gives us

$$s_2 = s_1$$

Solution

From Table A.7,

$$u_1 = 435.10 \text{ kJ/kg}, \qquad s_{T1}^0 = 7.5764 \text{ kJ/kg·K}$$

From Eq. 6.19,

$$s_2 - s_1 = 0 = (s_{T2}^0 - s_{T1}^0) - R \ln \frac{P_2}{P_1}$$

$$= (s_{T2}^0 - 7.5764) - 0.287 \ln \left(\frac{150}{400} \right)$$

$$s_{T2}^0 = 7.2949 \text{ kJ/kg·K}$$

From Table A.7,

$$T_2 = 457 \text{ K}, \qquad u_2 = 328.14 \text{ kJ/kg}$$

Therefore,

$$w = 435.10 - 328.14 = 106.96 \text{ kJ/kg}$$

The first of the three possibilities, constant specific heat, is also worth analyzing as a special case. In this instance, the result is Eq. 6.16 with the left side equal to zero, or

$$s_2 - s_1 = 0 = C_{p0} \ln \frac{T_2}{T_1} - R \ln \frac{P_2}{P_1}$$

This expression can also be written as

$$\ln\left(\frac{T_2}{T_1}\right) = \frac{R}{C_{p0}}\ln\left(\frac{P_2}{P_1}\right)$$

or

$$\frac{T_2}{T_1} = \left(\frac{P_2}{P_1}\right)^{R/C_{p0}} \qquad (6.20)$$

However,

$$\frac{R}{C_{p0}} = \frac{C_{p0} - C_{v0}}{C_{p0}} = \frac{k-1}{k} \qquad (6.21)$$

where k, the ratio of the specific heats, is defined as

$$k = \frac{C_{p0}}{C_{v0}} \qquad (6.22)$$

Equation (6.20) is now conveniently written as

$$\frac{T_2}{T_1} = \left(\frac{P_2}{P_1}\right)^{(k-1)/k} \qquad (6.23)$$

From this expression and the ideal-gas equation of state, it also follows that

$$\frac{T_2}{T_1} = \left(\frac{v_1}{v_2}\right)^{k-1} \qquad (6.24)$$

and

$$\frac{P_2}{P_1} = \left(\frac{v_1}{v_2}\right)^{k} \qquad (6.25)$$

From this last expression, we note that for this process

$$Pv^k = \text{constant} \qquad (6.26)$$

This is a special case of a polytropic process in which the polytropic exponent n is equal to the specific heat ratio k.

6.8 THE REVERSIBLE POLYTROPIC PROCESS FOR AN IDEAL GAS

When a gas undergoes a reversible process in which there is heat transfer, the process frequently takes place in such a manner that a plot of log P versus log V is a straight line, as shown in Fig. 6.12. For such a process PV^n is a constant.

A process having this relation between pressure and volume is called a polytropic process. An example is the expansion of the combustion gases in the cylinder of a water-cooled reciprocating engine. If the pressure and volume are measured during the expansion stroke of a polytropic process, as might be done with an engine indicator, and the logarithms

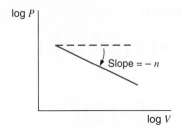

FIGURE 6.12
Example of a polytropic process.

of the pressure and volume are plotted, the result would be similar to the straight line in Fig. 6.12. From this figure it follows that

$$\frac{d \ln P}{d \ln V} = -n$$

$$d \ln P + n d \ln V = 0$$

If n is a constant (which implies a straight line on the log P versus log V plot), this equation can be integrated to give the following relation:

$$PV^n = \text{constant} = P_1 V_1^n = P_2 V_2^n \tag{6.27}$$

From this equation, the following relations can be written for a polytropic process:

$$\frac{P_2}{P_1} = \left(\frac{V_1}{V_2}\right)^n$$

$$\frac{T_2}{T_1} = \left(\frac{P_2}{P_1}\right)^{(n-1)/n} = \left(\frac{V_1}{V_2}\right)^{n-1} \tag{6.28}$$

For a control mass consisting of an ideal gas, the work done at the moving boundary during a reversible polytropic process can be derived (recall Eq. 3.21) from the relations

$$_1W_2 = \int_1^2 P\,dV \qquad \text{and} \qquad PV^n = \text{constant}$$

$$_1W_2 = \int_1^2 P\,dV = \text{constant} \int_1^2 \frac{dV}{V^n}$$

$$= \frac{P_2 V_2 - P_1 V_1}{1-n} = \frac{mR(T_2 - T_1)}{1-n} \tag{6.29}$$

for any value of n except $n = 1$.

The polytropic processes for various values of n are shown in Fig. 6.13 on P–v and T–s diagrams. The values of n for some familiar processes are

Isobaric process:	$n = 0,$	$P = \text{constant}$
Isothermal process:	$n = 1,$	$T = \text{constant}$
Isentropic process:	$n = k,$	$s = \text{constant}$
Isochoric process:	$n = \infty,$	$v = \text{constant}$

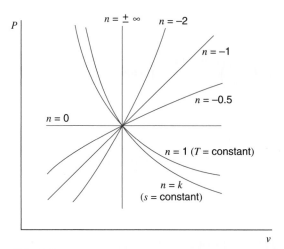
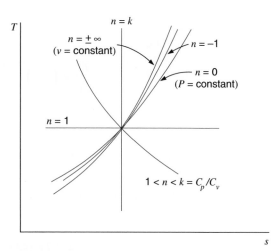

FIGURE 6.13 Polytropic process on P–v and T–s diagrams.

Example 6.7

In a reversible process, nitrogen is compressed in a cylinder from 100 kPa and 20 °C to 500 kPa. During this compression process, the relation between pressure and volume is $PV^{1.3} = $ constant. Calculate the work and heat transfer per kilogram, and show this process on P–v and T–s diagrams.

Control mass:	Nitrogen.
Initial state:	P_1, T_1; state 1 known.
Final state:	P_2.
Process:	Reversible, polytropic with exponent $n < k$.
Diagram:	Fig. 6.14.
Model:	Ideal gas, constant specific heat—value at 300 K.

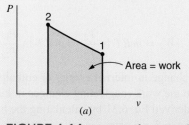
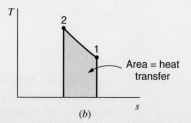

FIGURE 6.14 Diagram for Example 6.7.

Analysis

We need to find the boundary movement work. From Eq. 6.29, we have

$$_1W_2 = \int_1^2 P\,dV = \frac{P_2 V_2 - P_1 V_1}{1 - n} = \frac{mR(T_2 - T_1)}{1 - n}$$

The first law is

$$_1q_2 = u_2 - u_1 + _1w_2 = C_{v0}(T_2 - T_1) + _1w_2$$

Solution

From Eq. 6.28

$$\frac{T_2}{T_1} = \left(\frac{P_2}{P_1}\right)^{(n-1)/n} = \left(\frac{500}{100}\right)^{(1.3-1)/1.3} = 1.4498$$

$$T_2 = 293.2 \times 1.4498 = 425 \text{ K}$$

Then

$$_1w_2 = \frac{R(T_2 - T_1)}{1 - n} = \frac{0.2968(425 - 293.2)}{(1 - 1.3)} = -130.4 \text{ kJ/kg}$$

and from the energy equation,

$$_1q_2 = C_{v0}(T_2 - T_1) + {}_1w_2$$

$$= 0.745(425 - 293.2) - 130.4 = -32.2 \text{ kJ/kg}$$

The reversible isothermal process for an ideal gas is of particular interest. In this process

$$PV = \text{constant} = P_1V_1 = P_2V_2 \tag{6.30}$$

The work done at the boundary of a simple compressible mass during a reversible isothermal process can be found by integrating the equation

$$_1W_2 = \int_1^2 P \, dV$$

The integration is

$$_1W_2 = \int_1^2 P \, dV = \text{constant} \int_1^2 \frac{dV}{V} = P_1V_1 \ln \frac{V_2}{V_1} = P_1V_1 \ln \frac{P_1}{P_2} \tag{6.31}$$

or

$$_1W_2 = mRT \ln \frac{V_2}{V_1} = mRT \ln \frac{P_1}{P_2} \tag{6.32}$$

Because there is no change in internal energy or enthalpy in an isothermal process, the heat transfer is equal to the work (neglecting changes in kinetic and potential energy). Therefore, we could have derived Eq. 6.31 by calculating the heat transfer.

For example, using Eq. 6.7, we have

$$\int_1^2 T \, ds = {}_1q_2 = \int_1^2 du + \int_1^2 P \, dv$$

But $du = 0$ and $Pv = \text{constant} = P_1v_1 = P_2v_2$, such that

$$_1q_2 = \int_1^2 P \, dv = P_1v_1 \ln \frac{v_2}{v_1}$$

which yields the same result as Eq. 6.31.

In-Text Concept Questions

f. A liquid is compressed in a reversible adiabatic process. What is the change in T?

g. An ideal gas goes through a constant-T reversible heat addition process. How do the properties (v, u, h, s, P) change (up, down, or constant)?

h. Carbon dioxide is compressed to a smaller volume in a polytropic process with $n = 1.2$. How do the properties (u, h, s, P, T) change (up, down, or constant)?

6.9 ENTROPY CHANGE OF A CONTROL MASS DURING AN IRREVERSIBLE PROCESS

Consider a control mass that undergoes the cycles shown in Fig. 6.15. The cycle made up of the reversible processes A and B is a reversible cycle. Therefore, we can write

$$\oint \frac{\delta Q}{T} = \int_1^2 \left(\frac{\delta Q}{T}\right)_A + \int_2^1 \left(\frac{\delta Q}{T}\right)_B = 0$$

The cycle made up of the irreversible process C and the reversible process B is an irreversible cycle. Therefore, for this cycle the inequality of Clausius may be applied, giving the result

$$\oint \frac{\delta Q}{T} = \int_1^2 \left(\frac{\delta Q}{T}\right)_C + \int_2^1 \left(\frac{\delta Q}{T}\right)_B < 0$$

Subtracting the second equation from the first and rearranging, we have

$$\int_1^2 \left(\frac{\delta Q}{T}\right)_A > \int_1^2 \left(\frac{\delta Q}{T}\right)_C$$

Since path A is reversible, and since entropy is a property,

$$\int_1^2 \left(\frac{\delta Q}{T}\right)_A = \int_1^2 dS_A = \int_1^2 dS_C$$

Therefore,

$$\int_1^2 dS_C > \int_1^2 \left(\frac{\delta Q}{T}\right)_C$$

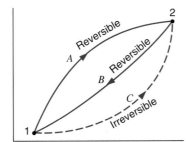

FIGURE 6.15
Entropy change of a control mass during an irreversible process.

As path C was arbitrary, the general result is

$$dS \geq \frac{\delta Q}{T}$$

$$S_2 - S_1 \geq \int_1^2 \frac{\delta Q}{T} \qquad (6.33)$$

In these equations, the equality holds for a reversible process and the inequality for an irreversible process.

This is one of the most important equations of thermodynamics. It is used to develop a number of concepts and definitions. In essence, this equation states the influence of irreversibility on the entropy of a control mass. Thus, if an amount of heat δQ is transferred to a control mass at temperature T in a reversible process, the change of entropy is given by the relation

$$dS = \left(\frac{\delta Q}{T}\right)_{\text{rev}}$$

If any irreversible effects occur while the amount of heat δQ is transferred to the control mass at temperature T, however, the change of entropy will be greater than for the reversible process. We would then write

$$dS > \left(\frac{\delta Q}{T}\right)_{\text{irr}}$$

Equation 6.33 holds when $\delta Q = 0$, when $\delta Q < 0$, and when $\delta Q > 0$. If δQ is negative, the entropy will tend to decrease as a result of the heat transfer. However, the influence of irreversibilities is still to increase the entropy of the mass, and from the absolute numerical perspective, we can still write for δQ

$$dS \geq \frac{\delta Q}{T}$$

6.10 ENTROPY GENERATION AND THE ENTROPY EQUATION

The conclusion from the previous considerations is that the entropy change in an irreversible process is larger than the change in a reversible process for the same δQ and T. This can be written out in a common form as an equality

$$dS = \frac{\delta Q}{T} + \delta S_{\text{gen}} \qquad (6.34)$$

provided that the last term is positive,

$$\delta S_{\text{gen}} \geq 0 \qquad (6.35)$$

The amount of entropy, δS_{gen}, is the entropy generation in the process due to irreversibilities occurring inside the system, a control mass for now but later extended to the more general control volume. This internal generation can be caused by the processes mentioned in Section 5.4, such as friction, unrestrained expansions, and the internal transfer of energy (redistribution) over a finite temperature difference. In addition to this internal entropy generation, external irreversibilities are possible by heat transfer over finite temperature differences as the δQ is transferred from a reservoir or by the mechanical transfer of work.

Equation 6.35 is then valid with the equal sign for a reversible process and the greater than sign for an irreversible process. Since the entropy generation is always positive and is the smallest in a reversible process, namely zero, we may deduce some limits for the heat transfer and work terms.

Consider a reversible process, for which the entropy generation is zero, and the heat transfer and work terms therefore are

$$\delta Q = T\,dS \qquad \text{and} \qquad \delta W = P\,dV$$

For an irreversible process with a nonzero entropy generation, the heat transfer from Eq. 6.34 becomes

$$\delta Q_{\text{irr}} = T\,dS - T\,\delta S_{\text{gen}}$$

and thus is smaller than that for the reversible case for the same change of state, dS. We also note that for the irreversible process, the work is no longer equal to $P\,dV$ but is smaller. Furthermore, since the first law is

$$\delta Q_{\text{irr}} = dU + \delta W_{\text{irr}}$$

and the property relation is valid,

$$T\,dS = dU + P\,dV$$

it is found that

$$\delta W_{\text{irr}} = P\,dV - T\,\delta S_{\text{gen}} \qquad (6.36)$$

showing that the work is reduced by an amount proportional to the entropy generation. For this reason the term $T\,\delta S_{\text{gen}}$ is often called *lost work*, although it is not a real work or energy quantity lost, but rather a lost opportunity to extract work.

Equation 6.34 can be integrated between the initial and final states to

$$S_2 - S_1 = \int_1^2 dS = \int_1^2 \frac{\delta Q}{T} + {}_1 S_{2\,\text{gen}} \qquad (6.37)$$

Thus, we have an expression for the change of entropy for an irreversible process as an equality, whereas in the previous section we had an inequality. In the limit of a reversible process, with a zero-entropy generation, the change in S expressed in Eq. 6.37 becomes identical to that expressed in Eq. 6.33 as the equal sign applies and the work term becomes $\int P\,dV$. Equation 6.37 is now the entropy balance equation for a control mass in the same form as the energy equation in Eq. 3.5, and it could include several subsystems. The equation can also be written in the general form

$$\Delta\text{ Entropy} = +\text{in} - \text{out} + \text{gen}$$

stating that we can generate but not destroy entropy. This is in contrast to energy, which we can neither generate nor destroy.

Some important conclusions can be drawn from Eqs. 6.34 to 6.37. First, there are two ways in which the entropy of a system can be increased—by transferring heat to it and by having an irreversible process. Since the entropy generation cannot be less than zero, there is only one way in which the entropy of a system can be decreased, and that is to transfer heat from the system. These changes are illustrated in a T–s diagram in Fig. 6.16 showing the halfplane into which the state moves due to a heat transfer or an entropy generation.

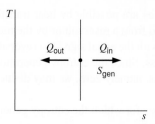

FIGURE 6.16
Change of entropy due
to heat transfer and
entropy generation.

Second, as we have already noted for an adiabatic process, $\delta Q = 0$, and therefore the increase in entropy is always associated with the irreversibilities.

Third, the presence of irreversibilities will cause the work to be smaller than the reversible work. This means less work out in an expansion process and more work into the control mass ($\delta W < 0$) in a compression process.

Finally, it should be emphasized that the change in s associated with the heat transfer is a transfer across the control surface, so a gain for the control volume is accompanied by a loss of the same magnitude outside the control volume. This is in contrast to the generation term that expresses all the entropy generated inside the control volume due to any irreversible process.

One other point concerning the representation of irreversible processes on P–v and T–s diagrams should be made. The work for an irreversible process is not equal to $\int P\,dV$, and the heat transfer is not equal to $\int T\,dS$. Therefore, the area underneath the path does not represent work and heat on the P–v and T–s diagrams, respectively. In fact, in many situations we are not certain of the exact state through which a system passes when it undergoes an irreversible process. For this reason, it is advantageous to show irreversible processes as dashed lines and reversible processes as solid lines. Thus, the area underneath the dashed line will never represent work or heat. Figure 6.17a shows an irreversible process, and, because the heat transfer and work for this process are zero, the area underneath the dashed line has no significance. Figure 6.17b shows the reversible process, and area 1–2–b–a–1 represents the work on the P–v diagram and the heat transfer on the T–s diagram.

In-Text Concept Questions

i. A substance has heat transfer out. Can you say anything about changes in s if the process is reversible? If it is irreversible?

j. A substance is compressed adiabatically, so P and T go up. Does that change s?

FIGURE 6.17
Reversible and
irreversible processes on
P–v and T–s diagrams.

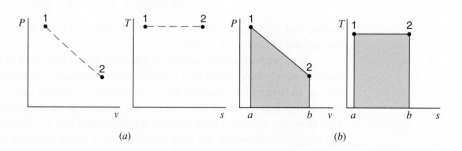

6.11 PRINCIPLE OF THE INCREASE OF ENTROPY

In the previous section, we considered irreversible processes in which the irreversibilities occurred inside the system or control mass. We also found that the entropy change of a control mass could be either positive or negative, since entropy can be increased by internal entropy generation and either increased or decreased by heat transfer, depending on the direction of that transfer. Now we would like to emphasize the difference between the energy and entropy equations and point out that energy is conserved but entropy is not.

Consider two mutually exclusive control volumes A and B with a common surface and their surroundings C such that they collectively include the whole world. Let some processes take place so that these control volumes exchange work and heat transfer as indicated in Fig. 6.18. Since a Q or W is transferred from one control volume to another, we only keep one symbol for each term and give the direction with the arrow. We will now write the energy and entropy equations for each control volume and then add them to see what the net effect is. As we write the equations, we do not try to memorize them, but just write them as

$$\text{Change} = +\text{in} - \text{out} + \text{generation}$$

and refer to the figure for the sign. We should know, however, that we cannot generate energy, but only entropy.

Energy:

$$(E_2 - E_1)_A = Q_a - W_a - Q_b + W_b$$

$$(E_2 - E_1)_B = Q_b - W_b - Q_c + W_c$$

$$(E_2 - E_1)_C = Q_c + W_a - Q_a - W_c$$

Entropy:

$$(S_2 - S_1)_A = \int \frac{\delta Q_a}{T_a} - \int \frac{\delta Q_b}{T_b} + S_{\text{gen } A}$$

$$(S_2 - S_1)_B = \int \frac{\delta Q_b}{T_b} - \int \frac{\delta Q_c}{T_c} + S_{\text{gen } B}$$

$$(S_2 - S_1)_C = \int \frac{\delta Q_c}{T_c} - \int \frac{\delta Q_a}{T_a} + S_{\text{gen } C}$$

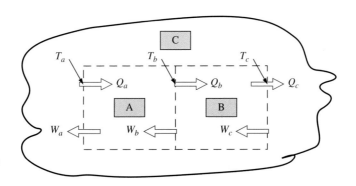

FIGURE 6.18 Total world divided into three control volumes.

Now we add all the energy equations to get the energy change for the total world:

$$(E_2 - E_1)_{\text{total}} = (E_2 - E_1)_A + (E_2 - E_1)_B + (E_2 - E_1)_C$$

$$= Q_a - W_a - Q_b + W_b + Q_b - W_b - Q_c + W_c + Q_c + W_a - Q_a - W_c$$

$$= 0 \tag{6.38}$$

and we see that total energy has not changed; that is, energy is conserved as all the right-hand-side transfer terms pairwise cancel out. The energy is not stored in the same form or place as it was before the process, but the total amount is the same. For entropy, we get something slightly different:

$$(S_2 - S_1)_{\text{total}} = (S_2 - S_1)_A + (S_2 - S_1)_B + (S_2 - S_1)_C$$

$$= \int \frac{\delta Q_a}{T_a} - \int \frac{\delta Q_b}{T_b} + S_{\text{gen }A} + \int \frac{\delta Q_b}{T_b} - \int \frac{\delta Q_c}{T_c} + S_{\text{gen }B}$$

$$+ \int \frac{\delta Q_c}{T_c} - \int \frac{\delta Q_a}{T_a} + S_{\text{gen }C}$$

$$= S_{\text{gen }A} + S_{\text{gen }B} + S_{\text{gen }C} \geq 0 \tag{6.39}$$

where all the transfer terms cancel, leaving only the positive entropy generation terms for each part of the total world. The total entropy increases and is then not conserved. Only if we have reversible processes in all parts of the world will the right-hand side become zero. This concept is referred to as the principle of the increase of entropy. Notice that if we add all the changes in entropy for the whole world from state 1 to state 2 we would get the total generation (increase), but we would not be able to specify where in the world the entropy was made. In order to get this more detailed information, we must make separate control volumes like A, B, and C and thus also evaluate all the necessary transfer terms so that we get the entropy generation by the balance of stored changes and transfers.

As an example of an irreversible process, consider a heat transfer process in which energy flows from a higher temperature domain to a lower temperature domain, as shown in Fig. 6.19. Let control volume A be a control mass at temperature T that receives a heat transfer of δQ from a surrounding control volume C at uniform temperature T_0. The transfer goes through the walls, control volume B, that separates domains A and C. Let us then analyze the incremental process from the point of view of control volume B, the walls,

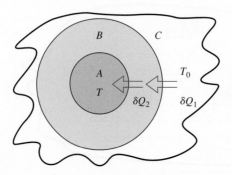

FIGURE 6.19 Heat transfer through a wall.

which do not have a change of state in time, but the state is nonuniform in space (it has T_0 on the outer side and T on the inner side).

Energy Eq.: $\quad dE = 0 = \delta Q_1 - \delta Q_2 \Rightarrow \delta Q_1 = \delta Q_2 = \delta Q$

Entropy Eq.: $\quad dS = 0 = \dfrac{\delta Q}{T_0} - \dfrac{\delta Q}{T} + \delta S_{\text{gen } B}$

So, from the energy equation, we find the two heat transfers to be the same, but realize that they take place at two different temperatures leading to an entropy generation as

$$\delta S_{\text{gen } B} = \frac{\delta Q}{T} - \frac{\delta Q}{T_0} = \delta Q\left(\frac{1}{T} - \frac{1}{T_0}\right) \geq 0 \qquad (6.40)$$

Since $T_0 > T$ for the heat transfer to move in the indicated direction, we see that the entropy generation is positive. Suppose the temperatures were reversed, so that $T_0 < T$. Then the parenthesis would be negative; to have a positive entropy generation, δQ must be negative, that is, move in the opposite direction. The direction of the heat transfer from a higher to a lower temperature domain is thus a logical consequence of the second law.

The principle of the increase of entropy (total entropy generation), Eq. 6.39, is illustrated by the following example.

Example 6.8

Suppose that 1 kg of saturated water vapor at 100 °C is condensed to a saturated liquid at 100 °C in a constant-pressure process by heat transfer to the surrounding air, which is at 25 °C. What is the net increase in entropy of the water plus surroundings?

Solution

For the control mass (water), from the steam tables, we obtain

$$\Delta S_{\text{c.m.}} = -ms_{fg} = -1 \times 6.0480 = -6.0480 \text{ kJ/K}$$

Concerning the surroundings, we have

$$Q_{\text{to surroundings}} = mh_{fg} = 1 \times 2257.0 = 2257 \text{ kJ}$$

$$\Delta S_{\text{surr}} = \frac{Q}{T_0} = \frac{2257}{298.15} = 7.5700 \text{ kJ/K}$$

$$\Delta S_{\text{gen total}} = \Delta S_{\text{c.m.}} + \Delta S_{\text{surr}} = -6.0480 + 7.5700 = 1.5220 \text{ kJ/K}$$

This increase in entropy is in accordance with the principle of the increase of entropy and tells us, as does our experience, that this process can take place.

It is interesting to note how this heat transfer from the water to the surroundings might have taken place reversibly. Suppose that an engine operating on the Carnot cycle received heat from the water and rejected heat to the surroundings, as shown in Fig. 6.20. The decrease in the entropy of the water is equal to the increase in the entropy of the surroundings.

$$\Delta S_{\text{c.m.}} = -6.0480 \text{ kJ/K}$$

$$\Delta S_{\text{surr}} = 6.0480 \text{ kJ/K}$$

$$Q_{\text{to surroundings}} = T_0 \Delta S = 298.15(6.0480) = 1803.2 \text{ kJ}$$

$$W = Q_H - Q_L = 2257 - 1803.2 = 453.8 \text{ kJ}$$

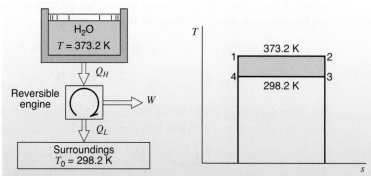

FIGURE 6.20 Reversible heat transfer with the surroundings.

Since this is a reversible cycle, the engine could be reversed and operated as a heat pump. For this cycle the work input to the heat pump would be 453.8 kJ.

 6.12 ENTROPY AS A RATE EQUATION

The second law of thermodynamics was used to write the balance of entropy in Eq. 6.34 for a variation and in Eq. 6.37 for a finite change. In some cases the equation is needed in a rate form so that a given process can be tracked in time. The rate form is also the basis for the development of the entropy balance equation in the general control volume analysis for an unsteady situation.

Take the incremental change in S from Eq. 6.34 and divide by δt. We get

$$\frac{dS}{\delta t} = \frac{1}{T}\frac{\delta Q}{\delta t} + \frac{\delta S_{\text{gen}}}{\delta t} \tag{6.41}$$

For a given control volume we may have more than one source of heat transfer, each at a certain surface temperature (semidistributed situation). Since we did not have to consider the temperature at which the heat transfer crossed the control surface for the energy equation, all the terms were written as a net heat transfer in a rate form in Eq. 3.3. Using this and a dot to indicate a rate, the final form for the entropy equation in the limit is

$$\frac{dS_{\text{c.m.}}}{dt} = \sum \frac{1}{T}\dot{Q} + \dot{S}_{\text{gen}} \tag{6.42}$$

expressing the rate of entropy change as due to the flux of entropy into the control mass from heat transfer and an increase due to irreversible processes inside the control mass. If only reversible processes take place inside the control volume, the rate of change of entropy is determined by the rate of heat transfer divided by the temperature terms alone.

Example 6.9

Consider an electric space heater that converts 1 kW of electric power into a heat flux of 1 kW delivered at 600 K from the hot wire surface. Let us look at the process of the energy conversion from electricity to heat transfer and find the rate of total entropy generation.

Control mass: The electric heater wire.

State: Constant wire temperature 600 K.

Analysis

The first and second laws of thermodynamics in rate form become

$$\frac{dE_{\text{c.m.}}}{dt} = \frac{dU_{\text{c.m.}}}{dt} = 0 = \dot{W}_{\text{el.in}} - \dot{Q}_{\text{out}}$$

$$\frac{dS_{\text{c.m.}}}{dt} = 0 = -\dot{Q}_{\text{out}}/T_{\text{surface}} + \dot{S}_{\text{gen}}$$

Notice that we neglected kinetic and potential energy changes in going from a rate of E to a rate of U. Then the left-hand side of the energy equation is zero since it is steady state and the right-hand side is electric work in minus heat transfer out. For the entropy equation the left-hand side is zero because of steady state and the right-hand side has a flux of entropy out due to heat transfer, and entropy is generated in the wire.

Solution

We now get the entropy generation as

$$\dot{S}_{\text{gen}} = \dot{Q}_{\text{out}}/T = 1\,\text{kW}/600\,\text{K} = 0.001\,67\,\text{kW/K}$$

Example 6.10

Consider a modern air conditioner using R-410A working in heat pump mode, as shown in Fig. 6.21. It has a COP of 4 with 10 kW of power input. The cold side is buried underground, where it is 8 °C, and the hot side is a house kept at 21 °C. For simplicity, assume that the cycle has a high temperature of 50 °C and a low temperature of −10 °C (recall Section 5.10). We would like to know where entropy is generated associated with the heat pump, assuming steady-state operation.

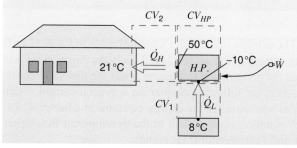

FIGURE 6.21 A heat pump for a house.

Let us look first at the heat pump itself, as in CV_{HP}, so from the COP

$$\dot{Q}_H = \beta_{HP} \times \dot{W} = 4 \times 10\,\text{kW} = 40\,\text{kW}$$

Energy Eq.: $\quad \dot{Q}_L = \dot{Q}_H - \dot{W} = 40\,\text{kW} - 10\,\text{kW} = 30\,\text{kW}$

Entropy Eq.: $\quad 0 = \dfrac{\dot{Q}_L}{T_{\text{low}}} - \dfrac{\dot{Q}_H}{T_{\text{high}}} + \dot{S}_{\text{gen}HP}$

$$\dot{S}_{\text{gen}HP} = \dfrac{\dot{Q}_H}{T_{\text{high}}} - \dfrac{\dot{Q}_L}{T_{\text{low}}} = \dfrac{40\,\text{kW}}{323\,\text{K}} - \dfrac{30\,\text{kW}}{263\,\text{K}} = 9.8\,\text{W/K}$$

Now consider CV_1 from the underground $8\,°\text{C}$ to the cycle $-10\,°\text{C}$.

Entropy Eq.: $\quad 0 = \dfrac{\dot{Q}_L}{T_L} - \dfrac{\dot{Q}_L}{T_{\text{low}}} + \dot{S}_{\text{gen}CV_1}$

$$\dot{S}_{\text{gen}CV_1} = \dfrac{\dot{Q}_L}{T_{\text{low}}} - \dfrac{\dot{Q}_L}{T_L} = \dfrac{30\,\text{kW}}{263\,\text{K}} - \dfrac{30\,\text{kW}}{281\,\text{K}} = 7.3\,\text{W/K}$$

And finally, consider CV_2 from the heat pump at $50\,°\text{C}$ to the house at $21\,°\text{C}$.

Entropy Eq.: $\quad 0 = \dfrac{\dot{Q}_H}{T_{\text{high}}} - \dfrac{\dot{Q}_H}{T_H} + \dot{S}_{\text{gen}CV_2}$

$$\dot{S}_{\text{gen}CV_2} = \dfrac{\dot{Q}_H}{T_H} - \dfrac{\dot{Q}_H}{T_{\text{high}}} = \dfrac{40\,\text{kW}}{294\,\text{K}} - \dfrac{40\,\text{kW}}{323\,\text{K}} = 12.2\,\text{W/K}$$

The total entropy generation rate becomes

$$\dot{S}_{\text{gen}TOT} = \dot{S}_{\text{gen}CV_1} + \dot{S}_{\text{gen}CV_2} + \dot{S}_{\text{gen}HP}$$

$$= \dfrac{\dot{Q}_L}{T_{\text{low}}} - \dfrac{\dot{Q}_L}{T_L} + \dfrac{\dot{Q}_H}{T_H} - \dfrac{\dot{Q}_H}{T_{\text{high}}} + \dfrac{\dot{Q}_H}{T_{\text{high}}} - \dfrac{\dot{Q}_L}{T_{\text{low}}}$$

$$= \dfrac{\dot{Q}_H}{T_H} - \dfrac{\dot{Q}_L}{T_L} = \dfrac{40\,\text{kW}}{294\,\text{K}} - \dfrac{30\,\text{kW}}{281\,\text{K}} = 29.3\,\text{W/K}$$

This last result is also obtained with a total control volume of the heat pump out to the $8\,°\text{C}$ and $21\,°\text{C}$ reservoirs that is the sum of the three control volumes shown. However, such an analysis would not be able to specify where the entropy is made; only the more detailed, smaller control volumes can provide this information.

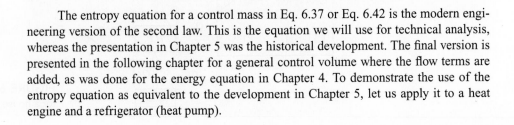

The entropy equation for a control mass in Eq. 6.37 or Eq. 6.42 is the modern engineering version of the second law. This is the equation we will use for technical analysis, whereas the presentation in Chapter 5 was the historical development. The final version is presented in the following chapter for a general control volume where the flow terms are added, as was done for the energy equation in Chapter 4. To demonstrate the use of the entropy equation as equivalent to the development in Chapter 5, let us apply it to a heat engine and a refrigerator (heat pump).

Consider an actual heat engine in a setup similar to Fig. 6.1 that operates in steady-state mode. The energy and entropy equations become

$$0 = \dot{Q}_H - \dot{Q}_L - \dot{W}_{HE} \tag{6.43}$$

$$0 = \frac{\dot{Q}_H}{T_H} - \frac{\dot{Q}_L}{T_L} + \dot{S}_{\text{gen}} \tag{6.44}$$

To express the work term as a fraction of the energy source $\dot{Q}_H$, we eliminate the heat transfer $\dot{Q}_L$ from the entropy equation as

$$\dot{Q}_L = \frac{T_L}{T_H} \dot{Q}_H + T_L \dot{S}_{\text{gen}}$$

Substitute this into the energy equation from which we get the work term as

$$\dot{W}_{HE} = \dot{Q}_H - \dot{Q}_L$$

$$= \dot{Q}_H - \frac{T_L}{T_H} \dot{Q}_H - T_L \dot{S}_{\text{gen}}$$

$$= \left(1 - \frac{T_L}{T_H}\right) \dot{Q}_H - T_L \dot{S}_{\text{gen}} \tag{6.45}$$

The result can be expressed as and related to the actual efficiency

$$\dot{W}_{HE} = \eta_{HE\,\text{carnot}} \dot{Q}_H - \text{loss}$$

$$= \eta_{HE\,\text{actual}} \dot{Q}_H \tag{6.46}$$

Before the result is discussed further, look at an actual refrigerator (heat pump) similar to Fig. 6.2 operating in steady-state mode. The energy and entropy equations become

$$0 = \dot{Q}_L - \dot{Q}_H + \dot{W}_{\text{ref.}} \tag{6.47}$$

$$0 = \frac{\dot{Q}_L}{T_L} - \frac{\dot{Q}_H}{T_H} + \dot{S}_{\text{gen}} \tag{6.48}$$

For the refrigerator, we want to express the heat transfer $\dot{Q}_L$ as a multiple of the work input, so use the entropy equation to solve for $\dot{Q}_H$ as

$$\dot{Q}_H = \frac{T_H}{T_L} \dot{Q}_L + T_H \dot{S}_{\text{gen}}$$

and substitute it into the energy equation

$$0 = \dot{Q}_L - \left[\frac{T_H}{T_L} \dot{Q}_L + T_H \dot{S}_{\text{gen}}\right] + \dot{W}_{\text{ref.}}$$

Now solve for $\dot{Q}_L$ to give

$$\dot{Q}_L = \frac{T_L}{T_H - T_L} \dot{W}_{\text{ref.}} - \frac{T_H T_L}{T_H - T_L} \dot{S}_{\text{gen}} \tag{6.49}$$

This result shows the Carnot COP and relates to the actual COP as

$$\dot{Q}_L = \beta_{\text{carnot}} \dot{W}_{\text{ref.}} - \text{loss}$$

$$= \beta_{\text{actual}} \dot{W}_{\text{ref.}} \tag{6.50}$$

From the results of the analysis of the heat engine and the refrigerator, we can conclude the following:

1. We get maximum benefit for a reversible process, $\dot{S}_{\text{gen}} = 0$, as Eq. 6.45 gives maximum $\dot{W}_{HE}$ for a given $\dot{Q}_H$ input, and for the refrigerator Eq. 6.49 gives maximum $\dot{Q}_L$ for a given $\dot{W}_{\text{ref.}}$ input.
2. For a reversible device, the analysis predicted the Carnot heat engine efficiency and the Carnot refrigerator COP.
3. For an actual device, the analysis shows that the decrease in performance (lower $\dot{W}_{HE}$ and $\dot{Q}_L$) is directly proportional to the entropy generation.

To predict the actual performance, the details of the processes must be known so the entropy generation can be found. This is very difficult to accomplish for any device so the manufacturers normally measure the performance over a range of operating conditions and then quote typically numbers for the efficiency or COP.

The application of the energy and entropy equations thus showed all the results that were presented in the historical development of the second law in Chapter 5, and this is the method that we will use in an engineering analysis of systems and devices.

6.13 SOME GENERAL COMMENTS ABOUT ENTROPY AND CHAOS

It is quite possible at this point that a student may have a good grasp of the material that has been covered and yet may have only a vague understanding of the significance of entropy. In fact, the question "What is entropy?" is frequently raised by students, with the implication that no one really knows! This section has been included in an attempt to give insight into the qualitative and philosophical aspects of the concept of entropy and to illustrate the broad application of entropy to many different disciplines.

First, we recall that the concept of energy arises from the first law of thermodynamics and the concept of entropy from the second law of thermodynamics. Actually, it is just as difficult to answer the question "What is energy?" as it is to answer the question "What is entropy?" However, since we regularly use the term *energy* and are able to relate this term to phenomena that we observe every day, the word *energy* has a definite meaning to us and thus serves as an effective vehicle for thought and communication. The word *entropy* could serve in the same capacity. If, when we observed a highly irreversible process (such as cooling coffee by placing an ice cube in it), we said, "That surely increases the entropy," we would soon be as familiar with the word *entropy* as we are with the word *energy*. In many cases, when we speak about higher efficiency, we are actually speaking about accomplishing a given objective with a smaller total increase in entropy.

A second point to be made regarding entropy is that in statistical thermodynamics, the property entropy is defined in terms of probability. Although this topic will not be examined in detail in this book, a few brief remarks regarding entropy and probability may prove helpful. From this point of view, the net increase in entropy that occurs during an irreversible process can be associated with a change of state from a less probable state to a more probable state. For instance, to use a previous example, one is more likely to find gas on both sides of the ruptured membrane in Fig. 5.15 than to find a gas on one side and a vacuum on the other. Thus, when the membrane ruptures, the direction of the process is from a less probable state to a more probable state, and associated with this process is an increase in entropy. Similarly, the more probable state is that a cup of coffee will be at the same temperature as its surroundings than at a higher (or lower) temperature. Therefore, as the coffee cools as the result of a transfer of heat to the surroundings, there is a change from a less probable to a more probable state, and associated with this is an increase in entropy.

To tie entropy a little closer to physics and to the level of disorder or chaos, let us consider a very simple system. Properties like U and S for a substance at a given state are averaged over many particles on the molecular level, so they (atoms and molecules) do not all exist in the same detailed quantum state. There are a number of different configurations possible for a given state that constitutes an uncertainty or chaos in the system. The number of possible configurations, w, is called the *thermodynamic probability*, and each of these is equally possible; this is used to define the entropy as

$$S = k \ln w \qquad (6.51)$$

where k is the Boltzmann constant, and it is from this definition that S is connected to the uncertainty or chaos. The larger the number of possible configurations is, the larger S is. For a given system, we would have to evaluate all the possible quantum states for kinetic energy, rotational energy, vibrational energy, and so forth, to find the equilibrium distribution and w. Without going into those details, which is the subject of statistical thermodynamics, a very simple example is used to illustrate the principle (Fig. 6.22).

Assume we have four identical objects that can only possess one kind of energy, namely, potential energy associated with elevation (the floor) in a tall building. Let the four objects have a combined 2 units of energy (floor height times mass times gravitation). How can this system be configured? We can have one object on the second floor and the remaining three on the ground floor, giving a total of 2 energy units. We could also have two objects on the first floor and two on the ground floor, again with a total of 2 energy units. These two configurations are equally possible, and we could therefore see the system 50 % of the time in one configuration and 50 % of the time in the other; we have some positive value of S.

Now let us add 2 energy units by heat transfer; that is done by giving the objects some energy that they share. Now the total energy is 4 units, and we can see the system in the following configurations (a–e):

Floor number:		0	1	2	3	4
Number of objects	a:	3				1
Number of objects	b:	2	1		1	
Number of objects	c:	2		2		
Number of objects	d:	1	2	1		
Number of objects	e:		4			

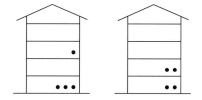

FIGURE 6.22

Illustration of energy distribution.

Now we have five different configurations ($w = 5$)—each equally possible—so we will observe the system 20 % of the time in each one, and we now have a larger value of S.

On the other hand, if we increase the energy by 2 units through work, it acts differently. Work is associated with the motion of a boundary, so now we pull in the building to make it higher and stretch it to be twice as tall; that is, the first floor has 2 energy units per object, and so forth, as compared with the original state. This means that we simply double the energy per object in the original configuration without altering the number of configurations, which stay at $w = 2$. In effect, S has not changed.

Floor number:		0	1	2	3	4
Number of objects	f:	3		1		
Number of objects	g:	2	2			

This example illustrates the profound difference between adding energy as a heat transfer changing S versus adding energy through a work term leaving S unchanged. In the first situation, we move a number of particles from lower energy levels to higher energy levels, thus changing the distribution and increasing the chaos. In the second situation, we do not move the particles between energy states, but we change the energy level of a given state, thus preserving the order and chaos.

SUMMARY

The inequality of Clausius and the property entropy (s) are modern statements of the second law. The final statement of the second law is the entropy balance equation that includes generation of entropy. All the results that were derived from the classical formulation of the second law in Chapter 5 can be rederived with the energy and entropy equations applied to the cyclic devices. For all reversible processes, entropy generation is zero and all real (irreversible) processes have positive entropy generation. How large the entropy generation is depends on the actual process.

Thermodynamic property relations for s are derived from consideration of a reversible process and lead to Gibbs relations. Changes in the property s are covered through general tables, approximations for liquids and solids, as well as ideal gases. Changes of entropy in various processes are examined in general together with special cases of polytropic processes. Just as reversible specific boundary work is the area below the process curve in a P–v diagram, the reversible heat transfer is the area below the process curve in a T–s diagram.

You should have learned a number of skills and acquired abilities from studying this chapter that will allow you to

- Know that Clausius inequality is an alternative statement of the second law.
- Know the relation between entropy and reversible heat transfer.
- Locate states in the tables involving entropy.
- Understand how a Carnot cycle looks in a T–s diagram.
- Know how different simple process curves look in a T–s diagram.
- Understand how to apply the entropy balance equation for a control mass.
- Recognize processes that generate entropy and where the entropy is made.
- Evaluate changes in s for liquids, solids, and ideal gases.
- Know the various property relations for a polytropic process in an ideal gas.
- Know the application of the unsteady entropy equation and what a flux of s is.

KEY CONCEPTS AND FORMULAS

Clausius inequality
$$\oint \frac{dQ}{T} \leq 0$$

Entropy
$$ds = \frac{dq}{T} + ds_{\text{gen}}; \qquad ds_{\text{gen}} \geq 0$$

Rate equation for entropy
$$\dot{S}_{\text{c.m.}} = \sum \frac{\dot{Q}_{\text{c.m.}}}{T} + \dot{S}_{\text{gen}}$$

Entropy equation
$$m(s_2 - s_1) = \int_1^2 \frac{\delta Q}{T} + {}_1 S_{2\,\text{gen}}; \qquad {}_1 S_{2\,\text{gen}} \geq 0$$

Total entropy change
$$\Delta S_{\text{net}} = \Delta S_{\text{cm}} + \Delta S_{\text{surr}} = S_{\text{gen}} \geq 0$$

Lost work
$$W_{\text{lost}} = \int T\, dS_{\text{gen}}$$

Actual boundary work
$$_1 W_2 = \int P\, dV - W_{\text{lost}}$$

Gibbs relations
$$T\, ds = du + P\, dv$$
$$T\, ds = dh - v\, dP$$

Solids, Liquids

$$v = \text{constant}, \qquad dv = 0$$

Change in s
$$s_2 - s_1 = \int \frac{du}{T} = \int C \frac{dT}{T} \approx C \ln \frac{T_2}{T_1}$$

Ideal Gas

Standard entropy
$$s_T^0 = \int_{T_0}^T \frac{C_{p0}}{T}\, dT \qquad \text{(Function of } T)$$

Change in s
$$s_2 - s_1 = s_{T2}^0 - s_{T1}^0 - R \ln \frac{P_2}{P_1} \quad \text{(Using Table A.7 or A.8)}$$

$$s_2 - s_1 = C_{p0} \ln \frac{T_2}{T_1} - R \ln \frac{P_2}{P_1} \quad \text{(For constant } C_p,\ C_v)$$

$$s_2 - s_1 = C_{v0} \ln \frac{T_2}{T_1} + R \ln \frac{v_2}{v_1} \quad \text{(For constant } C_p,\ C_v)$$

Ratio of specific heats
$$k = C_{p0}/C_{v0}$$

Polytropic processes
$$Pv^n = \text{constant}; \qquad PV^n = \text{constant}$$

$$\frac{P_2}{P_1} = \left(\frac{V_1}{V_2}\right)^n = \left(\frac{v_1}{v_2}\right)^n = \left(\frac{T_2}{T_1}\right)^{\frac{n}{n-1}}$$

$$\frac{T_2}{T_1} = \left(\frac{v_1}{v_2}\right)^{n-1} = \left(\frac{P_2}{P_1}\right)^{\frac{n-1}{n}}$$

$$\frac{v_2}{v_1} = \left(\frac{P_1}{P_2}\right)^{\frac{1}{n}} = \left(\frac{T_1}{T_2}\right)^{\frac{1}{n-1}}$$

Specific work
$$_1 w_2 = \frac{1}{1-n}(P_2 v_2 - P_1 v_1) = \frac{R}{1-n}(T_2 - T_1) \qquad n \neq 1$$

$$_1 w_2 = P_1 v_1 \ln \frac{v_2}{v_1} = RT_1 \ln \frac{v_2}{v_1} = RT_1 \ln \frac{P_1}{P_2} \qquad n = 1$$

The work is moving boundary work $w = \int P\, dv$

Identifiable processes	$n = 0$;	$P = $ constant;	Isobaric
	$n = 1$;	$T = $ constant;	Isothermal
	$n = k$;	$s = $ constant;	Isentropic
	$n = \pm\infty$;	$v = $ constant;	Isochoric or isometric

CONCEPT-STUDY GUIDE PROBLEMS

6.1 When a substance has completed a cycle, v, u, h, and s are unchanged. Did anything happen? Explain.

6.2 Assume a heat engine with a given Q_H. Can you say anything about Q_L if the engine is reversible? If it is irreversible?

6.3 $CV\ A$ is the mass inside a piston/cylinder; $CV\ B$ is that plus part of the wall out to a source of $_1Q_2$ at T_s. Write the entropy equation for the two control volumes, assuming no change of state of the piston mass or walls.

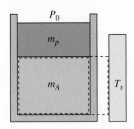

FIGURE P6.3

6.4 Consider the previous setup with the mass m_A and the piston/cylinder of mass m_p starting out at two different temperatures (Fig. P6.3). After a while, the temperature becomes uniform without any external heat transfer. Write the entropy equation storage term $(S_2 - S_1)$ for the total mass.

6.5 Water at $100\,°C$, quality $50\,\%$ in a rigid box is heated to $110\,°C$. How do the properties $(P, v, x, u,$ and $s)$ change (increase, stay about the same, or decrease)?

6.6 Liquid water at $20\,°C$, $100\,kPa$ is compressed in a piston/cylinder without any heat transfer to a pressure of $200\,kPa$. How do the properties $(T, v, u,$ and $s)$ change (increase, stay about the same, or decrease)?

6.7 A reversible process in a piston/cylinder is shown in Fig. P6.7. Indicate the storage change $u_2 - u_1$ and transfers $_1w_2$ and $_1q_2$ as positive, zero, or negative.

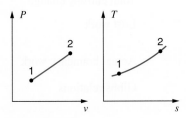

FIGURE P6.7

6.8 A reversible process in a piston/cylinder is shown in Fig. P6.8. Indicate the storage change $u_2 - u_1$ and transfers $_1w_2$ and $_1q_2$ as positive, zero, or negative.

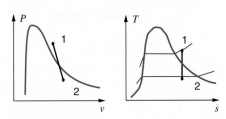

FIGURE P6.8

6.9 Air at $290\,K$, $100\,kPa$ in a rigid box is heated to $325\,K$. How do the properties $(P, v, u,$ and $s)$ change (increase, stay about the same, or decrease)?

6.10 Air at $20\,°C$, $100\,kPa$ is compressed in a piston/cylinder without any heat transfer to a pressure of $200\,kPa$. How do the properties $(T, v, u,$ and $s)$ change (increase, stay about the same, or decrease)?

6.11 Carbon dioxide is compressed to a smaller volume in a polytropic process with $n = 1.4$. How do the properties (u, h, s, P, T) change (up, down, or constant)?

6.12 Process A: Air at $300\,K$, $100\,kPa$ is heated to $310\,K$ at constant pressure. Process B: Air at $1300\,K$ is heated to $1310\,K$ at constant $100\,kPa$. Use the table below to compare the property changes.

Property	$\Delta_A > \Delta_B$	$\Delta_A \approx \Delta_B$	$\Delta_A < \Delta_B$
$\Delta = v_2 - v_1$			
$\Delta = h_2 - h_1$			
$\Delta = s_2 - s_1$			

6.13 Why do we write ΔS or $S_2 - S_1$, whereas we write $\int dQ/T$ and $_1S_{2gen}$?

6.14 A reversible heat pump has a flux of s entering as $\dot{Q}_L/T_L$. What can you say about the exit flux of s at T_H?

6.15 An electric baseboard heater receives 1500 W of electrical power that heats room air, which loses the same amount through the walls and windows. Specify exactly where entropy is generated in that process.

6.16 A 500 W electric space heater with a small fan inside heats air by blowing it over a hot electrical wire. For each control volume, (a) wire at T_{wire} only, (b) all the room air at T_{room}, and (c) total room plus the heater, specify the storage, entropy transfer terms, and entropy generation as rates (neglect any $\dot{Q}$ through the room walls or windows).

HOMEWORK PROBLEMS

Inequality of Clausius

6.17 Consider the steam power plant in Example 4.7 and assume an average T in the line between 1 and 2. Show that this cycle satisfies the inequality of Clausius.

6.18 A heat engine receives 6 kW from a 250 °C source and rejects heat at 30 °C. Examine each of three cases with respect to the inequality of Clausius.
   a. $\dot{W} = 6\,kW$
   b. $\dot{W} = 0\,kW$
   c. Carnot cycle

6.19 Use the inequality of Clausius to show that heat transfer from a warm space toward a colder space without work is a possible process; i.e., a heat engine with no work output.

6.20 Use the inequality of Clausius to show that heat transfer from a cold space toward a warmer space without work is an impossible process; i.e., a heat pump with no work input.

6.21 Let the steam power plant in Problem 5.26 have a temperature of 700 °C in the boiler and 40 °C during the heat rejection in the condenser. Does that satisfy the inequality of Clausius? Repeat the question for the cycle operated in reverse as a refrigerator.

Entropy of a Pure Substance

6.22 Determine the entropy for these states.
   a. Nitrogen, $P = 2000$ kPa, 120 K
   b. Nitrogen, 120 K, $v = 0.0050$ m³/kg
   c. R-410A, $T = 25$ °C, $v = 0.01$ m³/kg

6.23 Determine the missing property among P, T, s, and x for R-410A at
   a. $T = -20$ °C, $v = 0.1377$ m³/kg

   b. $T = 20$ °C, $v = 0.013\,77$ m³/kg
   c. $P = 400$ kPa, $s = 1.2108$ kJ/kg·K

6.24 Find the missing properties of P, v, s, and x for ammonia (NH_3) at
   a. $T = 65$ °C, $P = 600$ kPa
   b. $T = 20$ °C, $u = 800$ kJ/kg
   c. $T = 50$ °C, $v = 0.1185$ m³/kg

6.25 Find the entropy for the following water states and indicate each state on a T–s diagram relative to the two-phase region.
   a. 200 °C, $v = 0.02$ m³/kg
   b. 200 °C, 2000 kPa
   c. -2 °C, 100 kPa

6.26 Find the missing properties of P, v, s, and x for CO_2 and indicate each state on a T–s diagram relative to the two-phase region.
   a. -20 °C, 2000 kPa
   b. 20 °C, $s = 1.49$ kJ/kg·K
   c. -10 °C, $s = 1$ kJ/kg·K

6.27 Two kilograms of water at 120 °C with a quality of 25 % has its temperature raised 15 °C in a constant-volume process. What are the new quality and specific entropy?

6.28 Two kilograms of water at 400 kPa with a quality of 25 % has its temperature raised 20 °C in a constant-pressure process. What is the change in entropy?

6.29 Saturated liquid water at 20 °C is compressed to a higher pressure with constant temperature. Find the changes in u and s when the final pressure is
   a. 500 kPa
   b. 2000 kPa
   c. 20 000 kPa

6.30 Saturated vapor water at 250 °C is expanded to a lower pressure with constant temperature. Find the changes in u and s when the final pressure is
a. 100 kPa
b. 50 kPa
c. 10 kPa

Reversible Processes

6.31 In a Carnot engine with ammonia as the working fluid, the high temperature is 60 °C and as Q_H is received, the ammonia changes from saturated liquid to saturated vapor. The ammonia pressure at the low temperature is 190 kPa. Find T_L, the cycle thermal efficiency, the heat added per kilogram, and the entropy, s, at the beginning of the heat rejection process.

6.32 Consider a Carnot-cycle heat pump with R-410A as the working fluid. Heat is rejected from the R-410A at 35 °C, during which process the R-410A changes from saturated vapor to saturated liquid. The heat is transferred to the R-410A at 0 °C.
a. Show the cycle on a T–s diagram.
b. Find the quality of the R-410A at the beginning and end of the isothermal heat addition process at 0 °C.
c. Determine the COP for the cycle.

6.33 Do Problem 6.32 using refrigerant R-134a instead of R-410A.

6.34 Water is used as the working fluid in a Carnot-cycle heat engine, where it changes from saturated liquid to saturated vapor at 200 °C as heat is added. Heat is rejected in a constant-pressure process (also constant T) at 20 kPa. The heat engine powers a Carnot-cycle refrigerator that operates between −15 °C and +20 °C, shown in Fig. P6.34. Find the heat added to the water per kilogram of water. How much heat should be added to the water in the heat engine so that the refrigerator can remove 1 kJ from the cold space?

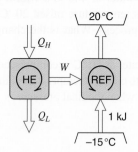

FIGURE P6.34

6.35 R-410A at 800 kPa and 160 °C is expanded in a piston/cylinder to 600 kPa, 40 °C in a reversible process. Find the sign for both the work and the heat transfer for this process.

6.36 A piston/cylinder compressor takes R-410A as saturated vapor 500 kPa and compresses it in a reversible adiabatic process to 3000 kPa. Find the final temperature and the specific compression work.

6.37 A piston/cylinder receives R-410A at 500 kPa and compresses it in a reversible adiabatic process to 1800 kPa, 60 °C. Find the initial temperature.

6.38 Compression and heat transfer bring carbon dioxide in a piston/cylinder from 1400 kPa, 20 °C to saturated vapor in an isothermal process. Find the specific heat transfer and the specific work.

6.39 A piston/cylinder maintaining constant pressure contains 0.1 kg saturated liquid water at 125 °C. It is now boiled to become saturated vapor in a reversible process. Find the work term and then the heat transfer from the energy equation. Find the heat transfer from the entropy equation; is it the same?

6.40 A piston/cylinder contains 0.5 kg of water at 200 kPa, 300 °C, and it now cools to 150 °C in an isobaric process. The heat goes into a heat engine that rejects heat to the ambient at 25 °C (shown in Fig. P6.40), and the whole process is assumed to be reversible. Find the heat transfer out of the water and the work given out by the heat engine.

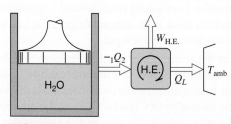

FIGURE P6.40

6.41 A cylinder fitted with a piston contains ammonia at 50 °C, 20 % quality with a volume of 1 L. The ammonia expands slowly, and during this process heat is transferred to maintain a constant temperature. The process continues until all the liquid is gone. Determine the work and heat transfer for this process.

6.42 Water in a piston/cylinder at 400 °C, 2000 kPa is expanded in a reversible adiabatic process. The specific work is measured to be 415.72 kJ/kg out. Find the final P and T and show the P–v and T–s diagrams for the process.

6.43 One kilogram of water at 300 °C expands against a piston in a cylinder until it reaches ambient pressure, 100 kPa, at which point the water has a quality of 90.2 %. It may be assumed that the expansion is reversible and adiabatic. What was the initial pressure in the cylinder, and how much work is done by the water?

6.44 Water at 800 kPa, 250 °C is brought to saturated vapor in a rigid container, shown in Fig. P6.44. Find the final T and the specific heat transfer in this isometric process.

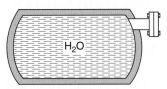

FIGURE P6.44

6.45 Estimate the specific heat transfer from the area in a T–s diagram and compare it to the correct value for the states and process in Problem 6.44.

6.46 A piston/cylinder has 2 kg of R-410A at 60 °C, 100 kPa that is compressed to 1000 kPa. The process happens so slowly that the temperature is constant. Find the heat transfer and work for the process, assuming it to be reversible.

6.47 A heavily insulated cylinder/piston contains ammonia at 1200 kPa, 60 °C. The piston is moved, expanding the ammonia in a reversible process until the temperature is −20 °C. During the process, 200 kJ of work is given out by the ammonia. What was the initial volume of the cylinder?

6.48 Water at 800 kPa, 250 °C is brought to saturated vapor in a piston/cylinder with an isothermal process. Find the specific work and heat transfer. Estimate the specific work from the area in the P–v diagram and compare it to the correct value.

6.49 A rigid, insulated vessel contains superheated vapor steam at 3 MPa, 400 °C. A valve on the vessel is opened, allowing steam to escape, as shown in Fig. P6.49. The overall process is irreversible, but the steam remaining inside the vessel goes through a reversible adiabatic expansion. Determine the fraction of steam that has escaped when the final state inside is saturated vapor.

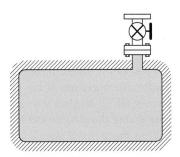

FIGURE P6.49

6.50 Water at 800 kPa, 200 °C is brought to saturated vapor in a piston/cylinder with an isobaric process. Find the specific work and heat transfer. Estimate the specific heat transfer from the area in a T–s diagram and compare it to the correct value.

Entropy of a Liquid or a Solid

6.51 Two 5 kg blocks of steel, one at 250 °C and the other at 25 °C, come in thermal contact. Find the final temperature and the change in entropy of the steel.

6.52 A rigid tank of 1.2 kg steel contains 1.5 kg of R-134a at 40 °C, 300 kPa. The tank is placed in a refrigerator that brings it to −20 °C. Find the process heat transfer and the combined steel and R-134a change in entropy.

6.53 A large slab of concrete, 5 m × 8 m × 0.3 m, is used as a thermal storage mass in a solar-heated house. If the slab cools overnight from 23 °C to 18 °C in an 18 °C house, what is the net entropy change associated with this process?

6.54 A foundry form box with 25 kg of 200 °C hot sand is dumped into a bucket with 50 L water at 15 °C. Assuming no heat transfer with the surroundings and no boiling away of liquid water, calculate the net entropy change for the mass.

6.55 Heat transfer to a block of 1.5 kg ice at −10 °C melts it to liquid at 20 °C in a kitchen. Find the entropy change of the water.

6.56 A piston/cylinder has constant pressure of 2000 kPa with water at 20 °C. It is now heated to 100 °C. Find the heat transfer and the entropy change using the steam tables. Repeat the calculation using constant specific heat and incompressibility.

6.57 A 4 L jug of milk at 25 °C is placed in your refrigerator, where it is cooled down to the refrigerator's inside constant temperature of 5 °C. Assume the milk has the property of liquid water and find the entropy change of the milk.

6.58 A 10 kg steel container is cured at 500 °C. Liquid water at 15 °C, 100 kPa is added to the container, so the final uniform temperature of the steel and the water becomes 60 °C. Neglect any water that might evaporate during the process and any air in the container. How much water should be added, and how much was the entropy changed?

6.59 A pan in an auto shop contains 5 L of engine oil at 20 °C, 100 kPa. Now 3 L of oil at 100 °C is mixed into the pan. Neglect any work term and find the final temperature and the entropy change.

6.60 A 5 kg aluminum radiator holds 2 kg of liquid R-134a at −10 °C. The setup is brought indoors and heated with 220 kJ. Find the final temperature and the change in entropy of the mass.

6.61 Find the total work the heat engine can give out as it receives energy from the rock bed as, described in Problem 5.51 (see Fig. P6.61). Hint: write the entropy balance equation for the control volume that is the combination of the rock bed and the heat engine.

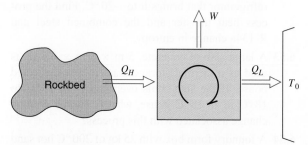

FIGURE P6.61

6.62 Consider Problem 6.51 if the two blocks of steel exchange energy through a heat engine similar to the setup in Problem 6.61. Find the work output of the heat engine.

6.63 Two kilograms of liquid lead initially at 400 °C is poured into a form. It then cools at constant pressure down to room temperature of 20 °C as heat is transferred to the room. The melting point of lead is

327 °C and the enthalpy change between the phases, h_{if}, is 24.6 kJ/kg. The specific heats are found in Tables A.3 and A.4. Calculate the net entropy change for the mass.

Entropy of Ideal Gases

6.64 Air inside a rigid tank is heated from 300 K to 350 K. Find the entropy increase $s_2 − s_1$. What is the entropy increase if the tank is heated from 1300 K to 1350 K?

6.65 A rigid tank contains 1 kg methane at 600 K, 1500 kPa. It is now cooled down to 300 K. Find the heat transfer and the change in entropy using ideal gas.

6.66 A piston/cylinder setup contains air at 100 kPa, 400 K that is compressed to a final pressure of 1000 kPa. Consider two different processes: (a) a reversible adiabatic process and (b) a reversible isothermal process. Show both processes in P–v and a T–s diagram. Find the final temperature and the specific work for both processes.

6.67 Prove that the two relations for changes in s, Eqs. 6.16 and 6.17, are equivalent once we assume constant specific heat. Hint: recall the relation for specific heat in Eq. 3.42.

6.68 A closed rigid container is filled with 1.5 kg water at 100 kPa, 55 °C, 1 kg of stainless steel and 0.5 kg of polyvinyl chloride, both at 20 °C and 0.1 kg of hot air at 400 K, 100 kPa. It is now left alone, with no external heat transfer and no water vaporizes. Find the final temperature and the change in entropy of the masses.

6.69 Water at 150 °C, 400 kPa is brought to 1200 °C in a constant-pressure process. Find the change in specific entropy using (a) the steam tables, (b) the ideal gas water Table A.8, and (c) the specific heat from Table A.5.

6.70 R-410A at 400 kPa is brought from 20 °C to 120 °C in a constant-pressure process. Evaluate the change in specific entropy using Table B.4 and using ideal gas with $C_p = 0.81$ kJ/kg·K.

6.71 R-410A at 300 kPa, 20 °C is brought to 220 °C in a constant-volume process. Evaluate the change in specific entropy using Table B.4 and using ideal gas with $C_v = 0.695$ kJ/kg·K.

6.72 Consider a small air pistol (Fig. P6.72) with a cylinder volume of 1 cm³ at 250 kPa, 27 °C. The bullet acts as a piston initially held by a trigger. The bullet is released, so the air expands in an adiabatic process. If the pressure should be 120 kPa as the bullet leaves the cylinder, find the final volume and the work done by the air.

FIGURE P6.72

6.73 Oxygen gas in a piston/cylinder at 300 K, 100 kPa with a volume of 0.1 m³ is compressed in a reversible adiabatic process to a final temperature of 700 K. Find the final pressure and volume using Table A.5 and repeat the process with Table A.8.

6.74 An insulated piston/cylinder setup contains carbon dioxide gas at 800 kPa, 300 K that is then compressed to 6 MPa in a reversible adiabatic process. Calculate the final temperature and the specific work using (a) ideal gas Tables A.8 and (b) using constant specific heats in Table A.5.

6.75 Extend the previous problem to solve using Table B.3.

6.76 A handheld pump for a bicycle (Fig. P6.76) has a volume of 25 cm³ when fully extended. You now press the plunger (piston) in while holding your thumb over the exit hole so that an air pressure of 300 kPa is obtained. The outside atmosphere is at P_0, T_0. Consider two cases: (1) it is done quickly (~1 s), and (2) it is done very slowly (~1 h).
   a. State the assumptions about the process for each case.
   b. Find the final volume and temperature for both cases.

FIGURE P6.76

6.77 Argon in a light bulb is at 90 kPa, 20 °C when it is turned on and electricity input now heats it to 75 °C. Find the entropy increase of the argon gas.

6.78 We wish to obtain a supply of cold helium gas by applying the following technique. Helium contained in a cylinder under ambient conditions, 100 kPa, 20 °C, is compressed in a reversible isothermal process to 600 kPa, after which the gas is expanded back to 100 kPa in a reversible adiabatic process.
   a. Show the process on a T–s diagram.
   b. Calculate the final temperature and the net work per kilogram of helium.

6.79 A 1 m³ insulated, rigid tank contains air at 800 kPa, 25 °C. A valve on the tank is opened, and the pressure inside quickly drops to 250 kPa, at which point the valve is closed. Assuming that the air remaining inside has undergone reversible adiabatic expansion, calculate the mass withdrawn during the process.

6.80 Two rigid, insulated tanks are connected with a pipe and valve. One tank has 0.5 kg air at 200 kPa, 300 K and the other has 0.75 kg air at 100 kPa, 400 K. The valve is opened, and the air comes to a single uniform state without any heat transfer. Find the final temperature and the change in entropy of the air.

6.81 Two rigid tanks, shown in Fig. P6.81, each contains 10 kg N₂ gas at 1000 K, 500 kPa. They are now thermally connected to a reversible heat pump, which heats one tank and cools the other, with no heat transfer to the surroundings. When one tank is heated to 1500 K, the process stops. Find the final (P, T) in both tanks and the work input to the heat pump, assuming constant heat capacities.

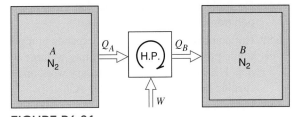

FIGURE P6.81

6.82 A rigid tank contains 4 kg air at 300 °C, 4 MPa that acts as the hot energy reservoir for a heat engine with its cold side at 20 °C, shown in Fig. P6.82. Heat transfer to the heat engine cools the air down in a reversible process to a final 20 °C and then stops. Find the final air pressure and the work output of the heat engine.

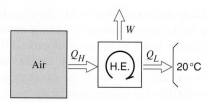

FIGURE P6.82

Polytropic Processes

6.83 An ideal gas having constant specific heat undergoes a reversible polytropic expansion with exponent $n = 1.4$. If the gas is carbon dioxide, will the heat transfer for this process be positive, negative, or zero?

6.84 A nitrogen gas goes through a polytropic process with $n = 1.3$ in a piston/cylinder arrangement. It starts out at 600 K, 600 kPa and ends at 900 K. Is the heat transfer positive, negative, or zero?

6.85 A cylinder/piston contains 1 kg methane gas at 200 kPa, 300 K. The gas is compressed reversibly to a pressure of 800 kPa. Calculate the work required if the process is adiabatic.

6.86 Do the previous problem but assume that the process is isothermal.

6.87 A piston/cylinder contains pure oxygen at 500 K, 600 kPa. The piston is moved to a volume such that the final temperature is 700 K in a polytropic process with exponent $n = 1.25$. Use ideal gas approximation and constant specific heat to find the final pressure, the specific work, and the heat transfer.

6.88 Do Problem 6.85 and assume that the process is polytropic with $n = 1.15$.

6.89 Hot combustion air at 1800 K expands in a polytropic process to a volume six times as large with $n = 1.3$. Find the specific boundary work and the specific heat transfer using Table A.7.

6.90 Helium in a piston/cylinder at 20 °C, 100 kPa is brought to 400 K in a reversible polytropic process with exponent $n = 1.25$. You may assume that helium is an ideal gas with constant specific heat. Find the final pressure and both the specific heat transfer and the specific work.

6.91 The power stroke in an internal combustion engine can be approximated with a polytropic expansion. Consider air in a cylinder volume of 0.2 L at 7 MPa, 1800 K, shown in Fig. P6.91. It now expands in a

reversible polytropic process with exponent $n = 1.5$ through a volume ratio of 10:1. Show this process on P–v and T–s diagrams, and calculate the work and heat transfer for the process.

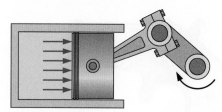

FIGURE P6.91

6.92 A cylinder/piston contains saturated vapor R-410A at 10 °C; the volume is 10 L. The R-410A is compressed to 2 MPa, 60 °C in a reversible (internally) polytropic process. Find the polytropic exponent n and calculate the work and heat transfer.

6.93 Air goes through a polytropic process with $n = 1.3$ in a piston/cylinder setup. It starts at 200 kPa, 300 K and ends with a pressure of 2000 kPa. Find the expansion ratio v_2/v_1, the specific work, and the specific heat transfer.

Entropy Generation

6.94 Consider a heat transfer of 100 kJ from 1500 K hot gases to a steel container at 750 K that has a heat transfer of the 100 kJ out to some air at 375 K. Determine the entropy generation in each of the control volumes indicated in Fig. P6.94.

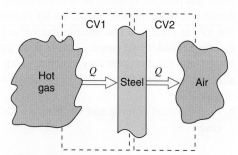

FIGURE P6.94

6.95 A rigid tank has 0.1 kg saturated vapor R-410A at 0 °C that is cooled to −20 °C by a −20 °C heat sink. Show the process in a T–s diagram; find the change in entropy of the R-410A, the heat sink, and the total entropy generation.

6.96 One kilogram of water at 500 °C and 1 kg saturated water vapor, both at 200 kPa, are mixed in a constant-pressure and adiabatic process. Find the final temperature and the entropy generation for the process.

6.97 A car uses an average power of 25 hp for a 1 h round trip. With a thermal efficiency of 35 %, how much fuel energy was used? What happened to all the energy? What change in entropy took place if we assume an ambient temperature of 20 °C?

6.98 A computer chip dissipates 2 kJ of electric work over time and rejects that work as heat transfer from its 50 °C surface to 25 °C air. How much entropy is generated in the chip? How much entropy, if any, is generated outside the chip?

6.99 An insulated cylinder/piston contains R-134a at 1 MPa, 50 °C, with a volume of 100 L. The R-134a expands, moving the piston until the pressure in the cylinder has dropped to 100 kPa. It is claimed that the R-134a does 190 kJ of work against the piston during the process. Is that possible?

6.100 The unrestrained expansion of the reactor water in Problem 3.81 has a final state in the two-phase region. Find the entropy generated in the process.

6.101 Heat transfer from a 20 °C kitchen to a block of 1.5 kg ice at −10 °C melts it to liquid at 20 °C. How much entropy is generated?

6.102 Ammonia is contained in a rigid sealed tank of unknown quality at 0 °C. When heated in boiling water to 100 °C, its pressure reaches 1200 kPa. Find the initial quality, the heat transfer to the ammonia, and the total entropy generation.

6.103 Water in a piston/cylinder is at 101 kPa, 25 °C and mass 0.5 kg. The piston rests on some stops, and the pressure should be 1000 kPa to float the piston. We now heat the water from a 200 °C reservoir, so the volume becomes five times the initial value. Find the total heat transfer and the entropy generation.

6.104 Do Problem 6.103, assuming the piston/cylinder is 1.5 kg of steel and has the same temperature as the water at any time.

6.105 A cylinder fitted with a movable piston contains water at 3 MPa, 50 % quality, at which point the volume is 20 L. The water now expands to 1.2 MPa as a result of receiving 600 kJ of heat from a large source at 300 °C. It is claimed that the water does 124 kJ of work during this process. Is this possible?

6.106 A piston/cylinder contains 1 kg water at 150 kPa, 20 °C. The piston is loaded, so the pressure is linear in volume. Heat is added from a 700 °C source until the water is at 1 MPa, 500 °C. Find the heat transfer and the entropy generation.

6.107 A closed, rigid container is filled with 1.5 kg water at 100 kPa, 55 °C, 1 kg of stainless steel and 0.5 kg of polyvinyl chloride, both at 20 °C and 0.1 kg of hot air at 400 K, 100 kPa. It is now left alone, with no external heat transfer, and no water vaporizes. Find the final temperature and the entropy generation for the process.

6.108 A cylinder/piston contains water at 200 kPa, 200 °C with a volume of 20 L. The piston is moved slowly, compressing the water to a pressure of 800 kPa. The loading on the piston is such that the product PV is a constant. Assuming that the room temperature is 20 °C, show that this process does not violate the second law.

6.109 A rigid steel tank of mass 2.5 kg contains 0.5 kg R-410A at 0 °C with a specific volume of 0.01 m³/kg. The system heats up to the room temperature, 25 °C. Find the process heat transfer and the entropy generation.

6.110 A piston/cylinder has ammonia at 2000 kPa, 80 °C with a volume of 0.1 m³. The piston is loaded with a linear spring and outside ambient is at 20 °C, shown in Fig. P6.110. The ammonia now cools down to 20 °C, at which point it has a quality of 15 %. Find the work, the heat transfer, and the total entropy generation in the process.

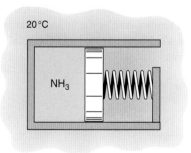

FIGURE P6.110

6.111 A 5 kg aluminum radiator holds 2 kg of liquid R-134a, both at −10 °C. The setup is brought indoors and heated with 220 kJ from a heat source at 100 °C. Find the total entropy generation for the process, assuming the R-134a remains a liquid.

6.112 A piston/cylinder of total 1 kg steel contains 0.5 kg ammonia at 1600 kPa, both masses at 120 °C. Some stops are placed so that a minimum volume is 0.02 m³, shown in Fig. P6.112. Now the whole system is cooled down to 30 °C by heat transfer to the ambient at 20 °C, and during the process the steel keeps the same temperature as the ammonia. Find the work, the heat transfer, and the total entropy generation in the process.

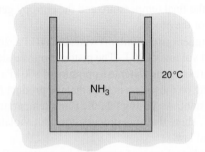

FIGURE P6.112

6.113 A piston/cylinder contains 0.1 kg water at 500 °C, 1000 kPa. The piston has a stop at half of the original volume, similar to Fig. P6.112. The water now cools to room temperature, 25 °C. Find the heat transfer and the entropy generation.

6.114 A cylinder/piston arrangement contains 10 g ammonia at 20 °C with a volume of 1 L. There are some stops, so if the piston is at the stops, the volume is 1.4 L. The ammonia is now heated to 200 °C by a 240 °C source. The piston and cylinder are made of 0.5 kg aluminum, and assume that the mass has the same temperature as the ammonia at any time. Find the total heat transfer and the total entropy generation.

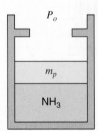

FIGURE P6.114

6.115 A cylinder/piston arrangement contains 0.1 kg R-410A of quality $x = 0.2534$ at −20 °C. Stops are mounted, so $V_{stop} = 3V_1$, similar to Fig. P6.114.

The system is now heated to the final temperature of 20 °C by a 60 °C source. Find the total entropy generation.

6.116 One kilogram of air at 300 K is mixed with 2 kg air at 400 K in a process at a constant 100 kPa and $Q = 0$. Find the final T and the entropy generation in the process.

6.117 Air in a rigid tank is at 900 K, 500 kPa, and it now cools to the ambient temperature of 300 K by heat loss to the ambient. Find the entropy generation.

6.118 Two rigid, insulated tanks are connected with a pipe and valve. One tank has 0.5 kg air at 200 kPa, 300 K and the other has 0.75 kg air at 100 kPa, 400 K. The valve is opened and the air comes to a single uniform state, without external heat transfer. Find the final T and P and the entropy generation.

6.119 One kilogram air at 100 kPa is mixed with 2 kg air at 200 kPa, both at 300 K, in a rigid, insulated tank. Find the final state (P, T) and the entropy generation in the process.

6.120 Argon in a light bulb is at 110 kPa, 80 °C. The light is turned off, so the argon cools to the ambient temperature of 20 °C. Disregard the glass and any other mass and find the specific entropy generation.

6.121 A rigid tank contains 2 kg air at 200 kPa and ambient temperature, 20 °C. An electric current now passes through a resistor inside the tank. After a total of 100 kJ of electrical work has crossed the boundary, the air temperature inside is 80 °C. Is this possible?

6.122 A spring-loaded piston/cylinder contains 1.5 kg air at 27 °C, 160 kPa. It is now heated in a process in which pressure is linear in volume, $P = A + BV$, to twice its initial volume, where it reaches 900 K. Find the work, the heat transfer, and the total entropy generation, assuming a source at 1000 K.

6.123 A constant pressure piston/cylinder contains 0.5 kg air at 300 K, 400 kPa. Assume the piston/cylinder has a total mass of 1 kg steel and is at the same temperature as the air at any time. The system is now heated to 1600 K by heat transfer from a 1700 K source. Find the entropy generation using constant specific heats for air.

6.124 Do Problem 6.123 using Table A.7.

6.125 Nitrogen at 200 °C, 300 kPa is in a piston/cylinder, volume 5 L, with the piston locked with a pin. The forces on the piston require a pressure inside of

200 kPa to balance it without the pin. The pin is removed, and the piston quickly comes to its equilibrium position without any heat transfer. Find the final P, T, and V and the entropy generation due to this partly unrestrained expansion.

6.126 One kilogram carbon dioxide at 100 kPa, 400 K is mixed with 2 kg carbon dioxide at 200 kPa, 2000 K, in a rigid, insulated tank. Find the final state (P, T) and the entropy generation in the process using constant specific heat from Table A.5.

6.127 Solve Problem 6.126 using Table A.8.

6.128 Nitrogen at 600 kPa, 127 °C is in a 0.5 m³ insulated tank connected to a pipe with a valve to a second insulated, initially empty tank of volume 0.25 m³, shown in Fig. P6.128. The valve is opened, and the nitrogen fills both tanks at a uniform state. Find the final pressure and temperature and the entropy generation this process causes. Why is the process irreversible?

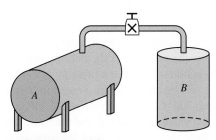

FIGURE P6.128

6.129 A cylinder/piston contains carbon dioxide at 1 MPa, 300 °C with a volume of 200 L. The total external force acting on the piston is proportional to V^3. This system is allowed to cool to room temperature, 20 °C. What is the total entropy generation for the process?

6.130 A cylinder/piston contains 100 L air at 110 kPa, 25 °C. The air is compressed in a reversible polytropic process to a final state of 800 kPa, 500 K. Assume the heat transfer is with the ambient at 25 °C and determine the polytropic exponent n and the final volume of the air. Find the work done by the air, the heat transfer, and the total entropy generation for the process.

Rates or Fluxes of Entropy

6.131 A room at 22 °C is heated electrically with 1500 W to keep a steady temperature. The outside ambient is at 5 °C. Find the flux of S ($=\dot{Q}/T$) into the room air, into the ambient, and the rate of entropy generation.

6.132 A mass of 3 kg nitrogen gas at 2000 K, $V = C$, cools with 500 W. What is dS/dt?

6.133 A heat pump (see Problem 5.43) should upgrade 5 MW of heat at 85 °C to heat delivered at 150 °C. For a reversible heat pump, what are the fluxes of entropy into and out of the heat pump?

6.134 Reconsider the heat pump in the previous problem and assume it has a COP of 2.5. What are the fluxes of entropy into and out of the heat pump and the rate of entropy generation inside it?

6.135 A heat pump with COP $= 4$ uses 1 kW of power input to heat a 25 °C room, drawing energy from the outside at 15 °C. Assume the high/low temperatures in the heat pump are 45 °C/0 °C. Find the total rates of entropy into and out of the heat pump, the rate from the outside at 15 °C, and the rate to the room at 25 °C.

6.136 An amount of power, say 1000 kW, comes from a furnace at 800 °C, going into water vapor at 400 °C. From the water, the power goes to a solid metal at 200 °C and then into some air at 70 °C. For each location, calculate the flux of s through a surface as ($\dot{Q}/T$). What makes the flux larger and larger?

6.137 Room air at 23 °C is heated by a 2000 W space heater with a surface filament temperature of 700 K, shown in Fig. P6.137. The room at steady state loses the power to the outside, which is at 7 °C. Find the rate(s) of entropy generation and specify where it is made.

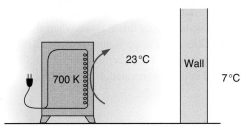

FIGURE P6.137

6.138 A car engine block receives 2 kW at its surface of 450 K from hot combustion gases at 1500 K. Near the cooling channel, the engine block transmits 2 kW out at its 400 K surface to the coolant flowing at 370 K. Finally, in the radiator, the coolant at 350 K delivers the 2 kW to air that is at 25 °C.

Find the rate of entropy generation inside the engine block, inside the coolant, and in the radiator/air combination.

6.139 A farmer runs a heat pump using 2 kW of power input. It keeps a chicken hatchery at a constant 30 °C while the room loses 15 kW to the colder outside ambient air at 10 °C. Find the COP of the heat pump, the rate of entropy generated in the heat pump and its heat exchangers, and the rate of entropy generated in the heat loss process.

Review Problems

6.140 An insulated cylinder/piston has an initial volume of 0.15 m^3 and contains steam at 400 kPa, 200 °C. The steam is expanded adiabatically, and the work output is measured very carefully to be 30 kJ. It is claimed that the final state of the water is in the two-phase (liquid and vapor) region. What is your evaluation of the claim?

6.141 The water in the two tanks of Problem 3.161 receives heat transfer from a reservoir at 300 °C. Find the total entropy generation due to this process.

6.142 A steel piston/cylinder of 1 kg contains 2.5 kg ammonia at 50 kPa, −20 °C. Now it is heated to 50 °C at constant pressure through the bottom of the cylinder from external hot gas at 200 °C, and we assume the steel has the same temperature as the ammonia. Find the heat transfer from the hot gas and the total entropy generation.

6.143 Water in a piston/cylinder, shown in Fig. P6.143, is at 1 MPa, 500 °C. There are two stops, a lower one at which $V_{min} = 1$ m^3 and an upper one at $V_{max} = 3$ m^3. The piston is loaded with a mass and outside atmosphere such that it floats when the pressure is 500 kPa. This setup is now cooled to 100 °C by rejecting heat to the surrounding at 20 °C. Find the total entropy generated in the process.

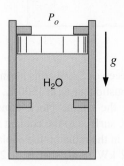

FIGURE P6.143

6.144 A piston/cylinder contains air at 300 K, 100 kPa. A reversible polytropic process with $n = 1.3$ brings the air to 500 K. Any heat transfer if it comes in is from a 325 °C reservoir, and if it goes out it is to the ambient at 300 K. Sketch the process in a P–v and a T–s diagram. Find the specific work and specific heat transfer in the process. Find the specific entropy generation (external to the air) in the process.

6.145 A closed tank, $V = 10$ L, containing 5 kg of water initially at 25 °C, is heated to 150 °C by a heat pump that is receiving heat from the surroundings at 25 °C. Assume that this process is reversible. Find the heat transfer to the water and the work input to the heat pump.

6.146 A resistor in a heating element is a total of 0.5 kg with specific heat of 0.8 kJ/kg·K. It is now receiving 500 W of electric power, so it heats from 20 °C to 160 °C. Neglect external heat loss and find the time the process took and the entropy generation.

6.147 Two tanks contain steam and they are both connected to a piston/cylinder, as shown in Fig. P6.147. Initially, the piston is at the bottom, and the mass of the piston is such that a pressure of 1.4 MPa below it will be able to lift it. Stream in A is 4 kg at 7 MPa, 700 °C and B has 2 kg at 3 MPa, 350 °C. The two valves are opened, and the water comes to a uniform state. Find the final temperature and the total entropy generation, assuming no heat transfer.

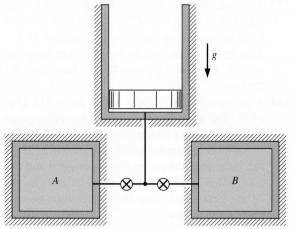

FIGURE P6.147

6.148 Assume the heat source in Problem 3.164 is at 300 °C in a setup similar to Fig. P6.147. Find the heat transfer and the entropy generation.

6.149 A device brings 2 kg ammonia from 150 kPa, $-20\,°C$ to 400 kPa, $80\,°C$ in a polytropic process. Find the polytropic exponent n, the work, and the heat transfer. Find the total entropy generated, assuming a source at $100\,°C$.

6.150 A rigid tank with 0.5 kg ammonia at 1600 kPa, $160\,°C$ is cooled in a reversible process by giving heat to a reversible heat engine that has its cold side at ambient temperature, $20\,°C$, shown in Fig. P6.150. The ammonia eventually reaches $20\,°C$ and the process stops. Find the heat transfer from the ammonia to the heat engine and the work output of the heat engine.

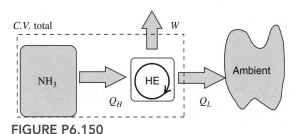

FIGURE P6.150

6.151 A piston/cylinder with constant loading of the piston contains 1 L water at 400 kPa, quality 15 %. It has some stops mounted, so the maximum possible volume is 11 L. A reversible heat pump extracting heat from the ambient at 300 K, 100 kPa heats the water to $300\,°C$. Find the total work and heat transfer for the water and the work input to the heat pump.

6.152 A small halogen light bulb receives electrical power of 50 W. The small filament is at 1000 K and gives out 20 % of the power as light and the rest as heat transfer to the gas, which is at 500 K; the glass is at 400 K. All the power is absorbed by the room walls at $25\,°C$. Find the rate of generation of entropy in the filament, in the total bulb including the glass, and in the total room including the bulb.

COMPUTER, DESIGN, AND OPEN-ENDED PROBLEMS

6.153 Write a computer program to solve Problem 6.54 using constant specific heat for both the sand and the liquid water. Let the amount and the initial temperatures be input variables.

6.154 Write a program to solve Problem 6.61 with the thermal storage rock bed in Problem 5.51. Let the size and temperatures be input variables so that the heat engine work output can be studied as a function of the system parameters.

6.155 Write a program to solve the following problem. One of the gases listed in Table A.6 undergoes a reversible adiabatic process in a cylinder from P_1, T_1 to P_2. We wish to calculate the final temperature and the work for the process by three methods:

a. Integrating the specific heat equation.
b. Assuming constant specific heat at temperature, T_1.
c. Assuming constant specific heat at the average temperature (by iteration).

6.156 Write a program to solve a problem similar to Problem 6.85, but instead of the ideal gas tables, use the formula for the specific heat as a function of temperature in Table A.6.

6.157 Write a program to study a general polytropic process in an ideal gas with constant specific heat. Take Problem 6.90 as an example.

6.158 Write a program to solve the general case of Problem 6.91, in which the initial state and the expansion ratio are input variables.

7 Entropy Equation for a Control Volume

In the preceding two chapters we discussed the second law of thermodynamics and the thermodynamic property entropy. As was done with the first-law analysis, we now consider the more general application of these concepts, the control volume analysis, and a number of cases of special interest. We will also discuss usual definitions of thermodynamic efficiencies.

7.1 THE SECOND LAW OF THERMODYNAMICS FOR A CONTROL VOLUME

The second law of thermodynamics can be applied to a control volume by a procedure similar to that used in Section 4.1, where the energy equation was developed for a control volume. We start with the second law expressed as a change of the entropy for a control mass in a rate form from Eq. 6.51,

$$\frac{dS_{\text{c.m.}}}{dt} = \sum \frac{\dot{Q}}{T} + \dot{S}_{\text{gen}} \tag{7.1}$$

to which we now will add the contributions from the mass flow rates into and out of the control volume. A simple example of such a situation is illustrated in Fig. 7.1. The flow of mass does carry an amount of entropy, s, per unit mass flowing, but it does not give rise to any other contributions. As a process may take place in the flow, entropy can be generated, but this is attributed to the space it belongs to (i.e., either inside or outside of the control volume).

The balance of entropy as an equation then states that the rate of change in total entropy inside the control volume is equal to the net sum of fluxes across the control surface plus the generation rate. That is,

$$\text{rate of change} = +\text{in} - \text{out} + \text{generation}$$

or

$$\frac{dS_{\text{c.v.}}}{dt} = \sum \dot{m}_i s_i - \sum \dot{m}_e s_e + \sum \frac{\dot{Q}_{\text{c.v.}}}{T} + \dot{S}_{\text{gen}} \tag{7.2}$$

These fluxes are mass flow rates carrying a level of entropy and the rate of heat transfer that takes place at a certain temperature (the temperature at the control surface). The

274

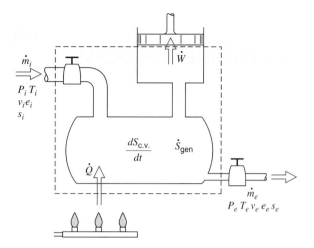

FIGURE 7.1 The entropy balance for a control volume on a rate form.

accumulation and generation terms cover the total control volume and are expressed in the lumped (integral form), so that

$$S_{c.v.} = \int \rho s \, dV = m_{c.v.}s = m_A s_A + m_B s_B + m_C s_C + \cdots$$

$$\dot{S}_{gen} = \int \rho \dot{s}_{gen} \, dV = \dot{S}_{gen. A} + \dot{S}_{gen. B} + \dot{S}_{gen. C} + \cdots \tag{7.3}$$

If the control volume has several different accumulation units with different fluid states and processes occurring in them, we may have to sum the various contributions over the different domains. If the heat transfer is distributed over the control surface, then an integral has to be done over the total surface area using the local temperature and rate of heat transfer per unit area, $(\dot{Q}/A)_{local}$, as

$$\sum \frac{\dot{Q}_{c.v.}}{T} = \int \frac{d\dot{Q}}{T} = \int_{surface} (\dot{Q}/A_{local})/T \, dA \tag{7.4}$$

These distributed cases typically require a much more detailed analysis, which is beyond the scope of the current presentation of the second law.

The generation term(s) in Eq. 7.2 from a summation of individual positive internal-irreversibility entropy-generation terms in Eq. 7.3 is (are) necessarily positive (or zero), such that an inequality is often written as

$$\frac{dS_{c.v.}}{dt} \geq \sum \dot{m}_i s_i - \sum \dot{m}_e s_e + \sum \frac{\dot{Q}_{c.v.}}{T} \tag{7.5}$$

Now the equality applies to internally reversible processes and the inequality to internally irreversible processes. The form of the second law in Eq. 7.2 or 7.5 is general, such that any particular case results in a form that is a subset (simplification) of this form. Examples of various classes of problems are illustrated in the following sections.

If there is no mass flow into or out of the control volume, it simplifies to a control mass and the equation for the total entropy reverts back to Eq. 6.51. Since that version of the second law has been covered in Chapter 6, here we will consider the remaining cases that were done for the energy equation in Chapter 4.

7.2 THE STEADY-STATE PROCESS AND THE TRANSIENT PROCESS

We now consider in turn the application of the second-law control volume equation, Eq. 7.2 or 7.5, to the two control volume model processes developed in Chapter 4.

Steady-State Process

For the steady-state process, which has been defined in Section 4.3, we conclude that there is no change with time of the entropy per unit mass at any point within the control volume, and therefore the first term of Eq. 7.2 equals zero. That is,

$$\frac{dS_{\text{c.v.}}}{dt} = 0 \tag{7.6}$$

so that, for the steady-state process,

$$\sum \dot{m}_e s_e - \sum \dot{m}_i s_i = \sum_{\text{c.s.}} \frac{\dot{Q}_{\text{c.v.}}}{T} + \dot{S}_{\text{gen}} \tag{7.7}$$

in which the various mass flows, heat transfer and entropy generation rates, and states are all constant with time.

If in a steady-state process there is only one area over which mass enters the control volume at a uniform rate and only one area over which mass leaves the control volume at a uniform rate, we can write

$$\dot{m}(s_e - s_i) = \sum_{\text{c.s.}} \frac{\dot{Q}_{\text{c.v.}}}{T} + \dot{S}_{\text{gen}} \tag{7.8}$$

and dividing the mass flow rate out gives

$$s_e = s_i + \sum \frac{q}{T} + s_{\text{gen}} \tag{7.9}$$

Since s_{gen} is always greater than or equal to zero, for an adiabatic process it follows that

$$s_e = s_i + s_{\text{gen}} \geq s_i \tag{7.10}$$

where the equality holds for a reversible adiabatic process.

Example 7.1

Steam enters a steam turbine at a pressure of 1 MPa, a temperature of 300 °C, and a velocity of 50 m/s. The steam leaves the turbine at a pressure of 150 kPa and a velocity of 200 m/s. Determine the work per kilogram of steam flowing through the turbine, assuming the process to be reversible and adiabatic.

Control volume:	Turbine.
Sketch:	Fig. 7.2.
Inlet state:	Fixed (Fig. 7.2).
Exit state:	P_e, $\mathbf{V}_e$ known.
Process:	Steady state, reversible, and adiabatic.
Model:	Steam tables.

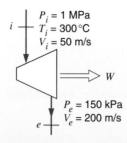

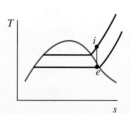

FIGURE 7.2 Sketch for Example 7.1.

Analysis

The continuity equation gives us

$$\dot{m}_e = \dot{m}_i = \dot{m}$$

From the energy equation, we have

$$h_i + \frac{\mathbf{V}_i^2}{2} = h_e + \frac{\mathbf{V}_e^2}{2} + w$$

and the second law is

$$s_e = s_i$$

Solution

From the steam tables, we get

$$h_i = 3051.2 \text{ kJ/kg}, \qquad s_i = 7.1228 \text{ kJ/kg·K}$$

The two properties known in the final state are pressure and entropy:

$$P_e = 0.15 \text{ MPa}, \qquad s_e = s_i = 7.1228 \text{ kJ/kg·K}$$

The quality and enthalpy of the steam leaving the turbine can be determined as follows:

$$s_e = 7.1228 = s_f + x_e s_{fg} = 1.4335 + x_e 5.7897$$

$$x_e = 0.9827$$

$$h_e = h_f + x_e h_{fg} = 467.1 + 0.9827 \times (2226.5)$$

$$= 2655.0 \text{ kJ/kg}$$

Therefore, the work per kilogram of steam for this isentropic process is found using the energy equation:

$$w = 3051.2 + \frac{50 \times 50}{2 \times 1000} - 2655.0 - \frac{200 \times 200}{2 \times 1000} = 377.5 \text{ kJ/kg}$$

Example 7.2

Consider the reversible adiabatic flow of steam through a nozzle. Steam enters the nozzle at 1 MPa and 300 °C, with a velocity of 30 m/s. The pressure of the steam at the nozzle exit is 0.3 MPa. Determine the exit velocity of the steam from the nozzle, assuming a reversible, adiabatic, steady-state process.

Control volume:	Nozzle.
Sketch:	Fig. 7.3.
Inlet state:	Fixed (Fig. 7.3).
Exit State:	P_e known.
Process:	Steady state, reversible, and adiabatic.
Model:	Steam tables.

Analysis

Because this is a steady-state process in which the work, heat transfer, and changes in potential energy are zero, we can write

Continuity Eq.: $\dot{m}_e = \dot{m}_i = \dot{m}$

Energy Eq.: $h_i + \dfrac{\mathbf{V}_i^2}{2} = h_e + \dfrac{\mathbf{V}_e^2}{2}$

Second law: $s_e = s_i$

Solution

From the steam tables, we have

$$h_i = 3051.2 \text{ kJ/kg}, \qquad s_i = 7.1228 \text{ kJ/kg·K}$$

The two properties that we know in the final state are entropy and pressure:

$$s_e = s_i = 7.1228 \text{ kJ/kg·K}, \qquad P_e = 0.3 \text{ MPa}$$

Therefore,

$$T_e = 159.1\,^\circ\text{C}, \qquad h_e = 2780.2 \text{ kJ/kg}$$

Substituting into the energy equation, we have

$$\frac{\mathbf{V}_e^2}{2} = h_i - h_e + \frac{\mathbf{V}_i^2}{2}$$

$$= 3051.2 - 2780.2 + \frac{30 \times 30}{2 \times 1000} = 271.5 \text{ kJ/kg}$$

$$\mathbf{V}_e = \sqrt{2000 \times 271.5} = 737 \text{ m/s}$$

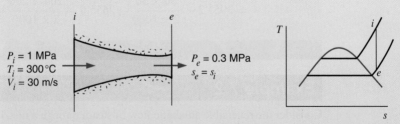

FIGURE 7.3 Sketch for Example 7.2.

Example 7.3

An inventor reports having a refrigeration compressor that receives saturated R-134a vapor at $-20\,°C$ and delivers the vapor at 1 MPa, $40\,°C$. The compression process is adiabatic. Does the process described violate the second law?

Control volume:	Compressor.
Inlet state:	Fixed (saturated vapor at T_i).
Exit state:	Fixed (P_e, T_e known).
Process:	Steady state, adiabatic.
Model:	R-134a tables.

Analysis

Because this is a steady-state adiabatic process, we can write the second law as

$$s_e = s_i + s_{gen}$$

Solution

From the R-134a tables, we read

$$s_e = 1.7148 \text{ kJ/kg·K}, \qquad s_i = 1.7395 \text{ kJ/kg·K}$$

Therefore, $s_e < s_i$, implying a negative entropy generation that is a violation of the second law and thus is impossible.

Example 7.4

An air compressor in a gas station (see Fig. 7.4) takes in a flow of ambient air at 100 kPa, 290 K and compresses it to 1000 kPa in a reversible adiabatic process. We want to know the specific work required and the exit air temperature.

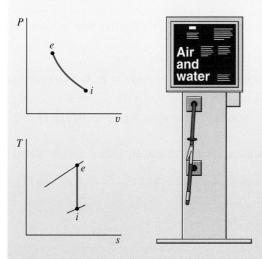

FIGURE 7.4 Diagram for Example 7.4.

Solution

C.V. air compressor, steady state, single flow through it, and assumed adiabatic $\dot{Q} = 0$.

Continuity Eq. 4.11: $\dot{m}_i = \dot{m}_e = \dot{m}$

Energy Eq. 4.12: $\dot{m}h_i = \dot{m}h_e + \dot{W}_C$

Entropy Eq. 7.8: $\dot{m}s_i + \dot{S}_{gen} = \dot{m}s_e$

Process: Reversible $\dot{S}_{gen} = 0$

Use constant specific heat from Table A.5, $C_{p0} = 1.004$ kJ/kg·K, $k = 1.4$. The entropy equation gives constant s, which gives the relation in Eq. 6.23:

$$s_i = s_e \Rightarrow T_e = T_i \left(\frac{P_e}{P_i} \right)^{\frac{k-1}{k}}$$

$$T_e = 290 \left(\frac{1000}{100} \right)^{0.2857} = 559.9 \text{ K}$$

The energy equation per unit mass gives the work term

$$w_c = h_i - h_e = C_{p0}(T_i - T_e) = 1.004 \times (290 - 559.9) = -271 \text{ kJ/kg}$$

Example 7.5

A desuperheater works by injecting liquid water into a flow of superheated steam. With 2 kg/s at 300 kPa, 200 °C, steam flowing in, what mass flow rate of liquid water at 20 °C should be added to generate saturated vapor at 300 kPa? We also want to know the rate of entropy generation in the process.

Solution

C.V. Desuperheater (see Fig. 7.5), no external heat transfer, and no work.

Continuity Eq. 4.9: $\dot{m}_1 + \dot{m}_2 = \dot{m}_3$

Energy Eq. 4.10: $\dot{m}_1 h_1 + \dot{m}_2 h_2 = \dot{m}_3 h_3 = (\dot{m}_1 + \dot{m}_2)h_3$

Entropy Eq. 7.7: $\dot{m}_1 s_1 + \dot{m}_2 s_2 + \dot{S}_{gen} = \dot{m}_3 s_3$

Process: $P = $ constant, $\dot{W} = 0$, and $\dot{Q} = 0$

All the states are specified (approximate state 2 with saturated liquid at 20 °C)

B.1.3: $h_1 = 2865.54 \dfrac{\text{kJ}}{\text{kg}}, \quad s_1 = 7.3115 \dfrac{\text{kJ}}{\text{kg·K}}; \quad h_3 = 2725.3 \dfrac{\text{kJ}}{\text{kg}}, \quad s_3 = 6.9918 \dfrac{\text{kJ}}{\text{kg·K}}$

B.1.2: $h_2 = 83.94 \dfrac{\text{kJ}}{\text{kg}}, \quad s_2 = 0.2966 \dfrac{\text{kJ}}{\text{kg·K}}$

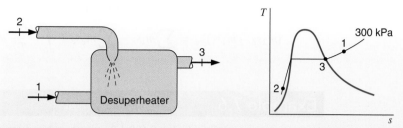

FIGURE 7.5 Sketch and diagram for Example 7.5.

Now we can solve for the flow rate $\dot{m}_2$ from the energy equation, having eliminated $\dot{m}_3$ by the continuity equation

$$\dot{m}_2 = \dot{m}_1 \frac{h_1 - h_3}{h_3 - h_2} = 2 \times \frac{2865.54 - 2725.3}{2725.3 - 83.94} = 0.1062 \,\text{kg/s}$$

$$\dot{m}_3 = \dot{m}_1 + \dot{m}_2 = 2.1062 \,\text{kg/s}$$

Generation is from the entropy equation

$$\dot{S}_{\text{gen}} = \dot{m}_3 s_3 - \dot{m}_1 s_1 - \dot{m}_2 s_2$$

$$= 2.1062 \times 6.9918 - 2 \times 7.3115 - 0.1062 \times 0.2966 = 0.072 \,\text{kW/K}$$

Transient Process

For the transient process, which was described in Section 4.6, the second law for a control volume, Eq. 7.2, can be written in the following form:

$$\frac{d}{dt}(ms)_{\text{c.v.}} = \sum \dot{m}_i s_i - \sum \dot{m}_e s_e + \sum \frac{\dot{Q}_{\text{c.v.}}}{T} + \dot{S}_{\text{gen}} \qquad (7.11)$$

If this is integrated over the time interval t, we have

$$\int_0^t \frac{d}{dt}(ms)_{\text{c.v.}} \, dt = (m_2 s_2 - m_1 s_1)_{\text{c.v.}}$$

$$\int_0^t \left(\sum \dot{m}_i s_i \right) dt = \sum m_i s_i, \quad \int_0^t \left(\sum \dot{m}_e s_e \right) dt = \sum m_e s_e, \quad \int_0^t \dot{S}_{\text{gen}} \, dt = {}_1 S_{2\text{gen}}$$

Therefore, for this period of time t, we can write the second law for the transient process as

$$(m_2 s_2 - m_1 s_1)_{\text{c.v.}} = \sum m_i s_i - \sum m_e s_e + \int_0^t \sum_{\text{c.s.}} \frac{\dot{Q}_{\text{c.v.}}}{T} \, dt + {}_1 S_{2\text{gen}} \qquad (7.12)$$

Since in this process the temperature is uniform throughout the control volume at any instant of time, the integral on the right reduces to

$$\int_0^t \sum_{\text{c.s.}} \frac{\dot{Q}_{\text{c.v.}}}{T} \, dt = \int_0^t \frac{1}{T} \sum_{\text{c.s.}} \dot{Q}_{\text{c.v.}} \, dt = \int_0^t \frac{\dot{Q}_{\text{c.v.}}}{T} \, dt$$

and therefore the second law for the transient process can be written

$$(m_2 s_2 - m_1 s_1)_{\text{c.v.}} = \sum m_i s_i - \sum m_e s_e + \int_0^t \frac{\dot{Q}_{\text{c.v.}}}{T} \, dt + {}_1 S_{2\text{gen}} \qquad (7.13)$$

Example 7.6

Assume an air tank has 40 L of 100 kPa air at ambient temperature $17\,^{\circ}\text{C}$. The adiabatic and reversible compressor is started so that it charges the tank up to a pressure of 1000 kPa and then it shuts off. We want to know how hot the air in the tank gets and the total amount of work required to fill the tank.

Solution

C.V. compressor and air tank in Fig. 7.6.

Continuity Eq. 4.20: $\quad m_2 - m_1 = m_{\text{in}}$

Energy Eq. 4.21: $\quad m_2 u_2 - m_1 u_1 = {}_1 Q_2 - {}_1 W_2 + m_{\text{in}} h_{\text{in}}$

Entropy Eq. 7.13: $\quad m_2 s_2 - m_1 s_1 = \int dQ/T + {}_1 S_{2\text{gen}} + m_{\text{in}} s_{\text{in}}$

Process: $\quad$ Adiabatic ${}_1 Q_2 = 0$, $\quad$ Process ideal ${}_1 S_{2\text{gen}} = 0$, $\quad s_1 = s_{\text{in}}$

$$\Rightarrow m_2 s_2 = m_1 s_1 + m_{\text{in}} s_{\text{in}} = (m_1 + m_{\text{in}}) s_1 = m_2 s_1 \Rightarrow s_2 = s_1$$

Constant $s \Rightarrow \quad$ Eq. 6.19 $\quad s_{T2}^0 = s_{T1}^0 + R \ln(P_2/P_i)$

$$s_{T2}^0 = 6.835\,21 + 0.287 \ln (10) = 7.496\,05 \text{ kJ/kg·K}$$

Interpolate in Table A.7 $\quad \Rightarrow T_2 = 555.7 \text{ K}, \quad u_2 = 401.49 \text{ kJ/kg}$

$$m_1 = P_1 V_1 / R T_1 = 100 \times 0.04/(0.287 \times 290) = 0.048\,06 \text{ kg}$$

$$m_2 = P_2 V_2 / R T_2 = 1000 \times 0.04/(0.287 \times 555.7) = 0.2508 \text{ kg}$$

$$\Rightarrow m_{\text{in}} = 0.2027 \text{ kg}$$

$${}_1 W_2 = m_{\text{in}} h_{\text{in}} + m_1 u_1 - m_2 u_2$$

$$= m_{\text{in}}(290.43) + m_1(207.19) - m_2(401.49) = -31.9 \text{ kJ}$$

Remark: The high final temperature makes the assumption of zero heat transfer poor. The charging process does not happen rapidly, so there will be a heat transfer loss. We need to know this to make a better approximation of the real process.

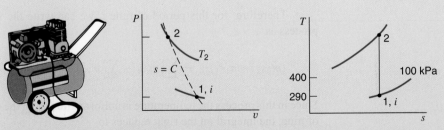

FIGURE 7.6 Sketch and diagram for Example 7.6.

In-Text Concept Questions

a. A reversible adiabatic flow of liquid water in a pump has increasing P. Is T increasing or decreasing?

b. A reversible adiabatic flow of air in a compressor has increasing P. Is T increasing or decreasing?

c. A compressor receives R-134a at $-10\,°C$, 200 kPa with an exit of 1200 kPa, $50\,°C$. What can you say about the process?

d. A flow of water at some velocity out of a nozzle is used to wash a car. The water then falls to the ground. What happens to the water state in terms of $\mathbf{V}$, T, and s?

7.3 THE STEADY-STATE SINGLE-FLOW PROCESS

An expression can be derived for the work in a steady-state, single-flow process that shows how the significant variables influence the work output. We have noted that when a steady-state process involves a single flow of fluid into and out of a control volume, the energy equation, Eq. 4.13, can be written as

$$q + h_i + \frac{1}{2}\mathbf{V}_i^2 + gZ_i = h_e + \frac{1}{2}\mathbf{V}_e^2 + gZ_e + w$$

The second law, Eq. 7.9, and recall Eq. 7.4, is

$$s_i + s_{\text{gen}} + \int \frac{\delta q}{T} = s_e$$

which we will write in a differential form as

$$\delta s_{\text{gen}} + \delta q/T = ds \qquad \Rightarrow \qquad \delta q = T\,ds - T\,\delta s_{\text{gen}}$$

To facilitate the integration and find q, we use the property relation, Eq. 6.8, and get

$$\delta q = T\,ds - T\,\delta s_{\text{gen}} = dh - v\,dP - T\,\delta s_{\text{gen}}$$

and we now have

$$q = \int_i^e \delta q = \int_i^e dh - \int_i^e v\,dP - \int_i^e T\,\delta s_{\text{gen}} = h_e - h_i - \int_i^e v\,dP - \int_i^e T\,\delta s_{\text{gen}}$$

This result is substituted into the energy equation, which we solve for work as

$$w = q + h_i - h_e + \frac{1}{2}(\mathbf{V}_i^2 - \mathbf{V}_e^2) + g(Z_i - Z_e)$$

$$= h_e - h_i - \int_i^e v\,dP - \int_i^e T\,\delta s_{\text{gen}} + h_i - h_e + \frac{1}{2}(\mathbf{V}_i^2 - \mathbf{V}_e^2) + g(Z_i - Z_e)$$

The enthalpy terms cancel, and the shaft work for a single flow going through an actual process becomes

$$w = -\int_i^e v\,dP + \frac{1}{2}(\mathbf{V}_i^2 - \mathbf{V}_e^2) + g(Z_i - Z_e) - \int_i^e T\,\delta s_{\text{gen}} \tag{7.14}$$

Several comments for this expression are in order:

1. We note that the last term always subtracts ($T > 0$ and $\delta s_{\text{gen}} \geq 0$), and we get the maximum work out for a reversible process where this term is zero. This is identical to the conclusion for the boundary work, Eq. 6.36, where it was concluded that any entropy generation reduces the work output. We do not write Eq. 7.14 because we expect to calculate the last integral for a process, but we show it to illustrate the effect of an entropy generation.

2. For a reversible process, the shaft work is associated with changes in pressure, kinetic energy, or potential energy either individually or in combination. When the pressure increases (pump or compressor) work tends to be negative, that is, we must have shaft work in, and when the pressure decreases (turbine), the work tends to be positive. The specific volume does not affect the sign of the work, but rather its magnitude, so a large amount of work will be involved when the specific volume is large (the fluid is a gas), whereas less work will take place when the specific volume is small (as for a liquid). When the flow reduces its kinetic energy (windmill) or potential energy (a dam and a turbine), we can extract the difference as work.

3. If the control volume does not have a shaft ($w = 0$), then the right-hand-side terms must balance out to zero. Any change in one of the terms must be accompanied by a net change of opposite sign in the other terms, and notice that the last term can only subtract. As an example, let us briefly look at a pipe flow with no changes in kinetic or potential energy. If the flow is considered reversible, then the last term is zero and the first term must be zero; that is, the pressure must be constant. Realizing the flow has some friction and is therefore irreversible, the first term must be positive (pressure is decreasing) to balance out the last term.

As mentioned in the comment above, Eq. 7.14 is useful to illustrate the work involved in a large class of flow processes such as turbines, compressors, and pumps in which changes in the kinetic and potential energies of the working fluid are small. The model process for these machines is then a reversible, steady-state process with no changes in kinetic or potential energy. The process is often also adiabatic, but this is not required for this expression, which reduces to

$$w = -\int_i^e v \, dP \tag{7.15}$$

From this result, we conclude that the shaft work associated with this type of process is given by the area shown in Fig. 7.7. It is important to note that this result applies to a very specific situation of a flow device and is very different from the boundary-type work $\int_1^2 P \, dv$ in a piston/cylinder arrangement. It was also mentioned in the comments that the

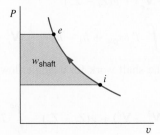

FIGURE 7.7 Shaft work from Eq. 7.15.

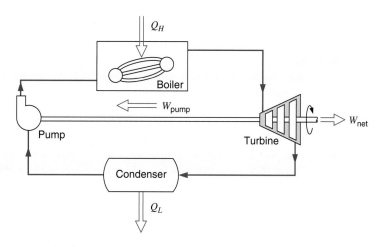

FIGURE 7.8 Simple steam power plant.

shaft work involved in this type of process is closely related to the specific volume of the fluid during the process. To amplify this point further, consider the simple steam power plant shown in Fig. 7.8. Suppose that this is a set of ideal components with no pressure drop in the piping, the boiler, or the condenser. Thus, the pressure increase in the pump is equal to the pressure decrease in the turbine. Neglecting kinetic and potential energy changes, the work done in each of these processes is given by Eq. 7.15. Since the pump handles liquid, which has a very small specific volume compared to that of the vapor that flows through the turbine, the power input to the pump is much less than the power output of the turbine. The difference is the net power output of the power plant.

This same line of reasoning can be applied qualitatively to actual devices that involve steady-state processes, even though the processes are not exactly reversible and adiabatic.

Example 7.7

Calculate the work per kilogram to pump water isentropically from 100 kPa, 30 °C to 5 MPa.

Control volume:	Pump.
Inlet state:	P_i, T_i known; state fixed.
Exit state:	P_e known.
Process:	Steady state, isentropic.
Model:	Steam tables.

Analysis

Since the process is steady state, reversible, and adiabatic, and because changes in kinetic and potential energies can be neglected, we have

$$\text{Energy Eq.:} \quad h_i = h_e + w$$

$$\text{Second law:} \quad s_e - s_i = 0$$

Solution

Since P_e and s_e are known, state e is fixed and therefore h_e is known and w can be found from the energy equation. However, the process is reversible and steady state, with negligible changes in kinetic and potential energies, so that Eq. 7.15 is also valid. Furthermore, since a liquid is being pumped, the specific volume will change very little during the process.

From the steam tables, $v_i = 0.001\ 004$ m^3/kg. Assuming that the specific volume remains constant and using Eq. 7.15, we have

$$-w = \int_i^e v\, dP = v(P_e - P_i) = 0.001\ 004 \times (5000 - 100) = 4.92\ \text{kJ/kg}$$

A simplified version of Eq. 7.14 arises when we consider a reversible flow of an incompressible fluid ($v = $ constant). The first integral is then readily done to give

$$w = -v(P_e - P_i) + \frac{1}{2}(\mathbf{V}_i^2 - \mathbf{V}_e^2) + g(Z_i - Z_e) \tag{7.16}$$

which is called the extended Bernoulli equation after Daniel Bernoulli, who wrote the equation for the zero work term, which then can be written

$$v P_i + \frac{1}{2}\mathbf{V}_i^2 + g Z_i = v P_e + \frac{1}{2}\mathbf{V}_e^2 + g Z_e \tag{7.17}$$

From this equation, it follows that the sum of flow work (Pv), kinetic energy, and potential energy is constant along a flow line. For instance, as the flow goes up, there is a corresponding reduction in the kinetic energy or pressure.

Example 7.8

Consider a nozzle used to spray liquid water. If the line pressure is 300 kPa and the water temperature is 20°C, how high a velocity can an ideal nozzle generate in the exit flow?

Analysis

For this single steady-state flow, we have no work or heat transfer, and since it is incompressible and reversible, the Bernoulli equation applies, giving

$$v P_i + \frac{1}{2}\mathbf{V}_i^2 + g Z_i = v P_i + 0 + 0 = v P_e + \frac{1}{2}\mathbf{V}_e^2 + g Z = v P_0 + \frac{1}{2}\mathbf{V}_e^2 + 0$$

and the exit kinetic energy becomes

$$\frac{1}{2}\mathbf{V}_e^2 = v(P_i - P_0)$$

Solution

We can now solve for the velocity using a value of $v = v_f = 0.001\ 002$ m^3/kg at 20°C from the steam tables.

$$\mathbf{V}_e = \sqrt{2v(P_i - P_0)} = \sqrt{2 \times 0.001\ 002(300 - 100)1000} = 20\ \text{m/s}$$

Notice the factor of 1000 used to convert from kPa to Pa for proper units.

As a final application of Eq. 7.14, we recall the reversible polytropic process for an ideal gas, discussed in Section 6.8 for a control mass process. For the steady-state process with no change in kinetic and potential energies, we have the relations

$$w = -\int_i^e v\, dP \quad\text{and}\quad Pv^n = \text{constant} = C^n$$

$$w = -\int_i^e v\, dP = -C \int_i^e \frac{dP}{P^{1/n}}$$

$$= -\frac{n}{n-1}(P_e v_e - P_i v_i) = -\frac{nR}{n-1}(T_e - T_i) \tag{7.18}$$

If the process is isothermal, then $n = 1$ and the integral becomes

$$w = -\int_i^e v\, dP = -\text{constant}\int_i^e \frac{dP}{P} = -P_i v_i \ln \frac{P_e}{P_i} \tag{7.19}$$

Note that the P–v and T–s diagrams of Fig. 6.13 are applicable to represent the slope of polytropic processes in this case as well.

These evaluations of the integral

$$\int_i^e v\, dP$$

may also be used in conjunction with Eq. 7.14 for instances in which kinetic and potential energy changes are not negligibly small.

In-Text Concept Questions

e. In a steady-state single flow, s is either constant or it increases. Is that true?

f. If a flow device has the same inlet and exit pressure, can shaft work be done?

g. A polytropic flow process with $n = 0$ might be which device?

7.4 PRINCIPLE OF THE INCREASE OF ENTROPY

The principle of the increase of entropy for a control mass analysis was discussed in Section 6.11. The same general conclusion is reached for a control volume analysis. This is demonstrated by the split of the whole world into a control volume A and its surroundings, control volume B, as shown in Fig. 7.9. Assume a process takes place in control volume A exchanging mass flows, energy, and entropy transfers with the surroundings. Precisely where the heat transfer enters control volume A, we have a temperature of T_A, which is not necessarily equal to the ambient temperature far away from the control volume.

First, let us write the entropy balance equation for the two control volumes:

$$\frac{dS_{CVA}}{dt} = \dot{m}_i s_i - \dot{m}_e s_e + \frac{\dot{Q}}{T_A} + \dot{S}_{\text{gen }A} \tag{7.20}$$

$$\frac{dS_{CVB}}{dt} = -\dot{m}_i s_i + \dot{m}_e s_e - \frac{\dot{Q}}{T_A} + \dot{S}_{\text{gen }B} \tag{7.21}$$

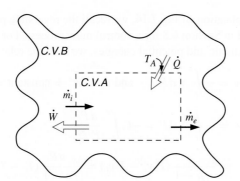

FIGURE 7.9 Entropy change for a control volume plus its surroundings.

and notice that the transfer terms are all evaluated right at the control volume surface. Now we will add the two entropy balance equations to find the net rate of change of S for the total world:

$$\frac{dS_{\text{net}}}{dt} = \frac{dS_{CVA}}{dt} + \frac{dS_{CVB}}{dt}$$

$$= \dot{m}_i s_i - \dot{m}_e s_e + \frac{\dot{Q}}{T_A} + \dot{S}_{\text{gen } A} - \dot{m}_i s_i + \dot{m}_e s_e - \frac{\dot{Q}}{T_A} + \dot{S}_{\text{gen } B}$$

$$= \dot{S}_{\text{gen } A} + \dot{S}_{\text{gen } B} \geq 0 \tag{7.22}$$

Here we notice that all the transfer terms cancel out, leaving only the positive generation terms for each part of the world. If no process takes place in the ambient environment, that generation term is zero. However, we also notice that for the heat transfer to move in the indicated direction, we must have $T_B \geq T_A$; that is, the heat transfer takes place over a finite temperature difference, so an irreversible process occurs in the surroundings. Such a situation is called an external irreversible process. This distinguishes it from any generation of s inside the control volume A, then called an internal irreversible process.

For this general control volume analysis, we arrive at the same conclusion as for the control mass situation—the entropy for the total world must increase or stay constant, $dS_{\text{net}}/dt \geq 0$, from Eq. 7.22. Only those processes that satisfy this equation can possibly take place; any process that would reduce the total entropy is impossible and will not occur.

Some other comments about the principle of the increase of entropy are in order. If we look at and evaluate changes in states for various parts of the world, we can find the net rate by the left-hand side of Eq. 7.22 and thus verify that it is positive for processes we consider. As we do this, we limit the focus to a control volume with a process occurring and the immediate ambient air affected by this process. Notice that the left-hand side sums the storage, but it does not explain where the entropy is made. If we want detailed information about where the entropy is made, we must make a number of control volume analyses and evaluate the storage and transfer terms for each control volume. Then the rate of generation is found from the balance, that is, from an equation like Eq. 7.22, and that must be positive or, at the least, zero. So, not only must the total entropy increase by the sum of the generation terms, but we also must have a positive or at least zero entropy generation in every conceivable control volume. This applies to very small (even differential dV) control volumes, so only processes that locally generate entropy (or let it stay constant) will happen; any process that locally would destroy entropy cannot take place. Remember, this does not preclude that entropy for some mass decreases as long as that is caused by a heat transfer (or net transfer by mass flow) out of that mass; that is, the negative storage is explained by a negative transfer term.

To further illustrate the principle of increase in entropy, consider the case of a steady state process with multiple flows as was done in Section 4.5 and the desuperheater shown in Ex. 7.5. Consider the mixing chamber in Fig. 7.5 with two inlet flows and a single exit flow operating in steady-state mode with no shaft work, and we neglect kinetic and potential energies. The energy and entropy equations for this case become

Energy Eq. 4.10: $$0 = \dot{m}_1 h_1 + \dot{m}_2 h_2 - \dot{m}_3 h_3 + \dot{Q} \tag{7.23}$$

Entropy Eq. 7.2: $$0 = \dot{m}_1 s_1 + \dot{m}_2 s_2 - \dot{m}_3 s_3 + \dot{Q}/T + \dot{S}_{\text{gen}}$$

As in the previous analysis, we can scale the equations with $\dot{m}_3$ to have the ratio $y = \dot{m}_1/\dot{m}_3$ and the other mass flow ratio is $1 - y = \dot{m}_2/\dot{m}_3$. The exit flow properties becomes

$$h_3 = y\, h_1 + (1 - y)h_2 + \widetilde{q} \tag{7.24}$$

$$s_3 = y\, s_1 + (1 - y)s_2 + \widetilde{q}/T + \widetilde{s}_{\text{gen}} \tag{7.25}$$

$$\widetilde{q} = \dot{Q}/\dot{m}_3; \qquad \widetilde{s}_{\text{gen}} = \dot{S}_{\text{gen}}/\dot{m}_3 \tag{7.26}$$

If the heat transfer is zero, the exit enthalpy becomes the mass flow weighted average of the two inlet enthalpies. However, the exit entropy becomes the mass flow weighted average of the two inlet entropies plus an amount due to the entropy generation. As the entropy generation is positive (minimum zero) the exit entropy is then larger resulting in a net increase of the entropy, which is stored in the surroundings.

In typical devices where several valves are used to control the flow rates, they introduce irreversible throttling processes besides having an irreversible mixing and possible heat transfer over finite temperature differences.

Example 7.9

Saturated vapor R-410A enters the uninsulated compressor of a home central air-conditioning system at $5\,°C$. The flow rate of refrigerant through the compressor is 0.08 kg/s, and the electrical power input is 3 kW. The exit state is $65\,°C$, 3000 kPa. Any heat transfer from the compressor is with the ambient environment at $30\,°C$. Determine the rate of entropy generation for this process.

Control volume:	Compressor out to ambient T_0.
Inlet state:	T_i, x_i known; state fixed.
Exit state:	P_e, T_e known; state fixed.
Process:	Steady-state, single fluid flow.
Model:	R-410A tables, B.4.

Analysis

Steady-state, single flow. Assume negligible changes in kinetic and potential energies.

$$\text{Continuity Eq.:} \quad \dot{m}_i = \dot{m}_e = \dot{m}$$

$$\text{Energy Eq.:} \quad 0 = \dot{Q}_{\text{c.v.}} + \dot{m}h_i - \dot{m}h_e - \dot{W}_{\text{c.v.}}$$

$$\text{Entropy Eq.:} \quad 0 = \dot{m}(s_i - s_e) + \frac{\dot{Q}_{\text{c.v.}}}{T_0} + \dot{S}_{\text{gen}}$$

Solution

From the R-410A tables, B.4, we get

$$h_i = 280.6 \text{ kJ/kg}, \qquad s_i = 1.0272 \text{ kJ/kg·K}$$

$$h_e = 307.8 \text{ kJ/kg}, \qquad s_e = 1.0140 \text{ kJ/kg·K}$$

From the energy equation,

$$\dot{Q}_{c.v.} = 0.08 \text{ kg/s} \times (307.8 - 280.6) \text{ kJ/kg} - 3.0 \text{ kW} = 2.176 - 3.0 = -0.824 \text{ kW}$$

From the entropy equation,

$$\dot{S}_{gen} = \dot{m}(s_e - s_i) - \frac{\dot{Q}_{c.v.}}{T_0}$$

$$= 0.08 \text{ kg/s} \times (1.0140 - 1.0272) \text{ kJ/kg·K} - (-0.824 \text{ kW}/303.2 \text{ K})$$

$$= -0.001\,06 + 0.002\,72 = +0.001\,66 \text{ kW/K}$$

Notice that the entropy generation also equals the storage effect in the surroundings.

Remark: In this process there are two sources of entropy generation: internal irreversibilities associated with the process taking place in the R-410A (compressor) and external irreversibilities associated with heat transfer across a finite temperature difference. Since we do not have the temperature at which the heat transfer leaves the R-410A, we cannot separate the two contributions.

7.5 ENGINEERING APPLICATIONS—EFFICIENCY

In Chapter 5 we noted that the second law of thermodynamics led to the concept of thermal efficiency for a heat engine cycle, namely,

$$\eta_{th} = \frac{W_{net}}{Q_H}$$

where W_{net} is the net work of the cycle and Q_H is the heat transfer from the high-temperature body.

In this chapter we have extended our application of the second law to control volume processes, and in Section 7.2 we considered several different types of devices. For steady-state processes, this included an ideal (reversible) turbine, compressor, and nozzle. We realize that actual devices of these types are not reversible, but the reversible models may in fact be very useful to compare with or evaluate the real, irreversible devices in making engineering calculations. This leads in each type of device to a component or machine process efficiency. For example, we might be interested in the efficiency of a turbine in a steam power plant or of the compressor in a gas turbine engine.

In general, we can say that to determine the efficiency of a machine in which a process takes place, we compare the actual performance of the machine under given conditions to the performance that would have been achieved in an ideal process. It is in the definition of this ideal process that the second law becomes a major consideration. For example, a steam turbine is intended to be an adiabatic machine. The only heat transfer is the unavoidable

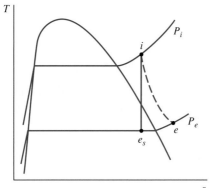
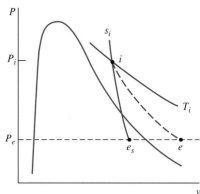

FIGURE 7.10 The process in a reversible adiabatic steam turbine and an actual turbine.

heat transfer that takes place between the given turbine and the surroundings. We also note that for a given steam turbine operating in a steady-state manner, the state of the steam entering the turbine and the exhaust pressure are fixed. Therefore, the ideal process is a reversible adiabatic process, which is an isentropic process, between the inlet state and the turbine exhaust pressure. In other words, the variables P_i, T_i, and P_e are the design variables—the first two because the working fluid has been prepared in prior processes to be at these conditions at the turbine inlet, while the exit pressure is fixed by the environment into which the turbine exhausts. Thus, the ideal turbine process would go from state i to state e_s, as shown in Fig. 7.10, whereas the real turbine process is irreversible, with the exhaust at a larger entropy at the real exit state e. Figure 7.10 shows typical states for a steam turbine, where state e_s is in the two-phase region, and state e may be as well, or may be in the superheated vapor region, depending on the extent of irreversibility of the real process. Denoting the work done in the real process i to e as w, and that done in the ideal, isentropic process from the same P_i, T_i to the same P_e as w_s, we define the efficiency of the turbine as

$$\eta_{\text{turbine}} = \frac{w}{w_s} = \frac{h_i - h_e}{h_i - h_{es}} \tag{7.27}$$

The same definition applies to a gas turbine, where all states are in the gaseous phase. Typical turbine efficiencies are 0.70–0.88, with large turbines usually having higher efficiencies than small ones.

Example 7.10

A steam turbine receives steam at a pressure of 1 MPa and a temperature of 300 °C. The steam leaves the turbine at a pressure of 15 kPa. The work output of the turbine is measured and is found to be 600 kJ/kg of steam flowing through the turbine. Determine the efficiency of the turbine.

Control volume:	Turbine.
Inlet state:	P_i, T_i known; state fixed.
Exit state:	P_e known.
Process:	Steady state.
Model:	Steam tables.

Analysis

The efficiency of the turbine is given by Eq. 7.27

$$\eta_{turbine} = \frac{w_a}{w_s}$$

Thus, to determine the turbine efficiency, we calculate the work that would be done in an isentropic process between the given inlet state and the final pressure. For this isentropic process, we have

Continuity Eq.: $\dot{m}_i = \dot{m}_e = \dot{m}$

Energy Eq.: $h_i = h_{es} + w_s$

Second law: $s_i = s_{es}$

Solution

From the steam tables, we get

$$h_i = 3051.2 \text{ kJ/kg}, \qquad s_i = 7.1228 \text{ kJ/kg·K}$$

Therefore, at $P_e = 15$ kPa,

$$s_{es} = s_i = 7.1228 = 0.7548 + x_{es}7.2536$$

$$x_{es} = 0.8779$$

$$h_{es} = 225.9 + 0.8779 \times (2373.1) = 2309.3 \text{ kJ/kg}$$

From the energy equation for the isentropic process,

$$w_s = h_i - h_{es} = 3051.2 - 2309.3 = 741.9 \text{ kJ/kg}$$

But, since

$$w_a = 600 \text{ kJ/kg}$$

we find that

$$\eta_{turbine} = \frac{w_a}{w_s} = \frac{600}{741.9} = 0.809 = 80.9\%$$

In connection with this example, it should be noted that to find the actual state e of the steam exiting the turbine, we need to analyze the real process taking place. For the real process

$$\dot{m}_i = \dot{m}_e = \dot{m}$$
$$h_i = h_e + w_a$$
$$s_e > s_i$$

Therefore, from the energy equation for the real process, we have

$$h_e = 3051.2 - 600 = 2451.2 \text{ kJ/kg}$$

$$2451.2 = 225.9 + x_e 2373.1$$

$$x_e = 0.9377$$

It is important to keep in mind that the turbine efficiency is defined in terms of an ideal, isentropic process from P_i and T_i to P_e, even when one or more of these variables is unknown. This is illustrated in the following example.

Example 7.11

Air enters a gas turbine at 1600 K and exits at 100 kPa, 830 K. The turbine efficiency is estimated to be 85 %. What is the turbine inlet pressure?

> *Control volume*:   Turbine.
> *Inlet state*: T_i known.
> *Exit state*: P_e, T_e known; state fixed.
> *Process*:   Steady state.
> *Model*:   Air tables, Table A.7.

Analysis

The efficiency, which is 85 %, is given by Eq. 7.27,

$$\eta_{\text{turbine}} = \frac{w}{w_s}$$

The energy equation for the real, irreversible process is

$$h_i = h_e + w$$

For the ideal, isentropic process from P_i, T_i to P_e, the energy equation is

$$h_i = h_{es} + w_s$$

and the second law is, from Eq. 6.19,

$$s_{es} - s_i = 0 = s_{Tes}^0 - s_{Ti}^0 - R \ln \frac{P_e}{P_i}$$

(Note that this equation is only for the ideal isentropic process and not for the real process, for which $s_e - s_i > 0$.)

Solution

From the air tables, Table A.7, at 1600 K, we get

$$h_i = 1757.3 \text{ kJ/kg}, \qquad s_{Ti}^0 = 8.6905 \text{ kJ/kg} \cdot \text{K}$$

From the air tables at 830 K (the actual turbine exit temperature),

$$h_e = 855.3 \text{ kJ/kg}$$

Therefore, from the energy equation for the real process,

$$w = 1757.3 - 855.3 = 902.0 \text{ kJ/kg}$$

Using the definition of turbine efficiency,

$$w_s = 902.0/0.85 = 1061.2 \text{ kJ/kg}$$

From the energy equation for the isentropic process,

$$h_{es} = 1757.3 - 1061.2 = 696.1 \text{ kJ/kg}$$

so that, from the air tables,

$$T_{es} = 683.7 \text{ K}, \qquad s^0_{Tes} = 7.7148 \text{ kJ/kg·K}$$

and the turbine inlet pressure is determined from

$$0 = 7.7148 - 8.6905 - 0.287 \ln \frac{100}{P_i}$$

or

$$P_i = 2995 \text{ kPa}$$

As was discussed in Section 4.4, unless specifically noted to the contrary, we normally assume compressors or pumps to be adiabatic. In this case the fluid enters the compressor at P_i and T_i, the condition at which it exists, and exits at the desired value of P_e, the reason for building the compressor. Thus, the ideal process between the given inlet state i and the exit pressure would be an isentropic process between state i and state e_s, as shown in Fig. 7.11, with a work input of w_s. The real process, however, is irreversible, and the fluid exits at the real state e with a larger entropy, and a larger amount of work input w is required. The compressor (or pump, in the case of a liquid) efficiency is defined as

$$\eta_{\text{comp}} = \frac{w_s}{w} = \frac{h_i - h_{es}}{h_i - h_e} \tag{7.28}$$

Typical compressor efficiencies are 0.70–0.88, with large compressors usually having higher efficiencies than small ones.

If an effort is made to cool a gas during compression by using a water jacket or fins, the ideal process is considered a reversible isothermal process, the work input for which is

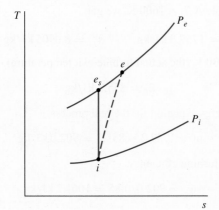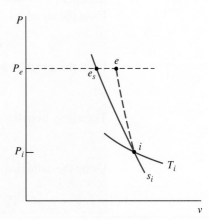

FIGURE 7.11 The compression process in an ideal and an actual adiabatic compressor.

w_T, compared to the larger work required w for the real compressor. The efficiency of the cooled compressor is then

$$\eta_{\text{cooled comp}} = \frac{w_T}{w} \tag{7.29}$$

Example 7.12

Air enters an automotive supercharger at 100 kPa, 300 K and is compressed to 150 kPa. The efficiency is 70 %. What is the required work input per kilogram of air? What is the exit temperature?

Control volume: Supercharger (compressor).

Inlet state: P_i, T_i known; state fixed.

Exit state: P_e known.

Process: Steady state.

Model: Ideal gas, 300 K specific heat, Table A.5.

Analysis

The efficiency, which is 70 %, is given by Eq. 7.28,

$$\eta_{\text{comp}} = \frac{w_s}{w}$$

The energy equation for the real, irreversible process is

$$h_i = h_e + w, \qquad w = C_{p0}(T_i - T_e)$$

For the ideal, isentropic process from P_i, T_i to P_e, the energy equation is

$$h_i = h_{es} + w_s, \qquad w_s = C_{p0}(T_i - T_{es})$$

and the second law is, from Eq. 6.23,

$$\frac{T_{es}}{T_i} = \left(\frac{P_e}{P_i}\right)^{(k-1)/k}$$

Solution

Using C_{p0} and k from Table A.5, from the second law, we get

$$T_{es} = 300\left(\frac{150}{100}\right)^{0.286} = 336.9\,\text{K}$$

From the energy equation for the isentropic process, we have

$$w_s = 1.004 \times (300 - 336.9) = -37.1\,\text{kJ/kg}$$

so that, from the efficiency, the real work input is

$$w = -37.1/0.70 = -53.0\,\text{kJ/kg}$$

and from the energy equation for the real process, the exit temperature is

$$T_e = 300 - \frac{-53.0}{1.004} = 352.8\,\text{K}$$

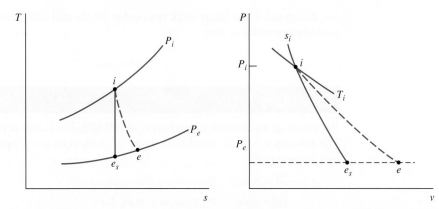

FIGURE 7.12 The ideal and actual processes in an adiabatic nozzle.

Our final example is that of nozzle efficiency. As discussed in Section 4.4, the purpose of a nozzle is to produce a high-velocity fluid stream, or in terms of energy, a large kinetic energy, at the expense of the fluid pressure. The design variables are the same as for a turbine: P_i, T_i, and P_e. A nozzle is usually assumed to be adiabatic, such that the ideal process is an isentropic process from state i to state e_s, as shown in Fig. 7.12, with the production of velocity $\mathbf{V}_{es}$. The real process is irreversible, with the exit state e having a larger entropy, and a smaller exit velocity $\mathbf{V}_e$. The nozzle efficiency is defined in terms of the corresponding kinetic energies,

$$\eta_{\text{nozz}} = \frac{\mathbf{V}_e^2/2}{\mathbf{V}_{es}^2/2} \tag{7.30}$$

Nozzles are simple devices with no moving parts. As a result, nozzle efficiency may be very high, typically 0.90 to 0.97.

In summary, to determine the efficiency of a device that carries out a process (rather than a cycle), we compare the actual performance to what would be achieved in a related but well-defined ideal process.

7.6 SUMMARY OF GENERAL CONTROL VOLUME ANALYSIS

One of the more important subjects to learn is the control volume formulation of the general laws (conservation of mass, momentum and energy, balance of entropy) and the specific laws that in the current presentation are given in an integral (mass averaged) form. The following steps show a systematic way to formulate a thermodynamic problem so that it does not become a formula chase, allowing you to solve general and even unfamiliar problems.

Formulation Steps

Step 1. Make a physical model of the system with components and illustrate all mass flows, heat flows, and work rates. Include an indication of forces like external pressures and gravitation.

Step 2. Define (i.e., choose) a control mass or control volume by placing a control surface that contains the substance you want to analyze. This choice is very important since the formulation will depend on it. Be sure that only those mass flows, heat fluxes, and work terms you want to analyze cross the control surface. Include as much of the system as you can to eliminate flows and fluxes that you don't want to enter the formulation. Number the states of the substance where it enters or leaves the control volume, and if it does not have the same state, label different parts of the system with storage.

Step 3. Write down the general laws for each of the chosen control volumes. For control volumes that have mass flows or heat and work fluxes between them, make certain that what leaves one control volume enters the other (i.e., have one term in each equation with an opposite sign). When the equations are written down, use the most general form and cancel terms that are not present. Only two forms of the general laws should be used in the formulation: (1) the original rate form (Eq. 7.2 for S) and (2) the time-integrated form (Eq. 7.12 for S), where now terms that are not present are canceled. It is very important to distinguish between storage terms (left-hand side) and flow terms.

Step 4. Write down the auxiliary or particular laws for whatever is inside each of the control volumes. The constitution for a substance is either written down or referenced to a table. The equation for a given process is normally easily written down. It is given by the way the system or devices are constructed and often is an approximation to reality. That is, we make a mathematical model of the real-world behavior.

Step 5. Finish the formulation by combining all the equations, but don't put in numbers yet. At this point, check which quantities are known and which are unknown. Here it is important to be able to find all the states of the substance and determine which two independent properties determine any given state. This task is most easily done by illustrating all the processes and states in a P–v, T–v, T–s, or similar diagram. These diagrams will also show what numbers to look up in the tables to determine where a given state is.

Step 6. The equations are now solved for the unknowns by writing all terms with unknown variables on one side and known terms on the other. It is usually easy to do this, but in some cases it may require an iteration technique to solve the equations (for instance, if you have a combined property of u, P, v, like $u + 1/2 Pv$ and not $h = u + Pv$). As you find the numerical values for different quantities, make sure they make sense and are within reasonable ranges.

SUMMARY

The second law of thermodynamics is extended to a general control volume with mass flow rates in or out for steady and transient processes. The vast majority of common devices and complete systems can be treated as nearly steady-state operations even if they have slower transients, as in a car engine or jet engine. Simplification of the entropy equation arises when applied to steady-state and single-flow devices like a turbine, nozzle, compressor, or pump. The second law and the Gibbs property relation are used to develop a general expression for reversible shaft work in a single flow that is useful in understanding the importance of the specific volume (or density) that influences the magnitude of the work. For a flow with no shaft work, consideration of the reversible process also leads to the derivation of the energy equation for an incompressible fluid as the Bernoulli equation. This covers the flows of liquids such as water or hydraulic fluid as well as air flow at low speeds, which can be considered incompressible for velocities less than a third of the speed of sound.

Many actual devices operate with some irreversibility in the processes that occur, so we also have entropy generation in the flow processes and the total entropy is always increasing. The characterization of performance of actual devices can be done with a comparison to a corresponding ideal device, giving efficiency as the ratio of two energy terms (work or kinetic energy).

You should have learned a number of skills and acquired abilities from studying this chapter that will allow you to

- Apply the second law to more general control volumes.
- Analyze steady-state, single-flow devices such as turbines, nozzles, compressors, and pumps, both reversible and irreversible.
- Know how to extend the second law to transient processes.
- Analyze complete systems as a whole or divide them into individual devices.
- Apply the second law to multiple-flow devices such as heat exchangers, mixing chambers, and turbines with several inlets and outlets.
- Recognize when you have an incompressible flow where you can apply the Bernoulli equation or the expression for reversible shaft work.
- Know when you can apply the Bernoulli equation and when you cannot.
- Know how to evaluate the shaft work for a polytropic process.
- Know how to apply the analysis to an actual device using an efficiency and identify the closest ideal approximation to the actual device.
- Know the difference between a cycle efficiency and a device efficiency.
- Have a sense of entropy as a measure of disorder or chaos.

KEY CONCEPTS AND FORMULAS		
Rate equation for entropy	rate of change = +in − out + generation $$\dot{S}_{\text{c.v.}} = \sum \dot{m}_i s_i - \sum \dot{m}_e s_e + \sum \frac{\dot{Q}_{\text{c.v.}}}{T} + \dot{S}_{\text{gen}}$$	
Steady-state single flow	$s_e = s_i + \int_i^e \frac{\delta q}{T} + s_{\text{gen}}$	
Reversible shaft work	$w = -\int_i^e v\,dP + \frac{1}{2}\mathbf{V}_i^2 - \frac{1}{2}\mathbf{V}_e^2 + gZ_i - gZ_e$	
Reversible heat transfer	$q = \int_i^e T\,ds = h_e - h_i - \int_i^e v\,dP$ (from the Gibbs relation)	
Bernoulli equation	$v(P_i - P_e) + \frac{1}{2}\mathbf{V}_i^2 - \frac{1}{2}\mathbf{V}_e^2 + gZ_i - gZ_e = 0$ (v = constant)	
Polytropic process work	$w = -\dfrac{n}{n-1}(P_e v_e - P_i v_i) = -\dfrac{nR}{n-1}(T_e - T_i) \qquad n \neq 1$ $w = -P_i v_i \ln \dfrac{P_e}{P_i} = -RT_i \ln \dfrac{P_e}{P_i} = RT_i \ln \dfrac{v_e}{v_i} \qquad n = 1$ The work is shaft work $w = -\int_i^e v\,dP$ and for ideal gas	
Isentropic efficiencies	$\eta_{\text{turbine}} = w_{Tac}/w_{Ts}$ (Turbine work is out) $\eta_{\text{compressor}} = w_{Cs}/w_{Cac}$ (Compressor work is in) $\eta_{\text{pump}} = w_{Ps}/w_{Pac}$ (Pump work is in) $\eta_{\text{nozzle}} = \frac{1}{2}\mathbf{V}_{ac}^2 \Big/ \frac{1}{2}\mathbf{V}_s^2$ (Kinetic energy is out)	

CONCEPT-STUDY GUIDE PROBLEMS

7.1 If we follow a mass element going through a reversible adiabatic flow process, what can we say about the change of state?

7.2 Which process will make the statement in In-Text Concept Question **e** true?

7.3 A reversible process in a steady flow with negligible kinetic and potential energy changes is shown in Fig. P7.3. Indicate the change $h_e - h_i$ and transfers w and q as positive, zero, or negative.

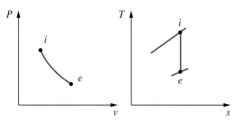

FIGURE P7.3

7.4 A reversible process in a steady flow of air with negligible kinetic and potential energy changes is shown

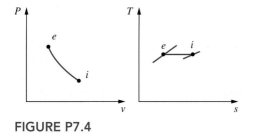

FIGURE P7.4

in Fig. P7.4. Indicate the change $h_e - h_i$ and transfers w and q as positive, zero, or negative.

7.5 A reversible steady isobaric flow has 1 kW of heat added with negligible changes in *KE* and *PE*; what is the work transfer?

7.6 An air compressor has a significant heat transfer out (review Example 7.4 to see how high *T* becomes if no heat transfer occurs). Is that good or should it be insulated?

7.7 Friction in a pipe flow causes a slight pressure decrease and a slight temperature increase. How does that affect entropy?

7.8 To increase the work out of a turbine for given inlet and exit pressures, how should the inlet state be changed?

7.9 An irreversible adiabatic flow of liquid water in a pump has a higher exit *P*. Is the exit *T* higher or lower?

7.10 The shaft work in a pump to increase the pressure is small compared to the shaft work in an air compressor for the same pressure increase. Why?

7.11 Liquid water is sprayed into the hot gases before they enter the turbine section of a large gas-turbine power plant. It is claimed that the larger mass flow rate produces more work. Is that the reason?

7.12 A tank contains air at 400 kPa, 300 K, and a valve opens up for flow out to the outside, which is at 100 kPa, 300 K. How does the state of the air that flows out change?

HOMEWORK PROBLEMS

Steady-State Reversible Single-Flow Processes

7.13 A turbine receives steam at 6 MPa, 600 °C with an exit pressure of 600 kPa. Assume the turbine is adiabatic and neglect kinetic energies. Find the exit temperature and the specific work.

7.14 A condenser receives R-410A at −20 °C and quality 80 %, with the exit flow being saturated liquid at −20 °C. Consider the cooling to be a reversible process and find the specific heat transfer from the entropy equation.

7.15 The exit nozzle in a jet engine receives air at 1200 K, 150 kPa with negligible kinetic energy. The exit

pressure is 80 kPa, and the process is reversible and adiabatic. Use constant specific heat at 300 K to find the exit velocity.

7.16 Do the previous problem using Table A.7.

7.17 A reversible adiabatic compressor receives 0.05 kg/s saturated vapor R-410A at 400 kPa and has an exit pressure of 1400 kPa. Neglect kinetic energies and find the exit temperature and the minimum power needed to drive the unit.

7.18 Nitrogen gas flowing in a pipe at 500 kPa, 200 °C, and at a velocity of 10 m/s, should be expanded in a nozzle to produce a velocity of 300 m/s.

Determine the exit pressure and cross-sectional area of the nozzle if the mass flow rate is 0.15 kg/s and the expansion is reversible and adiabatic.

7.19 A reversible isothermal expander (a turbine with heat transfer) has an inlet flow of carbon dioxide at 3 MPa, 80 °C and an exit flow at 1 MPa, 80 °C. Find the specific heat transfer from the entropy equation and the specific work from the energy equation assuming ideal gas.

7.20 Solve the previous problem using Table B.3.

7.21 A compressor in a commercial refrigerator receives R-410A at −25 °C and unknown quality. The exit is at 3000 kPa, 60 °C, and the process is assumed to be reversible and adiabatic. Neglect kinetic energies and find the inlet quality and the specific work.

7.22 A compressor brings a hydrogen gas flow at 280 K, 100 kPa up to a pressure of 1000 kPa in a reversible process. How hot is the exit flow and what is the specific work input?

7.23 Atmospheric air at −45 °C, 60 kPa enters the front diffuser of a jet engine, shown in Fig. P7.23, with a velocity of 900 km/h and a frontal area of 1 m². After adiabatic diffuser process, the velocity is 20 m/s. Find the diffuser exit temperature and the maximum pressure possible.

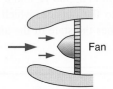

FIGURE P7.23

7.24 A compressor is surrounded by cold R-134a, so it works as an isothermal compressor. The inlet state is 0 °C, 100 kPa and the exit state is saturated vapor. Find the specific heat transfer and the specific work.

7.25 A flow of 2 kg/s saturated vapor R-410A at 500 kPa is heated at constant pressure to 60 °C. The heat is supplied by a heat pump that receives heat from the ambient at 300 K and work input, shown in Fig. P7.25. Assume everything is reversible and find the rate of work input.

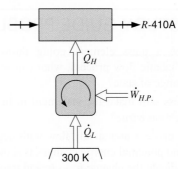

FIGURE P7.25

7.26 A flow of 2 kg/s hot exhaust air at 180 °C, 125 kPa supplies heat to a heat engine in a setup similar to that of the previous problem, with the heat engine rejecting heat to the ambient air at 290 K, and the air leaves the heat exchanger at 50 °C. Find the maximum possible power out of the heat engine.

7.27 A diffuser is a steady-state device in which a fluid flowing at high velocity is decelerated such that the pressure increases in the process. Air at 120 kPa, 30 °C enters a diffuser with a velocity of 200 m/s and exits with a velocity of 20 m/s. Assuming the process is reversible and adiabatic, what are the exit pressure and temperature of the air?

7.28 Air enters a turbine at 800 kPa, 1200 K and expands in a reversible adiabatic process to 100 kPa. Calculate the exit temperature and the specific work output using Table A.7 and repeat using constant specific heat from Table A.5.

7.29 An expander receives 0.5 kg/s air at 2000 kPa, 300 K with an exit state of 400 kPa, 300 K. Assume the process is reversible and isothermal. Find the rates of heat transfer and work, neglecting kinetic and potential energy changes.

7.30 A compressor receives air at 290 K, 95 kPa and shaft work of 5.5 kW from a gasoline engine. It should deliver a mass flow rate of 0.01 kg/s air to a pipeline. Find the maximum possible exit pressure of the compressor.

7.31 A reversible steady-state device receives a flow of 1 kg/s air at 400 K, 450 kPa and the air leaves at 600 K, 100 kPa. Heat transfer of 900 kW is added from a 1000 K reservoir, 50 kW is rejected at 350 K, and some heat transfer takes place at 500 K. Find the heat transferred at 500 K and the rate of work produced.

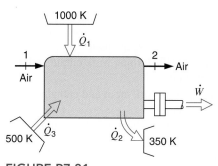

FIGURE P7.31

Multiple Devices and Cycles

7.32 A steam turbine in a power plant receives 5 kg/s steam at 3000 kPa, 500 °C. Twenty percent of the flow is extracted at 1000 kPa to a feedwater heater, and the remainder flows out at 200 kPa. Find the two exit temperatures and the turbine power output.

7.33 A reversible adiabatic compression of an air flow from 20 °C, 100 kPa to 200 kPa is followed by an expansion down to 100 kPa in an ideal nozzle. What are the two processes? How hot does the air get? What is the exit velocity?

7.34 One technique for operating a steam turbine in part-load power output is to throttle the steam to a lower pressure before it enters the turbine, as shown in Fig. P7.34. The steamline conditions are 2 MPa, 400 °C, and the turbine exhaust pressure is fixed at 10 kPa. Assuming the expansion inside the turbine is reversible and adiabatic, determine the specific turbine work for no throttling and the specific turbine work (part-load) if it is throttled to 500 kPa. Show both processes in a T–s diagram.

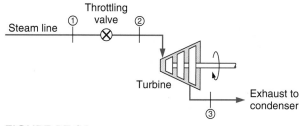

FIGURE P7.34

7.35 An adiabatic air turbine receives 1 kg/s air at 1500 K, 1.6 MPa and 2 kg/s air at 400 kPa, T_2 in a setup similar to that of Fig. P4.69 with an exit flow at 100 kPa. What should the temperature T_2 be so that the whole process can be reversible?

7.36 A turbocharger boosts the inlet air pressure to an automobile engine. It consists of an exhaust gas-driven turbine directly connected to an air compressor, as shown in Fig. P7.36. For a certain engine load the conditions are given in the figure. Assume that both the turbine and the compressor are reversible and adiabatic, having also the same mass flow rate. Calculate the turbine exit temperature and power output. Also find the compressor exit pressure and temperature.

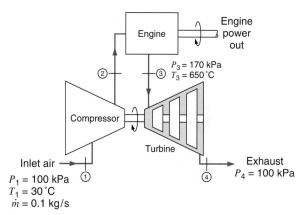

FIGURE P7.36

7.37 A flow of 5 kg/s water at 100 kPa, 30 °C should be delivered as steam at 1000 kPa, 350 °C to some application. We have a heat source at constant 500 °C. If the process should be reversible, how much heat transfer should we have?

7.38 A heat-powered portable air compressor consists of three components: (a) an adiabatic compressor, (b) a constant-pressure heater (heat supplied from an outside source), and (c) an adiabatic turbine (see Fig. P7.38). Ambient air enters the compressor at

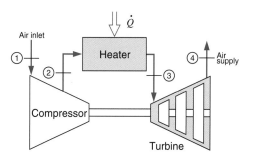

FIGURE P7.38

100 kPa, 300 K and is compressed to 600 kPa. All of the power from the turbine goes into the compressor, and the turbine exhaust is the supply of compressed air. If this pressure is required to be 200 kPa, what must the temperature be at the exit of the heater?

7.39 A two-stage compressor having an intercooler takes in air at 300 K, 100 kPa, and compresses it to 2 MPa, as shown in Fig. P7.39. The cooler then cools the air to 340 K, after which it enters the second stage, which has an exit pressure of 15 MPa. Both stages are adiabatic and reversible. Find q in the cooler, total specific work, and compare this to the work required with no intercooler.

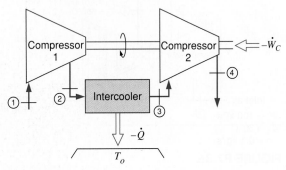

FIGURE P7.39

7.40 A certain industrial process requires a steady 0.75 kg/s supply of compressed air at 500 kPa at a maximum temperature of 30 °C, as shown in Fig. P7.40. This air is to be supplied by installing a compressor and an aftercooler. Local ambient conditions are 100 kPa, 20 °C. Using a reversible compressor, determine the power required to drive the compressor and the rate of heat rejection in the aftercooler.

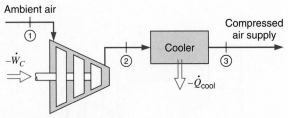

FIGURE P7.40

7.41 Consider a steam turbine power plant operating near critical pressure, as shown in Fig. P7.41. As a first approximation, it may be assumed that the turbine and the pump processes are reversible and adiabatic. Neglecting any changes in kinetic and potential energies, calculate

a. The specific turbine work output and the turbine exit state
b. The pump work input and enthalpy at the pump exit state
c. The thermal efficiency of the cycle

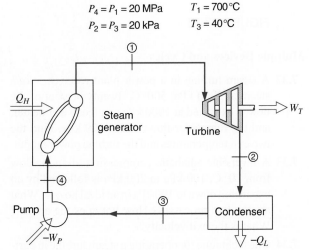

FIGURE P7.41

Transient Processes

7.42 A 10 m tall water tank with cross-sectional area 2 m² is on a tower, so the bottom is 5 m up from ground level and the top is open to the atmosphere. It is initially empty and then is filled by a pump taking water at ambient $T = 17\,°C$, 100 kPa from a small pond at ground level. Assume the process is reversible and find the total pump work.

7.43 A 0.5 m³ tank containing carbon dioxide at 300 K, 150 kPa is now filled from a supply of carbon dioxide at 300 K, 150 kPa by a compressor to a final tank pressure of 600 kPa. Assume the whole process is adiabatic and reversible. Find the final mass and temperature in the tank and the required work to the compressor.

7.44 A tank contains 1 kg of carbon dioxide at 6 MPa, 60 °C, and it is connected to a turbine with an exhaust at 1000 kPa. The carbon dioxide flows out of the tank and through the turbine until a final state in the tank of saturated vapor is reached. If the process is adiabatic and reversible, find the final mass in the tank and the turbine work output.

7.45 A supply line is supplied by an insulated compressor that takes in R-134a at 5 °C, quality of 96.5 %, and compresses it to 3 MPa in a reversible process. An insulated 2 m³ tank is charged with R-134a from the line, the tank is initially evacuated, and the valve is closed when the pressure inside the tank reaches 2 MPa. Calculate the total work input to the compressor to charge the tank.

7.46 An underground salt mine, 100 000 m³ in volume, contains air at 290 K, 100 kPa. The mine is used for energy storage, so the local power plant pumps it up to 2.1 MPa using outside air at 290 K, 100 kPa. Assume the pump is ideal and the process is adiabatic. Find the final mass and temperature of the air and the required pump work.

7.47 R-410A at 120 °C, 4 MPa is in an insulated tank, and flow is now allowed out to a turbine with a backup pressure of 800 kPa. The flow continues to a final tank pressure of 800 kPa, and the process stops. If the initial mass was 1 kg, how much mass is left in the tank and what is the turbine work, assuming a reversible process?

Reversible Shaft Work: Bernoulli Equation

7.48 A river flowing at 0.5 m/s across a 1 m high and 10 m wide area has a dam that creates an elevation difference of 3 m. How much energy can a turbine deliver per day if 80 % of the potential energy can be extracted as work?

7.49 How much liquid water at 15 °C can be pumped from 100 kPa to 300 kPa with a 5 kW motor?

7.50 A large storage tank contains saturated liquid nitrogen at ambient pressure, 100 kPa; it is to be pumped to 500 kPa and fed to a pipeline at the rate of 0.5 kg/s. How much power input is required for the pump, assuming it to be reversible?

7.51 Liquid water at 300 kPa, 15 °C flows in a garden hose with a small ideal nozzle. How high a velocity can be generated? If the water jet is directed straight up, how high will it go?

7.52 A wave comes rolling into the beach at 2 m/s horizontal velocity. Neglect friction and find how high up (elevation) on the beach the wave will reach.

7.53 An irrigation pump takes water from a river at 10 °C, 100 kPa and pumps it up to an open canal at a 50 m higher elevation. The pipe diameter into and out of the pump is 0.1 m, and the motor driving the pump is 5 hp. Neglect kinetic energies and friction and find the maximum possible mass flow rate.

7.54 Saturated R-410A at −10 °C is pumped/compressed to a pressure of 2.0 MPa at the rate of 0.5 kg/s in a reversible adiabatic process. Calculate the power required and the exit temperature for the two cases of inlet state of the R-410A:
a. quality of 100 %.
b. quality of 0 %.

7.55 The underwater bulb nose of a container ship has a velocity relative to the ocean water of 8 m/s. What is the pressure at the front stagnation point that is 2 m down from the water surface?

7.56 A small water pump at ground level has an inlet pipe down into a well at a depth H with the water at 100 kPa, 15 °C. The pump delivers water at 400 kPa to a building. The absolute pressure of the water must be at least twice the saturation pressure to avoid cavitation. What is the maximum depth this setup will allow?

7.57 A pump/compressor pumps a substance from 150 kPa, 10 °C to 1 MPa in a reversible adiabatic process. The exit pipe has a small crack so that a small amount leaks to the atmosphere at 100 kPa. If the substance is (a) water or (b) R-134a, find the temperature after compression and the temperature of the leak flow as it enters the atmosphere, neglecting kinetic energies.

7.58 A small pump is driven by a 2 kW motor with liquid water at 150 kPa, 10 °C entering. Find the maximum water flow rate you can get with an exit pressure of 1 MPa and negligible kinetic energies. The exit flow goes through a small hole in a spray nozzle out to the atmosphere at 100 kPa, as shown in Fig. P7.58. Find the spray velocity.

— Nozzle

FIGURE P7.58

7.59 A speedboat has a small hole in the front of the drive with the propeller that extends down into the water at a water depth of 0.4 m. Assuming we have a stagnation point at that hole when the boat is sailing with 40 km/h, what is the total pressure there?

7.60 A pipe in a small dam, of 0.5 m diameter, carries liquid water at 150 kPa, 20 °C with a flow rate of 2000 kg/s. The pipe runs to the bottom of the dam 15 m lower into a turbine with a pipe diameter of 0.35 m, as shown in Fig. P7.60. Assume no friction or heat transfer in the pipe and find the pressure of the turbine inlet. If the turbine exhausts to 100 kPa with negligible kinetic energy, what is the rate of work?

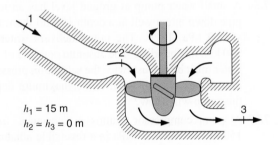

$h_1 = 15$ m
$h_2 = h_3 = 0$ m

FIGURE P7.60

7.61 Air flow at 100 kPa, 290 K, 100 m/s is directed toward a wall. At the wall, the flow stagnates (comes to zero velocity) without any heat transfer, as shown in Fig. P7.61. Find the stagnation pressure (a) assuming incompressible flow, (b) assuming adiabatic compression. Hint: T comes from the energy equation.

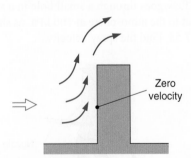

Zero velocity

FIGURE P7.61

7.62 A flow of air at 100 kPa, 300 K enters a device and goes through a polytropic process with $n = 1.3$ before it exits at 600 K. Find the exit pressure, the specific work, and the heat transfer using constant specific heats.

7.63 Solve the previous problem but use the air tables (Table A.7).

7.64 Helium gas enters a steady-flow expander at 800 kPa, 300 °C and exits at 120 kPa. The expansion process can be considered a reversible polytropic process with exponent $n = 1.3$. Calculate the mass flow rate for 150 kW power output from the expander.

7.65 A flow of 4 kg/s ammonia goes through a device in a polytropic process with an inlet state of 150 kPa, -20 °C and an exit state of 400 kPa, 60 °C. Find the polytropic exponent n, the specific work, and the heat transfer.

7.66 Calculate the air temperature and pressure at the stagnation point right in front of a meteorite entering the atmosphere (-50 °C, 50 kPa) with a velocity of 2000 m/s. Do this assuming air is incompressible at the given state and repeat for air being a compressible substance going through an adiabatic compression.

Irreversible Flow Processes

Steady Flow Processes

7.67 Consider a steam turbine with inlet 2 MPa, 350 °C and an exhaust flow as saturated vapor, 100 kPa. There is a heat loss of 6 kJ/kg to the ambient. Is the turbine possible?

7.68 A large condenser in a steam power plant dumps 15 MW by condensing saturated water vapor at 45 °C to saturated liquid. What is the water flow rate and the entropy generation rate with an ambient at 25 °C?

7.69 R-410A at -5 °C, 700 kPa is throttled, so it becomes cold at -30 °C. What is exit P and the specific entropy generation?

7.70 A compressor in a commercial refrigerator receives R-410A at -25 °C and $x = 1$. The exit is at 1000 kPa, 40 °C. Is this compressor possible?

7.71 R-134a at 30 °C, 800 kPa is throttled in a steady flow to a lower pressure, so it comes out at -10 °C. What is the specific entropy generation?

7.72 Analyze the steam turbine described in Problem 4.67. Is it possible?

7.73 Two flowstreams of water, one at 0.6 MPa, saturated vapor, and the other at 0.6 MPa, 600 °C, mix adiabatically in a steady-flow process to produce a single flow out at 0.6 MPa, 400 °C. Find the total entropy generation for this process.

7.74 A geothermal supply of hot water at 500 kPa, 150 °C is fed to an insulated flash evaporator at the rate of 1.5 kg/s. A stream of saturated liquid at 200 kPa is drained from the bottom of the chamber, as shown in Fig. P7.74, and a stream of saturated vapor at 200 kPa is drawn from the top and fed to a turbine. Find the rate of entropy generation in the flash evaporator.

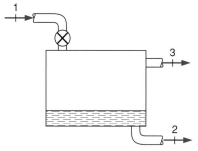

FIGURE P7.74

7.75 A steam turbine has an inlet of 2 kg/s water at 1000 kPa, 400 °C with velocity of 15 m/s. The exit is at 100 kPa, 150 °C and very low velocity. Find the power produced and the rate of entropy generation.

7.76 A factory generates compressed air from ambient 100 kPa, 17 °C by compression to 1000 kPa, 600 K, after which it cools in a constant-pressure cooler to 300 K by heat transfer to the ambient. Find the specific entropy generation in the compressor and in the cooler operation.

7.77 A mixing chamber receives 5 kg/min ammonia as saturated liquid at −20 °C from one line and ammonia at 40 °C, 250 kPa from another line through a valve. The chamber also receives 325 kJ/min energy as heat transferred from a 40 °C reservoir, as shown in Fig. P7.77. This should produce saturated ammonia vapor at −20 °C in the exit line. What is the mass flow rate in the second line, and what is the total entropy generation in the process?

7.78 Carbon dioxide at 300 K, 200 kPa is brought through a steady-flow device, where it is heated to 600 K by a 700 K reservoir in a constant-pressure process. Find the specific work, specific heat transfer, and specific entropy generation.

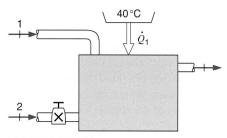

FIGURE P7.77

7.79 Methane at 1 MPa, 300 K is throttled through a valve to 200 kPa. Assume no change in the kinetic energy and ideal gas behavior. What is the specific entropy generation?

7.80 A dual fluid heat exchanger has 5 kg/s water enter at 40 °C, 150 kPa and leave at 10 °C, 150 kPa. The other fluid is glycol coming in at −10 °C, 160 kPa and leaving at 10 °C, 160 kPa. Find the mass flow rate of glycol and the rate of entropy generation.

7.81 Two flows of air are both at 200 kPa; one has 2 kg/s at 400 K, and the other has 1 kg/s at 290 K. The two flows are mixed together in an insulated box to produce a single exit flow at 200 kPa. Find the exit temperature and the total rate of entropy generation.

7.82 A condenser in a power plant receives 5 kg/s steam at 15 kPa, quality 90 % and rejects the heat to cooling water with an average temperature of 17 °C. Find the power given to the cooling water in this constant-pressure process, shown in Fig. P7.82, and the total rate of entropy generation when the condenser exit is saturated liquid.

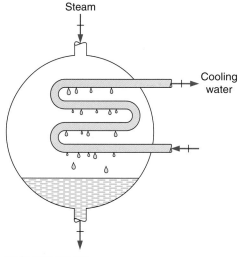

FIGURE P7.82

7.83 A large supply line has a steady flow of R-410A at 1000 kPa, 60 °C. It is used in three different adiabatic devices shown in Fig. P7.83: a throttle flow, an ideal nozzle, and an ideal turbine. All the exit flows are at 300 kPa. Find the exit temperature and specific entropy generation for each device and the exit velocity of the nozzle.

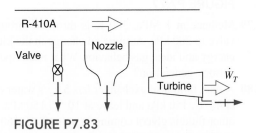

FIGURE P7.83

7.84 A two-stage compressor takes nitrogen in at 20 °C, 150 kPa and compresses it to 600 kPa, 450 K. Then it flows through an intercooler, where it cools to 320 K, and the second stage compresses it to 3000 kPa, 530 K. Find the specific entropy generation in each of the two compressor stages.

7.85 The intercooler in the previous problem uses cold liquid water to cool the nitrogen. The nitrogen flow is 0.1 kg/s, and the liquid water inlet is 20 °C and is setup to flow in the opposite direction from the nitrogen, so the water leaves at 35 °C. Find the flow rate of the water and the entropy generation in this intercooler.

7.86 A counterflowing heat exchanger has one line with 2 kg/s at 125 kPa, 1000 K entering, and the air is leaving at 100 kPa, 400 K. The other line has 0.5 kg/s water coming in at 200 kPa, 20 °C and leaving at 200 kPa. What is the exit temperature of the water and the total rate of entropy generation?

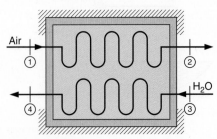

FIGURE P7.86

7.87 A large supply line has a steady air flow at 500 K, 200 kPa. It is used in the three different adiabatic devices shown in Fig. P7.83. All the exit flows are at 100 kPa. Find the exit temperature and specific entropy generation for each device and the exit velocity of the nozzle.

7.88 In a heat-driven refrigerator with ammonia as the working fluid, a turbine with inlet conditions of 2.0 MPa, 70 °C is used to drive a compressor with inlet saturated vapor at −20 °C. The exhausts, both at 1.2 MPa, are then mixed together, as shown in Fig. P7.88. The ratio of the mass flow rate to the turbine to the total exit flow was measured to be 0.62. Can this be true?

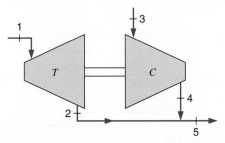

FIGURE P7.88

7.89 Repeat Problem 7.87 for the throttle and the nozzle when the inlet air temperature is 2000 K and use the air tables.

7.90 Carbon dioxide used as a natural refrigerant flows through a cooler at 10 MPa, which is supercritical, so no condensation occurs. The inlet is at 200 °C and the exit is at 40 °C. Assume the heat transfer is to the ambient at 20 °C and find the specific entropy generation.

7.91 A steam turbine in a power plant receives steam at 3000 kPa, 500 °C. The turbine has two exit flows; one is 20 % of the flow at 1000 kPa, 350 °C to a feedwater heater, and the remainder flows out at 200 kPa, 200 °C. Find the specific turbine work and the specific entropy generation, both per kilogram flow in.

7.92 One type of feedwater heater for preheating the water before entering a boiler operates on the principle of mixing the water with steam that has been bled from the turbine. For the states shown in Fig. P7.92, calculate the rate of net entropy increase for the process, assuming the process to be steady flow and adiabatic.

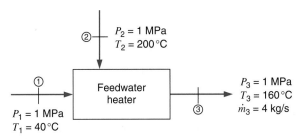

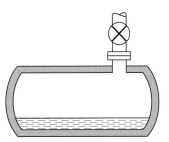

$P_1 = 1$ MPa
$T_1 = 40\,°C$

FIGURE P7.92

7.93 A co-flowing (same direction) heat exchanger, shown in Fig. P7.93, has one line with 0.5 kg/s oxygen at 17 °C, 200 kPa entering, and the other line has 0.6 kg/s nitrogen at 150 kPa, 500 K entering. The heat exchanger is very long, so the two flows exit at the same temperature. Use constant heat capacities and find the exit temperature and the total rate of entropy generation.

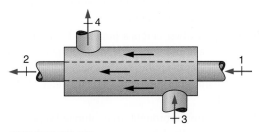

FIGURE P7.93

7.94 A supply of 5 kg/s ammonia at 500 kPa, 20 °C is needed. Two sources are available; one is saturated liquid at 20 °C, and the other is at 500 kPa, 120 °C. Flows from the two sources are fed through valves to an insulated mixing chamber, which then produces the desired output state. Find the two source mass flow rates and the total rate of entropy generation by this setup.

Transient Flow Processes

7.95 Calculate the specific entropy generated in the filling process given in Example 4.10.

7.96 A 1 m³ rigid tank contains 100 kg R-410A at a temperature of 15 °C, as shown in Fig. P7.96. A valve on top of the tank is opened, and saturated vapor is throttled to ambient pressure, 100 kPa, and flows to a collector system. During the process, the temperature inside the tank remains at 15 °C by heat transfer from the 20 °C ambient. The valve is closed when no more liquid remains inside. Calculate the heat transfer to the tank and total entropy generation in the process.

FIGURE P7.96

7.97 An initially empty 0.1 m³ cannister is filled with R-410A from a line flowing saturated liquid at −5 °C. This is done quickly such that the process is adiabatic. Find the final mass, liquid, and vapor volumes, if any, in the cannister. Is the process reversible?

7.98 A 1 L can of R-134a is at room temperature, 20 °C, with a quality of 50 %. A leak in the top valve allows vapor to escape and heat transfer from the room takes place, so we reach a final state of 5 °C with a quality of 100 %. Find the mass that escaped, the heat transfer, and the entropy generation, not including that made in the valve.

7.99 A cook filled a pressure cooker with 3 kg water at 20 °C and a small amount of air and forgot about it. The pressure cooker has a vent valve, so if $P > 200$ kPa, steam escapes to maintain a pressure of 200 kPa. How much entropy was generated in the throttling of the steam through the vent to 100 kPa when 60 % of the original mass escaped?

7.100 A 10 m tall pipe with a 0.1 m diameter is filled with liquid water at 20 °C. It is open at the top to the atmosphere, 100 kPa, and a small nozzle is mounted in the bottom. The water is now let out through the nozzle splashing out to the ground until the pipe is empty. Find the water's initial exit velocity, the average kinetic energy in the exit flow, and the total entropy generation for the process.

7.101 A 200 L insulated tank contains nitrogen gas at 200 kPa, 300 K. A line with nitrogen at 500 K,

500 kPa adds 40 % more mass to the tank with a flow through a valve. Use constant specific heats to find the final temperature and the entropy generation.

7.102 A 200 L insulated tank contains nitrogen gas at 200 kPa, 300 K. A line with nitrogen at 1500 K, 1000 kPa adds 40 % more mass to the tank with a flow through a valve. Use Table A.8 to find the final temperature and the entropy generation.

7.103 An insulated piston/cylinder contains 0.1 m³ air at 250 kPa, 300 K and it maintains constant pressure. More air flows in through a valve from a line at 300 kPa, 400 K, so the volume increases 60 %. Use constant specific heats to solve for the final temperature and the total entropy generation.

7.104 A balloon is filled with air from a line at 200 kPa, 300 K to a final state of 110 kPa, 300 K with a mass of 0.1 kg air. Assume the pressure is proportional to the balloon volume as $P = 100$ kPa $+ CV$. Find the heat transfer to/from the ambient at 300 K and the total entropy generation.

Device Efficiency

7.105 A steam turbine inlet is at 1200 kPa, 400 °C. The exit is at 200 kPa. What is the lowest possible exit temperature? Which efficiency does that correspond to?

7.106 A steam turbine inlet is at 1200 kPa, 400 °C. The exit is at 200 kPa. What is the highest possible exit temperature? Which efficiency does that correspond to?

7.107 A steam turbine inlet is at 1200 kPa, 400 °C. The exit is at 200 kPa, 200 °C. What is the isentropic efficiency?

7.108 A compressor in a commercial refrigerator receives R-410A at −25 °C and $x = 1$. The exit is at 1400 kPa, 60 °C. Neglect kinetic energies and find the isentropic compressor efficiency.

7.109 The exit velocity of a nozzle is 500 m/s. If $\eta_{nozzle} = 0.88$, what is the ideal exit velocity?

7.110 An emergency drain pump, shown in Fig. P7.110, should be able to pump 0.1 m³/s liquid water at 15 °C, 10 m vertically up, delivering it with a velocity of 20 m/s. It is estimated that the pump, pipe, and nozzle have a combined isentropic efficiency expressed for the pump as 60 %. How much power is needed to drive the pump?

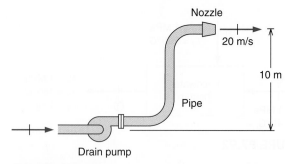

FIGURE P7.110

7.111 Find the isentropic efficiency of the R-134a compressor in Example 4.8.

7.112 A gas turbine with air flowing in at 1200 kPa, 1200 K has an exit pressure of 200 kPa and an isentropic efficiency of 87 %. Find the exit temperature.

7.113 A gas turbine with air flowing in at 1200 kPa, 1200 K has an exit pressure of 200 kPa. Find the lowest possible exit temperature. Which efficiency does that correspond to?

7.114 Liquid water enters a pump at 15 °C, 100 kPa and exits at a pressure of 5 MPa. If the isentropic efficiency of the pump is 75 %, determine the enthalpy (steam table reference) of the water at the pump exit.

7.115 Ammonia is brought from saturated vapor at 300 kPa to 1400 kPa, 140 °C in a steady-flow adiabatic compressor. Find the compressor specific work, entropy generation, and its isentropic efficiency.

7.116 Find the isentropic efficiency of the nozzle in Example 4.4.

7.117 A centrifugal compressor takes in ambient air at 100 kPa, 17 °C and discharges it at 450 kPa. The compressor has an isentropic efficiency of 80 %. What is your best estimate for the discharge temperature?

7.118 A refrigerator uses carbon dioxide that is brought from 1 MPa, −20 °C to 6 MPa using 2.2 kW power input to the compressor with a flow rate of 0.02 kg/s. Find the compressor exit temperature and its isentropic efficiency.

7.119 Redo Problem 7.36, assuming the compressor and turbine in the turbocharger both have isentropic efficiency of 85 %.

7.120 A pump receives water at 100 kPa, 15 °C, and a power input of 1.5 kW. The pump has an isentropic efficiency of 75%, and it should flow 1.2 kg/s delivered at 30 m/s exit velocity. How high an exit pressure can the pump produce?

7.121 A turbine receives air at 1500 K, 1000 kPa and expands it to 100 kPa. The turbine has an isentropic efficiency of 85%. Find the actual turbine exit air temperature and the specific entropy increase in the actual turbine using Table A.7.

7.122 Carbon dioxide enters an adiabatic compressor at 100 kPa, 300 K and exits at 1000 kPa, 520 K. Find the compressor efficiency and the entropy generation for the process.

7.123 A small air turbine with an isentropic efficiency of 80% should produce 270 kJ/kg of work. The inlet temperature is 1000 K and it exhausts to the atmosphere. Find the required inlet pressure and the exhaust temperature.

7.124 A compressor in an industrial air-conditioner compresses ammonia from a state of saturated vapor at 200 kPa to a pressure 800 kPa. At the exit, the temperature is measured to be 120 °C and the mass flow rate is 0.5 kg/s. What is the required motor size for this compressor and what is its isentropic efficiency?

7.125 Repeat Problem 7.41, assuming the turbine and the pump each has an isentropic efficiency of 85%.

7.126 A nozzle in a high-pressure liquid water sprayer has an area of 0.5 cm^2. It receives water at 350 kPa, 20 °C, and the exit pressure is 100 kPa. Neglect the inlet kinetic energy and assume a nozzle isentropic efficiency of 85%. Find the ideal nozzle exit velocity and the actual nozzle mass flow rate.

7.127 Air flows into an insulated nozzle at 1 MPa, 1200 K with 15 m/s and a mass flow rate of 2 kg/s. It expands to 650 kPa, and the exit temperature is 1100 K. Find the exit velocity and the nozzle efficiency.

7.128 A nozzle is required to produce a flow of air at 200 m/s at 20 °C, 100 kPa. It is estimated that the nozzle has an isentropic efficiency of 92%. What nozzle inlet pressure and temperature are required, assuming the inlet kinetic energy is negligible?

7.129 A water-cooled air compressor takes air in at 20 °C, 90 kPa and compresses it to 500 kPa. The isothermal efficiency is 88% and the actual compressor has the same heat transfer as the ideal one. Find the specific compressor work and the exit temperature.

Review Problems

7.130 A flow of saturated liquid R-410A at 200 kPa in an evaporator is brought to a state of superheated vapor at 200 kPa, 20 °C. Assuming the process is reversible, find the specific heat transfer and specific work.

7.131 A co-flowing heat exchanger has one line with 2 kg/s saturated water vapor at 100 kPa entering. The other line is 1 kg/s air at 200 kPa, 1200 K. The heat exchanger is very long, so the two flows exit at the same temperature. Find the exit temperature by trial and error. Calculate the rate of entropy generation.

7.132 Air at 100 kPa, 17 °C is compressed to 400 kPa, after which it is expanded through a nozzle back to the atmosphere. The compressor and the nozzle both have an isentropic efficiency of 90% and are adiabatic. The kinetic energy into and out of the compressor can be neglected. Find the compressor work and its exit temperature and find the nozzle exit velocity.

7.133 A vortex tube has an air inlet flow at 20 °C, 200 kPa and two exit flows of 100 kPa, one at 0 °C and the other at 40 °C, as shown in Fig. P7.133. The tube has no external heat transfer and no work, and all the flows are steady and have negligible kinetic energy. Find the fraction of the inlet flow that comes out at 0 °C. Is this setup possible?

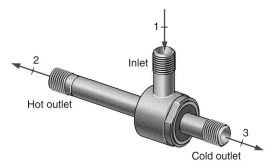

FIGURE P7.133

7.134 A stream of ammonia enters a steady-flow device at 100 kPa, 50 °C at the rate of 1 kg/s. Two streams exit the device at equal mass flow rates; one is at 200 kPa, 50 °C, and the other is as saturated liquid at 10 °C. It is claimed that the device operates in a room at 25 °C on an electrical power input of 250 kW. Is this possible?

7.135 A certain industrial process requires a steady 0.5 kg/s supply of compressed air at 500 kPa at a maximum temperature of 30 °C. This air is to be supplied by installing a compressor and an aftercooler; see Fig. P7.40. Local ambient conditions are 100 kPa, 20 °C. Using an isentropic compressor efficiency of 80 %, determine the power required to drive the compressor and the rate of heat rejection in the aftercooler.

7.136 Carbon dioxide flows through a device, entering at 300 K, 200 kPa and leaving at 500 K. The process is steady state polytropic with $n = 3.8$, and heat transfer comes from a 600 K source. Find the specific work, specific heat transfer, and specific entropy generation due to this process.

7.137 A flow of nitrogen, 0.1 kg/s, comes out of a compressor stage at 500 kPa, 500 K and is now cooled to 310 K in a counterflowing intercooler by liquid water at 125 kPa, 15 °C that leaves at 22 °C. Find the flow rate of water and the total rate of entropy generation.

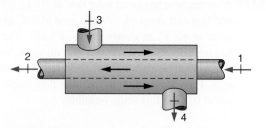

FIGURE P7.137

7.138 An initially empty spring-loaded piston/cylinder requires 100 kPa to float the piston. A compressor with a line and valve now charges the cylinder with water to a final pressure of 1.4 MPa, at which point the volume is 0.6 m³, state 2. The inlet condition to the reversible adiabatic compressor is saturated vapor at 100 kPa. After charging, the valve is closed and the water eventually cools to room temperature, 20 °C, state 3. Find the final mass of water, the piston work from 1 to 2, the required compressor work, and the final pressure, $P3$.

7.139 Consider the scheme shown in Fig. P7.139 for producing fresh water from salt water. The conditions are as shown in the figure. Assume that the properties of salt water are the same as those of pure water, and that the pump is reversible and adiabatic.

a. Determine the ratio (m_7/m_1), the fraction of salt water purified.
b. Determine the input quantities, w_P and q_H.
c. Make a second law analysis of the overall system.

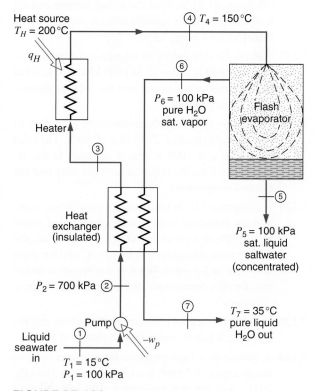

FIGURE P7.139

7.140 A rigid 1.0 m³ tank contains water initially at 120 °C, with 50 % liquid and 50 % vapor by volume. A pressure-relief valve on top of the tank is set to 1.0 MPa (the tank pressure cannot exceed 1.0 MPa; water will be discharged instead). Heat is now transferred to the tank from a 200 °C heat source until the tank contains saturated vapor at 1.0 MPa. Calculate the heat transfer to the tank and show that this process does not violate the second law.

7.141 A horizontal insulated cylinder has a frictionless piston held against stops by an external force of 500 kN, as shown in Fig. P7.141. The piston cross-sectional area is 0.5 m², and the initial volume is 0.25 m³. Argon gas in the cylinder is at 200 kPa, 100 °C. A valve is now opened to a line flowing argon at 1.2 MPa, 200 °C, and gas flows in until the cylinder pressure just balances the external force, at which point the valve is closed. Use constant specific heat to verify that the final temperature is 645 K, and find the total entropy generation.

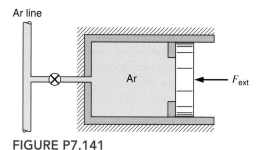

FIGURE P7.141

7.142 Supercharging of an engine is used to increase the inlet air density so that more fuel can be added, the result of which is increased power output. Assume that ambient air, 100 kPa and 27 °C, enters the supercharger at a rate of 250 L/s. The supercharger (compressor) has an isentropic efficiency of 75 % and uses 20 kW of power input. Assume that the ideal and actual compressors have the same exit pressure. Find the ideal specific work and verify that the exit pressure is 175 kPa. Find the percent-

age increase in air density entering the engine due to the supercharger and the entropy generation.

7.143 A certain industrial process requires a steady 0.5 kg/s of air at 200 m/s at the condition of 150 kPa, 300 K, as shown in Fig. P7.143. This air is to be the exhaust from a specially designed turbine whose inlet pressure is 400 kPa. The turbine process may be assumed to be reversible and polytropic, with polytropic exponent $n = 1.20$.
a. What is the turbine inlet temperature?
b. What are the power output and heat transfer rate for the turbine?
c. Calculate the rate of net entropy increase if the heat transfer comes from a source at a temperature 100 °C higher than the turbine inlet temperature.

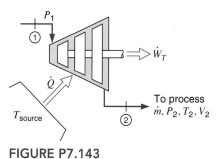

FIGURE P7.143

COMPUTER, DESIGN, AND OPEN-ENDED PROBLEMS

7.144 Use the menu-driven software to get the properties for the calculation of the isentropic efficiency of the pump in the steam power plant of Problem 4.92.

7.145 Write a program to solve the general case of Problem 7.23, in which the states, velocities, and area are input variables. Use a constant specific heat and find the diffuser exit area, temperature, and pressure.

7.146 Write a program to solve Problem 7.138 in which the inlet and exit flow states are input variables. Use a constant specific heat, and let the program calculate the split of the mass flow and the overall entropy generation.

7.147 Small gasoline engine or electric motor-driven air compressors are used to supply compressed air to power tools, machine shops, and so on. The compressor charges air into a tank that acts as a storage

buffer. Find examples of these and discuss their sizes in terms of tank volume, charging pressure, engine, or motor power. Also, find the time it will take to charge the system from startup and its continuous supply capacity.

7.148 A reversible adiabatic compressor receives air at the state of the surroundings, 20 °C, 100 kPa. It should compress the air to a pressure of 1.2 MPa in two stages with a constant-pressure intercooler between the two stages. Investigate the work input as a function of the pressure between the two stages, assuming the intercooler brings the air down to 50 °C.

7.149 (Adv.) Investigate the optimal pressure, P_2, for a constant-pressure intercooler between two stages in a compressor. Assume that the compression process in each stage follows a polytropic process and that the intercooler brings the substance to the original inlet temperature, T_1. Show that the minimal work for the combined stages arises when

$$P_2 = (P_1 P_3)^{1/2}$$

where P_3 is the final exit pressure.

7.150 (Adv.) Reexamine the previous problem when the intercooler cools the substance to a temperature, $T_2 > T_1$, due to finite heat-transfer rates. What is the effect of having isentropic efficiencies for the

compressor stages of less than 100 % on the total work and selection of P_2?

7.151 Consider a geothermal supply of hot water available as saturated liquid at $P_1 = 1.5$ MPa. The liquid is to be flashed (throttled) to some lower pressure, P_2. The saturated liquid and saturated vapor at this pressure are separated, and the vapor is expanded through a reversible adiabatic turbine to the exhaust pressure, $P_3 = 10$ kPa. Study the turbine power output per unit initial mass, m_1, as a function of the pressure, P_2.

Exergy

8

The previous chapters presented the basic set of general laws for a control volume and applied them to thermal science problems involving processes with storage of energy and transfer of energy by flow of mass or as work and heat transfers. We now turn to the first extension of these principles, which involves additional considerations of processes and system characteristics based on advanced use of the energy and entropy equations. We would like to know the general limitations for the operation of systems and devices so that we can design them for optimal efficiency with a minimal use of resources to accomplish a certain task.

8.1 EXERGY, REVERSIBLE WORK, AND IRREVERSIBILITY

We introduced the reversible boundary work for a control mass in Chapter 6 and, single-flow reversible shaft work in Chapter 7. A different kind of comparison to a reversible device was done with the efficiency introduced for simple devices like a turbine, compressor, or nozzle. This efficiency compared the desired output of an actual device with the output from a similar reversible device, and the output was measured in energy. We will now develop a more general concept to use in the evaluation of actual system and devices.

Before we show the specific analysis, we define the concept in words and look at some simple situations in which we can do an evaluation of it. The concept of exergy is defined as the possible work we can extract from a given physical setup when it is allowed to interact with the ambient and the process end state is at P_0, T_0.

$$\Phi = W_{\text{out}} \text{ given an ambient } P_0, T_0$$

This is closely related to reversible work, as we will illustrate with some examples. Later in the chapter, a more precise definition of the property exergy will be given.

We start with a simple situation shown in Fig. 8.1a, in which there is an energy source Q as a heat transfer from a very large constant-temperature reservoir. How much work is it possible to extract from this system? From the description in Chapter 5 and the discussion in Chapter 6, we know that the maximum work out is obtained from a reversible heat engine. As it is allowed to interact with the ambient, we will let the ambient be the other energy reservoir that is at a constant temperature, T_0. Since the two reservoirs are at constant temperatures, the heat engine must be operating in a Carnot cycle and we therefore get the work as

$$\text{Energy:} \quad W_{\text{rev HE}} = Q - Q_0$$

$$\text{Entropy:} \quad 0 = \frac{Q}{T} - \frac{Q_0}{T_0}$$

313

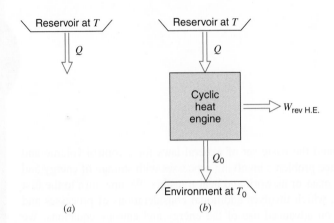

FIGURE 8.1
Constant-temperature
energy source.

so that

$$\Phi_{QT} = W_{\text{rev HE}} = Q\left(1 - \frac{T_0}{T}\right) \tag{8.1}$$

It is only a fraction of the heat transfer that can be available as work, and that fraction is the exergy value of Q, which equals the Carnot heat engine efficiency times Q. The split is shown in the T–S diagram in Fig. 8.2 with the total shaded area as Q. The portion of Q that is below T_0 cannot be converted into work by the heat engine and must be discarded as the unavailable part of Q.

   Let us next consider the same situation, except that the heat transfer Q is available from a constant-pressure source, for example, a simple heat exchanger, as shown in Fig. 8.3a. The Carnot cycle must now be replaced by a sequence of such engines, with the result shown in Fig. 8.3b. The only difference between the first and second examples is that the second includes an integral, which corresponds to ΔS.

$$\Delta S = \int \frac{\delta Q_{\text{rev}}}{T} = \frac{Q_0}{T_0} \tag{8.2}$$

Substituting into the first law, we have

$$\Phi_{QT} = W_{\text{rev HE}} = Q - T_0\,\Delta S \tag{8.3}$$

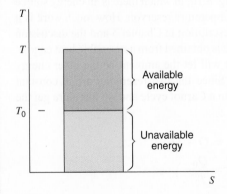

FIGURE 8.2
T–S diagram for a
constant-temperature
energy source.

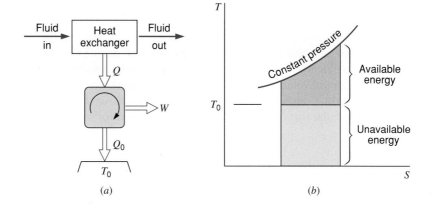

FIGURE 8.3
Changing-temperature
energy source.

Note that this ΔS quantity does not include the standard sign convention. It corresponds to the amount of change of entropy shown in Fig. 8.3*b*. Equation 8.2 specifies the available portion of the quantity Q. The portion unavailable for producing work in this circumstance lies below T_0 in Fig. 8.3*b*.

In the preceding paragraphs we examined a simple cyclic heat engine receiving energy from different sources. We will now analyze real irreversible processes occurring in a general control volume.

Consider the actual control volume shown in Fig. 8.4 with mass and energy transfers including storage effects. For this control volume the continuity equation is Eq. 4.1, the energy equation from Eq. 4.7, and the entropy equation from Eq. 7.2.

$$\frac{dm_{\text{c.v.}}}{dt} = \sum \dot{m}_i - \sum \dot{m}_e \tag{8.4}$$

$$\frac{dE_{\text{e.v}}}{dt} = \sum \dot{Q}_j + \sum \dot{m}_i h_{\text{tot } i} - \sum \dot{m}_e h_{\text{tot } e} - \dot{W}_{\text{c.v. ac}} \tag{8.5}$$

$$\frac{dS_{\text{c.v.}}}{dt} = \sum \frac{\dot{Q}_j}{T_j} + \sum \dot{m}_i s_i - \sum \dot{m}_e s_e + \dot{S}_{\text{gen ac}} \tag{8.6}$$

We wish to establish a quantitative measure in energy terms of the extent or degree to which this actual process is irreversible. This is done by comparison to a similar control volume that only includes reversible processes, which is the ideal counterpart to the actual control volume. The ideal control volume is identical to the actual control volume in as many aspects as possible. It has the same storage effect (left-hand side of the equations), the same heat transfers $\dot{Q}_j$ at T_j, and the same flows $\dot{m}_i$, $\dot{m}_e$ at the same states, so the first

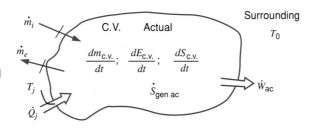

FIGURE 8.4 An actual
control volume that
includes irreversible
processes.

four terms in Eqs. 8.5 and 8.6 are the same. What is different? Since it must be reversible, the entropy generation term is zero, whereas the actual one in Eq. 8.6 is positive. The last term in Eq. 8.6 is substituted with a reversible positive flux of S, and the only reversible process that can increase entropy is a heat transfer in, so we allow one, $\dot{Q}_0^{\text{rev}}$, from the ambient at T_0. This heat transfer must also be present in the energy equation for the ideal control volume together with a reversible work term, both of which replace the actual work term. Comparing only the last terms in Eqs. 8.5 and 8.6 for the actual control volume to the similar part of the equations for the ideal control volume gives

Actual C.V. terms   Ideal C.V. terms

$$\dot{S}_{\text{gen ac}} = \frac{\dot{Q}_0^{\text{rev}}}{T_0} \tag{8.7}$$

$$-\dot{W}_{\text{c.v. ac}} = \dot{Q}_0^{\text{rev}} - \dot{W}^{\text{rev}} \tag{8.8}$$

From the equality of the entropy generation to the entropy flux in Eq. 8.7, we get

$$\dot{Q}_0^{\text{rev}} = T_0 \dot{S}_{\text{gen ac}} \tag{8.9}$$

and the reversible work from Eq. 8.8 becomes

$$\dot{W}^{\text{rev}} = \dot{W}_{\text{c.v. ac}} + \dot{Q}_0^{\text{rev}} \tag{8.10}$$

Notice that the ideal control volume has heat transfer from the ambient even if the actual control volume is adiabatic, and only if the actual control volume process is reversible is this heat transfer zero and the two control volumes identical.

    To see the reversible work as a result of all the flows and fluxes in the actual control volume, we solve for the entropy generation rate in Eq. 8.6 and substitute it into Eq. 8.9 and the result into Eq. 8.10. The actual work is found from the energy equation Eq. 8.5 and substituted into Eq. 8.10, giving the final result for the reversible work. Following this, we get

$$\dot{W}^{\text{rev}} = \dot{W}_{\text{c.v. ac}} + \dot{Q}_0^{\text{rev}}$$

$$= \sum \dot{Q}_j + \sum \dot{m}_i h_{\text{tot} i} - \sum \dot{m}_e h_{\text{tot} e} - \frac{dE_{\text{c.v.}}}{dt}$$

$$+ T_0 \left[\frac{dS_{\text{c.v.}}}{dt} - \sum \frac{\dot{Q}_j}{T_j} - \sum \dot{m}_i s_i + \sum \dot{m}_e s_e \right]$$

Now combine similar terms and rearrange to become

$$\dot{W}^{\text{rev}} = \sum \left(1 - \frac{T_0}{T_j}\right) \dot{Q}_j$$

$$+ \sum \dot{m}_i (h_{\text{tot} i} - T_0 s_i) - \sum \dot{m}_e (h_{\text{tot} e} - T_0 s_e)$$

$$- \left[\frac{dE_{\text{c.v.}}}{dt} - T_0 \frac{dS_{\text{c.v.}}}{dt} \right] \tag{8.11}$$

The contributions from the heat transfers appear to be independent, each producing work as if the heat transfer goes to a Carnot heat engine with low temperature T_0. Each flow makes a unique contribution, and the storage effect is expressed in the last parenthesis. This result represents the theoretical upper limit for the rate of work that can be produced by a general control volume, and it can be compared to the actual work and thus provide the measure by which the actual control volume system(s) can be evaluated. The difference between this reversible work and the actual work is called the *irreversibility* $\dot{I}$, as

$$\dot{I} = \dot{W}^{\text{rev}} - \dot{W}_{\text{c.v. ac}} \tag{8.12}$$

and since this represents the difference between what is theoretically possible and what actually is produced, it is also called *lost work*. Notice that the energy is not lost. Energy is conserved; it is a lost opportunity to convert some other form of energy into work. We can also express the irreversibility in a different form by using Eqs. 8.9 and 8.10:

$$\dot{I} = \dot{W}^{\text{rev}} - \dot{W}_{\text{c.v. ac}} = \dot{Q}_0^{\text{rev}} = T_0 \dot{S}_{\text{gen ac}} \tag{8.13}$$

From this we see that the irreversibility is directly proportional to the entropy generation but is expressed in energy units, and this requires a fixed and known reference temperature T_0 to be generally useful. Notice how the reversible work is higher than the actual work by the positive irreversibility. If the device is like a turbine or is the expansion work in the piston/cylinder of an engine, the actual work is positive out and the reversible work is then larger, so more work could be produced in a reversible process. On the other hand, if the device requires work input, the actual work is negative, as in a pump or compressor, the reversible work is higher which is closer to zero, and thus the reversible device requires less work input. These conditions are illustrated in Fig. 8.5, with the positive actual work as case 1 and the negative actual work as case 2.

The subsequent examples will illustrate the concepts of reversible work and irreversibility for the simplifying cases of steady-state processes, the control mass process, and the transient process. These situations are all special cases of the general theory shown above.

The Steady-State Process

Consider now a typical steady single-flow device involving heat transfer and actual work. For a single flow, the continuity equations simplify to state the equality of the mass flow rates in and out (recall Eq. 4.11). For this case, the reversible work in Eq. 8.11 is divided

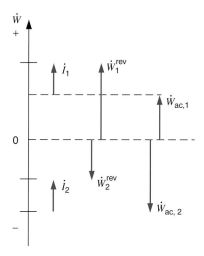

FIGURE 8.5 The actual and reversible rates of work.

with the mass flow rate to express the reversible specific work as

$$w^{\text{rev}} = \dot{W}^{\text{rev}}/\dot{m} = \sum \left(1 - \frac{T_0}{T_j}\right) q_j + (h_{\text{tot}\,i} - T_0 s_i) - (h_{\text{tot}\,e} - T_0 s_e) \qquad (8.14)$$

and with steady state, the last term in Eq. 8.11 drops out. For these cases, the irreversibility in Eqs. 8.12 and 8.13 is expressed as a specific irreversibility:

$$i = \dot{I}/\dot{m} = w^{\text{rev}} - w_{\text{c.v. ac}} = q_0^{\text{rev}} = T_0 s_{\text{gen ac}}$$

$$= T_0 \left[s_e - s_i - \sum \frac{q_j}{T_j} \right] \qquad (8.15)$$

The following examples will illustrate the reversible work and the irreversibility for a heat exchanger and a compressor with a heat loss.

Example 8.1

A feedwater heater has 5 kg/s water at 5 MPa and 40 °C flowing through it, being heated from two sources, as shown in Fig. 8.6. One source adds 900 kW from a 100 °C reservoir, and the other source transfers heat from a 200 °C reservoir such that the water exit condition is 5 MPa, 180 °C. Find the reversible work and the irreversibility.

Control volume:	Feedwater heater extending out to the two reservoirs.
Inlet state:	P_i, T_i known; state fixed.
Exit state:	P_e, T_e known; state fixed.
Process:	Constant-pressure heat addition with no change in kinetic or potential energy.
Model:	Steam tables.

Analysis

This control volume has a single inlet and exit flow with two heat-transfer rates coming from reservoirs different from the ambient surroundings. There is no actual work or actual heat transfer with the surroundings at 25 °C. For the actual feedwater heater, the energy equation becomes

$$h_i + q_1 + q_2 = h_e$$

The reversible work for the given change of state is, from Eq. 8.14, with heat transfer q_1 from reservoir T_1 and heat transfer q_2 from reservoir T_2,

$$w^{\text{rev}} = T_0(s_e - s_i) - (h_e - h_i) + q_1\left(1 - \frac{T_0}{T_1}\right) + q_2\left(1 - \frac{T_0}{T_2}\right)$$

From Eq. 8.15, since the actual work is zero, we have

$$i = w^{\text{rev}} - w = w^{\text{rev}}$$

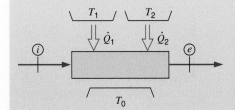

FIGURE 8.6 The feedwater heater for Example 8.1.

Solution

From the steam tables, the inlet and exit state properties are

$$h_i = 171.95 \text{ kJ/kg}, \qquad s_i = 0.5705 \text{ kJ/kg·K}$$

$$h_e = 765.24 \text{ kJ/kg}, \qquad s_e = 2.1341 \text{ kJ/kg·K}$$

The second heat transfer is found from the energy equation as

$$q_2 = h_e - h_i - q_1 = 765.24 - 171.95 - 900/5 = 413.29 \text{ kJ/kg}$$

The reversible work is

$$w^{\text{rev}} = T_0(s_e - s_i) - (h_e - h_i) + q_1\left(1 - \frac{T_0}{T_1}\right) + q_2\left(1 - \frac{T_0}{T_2}\right)$$

$$= 298.2(2.1341 - 0.5705) - (765.24 - 171.95)$$

$$+ 180\left(1 - \frac{298.2}{373.2}\right) + 413.29\left(1 - \frac{298.2}{473.2}\right)$$

$$= 466.27 - 593.29 + 36.17 + 152.84 = 62.0 \text{ kJ/kg}$$

The irreversibility is

$$i = w^{\text{rev}} = 62.0 \text{ kJ/kg}$$

Example 8.2

Consider an air compressor that receives ambient air at 100 kPa and 25 °C. It compresses the air to a pressure of 1 MPa, where it exits at a temperature of 540 K. Since the air and compressor housing are hotter than the ambient surroundings, 50 kJ per kilogram air flowing through the compressor are lost. Find the reversible work and the irreversibility in the process.

Control volume:	The air compressor.
Sketch:	Fig. 8.7.
Inlet state:	P_i, T_i known; state fixed.
Exit state:	P_e, T_e known; state fixed.
Process:	Nonadiabatic compression with no change in kinetic or potential energy.
Model:	Ideal gas.

Analysis

This steady-state process has a single inlet and exit flow, so all quantities are determined on a mass basis as specific quantities. From the ideal gas air tables, we obtain

$$h_i = 298.6 \text{ kJ/kg}, \qquad s^0_{T_i} = 6.8631 \text{ kJ/kg·K}$$

$$h_e = 544.7 \text{ kJ/kg}, \qquad s^0_{T_e} = 7.4664 \text{ kJ/kg·K}$$

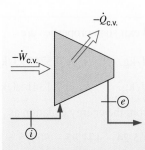

FIGURE 8.7 Illustration for Example 8.2.

so the energy equation for the actual compressor gives the work as

$$q = -50 \text{ kJ/kg}$$

$$w = h_i - h_e + q = 298.6 - 544.7 - 50 = -296.1 \text{ kJ/kg}$$

The reversible work for the given change of state is, from Eq. 8.14, with $T_j = T_0$

$$w^{\text{rev}} = T_0(s_e - s_i) - (h_e - h_i) + q\left(1 - \frac{T_0}{T_0}\right)$$

$$= 298.2(7.4664 - 6.8631 - 0.287 \ln 10) - (544.7 - 298.6) + 0$$

$$= -17.2 - 246.1 = -263.3 \text{ kJ/kg}$$

From Eq. 8.15, we get

$$i = w^{\text{rev}} - w$$

$$= -263.3 - (-296.1) = 32.8 \text{ kJ/kg}$$

The expression for the reversible work includes the kinetic and potential energies in the total enthalpy for the flow terms. In many devices these terms are negligible, so the total enthalpy reduces to the thermodynamic property enthalpy. For devices such as nozzles and diffusers, the kinetic energy terms are important, whereas for longer pipes and channel flows that run through different elevations, the potential energy becomes important and must be included in the formulation. There are also steady-state processes involving more than one fluid stream entering or exiting the control volume. In such cases, it is necessary to use the original expression for the rate of work in Eq. 8.11 and drop only the last term.

The Control Mass Process

For a control mass we do not have a flow of mass in or out, so the reversible work is

$$\dot{W}^{\text{rev}} = \sum \left(1 - \frac{T_0}{T_j}\right)\dot{Q}_j - \left[\frac{dE_{\text{c.v.}}}{dt} - T_0\frac{dS_{\text{c.v.}}}{dt}\right] \tag{8.16}$$

showing the effects of heat transfers and storage changes. In most applications, we look at processes that bring the control mass from an initial state 1 to a final state 2, so Eq. 8.16 is integrated in time to give

$$_1W_2^{\text{rev}} = \sum \left(1 - \frac{T_0}{T_j}\right)_1Q_{2j} - [E_2 - E_1 - T_0(S_2 - S_1)] \tag{8.17}$$

and similarly, the irreversibility from Eq. 8.13 integrated in time becomes

$$_1I_2 = {_1W_2^{rev}} - {_1W_{2\,ac}} = T_0\,{_1S_{2\,gen\,ac}}$$
$$= T_0(S_2 - S_1) - \sum \frac{T_0}{T_j}\,{_1Q_{2\,j}} \tag{8.18}$$

where the last equality has substituted the entropy generation from the entropy equation as Eq. 6.14 or Eq. 8.6 integrated in time.

For many processes, the changes in kinetic and potential energies are negligible, so the energy change $E_2 - E_1$ becomes $U_2 - U_1$, used in Eq. 8.17.

Example 8.3

An insulated rigid tank is divided into two parts, A and B, by a diaphragm. Each part has a volume of 1 m³. Initially, part A contains water at room temperature, 20 °C, with a quality of 50 %, while part B is evacuated. The diaphragm then ruptures and the water fills the total volume. Determine the reversible work for this change of state and the irreversibility of the process.

> *Control mass*:   Water
> *Initial state*: T_1, x_1 known; state fixed.
> *Final state*: V_2 known.
> *Process*:   Adiabatic, no change in kinetic or potential energy.
> *Model*:   Steam tables.

Analysis

There is a boundary movement for the water, but since it occurs against no resistance, no work is done. Therefore, the first law reduces to

$$m(u_2 - u_1) = 0$$

From Eq. 8.17 with no change in internal energy and no heat transfer,

$$_1W_2^{rev} = T_0(S_2 - S_1) = T_0 m(s_2 - s_1)$$

From Eq. 8.18

$$_1I_2 = {_1W_2^{rev}} - {_1W_2} = {_1W_2^{rev}}$$

Solution

From the steam tables at state 1,

$$u_1 = 1243.5\ \text{kJ/kg} \qquad v_1 = 28.895\ \text{m}^3/\text{kg} \qquad s_1 = 4.4819\ \text{kJ/kg·K}$$

Therefore,

$$v_2 = V_2/m = 2 \times v_1 = 57.79\ \text{m}^3/\text{kg} \qquad u_2 = u_1 = 1243.5\ \text{kJ/kg}$$

These two independent properties, v_2 and u_2, fix state 2. The final temperature T_2 must be found by trial and error in the steam tables.

> For $\quad T_2 = 5\,°C \quad$ and $\quad v_2 \Rightarrow x = 0.3928, \qquad u = 948.5\ \text{kJ/kg}$
> For $\quad T_2 = 10\,°C \quad$ and $\quad v_2 \Rightarrow x = 0.5433, \qquad u = 1317\ \text{kJ/kg}$

so the final interpolation in u gives a temperature of $9\,°C$. If the software is used, the final state is interpolated to be

$$T_2 = 9.1\,°C \qquad x_2 = 0.513 \qquad s_2 = 4.644\ \text{kJ/kg·K}$$

with the given u and v. Since the actual work is zero, we have

$$_1I_2 = {}_1W_2^{\text{rev}} = T_0(V_1/v_1)(s_2 - s_1)$$
$$= 293.2(1/28.895)(4.644 - 4.4819) = 1.645\ \text{kJ}$$

The Transient Process

The transient process has a change in the control volume from state 1 to state 2, as for the control mass, together with possible mass flow in at state i and/or out at state e. The instantaneous rate equations in Eq. 8.11 for the work and Eq. 8.13 for the irreversibility are integrated in time to yield

$$_1W_2^{\text{rev}} = \sum\left(1 - \frac{T_0}{T_j}\right){}_1Q_{2\,j} + \sum m_i(h_{\text{tot}\,i} - T_0 s_i) - \sum m_e(h_{\text{tot}\,e} - T_0 s_e)$$
$$- [m_2 e_2 - m_1 e_1 - T_0(m_2 s_2 - m_1 s_1)] \tag{8.19}$$

$$_1I_2 = {}_1W_2^{\text{rev}} - {}_1W_{2\,\text{ac}} = T_0\,{}_1S_{2\,\text{gen ac}}$$

$$= T_0\left[(m_2 s_2 - m_1 s_1) + \sum m_e s_e - \sum m_i s_i - \sum \frac{1}{T_j}{}_1Q_{2\,j}\right] \tag{8.20}$$

where the last expression substituted the entropy generation term (integrated in time) from the entropy equation, Eq. 8.6.

Example 8.4

A $1\ \text{m}^3$ rigid tank, Fig. 8.8, contains ammonia at 200 kPa and ambient temperature $20\,°C$. The tank is connected with a valve to a line flowing saturated liquid ammonia at $-10\,°C$. The valve is opened, and the tank is charged quickly until the flow stops and the valve is closed. As the process happens very quickly, there is no heat transfer. Determine the final mass in the tank and the irreversibility in the process.

Control volume:	The tank and the valve.
Initial state:	T_1, P_1 known; state fixed.
Inlet state:	T_i, x_i known; state fixed.
Final state:	$P_2 = P_{\text{line}}$ known.
Process:	Adiabatic, no kinetic or potential energy change.
Model:	Ammonia tables.

Analysis

Since the line pressure is higher than the initial pressure inside the tank, flow is going into the tank and the flow stops when the tank pressure has increased to the line pressure.

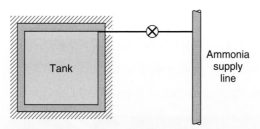

FIGURE 8.8 Ammonia tank and line for Example 8.4.

The continuity, energy, and entropy equations are

$$m_2 - m_1 = m_i$$

$$m_2 u_2 - m_1 u_1 = m_i h_i = (m_2 - m_1) h_i$$

$$m_2 s_2 - m_1 s_1 = m_i s_i + {}_1 S_{2\text{gen}}$$

where kinetic and potential energies are zero for the initial and final states and neglected for the inlet flow.

Solution

From the ammonia tables, the initial and line state properties are

$$v_1 = 0.6995 \, \text{m}^3/\text{kg} \quad u_1 = 1369.5 \, \text{kJ/kg} \quad s_1 = 5.927 \, \text{kJ/kg·K}$$

$$h_i = 134.41 \, \text{kJ/kg} \quad s_i = 0.5408 \, \text{kJ/kg·K}$$

The initial mass is therefore

$$m_1 = V/v_1 = 1/0.6995 = 1.4296 \, \text{kg}$$

It is observed that only the final pressure is known, so one property is needed. The unknowns are the final mass and final internal energy in the energy equation. Since only one property is unknown, the two quantities are not independent. From the energy equation, we have

$$m_2(u_2 - h_i) = m_1(u_1 - h_i)$$

from which it is seen that $u_2 > h_i$ and the state therefore is two-phase or superheated vapor. Assume that the state is two phase; then

$$m_2 = V/v_2 = 1/(0.001\,534 + x_2 \times 0.416\,84)$$

$$u_2 = 133.964 + x_2 \times 1175.257$$

so the energy equation is

$$\frac{133.964 + x_2 \times 1175.257 - 134.41}{0.001\,534 + x_2 \times 0.041\,684} = 1.4296(1369.5 - 134.41) = 1765.67 \, \text{kJ}$$

This equation is solved for the quality and the rest of the properties to give

$$x_2 = 0.007\,182 \quad v_2 = 0.004\,527\,6 \, \text{m}^3/\text{kg} \quad s_2 = 0.5762 \, \text{kJ/kg}$$

Now the final mass and the irreversibility are found:

$$m_2 = V/v_2 = 1/0.004\,527\,6 = 220.87\,\text{kg}$$

$$_1S_{2\,\text{gen}} = m_2s_2 - m_1s_1 - m_is_i = 127.265 - 8.473 - 118.673 = 0.119\,\text{kJ/K}$$

$$I_{\text{c.v.}} = T_0\,_1S_{2\text{gen}} = 293.15 \times 0.119 = 34.885\,\text{kJ}$$

In-Text Concept Questions

a. Can any energy transfer as heat transfer be 100% available?

b. Is electrical work 100% available?

c. A nozzle involves no actual work; how should you then interpret the reversible work?

d. If an actual control volume process is reversible, what can you say about the work term?

e. Can entropy change in a control volume process that is reversible?

8.2 EXERGY AND SECOND-LAW EFFICIENCY

What is the maximum reversible work that can be done by a given mass in a given state? In the previous section, we developed expressions for the reversible work for a given change of state for a control mass and control volume undergoing specific types of processes. For any given case, what final state will give the maximum reversible work?

The answer to this question is that, for any type of process, when the mass comes into equilibrium with the environment, no spontaneous change of state will occur and the mass will be incapable of doing any work. Therefore, if a mass in a given state undergoes a completely reversible process until it reaches a state in which it is in equilibrium with the environment, the maximum reversible work will have been done by the mass. In this sense, we refer to the exergy at the original state in terms of the potential for achieving the maximum possible work by the mass.

If a control mass is in equilibrium with the surroundings, it must certainly be in pressure and temperature equilibrium with the surroundings, that is, at pressure P_0 and temperature T_0. It must also be in chemical equilibrium with the surroundings, which implies that no further chemical reaction will take place. Equilibrium with the surroundings also requires that the system have zero velocity and minimum potential energy. Similar requirements can be set forth regarding electrical and surface effects if these are relevant to a given problem.

The same general remarks can be made about a quantity of mass that undergoes a steady-state process. With a given state for the mass entering the control volume, the reversible work will be maximum when this mass leaves the control volume in equilibrium with the surroundings. This means that as the mass leaves the control volume, it must be at the pressure and temperature of the surroundings, be in chemical equilibrium with the surroundings, and have minimum potential energy and zero velocity. (The mass leaving the control volume must of necessity have some velocity, but it can be made to approach zero.)

Let us consider the exergy from the different types of processes and situations that can arise and start with the expression for the reversible work in Eq. 8.11. For that expression, we recognized separate contributions to the reversible work as one from heat transfer, another one from the mass flows, and finally, a contribution from the storage effect that is a change of state of the substance inside the control volume. We will now measure the exergy as the maximum work we can get out relative to the surroundings.

Starting with the heat transfer, we see that the contributions to the reversible work from these terms relative to the surroundings at T_0 are

$$\dot{\Phi}_q = \sum \left(1 - \frac{T_0}{T_j}\right)\dot{Q}_j \qquad (8.21)$$

which was the result we found in Eq. 8.1. This is now labeled as a rate of exergy $\dot{\Phi}_q$ that equals the possible reversible work that can be extracted from the heat transfers; as such, this is the value of the heat transfers expressed in work. We notice that if the heat transfers come at a higher temperature T_j, the value (exergy) increases and we could extract a larger fraction of the heat transfers as work. This is sometimes expressed as a higher quality of the heat transfer. One limit is an infinite high temperature ($T_j \to \infty$), for which the heat transfer is 100 % exergy, and another limit is $T_j = T_0$, for which the heat transfer has zero exergy.

Shifting attention to the flows and the exergy associated with those terms, we like to express the exergy for each flow separately and use the surroundings as a reference for thermal energy as well as kinetic and potential energy. Having a flow at some state that goes through a reversible process will result in the maximum possible work out when the fluid leaves in equilibrium with the surroundings. The fluid is in equilibrium with the surroundings when it approaches the *dead state* that has the smallest possible energy where $T = T_0$ and $P = P_0$, with zero velocity and reference elevation Z_0 (normally zero at standard sea level). Assuming this is the case, a single flow into a control volume without the heat transfer and an exit state that is the dead state give a specific reversible work from Eq. 8.14 with the symbol ψ representing a flow exergy as

$$
\begin{aligned}
\psi &= (h_{\text{tot}} - T_0 s) - (h_{\text{tot }0} - T_0 s_0) \\
&= \left(h - T_0 s + \frac{1}{2}\mathbf{V}^2 + gZ\right) - (h_0 - T_0 s_0 + gZ_0)
\end{aligned} \qquad (8.22)
$$

where we have written out the total enthalpy to show the kinetic and potential energy terms explicitly. A flow at the ambient dead state therefore has an exergy of zero, whereas most flows are at different states in and out. A single steady flow has terms in specific exergy as

$$
\begin{aligned}
\psi_i - \psi_e &= [(h_{\text{tot }i} - T_0 s_i) - (h_0 - T_0 s_0 + gZ_0)] - [(h_{\text{tot }e} - T_0 s_e) - (h_0 - T_0 s_0 + gZ_0)] \\
&= (h_{\text{tot }i} - T_0 s_i) - (h_{\text{tot }e} - T_0 s_e)
\end{aligned} \qquad (8.23)
$$

so the constant offset disappears when we look at differences in exergies. The last expression for the change in exergy is identical to the two terms in Eq. 8.14 for the reversible work, so we see that the reversible work from a single steady-state flow equals the decrease in exergy of the flow.

The reversible work from a storage effect due to a change of state in the control volume can also be used to find an exergy. In this case, the volume may change, and some work is exchanged with the ambient, which is not available as useful work. Starting with the rate form, where we have a rate of volume change $\dot{V}$, the work done against the surroundings is

$$\dot{W}_{\text{surr}} = P_0 \dot{V} \tag{8.24}$$

so the maximum available rate of work from the storage terms in Eq. 8.11 becomes

$$\dot{W}_{\text{avail}}^{\text{max}} = \dot{W}_{\text{storage}}^{\text{rev}} - \dot{W}_{\text{surr}}$$

$$= -\left[\frac{dE_{\text{c.v.}}}{dt} - T_0 \frac{dS_{\text{c.v.}}}{dt} \right] - P_0 \dot{V} \tag{8.25}$$

Integrating this from a given state to the final state (being the dead ambient state) gives the exergy as

$$\Phi = -[E_0 - E - T_0(S_0 - S) + P_0(V_0 - V)]$$

$$= (E - T_0 S) - (E_0 - T_0 S_0) + P_0(V - V_0)$$

$$\dot{\Phi}_{\text{c.v.}} = \frac{dE_{\text{c.v.}}}{dt} - T_0 \frac{dS_{\text{c.v.}}}{dt} + P_0 \dot{V} \tag{8.26}$$

so the maximum available rate of work, Eq. 8.25, is the negative rate of change of stored exergy, Eq. 8.26. For a control mass the specific exergy becomes, after dividing with mass m,

$$\phi = (e - T_0 s + P_0 v) - (e_0 - T_0 s_0 + P_0 v_0) \tag{8.27}$$

As we did for the flow terms, we often look at differences between two states as

$$\phi_2 - \phi_1 = (e_2 - T_0 s_2 + P_0 v_2) - (e_1 - T_0 s_1 + P_0 v_1) \tag{8.28}$$

where the constant offset (the last parenthesis in Eq. 8.27) drops out.

Now that we have developed the expressions for the exergy associated with the different energy terms, we can write the final expression for the relation between the actual rate of work, the reversible rate of work, and the various exergies. The reversible work from Eq. 8.11, with the right-hand-side terms expressed with the exergies, becomes

$$\dot{W}^{\text{rev}} = \dot{\Phi}_q + \sum \dot{m}_i \psi_i - \sum \dot{m}_e \psi_e - \dot{\Phi}_{\text{c.v.}} + P_0 \dot{V} \tag{8.29}$$

and then the actual work from Eqs. 8.9 and 8.10 becomes

$$\dot{W}_{\text{c.v. ac}} = \dot{W}^{\text{rev}} - \dot{Q}_0^{\text{rev}} = \dot{W}^{\text{rev}} - \dot{I} \tag{8.30}$$

From this last expression, we see that the irreversibility destroys part of the potential work from the various types of exergy expressed in Eq. 8.29. These two equations can then be written out for all the special cases that we considered earlier, such as the control mass process, the steady single flow, and the transient process.

The less the irreversibility associated with a given change of state, the greater the amount of work that will be done (or the smaller the amount of work that will be required). This relation is significant for at least two reasons. The first is that exergy is one of our natural resources. This exergy is found in such forms as oil reserves, coal reserves, and uranium reserves. Suppose we wish to accomplish a given objective that requires a certain amount of work. If this work is produced reversibly while drawing on one of the exergy reserves,

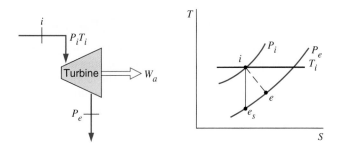

FIGURE 8.9
Irreversible turbine.

the decrease in exergy is exactly equal to the reversible work. However, since there are irreversibilities in producing this required amount of work, the actual work will be less than the reversible work, and the decrease in exergy will be greater (by the amount of the irreversibility) than if this work had been produced reversibly. Thus, the more irreversibilities we have in all our processes, the greater will be the decrease in our exergy reserves.[1] The conservation and effective use of these exergy reserves is an important responsibility for all of us.

The second reason that it is desirable to accomplish a given objective with the smallest irreversibility is an economic one. Work costs money, and in many cases a given objective can be accomplished at less cost when the irreversibility is less. It should be noted, however, that many factors enter into the total cost of accomplishing a given objective, and an optimization process that considers many factors is often necessary to arrive at the most economical design. For example, in a heat-transfer process, the smaller the temperature difference across which the heat is transferred, the less the irreversibility. However, for a given rate of heat transfer, a smaller temperature difference will require a larger (and therefore more expensive) heat exchanger. These various factors must all be considered in developing the optimum and most economical design.

In many engineering decisions, other factors, such as the impact on the environment (for example, air pollution and water pollution) and the impact on society must be considered in developing the optimum design.

Along with the increased use of exergy analysis in recent years, a term called second-law efficiency has come into more common use. This term refers to comparison of the desired output of a process with the cost, or input, in terms of the exergy. Thus, the isentropic turbine efficiency defined by Eq. 7.27 as the actual work output divided by the work for a hypothetical isentropic expansion from the same inlet state to the same exit pressure might well be called a *first-law efficiency*, in that it is a comparison of two energy quantities. The second-law efficiency, as just described, would be the actual work output of the turbine divided by the decrease in exergy from the same inlet state to the same exit state. For the turbine shown in Fig. 8.9, the second-law efficiency is

$$\eta_{\text{2nd law}} = \frac{w_a}{\psi_i - \psi_e} \tag{8.31}$$

In this sense, this concept provides a rating or measure of the real process in terms of the actual change of state and is simply another convenient way of utilizing the concept of exergy. In a similar manner, the second-law efficiency of a pump or compressor is the ratio of the increase in exergy to the work input to the device.

---

[1]In many popular talks, reference is made to our energy reserves. From a thermodynamic point of view, *exergy reserves* would be a much more acceptable term. There is much energy in the atmosphere and the ocean but relatively little exergy.

Example 8.5

An insulated steam turbine (Fig. 8.10), receives 30 kg of steam per second at 3 MPa, 350 °C. At the point in the turbine where the pressure is 0.5 MPa, steam is bled off for processing equipment at the rate of 5 kg/s. The temperature of this steam is 200 °C. The balance of the steam leaves the turbine at 15 kPa, 90 % quality. Determine the exergy per kilogram of the steam entering and at both points at which steam leaves the turbine, the isentropic efficiency and the second-law efficiency for this process.

Control volume:	Turbine.
Inlet state:	P_1, T_1 known; state fixed.
Exit state:	P_2, T_2 known; P_3, x_3 known; both states fixed.
Process:	Steady state.
Model:	Steam tables.

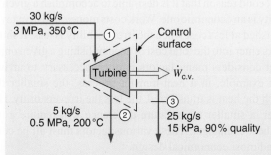

FIGURE 8.10 Sketch for Example 8.5.

Analysis

The exergy at any point for the steam entering or leaving the turbine is given by Eq. 8.22;

$$\psi = (h - h_0) - T_0(s - s_0) + \frac{\mathbf{V}^2}{2} + g(Z - Z_0)$$

Since there are no changes in kinetic and potential energy in this problem, this equation reduces to

$$\psi = (h - h_0) - T_0(s - s_0)$$

For the ideal isentropic turbine,

$$\dot{W}_s = \dot{m}_1 h_1 - \dot{m}_2 h_{2s} - \dot{m}_3 h_{3s}$$

For the actual turbine,

$$\dot{W} = \dot{m}_1 h_1 - \dot{m}_2 h_2 - \dot{m}_3 h_3$$

Solution

At the pressure and temperature of the surroundings, 0.1 MPa, 25 °C, the water is a slightly compressed liquid, and the properties of the water are essentially equal to those for saturated liquid at 25 °C.

$$h_0 = 104.9 \text{ kJ/kg} \qquad s_0 = 0.3674 \text{ kJ/kg·K}$$

From Eq. 8.22

$$\psi_1 = (3115.3 - 104.9) - 298.15(6.7428 - 0.3674) = 1109.6 \text{ kJ/kg}$$

$$\psi_2 = (2855.4 - 104.9) - 298.15(7.0592 - 0.3674) = 755.3 \text{ kJ/kg}$$

$$\psi_3 = (2361.8 - 104.9) - 298.15(7.2831 - 0.3674) = 195.0 \text{ kJ/kg}$$

$$\dot{m}_1\psi_1 - \dot{m}_2\psi_2 - \dot{m}_3\psi_3 = 30(1109.6) - 5(755.3) - 25(195.0) = 24\,637 \text{ kW}$$

For the ideal isentropic turbine,

$$s_{2s} = 6.7428 = 1.8606 + x_{2s} \times 4.906, \qquad x_{2s} = 0.9842$$

$$h_{2s} = 640.2 + 0.9842 \times 2108.5 = 2715.4$$

$$s_{3s} = 6.7428 = 0.7549 + x_{3s} \times 7.2536, \qquad x_{3s} = 0.8255$$

$$h_{3s} = 225.9 + 0.8255 \times 2373.1 = 2184.9$$

$$\dot{W}_s = 30(3115.3) - 5(2715.4) - 25(2184.9) = 25\,260 \text{ kW}$$

For the actual turbine,

$$\dot{W} = 30(3115.3) - 5(2855.4) - 25(2361.8) = 20\,137 \text{ kW}$$

The isentropic efficiency is

$$\eta_s = \frac{20\,137}{25\,260} = 0.797$$

and the second-law efficiency is

$$\eta_{\text{2nd law}} = \frac{20\,137}{24\,637} = 0.817$$

For a device that does not involve the production or the input of work, the definition of second-law efficiency refers to the accomplishment of the goal of the process relative to the process input in terms of exergy changes or transfers. For example, in a heat exchanger, energy is transferred from a high-temperature fluid stream to a low-temperature fluid stream, as shown in Fig. 8.11, in which case the second-law efficiency is defined as

$$\eta_{\text{2nd law}} = \frac{\dot{m}_1(\psi_2 - \psi_1)}{\dot{m}_3(\psi_3 - \psi_4)} \tag{8.32}$$

The previous expressions for the second-law efficiency can be presented by a single expression. First, notice that the actual work from Eq. 8.30 is

$$\dot{W}_{\text{c.v.}} = \dot{\Phi}_{\text{source}} - \dot{I}_{\text{c.v.}} = \dot{\Phi}_{\text{source}} - T\dot{S}_{\text{gen c.v.}} \tag{8.33}$$

FIGURE 8.11
A two-fluid heat exchanger.

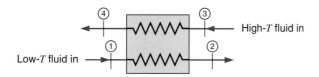

where $\dot{\Phi}_{source}$ is the total rate of exergy supplied from all sources: flows, heat transfers, and work inputs. In other words, the outgoing exergy, $\dot{W}_{c.v.}$, equals the incoming exergy less the irreversibility. Then, for all cases we may write

$$\eta_{2nd\ law} = \frac{\dot{\Phi}_{wanted}}{\dot{\Phi}_{source}} = \frac{\dot{\Phi}_{source} - \dot{I}_{c.v.}}{\dot{\Phi}_{source}} \tag{8.34}$$

and the wanted quantity is then expressed as exergy, whether it is actually a work term or a heat transfer. We can verify that this covers the turbine, Eq. 8.31, the pump or compressor, where work input is the source, and the heat exchanger efficiency in Eq. 8.32.

Example 8.6

In a boiler, heat is transferred from the products of combustion to the steam. The temperature of the products of combustion decreases from 1100 °C to 550 °C, while the pressure remains constant at 0.1 MPa. The average constant-pressure specific heat of the products of combustion is 1.09 kJ/kg·K. The water enters at 0.8 MPa, 150 °C, and leaves at 0.8 MPa, 250 °C. Determine the second-law efficiency for this process and the irreversibility per kilogram of water evaporated.

Control volume:	Overall heat exchanger.
Sketch:	Fig. 8.12.
Inlet states:	Both known, given in Fig. 8.12.
Exit states:	Both known, given in Fig. 8.12.
Process:	Overall, adiabatic.
Diagram:	Fig. 8.13.
Model:	Products—ideal gas, constant specific heat. Water—steam tables.

Analysis

For the products, the entropy change for this constant-pressure process is

$$(s_e - s_i)_{prod} = C_{po} \ln \frac{T_e}{T_i}$$

For this control volume, we can write the following governing equations: Continuity equation:

$$(\dot{m}_i)_{H_2O} = (\dot{m}_e)_{H_2O} \tag{a}$$

$$(\dot{m}_i)_{prod} = (\dot{m}_e)_{prod} \tag{b}$$

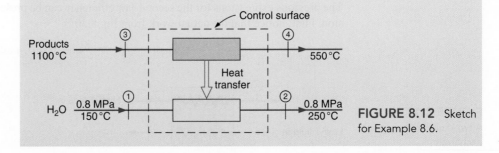

FIGURE 8.12 Sketch for Example 8.6.

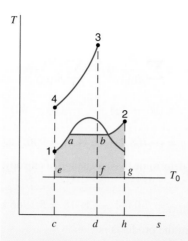

FIGURE 8.13 Temperature–S diagram for Example 8.6.

Energy equation (a steady-state process):

$$(\dot{m}_i h_i)_{H_2O} + (\dot{m}_i h_i)_{prod} = (\dot{m}_e h_e)_{H_2O} + (\dot{m}_e h_e)_{prod} \qquad (c)$$

Entropy equation (the process is adiabatic for the control volume shown):

$$(\dot{m}_e s_e)_{H_2O} + (\dot{m}_e s_e)_{prod} = (\dot{m}_i s_i)_{H_2O} + (\dot{m}_i s_i)_{prod} + \dot{S}_{gen}$$

Solution

From Eqs. a, b, and c, we can calculate the ratio of the mass flow of products to the mass flow of water.

$$\dot{m}_{prod}(h_i - h_e)_{prod} = \dot{m}_{H_2O}(h_e - h_i)_{H_2O}$$

$$\frac{\dot{m}_{prod}}{\dot{m}_{H_2O}} = \frac{(h_e - h_i)_{H_2O}}{(h_i - h_e)_{prod}} = \frac{2950 - 632.2}{1.09(1100 - 550)} = 3.866$$

The increase in exergy of the water is, per kilogram of water,

$$\psi_2 - \psi_1 = (h_2 - h_1) - T_0(s_2 - s_1)$$

$$= (2950 - 632.2) - 298.15(7.0384 - 1.8418)$$

$$= 768.4 \text{ kJ/kg}$$

The decrease in exergy of the products, per kilogram of water, is

$$\frac{\dot{m}_{prod}}{\dot{m}_{H_2O}}(\psi_3 - \psi_4) = \frac{\dot{m}_{prod}}{\dot{m}_{H_2O}}[(h_3 - h_4) - T_0(s_3 - s_4)]$$

$$= 3.866\left[1.09(1100 - 550) - 298.15\left(1.09 \ln\frac{1373.15}{823.15}\right)\right]$$

$$= 1674.7 \text{ kJ/kg}$$

Therefore, the second-law efficiency is, from Eq. 8.32,

$$\eta_{2nd\ law} = \frac{768.4}{1674.7} = 0.459$$

From Eq. 8.30, $\dot{I} = \dot{W}^{\mathrm{rev}}$, and Eq. 8.29, the process irreversibility per kilogram of water is

$$\frac{\dot{I}}{\dot{m}_{H_2O}} = \sum_i \frac{\dot{m}_i}{\dot{m}_{H_2O}}\psi_i - \sum_2 \frac{\dot{m}_e}{\dot{m}_{H_2O}}\psi_e$$

$$= (\psi_1 - \psi_2) + \frac{\dot{m}_{\mathrm{prod}}}{\dot{m}_{H_2O}}(\psi_3 - \psi_4)$$

$$= (-768.4 + 1674.7) = 906.3 \text{ kJ/kg}$$

It is also of interest to determine the net change of entropy. The change in the entropy of the water is

$$(s_2 - s_1)_{H_2O} = 7.0384 - 1.8418 = 5.1966 \text{ kJ/kg·K}$$

The change in the entropy of the products is

$$\frac{\dot{m}_{\mathrm{prod}}}{\dot{m}_{H_2O}}(s_4 - s_3)_{\mathrm{prod}} = -3.866\left(1.09 \ln\frac{1373.15}{823.15}\right) = -2.1564 \text{ kJ/kg·K}$$

Thus, there is a net increase in entropy during the process. The irreversibility could also have been calculated from Eqs. 8.6 and 8.13:

$$\dot{I} = \sum \dot{m}_e T_0 s_e - \sum \dot{m}_i T_0 s_i = T_0 \dot{S}_{\mathrm{gen}}$$

$$\frac{\dot{I}}{\dot{m}_{H_2O}} = T_0(s_2 - s_1)_{H_2O} + T_0\frac{\dot{m}_{\mathrm{prod}}}{\dot{m}_{H_2O}}(s_4 - s_3)_{\mathrm{prod}}$$

$$= 298.15(5.1966) + 298.15(-2.1564)$$

$$= 906.3 \text{ kJ/kg}$$

These two processes are shown on the T–s diagram of Fig. 8.13. Line 3–4 represents the process for the 3.866 kg of products. Area 3–4–c–d–3 represents the heat transferred from the 3.866 kg of products of combustion, and area 3–4–e–f–3 represents the decrease in exergy of these products. Area 1–a–b–2–h–c–1 represents the heat transferred to the water, and this is equal to area 3–4–c–d–3, which represents the heat transferred from the products of combustion. Area 1–a–b–2–g–e–1 represents the increase in exergy of the water. The difference between area 3–4–e–f–3 and area 1–a–b–2–g–e–1 represents the net decrease in exergy. It is readily shown that this net change is equal to area f–g–h–d–f, or $T_0(\Delta s)_{\mathrm{net}}$. Since the actual work is zero, this area also represents the irreversibility, which agrees with our calculation.

8.3 EXERGY BALANCE EQUATION

The previous treatment of exergy in different situations was done separately for the steady-flow, control mass, and transient processes. For each case, an actual process was compared to an ideal counterpart, which led to the reversible work and the irreversibility. When the reference was made with respect to the ambient state, we found the flow exergy, ψ, in Eq. 8.22, and the no-flow exergy, ϕ, in Eq. 8.27. We want to show that these forms of exergy are consistent with one another. The whole concept is unified by a formulation of the exergy for a general control volume from which we will recognize all the previous forms of exergy as special cases of the more general form.

In this analysis, we start out with the definition of exergy, $\Phi = m\phi$, as the maximum available work at a given state of a mass from Eq. 8.27, as

$$\Phi = m\phi = m(e - e_0) + P_0 m(v - v_0) - T_0 m(s - s_0) \qquad (8.35)$$

Here, subscript "0" refers to the ambient state with zero kinetic energy, the dead state, from which we take our reference. Because the properties at the reference state are constants, the rate of change for Φ becomes

$$
\begin{aligned}
\frac{d\Phi}{dt} &= \frac{dme}{dt} - e_0 \frac{dm}{dt} + P_0 \frac{dV}{dt} - P_0 v_0 \frac{dm}{dt} - T_0 \frac{dms}{dt} + T_0 s_0 \frac{dm}{dt} \\
&= \frac{dme}{dt} + P_0 \frac{dV}{dt} - T_0 \frac{dms}{dt} - (h_0 - T_0 s_0)\frac{dm}{dt} \qquad (8.36)
\end{aligned}
$$

and we used, $h_0 = e_0 + P_0 v_0$, to shorten the expression. Now we substitute the rate of change of mass from the continuity equation, Eq. 4.1,

$$\frac{dm}{dt} = \sum \dot{m}_i - \sum \dot{m}_e$$

the rate of change of total energy from the energy equation, Eq. 4.8,

$$\frac{dE}{dt} = \frac{dme}{dt} = \sum \dot{Q}_{\text{c.v.}} - \dot{W}_{\text{c.v.}} + \sum \dot{m}_i h_{\text{tot } i} - \sum \dot{m}_e h_{\text{tot } e}$$

and the rate of change of entropy from the entropy equation, Eq. 7.2,

$$\frac{dS}{dt} = \frac{dms}{dt} = \sum \dot{m}_i s_i - \sum \dot{m}_e s_e + \sum \frac{\dot{Q}_{\text{c.v.}}}{T} + \dot{S}_{\text{gen}}$$

into the rate of exergy equation, Eq. 8.36. When that is done, we get

$$
\begin{aligned}
\frac{d\Phi}{dt} &= \sum \dot{Q}_{\text{c.v.}} - \dot{W}_{\text{c.v.}} + \sum \dot{m}_i h_{\text{tot } i} - \sum \dot{m}_e h_{\text{tot } e} + P_0 \frac{dV}{dt} \\
&\quad - T_0 \sum \dot{m}_i s_i + T_0 \sum \dot{m}_e s_e - \sum T_0 \frac{\dot{Q}_{\text{c.v.}}}{T} - T_0 \dot{S}_{\text{gen}} \\
&\quad - (h_0 - T_0 s_0)\left[\sum \dot{m}_i - \sum \dot{m}_e\right] \qquad (8.37)
\end{aligned}
$$

Now collect the terms relating to the heat transfer together and those relating to the flow together and group them as

$$
\begin{aligned}
\frac{d\Phi}{dt} &= \sum \left(1 - \frac{T_0}{T}\right) \dot{Q}_{\text{c.v.}} && \text{Transfer by heat at } T \\
&\quad - \dot{W}_{\text{c.v.}} + P_0 \frac{dV}{dt} && \text{Transfer by shaft/boundary work} \\
&\quad + \sum \dot{m}_i \psi_i - \sum \dot{m}_e \psi_e && \text{Transfer by flow} \\
&\quad - T_0 \dot{S}_{\text{gen}} && \text{Exergy destruction} \qquad (8.38)
\end{aligned}
$$

The final form of the exergy balance equation is identical to the equation for the reversible work, Eq. 8.29, where the reversible work is substituted for by the actual work and the irreversibility from Eq. 8.30 and rearranged to solve for the storage term, $\dot{\Phi}_{c.v.}$. The rate equation for exergy can be stated verbally, like all the other balance equations:

$$\text{Rate of exergy storage} = \text{Transfer by heat} + \text{Transfer by shaft/boundary work}$$
$$+ \text{Transfer by flow} - \text{Exergy destruction}$$

and we notice that all the transfers take place with some surroundings and thus do not add up to any net change when the total world is considered. Only the exergy destruction due to entropy generation lowers the overall exergy level, and we can thus identify the regions in space where this occurs as the locations that have entropy generation. The exergy destruction is identical to the previously defined term, irreversibility.

Example 8.7

Let us look at the flows and fluxes of exergy for the feedwater heater in Example 8.1. The feedwater heater has a single flow, two heat transfers, and no work involved. When we do the balance of terms in Eq. 8.38 and evaluate the flow exergies from Eq. 8.22, we need the reference properties (take saturated liquid instead of 100 kPa at 25 °C):

$$\text{Table B.1.1: } h_0 = 104.87 \text{ kJ/kg}, \qquad s_0 = 0.3673 \text{ kJ/kg·K}$$

The flow exergies become

$$\psi_i = h_{\text{tot } i} - h_0 - T_0(s_i - s_0)$$
$$= 171.97 - 104.87 - 298.2 \times (0.5705 - 0.3687) = 6.92 \text{ kJ/kg}$$
$$\psi_e = h_{\text{tot } e} - h_0 - T_0(s_e - s_0)$$
$$= 765.25 - 104.87 - 298.2 \times (2.1341 - 0.3687) = 133.94 \text{ kJ/kg}$$

and the exergy fluxes from each of the heat transfers are

$$\left(1 - \frac{T_0}{T_1}\right)q_1 = \left(1 - \frac{298.2}{373.2}\right)180 = 36.17 \text{ kJ/kg}$$

$$\left(1 - \frac{T_0}{T_2}\right)q_2 = \left(1 - \frac{298.2}{473.2}\right)413.28 = 152.84 \text{ kJ/kg}$$

The destruction of exergy is then the balance ($w = 0$) of Eq. 8.38 as

$$T_0 s_{\text{gen}} = \sum \left(1 - \frac{T_0}{T}\right)q_{c.v.} + \psi_i - \psi_e$$

$$= 36.17 + 152.84 + 6.92 - 133.94 = 62.0 \text{ kJ/kg}$$

We can now express the heater's second-law efficiency as

$$\eta_{\text{2nd law}} = \frac{\dot{\Phi}_{\text{source}} - \dot{I}_{c.v.}}{\dot{\Phi}_{\text{source}}} = \frac{36.17 + 152.84 - 62.0}{36.17 + 152.84} = 0.67$$

The exergy fluxes are shown in Fig. 8.14, and the second-law efficiency shows that there is a potential for improvement. We should lower the temperature difference between the source and the water flow by adding more energy from the low-temperature source, thus decreasing the irreversibility.

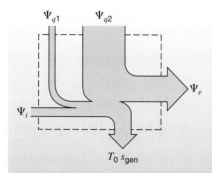

FIGURE 8.14 Fluxes, flows, and destruction of exergy in the feedwater heater.

Example 8.8

Assume a 500 W heating element in a stove with an element surface temperature of 1000 K. On top of the element is a ceramic top with a top surface temperature of 500 K, both shown in Fig. 8.15. Let us disregard any heat transfer downward, and follow the flux of exergy, and find the exergy destruction in the process.

Solution

Take just the heating element as a control volume in steady state with electrical work going in and heat transfer going out.

$$\text{Energy Eq.:} \qquad 0 = \dot{W}_{\text{electrical}} - \dot{Q}_{\text{out}}$$

$$\text{Entropy Eq.:} \qquad 0 = -\frac{\dot{Q}_{\text{out}}}{T_{\text{surf}}} + \dot{S}_{\text{gen}}$$

$$\text{Exergy Eq.:} \qquad 0 = -\left(1 - \frac{T_0}{T}\right)\dot{Q}_{\text{out}} - (-\dot{W}_{\text{electrical}}) - T_0\dot{S}_{\text{gen}}$$

From the balance equations, we get

$$\dot{Q}_{\text{out}} = \dot{W}_{\text{electrical}} = 500 \text{ W}$$

$$\dot{S}_{\text{gen}} = \dot{Q}_{\text{out}}/T_{\text{surf}} = 500 \text{ W}/1000 \text{ K} = 0.5 \text{ W/K}$$

$$\dot{\Phi}_{\text{destruction}} = T_0\dot{S}_{\text{gen}} = 298.15 \text{ K} \times 0.5 \text{ W/K} = 149 \text{ W}$$

$$\dot{\Phi}_{\text{transfer out}} = \left(1 - \frac{T_0}{T}\right)\dot{Q}_{\text{out}} = \left(1 - \frac{298.15}{1000}\right)500 = 351 \text{ W}$$

so the heating element receives 500 W of exergy flux, destroys 149 W, and gives out the balance of 351 W with the heat transfer at 1000 K.

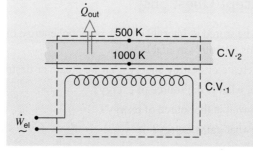

FIGURE 8.15 The electric heating element and ceramic top of a stove.

Take a second control volume from the heating element surface to the ceramic stove top. Here heat transfer comes in at 1000 K and leaves at 500 K with no work involved.

Energy Eq.: $\qquad 0 = \dot{Q}_{\text{in}} - \dot{Q}_{\text{out}}$

Entropy Eq.: $\qquad 0 = \dfrac{\dot{Q}_{\text{in}}}{T_{\text{surf}}} - \dfrac{\dot{Q}_{\text{out}}}{T_{\text{top}}} + \dot{S}_{\text{gen}}$

Exergy Eq.: $\qquad 0 = \left(1 - \dfrac{T_0}{T_{\text{surf}}}\right)\dot{Q}_{\text{in}} - \left(1 - \dfrac{T_0}{T_{\text{top}}}\right)\dot{Q}_{\text{out}} - T_0\dot{S}_{\text{gen}}$

From the energy equation, we see that the two heat transfers are equal, and the entropy generation then becomes

$$\dot{S}_{\text{gen}} = \frac{\dot{Q}_{\text{out}}}{T_{\text{top}}} - \frac{\dot{Q}_{\text{in}}}{T_{\text{surf}}} = 500\left(\frac{1}{500} - \frac{1}{1000}\right)\text{W/K} = 0.5 \text{ W/K}$$

The terms in the exergy equation become

$$0 = \left(1 - \frac{298.15}{1000}\right)500 \text{ W} - \left(1 - \frac{298.15}{500}\right)500 \text{ W} - 298.15 \text{ K} \times 0.5 \text{ W/K}$$

or

$$0 = 351 \text{ W} - 202 \text{ W} - 149 \text{ W}$$

This means that the top layer receives 351 W of exergy from the electric heating element and gives out 202 W from the top surface, having destroyed 149 W of exergy in the process. The flow of exergy and its destruction are illustrated in Fig. 8.16.

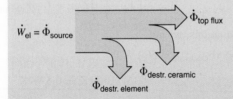

$\dot{W}_{\text{el}} = \dot{\Phi}_{\text{source}}$

$\dot{\Phi}_{\text{top flux}}$

$\dot{\Phi}_{\text{destr. ceramic}}$

$\dot{\Phi}_{\text{destr. element}}$

FIGURE 8.16 The fluxes and destruction terms of exergy.

In-Text Concept Questions

f. Energy can be stored as internal energy, potential energy, or kinetic energy. Are those energy forms all 100 % available?

g. We cannot create or destroy energy. Can we create or destroy exergy?

h. In a turbine, what is the source of exergy?

i. In a pump, what is the source of exergy?

j. In a pump, what gains exergy?

8.4 ENGINEERING APPLICATIONS

The most important application of the concept of exergy is to analyze single devices and complete systems with respect to the energy transfers, as well as the exergy transfers and destruction. Consideration of the energy terms leads to a first-law efficiency as a conversion efficiency for heat engines or a device efficiency measuring the actual device relative to a corresponding reversible device. Focusing on the exergy instead of the energy leads to a second-law efficiency for devices, as shown in Eqs. 8.31–8.34. These second-law efficiencies are generally larger than the first-law efficiency, as they express the operation of the actual device relative to what is theoretically possible with the same inlet and exit states as in the actual device. This is different from the first-law efficiency, where the ideal device used in the comparison does not have the same exit or end state as the actual device.

These efficiencies are used as guidelines for the evaluation of actual devices and systems such as pumps, compressors, turbines, and nozzles, to mention a few common devices. Such comparisons rely on experience with respect to the judgment of the result; i.e., is a second-law efficiency of 85 % considered good enough? This might be excellent for a compressor generating a very high pressure but not good enough for one that creates a moderately high pressure, and it is too low for a nozzle to be considered good.

Besides using a second-law efficiency for devices, as previously shown, we can use it for complete cycle systems such as heat engines or heat pumps. Consider a simple heat engine that gives out actual work from a high-temperature heat transfer with a first-law efficiency that is an energy conversion efficiency

$$W_{HE} = \eta_{HE\,I}Q_H$$

What, then, is the second-law efficiency? We basically form the same relation but express it in terms of exergy rather than energy and recall that work is 100 % exergy:

$$W_{HE} = \eta_{HE\,II}\Phi_H = \eta_{HE\,II}\left(1 - \frac{T_0}{T_H}\right)Q_H \qquad (8.39)$$

A second-law efficiency for a heat pump would be the ratio of exergy gained Φ_H (or $\Phi_H - \Phi_L$ if the low-temperature Φ_L is important) and the exergy from the source, which is the work input as

$$\eta_{HP\,II} = \frac{\Phi_H}{W_{HP}} = \left(1 - \frac{T_0}{T_H}\right)Q_H/W_{HP} \qquad (8.40)$$

A similar but slightly different measure of performance is to look at the exergy destruction term(s), either absolute or relative to the exergy input from the source. Consider a more complex system such as a complete steam power plant with several devices; look at Problem 4.92 for an example. If we do the analysis of every component and find the exergy destruction in all parts of the system, we would then use those findings to guide us in deciding where we should spend engineering effort to improve the system. Look at the system parts that have the largest exergy destruction first and try to reduce that by altering the system design and operating conditions. For the power plant, for instance, try to lower the temperature differences in the heat exchangers (recall Examples 8.1 and 8.7), reduce the pressure and heat loss in the piping, and ensure that the turbine is operating in its optimal range, to mention just a few of the more important places that have exergy destruction.

In the steam condenser, a large amount of energy is rejected to the surroundings but very little exergy is destroyed or lost, so the consideration of energy is misleading; the flows and fluxes of exergy provide a much better impression of the importance for the overall performance.

In-Text Concept Questions

k. In a heat engine, what is the source of exergy?

l. In a heat pump, what is the source of exergy?

m. In Eq. 8.39 for the heat engine, the source of exergy was written as a heat transfer. What does the expression look like if the source is a flow of hot gas being cooled down as it gives energy to the heat engine?

SUMMARY

Work out of a Carnot-cycle heat engine is the available energy in the heat transfer from the hot source; the heat transfer to the ambient air is unavailable. When an actual device is compared to an ideal device with the same flows and states in and out, we get to the concepts of reversible work and exergy. The reversible work is the maximum work we can get out of a given set of flows and heat transfers or, alternatively, the minimum work we have to put into the device. The comparison between the actual work and the theoretical maximum work gives a second-law efficiency. When exergy is used, the second-law efficiency can also be used for devices that do not involve shaftwork such as heat exchangers. In that case, we compare the exergy given out by one flow to the exergy gained by the other flow, giving a ratio of exergies instead of energies used for the first-law efficiency. Any irreversibility (entropy generation) in a process destroys exergy and is undesirable. The concept of available work can be used to give a general definition of exergy as being the reversible work minus the work that must go to the ambient air. From this definition, we can construct the exergy balance equation and apply it to different control volumes. From a design perspective, we can then focus on the flows and fluxes of exergy and improve the processes that destroy exergy.

You should have learned a number of skills and acquired abilities from studying this chapter that will allow you to

- Understand the concept of available energy.
- Understand that energy and exergy are different concepts.
- Be able to conceptualize the ideal counterpart to an actual system and find the reversible work and heat transfer in the ideal system.
- Understand the difference between a first-law and a second-law efficiency.
- Relate the second-law efficiency to the transfer and destruction of exergy.
- Be able to look at flows (fluxes) of exergy.
- Determine irreversibilities as the destruction of exergy.
- Know that destruction of exergy is due to entropy generation.
- Know that transfers of exergy do not change total or net exergy in the world.
- Know that the exergy equation is based on the energy and entropy equations and thus does not add another law.

KEY CONCEPTS AND FORMULAS

Available work from heat

$$W = Q\left(1 - \frac{T_0}{T_H}\right)$$

Reversible flow work with extra q_0^{rev} from ambient at T_0 and q in at T_H

$$q_0^{\text{rev}} = T_0(s_e - s_i) - q\frac{T_0}{T_H}$$

$$w^{\text{rev}} = h_i - h_e - T_0(s_i - s_e) + q\left(1 - \frac{T_0}{T_H}\right)$$

Flow irreversibility

$$i = w^{\text{rev}} - w = q_0^{\text{rev}} = T_0\dot{S}_{\text{gen}}/\dot{m} = T_0 s_{\text{gen}}$$

Reversible work C.M.

$$_1W_2^{\text{rev}} = T_0(S_2 - S_1) - (U_2 - U_1) + {_1}Q_2\left(1 - \frac{T_0}{T_H}\right)$$

Irreversibility C.M.

$$_1I_2 = T_0(S_2 - S_1) - {_1}Q_2\frac{T_0}{T_H} = T_0\,{_1}S_{2\,\text{gen}}$$

Second-law efficiency

$$\eta_{\text{2nd law}} = \frac{\dot{\Phi}_{\text{wanted}}}{\dot{\Phi}_{\text{source}}} = \frac{\dot{\Phi}_{\text{source}} - \dot{I}_{\text{c.v.}}}{\dot{\Phi}_{\text{source}}}$$

Exergy, flow

$$\psi = [h - T_0 s + \tfrac{1}{2}\mathbf{V}^2 + gZ] - [h_0 - T_0 s_0 + gZ_0]$$

Exergy, stored

$$\phi = (e - e_0) + P_0(v - v_0) - T_0(s - s_0); \quad \Phi = m\phi$$

Exergy transfer by heat

$$\phi_{\text{transfer}} = q\left(1 - \frac{T_0}{T_H}\right)$$

Exergy transfer by flow

$$\phi_{\text{transfer}} = h_{\text{tot}\ i} - h_{\text{tot}\ e} - T_0(s_i - s_e)$$

Exergy rate Eq.

$$\frac{d\Phi}{dt} = \sum\left(1 - \frac{T_0}{T}\right)\dot{Q}_{\text{c.v.}} - \dot{W}_{\text{c.v.}} + P_0\frac{dV}{dt}$$

$$+ \sum \dot{m}_i \psi_i - \sum \dot{m}_e \psi_e - T_0\dot{S}_{\text{gen}}$$

Exergy Eq. C.M. ($\Phi = m\phi$)

$$\Phi_2 - \Phi_1 = \left(1 - \frac{T_0}{T_H}\right){_1}Q_2 - {_1}W_2$$

$$+ P_0(V_2 - V_1) - {_1}I_2$$

CONCEPT-STUDY GUIDE PROBLEMS

8.1 Why does the reversible C.V. counterpart to the actual C.V. have the same storage and flow terms?

8.2 Can one of the heat transfers in Eqs. 8.5 and 8.6 be to or from the ambient air?

8.3 Is all the energy in the ocean available?

8.4 Does a reversible process change the exergy if there is no work involved?

8.5 Is the reversible work between two states the same as ideal work for the device?

8.6 When is the reversible work the same as the isentropic work?

8.7 If I heat some cold liquid water to T_0, do I increase its exergy?

8.8 Are reversible work and availability (exergy) connected?

8.9 Consider, the availability (exergy) associated with a flow. The total exergy is based on the thermodynamic state and the kinetic and potential energies. Can they all be negative?

8.10 Verify that Eq. 8.29 reduces to Eq. 8.14 for a steady-state process.

8.11 What is the second-law efficiency of a Carnot heat engine?

8.12 What is the second-law efficiency of a reversible heat engine?

8.13 For a nozzle, what is the output and input (source) expressed in exergies?

8.14 Is the exergy equation independent of the energy and entropy equations?

8.15 Use the exergy balance equation to find the efficiency of a steady-state Carnot heat engine operating between two fixed temperature reservoirs.

HOMEWORK PROBLEMS

Exergy, Reversible Work

8.16 A control mass gives out 10 kJ of energy in the form of
a. Electrical work from a battery.
b. Mechanical work from a spring.
c. Heat transfer at $500\,°C$.
Find the change in exergy of the control mass for each of the three cases.

8.17 A fraction of some power to a motor (1), 2 kW, is turned into heat transfer at 500 K (2) and then it dissipates in the ambient at 300 K (3). Give the rates of exergy along the process 1–2–3.

8.18 A heat engine receives 5 kW at 800 K and 10 kW at 1000 K, rejecting energy by heat transfer at 600 K. Assume it is reversible and find the power output. How much power could be produced if it could reject energy at $T_0 = 298$ K?

8.19 A household refrigerator has a freezer at T_F and a cold space at T_C from which energy is removed and rejected to the ambient at T_A, as shown in Fig. P8.19. Assume that the rate of heat transfer

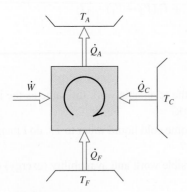

FIGURE P8.19

from the cold space, $\dot{Q}_C$, is the same as from the freezer, $\dot{Q}_F$, and find an expression for the

minimum power into the heat pump. Evaluate this power when $T_A = 20\,°C$, $T_C = 5\,°C$, $T_F = -10\,°C$, and $\dot{Q}_F = 3$ kW.

8.20 An air compressor takes air in at the state of the surroundings, 100 kPa, 300 K. The air exits at 400 kPa, 480 K using 100 kW of power. Determine the minimum compressor work input.

8.21 The compressor in a refrigerator takes refrigerant R-134a in at 100 kPa, $-20\,°C$ and compresses it to 1 MPa, $70\,°C$. With the room at $20\,°C$, find the minimum compressor work.

8.22 Calculate the reversible work out of the two-stage turbine shown in Problem 4.68, assuming the ambient is at $25\,°C$. Compare this to the actual work, which was found to be 18.08 MW.

8.23 A steam turbine receives steam at 6 MPa, $800\,°C$. It has a heat loss of 49.7 kJ/kg and an isentropic efficiency of 90 %. For an exit pressure of 15 kPa and surroundings at $20\,°C$, find the actual work and the reversible work between the inlet and the exit.

8.24 A compressor in a refrigerator receives R-410A at 150 kPa, $-40\,°C$, and brings it up to 600 kPa, using an actual specific work of 58.65 kJ/kg in adiabatic compression. Find the specific reversible work.

8.25 Air flows through a constant-pressure heating device, shown in Fig. P8.25. It is heated up in a reversible process with a work input of 200 kJ/kg

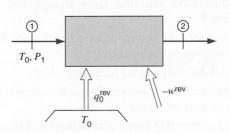

FIGURE P8.25

air flowing. The device exchanges heat with the ambient at 300 K. The air enters at 300 K, 400 kPa. Assuming constant specific heat, develop an expression for the exit temperature and solve for it by iterations.

8.26 An adiabatic and reversible air compressor takes air in at 100 kPa, 310 K. The air exits at 600 kPa at the rate of 0.4 kg/s. Determine the minimum compressor work input and repeat for an inlet at 295 K instead. Why is the work less for a lower inlet T?

8.27 An air flow of 5 kg/min at 1500 K, 125 kPa goes through a constant-pressure heat exchanger, giving energy to a heat engine, shown in Fig. P8.27. The air exits at 500 K, and the ambient is at 298 K, 100 kPa. Find the rate of heat transfer delivered to the engine and the power the engine can produce.

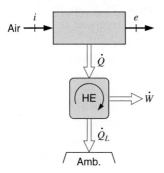

FIGURE P8.27

8.28 Water at 800 °C, 10 MPa is flowing through a heat exchanger giving off energy to come out as saturated liquid water at 10 MPa in a steady-flow process. Find the specific heat transfer and the specific flow exergy the water has delivered.

8.29 A rock bed consists of 6000 kg granite and is at 70 °C. A small house with a lumped mass of 12 000 kg wood and 1000 kg iron is at 15 °C. They are now brought to a uniform final temperature with no external heat transfer by connecting the house and the rock bed through some heat engines. If the process is reversible, find the final temperature and the work done in the process.

8.30 A constant-pressure piston/cylinder has 1 kg of saturated liquid water at 100 kPa. A rigid tank contains air at 1200 K, 1000 kPa. They are now thermally connected by a reversible heat engine cooling the air tank and boiling the water to saturated vapor. Find the required amount of air and the work out of the heat engine.

8.31 A basement is flooded with 16 m³ of water at 15 °C. It is pumped out with a small pump driven by a 0.75 kW electric motor. The hose can reach 8 m vertically up, and to ensure that the water can flow over the edge of a dike, it should have a velocity of 20 m/s at that point generated by a nozzle (see Fig. P8.31). Find the maximum flow rate you can get and how fast the basement can be emptied.

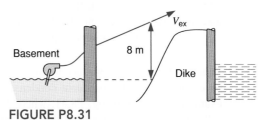

FIGURE P8.31

Irreversibility

8.32 A 20 °C room is heated with a 2500 W electric baseboard heater. What is the rate of irreversibility?

8.33 A refrigerator removes 1.5 kW from the cold space at −10 °C, using 750 W of power input while it rejects heat to the kitchen at 25 °C. Find the rate of irreversibility.

8.34 Calculate the irreversibility for the condenser in Problem 7.82, assuming an ambient temperature of 17 °C.

8.35 A throttle process is an irreversible process. Assume that an air flow at 1000 kPa, 400 K runs through a valve out to ambient 100 kPa. Find the reversible work and irreversibility, assuming an ambient temperature of 25 °C.

8.36 A compressor in a refrigerator receives R-410A at 150 kPa, −20 °C, and brings it up to 600 kPa, 60 °C in an adiabatic compression. Find the specific work, reversible work, entropy generation, and irreversibility.

8.37 A constant-pressure piston/cylinder contains 2 kg of water at 5 MPa and 100 °C. Heat is added from a reservoir at 600 °C to the water until it reaches 600 °C. Find the total irreversibility in the process.

8.38 A constant "flow" of steel parts at 2 kg/s at 20 °C goes into a furnace, where the parts are heat treated to 900 °C by a source at an average of 1250 K. Find the reversible work and the irreversibility in this process.

8.39 An air compressor receives atmospheric air at $T_0 = 17\,°C$, 100 kPa, and compresses it to 1400 kPa. The compressor has an isentropic efficiency of 88%, and it loses energy by heat transfer to the atmosphere as 10% of the isentropic work. Find the actual exit temperature and the reversible work.

8.40 Two flows of air, both at 400 kPa, mix in an insulated mixing chamber. One flow is 1 kg/s at 1500 K and the other is 2 kg/s at 300 K. Find the irreversibility in the process per kilogram of air flowing out.

8.41 Fresh water can be produced from salt water by evaporation and subsequent condensation. An example is shown in Fig. P8.41, where 150 kg/s salt water, state 1, comes from the condenser in a large power plant. The water is throttled to the saturated pressure in the flash evaporator and the vapor, state 2, is then condensed by cooling with sea water. As the evaporation takes place below atmospheric pressure, pumps must bring the liquid water flows back up to P_0. Assume that the salt water has the same properties as pure water, the ambient is at $20\,°C$, and there are no external heat transfers. With the states as shown in the table below, find the irreversibility in the throttling valve and in the condenser.

State	1	2	5	7	8
T [°C]	30	25	23	17	20
h [kJ/kg]	125.77	2547.2	96.5	71.37	83.96
s [kJ/kg·K]	0.4369	8.558	0.3392	0.2535	0.2966

8.42 A rock bed consists of 6000 kg granite and is at $70\,°C$. A small house with a lumped mass of 12 000 kg wood and 1000 kg iron is at $15\,°C$. They are now brought to a uniform final temperature by circulating water between the rock bed and the house. Find the final temperature and the irreversibility of the process, assuming an ambient of $15\,°C$.

8.43 A computer CPU chip consists of 50 g silicon, 20 g copper, and 50 g polyvinyl chloride (plastic). It now heats from ambient, $25\,°C$, to $70\,°C$ in an adiabatic process as the computer is turned on. Find the amount of irreversibility.

8.44 R-134a is flowed into an insulated 0.2 m³ initially empty container from a line at 500 kPa, saturated vapor until the flow stops by itself. Find the final mass and temperature in the container and the total irreversibility in the process.

8.45 The water cooler in Problem 5.22 operates steady state. Find the rate of exergy destruction (irreversibility).

8.46 Air enters the turbocharger compressor (see Fig. P8.46) of an automotive engine at 100 kPa, $30\,°C$ and exits at 200 kPa. The air is cooled by $50\,°C$ in an intercooler before entering the engine. The isentropic efficiency of the compressor is 75%. Determine the temperature of the air entering the engine and the irreversibility of the compression-cooling process.

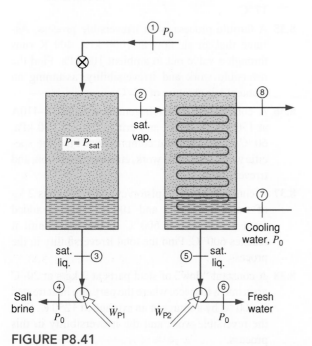

FIGURE P8.41

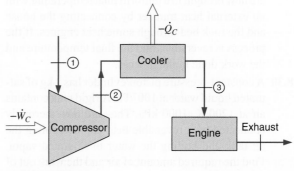

FIGURE P8.46

8.47 A constant-pressure piston/cylinder has 1 kg of saturated liquid water at 100 kPa. A rigid tank contains air at 1000 K, 1000 kPa. They are now thermally connected by conduction through the walls cooling the air tank and bringing the water to saturated vapor. Find the required amount of air and the irreversibility of the process, assuming no external heat transfer.

Exergy

8.48 The automatic transmission in a car receives 25 kW of shaft work and gives out 23 kW to the drive shaft. The balance is dissipated in the hydraulic fluid and metal casing, all at 45 °C, which, in turn, transmit it to the outer atmosphere at 20 °C. Find all the exergy transfer rates.

8.49 A heat engine receives 1 kW heat transfer at 1000 K and gives out 500 W as work, with the rest as heat transfer to the ambient at 25 °C. What are the fluxes of exergy in and out?

8.50 In a refrigerator, 1 kW is removed from the −10 °C cold space and 1.3 kW is moved into the 30 °C warm space. Find the exergy fluxes, including the direction associated with the two heat transfers.

8.51 A heat pump has a COP of 3 using a power input of 3 kW. Its low temperature is T_0 and its high temperature is 80 °C, with an ambient at T_0. Find the fluxes of exergy associated with the energy fluxes in and out.

8.52 A flow of air at 1000 kPa, 300 K is throttled to 500 kPa. What is the irreversibility? What is the drop in flow exergy?

8.53 A power plant has an overall thermal efficiency of 40 %, receiving 100 MW of heat transfer from hot gases at an average of 1300 K, and rejects heat transfer at 50 °C from the condenser to a river at ambient temperature, 20 °C. Find the rate of both energy and exergy (a) from the hot gases and (b) from the condenser.

8.54 Find the change in exergy from inlet to exit of the condenser in Problem 7.41.

8.55 A steady-flow device receives R-410A at 40 °C, 800 kPa and it exits at 40 °C, 100 kPa. Assume a reversible isothermal process. Find the change in specific exergy.

8.56 Consider the springtime melting of ice in the mountains, which provides cold water running in a river

at 2 °C while the air temperature is 20 °C. What is the exergy of the water relative to the ambient temperature?

8.57 Nitrogen flows in a pipe with a velocity of 300 m/s at 500 kPa, 300 °C. What is its exergy with respect to an ambient at 100 kPa, 20 °C?

8.58 Compressed air for machines and tools in a plant is generated by a central compressor receiving air at 100 kPa, 300 K, 1 kg/s and delivering it at 600 kPa to a buffer tank and a distribution pipe. After flowing through the tank and pipe, the air is at the ambient 300 K, but 600 kPa at its point of use. Assume a reversible adiabatic compressor and find the compressor exit temperature and the increase in air exergy through the compressor.

8.59 For the air system in the previous problem, find the increase in the air exergy from the inlet to the point of use. How much exergy was lost in the flow after the compressor exit?

8.60 A geothermal source provides 10 kg/s of hot water at 500 kPa, 145 °C, flowing into a flash evaporator that separates vapor and liquid at 250 kPa. Find the three fluxes of exergy (inlet and two outlets) and the irreversibility rate.

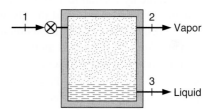

FIGURE P8.60

8.61 An air compressor is used to charge an initially empty 200 L tank with air up to 5 MPa. The air inlet to the compressor is at 100 kPa, 17 °C and the compressor's isentropic efficiency is 80 %. Find the total compressor work and the change in exergy of the air.

8.62 Find the exergy at all four states in the power plant in Problem 7.41 with an ambient at 298 K.

8.63 An electric stove has one heating element at 300 °C producing 750 W of electric power. It transfers 90 % of the power to 1 kg water in a kettle initially at 20 °C, 100 kPa; the other 10 % leaks to the room air. The water at a uniform T is brought to the boiling point. At the start of the process, what is the rate of exergy transfer by (a) electrical input, (b) from the heating element, and (c) into the water at T_{water}?

8.64 A wooden bucket (2 kg) with 10 kg hot liquid water, both at 85 °C, is lowered 400 m down into a mine shaft. What is the exergy of the bucket and water with respect to the surface ambient at 20 °C?

8.65 A flow of 0.1 kg/s hot water at 80 °C is mixed with a flow of 0.2 kg/s cold water at 20 °C in a shower fixture. What is the rate of exergy destruction (irreversibility) for this process?

8.66 A 1 kg block of copper at 350 °C is quenched in a 10 kg oil bath initially at the ambient temperature of 20 °C. Calculate the final uniform temperature (no heat transfer to/from the ambient) and the change of exergy of the system (copper and oil).

8.67 A 200 L insulated tank contains nitrogen gas at 200 kPa, 300 K. A line with nitrogen at 500 K, 500 kPa adds 40% more mass to the tank with a flow through a valve. Use constant specific heats to find the final temperature and the exergy destruction.

8.68 A 10 kg iron disk brake on a car is initially at 10 °C. Suddenly the brake pad hangs up, increasing the brake temperature by friction to 110 °C while the car maintains constant speed. Find the change in exergy of the disk and the energy depletion of the car's gas tank due to this process alone. Assume that the engine has a thermal efficiency of 35%.

8.69 Water as saturated liquid at 200 kPa goes through a constant-pressure heat exchanger, as shown in Fig. P8.69. The heat input is supplied from a

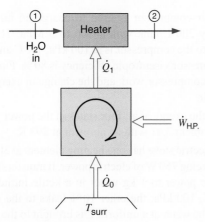

FIGURE P8.69

reversible heat pump extracting heat from the surroundings at 17 °C. The water flow rate is 2 kg/min and the whole process is reversible; that is, there is no overall net entropy change. If the heat pump receives 40 kW of work, find the water exit state and the increase in specific exergy of the water.

Exergy Balance Equation

8.70 Apply the exergy equation to solve Problem 8.32 with $T_0 = 20$ °C.

8.71 Estimate some reasonable temperatures to use and find all the fluxes of exergy in the refrigerator in Example 5.2.

8.72 Find the specific flow exergy into and out of the steam turbine in Example 7.1, assuming an ambient at 293 K. Use the exergy balance equation to find the reversible specific work. Does this calculation of specific work depend on T_0?

8.73 Apply the exergy equation to solve Problem 8.33.

8.74 Evaluate the steady-state exergy fluxes due to a heat transfer of 50 W through a wall with 600 K on one side and 400 K on the other side. What is the exergy destruction in the wall?

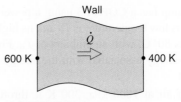

FIGURE P8.74

8.75 Apply the exergy equation to find the exergy destruction for Problem 8.49.

8.76 A flow of 1 kg/s of air at 300 K is mixed with a flow of 2 kg/s air at 1200 K in an insulated pipe junction at a pressure of 100 kPa. Find the exit temperature and the exergy destruction.

8.77 The condenser in a power plant cools 10 kg/s water at 10 kPa, quality 90%, so it comes out as saturated liquid at 10 kPa. The cooling is done by ocean water coming in at ambient 15 °C and returned to the ocean at 20 °C. Find the transfer out of the water and the transfer into the ocean water of both energy and exergy (four terms).

8.78 Consider the automobile engine in Example 5.1 and assume the fuel energy is delivered at a constant 1500 K. Of the 70 % of the energy that is lost, 40 % is exhaust flow at 900 K and the remainder is 30 % heat transfer to the walls at 450 K that goes on to the coolant fluid at 370 K, finally ending up in atmospheric air at ambient 20 °C. Find all the energy and exergy flows for this heat engine. Also find the exergy destruction and where that is done.

8.79 A piston/cylinder has forces on the piston, so it maintains constant pressure. It contains 2 kg of ammonia at 800 kPa, 40 °C and is now heated to 100 °C by a reversible heat engine that receives heat from a 200 °C source. Find the work out of the heat engine using the exergy balance equation.

8.80 A disk brake of 2 kg steel and 1 kg brake pads is at 20 °C. The brakes are now stopping a car, so they dissipate energy by friction and heat up to $T_2 = 200$ °C. Assume the brake pads have specific heat of 0.6 kJ/kg·K. After this process, the disk and pads now slowly cool to the ambient, 20 °C $= T_3$. Find the exergy destruction in the braking process (1 to 2) and in the cooling process (2 to 3).

8.81 A small house kept at 20 °C inside loses 12 kW to the outside ambient at 0 °C. A heat pump is used to help heat the house together with possible electrical heat. The heat pump is driven by a motor of 2.5 kW, and it has a COP that is one-quarter that of a Carnot heat pump unit. Find the actual heat pump COP and the exergy destruction in the whole process.

8.82 A farmer runs a heat pump using 3 kW of power input. It keeps a chicken hatchery at a constant 30 °C, while the room loses 10 kW to the colder outside ambient at 10 °C. Find the COP of the heat pump, the rate of exergy destruction in the heat pump and its heat exchangers, and the rate of exergy destruction in the heat loss process.

Device Second-Law Efficiency

8.83 A heat engine receives 1 kW heat transfer at 1000 K and gives out 400 W as work, with the rest as heat transfer to the ambient. Find its first- and second-law efficiencies.

8.84 A heat exchanger increases the exergy of 3 kg/s water by 1650 kJ/kg using 10 kg/s air coming in at 1400 K and leaving with 600 kJ/kg less exergy. What are the irreversibility and the second-law efficiency?

8.85 Find the second-law efficiency of the heat pump in Problem 8.51.

8.86 A steam turbine inlet is at 1200 kPa, 500 °C. The actual exit is at 300 kPa, with actual work of 407 kJ/kg. What is its second-law efficiency?

8.87 Find the isentropic efficiency and the second-law efficiency for the compressor in Problem 8.24.

8.88 A steam turbine has inlet at 4 MPa, 500 °C and actual exit of 200 kPa, 150 °C. Find its first-law (isentropic) and its second-law efficiencies.

8.89 Find the second-law efficiency for the compressed air system in Problem 8.58. Consider the total system from the inlet to the final point of use.

8.90 A turbine receives steam at 3000 kPa, 500 °C and has two exit flows, one at 1000 kPa, 350 °C with 20 % of the flow and the remainder at 200 kPa, 200 °C. Find the isentropic and second-law efficiencies.

8.91 A heat engine operating in an environment at 298 K produces 5 kW of power output with a first-law efficiency of 50 %. It has a second-law efficiency of 80 % and $T_L = 310$ K. Find all the energy and exergy transfers in and out.

8.92 Air flows into a heat engine at ambient conditions of 100 kPa, 300 K, as shown in Fig. P8.92. Energy is supplied as 1200 kJ per kilogram of air from a 1500 K source, and in some part of the process a heat transfer loss of 300 kJ per kilogram of air air occurs at 750 K. The air leaves the engine at 100 kPa, 800 K. Find the first- and second-law efficiencies.

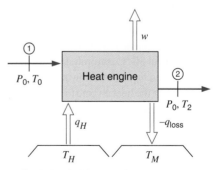

FIGURE P8.92

8.93 Air enters a compressor at ambient conditions of 100 kPa, 300 K, and exits at 800 kPa. If the isentropic compressor efficiency is 70 %, what is the second-law efficiency of the compressor process?

8.94 Consider that in the previous problem, after the compressor, the air flows in a pipe to an air tool, and at that point the temperature has dropped to ambient at 300 K and air pressure of 750 kPa. What is the second-law efficiency for the total system?

8.95 Use the exergy equation to analyze the compressor in Example 4.8 to find its second-law efficiency, assuming an ambient at 20 °C.

8.96 Calculate the second-law efficiency of the co-flowing heat exchanger in Problem 7.93 with an ambient at 17 °C.

8.97 An air compressor receives air at 290 K, 100 kPa and brings it up to a higher pressure in an adiabatic process. The actual specific work is 210 kJ/kg, and the isentropic efficiency is 82 %. Find the exit pressure and the second-law efficiency.

8.98 A heat exchanger brings 10 kg/s water from 100 °C up to 500 °C at 2000 kPa using air coming in at 1400 K and leaving at 460 K. What is the second-law efficiency?

8.99 Calculate the second-law efficiency of the counter-flowing heat exchanger in Problem 7.86 with an ambient at 20 °C.

8.100 A steam turbine receives 5 kg/s steam at 400 °C, 10 MPa. One flow of 0.8 kg/s is extracted at 3 MPa as saturated vapor, and the remainder runs out at 1500 kPa with a quality of 0.975. Find the second-law efficiency of the turbine.

Additional problems with applications of exergy related to cycles are found in Chapters 9 and 10.

Review Problems

8.101 The high-temperature heat source for a cyclic heat engine is a steady-flow heat exchanger where R-134a enters at 80 °C, saturated vapor, and exits at 80 °C, saturated liquid, at a flow rate of 5 kg/s. Heat is rejected from the heat engine to a steady-flow heat exchanger, where air enters at 150 kPa and ambient temperature 20 °C and exits at 125 kPa, 70 °C. The rate of irreversibility for the overall process is 175 kW. Calculate the mass flow rate of the air and the thermal efficiency of the heat engine.

8.102 A cylinder with a linear spring-loaded piston contains carbon dioxide gas at 2 MPa with a volume of 50 L. The device is made of aluminum and has a mass of 4 kg. Everything (aluminum and gas) is initially at 200 °C. By heat transfer, the whole system

cools to the ambient temperature of 25 °C, at which point the gas pressure is 1.5 MPa. Find the exergy at the initial and final states and the destruction of exergy in the process.

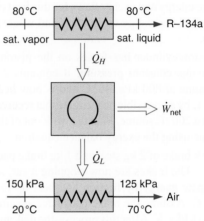

FIGURE P8.101

8.103 A two-stage compressor takes nitrogen in at 20 °C, 150 kPa and compresses it to 600 kPa, 450 K. Then it flows through an intercooler, where it cools to 320 K, and the second stage compresses it to 3000 kPa, 530 K. Find the specific exergy increase and the specific exergy destruction in each of the two compressor stages.

8.104 The intercooler in the previous problem uses cold liquid water to cool the nitrogen. The nitrogen flow is 0.1 kg/s, and the liquid water inlet is 20 °C and is setup to flow in the opposite direction from the nitrogen, so the water leaves at 35 °C. Find the flow rate of the water and the exergy destruction in this intercooler.

8.105 Find the irreversibility in the cooling process of the glass sheet in Problem 4.115.

8.106 Air in a piston/cylinder arrangement is at 110 kPa, 25 °C, with a volume of 50 L. It goes through a reversible polytropic process to a final state of 700 kPa, 500 K and exchanges heat with the ambient at 25 °C through a reversible device. Find the total work (including that of the external device) and the heat transfer from the ambient.

8.107 Consider the irreversible process in Problem 6.147. Assume that the process could be done reversibly by adding heat engines/pumps between tanks A and B and the cylinder. The total system is insulated, so there is no heat transfer to or from the ambient. Find the final state, the work given out to the piston, and the total work to or from the heat engines/pumps.

8.108 Consider the heat engine in Problem 8.92. The exit temperature was given as 800 K, but what are the theoretical limits for this temperature? Find the lowest and highest temperatures, assuming the heat transfers are as given. For each case, give the first- and second-law efficiencies.

8.109 A small air gun has 1 cm³ of air at 250 kPa, 27 °C. The piston is a bullet of mass 20 g. What is the highest potential velocity with which the bullet can leave?

8.110 Consider the nozzle in Problem 7.126. What is the second-law efficiency for the nozzle?

8.111 Consider the light bulb in Problem 6.152. What are the fluxes of exergy at the various locations mentioned? What is the exergy destruction in the filament, the entire bulb including the glass, and the entire room including the bulb? The light does not affect the gas or the glass in the bulb, but it becomes absorbed by the walls of the room.

8.112 Air enters a steady-flow turbine at 1600 K and exhausts to the atmosphere at 1000 K. The second-law efficiency is 85 %. What is the turbine inlet pressure?

8.113 A piston/cylinder arrangement has a load on the piston, so it maintains constant pressure. It contains 1 kg of steam at 500 kPa, 50 % quality. Heat from a reservoir at 700 °C brings the steam to 600 °C. Find the second-law efficiency for this process. Note that no formula is given for this particular case, so determine a reasonable expression for it.

8.114 A jet of air at 200 m/s flows at 25 °C, 100 kPa toward a wall, where the jet flow stagnates and leaves at very low velocity. Consider the process to be adiabatic and reversible. Use the exergy equation and the second law to find the stagnation temperature and pressure.

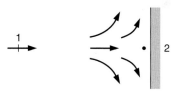

FIGURE P8.114

COMPUTER, DESIGN, AND OPEN-ENDED PROBLEMS

8.115 Use software to determine the properties of water as needed and calculate the second law efficiency of the low-pressure turbine in Problem 4.95.

8.116 The maximum power a windmill can possibly extract from the wind is

$$\dot{W} = \frac{16}{27}\rho A \mathbf{V} \frac{1}{2} \mathbf{V}^2 = \frac{16}{27}\dot{m}_{air} \times KE$$

Water flowing through Hoover Dam (see Problem 4.38) produces $\dot{W} = 0.8\dot{m}_{water}\,gh$. Burning 1 kg of coal gives 24 000 kJ delivered at 900 K to a heat engine. Find other examples in the literature and from problems in the previous chapters with steam and gases into turbines. List the availability (exergy) for a flow of 1 kg/s of substance with the above examples. Use a reasonable choice for

the values of the parameters and do the necessary analysis.

8.117 Determine the amount of power saved when the inlet temperature to an air compressor is lowered, as in Problem 8.26. Plot a graph of the required power versus the inlet temperature in the range 290 K to 310 K.

8.118 Consider the air compressor in Problem 8.93 and assume it uses 90 kW of power. Investigate the required power if the exit pressure can be reduced; plot the power for a range of exit pressures from 800 kPa down to 600 kPa.

8.119 Reconsider the use of the geothermal energy as discussed in Problem 4.99. The analysis that was done and the original problem statement specified the turbine exit state as 10 kPa, 90 % quality. Reconsider this problem with an adiabatic turbine having an isentropic efficiency of 85 % and an exit pressure of 10 kPa. Include a second-law analysis and discuss the changes in exergy. Describe another way of using the geothermal energy and make appropriate calculations.

8.120 Consider the nuclear power plant shown in Problem 4.95. Select one feedwater heater and one pump and analyze their performance. Check the energy balances and do the second-law analysis. Determine the change of exergy in all the flows and discuss measures of performance for both the pump and the feedwater heater.

Power and Refrigeration Systems—With Phase Change

9

Some power plants, such as the simple steam power plant, which we have considered several times, operate in a cycle. That is, the working fluid undergoes a series of processes and finally returns to the initial state. In other power plants, such as the internal-combustion engine and the gas turbine, the working fluid does not go through a thermodynamic cycle, even though the engine itself may operate in a mechanical cycle. In this instance, the working fluid has a different composition or is in a different state at the conclusion of the process than it had or was in at the beginning. Such equipment is sometimes said to operate on an *open cycle* (the word *cycle* is a misnomer), whereas the steam power plant operates on a *closed cycle*. The same distinction between open and closed cycles can be made regarding refrigeration devices. For both the open- and closed-cycle apparatus, however, it is advantageous to analyze the performance of an idealized closed cycle similar to the actual cycle. Such a procedure is particularly advantageous for determining the influence of certain variables on performance. For example, the spark-ignition internal-combustion engine is usually approximated by the Otto cycle. From an analysis of the Otto cycle, we conclude that increasing the compression ratio increases the efficiency. This is also true for the actual engine, even though the Otto-cycle efficiencies may deviate significantly from the actual efficiencies.

This chapter and the next are concerned with these idealized cycles for both power and refrigeration apparatus. This chapter focuses on systems with phase change, that is, systems utilizing condensing working fluids, while Chapter 10 deals with gaseous working fluids, where there is no change of phase. In both chapters, an attempt will be made to point out how the processes in the actual apparatus deviate from the ideal. Consideration is also given to certain modifications of the basic cycles that are intended to improve performance. These modifications include the use of devices such as regenerators, multistage compressors and expanders, and intercoolers. Various combinations of these types of systems and also special applications, such as cogeneration of electrical power and energy, combined cycles, topping and bottoming cycles, and binary cycle systems, are also discussed in these chapters and in the chapter-end problems.

9.1 INTRODUCTION TO POWER SYSTEMS

In introducing the second law of thermodynamics in Chapter 5, we considered cyclic heat engines consisting of four separate processes. We noted that these engines can be operated as steady-state devices involving shaft work, as shown in Fig. 5.18, or as cylinder/piston devices involving boundary-movement work, as shown in Fig. 5.19. The former may have a working fluid that changes phase during the processes in the cycle or may have a single-phase working fluid throughout. The latter type would normally have a gaseous working fluid throughout the cycle.

For a reversible steady-state process involving negligible kinetic and potential energy changes, the shaft work per unit mass is given by Eq. 7.15,

$$w = -\int v\,dP$$

For a reversible process involving a simple compressible substance, the boundary movement work per unit mass is given by Eq. 3.17,

$$w = \int P\,dv$$

The areas represented by these two integrals are shown in Fig. 9.1. It is of interest to note that, in the former case, there is no work involved in a constant-pressure process, while in the latter case, there is no work involved in a constant-volume process.

Let us now consider a power system consisting of four steady-state processes, as in Fig. 5.18. We assume that each process is internally reversible and has negligible changes in kinetic and potential energies, which results in the work for each process being given by Eq. 7.15. For convenience of operation, we will make the two heat-transfer processes (boiler and condenser) constant-pressure processes, such that those are simple heat exchangers involving no work. Let us also assume that the turbine and pump processes are both adiabatic and are therefore isentropic processes. Thus, the four processes comprising the cycle are as shown in Fig. 9.2. Note that if the entire cycle takes place inside the two-phase liquid–vapor dome, the resulting cycle is the Carnot cycle, since the two constant-pressure processes are also isothermal. Otherwise, this cycle is not a Carnot cycle. In either case, we find that the

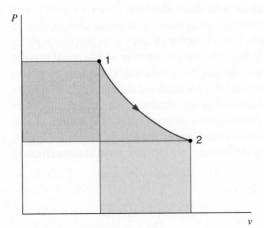

FIGURE 9.1

Comparison of shaft work and boundary-movement work.

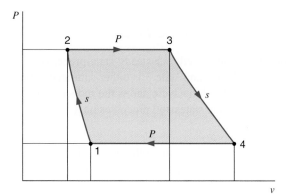

FIGURE 9.2 Four-process power cycle.

net work output for this power system is given by

$$w_{\text{net}} = -\int_1^2 v\,dP + 0 - \int_3^4 v\,dP + 0 = -\int_1^2 v\,dP + \int_4^3 v\,dP$$

and, since $P_2 = P_3$ and $P_1 = P_4$, we find that the system produces a net work output because the specific volume is larger during the expansion from 3 to 4 than it is during the compression from 1 to 2. This result is also evident from the areas $-\int v\,dP$ in Fig. 9.2. We conclude that it would be advantageous to have this difference in specific volume be as large as possible, as, for example, the difference between a vapor and a liquid.

If the four-process cycle shown in Fig. 9.2 were accomplished in a cylinder/piston system involving boundary-movement work, then the net work output for this power system would be given by

$$w_{\text{net}} = \int_1^2 P\,dv + \int_2^3 P\,dv + \int_3^4 P\,dv + \int_4^1 P\,dv$$

and from these four areas in Fig. 9.2, we note that the pressure is higher during any given change in volume in the two expansion processes than in the two compression processes, resulting in a net positive area and a net work output.

For either of the two cases just analyzed, it is noted from Fig. 9.2 that the net work output of the cycle is equal to the area enclosed by the process lines 1–2–3–4–1, and this area is the same for both cases, even though the work terms for the four individual processes are different for the two cases.

In this chapter we will consider the first of the two cases examined above, steady-state flow processes involving shaft work, utilizing condensing working fluids, such that the difference in the $-\int v\,dP$ work terms between the expansion and compression processes is a maximum. Then, in Chapter 10, we will consider systems utilizing gaseous working fluids for both cases, steady-state flow systems with shaft work terms and piston/cylinder systems involving boundary-movement work terms.

In the next several sections, we consider the Rankine cycle, which is the ideal four-steady-state process cycle shown in Fig. 9.2, utilizing a phase change between vapor and liquid to maximize the difference in specific volume during expansion and compression. This is the idealized model for a steam power plant system.

9.2 THE RANKINE CYCLE

We now consider the idealized four-steady-state-process cycle shown in Fig. 9.2, in which state 1 is saturated liquid and state 3 is either saturated vapor or superheated vapor. This system is termed the Rankine cycle and is the model for the simple steam power plant. It is convenient to show the states and processes on a *T–s* diagram, as given in Fig. 9.3. The four processes are:

1–2: Reversible adiabatic pumping process in the pump

2–3: Constant-pressure transfer of heat in the boiler

3–4: Reversible adiabatic expansion in the turbine (or another prime mover such as a steam engine)

4–1: Constant-pressure transfer of heat in the condenser

As mentioned earlier, the Rankine cycle also includes the possibility of superheating the vapor, as cycle 1–2–3′–4′–1.

If changes of kinetic and potential energy are neglected, heat transfer and work may be represented by various areas on the *T–s* diagram. The heat transferred to the working fluid is represented by area *a–2–2′–3–b–a* and the heat transferred from the working fluid by area *a–1–4–b–a*. From the energy equation we conclude that the area representing the work is the difference between these two areas—area 1–2–2′–3–4–1. The thermal efficiency is defined by the relation

$$\eta_{th} = \frac{w_{net}}{q_H} = \frac{\text{area } 1\text{–}2\text{–}2'\text{–}3\text{–}4\text{–}1}{\text{area } a\text{–}2\text{–}2'\text{–}3\text{–}b\text{–}a} \tag{9.1}$$

For analyzing the Rankine cycle, it is helpful to think of efficiency as depending on the average temperature at which heat is supplied and the average temperature at which heat is rejected. Any changes that increase the average temperature at which heat is supplied or decrease the average temperature at which heat is rejected will increase the Rankine-cycle efficiency.

In analyzing the ideal cycles in this chapter, the changes in kinetic and potential energies from one point in the cycle to another are neglected. In general, this is a reasonable assumption for the actual cycles.

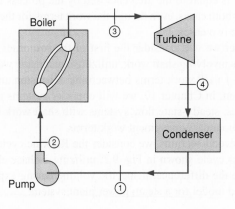
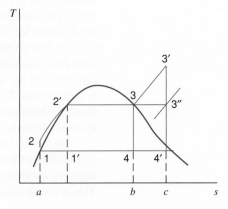

FIGURE 9.3 Simple steam power plant that operates on the Rankine cycle.

It is readily evident that the Rankine cycle has lower efficiency than a Carnot cycle with the same maximum and minimum temperatures as a Rankine cycle because the average temperature between 2 and $2'$ is less than the temperature during evaporation. We might well ask, why choose the Rankine cycle as the ideal cycle? Why not select the Carnot cycle $1'$–$2'$–3–4–$1'$? At least two reasons can be given. The first reason concerns the pumping process. State $1'$ is a mixture of liquid and vapor. Great difficulties are encountered in building a pump that will handle the mixture of liquid and vapor at $1'$ and deliver saturated liquid at $2'$. It is much easier to condense the vapor completely and handle only liquid in the pump; the Rankine cycle is based on this fact. The second reason concerns superheating the vapor. In the Rankine cycle the vapor is superheated at constant pressure, process 3–$3'$. In the Carnot cycle all the heat transfer is at constant temperature, and therefore the vapor is superheated in process 3–$3''$. Note, however, that during this process the pressure is dropping, which means that the heat must be transferred to the vapor as it undergoes an expansion process in which work is done. This heat transfer is also very difficult to achieve in practice. Thus, the Rankine cycle is the ideal cycle that can be approximated in practice. In the following sections, we will consider some variations on the Rankine cycle that enable it to approach more closely the efficiency of the Carnot cycle.

Before we discuss the influence of certain variables on the performance of the Rankine cycle, we will study an example.

Example 9.1

Determine the efficiency of a Rankine cycle using steam as the working fluid in which the condenser pressure is 10 kPa. The boiler pressure is 2 MPa. The steam leaves the boiler as saturated vapor.

In solving Rankine-cycle problems, we let w_p denote the work into the pump per kilogram of fluid flowing and q_L denote the heat rejected from the working fluid per kilogram of fluid flowing.

To solve this problem we consider, in succession, a control surface around the pump, the boiler, the turbine, and the condenser. For each, the thermodynamic model is the steam tables, and the process is steady state with negligible changes in kinetic and potential energies. First, consider the pump:

Control volume: Pump.
Inlet state: P_1 known, saturated liquid; state fixed.
Exit state: P_2 known.

Analysis

$$\text{Energy Eq.:} \quad w_p = h_2 - h_1$$
$$\text{Entropy Eq.:} \quad s_2 = s_1$$

and so

$$h_2 - h_1 = \int_1^2 v\, dP$$

Solution

Assuming the liquid to be incompressible, we have

$$w_p = v(P_2 - P_1) = 0.001\,01 \text{ m}^3/\text{kg} \times (2000 - 10)\text{ kPa} = 2.0 \text{ kJ/kg}$$
$$h_2 = h_1 + w_p = 191.8 + 2.0 = 193.8 \text{ kJ/kg}$$

Now consider the boiler:

Control volume: Boiler.

Inlet state: P_2, h_2 known; state fixed.

Exit state: P_3 known, saturated vapor; state fixed.

Analysis

Energy Eq.: $q_H = h_3 - h_2$

Solution

Substituting, we obtain

$$q_H = h_3 - h_2 = 2799.5 - 193.8 = 2605.7 \text{ kJ/kg}$$

Turning to the turbine next, we have:

Control volume: Turbine.

Inlet state: State 3 known (above).

Exit state: P_4 known.

Analysis

Energy Eq.: $w_t = h_3 - h_4$

Entropy Eq.: $s_3 = s_4$

Solution

We can determine the quality at state 4 as follows:

$$s_3 = s_4 = 6.3409 = 0.6493 + x_4 \, 7.5009, \qquad x_4 = 0.7588$$

$$h_4 = 191.8 + 0.7588(2392.8) = 2007.5 \text{ kJ/kg}$$

$$w_t = 2799.5 - 2007.5 = 792.0 \text{ kJ/kg}$$

Finally, we consider the condenser.

Control volume: Condenser.

Inlet state: State 4 known (as given).

Exit state: State 1 known (as given).

Analysis

Energy Eq.: $q_L = h_4 - h_1$

Solution

Substituting, we obtain

$$q_L = h_4 - h_1 = 2007.5 - 191.8 = 1815.7 \text{ kJ/kg}$$

We can now calculate the thermal efficiency:

$$\eta_{\text{th}} = \frac{w_{\text{net}}}{q_H} = \frac{q_H - q_L}{q_H} = \frac{w_t - w_p}{q_H} = \frac{792.0 - 2.0}{2605.7} = 30.3\,\%$$

We could also write an expression for thermal efficiency in terms of properties at various points in the cycle:

$$\eta_{th} = \frac{(h_3 - h_2) - (h_4 - h_1)}{h_3 - h_2} = \frac{(h_3 - h_4) - (h_2 - h_1)}{h_3 - h_2}$$

$$= \frac{2605.7 - 1815.7}{2605.7} = \frac{792.0 - 2.0}{2605.7} = 30.3\%$$

9.3 EFFECT OF PRESSURE AND TEMPERATURE ON THE RANKINE CYCLE

Let us first consider the effect of exhaust pressure and temperature on the Rankine cycle. This effect is shown on the T–s diagram of Fig. 9.4. Let the exhaust pressure drop from P_4 to P_4' with the corresponding decrease in temperature at which heat is rejected. The net work is increased by area 1–4–4'–1'–2'–2–1 (shown by the shading). The heat transferred to the steam is increased by area a'–2'–2–a–a'. Since these two areas are approximately equal, the net result is an increase in cycle efficiency. This is also evident from the fact that the average temperature at which heat is rejected is decreased. Note, however, that lowering the back pressure causes the moisture content of the steam leaving the turbine to increase. This is a significant factor because if the moisture in the low-pressure stages of the turbine exceeds about 10 %, not only is there a decrease in turbine efficiency, but erosion of the turbine blades may also be a very serious problem.

Next, consider the effect of superheating the steam in the boiler, as shown in Fig. 9.5. We see that the work is increased by area 3–3'–4'–4–3, and the heat transferred in the boiler is increased by area 3–3'–b'–b–3. Since the ratio of these two areas is greater than the ratio of net work to heat supplied for the rest of the cycle, it is evident that for given pressures, superheating the steam increases the Rankine-cycle efficiency. This increase in efficiency would also follow from the fact that the average temperature at which heat is transferred to the steam is increased. Note also that when the steam is superheated, the quality of the steam leaving the turbine increases.

Finally, the influence of the maximum pressure of the steam must be considered, and this is shown in Fig. 9.6. In this analysis the maximum temperature of the steam, as well as

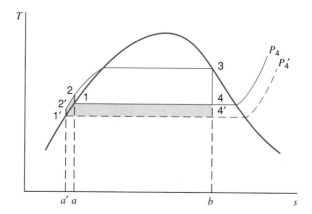

FIGURE 9.4 Effect of exhaust pressure on Rankine-cycle efficiency.

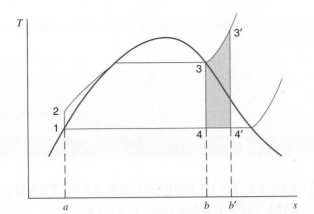

FIGURE 9.5 Effect of superheating on Rankine-cycle efficiency.

the exhaust pressure, is held constant. The heat rejected decreases by area $b'–4'–4–b–b'$. The net work increases by the amount of the single cross-hatching and decreases by the amount of the double cross-hatching. Therefore, the net work tends to remain the same but the heat rejected decreases, and hence the Rankine-cycle efficiency increases with an increase in maximum pressure. Note that in this instance, too, the average temperature at which heat is supplied increases with an increase in pressure. The quality of the steam leaving the turbine decreases as the maximum pressure increases.

To summarize this section, we can say that the net work and the efficiency of the Rankine cycle can be increased by lowering the condenser pressure, by increasing the pressure during heat addition, and by superheating the steam. The quality of the steam leaving the turbine is increased by superheating the steam and decreased by lowering the exhaust pressure and by increasing the pressure during heat addition. These effects are shown in Figs. 9.7 and 9.8.

In connection with these considerations, we note that the cycle is modeled with four known processes (two isobaric and two isentropic) between the four states with a total of eight properties. Assuming state 1 is saturated liquid ($x_1 = 0$), we have three $(8 − 4 − 1)$ parameters to determine. The operating conditions are physically controlled by the high pressure generated by the pump, $P_2 = P_3$, the superheat to T_3 (or $x_3 = 1$ if none), and the condenser temperature T_1, which is a result of the amount of heat transfer that takes place.

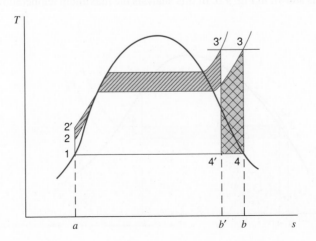

FIGURE 9.6 Effect of boiler pressure on Rankine-cycle efficiency.

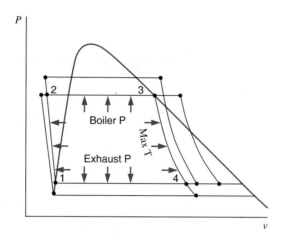

FIGURE 9.7 Effect of pressure and temperature on Rankine-cycle work.

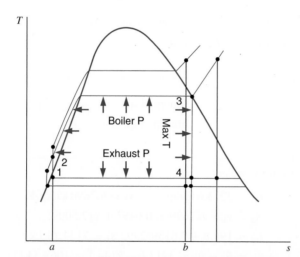

FIGURE 9.8 Effect of pressure and temperature on Rankine-cycle efficiency.

Example 9.2

In a Rankine cycle, steam leaves the boiler and enters the turbine at 4 MPa and 400 °C. The condenser pressure is 10 kPa. Determine the cycle efficiency.

To determine the cycle efficiency, we must calculate the turbine work, the pump work, and the heat transfer to the steam in the boiler. We do this by considering a control surface around each of these components in turn. In each case the thermodynamic model is the steam tables, and the process is steady state with negligible changes in kinetic and potential energies.

    Control volume: Pump.
        Inlet state: P_1 known, saturated liquid; state fixed.
        Exit state: P_2 known.

Analysis

$$\text{Energy Eq.:} \quad w_p = h_2 - h_1$$
$$\text{Entropy Eq.:} \quad s_2 = s_1$$

Since $s_2 = s_1$,

$$h_2 - h_1 = \int_1^2 v \, dP = v(P_2 - P_1)$$

Solution

Substituting, we obtain

$$w_p = v(P_2 - P_1) = (0.001\ 01)(4000 - 10) = 4.0 \text{ kJ/kg}$$
$$h_1 = 191.8 \text{ kJ/kg}$$
$$h_2 = 191.8 + 4.0 = 195.8 \text{ kJ/kg}$$

For the turbine we have:

Control volume:	Turbine.
Inlet state:	P_3, T_3 known; state fixed.
Exit state:	P_4 known.

Analysis

$$\text{Energy Eq.:} \quad w_t = h_3 - h_4$$
$$\text{Entropy Eq.:} \quad s_4 = s_3$$

Solution

Upon substitution we get

$$h_3 = 3213.6 \text{ kJ/kg}, \qquad s_3 = 6.7690 \text{ kJ/kg·K}$$
$$s_3 = s_4 = 6.7690 = 0.6493 + x_4 7.5009, \qquad x_4 = 0.8159$$
$$h_4 = 191.8 + 0.8159(2392.8) = 2144.1 \text{ kJ/kg}$$
$$w_t = h_3 - h_4 = 3213.6 - 2144.1 = 1069.5 \text{ kJ/kg}$$
$$w_{\text{net}} = w_t - w_p = 1069.5 - 4.0 = 1065.5 \text{ kJ/kg}$$

Finally, for the boiler we have:

Control volume:	Boiler.
Inlet state:	P_2, h_2 known; state fixed.
Exit state:	State 3 fixed (as given).

Analysis

$$\text{Energy Eq.:} \quad q_H = h_3 - h_2$$

Solution

Substituting gives

$$q_H = h_3 - h_2 = 3213.6 - 195.8 = 3017.8 \text{ kJ/kg}$$
$$\eta_{\text{th}} = \frac{w_{\text{net}}}{q_H} = \frac{1065.5}{3017.8} = 35.3 \%$$

The net work could also be determined by calculating the heat rejected in the condenser, q_L, and noting, from the first law, that the net work for the cycle is equal to the net heat transfer. Considering a control surface around the condenser, we have

$$q_L = h_4 - h_1 = 2144.1 - 191.8 = 1952.3 \text{ kJ/kg}$$

Therefore,

$$w_{net} = q_H - q_L = 3017.8 - 1952.3 = 1065.5 \text{ kJ/kg}$$

9.4 THE REHEAT CYCLE

In the previous section, we noted that the efficiency of the Rankine cycle could be increased by increasing the pressure during the addition of heat. However, the increase in pressure also increases the moisture content of the steam in the low-pressure end of the turbine. The reheat cycle has been developed to take advantage of the increased efficiency with higher pressures and yet avoid excessive moisture in the low-pressure stages of the turbine. This cycle is shown schematically and on a T–s diagram in Fig. 9.9. The unique feature of this cycle is that the steam is expanded to some intermediate pressure in the turbine and is then reheated in the boiler, after which it expands in the turbine to the exhaust pressure. It is evident from the T–s diagram that there is very little gain in efficiency from reheating the steam, because the average temperature at which heat is supplied is not greatly changed. The chief advantage is in decreasing to a safe value the moisture content in the low-pressure stages of the turbine. If metals could be found that would enable us to superheat the steam to 3′, the simple Rankine cycle would be more efficient than the reheat cycle, and there would be no need for the reheat cycle.

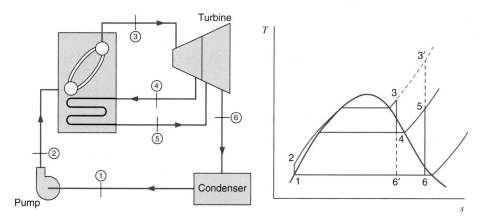

FIGURE 9.9
The ideal reheat cycle.

Example 9.3

Consider a reheat cycle utilizing steam. Steam leaves the boiler and enters the turbine at 4 MPa, 400 °C. After expansion in the turbine to 400 kPa, the steam is reheated to 400 °C and then expanded in the low-pressure turbine to 10 kPa. Determine the cycle efficiency.

For each control volume analyzed, the thermodynamic model is the steam tables, the process is steady state, and changes in kinetic and potential energies are negligible.

For the high-pressure turbine,

Control volume: High-pressure turbine.

Inlet state: P_3, T_3 known; state fixed.

Exit state: P_4 known.

Analysis

$$\text{Energy Eq.:} \quad w_{h-p} = h_3 - h_4$$
$$\text{Entropy Eq.:} \quad s_3 = s_4$$

Solution

Substituting,

$$h_3 = 3213.6, \qquad s_3 = 6.7690$$
$$s_4 = s_3 = 6.7690 = 1.7766 + x_4 5.1193, \qquad x_4 = 0.9752$$
$$h_4 = 604.7 + 0.9752(2133.8) = 2685.6 \text{ kJ/kg}$$

For the low-pressure turbine,

Control volume: Low-pressure turbine.

Inlet state: P_5, T_5 known; state fixed.

Exit state: P_6 known.

Analysis

$$\text{Energy Eq.:} \quad w_{l-p} = h_5 - h_6$$
$$\text{Entropy Eq.:} \quad s_5 = s_6$$

Solution

Upon substituting,

$$h_5 = 3273.4 \text{ kJ/kg} \qquad s_5 = 7.8985 \text{ kJ/kg·K}$$
$$s_6 = s_5 = 7.8985 = 0.6493 + x_6 7.5009, \qquad x_6 = 0.9664$$
$$h_6 = 191.8 + 0.9664(2392.8) = 2504.3 \text{ kJ/kg}$$

For the overall turbine, the total work output w_t is the sum of w_{h-p} and w_{l-p}, so that

$$w_t = (h_3 - h_4) + (h_5 - h_6)$$
$$= (3213.6 - 2685.6) + (3273.4 - 2504.3)$$
$$= 1297.1 \text{ kJ/kg}$$

For the pump,

> *Control volume*:   Pump.
>> *Inlet state*: P_1 known, saturated liquid; state fixed.
>> *Exit state*: P_2 known.

Analysis

$$\text{Energy Eq.:} \quad w_p = h_2 - h_1$$
$$\text{Entropy Eq.:} \quad s_2 = s_1$$

Since $s_2 = s_1$,

$$h_2 - h_1 = \int_1^2 v\, dP = v(P_2 - P_1)$$

Solution

Substituting,

$$w_p = v(P_2 - P_1) = (0.001\,01)(4000 - 10) = 4.0 \text{ kJ/kg}$$
$$h_2 = 191.8 + 4.0 = 195.8 \text{ kJ/kg}$$

Finally, for the boiler

> *Control volume*:   Boiler.
>> *Inlet states*:   States 2 and 4 both known (above).
>> *Exit states*:   States 3 and 5 both known (as given).

Analysis

$$\text{Energy Eq.:} \quad q_H = (h_3 - h_2) + (h_5 - h_4)$$

Solution

Substituting,

$$q_H = (h_3 - h_2) + (h_5 - h_4)$$
$$= (3213.6 - 195.8) + (3273.4 - 2685.6) = 3605.6 \text{ kJ/kg}$$

Therefore,

$$w_{\text{net}} = w_t - w_p = 1297.1 - 4.0 = 1293.1 \text{ kJ/kg}$$

$$\eta_{\text{th}} = \frac{w_{\text{net}}}{q_H} = \frac{1293.1}{3605.6} = 35.9\,\%$$

By comparing this example with Example 9.2, we find that through reheating the gain in efficiency is relatively small, but the moisture content of the vapor leaving the turbine is decreased from 18.4 % to 3.4 %.

9.5 THE REGENERATIVE CYCLE AND FEEDWATER HEATERS

Another important variation from the Rankine cycle is the regenerative cycle, which uses feedwater heaters. The basic concepts of this cycle can be demonstrated by considering the Rankine cycle without superheat, as shown in Fig. 9.10. During the process between states 2 and 2′, the working fluid is heated while in the liquid phase, and the average temperature of the working fluid is much lower than during the vaporization process 2′–3. The process between states 2 and 2′ causes the average temperature at which heat is supplied in the Rankine cycle to be lower than in the Carnot cycle 1′–2′–3–4–1′. Consequently, the efficiency of the Rankine cycle is lower than that of the corresponding Carnot cycle. In the regenerative cycle the working fluid enters the boiler at some state between 2 and 2′; consequently, the average temperature at which heat is supplied is higher.

Consider first an idealized regenerative cycle, as shown in Fig. 9.11. The unique feature of this cycle compared to the Rankine cycle is that after leaving the pump, the liquid

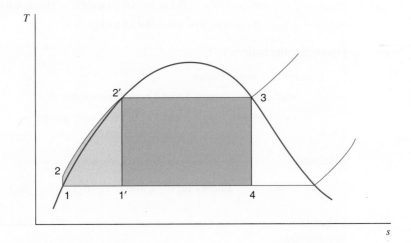

FIGURE 9.10 T–s diagram showing the relationships between Carnot-cycle efficiency and Rankine-cycle efficiency.

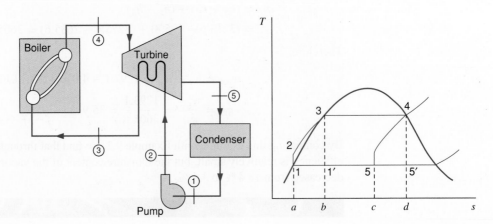

FIGURE 9.11 The ideal regenerative cycle.

circulates around the turbine casing, counterflow to the direction of vapor flow in the turbine. Thus, it is possible to transfer to the liquid flowing around the turbine the heat from the vapor as it flows through the turbine. Let us assume for the moment that this is a reversible heat transfer; that is, at each point the temperature of the vapor is only infinitesimally higher than the temperature of the liquid. In this instance, line 4–5 on the T–s diagram of Fig. 9.11, which represents the states of the vapor flowing through the turbine, is exactly parallel to line 1–2–3, which represents the pumping process (1–2) and the states of the liquid flowing around the turbine. Consequently, areas 2–3–b–a–2 and 5–4–d–c–5 are not only equal but congruous, and these areas, respectively, represent the heat transferred to the liquid and from the vapor. Heat is also transferred to the working fluid at constant temperature in process 3–4, and area 3–4–d–b–3 represents this heat transfer. Heat is transferred from the working fluid in process 5–1, and area 1–5–c–a–1 represents this heat transfer. This area is exactly equal to area $1'$–$5'$–d–b–$1'$, which is the heat rejected in the related Carnot cycle $1'$–3–4–$5'$–$1'$. Thus, the efficiency of this idealized regenerative cycle is exactly equal to the efficiency of the Carnot cycle with the same heat supply and heat rejection temperatures.

Obviously, this idealized regenerative cycle is impractical. First, it would be impossible to effect the necessary heat transfer from the vapor in the turbine to the liquid feedwater. Furthermore, the moisture content of the vapor leaving the turbine increases considerably as a result of the heat transfer. The disadvantage of this was noted previously. The practical regenerative cycle extracts some of the vapor after it has partially expanded in the turbine and uses feedwater heaters (FWH), as shown in Fig. 9.12.

Steam enters the turbine at state 5. After expansion to state 6, some of the steam is extracted and enters the FWH. The steam that is not extracted is expanded in the turbine to state 7 and is then condensed in the condenser. This condensate is pumped into the FWH, where it mixes with the steam extracted from the turbine. The proportion of steam extracted is just sufficient to cause the liquid leaving the FWH to be saturated at state 3. Note that the liquid has not been pumped to the boiler pressure, but only to the intermediate pressure corresponding to state 6. Another pump is required to pump the liquid leaving the FWH boiler pressure. The significant point is that the average temperature at which heat is supplied has been increased.

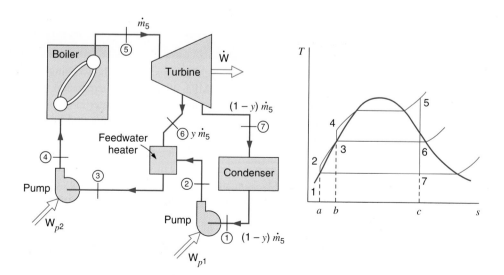

FIGURE 9.12

Regenerative cycle with an open FWH.

Consider a control volume around the open FWH in Fig. 9.12. The conservation of mass requires

$$\dot{m}_2 + \dot{m}_6 = \dot{m}_3$$

satisfied with the extraction fraction as

$$y = \dot{m}_6/\dot{m}_5 \tag{9.2}$$

so

$$\dot{m}_7 = (1 - y)\dot{m}_5 = \dot{m}_1 = \dot{m}_2$$

The energy equation with no external heat transfer and no work becomes

$$\dot{m}_2 h_2 + \dot{m}_6 h_6 = \dot{m}_3 h_3 \tag{9.3}$$

into which we substitute the mass flow rates ($\dot{m}_3 = \dot{m}_5$) as

$$(1 - y)\dot{m}_5 h_2 + y\dot{m}_5 h_6 = \dot{m}_5 h_3 \tag{9.4}$$

We take state 3 as the limit of saturated liquid (we do not want to heat it further, as it would move into the two-phase region and damage the pump $P2$) and then solve for y:

$$y = \frac{h_3 - h_2}{h_6 - h_2} \tag{9.5}$$

This establishes the maximum extraction fraction we should take out at this extraction pressure.

This cycle is somewhat difficult to show on a T–s diagram because the masses of steam flowing through the various components vary. The T–s diagram of Fig. 9.12 simply shows the state of the fluid at the various points.

Area 4–5–c–b–4 in Fig. 9.12 represents the heat transferred per kilogram of working fluid. Process 7–1 is the heat rejection process, but since not all the steam passes through the condenser, area 1–7–c–a–1 represents the heat transfer per kilogram flowing through the condenser, which does not represent the heat transfer per kilogram of working fluid entering the turbine. Between states 6 and 7, only part of the steam is flowing through the turbine. The example that follows illustrates the calculations for the regenerative cycle.

Example 9.4

Consider a regenerative cycle using steam as the working fluid. Steam leaves the boiler and enters the turbine at 4 MPa, 400 °C. After expansion to 400 kPa, some of the steam is extracted from the turbine to heat the feedwater in an open FWH. The pressure in the FWH is 400 kPa, and the water leaving it is saturated liquid at 400 kPa. The steam not extracted expands to 10 kPa. Determine the cycle efficiency.

The line diagram and T–s diagram for this cycle are shown in Fig. 9.12.

As in previous examples, the model for each control volume is the steam tables, the process is steady state, and kinetic and potential energy changes are negligible.

From Examples 9.2 and 9.3, we have the following properties:

$$h_5 = 3213.6 \qquad h_6 = 2685.6 \text{ kJ/kg}$$

$$h_7 = 2144.1 \qquad h_1 = 191.8 \text{ kJ/kg}$$

For the low-pressure pump,

> *Control volume*:   Low-pressure pump.
> *Inlet state*: P_1 known, saturated liquid; state fixed.
> *Exit state*: P_2 known.

Analysis

$$\text{Energy Eq.:} \quad w_{p1} = h_2 - h_1$$
$$\text{Entropy Eq.:} \quad s_2 = s_1$$

Therefore,

$$h_2 - h_1 = \int_1^2 v\, dP = v(P_2 - P_1)$$

Solution

Substituting,

$$w_{p1} = v(P_2 - P_1) = (0.001\,01)(400 - 10) = 0.4\,\text{kJ/kg}$$
$$h_2 = h_1 + w_p = 191.8 + 0.4 = 192.2\,\text{kJ/kg}$$

For the turbine,

> *Control volume*:   Turbine.
> *Inlet state*: P_5, T_5 known; state fixed.
> *Exit state*: P_6 known; P_7 known.

Analysis

$$\text{Energy Eq.:} \quad w_t = (h_5 - h_6) + (1 - y)(h_6 - h_7)$$
$$\text{Entropy Eq.:} \quad s_5 = s_6 = s_7$$

Solution

From the second law, the values for h_6 and h_7 given previously were calculated in Examples 9.2 and 9.3.

For the FWH,

> *Control volume*:   FWH.
> *Inlet states*:   States 2 and 6 both known (as given).
> *Exit state*: P_3 known, saturated liquid; state fixed.

Analysis

$$\text{Energy Eq.:} \quad y(h_6) + (1 - y)h_2 = h_3$$

Solution

After substitution,

$$y(2685.6) + (1 - y)(192.2) = 604.7$$
$$y = 0.1654$$

We can now calculate the turbine work.

$$w_t = (h_5 - h_6) + (1 - y)(h_6 - h_7)$$
$$= (3213.6 - 2685.6) + (1 - 0.1654)(2685.6 - 2144.1)$$
$$= 979.9 \text{ kJ/kg}$$

For the high-pressure pump,

Control volume: High-pressure pump.

Inlet state: State 3 known (as given).

Exit state: P_4 known.

Analysis

$$\text{Energy Eq.:} \quad w_{p2} = h_4 - h_3$$
$$\text{Entropy Eq.:} \quad s_4 = s_3$$

Solution

Substituting,

$$w_{p2} = v(P_4 - P_3) = (0.001\ 084)(4000 - 400) = 3.9 \text{ kJ/kg}$$
$$h_4 = h_3 + w_{p2} = 604.7 + 3.9 = 608.6 \text{ kJ/kg}$$

Therefore,

$$w_{\text{net}} = w_t - (1 - y)w_{p1} - w_{p2}$$
$$= 979.9 - (1 - 0.1654)(0.4) - 3.9 = 975.7 \text{ kJ/kg}$$

Finally, for the boiler,

Control volume: Boiler.

Inlet state: P_4, h_4 known (as given); state fixed.

Exit state: State 5 known (as given).

Analysis

$$\text{Energy Eq.:} \quad q_H = h_5 - h_4$$

Solution

Substituting,

$$q_H = h_5 - h_4 = 3213.6 - 608.6 = 2605.0 \text{ kJ/kg}$$

$$\eta_{\text{th}} = \frac{w_{\text{net}}}{q_H} = \frac{975.7}{2605.0} = 37.5\,\%$$

Note the increase in efficiency over the efficiency of the Rankine cycle in Example 9.2.

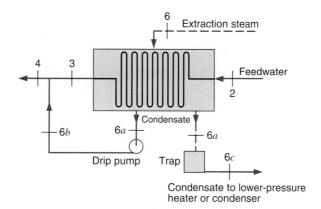

FIGURE 9.13

Schematic arrangement for a closed FWH.

Up to this point, the discussion and examples have tacitly assumed that the extraction steam and feedwater are mixed in the FWH. Another frequently used type of FWH, known as a closed feedwater heater, is one in which the steam and feedwater do not mix. Rather, heat is transferred from the extracted steam as it condenses on the outside of tubes while the feedwater flows through the tubes. In a closed heater, a schematic sketch of which is shown in Fig. 9.13, the steam and feedwater may be at considerably different pressures. The condensate may be pumped into the feedwater line, or it may be removed through a trap to a lower-pressure heater or to the condenser. (A trap is a device that permits liquid but not vapor to flow to a region of lower pressure.)

Let us analyze the closed FWH in Fig. 9.13 when a trap with a drain to the condenser is used. We assume we can heat the feedwater up to the temperature of the condensing extraction flow, that is, $T_3 = T_4 = T_{6a}$, as there is no drip pump. Conservation of mass for the feedwater heater is

$$\dot{m}_4 = \dot{m}_3 = \dot{m}_2 = \dot{m}_5; \qquad \dot{m}_6 = y\dot{m}_5 = \dot{m}_{6a} = \dot{m}_{6c}$$

Notice that the extraction flow is added to the condenser, so the flow rate at state 2 is the same as at state 5. The energy equation is

$$\dot{m}_5 h_2 + y\dot{m}_5 h_6 = \dot{m}_5 h_3 + y\dot{m}_5 h_{6a} \qquad (9.6)$$

which we can solve for y as

$$y = \frac{h_3 - h_2}{h_6 - h_{6a}} \qquad (9.7)$$

Open FWHs have the advantages of being less expensive and having better heat-transfer characteristics than closed FWHs. They have the disadvantage of requiring a pump to handle the feedwater between each heater.

In many power plants a number of extraction stages are used, though rarely more than five. The number is, of course, determined by economics. It is evident that using a very large number of extraction stages and FWHs allows the cycle efficiency to approach that of the idealized regenerative cycle of Fig. 9.11, where the feedwater enters the boiler as saturated liquid at the maximum pressure. In practice, however, this cannot be economically justified because the savings effected by the increase in efficiency would be more than offset by the cost of additional equipment (FWHs, piping, and so forth).

A *typical arrangement* of the main components in an actual power plant is shown in Fig. 9.14. Note that one open FWH is a deaerating FWH; this heater has the dual purpose

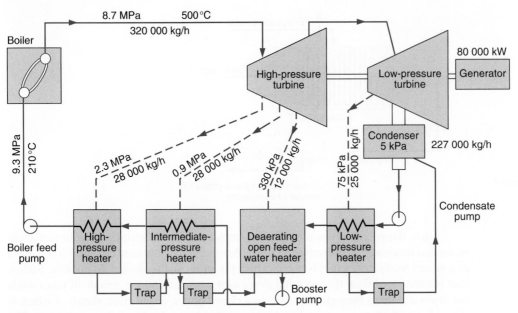

FIGURE 9.14 Arrangement of heaters in an actual power plant utilizing regenerative FWHs.

of heating and removing the air from the feedwater. Unless the air is removed, excessive corrosion occurs in the boiler. Note also that the condensate from the high-pressure heater drains (through a trap) to the intermediate heater, and the condensate from the intermediate heater drains to the deaerating FWH. The condensate from the low-pressure heater drains to the condenser.

Many actual power plants combine one reheat stage with a number of extraction stages. The principles already considered are readily applied to such a cycle.

9.6 DEVIATION OF ACTUAL CYCLES FROM IDEAL CYCLES

Before we leave the matter of vapor power cycles, a few comments are in order regarding the ways in which an actual cycle deviates from an ideal cycle. The most important of these losses are due to the turbine, the pump(s), the pipes, and the condenser. These losses are discussed next.

Turbine Losses

Turbine losses, as described in Section 7.5, represent by far the largest discrepancy between the performance of a real cycle and a corresponding ideal Rankine-cycle power plant. The large positive turbine work is the principal number in the numerator of the cycle thermal efficiency and is directly reduced by the factor of the isentropic turbine efficiency. Turbine losses are primarily those associated with the flow of the working fluid through the turbine blades and passages, with heat transfer to the surroundings also being a loss but of secondary importance. The turbine process might be represented as shown in Fig. 9.15, where state 4_s is the state after an ideal isentropic turbine expansion and state 4 is the actual state leaving

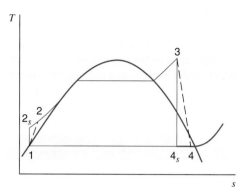

FIGURE 9.15 *T–s* diagram showing the effect of turbine and pump inefficiencies on cycle performance.

the turbine following an irreversible process. The turbine governing procedures may also cause a loss in the turbine, particularly if a throttling process is used to govern the turbine operation.

Pump Losses

The losses in the pump are similar to those in the turbine and are due primarily to the irreversibilities with the fluid flow. Pump efficiency was discussed in Section 7.5, and the ideal exit state 2_s and real exit state 2 are shown in Fig. 9.15. Pump losses are much smaller than those of the turbine, since the associated work is far smaller.

Piping Losses

Pressure drops caused by frictional effects and heat transfer to the surroundings are the most important piping losses. Consider, for example, the pipe connecting the turbine to the boiler. If only frictional effects occur, states *a* and *b* in Fig. 9.16 would represent the states of the steam leaving the boiler and entering the turbine, respectively. Note that the frictional effects cause an increase in entropy. Heat transferred to the surroundings at constant pressure can be represented by process *bc*. This effect decreases entropy. Both the pressure drop and heat transfer decrease the exergy of the steam entering the turbine. The irreversibility of this process can be calculated by the methods outlined in Chapter 8.

A similar loss is the pressure drop in the boiler. Because of this pressure drop, the water entering the boiler must be pumped to a higher pressure than the desired steam pressure leaving the boiler, which requires additional pump work.

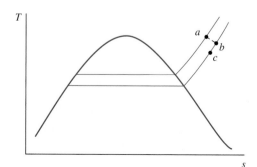

FIGURE 9.16 *T–s* diagram showing the effect of losses between the boiler and turbine.

Condenser Losses

The losses in the condenser are relatively small. One of these minor losses is the cooling below the saturation temperature of the liquid leaving the condenser. This represents a loss because additional heat transfer is necessary to bring the water to its saturation temperature.

The influence of these losses on the cycle is illustrated in the following example, which should be compared to Example 9.2.

Example 9.5

A steam power plant operates on a cycle with pressures and temperatures as designated in Fig. 9.17. The efficiency of the turbine is 86 %, and the efficiency of the pump is 80 %. Determine the thermal efficiency of this cycle.

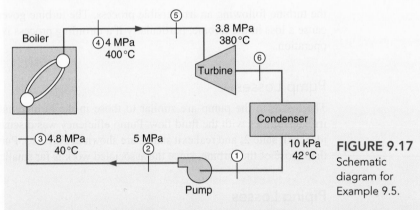

FIGURE 9.17
Schematic diagram for Example 9.5.

As in previous examples, for each control volume the model used is the steam tables, and each process is steady state with no changes in kinetic or potential energy. This cycle is shown on the T–s diagram of Fig. 9.18.

Control volume: Turbine.

Inlet state: P_5, T_5 known; state fixed.

Exit state: P_6 known.

Analysis

Energy Eq.: $w_t = h_5 - h_6$

Entropy Eq.: $s_{6s} = s_5$

The efficiency is

$$\eta_t = \frac{w_t}{h_5 - h_{6s}} = \frac{h_5 - h_6}{h_5 - h_{6s}}$$

Solution

From the steam tables, we get

$$h_5 = 3169.1 \text{ kJ/kg}, \qquad s_5 = 6.7235 \text{ kJ/kg} \cdot \text{K}$$

$$s_{6s} = s_5 = 6.7235 = 0.6493 + x_{6s}7.5009, \qquad x_{6s} = 0.8098$$

$$h_{6s} = 191.8 + 0.8098(2392.8) = 2129.5 \text{ kJ/kg}$$

$$w_t = \eta_t(h_5 - h_{6s}) = 0.86(3169.1 - 2129.5) = 894.1 \text{ kJ/kg}$$

For the pump, we have:

Control volume:

Pump.

Inlet state: P_1, T_1 known; state fixed.

Exit state: P_2 known.

Analysis

$$\text{Energy Eq.:} \quad w_p = h_2 - h_1$$
$$\text{Entropy Eq.:} \quad s_{2s} = s_1$$

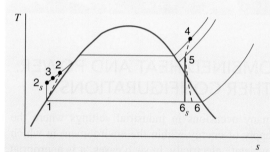

FIGURE 9.18 *T–s* diagram for Example 9.5.

The pump efficiency is

$$\eta_p = \frac{h_{2s} - h_1}{w_p} = \frac{h_{2s} - h_1}{h_2 - h_1}$$

Since $s_{2s} = s_1$,

$$h_{2s} - h_1 = v(P_2 - P_1)$$

Therefore,

$$w_p = \frac{h_{2s} - h_1}{\eta_p} = \frac{v(P_2 - P_1)}{\eta_p}$$

Solution

Substituting, we obtain

$$w_p = \frac{v(P_2 - P_1)}{\eta_p} = \frac{(0.001\,009)(5000 - 10)}{0.80} = 6.3 \text{ kJ/kg}$$

Therefore,

$$w_{net} = w_t - w_p = 894.1 - 6.3 = 887.8 \text{ kJ/kg}$$

Finally, for the boiler:

Control volume: Boiler.

Inlet state: P_3, T_3 known; state fixed.

Exit state: P_4, T_4 known, state fixed.

Analysis

$$\text{Energy Eq.:} \quad q_H = h_4 - h_3$$

Solution

Substitution gives

$$q_H = h_4 - h_3 = 3213.6 - 171.8 = 3041.8 \text{ kJ/kg}$$

$$\eta_{th} = \frac{887.8}{3041.8} = 29.2\%$$

This result compares to the Rankine efficiency of 35.3% for the similar cycle of Example 9.2.

9.7 COMBINED HEAT AND POWER: OTHER CONFIGURATIONS

There are many occasions in industrial settings where the need arises for a specific source or supply of energy within the environment in which a steam power plant is being used to generate electricity. In such cases, it is appropriate to consider supplying this source of energy in the form of steam that has already been expanded through the high-pressure section of the turbine in the power plant cycle, thereby eliminating the construction and use of a second boiler or other energy source. Such an arrangement is shown in Fig. 9.19, in which the turbine is tapped at some intermediate pressure to furnish the necessary amount of process steam required for the particular energy need—perhaps to operate a special process in the plant, or in many cases simply for the purpose of space heating the facilities.

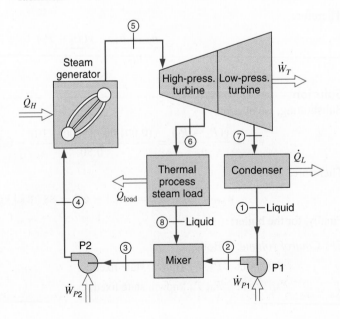

FIGURE 9.19
Example of a cogeneration system.

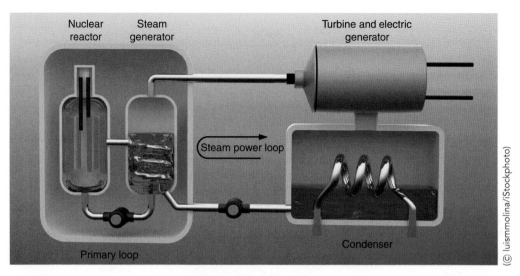

FIGURE 9.20 Schematic diagram of a shipboard nuclear propulsion system.

This type of application is called *combined heat and power* (CHP); sometimes it is also called *cogeneration*, where it refers to the generation of a by-product while making the main product. In some instances the steam is the main product and the electricity (work) is the by-product typical of factories and smaller units, whereas in other cases the electricity is the main product and the steam is a by-product, as for major utility companies. For instance, the power plant shown in Figs. 1.1 and 1.2 generates electricity for the grid and, in addition, produces warm water that is distributed in pipes below the streets in the nearby city for heating purposes, a setup that is called *district heating*. This is economically feasible only if the population density is high enough and the distribution distances are short enough.

The basic Rankine cycle can also be used for a number of other applications where special considerations are required. Most major power plants use coal as the fuel because of cost, but other fuels are possible. Figure 9.20 shows an example of a power plant driven by a nuclear reactor for a submarine. The benefit in this situation is having a power source that does not need frequent refueling, requires a modest amount of space, and does not need air for combustion. There are, of course, extra safety precautions that must be taken to operate such a plant.

Other alternative energy sources can be used to operate a power plant. For very-low-temperature sources like unconcentrated solar power or low-temperature waste heat, a cycle can be made with substances other than water that will boil at a much lower temperature and still be condensed at ambient temperature. Such "bottom" temperature cycles are described in the following chapter for combined systems.

Since a higher temperature of the source will improve the efficiency of the cycle, a source like solar power can be concentrated by a collection system of mirrors called *heliostats* that will track the sun to focus the light beams at a fixed location. Due to the limited amount of time with power input from the sun, an energy storage system can extend the operating period of the power plant, greatly improving the utilization of the invested resources. This is the principle behind a new solar power plant being built in Nevada that uses molten salt to transfer the energy from the solar collector to the Rankine-cycle parts of the plant, as shown in Fig. 9.21. This system will allow power to be produced during the

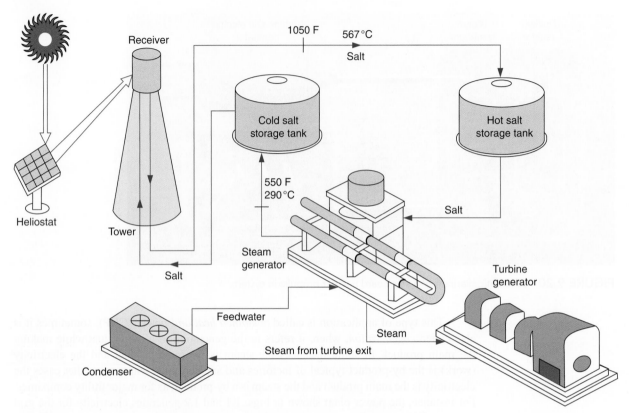

FIGURE 9.21 A facility being built in Tonopah, Nevada, for power generation.

day and continue after sunset to about midnight, thus providing a time buffer between the energy collection and the use of the energy.

In-Text Concept Questions

a. Consider a Rankine cycle without superheat. How many single properties are needed to determine the cycle? Repeat the answer for a cycle with superheat.

b. Which component determines the high pressure in a Rankine cycle? What factor determines the low pressure?

c. What is the difference between an open and a closed FWH?

d. In a cogenerating power plant, what is cogenerated?

9.8 INTRODUCTION TO REFRIGERATION SYSTEMS

In Section 9.1, we discussed cyclic heat engines consisting of four separate processes, either steady-state or piston/cylinder boundary-movement work devices. We further allowed for a working fluid that changes phase or for one that remains in a single phase throughout the cycle. We then considered a power system comprised of four reversible steady-state

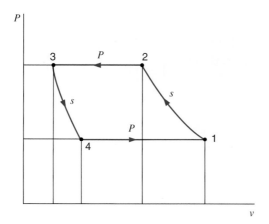

FIGURE 9.22
Four-process
refrigeration cycle.

processes, two of which were constant-pressure heat-transfer processes, for simplicity of equipment requirements, since these two processes involve no work. It was further assumed that the other two work-involved processes were adiabatic and therefore isentropic. The resulting power cycle appeared as in Fig. 9.2.

We now consider the basic ideal refrigeration system cycle in exactly the same terms as those described earlier, except that each process is the reverse of that in the power cycle. The result is the ideal cycle shown in Fig. 9.22. Note that if the entire cycle takes place inside the two-phase liquid–vapor dome, the resulting cycle is, as with the power cycle, the Carnot cycle, since the two constant-pressure processes are also isothermal. Otherwise, this cycle is not a Carnot cycle. It is also noted, as before, that the net work input to the cycle is equal to the area enclosed by the process lines 1–2–3–4–1, independently of whether the individual processes are steady state or cylinder/piston boundary movement.

In the next section, we make one modification to this idealized basic refrigeration system cycle in presenting and applying the model of refrigeration and heat pump systems.

 9.9 THE VAPOR-COMPRESSION REFRIGERATION CYCLE

In this section, we consider the ideal refrigeration cycle for a working substance that changes phase during the cycle, in a manner equivalent to that done with the Rankine power cycle in Section 9.2. In doing so, we note that state 3 in Fig. 9.22 is saturated liquid at the condenser temperature and state 1 is saturated vapor at the evaporator temperature. This means that the isentropic expansion process from 3–4 will be in the two-phase region, and the substance there will be mostly liquid. As a consequence, there will be very little work output from this process, so it is not worth the cost of including this piece of equipment in the system. We therefore replace the turbine with a throttling device, usually a valve or a length of small-diameter tubing, by which the working fluid is throttled from the high-pressure to the low-pressure side. The resulting cycle become the ideal model for a vapor-compression refrigeration system, which is shown in Fig. 9.23. Saturated vapor at low pressure enters the compressor and undergoes reversible adiabatic compression, process 1–2. Heat is then rejected at constant pressure in process 2–3, and the working fluid exits the condenser as saturated liquid. An adiabatic throttling process, 3–4, follows, and the working fluid is then evaporated at constant pressure, process 4–1, to complete the cycle.

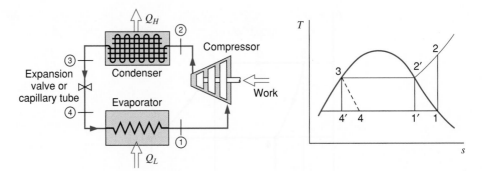

FIGURE 9.23 The ideal vapor-compression refrigeration cycle.

The similarity of this cycle to the reverse of the Rankine cycle has already been noted. We also note the difference between this cycle and the ideal Carnot cycle, in which the working fluid always remains inside the two-phase region, $1'$–$2'$–3–$4'$–$1'$. It is much more expedient to have a compressor handle-only vapor than a mixture of liquid and vapor, as would be required in process $1'$–$2'$ of the Carnot cycle. It is virtually impossible to compress, at a reasonable rate, a mixture such as that represented by state $1'$ and still maintain equilibrium between liquid and vapor. The other difference, that of replacing the turbine by the throttling process, has already been discussed.

The standard vapor-compression refrigeration cycle has four known processes (one isentropic, two isobaric, and one isenthalpic) between the four states with eight properties. It is assumed that state 3 is saturated liquid and state 1 is saturated vapor, so there are two $(8 - 4 - 2)$ parameters that determine the cycle. The compressor generates the high pressure, $P_2 = P_3$, and the heat transfer between the evaporator and the cold space determines the low temperature, $T_4 = T_1$.

The system described in Fig. 9.23 can be used for either of two purposes. The first use is as a refrigeration system, in which case it is desired to maintain a space at a low temperature T_1 relative to the ambient temperature T_3. (In a real system, it would be necessary to allow a finite temperature difference in both the evaporator and condenser to provide a finite rate of heat transfer in each.) Thus, the reason for building the system in this case is the quantity q_L. The measure of performance of a refrigeration system is given in terms of the coefficient of performance (COP), β, which was defined in Chapter 5 as

$$\beta = \frac{q_L}{w_c} \tag{9.8}$$

The second use of this system described in Fig. 9.23 is as a heat pump system, in which case it is desired to maintain a space at a temperature T_3 above that of the ambient (or other source) T_1. In this case, the reason for building the system is the quantity q_H, and the COP for the heat pump, β', is now

$$\beta' = \frac{q_H}{w_c} \tag{9.9}$$

Refrigeration systems and heat pump systems are, of course, different in terms of design variables, but the analysis of the two is the same. When we discuss refrigerators in this and the following two sections, it should be kept in mind that the same comments generally apply to heat pump systems as well.

Example 9.6

Consider an ideal refrigeration cycle that uses R-134a as the working fluid. The temperature of the refrigerant in the evaporator is $-20\,°C$, and in the condenser it is $40\,°C$. The refrigerant is circulated at the rate of 0.03 kg/s. Determine the COP and the capacity of the plant in rate of refrigeration.

The diagram for this example is shown in Fig. 9.23. For each control volume analyzed, the thermodynamic model is as exhibited in the R-134a tables. Each process is steady state, with no changes in kinetic or potential energy.

Control volume: Compressor.

Inlet state: T_1 known, saturated vapor; state fixed.

Exit state: P_2 known (saturation pressure at T_3).

Analysis

$$\text{Energy Eq.:} \quad w_c = h_2 - h_1$$
$$\text{Entropy Eq.:} \quad s_2 = s_1$$

Solution

At $T_3 = 40\,°C$,

$$P_g = P_2 = 1017\,\text{kPa}$$

From the R-134a tables, we get

$$h_1 = 386.1\,\text{kJ/kg}, \quad s_1 = 1.7395\,\text{kJ/kg}$$

Therefore,

$$s_2 = s_1 = 1.7395\,\text{kJ/kg·K}$$

so that by suitable interpolation in Table B.5

$$T_2 = 47.7\,°C \quad \text{and} \quad h_2 = 428.4\,\text{kJ/kg}$$
$$w_c = h_2 - h_1 = 428.4 - 386.1 = 42.3\,\text{kJ/kg}$$

Control volume: Expansion valve.

Inlet state: T_3 known, saturated liquid; state fixed.

Exit state: T_4 known.

Analysis

$$\text{Energy Eq.:} \quad h_3 = h_4$$
$$\text{Entropy Eq.:} \quad s_3 + s_{\text{gen}} = s_4$$

Solution

Numerically, we have

$$h_4 = h_3 = 256.5\,\text{kJ/kg}$$

Control volume: Evaporator.

Inlet state: State 4 known (as given).

Exit state: State 1 known (as given).

Analysis

$$\text{Energy Eq.:} \quad q_L = h_1 - h_4$$

Solution

Substituting, we have

$$q_L = h_1 - h_4 = 386.1 - 256.5 = 129.6 \text{ kJ/kg}$$

Therefore,

$$\beta = \frac{q_L}{w_c} = \frac{129.6}{42.3} = 3.064$$

Refrigeration capacity $= 129.6 \times 0.03 = 3.89 \text{ kW}$

9.10 WORKING FLUIDS FOR VAPOR-COMPRESSION REFRIGERATION SYSTEMS

A much larger number of working fluids (refrigerants) are utilized in vapor-compression refrigeration systems than in vapor power cycles. Ammonia and sulfur dioxide were important in the early days of vapor-compression refrigeration, but both are highly toxic and therefore dangerous substances. For many years, the principal refrigerants have been the halogenated hydrocarbons, which are marketed under the trade names Freon and Genatron. For example, dichlorodifluoromethane (CCl_2F_2) is known as Freon-12 and Genatron-12, and therefore as refrigerant-12 or R-12. This group of substances, known commonly as *chlorofluorocarbons* (CFCs), are chemically very stable at ambient temperature, especially those lacking any hydrogen atoms. This characteristic is necessary for a refrigerant working fluid. This same characteristic, however, has devastating consequences if the gas, having leaked from an appliance into the atmosphere, spends many years slowly diffusing upward into the stratosphere. There it is broken down, releasing chlorine, which destroys the protective ozone layer of the stratosphere. It is therefore of overwhelming importance to us all to eliminate completely the widely used but life-threatening CFCs, particularly R-11 and R-12, and to develop suitable and acceptable replacements. The CFCs containing hydrogen (often termed *hydrochlorofluorocarbons* [HCFCs]), such as R-22, have shorter atmospheric lifetimes and therefore are not as likely to reach the stratosphere before being broken down and rendered harmless. The most desirable fluids, called *hydrofluorocarbons* (HFCs), contain no chlorine at all, but they do contribute to the atmospheric greenhouse gas effect in a manner similar to, and in some cases to a much greater extent than, carbon dioxide. The sale of refrigerant fluid R-12, which has been widely used in refrigeration systems, has already been banned in many countries, and R-22, used in air-conditioning systems, is scheduled to be banned in the near future. Some alternative refrigerants, several of which are mixtures of different fluids, and therefore are not pure substances, are listed in Table 9.1.

There are two important considerations when selecting refrigerant working fluids: the temperature at which refrigeration is needed and the type of equipment to be used.

As the refrigerant undergoes a change of phase during the heat-transfer process, the pressure of the refrigerant is the saturation pressure during the heat supply and heat

TABLE 9.1

Refrigerants and New Replacements

Old refrigerant	R-11	R-12	R-13	R-22	R-502	R-503
Alternative refrigerant	R-123 R-245fa	R-134a R-152a R-401a	R-23 (low T) CO_2 R-170 (ethane)	NH_3 R-410A	R-404a R-407a R-507a	R-23 (low T) CO_2

rejection processes. Low pressures mean large specific volumes and correspondingly large equipment. High pressures mean smaller equipment, but it must be designed to withstand higher pressure. In particular, the pressures should be well below the critical pressure. For extremely low-temperature applications, a binary fluid system may be used by cascading two separate systems.

The type of compressor used has a particular bearing on the refrigerant. Reciprocating compressors are best adapted to low specific volumes, which means higher pressures, whereas centrifugal compressors are most suitable for low pressures and high specific volumes.

It is also important that the refrigerants used in domestic appliances be nontoxic. Other beneficial characteristics, in addition to being environmentally acceptable, are miscibility with compressor oil, dielectric strength, stability, and low cost. Refrigerants, however, have an unfortunate tendency to cause corrosion. For given temperatures during evaporation and condensation, not all refrigerants have the same COP for the ideal cycle. It is, of course, desirable to use the refrigerant with the highest COP, other factors permitting.

9.11 DEVIATION OF THE ACTUAL VAPOR-COMPRESSION REFRIGERATION CYCLE FROM THE IDEAL CYCLE

The actual refrigeration cycle deviates from the ideal cycle primarily because of pressure drops associated with fluid flow and heat transfer to or from the surroundings. The actual cycle might approach the one shown in Fig. 9.24.

The vapor entering the compressor will probably be superheated. During the compression process, there are irreversibilities and heat transfer either to or from the surroundings, depending on the temperature of the refrigerant and the surroundings. Therefore, the

FIGURE 9.24

The actual vapor-compression refrigeration cycle.

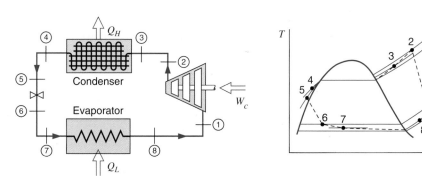

entropy might increase or decrease during this process, for the irreversibility and the heat transferred to the refrigerant cause an increase in entropy, and the heat transferred from the refrigerant causes a decrease in entropy. These possibilities are represented by the two dashed lines 1–2 and 1–2′. The pressure of the liquid leaving the condenser will be less than the pressure of the vapor entering, and the temperature of the refrigerant in the condenser will be somewhat higher than that of the surroundings to which heat is being transferred. Usually, the temperature of the liquid leaving the condenser is lower than the saturation temperature. It might drop somewhat more in the piping between the condenser and the expansion valve. This represents a gain, however, because as a result of this heat transfer the refrigerant enters the evaporator with a lower enthalpy, which permits more heat to be transferred to the refrigerant in the evaporator.

There is some drop in pressure as the refrigerant flows through the evaporator. It may be slightly superheated as it leaves the evaporator, and through heat transferred from the surroundings, its temperature will increase in the piping between the evaporator and the compressor. This heat transfer represents a loss because it increases the work of the compressor, since the fluid entering it has an increased specific volume.

Example 9.7

A refrigeration cycle utilizes R-134a as the working fluid. The following are the properties at various points of the cycle designated in Fig. 9.24:

$$P_1 = 125 \text{ kPa}, \qquad T_1 = -10\,°\text{C}$$
$$P_2 = 1.2 \text{ MPa}, \qquad T_2 = 100\,°\text{C}$$
$$P_3 = 1.19 \text{ MPa}, \qquad T_3 = 80\,°\text{C}$$
$$P_4 = 1.16 \text{ MPa}, \qquad T_4 = 45\,°\text{C}$$
$$P_5 = 1.15 \text{ MPa}, \qquad T_5 = 40\,°\text{C}$$
$$P_6 = P_7 = 140 \text{ kPa}, \qquad x_6 = x_7$$
$$P_8 = 130 \text{ kPa}, \qquad T_8 = -20\,°\text{C}$$

The heat transfer from R-134a during the compression process is 4 kJ/kg. Determine the COP of this cycle.

For each control volume, the R-134a tables are the model. Each process is steady state, with no changes in kinetic or potential energy.

As before, we break the process down into stages, treating the compressor, the throttling value and line, and the evaporator, in turn.

Control volume: Compressor.
Inlet state: P_1, T_1 known; state fixed.
Exit state: P_2, T_2 known; state fixed.

Analysis

From the energy equation, we have

$$q + h_1 = h_2 + w$$
$$w_c = -w = h_2 - h_1 - q$$

Solution

From the R-134a tables, we read

$$h_1 = 394.9 \text{ kJ/kg}, \qquad h_2 = 480.9 \text{ kJ/kg}$$

Therefore,

$$w_c = 480.9 - 394.9 - (-4) = 90.0 \text{ kJ/kg}$$

Control volume:	Throttling valve plus line.
Inlet state:	P_5, T_5 known; state fixed.
Exit state:	$P_7 = P_6$ known, $x_7 = x_6$.

Analysis

$$\text{Energy Eq.:} \quad h_5 = h_6$$

Since $x_7 = x_6$, it follows that $h_7 = h_6$.

Solution

Numerically, we obtain

$$h_5 = h_6 = h_7 = 256.4 \text{ kJ/kg}$$

Control volume:	Evaporator.
Inlet state:	P_7, h_7 known (above).
Exit state:	P_8, T_8 known; state fixed.

Analysis

$$\text{Energy Eq.:} \quad q_L = h_8 - h_7$$

Solution

Substitution gives

$$q_L = h_8 - h_7 = 386.6 - 256.4 = 130.2 \text{ kJ/kg}$$

Therefore,

$$\beta = \frac{q_L}{w_c} = \frac{130.2}{90.0} = 1.44$$

In-Text Concept Questions

e. A refrigerator in my 20 °C kitchen uses R-134a, and I want to make ice cubes at −5 °C. What is the minimum high P and the maximum low P it can use?

f. How many parameters are needed to completely determine a standard vapor-compression refrigeration cycle?

9.12 REFRIGERATION CYCLE CONFIGURATIONS

The basic refrigeration cycle can be modified for special applications and to increase the COP. For larger temperature differences, an improvement in performance is achieved with a two-stage compression with dual loops shown in Fig. 9.25. This configuration can be used when the temperature between the compressor stages is too low to use a two-stage compressor with intercooling (see Fig. P7.39), as there is no cooling medium with such a low temperature. The lowest-temperature compressor then handles a smaller flow rate at the very large specific volume, which means large specific work, and the net result increases the COP.

A regenerator can be used for the production of liquids from gases done in a Linde-Hampson process, as shown in Fig. 9.26, which is a simpler version of the liquid oxygen plant shown in Fig. 9.27. The regenerator cools the gases further before the throttle process, and the cooling is provided by the cold vapor that flows back to the compressor. The compressor is typically a multistage piston/cylinder type, with intercooling between the stages to reduce the compression work, and it approaches isothermal compression.

Finally, the temperature range may be so large that two different refrigeration cycles must be used with two different substances stacking (temperature-wise) one cycle on top of the other cycle, called a *cascade refrigeration system*, shown in Fig. 9.28. In this system, the evaporator in the higher-temperature cycle absorbs heat from the condenser in the lower-temperature cycle, requiring a temperature difference between the two. This dual fluid heat

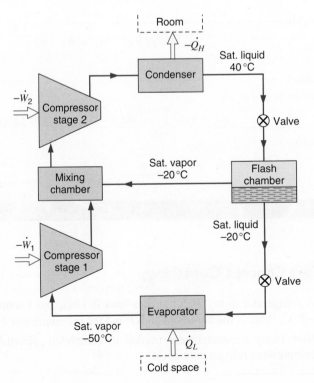

FIGURE 9.25

A two-stage compression dual-loop refrigeration system.

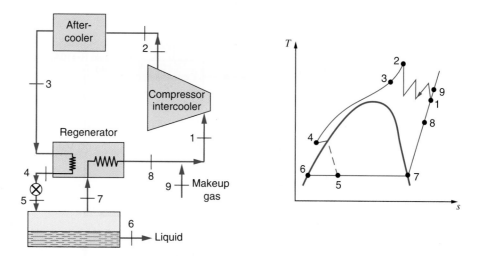

FIGURE 9.26

A Linde-Hampson system for liquefaction of gases.

exchanger couples the mass flow rates in the two cycles through the energy balance with no external heat transfer. The net effect is to lower the overall compressor work and increase the cooling capacity compared to a single-cycle system. A special low-temperature refrigerant like R-23 or a hydrocarbon is needed to produce thermodynamic properties suitable for the temperature range, including viscosity and conductivity.

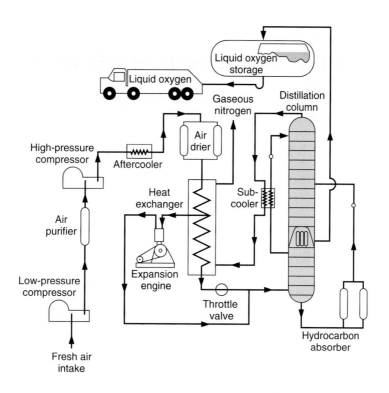

FIGURE 9.27

A simplified diagram of a liquid oxygen plant.

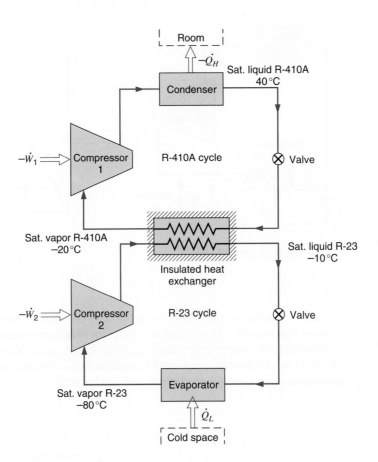

FIGURE 9.28
A two-cycle cascade
refrigeration system.

9.13 THE ABSORPTION REFRIGERATION CYCLE

The absorption refrigeration cycle differs from the vapor-compression cycle in the manner in which compression is achieved. In the absorption cycle, low-pressure vapor is absorbed in water, and the liquid solution is pumped to a high pressure by a liquid pump. Figure 9.29 shows a schematic arrangement of the essential elements of such a system using ammonia.

The low-pressure ammonia vapor leaving the evaporator enters the absorber, where it is absorbed in the weak ammonia solution. This process takes place at a temperature slightly higher than that of the surroundings. Heat must be transferred to the surroundings during this process. The strong ammonia solution is then pumped through a heat exchanger to the generator, where a higher pressure and temperature are maintained. Under these conditions, ammonia vapor is driven from the solution as heat is transferred from a high-temperature source. The ammonia vapor goes to the condenser, where it is condensed, as in a vapor-compression system, and then to the expansion valve and evaporator. The weak ammonia solution is returned to the absorber through the heat exchanger.

The distinctive feature of the absorption system is that very little work input is required because the pumping process involves a liquid. This follows from the fact that for a reversible steady-state process with negligible changes in kinetic and potential energy, the work is equal to $-\int v\, dP$ and the specific volume of the liquid is much less than the specific volume of

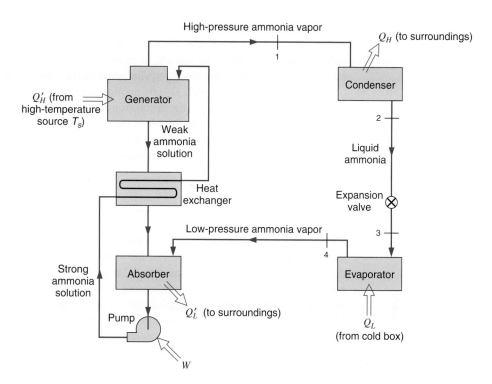

FIGURE 9.29
An ammonia absorption
refrigeration cycle.

the vapor. However, a relatively high-temperature source of heat must be available ($100°$ to $200\,°C$).

All the parts of the cycle to the left of states 4 and 1 in Fig. 9.29 constitute a substitute for the compressor and work input in the standard refrigeration cycle in Fig. 9.23. It works as a heat engine with a small amount of work input to the pump, so the total work available to drive the refrigeration part is

$$\dot{W}_{\text{in}} = \dot{W}_P + \eta_{\text{HE}}\,\dot{Q}'_H \tag{9.10}$$

The upper limit of the heat engine efficiency is the Carnot cycle efficiency as

$$\eta_{\text{HE}} \leq 1 - \frac{T_0}{T_s}$$

and with the standard definition of the COP, Eq. 9.8 or Eq. 5.2, we have

$$\dot{Q}_L = \text{COP}\,\dot{W}_{\text{in}} = \text{COP}\,(\dot{W}_P + \eta_{\text{HE}}\,\dot{Q}'_H) \tag{9.11}$$

Since the major energy input to this cycle is from the heat source, it is common to use the heat transfer ratio for a modified COP as

$$\text{COP}' = \beta_{\text{absorption ref.}} = \frac{\dot{Q}_L}{\dot{Q}'_H}$$
$$= \text{COP}\,(\eta_{\text{HE}} + \dot{W}_P/\dot{Q}'_H) \tag{9.12}$$

and as the pump work is very small, it is often neglected. The COP' ratio is usually less than one for these absorption refrigeration cycles.

There is more equipment in an absorption system than in a vapor-compression system, and it can usually be economically justified only when a suitable source of heat is available that would otherwise be wasted. In recent years, the absorption cycle has been given increased attention in connection with alternative energy sources, for example, solar energy or geothermal energy. It should also be pointed out that other working fluid combinations have been used successfully in the absorption cycle, one being lithium bromide in water.

The absorption cycle reemphasizes the important principle that since the shaft work in a reversible steady-state process with negligible changes in kinetic and potential energies is given by $-\int v\,dP$, a compression process should take place with the smallest possible specific volume.

SUMMARY

The standard power-producing cycle and refrigeration cycle for fluids with phase change during the cycle are presented. The Rankine cycle and its variations represent a steam power plant, which generates most of the world production of electricity. The heat input can come from combustion of fossil fuels, a nuclear reactor, solar radiation, or any other heat source that can generate a temperature high enough to boil water at high pressure. In low- or very-high-temperature applications, working fluids other than water can be used. Modifications to the basic cycle such as reheat, closed, and open FWHs are covered, together with applications where the electricity is cogenerated with a base demand for process steam.

Standard refrigeration systems are covered by the vapor-compression refrigeration cycle. Applications include household and commercial refrigerators, air-conditioning systems, and heat pumps, as well as lower-temperature-range special-use installations. As a special case, we briefly discuss the ammonia absorption cycle.

For combinations of cycles, see Section 10.12.

You should have learned a number of skills and acquired abilities from studying this chapter that will allow you to

- Apply the general laws to control volumes with several devices forming a complete system.
- Know how common power-producing devices work.
- Know how simple refrigerators and heat pumps work.
- Know that no cycle devices operate in Carnot cycles.
- Know that real devices have lower efficiencies/COP than ideal cycles.
- Understand the most influential parameters for each type of cycle.
- Understand the importance of the component efficiency for the overall cycle efficiency or COP.
- Know that most real cycles have modifications to the basic cycle setup.
- Know that many of these devices affect our environment.

KEY CONCEPTS AND FORMULAS

Rankine Cycle

Efficiency	$\eta_{th} = \dfrac{w_{net}}{q_H} = \dfrac{w_t - w_p}{q_H}$
Superheat	$\Delta T_{superheat} = T_{3'} - T_3$ (Fig. 9.5)
Reheat	Heat water after some turbine section
Open feedwater heater	Feedwater mixed with extraction steam, exit as saturated liquid

Extraction fraction	$y = \dot{m}_6/\dot{m}_5 = \dfrac{h_3 - h_2}{h_6 - h_2}$ (Fig. 9.12)
Closed feedwater heater	Feedwater heated by extraction steam, no mixing
Extraction fraction with trap	$y = \dfrac{h_3 - h_2}{h_6 - h_{6a}}$ (Fig. 9.13)
Deaerating FWH	Open FWM operating at P_{atm} to vent gas out
Cogeneration	Turbine power is cogenerated with a desired steam supply

Refrigeration Cycle ────────────────────

Coefficient of performance $\text{COP} = \beta_{\text{REF}} = \dfrac{\dot{Q}_L}{\dot{W}_c} = \dfrac{q_L}{w_c} = \dfrac{h_1 - h_3}{h_2 - h_1}$

CONCEPT-STUDY GUIDE PROBLEMS

9.1 Is a steam power plant running in a Carnot cycle? Name the four processes.

9.2 Raising the boiler pressure in a Rankine cycle for fixed superheat and condenser temperatures, in what direction do these change: turbine work, pump work and turbine exit T or x?

9.3 For other properties fixed in a Rankine cycle, raising the condenser temperature causes changes in which work and heat transfer terms?

9.4 Mention two benefits of a reheat cycle.

9.5 What is the benefit of the moisture separator in the power plant of Problem 4.95?

9.6 Instead of using the moisture separator in Problem 4.95, what could have been done to remove any liquid in the flow?

9.7 Can the energy removed in a power plant condenser be useful?

9.8 If the district heating system (see Fig. 1.1) should supply hot water at $90\,^\circ\text{C}$, what is the lowest possible condenser pressure with water as the working substance?

9.9 What is the mass flow rate through the condensate pump in Fig. 9.14?

9.10 A heat pump for a $20\,^\circ\text{C}$ house uses R-410A, and the outside temperature is $-5\,^\circ\text{C}$. What is the minimum high P and the maximum low P it can use?

9.11 A heat pump uses carbon dioxide, and it must condense at a minimum of $22\,^\circ\text{C}$ and receives energy from the outside on a winter day at $-10\,^\circ\text{C}$. What restrictions does that place on the operating pressures?

9.12 Since any heat transfer is driven by a temperature difference, how does that affect all the real cycles relative to the ideal cycles?

HOMEWORK PROBLEMS

Rankine Cycles, Power Plants

Simple Cycles

9.13 A steam power plant, as shown in Fig. 9.3 operating in a Rankine cycle, has saturated vapor at 3.0 MPa leaving the boiler. The turbine exhausts to the condenser operating at 10 kPa. Find the specific work and heat transfer in each of the ideal components and the cycle efficiency.

9.14 Consider a solar-energy-powered ideal Rankine cycle that uses water as the working fluid. Saturated vapor leaves the solar collector at 1500 kPa, and the condenser pressure is 15 kPa. Determine the thermal efficiency of this cycle.

9.15 The power plant in the previous problem is augmented with a natural gas burner to superheat the water to $300\,^\circ\text{C}$ before entering the turbine. Find the cycle efficiency with this configuration and the specific heat transfer added by the natural gas burner.

9.16 A utility runs a Rankine cycle with a water boiler at 3.0 MPa, and the cycle has the highest and lowest temperatures of 450 °C and 60 °C, respectively. Find the plant efficiency and the efficiency of a Carnot cycle with the same temperatures.

9.17 The power plant in the previous problem has too low a quality in the low-pressure turbine section, so the plant wants to increase the superheat. What should the superheat be so that the quality of the water in the turbine stays above 92 %?

9.18 A power plant for a polar expedition uses ammonia, which is heated to 80 °C at 1000 kPa in the boiler, and the condenser is maintained at −10 °C. Find the cycle efficiency.

9.19 A Rankine cycle with R-410A has the boiler at 3 MPa superheating to 180 °C, and the condenser operates at 800 kPa. Find all four energy transfers and the cycle efficiency.

9.20 A supply of geothermal hot water is used as the energy source in an ideal Rankine cycle with R-134a, as shown in Fig. P9.20. Saturated vapor R-134a leaves the boiler at a temperature of 85 °C, and the condenser temperature is 40 °C. Calculate the thermal efficiency of this cycle.

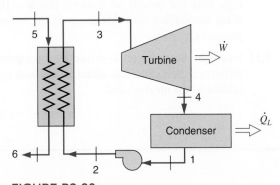

FIGURE P9.20

9.21 Do Problem 9.20 with R-410A as the working fluid and the boiler exit at 4000 kPa, 70 °C.

9.22 A low-temperature power plant operates with R-410A maintaining −20 °C in the condenser and a high pressure of 3 MPa with superheat. Find the temperature out of the boiler/superheater so that

the turbine exit temperature is 40 °C, and find the overall cycle efficiency.

9.23 Do Problem 9.20 with ammonia as the working fluid.

9.24 Geothermal water can be used directly as a source to a steam turbine. Consider 10 kg/s water at 500 kPa, 150 °C brought to a flash chamber, where it is throttled to 200 kPa, as shown in Fig. P9.24. From the chamber, saturated vapor at 200 kPa flows to the turbine with an exit at 10 kPa. From state 4, it is cooled in a condenser and pumped back into the ground. Determine the quality at the turbine exit and the power that can be obtained from the turbine.

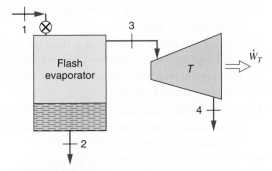

FIGURE P9.24

9.25 Some geothermal locations have higher pressure and temperature, so assume the geothermal power plant as in Problem 9.24 but with a supply as saturated liquid at 250 °C. For this case, the flash chamber operates at 1000 kPa. For a supply of 10 kg/s, determine the mass flow that goes through the turbine and the power output.

9.26 With a higher supply pressure and temperature of the geothermal source, it is possible to have two flash evaporators, as shown in Fig. P9.26. Assume the supply is saturated water at 250 °C and the first chamber flashes to 2000 kPa and saturated liquid at state 2 is flashed to 500 kPa, with the saturated vapor out added to the turbine, which has an exit state of 10 kPa with quality 78 %. For a supply of 10 kg/s, determine the mass flows at states 3 and 5 into the turbine and the total power output.

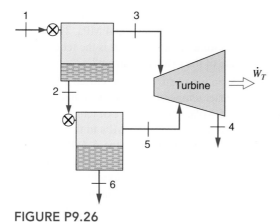

FIGURE P9.26

9.27 A coal-fired power plant produces 25 kg/s steam at 3 MPa, 600 °C in the boiler. It cools the condenser with ocean water coming in at 15 °C and returned at 23 °C, so the condenser exit is at 50 °C. Find the net power output and the required mass flow rate of ocean water.

9.28 Consider an ideal Rankine cycle using water with a high-pressure side of the cycle at a supercritical pressure. Such a cycle has the potential advantage of minimizing local temperature differences between the fluids in the steam generator, such as the instance in which the high-temperature energy source is the hot exhaust gas from a gas-turbine engine. Calculate the thermal efficiency of the cycle if the state entering the turbine is 30 MPa, 550 °C and the condenser pressure is 10 kPa. What is the steam quality at the turbine exit?

Reheat Cycles

9.29 The power plant in Problem 9.16 has too low a quality in the low-pressure turbine section, so the plant wants to apply reheat at 800 kPa. What should the superheat for the reheat be so that the turbine quality stays above 95 %?

9.30 Consider the supercritical cycle in Problem 9.28 and assume that the turbine first expands to 3 MPa, followed by a reheat to 500 °C, with a further expansion in the low-pressure turbine to 10 kPa. Find the combined specific turbine work and the total specific heat transfer in the boiler.

9.31 A small power plant produces steam at 3 MPa, 600 °C in the boiler. It keeps the condenser at 45 °C

by transfer of 10 MW out as heat transfer. The first turbine section expands to 500 kPa, and then flow is reheated followed by expansion in the low-pressure turbine. Find the reheat temperature so that the turbine output is saturated vapor. For this reheat, find the total turbine power output and the boiler heat transfer.

9.32 Consider an ideal steam reheat cycle where steam enters the high-pressure turbine at 3.0 MPa, 400 °C and then expands to 0.8 MPa. It is then reheated to 350 °C and expands to 10 kPa in the low-pressure turbine. Calculate the cycle thermal efficiency and the moisture content of the steam leaving the low-pressure turbine.

9.33 The reheat pressure affects the operating variables and thus the turbine performance. Repeat Problem 9.31 twice, using 0.6 and 1.0 MPa for the reheat pressure.

Open FWHs

9.34 An open FWH receives steam at 1 MPa, 200 °C from the turbine and 1 MPa, 100 °C water from the feedwater line. Find the required fraction of the extraction flow in the turbine.

9.35 A power plant for a polar expedition uses ammonia and the boiler exit is 80 °C, 1000 kPa, and the condenser operates at −15 °C. A single open FWH operates at 400 kPa with an exit state of saturated liquid. Find the mass fraction extracted in the turbine.

9.36 Find the cycle efficiency for the cycle in Problem 9.35.

9.37 A steam power plant has high and low pressures of 20 MPa and 15 kPa, respectively, and one open FWH operating at 1 MPa with the exit as saturated liquid. The maximum temperature is 800 °C, and the turbine has a total power output of 5 MW. Find the fraction of the extraction flow to the FWH and the total condenser heat transfer rate.

9.38 A low-temperature power plant operates with R-410A maintaining −20 °C in the condenser, a high pressure of 3 MPa with superheat to 80 °C. There is one open FWH operating at 800 kPa with an exit as saturated liquid at 0 °C. Find the extraction fraction of the flow out of the turbine and the turbine work per unit mass flowing through the boiler.

9.39 A Rankine cycle operating with ammonia is heated by some low-temperature source, so the highest T is 120 °C at a pressure of 5000 kPa. Its low pressure is 1003 kPa, and it operates with one open FWH at 2033 kPa. The total flow rate is 5 kg/s. Find the extraction flow rate to the FWH, assuming its outlet state is saturated liquid at 2033 kPa. Find the total power to the two pumps.

9.40 A steam power plant operates with a boiler output of 20 kg/s steam at 2 MPa, 600 °C. The condenser operates at 50 °C, dumping energy into a river that has an average temperature of 20 °C. There is one open FWH with extraction from the turbine at 600 kPa, and its exit is saturated liquid. Find the mass flow rate of the extraction flow. If the river water should not be heated more than 5 °C, how much water should be pumped from the river to the heat exchanger (condenser)?

9.41 In a nuclear power plant the reactor transfers the heat to a flow of liquid sodium, which in a heat exchanger transfers the heat to boiling water. Saturated vapor steam at 5 MPa exits this heat exchanger and is then superheated to 600 °C in an external gas-fired superheater. The steam enters the turbine, which has an extraction at 0.4 MPa to an open FWH, and the condenser pressure is 10 kPa. Determine the heat transfer in the reactor and in the superheater to produce a net power output of 5 MW.

9.42 Consider an ideal steam regenerative cycle in which steam enters the turbine at 3.0 MPa, 400 °C and exhausts to the condenser at 10 kPa. Steam is extracted from the turbine at 0.8 MPa for an open FWH. The feedwater leaves the heater as saturated liquid. The appropriate pumps are used for the water leaving the condenser and the FWH. Calculate the thermal efficiency of the cycle and the specific net work.

Closed FWHs

9.43 Write the analysis (continuity and energy equations) for the closed FWH with a drip pump, as shown in Fig. 9.13. Assume that the control volume has state 4 out, so it includes the drip pump. Find the equation for the extraction fraction.

9.44 A closed FWH in a regenerative steam power cycle heats 20 kg/s of water from 140 °C, 20 MPa to 200 °C, 20 MPa. The extraction steam from the turbine enters the heater at 4 MPa, 275 °C and leaves as saturated liquid. What is the required mass flow rate of the extraction steam?

9.45 A power plant with one closed FWH has a condenser temperature of 45 °C, a maximum pressure of 5 MPa, and a boiler exit temperature of 900 °C. Extraction steam at 1 MPa to the FWH condenses and is pumped up to the 5 MPa feedwater line, where all the water goes to the boiler at 200 °C. Find the fraction of extraction steam flow and the two specific pump work inputs.

9.46 Do Problem 9.37 with a closed FWH instead of an open FWH and a drip pump to add the extraction flow to the feedwater line at 20 MPa. Assume the temperature is 175 °C after the drip pump flow is added to the line. One main pump brings the water to 20 MPa from the condenser.

9.47 Repeat Problem 9.42, but assume a closed FWH instead of an open FWH. A single pump is used to pump the water leaving the condenser up to the boiler pressure of 3.0 MPa. Condensate from the FWH is going through a drip pump and is added to the feedwater line, so state 4 is at T_6.

9.48 Repeat Problem 9.42, but assume a closed FWH instead of an open FWH. A single pump is used to pump the water leaving the condenser up to the boiler pressure of 3.0 MPa. Condensate from the FWH is drained through a trap to the condenser.

9.49 Assume that the power plant in Problem 9.39 has one closed FWH instead of the open FWH. The extraction flow out of the FWH is saturated liquid at 2033 kPa being dumped into the condenser, and the feedwater is heated to 50 °C. Find the extraction flow rate and the total turbine power output.

9.50 Assume a variation of the cycle in Problem 9.42 with a closed FWH at 0.8 MPa and one open FWH at 100 kPa. A pump is used to bring the water leaving the condenser up to 100 kPa for an open FWH, and a second pump brings the feedwater up to 3.0 MPa, as shown in Fig. P9.50. Condensate from the closed FWH is drained through a trap to the open FWH. Calculate the thermal efficiency of the cycle and the specific net work.

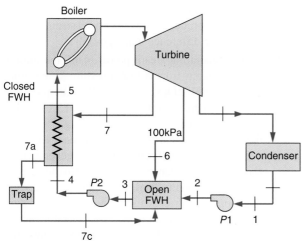

FIGURE P9.50

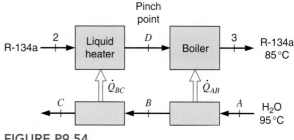

FIGURE P9.54

Nonideal Cycles

9.51 A steam power cycle has a high pressure of 3.0 MPa and a condenser exit temperature of 45 °C. The turbine efficiency is 85 %, and other cycle components are ideal. If the boiler superheats to 800 °C, find the cycle thermal efficiency.

9.52 A Rankine cycle with water superheats to 500 °C at 3 MPa in the boiler, and the condenser operates at 100 °C. All components are ideal except the turbine, which has an exit state measured to be saturated vapor at 100 °C. Find the cycle efficiency with (a) an ideal turbine and (b) the actual turbine.

9.53 For the steam power plant described in Problem 9.13, assume the isentropic efficiencies of the turbine and pump are 85 % and 80 %, respectively. Find the components' specific work and heat transfers and the cycle efficiency.

9.54 Consider the boiler in Problem 9.20, where the geothermal hot water brings the R-134a to saturated vapor. Assume a counterflowing heat exchanger arrangement (Fig. P9.54). The geothermal water temperature should be equal to or greater than that of the R-134a temperature at any location inside the heat exchanger. The point with the smallest temperature difference between the source and the working fluid is called the *pinch point*. If 2 kg/s of geothermal water is available at 95 °C, what is the maximum power output of this cycle for R-134a as the working fluid? (Hint: split the heat exchanger C.V. into two so that the pinch point with $\Delta T = 0$, $T = 85$ °C appears.)

9.55 Do the previous problem with ammonia as the working fluid.

9.56 A concentrated solar power plant receives the energy from molten salt coming in at 560 °C and leaving at 300 °C in a counterflow heat exchanger where the water comes in at 3 MPa, 60 °C and leaves at 450 °C, 3 MPa. The molten salt has 5 kg/s flow with $C_p = 1.5$ kJ/kg·K. What is the possible water flow rate, the rate of energy transfer, and the rate of entropy generation?

9.57 A steam power plant operates with a high pressure of 4 MPa and has a boiler exit of 600 °C receiving heat from a 700 °C source. The ambient at 20 °C provides cooling to maintain the condenser at 60 °C. All components are ideal except for the turbine, which has an isentropic efficiency of 92 %. Find the ideal and actual turbine exit qualities. Find the actual specific work and specific heat transfer in all four components.

9.58 For the previous problem, also find the specific entropy generation in the boiler heat source setup.

9.59 Consider the power plant in Problem 9.35. Assume that the high temperature source is a flow of liquid water at 120 °C into a heat exchanger at constant pressure, 300 kPa, and that the water leaves at 90 °C. Assume that the condenser rejects heat to the ambient, which is at −20 °C. List all the places that have entropy generation and find the entropy generated in the boiler heat exchanger per kilogram of ammonia flowing.

9.60 Repeat Problem 9.37, assuming the turbine has an isentropic efficiency of 85 %.

9.61 Steam leaves a power plant steam generator at 3.5 MPa, 400 °C ($h_1 = 3222.3$ kJ/kg, $s_1 = 6.8405$ kJ/kg·K) and enters the turbine at 3.4 MPa, 375 °C ($h_2 = 3165.7$ kJ/kg, $s_2 = 6.7675$ kJ/kg·K).

The isentropic turbine efficiency is 88%, and the turbine exhaust pressure is 10 kPa. Condensate leaves the condenser and enters the pump at 35 °C, 10 kPa. The isentropic pump efficiency is 80%, and the discharge pressure is 3.7 MPa. The feedwater enters the steam generator at 3.6 MPa, 30 °C ($h = 129.0$ kJ/kg). Calculate the thermal efficiency of the cycle and the entropy generation for the process in the line between the steam generator exit and the turbine inlet, assuming an ambient temperature of 25 °C.

9.62 Find the entropy generation per unit mass leaving the open FWH in Problem 9.34.

9.63 Find the rate of entropy generation in the closed FWH in Problem 9.44.

Combined Heat and Power

9.64 A cogenerating steam power plant, as in Fig. 9.19, operates with a boiler output of 25 kg/s steam at 7 MPa, 500 °C. The condenser operates at 7.5 kPa, and the process heat is extracted as 5 kg/s from the turbine at 500 kPa, state 6 and after use is returned as saturated liquid at 100 kPa, state 8. Assume all components are ideal and find the temperature after pump 1, the total turbine output, and the total process heat transfer.

9.65 A steam power plant has 3 MPa, 500 °C into the turbine, and to have the condenser itself deliver the process heat, it is run at 101 kPa. How much net power as work is produced for process heat of 13 MW?

9.66 A 15 kg/s steady supply of saturated-vapor steam at 500 kPa is required for drying a wood pulp slurry in a paper mill (see Fig. P9.66). It is decided to supply this steam by cogeneration; that is, the steam supply will be the exhaust from a steam turbine. Water at 20 °C, 100 kPa is pumped to a pressure of 5 MPa and then fed to a steam generator with an exit at 400 °C. What is the additional heat transfer rate to

the steam generator beyond what would have been required to produce only the desired steam supply? What is the difference in net power?

9.67 In a cogenerating steam power plant, the turbine receives steam from a high-pressure steam drum ($h = 3445.9$ kJ/kg, $s = 6.9108$ kJ/kg·K) and a low-pressure steam drum ($h = 2855.4$ kJ/kg, $s = 7.0592$ kJ/kg·K), as shown in Fig. P9.67. For the turbine calculation, assume a mass-weighted average entropy and neglect entropy generation by mixing. The condenser is made as two closed heat exchangers used to heat water running in a separate loop for district heating. The high-temperature heater adds 30 MW and the low-temperature heater adds 31 MW to the district heating water flow. Find the power cogenerated by the turbine and the temperature in the return line to the deaerator.

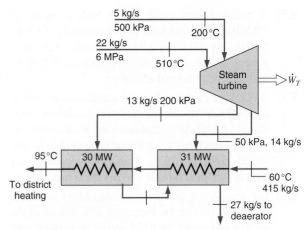

FIGURE P9.67

9.68 A boiler delivers steam at 10 MPa, 550 °C to a two-stage turbine, as shown in Fig. 9.19. After the first stage, 25% of the steam is extracted at 1.4 MPa for a process application and returned at 1 MPa, 90 °C to the feedwater line. The remainder of the steam continues through the low-pressure turbine stage, which exhausts to the condenser at 10 kPa. One pump brings the feedwater to 1 MPa and a second pump brings it to 10 MPa. Assume all components are ideal. If the process application requires 5 MW of power, how much power can then be cogenerated by the turbine?

9.69 A smaller power plant produces 25 kg/s steam at 3 MPa, 600 °C in the boiler. It cools the con-

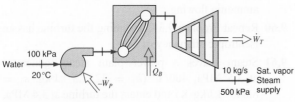

FIGURE P9.66

denser to an exit temperature of 45 °C, and the cycle is shown in Fig. P9.69. An extraction is done at 500 kPa to an open FWH; in addition, a steam supply of 5 kg/s is taken out and not returned. The missing 5 kg/s water is added to the FWH from a 20 °C, 500 kPa source. Find the extraction flow rate needed to cover both the FWH and the steam supply. Find the total turbine power output.

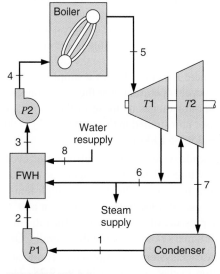

FIGURE P9.69

Refrigeration Cycles

9.70 A refrigeration cycle, as in Fig. 9.23, can be used for cooling or for heating purposes using one of the two heat exchangers. Suppose a refrigerator should cool meat at −10 °C in a 30 °C kitchen. What are the minimum high and maximum low pressures in the cycle if the working substance is (a) R-134a or (b) R-410A?

9.71 A refrigerator with R-134a as the working fluid has a minimum temperature of −10 °C and a maximum pressure of 2 MPa. Assume an ideal refrigeration cycle, as in Fig. 9.23. Find the specific heat transfer from the cold space and that to the hot space, and the COP.

9.72 Repeat the previous problem with R-410A as the working fluid. Will that work in an ordinary kitchen?

9.73 The natural refrigerant carbon dioxide has a fairly low critical temperature. Find the high temperature,

the condensing temperature, and the COP if it is used in a standard cycle with high and low pressures of 6 MPa and 3 MPa, respectively.

9.74 A refrigerator receives 500 W of electrical power to the compressor driving the cycle flow of R-134a. The refrigerator operates with a condensing temperature of 40 °C and a low temperature of −5 °C. Find the COP for the cycle and its cooling capacity.

9.75 A new air conditioner using R-410A is used in heat pump mode. The high pressure is 2000 kPa and the low pressure is 400 kPa. It warms a house at 20 °C, driven by an electric power input of 2.5 kW in an ambient at −5 °C. Find the COP, the heating rate, and the rate of entropy generation for the heat pump.

9.76 A heat pump for heat upgrade uses ammonia with a low temperature of 25 °C and a high pressure of 5000 kPa. If it receives 1 MW of shaft work, what is the rate of heat transfer at the high temperature?

9.77 Reconsider the heat pump in the previous problem. Assume the compressor is split into two. First, compress to 2000 kPa. Then take heat transfer out at constant P to reach saturated vapor and then compress to 5000 kPa. Find the two rates of heat transfer, at 2000 kPa and at 5000 kPa, for a total of 1 MW shaft work input.

9.78 A car air conditioner operating with R-134a as the working fluid has a minimum temperature of −10 °C and a maximum pressure of 1 MPa. The actual adiabatic compressor exit temperature is 50 °C. Assume no pressure loss in the heat exchangers. Find the specific heat transfer from the cold space and that to the hot space, the COP, and the isentropic efficiency of the compressor.

9.79 An air conditioner in the airport of Timbuktu runs a cooling system using R-410A with a high pressure of 1800 kPa and a low pressure of 200 kPa. It should cool the desert air at 45 °C down to 15 °C. Find the cycle COP. Will the system work?

9.80 Consider an ideal heat pump that has a condenser temperature of 50 °C and an evaporator temperature of 0 °C. Determine the COP of this heat pump for the working fluids R-134a and ammonia.

9.81 A refrigerator in a meat warehouse must maintain a low temperature of −15 °C, and the outside temperature is 20 °C. It uses ammonia as the refrigerant, which must remove 5 kW from the cold space. Find the flow rate of the ammonia needed, assuming a standard vapor compression refrigeration cycle with a condenser at 20 °C.

9.82 The air conditioner in an automobile uses R-134a and the compressor power input is 1.5 kW, bringing the R-134a from 201.7 kPa to 1200 kPa by compression. The cold space is a heat exchanger that cools atmospheric air from the outside at 30 °C down to 10 °C and blows it into the car. What is the mass flow rate of the R-134a, and what is the low-temperature heat transfer rate? What is the mass flow rate of air at 10 °C?

9.83 A refrigerator in a laboratory uses R-134a as the working substance. The high pressure is 1200 kPa, the low pressure is 165 kPa, and the compressor is reversible. It should remove 500 W from a specimen currently at −10 °C (not equal to T_L in the cycle) that is inside the refrigerated space. Find the cycle COP and the electrical power required.

9.84 Consider the previous problem and find the two rates of entropy generation in the process and where they occur.

9.85 A refrigerator using R-134a is located in a 20 °C room. Consider the cycle to be ideal, except that the compressor is neither adiabatic nor reversible. Saturated vapor at −20 °C enters the compressor, and the R-134a exits the compressor at 50 °C. The condenser temperature is 40 °C. The mass flow rate of refrigerant around the cycle is 0.2 kg/s, and the COP is measured and found to be 2.3. Find the power input to the compressor and the rate of entropy generation in the compressor process.

9.86 A small heat pump unit is used to heat water for a hot-water supply. Assume that the unit uses ammonia and operates on the ideal refrigeration cycle. The evaporator temperature is 15 °C and the condenser temperature is 60 °C. If the amount of hot water needed is 0.1 kg/s, determine the amount of energy saved by using the heat pump instead of directly heating the water from 15 °C to 60 °C.

Extended Refrigeration Cycles

9.87 One means of improving the performance of a refrigeration system that operates over a wide temperature range is to use a two-stage compressor. Consider an ideal refrigeration system of this type that uses R-410A as the working fluid, as shown in

Fig. 9.25. Saturated liquid leaves the condenser at 40 °C and is throttled to −20 °C. The liquid and vapor at this temperature are separated, and the liquid is throttled to the evaporator temperature, −50 °C. Vapor leaving the evaporator is compressed to the saturation pressure corresponding to −20 °C, after which it is mixed with the vapor leaving the flash chamber. It may be assumed that both the flash chamber and the mixing chamber are well insulated to prevent heat transfer from the ambient. Vapor leaving the mixing chamber is compressed in the second stage of the compressor to the saturation pressure corresponding to the condenser temperature, 40 °C. Determine the COP of the system. Compare it to the COP of a simple ideal refrigeration cycle operating over the same condenser and evaporator ranges as those of the two-stage compressor unit studied in this problem ($T_2 = 90.7$ °C, $h_2 = 350.35$ kJ/kg).

9.88 A cascade system with one refrigeration cycle operating with R-410A has an evaporator at −40 °C and a high pressure of 1200 kPa. The high-temperature cycle uses R-134a with an evaporator at 0 °C and a high pressure of 1200 kPa. Find the ratio of the two cycle's mass flow rates and the overall COP.

9.89 A cascade system is composed of two ideal refrigeration cycles, as shown in Fig. 9.28. The high-temperature cycle uses R-410A. Saturated liquid leaves the condenser at 40 °C, and saturated vapor leaves the heat exchanger at −20 °C. The low-temperature cycle uses a different refrigerant, R-23. Saturated vapor leaves the evaporator at −80 °C, $h = 330$ kJ/kg, and saturated liquid leaves the heat exchanger at −10 °C, $h = 185$ kJ/kg. R-23 out of the compressor has $h = 405$ kJ/kg. Calculate the ratio of the mass flow rates through the two cycles and the COP of the system.

R-410A	T, °C	P	h	s
1′	−20	0.400	271.89	1.0779
2′	71	2.421	322.61	1.0779
3′	40	2.421	124.09	
4′	−20		124.09	

9.90 A split evaporator is used to provide cooling of the refrigerator section and separate cooling of the freezer section, as shown in Fig. P9.90. Assume constant pressure in the two evaporators. How does the COP $= (Q_{L1} + Q_{L2})/W$ compare to that of a refrigerator with a single evaporator at the lowest temperature?

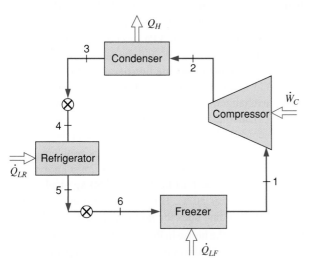

FIGURE P9.90

9.91 A refrigerator using R-410A is powered by a small, natural gas-fired heat engine with a thermal efficiency of 25 %, as shown in Fig. P9.91. The R-410A condenses at 40 °C and evaporates at −20 °C, and the cycle is standard. Find the two specific heat transfers in the refrigeration cycle. What is the overall COP as Q_L/Q_1?

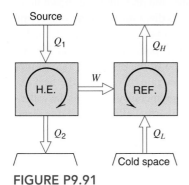

FIGURE P9.91

Ammonia Absorption Cycles

9.92 Notice that the configuration in Fig. 9.29 has the left-hand side column of devices substitute for a compressor in the standard cycle. What is an expression for the equivalent work output from the left-hand-side devices, assuming they are reversible and the high and low temperatures are constant, as a function of the pump work W and the two temperatures?

9.93 As explained in the previous problem, the ammonia absorption cycle is very similar to the setup sketched in Problem 9.91. Assume the heat engine has an efficiency of 30 % and the COP of the refrigeration cycle is 3.0. What is the ratio of the cooling to the heating heat transfer Q_L/Q_1?

9.94 Give an estimate for the overall COP′ of an ammonia absorption cycle used as a chiller to cool water to 5 °C in a 25 °C ambient when the small pump work is neglected. A heat source is available at 100 °C. Also find the efficiency of the heat engine part and the COP of the refrigeration cycle part.

9.95 Consider a small ammonia absorption refrigeration cycle that is powered by solar energy and is to be used as an air conditioner. Saturated vapor ammonia leaves the generator at 50 °C, and saturated vapor leaves the evaporator at 10 °C. If 3000 kJ of heat is required in the generator (solar collector) per kilogram of ammonia vapor generated, determine the overall performance of this system.

9.96 The performance of an ammonia absorption cycle refrigerator is to be compared with that of a similar vapor-compression system. Consider an absorption system having an evaporator temperature of −10 °C and a condenser temperature of 50 °C. The generator temperature in this system is 150 °C. In this cycle, 0.42 kJ is transferred to the ammonia in the evaporator for each kilojoule transferred from the high-temperature source to the ammonia solution in the generator. To make the comparison, assume that a reservoir is available at 150 °C, and that heat is transferred from this reservoir to a reversible engine that rejects heat to the surroundings at 25 °C. This work is then used to drive an ideal vapor-compression system with ammonia as the refrigerant. Compare the amount of refrigeration that can be achieved per kilojoule from the high-temperature source with the 0.42 kJ that can be achieved in the absorption system.

Exergy Concepts

9.97 If we neglect the external irreversibilities due to the heat transfers over finite temperature differences in a power plant, how would you define its second-law efficiency?

9.98 Find the exergy of the water at all four states in the Rankine cycle described in Problem 9.16. Assume that the high-temperature source is 500 °C and the low-temperature reservoir is at the ambient 25 °C. Determine the flow of exergy into or out of the reservoirs per kilogram of steam flowing in the cycle. What is the overall cycle second-law efficiency?

9.99 A condenser is maintained at 60 °C by cooling it with atmospheric air coming in at 20 °C and leaving at 35 °C. The condenser must reject 25 MW from the water to the air. Find the flow rate of air and the second-law efficiency of the heat exchanger.

9.100 Find the flows and fluxes of exergy in the condenser of Problem 9.27. Use them to determine the second-law efficiency, $T_0 = 15$ °C.

9.101 Find the second-law efficiency for the open FWH in Problem 9.34.

9.102 The power plant using ammonia in Problem 9.59 has a flow of liquid water at 120 °C, 300 kPa as a heat source; the water leaves the heat exchanger at 90 °C. Find the second-law efficiency of this heat exchanger.

9.103 A concentrated solar power plant receives the energy from molten salt coming in at 560 °C and leaving at 300 °C in a counterflow heat exchanger where the water comes in at 3 MPa, 60 °C and leaves at 450 °C, 3 MPa. The molten salt has 5 kg/s flow with $C_p = 1.5$ kJ/kg·K. What is the possible water flow rate and the rate of energy transfer? Find the second-law efficiency of this heat exchanger.

9.104 What is the second-law efficiency of the heat pump in Problem 9.76?

9.105 The condenser in a refrigerator receives R-134a at 700 kPa, 50 °C, and it exits as saturated liquid at 25 °C. The flow rate is 0.1 kg/s, and the condenser has air flowing in at ambient 15 °C and leaving at 35 °C. Find the minimum flow rate of air and the heat exchanger second-law efficiency.

9.106 A new air conditioner using R-410A is used in heat pump mode. The high pressure is 2000 kPa and the low pressure is 400 kPa. It warms a house at 20 °C driven by an electric power input of 2 kW in an ambient at −5 °C. Find the destruction of exergy in four places: (1) inside the heat pump, (2)

in the high-T heat exchanger, (3) in the low-T heat exchanger, and (4) in the house walls/windows, and so on, that separate the inside from the outside of the house.

9.107 An air conditioner using R-410A is used in cooling mode. The high pressure is 3000 kPa and the low pressure is 800 kPa. It cools a house at 20 °C with a rate of 12 kW, and the outside ambient is at 35 °C. Find the destruction of exergy in four places: (1) inside the refrigerator, (2) in the high-T heat exchanger, (3) in the low-T heat exchanger, and (4) in the house walls/windows, and so on, that separate the inside from the outside of the house.

9.108 Assume the house in the previous problem has a combined 12 000 kg hard wood, 2500 kg gypsum plates ($C_p = 1$ kJ/kg·K), and 750 kg steel, all of which is at 20 °C. If the air conditioner is turned off, how fast does the house heat up (°C/s)?

9.109 Assume the house in Problem 9.106 has a combined 12 000 kg hard wood, 2500 kg gypsum plates ($C_p = 1$ kJ/kg·K), and 750 kg steel, all of which is at 20 °C. If the heat pump is turned off, how fast does the house cool down (°C/s)?

Combined Cycles

See Section 10.12 for text and figures.

9.110 A binary system power plant uses mercury for the high-temperature cycle and water for the low-temperature cycle, as shown in Fig. 10.23. The temperatures and pressures are shown in the corresponding T–s diagram. The maximum temperature in the steam cycle is where the steam leaves the superheater at point 4, where it is 500 °C. Determine the ratio of the mass flow rate of mercury to the mass flow rate of water in the heat exchanger that condenses mercury and boils the water, and determine the thermal efficiency of this ideal cycle.

The following saturation properties for mercury are known:

P, MPa	T_g, °C	h_f, kJ/kg	h_g, kJ/kg	s_f, kJ/kg·K	s_g, kJ/kg·K
0.04	309	42.21	335.64	0.1034	0.6073
1.60	562	75.37	364.04	0.1498	0.4954

9.111 A Rankine steam power plant should operate with a high pressure of 3 MPa and a low pressure of

10 kPa, and the boiler exit temperature should be 500 °C. The available high-temperature source is the exhaust of 175 kg/s air at 600 °C from a gas turbine. If the boiler operates as a counterflowing heat exchanger in which the temperature difference at the pinch point is 20 °C, find the maximum water mass flow rate possible and the air exit temperature.

9.112 Consider an ideal, dual-loop heat-powered refrigeration cycle using R-134a as the working fluid, as shown in Fig. P9.112. Saturated vapor at 90 °C leaves the boiler and expands in the turbine to the condenser pressure. Saturated vapor at −15 °C leaves the evaporator and is compressed to the condenser pressure. The ratio of the flows through the two loops is such that the turbine produces just enough power to drive the compressor. The two exiting streams mix together and enter the condenser. Saturated liquid leaving the condenser at 45 °C is then separated into two streams in the necessary proportions. Determine the ratio of the mass flow rate through the power loop to that through the refrigeration loop. Also find the performance of the cycle in terms of the ratio Q_L/Q_H.

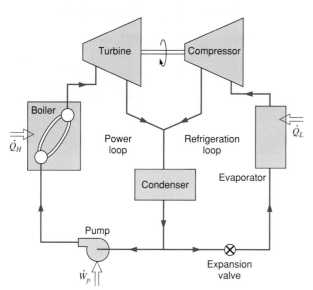

FIGURE P9.112

9.113 For Problem 9.111, determine the change in exergy of the water flow and that of the air flow. Use these values to determine the second-law efficiency of the boiler heat exchanger.

Review Problems

9.114 A simple steam power plant is said to have the following four states: 1: (20 °C, 100 kPa), 2: (25 °C, 1 MPa), 3: (1000 °C, 1 MPa), 4: (250 °C, 100 kPa), with an energy source at 1100 °C, and it rejects energy to a 0 °C ambient. Is this cycle possible? Are any of the devices impossible?

9.115 A supercritical power plant has a high pressure of 30 MPa, and the boiler heats the water to 500 °C with 45 °C in the condenser. To avoid a quality in the turbine of less than 92 %, determine the pressure(s) at which to make reheat (look only at the P's listed in Table B.1; do not interpolate between them) if the reheat takes it up to 400 °C.

9.116 A dairy farmer needs to heat a 0.1 kg/s flow of milk from room temperature, 20 °C, to 60 °C in order to pasteurize it and then send the flow through a cooler, bringing it to 10 °C for storage. He wants to buy a heat pump to do this and finds one using R-134a with a high pressure of 3 MPa and a low pressure of 300 kPa. The farmer is very clever and uses the heat pump to heat the milk in a heat exchanger, after which the milk flows through the evaporator's heat exchanger to cool it. Find the power required to run the heat pump so that it can do both the heating and the cooling, assuming milk has the properties of water. Is there any excess heating or cooling capacity? What is the total rate of exergy destruction attained by running this system?

9.117 Consider an ideal combined reheat and regenerative cycle in which steam enters the high-pressure turbine at 3.0 MPa, 400 °C, and is extracted to an open FWH at 0.8 MPa with exit as saturated liquid. The remainder of the steam is reheated to 400 °C at this pressure, 0.8 MPa, and is fed to the low-pressure turbine. The condenser pressure is 10 kPa. Calculate the thermal efficiency of the cycle and the net work per kilogram of steam.

9.118 An industrial application has the following steam requirement: one 10 kg/s stream at a pressure of 0.5 MPa and one 5 kg/s stream at 1.4 MPa (both saturated or slightly superheated vapor). It is obtained by cogeneration, whereby a high-pressure boiler supplies steam at 10 MPa, 500 °C to a reversible turbine. The required amount is withdrawn at 1.4 MPa, and the remainder is expanded in the low-pressure end of the turbine to 0.5 MPa, providing the second required steam flow.

a. Determine the power output of the turbine and the heat transfer rate in the boiler.

b. Compute the rates needed if the steam is generated in a low-pressure boiler without cogeneration. Assume that for each, 20 °C liquid water is pumped to the required pressure and fed to a boiler.

9.119 A jet ejector, a device with no moving parts, functions as the equivalent of a coupled turbine-compressor unit as shown in Fig. P9.119a. Thus, the turbine-compressor in the dual-loop cycle of Fig. P9.112 could be replaced by a jet ejector. The primary stream of the jet ejector enters from the boiler, the secondary stream enters from the evaporator, and the discharge flows to the condenser. Alternatively, a jet ejector may be used with water as the working fluid. The purpose of the device is to chill water, usually for an air-conditioning system. In this application, the physical setup is as shown in Fig. P9.119b. Using the data given in the diagram, evaluate the performance of this cycle in terms of the ratio Q_L/Q_H. Do this assuming an ideal cycle, and repeat assuming an ejector efficiency of 20 %.

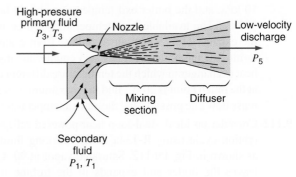

FIGURE P9.119a

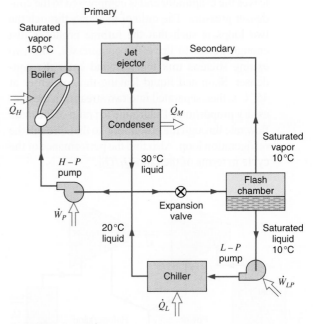

FIGURE 9.119b

COMPUTER, DESIGN, AND OPEN-ENDED PROBLEMS

9.120 Use the software for the properties to consider the moisture separator in Problem 4.95. Steam comes in at state 3 and leaves as liquid, state 9, with the rest, at state 4, going to the low-pressure turbine. Assume no heat transfer and find the total entropy generation and irreversibility in the process.

9.121 The effect of evaporator temperature on the COP of a heat pump is to be studied. Consider an ideal cycle with R-134a as the working fluid and a con-denser temperature of 40 °C. Plot a curve for the COP versus the evaporator temperature for temperatures from +15 to −25 °C.

9.122 A hospital requires 2 kg/s steam at 200 °C, 125 kPa for sterilization purposes, and space heating requires 15 kg/s hot water at 90 °C, 100 kPa. Both of these requirements are provided by the hospital's steam power plant. Discuss some arrangement that will accomplish this.

9.123 Investigate the maximum power out of a steam power plant with operating conditions, as in Problem 9.28. The energy source is 100 kg/s combustion products (air) at 125 kPa, 1200 K. Make sure the air temperature is higher than the water temperature throughout the boiler.

9.124 Use the computer software to solve Problem 9.71 with R-12 as the working substance.

9.125 Use the computer software to solve Problem 9.81 with R-12 as the working substance.

9.126 Use the computer software to solve Problem 9.20 with R-22 as the working substance.

9.127 Use the computer software to solve Problem 9.54 with R-22 as the working substance.

Consider also Problem 8.119.

9.128 Consider the high-pressure closed FWH in the nuclear power plant described in Problem 4.95. Determine its second-law efficiency use CATT3.

10 Power and Refrigeration Systems—Gaseous Working Fluids

In the previous chapter, we studied power and refrigeration systems that utilize condensing working fluids, in particular those involving steady-state flow processes with shaft work. It was noted that condensing working fluids have the maximum difference in the $-\int v\,dP$ work terms between the expansion and compression processes. In this chapter, we continue to study power and refrigeration systems involving steady-state flow processes, but those with gaseous working fluids throughout, recognizing that the difference in expansion and compression work terms is considerably smaller. We then study power cycles for piston/cylinder systems involving boundary-movement work. We conclude the chapter by examining combined cycle system arrangements.

We begin the chapter by introducing the concept of the air-standard cycle, the basic model to be used with gaseous power systems.

10.1 AIR-STANDARD POWER CYCLES

In Section 9.1, we considered idealized four-process cycles, including both steady-state-process and piston/cylinder boundary-movement cycles. The question of phase-change cycles and single-phase cycles was also mentioned. We then examined the Rankine power plant cycle in detail, the idealized model of a phase-change power cycle. However, many work-producing devices (engines) utilize a working fluid that is always a gas. The spark-ignition automotive engine is a familiar example, as are the diesel engine and the conventional gas turbine. In all these engines there is a change in the composition of the working fluid, because during combustion it changes from air and fuel to combustion products. For this reason, these engines are called *internal-combustion engines*. In contrast, the steam power plant may be called an *external-combustion engine*, because heat is transferred from the products of combustion to the working fluid. External-combustion engines using a gaseous working fluid (usually air) have been built. To date they have had only limited application, but use of the gas-turbine cycle in conjunction with a nuclear reactor has been investigated extensively. Other external-combustion engines are currently receiving serious attention in an effort to combat air pollution.

Because the working fluid does not go through a complete thermodynamic cycle in the engine (even though the engine operates in a mechanical cycle), the internal-combustion engine operates on the so-called open cycle. However, for analyzing internal-combustion engines, it is advantageous to devise closed cycles that closely approximate the open cycles. One such approach is the air-standard cycle, which is based on the following assumptions:

1. A fixed mass of air is the working fluid throughout the entire cycle, and the air is always an ideal gas. Thus, there is no inlet process or exhaust process.
2. The combustion process is replaced by a process transferring heat from an external source.
3. The cycle is completed by heat transfer to the surroundings (in contrast to the exhaust and intake process of an actual engine).
4. All processes are internally reversible.
5. An additional assumption is often made that air has a constant specific heat, evaluated at 300 K, called *cold air* properties, recognizing that this is not the most accurate model.

The principal value of the air-standard cycle is to enable us to examine qualitatively the influence of a number of variables on performance. The quantitative results obtained from the air-standard cycle, such as efficiency and mean effective pressure, will differ from those of the actual engine. Our emphasis, therefore, in our consideration of the air-standard cycle will be primarily on the qualitative aspects.

10.2 THE BRAYTON CYCLE

In discussing idealized four-steady-state-process power cycles in Section 9.1, a cycle involving two constant-pressure and two isentropic processes was examined, and the results were shown in Fig. 9.2. This cycle used with a condensing working fluid is the Rankine cycle, but when used with a single-phase, gaseous working fluid it is termed the Brayton cycle. The air-standard Brayton cycle is the ideal cycle for the simple gas turbine. The simple open-cycle gas turbine utilizing an internal-combustion process and the simple closed-cycle gas turbine, which utilizes heat-transfer processes, are both shown schematically in Fig. 10.1. The air-standard Brayton cycle is shown on the *P–v* and *T–s* diagrams of Fig. 10.2.

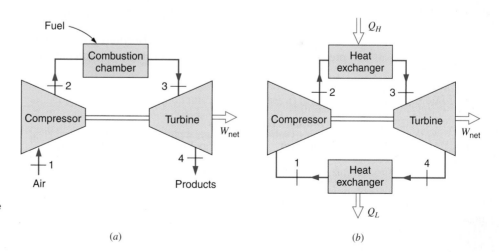

FIGURE 10.1 A gas turbine operating on the Brayton cycle. (*a*) Open cycle. (*b*) Closed cycle.

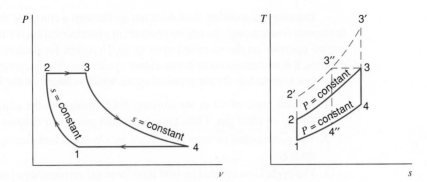

FIGURE 10.2 The air-standard Brayton cycle.

The analysis of the Brayton cycle is done with a control volume around each of the four devices shown in Fig. 10.1b, and the results for the energy and entropy equations are shown in Table 10.1. From the analysis, we can find the overall conversion efficiency for the cycle as

$$\eta_{th} = 1 - \frac{q_L}{q_H} = 1 - \frac{h_4 - h_1}{h_3 - h_2} \approx 1 - \frac{C_p(T_4 - T_1)}{C_p(T_3 - T_2)} = 1 - \frac{T_1(T_4/T_1 - 1)}{T_2(T_3/T_2 - 1)}$$

From the ideal cycle, we know that the pressure increase in the compressor equals the pressure decrease in the turbine, so

$$\frac{P_3}{P_4} = \frac{P_2}{P_1}$$

and from the two isentropic processes we get the power relations as

$$\frac{P_2}{P_1} = \left(\frac{T_2}{T_1}\right)^{k/(k-1)} = \frac{P_3}{P_4} = \left(\frac{T_3}{T_4}\right)^{k/(k-1)}$$

$$\frac{T_3}{T_4} = \frac{T_2}{T_1} \quad \therefore \quad \frac{T_3}{T_2} = \frac{T_4}{T_1} \quad \text{and} \quad \frac{T_3}{T_2} - 1 = \frac{T_4}{T_1} - 1$$

The cycle efficiency thus becomes

$$\eta_{th} = 1 - \frac{T_1}{T_2} = 1 - \frac{1}{(P_2/P_1)^{(k-1)/k}} \tag{10.1}$$

The efficiency of the air-standard Brayton cycle is therefore a function of the isentropic pressure ratio. The fact that efficiency increases with pressure ratio is evident from the T–s

TABLE 10.1
The Brayton Cycle Processes

Component	Energy Eq.	Entropy Eq.	Process
Compressor	$0 = h_1 + w_C - h_2$	$0 = s_1 - s_2 + (0/T) + 0$	$q = 0, s_1 = s_2$
Combustion	$0 = h_2 - h_3 + q_H$	$0 = s_2 - s_3 + \int dq/T + 0$	$P_3 = P_2 = C$
Turbine	$0 = h_3 - h_4 - w_T$	$0 = s_3 - s_4 + (0/T) + 0$	$q = 0, s_3 = s_4$
Heat exchanger	$0 = h_4 - h_1 - q_L$	$0 = s_4 - s_1 - \int dq/T + 0$	$P_4 = P_1 = C$

diagram of Fig. 10.2 because increasing the pressure ratio changes the cycle from 1–2–3–4–1 to 1–2′–3′–4–1. The latter cycle has a greater heat supply and the same heat rejected as the original cycle; therefore, it has greater efficiency. Note that the latter cycle has a higher maximum temperature, $T_{3'}$, than the original cycle, T_3. In the actual gas turbine, the allowable maximum temperature of the gas entering the turbine is determined by material considerations. Therefore, if we fix the temperature T_3 and increase the pressure ratio, the resulting cycle is 1–2′–3″–4″–1. This cycle would have a higher efficiency than the original cycle, but the heat transfer and work per kilogram of working fluid are thereby changed.

With the advent of nuclear reactors, the closed-cycle gas turbine has become more important. Heat is transferred, either directly or via a second fluid, from the fuel in the nuclear reactor to the working fluid in the gas turbine. Heat is rejected from the working fluid to the surroundings.

The actual gas-turbine engine differs from the ideal cycle primarily because of irreversibilities in the compressor and turbine, and because of pressure drop in the flow passages and combustion chamber (or in the heat exchanger of a closed-cycle turbine). Thus, the state points in a simple open-cycle gas turbine might be as shown in Fig. 10.3.

The efficiencies of the compressor and turbine are defined in relation to isentropic processes. With the states designated as in Fig. 10.3, the definitions of compressor and turbine efficiencies are

$$\eta_{\text{comp}} = \frac{h_{2s} - h_1}{h_2 - h_1} \tag{10.2}$$

$$\eta_{\text{turb}} = \frac{h_3 - h_4}{h_3 - h_{4s}} \tag{10.3}$$

Another important feature of the Brayton cycle is the large amount of compressor work (also called back work) compared to turbine work. Thus, the compressor might require 40 % to 80 % of the output of the turbine. This is particularly important when the actual cycle is considered because the effect of the losses is to require a larger amount of compression work from a smaller amount of turbine work. Thus, the overall efficiency drops very rapidly with a decrease in the efficiencies of the compressor and turbine. In fact, if these efficiencies

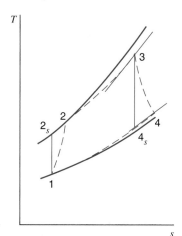

FIGURE 10.3 Effect of inefficiencies on the gas-turbine cycle.

drop below about 60 %, all the work of the turbine will be required to drive the compressor, and the overall efficiency will be zero. This is in sharp contrast to the Rankine cycle, where only 1 % or 2 % of the turbine work is required to drive the pump. This demonstrates the inherent advantage of the cycle utilizing a condensing working fluid, such that a much larger difference in specific volume between the expansion and compression processes is utilized effectively.

Example 10.1

In an air-standard Brayton cycle, the air enters the compressor at 0.1 MPa and 15 °C. The pressure leaving the compressor is 1.0 MPa, and the maximum temperature in the cycle is 1100 °C. Determine

1. The pressure and temperature at each point in the cycle.
2. The compressor work, turbine work, and cycle efficiency.

For each control volume analyzed, the model is ideal gas with constant specific heat, at 300 K, and each process is steady state with no kinetic or potential energy changes. The diagram for this example is Fig. 10.2.

We consider the compressor, the turbine, and the high-temperature and low-temperature heat exchangers in turn.

Control volume: Compressor.
Inlet state: P_1, T_1 known; state fixed.
Exit state: P_2 known.

Analysis

$$\text{Energy Eq.:} \quad w_c = h_2 - h_1$$

(Note that the compressor work w_c is here defined as work input to the compressor.)

$$\text{Entropy Eq.:} \quad s_2 = s_1 \Rightarrow \frac{T_2}{T_1} = \left(\frac{P_2}{P_1}\right)^{(k-1)/k}$$

Solution

Solving for T_2, we get

$$T_2 = T_1 \left(\frac{P_2}{P_1}\right)^{(k-1)/k} = 288.2 \times 10^{0.286} = 556.8 \text{ K}$$

Therefore,

$$w_c = h_2 - h_1 = C_p(T_2 - T_1)$$

$$= 1.004 \frac{\text{kJ}}{\text{kg·K}}(556.8 - 288.2) \text{ K} = 269.5 \text{ kJ/kg}$$

Consider the turbine next.

Control volume: Turbine.
Inlet state: $P_3 \ (= P_2)$ known, T_3 known, state fixed.
Exit state: $P_4 \ (= P_1)$ known.

Analysis

$$\text{Energy Eq.:} \quad w_t = h_3 - h_4$$

$$\text{Entropy Eq.:} \quad s_3 = s_4 \Rightarrow \frac{T_3}{T_4} = \left(\frac{P_3}{P_4}\right)^{(k-1)/k}$$

Solution

Solving for T_4, we get

$$T_4 = T_3(P_4/P_3)^{(k-1)/k} = 1373.2 \times 0.1^{0.286} = 710.8 \text{ K}$$

Therefore,

$$w_t = h_3 - h_4 = C_p(T_3 - T_4)$$
$$= 1.004(1373.2 - 710.8) = 664.7 \text{ kJ/kg}$$
$$w_{\text{net}} = w_t - w_c = 664.7 - 269.5 = 395.2 \text{ kJ/kg}$$

Now we turn to the heat exchangers.

Control volume: High-temperature heat exchanger.
     Inlet state: State 2 fixed (as given).
     Exit state: State 3 fixed (as given).

Analysis

$$\text{Energy Eq.:} \quad q_H = h_3 - h_2 = C_p(T_3 - T_2)$$

Solution

Substitution gives

$$q_H = h_3 - h_2 = C_p(T_3 - T_2) = 1.004(1373.2 - 556.8) = 819.3 \text{ kJ/kg}$$

Control volume: Low-temperature heat exchanger.
     Inlet state: State 4 fixed (above).
     Exit state: State 1 fixed (above).

Analysis

$$\text{Energy Eq.:} \quad q_L = h_4 - h_1 = C_p(T_4 - T_1)$$

Solution

Upon substitution we have

$$q_L = h_4 - h_1 = C_p(T_4 - T_1) = 1.004(710.8 - 288.2) = 424.1 \text{ kJ/kg}$$

Therefore,

$$\eta_{\text{th}} = \frac{w_{\text{net}}}{q_H} = \frac{395.2}{819.3} = 48.2\%$$

This may be checked by using Eq. 10.1.

$$\eta_{th} = 1 - \frac{1}{(P_2/P_1)^{(k-1)/k}} = 1 - \frac{1}{10^{0.286}} = 48.2\%$$

Example 10.2

Consider a gas turbine with air entering the compressor under the same conditions as in Example 10.1 and leaving at a pressure of 1.0 MPa. The maximum temperature is $1100\,^\circ C$. Assume a compressor efficiency of 80 %, a turbine efficiency of 85 %, and a pressure drop between the compressor and turbine of 15 kPa. Determine the compressor work, turbine work, and cycle efficiency.

As in the previous example, for each control volume the model is ideal gas with constant specific heat, at 300 K, and each process is steady state with no kinetic or potential energy changes. In this example, the diagram is Fig. 10.3.

We consider the compressor, the turbine, and the high-temperature heat exchanger in turn.

Control volume: Compressor.

Inlet state: P_1, T_1 known; state fixed.

Exit state: P_2 known.

Analysis

Energy Eq. real process: $w_c = h_2 - h_1$

Entropy Eq. ideal process: $s_{2_s} = s_1 \Rightarrow \dfrac{T_{2_s}}{T_1} = \left(\dfrac{P_2}{P_1}\right)^{(k-1)/k}$

In addition,

$$\eta_c = \frac{h_{2_s} - h_1}{h_2 - h_1} = \frac{T_{2_s} - T_1}{T_2 - T_1}$$

Solution

Solving for T_{2_s}, we get

$$\left(\frac{P_2}{P_1}\right)^{(k-1)/k} = \frac{T_{2_s}}{T_1} = 10^{0.286} = 1.932, \qquad T_{2_s} = 556.8 \text{ K}$$

The efficiency is

$$\eta_c = \frac{h_{2_s} - h_1}{h_2 - h_1} = \frac{T_{2_s} - T_1}{T_2 - T_1} = \frac{556.8 - 288.2}{T_2 - T_1} = 0.80$$

Therefore,

$$T_2 - T_1 = \frac{556.8 - 288.2}{0.80} = 335.8, \qquad T_2 = 624.0 \text{ K}$$

$$w_c = h_2 - h_1 = C_p(T_2 - T_1)$$

$$= 1.004(624.0 - 288.2) = 337.0 \text{ kJ/kg}$$

Control volume: Turbine.
Inlet state: P_3 (P_2 – drop) known, T_3 known; state fixed.
Exit state: P_4 known.

Analysis

Energy Eq. real process: $w_c = h_3 - h_4$

Entropy Eq. ideal process: $s_{4_s} = s_3 \Rightarrow \dfrac{T_3}{T_{4_s}} = \left(\dfrac{P_3}{P_4}\right)^{(k-1)/k}$

In addition,

$$\eta_t = \frac{h_3 - h_4}{h_3 - h_{4_s}} = \frac{T_3 - T_4}{T_3 - T_{4_s}}$$

Solution

Substituting numerical values, we obtain

$$P_3 = P_2 - \text{pressure drop} = 1.0 - 0.015 = 0.985 \text{ MPa}$$

$$\left(\frac{P_3}{P_4}\right)^{(k-1)/k} = \frac{T_3}{T_{4_s}} = 9.85^{0.286} = 1.9236, \qquad T_{4_s} = 713.9 \text{ K}$$

$$\eta_t = \frac{h_3 - h_4}{h_3 - h_{4_s}} = \frac{T_3 - T_4}{T_3 - T_{4_s}} = 0.85$$

$$T_3 - T_4 = 0.85(1373.2 - 713.9) = 560.4 \text{ K}$$

$$T_4 = 812.8 \text{ K}$$

$$w_t = h_3 - h_4 = C_p(T_3 - T_4)$$

$$= 1.004(1373.2 - 812.8) = 562.4 \text{ kJ/kg}$$

$$w_{\text{net}} = w_t - w_c = 562.4 - 337.0 = 225.4 \text{ kJ/kg}$$

Finally, for the heat exchanger:

Control volume: High-temperature heat exchanger.
Inlet state: State 2 fixed (as given).
Exit state: State 3 fixed (as given).

Analysis

Energy Eq.: $q_H = h_3 - h_2$

Solution

Substituting, we have

$$q_H = h_3 - h_2 = C_p(T_3 - T_2)$$

$$= 1.004(1373.2 - 624.0) = 751.8 \text{ kJ/kg}$$

so that

$$\eta_{\text{th}} = \frac{w_{\text{net}}}{q_H} = \frac{225.4}{751.8} = 30.0\%$$

The following comparisons can be made between Examples 10.1 and 10.2.

	w_c	w_t	w_{net}	q_H	η_{th}
Example 10.1 (Ideal)	269.5	664.7	395.2	819.3	48.2
Example 10.2 (Actual)	337.0	562.4	225.4	751.8	30.0

As stated previously, the irreversibilities decrease the turbine work and increase the compressor work. Since the net work is the difference between these two, it decreases very rapidly as compressor and turbine efficiencies decrease. The development of highly efficient compressors and turbines is therefore an important aspect of the development of gas turbines.

Note that in the ideal cycle (Example 10.1), about 41 % of the turbine work is required to drive the compressor and 59 % is delivered as net work. In the actual turbine (Example 10.2), 60 % of the turbine work is required to drive the compressor and 40 % is delivered as net work. Thus, if the net power of this unit is to be 10 000 kW, a 25 000 kW turbine and a 15 000 kW compressor are required. This result demonstrates that a gas turbine has a high back-work ratio.

10.3 THE SIMPLE GAS-TURBINE CYCLE WITH A REGENERATOR

The efficiency of the gas-turbine cycle may be improved by introducing a regenerator. The simple open-cycle gas-turbine cycle with a regenerator is shown in Fig. 10.4, and the corresponding ideal air-standard cycle with a regenerator is shown on the P–v and T–s diagrams. In cycle 1–2–x–3–4–y–1, the temperature of the exhaust gas leaving the turbine in state 4 is higher than the temperature of the gas leaving the compressor. Therefore, heat can be transferred from the exhaust gases to the high-pressure gases leaving the compressor. If this is done in a counterflow heat exchanger (a regenerator), the temperature of the high-pressure gas leaving the regenerator, T_x, may, in the ideal case, have a temperature equal to T_4, the temperature of the gas leaving the turbine. Heat transfer from the external source is necessary only to increase the temperature from T_x to T_3. Area x–3–d–b–x represents the heat transferred, and area y–1–a–c–y represents the heat rejected.

The influence of pressure ratio on the simple gas-turbine cycle with a regenerator is shown by considering cycle 1–2′–3′–4–1. In this cycle the temperature of the exhaust gas

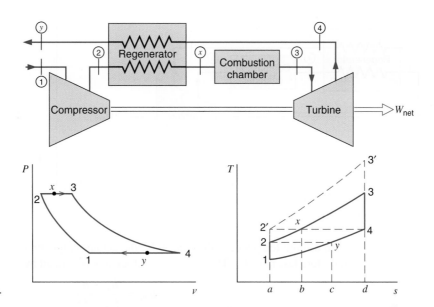

FIGURE 10.4 The ideal regenerative cycle.

leaving the turbine is just equal to the temperature of the gas leaving the compressor; therefore, utilizing a regenerator is not possible. This can be shown more exactly by determining the efficiency of the ideal gas-turbine cycle with a regenerator.

The efficiency of this cycle with regeneration is found as follows, where the states are as given in Fig. 10.4.

$$\eta_{th} = \frac{w_{net}}{q_H} = \frac{w_t - w_c}{q_H}$$

$$q_H \cong C_p(T_3 - T_x)$$

$$w_t \cong C_p(T_3 - T_4)$$

But for an ideal regenerator, $T_4 = T_x$, and therefore $q_H = w_t$. Consequently,

$$\eta_{th} = 1 - \frac{w_c}{w_t} \cong 1 - \frac{C_p(T_2 - T_1)}{C_p(T_3 - T_4)}$$

$$= 1 - \frac{T_1(T_2/T_1 - 1)}{T_3(1 - T_4/T_3)} = 1 - \frac{T_1[(P_2/P_1)^{(k-1)/k} - 1]}{T_3[1 - (P_1/P_2)^{(k-1)/k}]}$$

$$\eta_{th} = 1 - \frac{T_1}{T_3}\left(\frac{P_2}{P_1}\right)^{(k-1)/k} = 1 - \frac{T_2}{T_3}$$

Thus, for the ideal cycle with regeneration, the thermal efficiency depends not only on the pressure ratio but also on the ratio of the minimum to the maximum temperature. We note that, in contrast to the Brayton cycle, the efficiency decreases with an increase in pressure ratio.

The effectiveness or efficiency of a regenerator is given by the regenerator efficiency, which can best be defined by reference to Fig. 10.5. State x represents the high-pressure gas leaving the regenerator. In the ideal regenerator, there would be only an infinitesimal temperature difference between the two streams, and the high-pressure gas would leave the regenerator at temperature T'_x, and $T'_x = T_4$. In an actual regenerator, which must operate

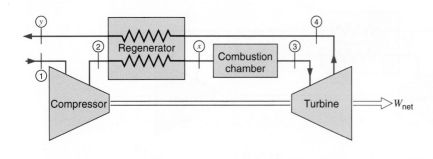

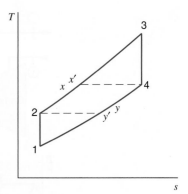

FIGURE 10.5 *T–s* diagram illustrating the definition of regenerator efficiency.

with a finite temperature difference T_x, the actual temperature leaving the regenerator is therefore less than T'_x. The regenerator efficiency is defined by

$$\eta_{\text{reg}} = \frac{h_x - h_2}{h'_x - h_2} \tag{10.4}$$

If the specific heat is assumed to be constant, the regenerator efficiency is also given by the relation

$$\eta_{\text{reg}} = \frac{T_x - T_2}{T'_x - T_2}$$

A higher efficiency can be achieved by using a regenerator with a greater heat-transfer area. However, this also increases the pressure drop, which represents a loss, and both the pressure drop and the regenerator efficiency must be considered in determining which regenerator gives maximum thermal efficiency for the cycle. From an economic point of view, the cost of the regenerator must be weighed against the savings that can be effected by its use.

Example 10.3

If an ideal regenerator is incorporated into the cycle of Example 10.1, determine the thermal efficiency of the cycle.

The diagram for this example is Fig. 10.5. Values are from Example 10.1. Therefore, for the analysis of the high-temperature heat exchanger (combustion chamber), from the energy equation, we have

$$q_H = h_3 - h_x$$

so that the solution is

$$T_x = T_4 = 710.8 \text{ K}$$

$$q_H = h_3 - h_x = C_p(T_3 - T_x) = 1.004(1373.2 - 710.8) = 664.7 \text{ kJ/kg}$$

$$w_{\text{net}} = 395.2 \text{ kJ/kg (from Example 10.1)}$$

$$\eta_{\text{th}} = \frac{395.2}{664.7} = 59.5 \%$$

10.4 GAS-TURBINE POWER CYCLE CONFIGURATIONS

The Brayton cycle, being the idealized model for the gas-turbine power plant, has a reversible, adiabatic compressor and a reversible, adiabatic turbine. In the following example, we consider the effect of replacing these components with reversible, isothermal processes.

Example 10.4

An air-standard power cycle has the same states given in Example 10.1. In this cycle, however, the compressor and turbine are both reversible, isothermal processes. Calculate the compressor work and the turbine work, and compare the results with those of Example 10.1.

    Control volumes: Compressor, turbine.

Analysis

For each reversible, isothermal process, from Eq. 7.19:

$$w = -\int_{i}^{e} v \, dP = -P_i v_i \ln \frac{P_e}{P_i} = -RT_i \ln \frac{P_e}{P_i}$$

Solution

For the compressor,

$$w = -0.287 \text{ kJ/kg·K} \times 288.2 \text{ K} \times \ln 10 = -190.5 \text{ kJ/kg}$$

compared with -269.5 kJ/kg in the adiabatic compressor.
    For the turbine,

$$w = -0.287 \times 1373.2 \times \ln 0.1 = +907.5 \text{ kJ/kg}$$

compared with $+664.7$ kJ/kg in the adiabatic turbine.

It is found that the isothermal process would be preferable to the adiabatic process in both the compressor and turbine. The resulting cycle, called the Ericsson cycle, consists of two reversible, constant-pressure processes and two reversible, constant-temperature processes. The reason the actual gas turbine does not attempt to emulate this cycle rather than the Brayton cycle is that the compressor and turbine processes are both high-flow-rate processes involving work-related devices in which it is not practical to attempt to transfer large quantities of heat. As a consequence, the processes tend to be essentially adiabatic, so that this becomes the process in the model cycle.

    There is a modification of the Brayton/gas turbine cycle that tends to change its performance in the direction of the Ericsson cycle. This modification is to use multiple stages of compression with intercooling and multiple stages of expansion with reheat.

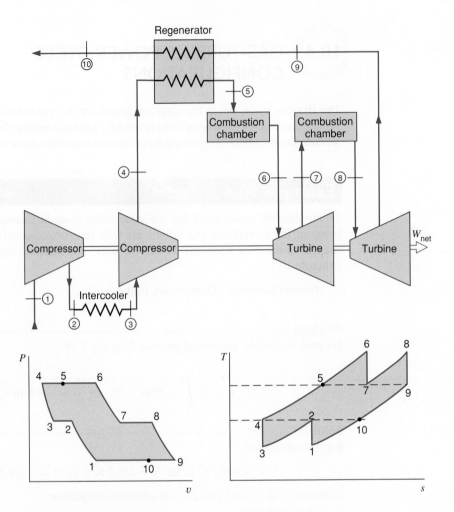

FIGURE 10.6 The ideal gas-turbine cycle utilizing intercooling, reheat, and a regenerator.

Such a cycle with two stages of compression and expansion, and also incorporating a regenerator, is shown in Fig. 10.6. The air-standard cycle is given on the corresponding T–s diagram. It may be shown that for this cycle, the maximum efficiency is obtained if equal pressure ratios are maintained across the two compressors and the two turbines. In this ideal cycle, it is assumed that the temperature of the air leaving the intercooler, T_3, is equal to the temperature of the air entering the first stage of compression, T_1, and that the temperature after reheating, T_8, is equal to the temperature entering the first turbine, T_6. Furthermore, in the ideal cycle it is assumed that the temperature of the high-pressure air leaving the regenerator, T_5, is equal to the temperature of the low-pressure air leaving the turbine, T_9.

From the discussion of the expression for a steady-state shaft work Eq. 7.14

$$w = -\int v\, dP + \Delta ke + \Delta pe - \text{loss}$$

we recognized that the work will be less if the specific volume is smaller for a given change in pressure. This fact is used in the application of *intercoolers* used in many compression processes where we need a high-pressure gas but not necessarily at a high temperature.

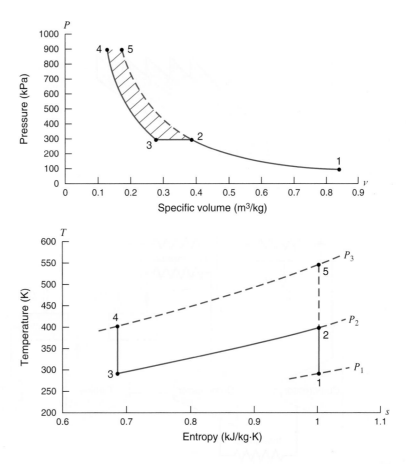

FIGURE 10.7
A compressor with an intercooler.

By cooling a gas at constant pressure, as in a heat exchanger, the specific volume is reduced and any subsequent compression can then be done with a lower work input. Consider the reversible compression process between an initial state 1 and a final state 4 shown in Fig. 10.6, which requires an amount of work equal to the area under the curve, as seen from the P axis in the P–v diagram. The flow is taken out at an intermediate pressure at state 2 and is cooled to the original inlet temperature before the compression to the final pressure. The whole process is illustrated in Fig. 10.7 in both the P–v and T–s diagrams. If the process is done without the intercooler, it follows the path 1–2–5, which requires a larger amount of work since the specific volume is larger for the last part of the process. The work input difference corresponds to the area enclosed by the curves 2–3–4–5–2, shown as crosshatched in Fig. 10.7.

If a large number of compression and expansion stages are used, it is evident that the Ericsson cycle is approached. This is shown in Fig. 10.8. In practice, the economical limit to the number of stages is usually two or three. The turbine and compressor losses and pressure drops that have already been discussed would be involved in any actual unit employing this cycle.

The turbines and compressors using this cycle can be utilized in a variety of ways. Two possible arrangements for closed cycles are shown in Fig. 10.9. One advantage frequently sought in a given arrangement is ease of control of the unit under various loads. Detailed discussion of this point, however, is beyond the scope of this book.

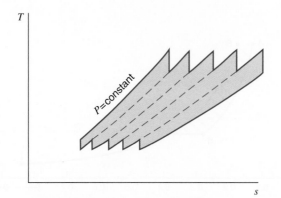

FIGURE 10.8
T–s diagram that shows how the gas-turbine cycle with many stages approaches the Ericsson cycle.

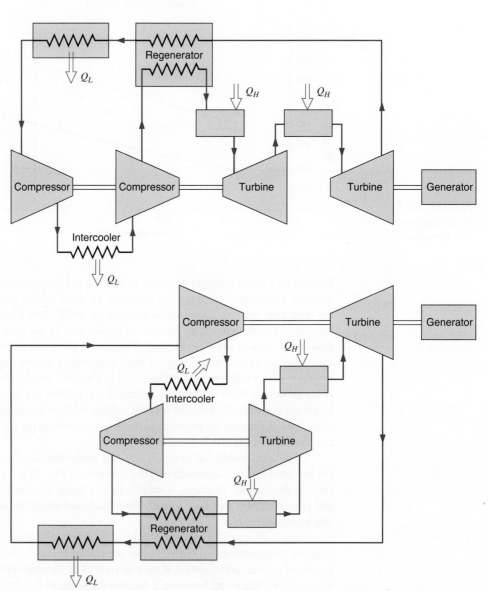

FIGURE 10.9
Some arrangements of components that may be utilized in stationary gas-turbine power plants.

10.5 THE AIR-STANDARD CYCLE FOR JET PROPULSION

The next air-standard power cycle we consider is utilized in jet propulsion. In this cycle, the work done by the turbine is just sufficient to drive the compressor. The gases are expanded in the turbine to a pressure for which the turbine work is just equal to the compressor work. The exhaust pressure of the turbine will then be greater than that of the surroundings, and the gas can be expanded in a nozzle to the pressure of the surroundings. Since the gases leave at a high velocity, the change in momentum that the gases undergo gives a thrust to the aircraft in which the engine is installed. A jet engine is shown in Fig. 10.10, and the air-standard cycle for this situation is shown in Fig. 10.11. The principles governing this cycle follow from the analysis of the Brayton cycle plus that for a reversible, adiabatic nozzle.

Example 10.5

Consider an ideal jet propulsion cycle in which air enters the compressor at 0.1 MPa and 15 °C. The pressure leaving the compressor is 1.0 MPa, and the maximum temperature is 1100 °C. The air expands in the turbine to a pressure at which the turbine work is just equal to the compressor work. On leaving the turbine, the air expands in a nozzle to 0.1 MPa. The process is reversible and adiabatic. Determine the velocity of the air leaving the nozzle.

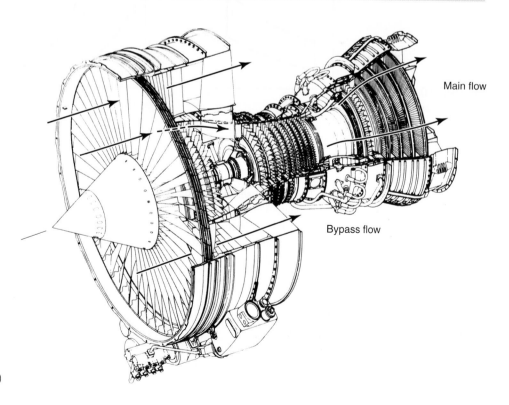

FIGURE 10.10

A turbofan jet engine. (Adapted from General Electric Aircraft Engines.)

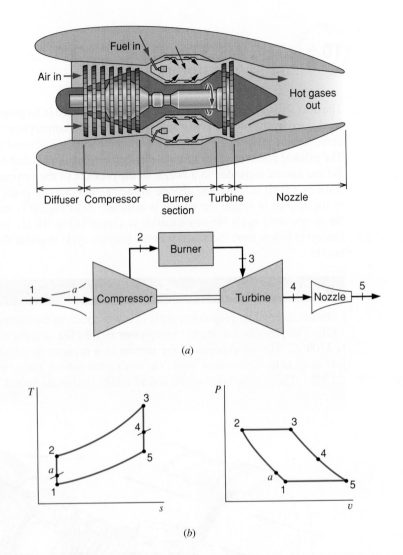

FIGURE 10.11
The ideal gas-turbine
cycle for a jet engine.

(b)

The model used is ideal gas with constant specific heat, at 300 K, and each process is steady state with no potential energy change. The only kinetic energy change occurs in the nozzle. The diagram is shown in Fig. 10.11.

The compressor analysis is the same as in Example 10.1. From the results of that solution, we have

$$P_1 = 0.1 \text{ MPa}, \qquad T_1 = 288.2 \text{ K}$$

$$P_2 = 1.0 \text{ MPa}, \qquad T_2 = 556.8 \text{ K}$$

$$w_c = 269.5 \text{ kJ/kg}$$

The turbine analysis is also the same as in Example 10.1. Here, however,

$$P_3 = 1.0 \, \text{MPa}, \qquad T_3 = 1373.2 \, \text{K}$$

$$w_c = w_t = C_p(T_3 - T_4) = 269.5 \, \text{kJ/kg}$$

$$T_3 - T_4 = \frac{269.5}{1.004} = 268.6 \, \text{K}, \qquad T_4 = 1104.6 \, \text{K}$$

so that

$$P_4 = P_3(T_4/T_3)^{k/(k-1)}$$

$$= 1.0 \, \text{MPa} \times (1104.6/1373.2)^{3.5} = 0.4668 \, \text{MPa}$$

Control volume:	Nozzle.
Inlet state:	State 4 fixed (above).
Exit state:	P_5 known.

Analysis

$$\text{Energy Eq.:} \quad h_4 = h_5 + \frac{\mathbf{V}_5^2}{2}$$

$$\text{Entropy Eq.:} \quad s_4 = s_5 \Rightarrow T_5 = T_4(P_5/P_4)^{(k-1)/k}$$

Solution

Since P_5 is 0.1 MPa, from the second law, we find that $T_5 = 710.8$ K. Then

$$\mathbf{V}_5^2 = 2C_{p0}(T_4 - T_5)$$

$$\mathbf{V}_5^2 = 2 \times 1000 \, \frac{\text{J}}{\text{kJ}} \times 1.004 \, \frac{\text{kJ}}{\text{kg}} \times (1104.6 - 710.8) \, \text{K}$$

$$\mathbf{V}_5 = 889 \, \text{m/s}$$

In-Text Concept Questions

a. The Brayton cycle has the same four processes as the Rankine cycle, but the T–s and P–v diagrams look very different; why is that?

b. Is it always possible to add a regenerator to the Brayton cycle? What happens when the pressure ratio is increased?

c. Why would you use an intercooler between compressor stages?

d. The jet engine does not produce shaft work; how is power produced?

10.6 THE AIR-STANDARD REFRIGERATION CYCLE

If we consider the original ideal four-process refrigeration cycle of Fig. 10.12 with a non-condensing (gaseous) working fluid, then the work output during the isentropic expansion process is not negligibly small, as was the case with a condensing working fluid. Therefore, we retain the turbine in the four-steady-state-process ideal air-standard refrigeration cycle shown in Fig. 10.12. This cycle is seen to be the reverse Brayton cycle, and it is used in practice in the liquefaction of air (see Fig. 9.26 for the Linde-Hampson system) and other gases and also in certain special situations that require refrigeration, such as aircraft cooling systems. After compression from state 1 to 2, the air is cooled as heat is transferred to the surroundings at temperature T_0. The air is then expanded in process 3–4 to the pressure entering the compressor, and the temperature drops to T_4 in the expander. Heat is then transferred to the air until temperature T_L is reached. The work for this cycle is represented by area 1–2–3–4–1, and the refrigeration effect is represented by area 4–1–b–a–4. The coefficient of performance (COP) is the ratio of these two areas.

The COP of the air-standard refrigeration cycle involves the net work between the compressor and expander work terms, and it becomes

$$\beta = \frac{q_L}{w_{\text{net}}} = \frac{q_L}{w_C - w_E} = \frac{h_1 - h_4}{h_2 - h_1 - (h_3 - h_4)} \approx \frac{C_p(T_1 - T_4)}{C_p(T_2 - T_1) - C_p(T_3 - T_4)}$$

Using a constant specific heat to evaluate the differences in enthalpies and writing the power relations for the two isentropic processes, we get

$$\frac{P_2}{P_1} = \left(\frac{T_2}{T_1}\right)^{k/(k-1)} = \frac{P_3}{P_4} = \left(\frac{T_3}{T_4}\right)^{k/(k-1)}$$

and

$$\beta = \frac{T_1 - T_4}{T_2 - T_1 - T_3 + T_4} = \frac{1}{\dfrac{T_2}{T_1}\dfrac{1 - T_3/T_2}{1 - T_4/T_1} - 1} = \frac{1}{\dfrac{T_2}{T_1} - 1}$$

$$= \frac{1}{r_p^{(k-1)/k} - 1} \tag{10.5}$$

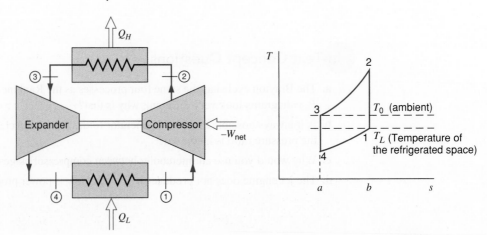

FIGURE 10.12
The air-standard refrigeration cycle.

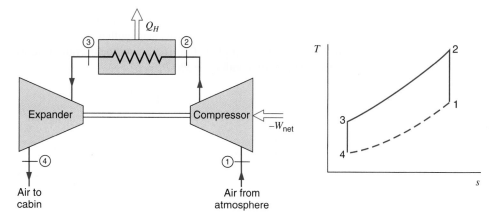

FIGURE 10.13
An air-refrigeration cycle that might be utilized for aircraft cooling.

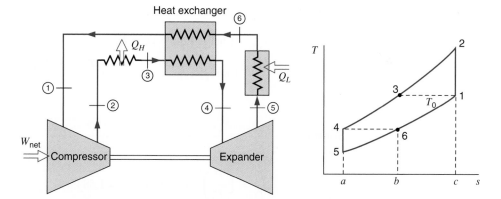

FIGURE 10.14
The air-refrigeration cycle utilizing a heat exchanger.

Here we used $T_3/T_2 = T_4/T_1$ with the pressure ratio $r_p = P_2/P_1$, and we have a result similar to that of the other cycles. The refrigeration cycle is a Brayton cycle with the flow in the reverse direction giving the same relations between the properties.

In practice, this cycle has been used to cool aircraft in an open cycle; a simplified form is shown in Fig. 10.13. Upon leaving the expander, the cool air is blown directly into the cabin, thus providing the cooling effect where needed.

When counterflow heat exchangers are incorporated, very low temperatures can be obtained. This is essentially the cycle used in low-pressure air liquefaction plants and in other liquefaction devices such as the Collins helium liquefier. The ideal cycle is as shown in Fig. 10.14. Because the expander operates at very low temperature, the designer is faced with unique problems in providing lubrication and choosing materials.

Example 10.6

Consider the simple air-standard refrigeration cycle of Fig. 10.12. Air enters the compressor at 0.1 MPa and $-20\,°C$ and leaves at 0.5 MPa. Air enters the expander at $15\,°C$. Determine

1. The COP for this cycle.
2. The rate at which air must enter the compressor to provide 1 kW of refrigeration.

For each control volume in this example, the model is ideal gas with constant specific heat, at 300 K, and each process is steady state with no kinetic or potential energy changes. The diagram for this example is Fig. 10.12, and the overall cycle was considered, resulting in a COP in Eq. 10.5 with $r_p = P_2/P_1 = 5$.

$$\beta = \left[r_p^{(k-1)/k} - 1\right]^{-1}$$

$$= [5^{0.286} - 1]^{-1} = 1.711$$

Control volume: Expander.
Inlet state: $P_3 (= P_2)$ known, T_3, known; state fixed.
Exit state: $P_4 (= P_1)$ known.

Analysis

Energy Eq.: $w_t = h_3 - h_4$

Entropy Eq.: $s_3 = s_4 \Rightarrow \dfrac{T_3}{T_4} = \left(\dfrac{P_3}{P_4}\right)^{(k-1)/k}$

Solution
Therefore,

$$\frac{T_3}{T_4} = \left(\frac{P_3}{P_4}\right)^{(k-1)/k} = 5^{0.286} = 1.5845, \qquad T_4 = 181.9 \text{ K}$$

Control volume: Low-temperature heat exchanger.
Inlet state: State 4 known (as given).
Exit state: State 1 known (as given).

Analysis

Energy Eq.: $q_L = h_1 - h_4$

Solution
Substituting, we obtain

$$q_L = h_1 - h_4 = C_p(T_1 - T_4) = 1.004 \times (253.2 - 181.9) = 71.6 \text{ kJ/kg}$$

To provide 1 kW of refrigeration capacity, we have

$$\dot{m} = \frac{\dot{Q}_L}{q_L} = \frac{1}{71.6} \frac{\text{kW}}{\text{kJ/kg}} = 0.014 \text{ kg/s}$$

10.7 RECIPROCATING ENGINE POWER CYCLES

In Section 9.1, we discussed power cycles incorporating either steady-state processes or piston/cylinder boundary work processes. In that section, it was noted that for the steady-state process, there is no work in a constant-pressure process. Each of the steady-state power cycles presented in subsequent sections of that chapter and to this point in the present chapter incorporated two constant-pressure heat transfer processes. It should now be noted that in a boundary-work process, $\int P\, dv$, there is no work in a constant-volume process. In the next four sections, we will present ideal air-standard power cycles for piston/cylinder boundary-work processes, each example of which includes either one or two constant-volume heat transfer processes.

Before we describe the reciprocating engine cycles, we want to present a few common definitions and terms. Car engines typically have four, six, or eight cylinders, each with a diameter called *bore B*. The piston is connected to a crankshaft, as shown in Fig. 10.15, and as it rotates, changing the crank angle, θ, the piston moves up or down with a stroke.

$$S = 2R_{\text{crank}} \tag{10.6}$$

This gives a displacement for all cylinders as

$$V_{\text{displ}} = N_{\text{cyl}}(V_{\text{max}} - V_{\text{min}}) = N_{\text{cyl}}A_{\text{cyl}}S \tag{10.7}$$

which is the main characterization of the engine size. The ratio of the largest to the smallest volume is the compression ratio

$$r_v = CR = V_{\text{max}}/V_{\text{min}} \tag{10.8}$$

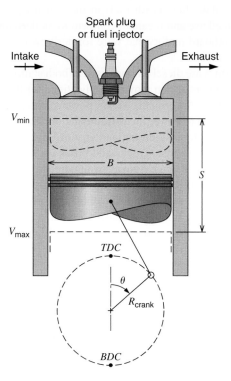

FIGURE 10.15
The piston/cylinder configuration for an internal-combustion engine.

and both of these characteristics are fixed with the engine geometry. The net specific work in a complete cycle is used to define a mean effective pressure

$$w_{net} = \oint P \, dv \equiv P_{meff}(v_{max} - v_{min}) \tag{10.9}$$

or net work per cylinder per cycle

$$W_{net} = m w_{net} = P_{meff}(V_{max} - V_{min}) \tag{10.10}$$

We now use this to find the rate of work (power) for the whole engine as

$$\dot{W} = N_{cyl} m w_{net} \frac{RPM}{60} = P_{meff} V_{displ} \frac{RPM}{60} \tag{10.11}$$

where RPM is revolutions per minute. This result should be corrected with a factor $\frac{1}{2}$ for a four-stroke engine, where two revolutions are needed for a complete cycle to also accomplish the intake and exhaust strokes.

Most engines are four-stroke engines where the following processes occur; the piston motion and crank position refer to Fig. 10.15.

Notice how the intake and the exhaust process each takes one whole stroke of the piston, so two revolutions with four strokes are needed for the complete cycle. In a two-stroke engine, the exhaust flow starts before the expansion is completed and the intake flow overlaps in time with part of the exhaust flow and continues into the compression stroke. This reduces the effective compression and expansion processes, but there is power output in every revolution and the total power is nearly twice the power of the same-size four-stroke engine. Two-stroke engines are used as large diesel engines in ships and as small gasoline engines for lawnmowers and handheld power tools like weed cutters. Because of potential cross-flow from the intake flow (with fuel) to the exhaust port, the two-stroke gasoline engine has seen reduced use and it cannot conform to modern low-emission requirements. For instance, most outboard motors that were formerly two-stroke engines are now made as four-stroke engines.

The largest engines are diesel engines used in both stationary applications as primary or backup power generators and in moving applications for the transportation industry, as in locomotives and ships. An ordinary steam power plant cannot start by itself and thus could have a diesel engine to power its instrumentation and control systems, and so on, to make a cold start. A remote location on land or a drilling platform at sea also would use a diesel engine as a power source. Trucks and buses use diesel engines due to their high efficiency and durability; they range from a few hundred to perhaps 500 hp. Ships use diesel engines running at 100–180 RPM, so they do not need a gearbox to the propeller (these engines can even reverse and run backward without a gearbox). The world's biggest engine is a two-stroke diesel engine with 25 m^3 displacement volume and 14 cylinders, giving a maximum of 105 000 hp, used in a modern container ship.

10.8 THE OTTO CYCLE

The air-standard Otto cycle is an ideal cycle that approximates a spark-ignition internal-combustion engine. This cycle is shown on the P–v and T–s diagrams of Fig. 10.16 and the processes are listed in Table 10.2. Process 1–2 is an isentropic compression of the air

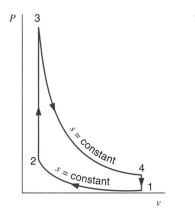 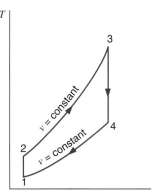

FIGURE 10.16 The air-standard Otto cycle.

as the piston moves from bottom dead center (BDC) to top dead center (TDC). Heat is then added at constant volume while the piston is momentarily at rest at TDC. This process corresponds to the ignition of the fuel–air mixture by the spark and the subsequent burning in the actual engine. Process 3–4 is an isentropic expansion, and process 4–1 is the rejection of heat from the air while the piston is at BDC.

Process, Piston Motion	Crank Position, Crank Angle	Property Variation
Intake, 1 S	TDC to BDC, 0–180 deg.	$P \approx C$, $V \nearrow$, flow in
Compression, 1 S	BDC to TDC, 180–360 deg.	$V \searrow$, $P \nearrow$, $T \nearrow$, $Q = 0$
Ignition and combustion	fast ∼ TDC, 360 deg.	$V = C$, Q in, $P \nearrow$, $T \nearrow$
Expansion, 1 S	TDC to BDC, 360–540 deg.	$V \nearrow$, $P \searrow$, $T \searrow$, $Q = 0$
Exhaust, 1 S	BDC to TDC, 540–720 deg.	$P \approx C$, $V \searrow$, flow out

The thermal efficiency of this cycle is found as follows, assuming constant specific heat of air:

$$\eta_{\text{th}} = \frac{q_H - q_L}{q_H} = 1 - \frac{q_L}{q_H} \approx 1 - \frac{C_v(T_4 - T_1)}{C_v(T_3 - T_2)}$$

$$= 1 - \frac{T_1(T_4/T_1 - 1)}{T_2(T_3/T_2 - 1)}$$

We note further that

$$\frac{T_2}{T_1} = \left(\frac{V_1}{V_2}\right)^{k-1} = \left(\frac{V_4}{V_3}\right)^{k-1} = \frac{T_3}{T_4}$$

TABLE 10.2
The Otto Cycle Processes

Process	Energy Eq.	Entropy Eq.	Process Eq.
Compression	$u_2 - u_1 = -_1 w_2$	$s_2 - s_1 = (0/T) + 0$	$q = 0, s_1 = s_2$
Combustion	$u_3 - u_2 = q_H$	$s_3 - s_2 = \int dq_H/T + 0$	$v_3 = v_2 = C$
Expansion	$u_4 - u_3 = -_3 w_4$	$s_4 - s_3 = (0/T) + 0$	$q = 0, s_3 = s_4$
Heat rejection	$u_1 - u_4 = -q_L$	$s_1 - s_4 = -\int dq_L/T + 0$	$v_4 = v_1 = C$

Therefore,

$$\frac{T_3}{T_2} = \frac{T_4}{T_1}$$

and

$$\eta_{th} = 1 - \frac{T_1}{T_2} = 1 - (r_v)^{1-k} = 1 - \frac{1}{r_v^{k-1}} \qquad (10.12)$$

where

$$r_v = \text{compression ratio} = \frac{V_1}{V_2} = \frac{V_4}{V_3}$$

It is important to note that the efficiency of the air-standard Otto cycle is a function only of the compression ratio and that the efficiency is increased by increasing the compression ratio. Figure 10.17 shows a plot of the air-standard cycle thermal efficiency versus compression ratio. It is also true that the efficiency of an actual spark-ignition engine can be increased by increasing the compression ratio. The trend toward higher compression ratios is prompted by the effort to obtain higher thermal efficiency. In the actual engine, there is an increased tendency for the fuel to detonate as the compression ratio is increased. After detonation the fuel burns rapidly, and strong pressure waves present in the engine cylinder give rise to the so-called spark knock. Therefore, the maximum compression ratio that can be used is fixed by the fact that detonation must be avoided. Advances over the years in compression ratios in actual engines were originally made possible by developing fuels with better antiknock characteristics, primarily through the addition of tetraethyl lead. More recently, however, nonleaded gasolines with good antiknock characteristics have been developed in an effort to reduce atmospheric contamination.

Some of the most important ways in which the actual open-cycle spark-ignition engine deviates from the air-standard cycle are as follows:

1. The specific heats of the actual gases increase with an increase in temperature.

2. The combustion process replaces the heat-transfer process at high temperature, and combustion is not instantaneous or fully complete.

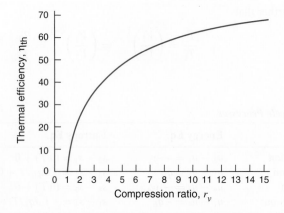

FIGURE 10.17
Thermal efficiency of the Otto cycle as a function of compression ratio.

3. Each mechanical cycle of the engine involves an inlet and an exhaust process and, because of the pressure drop through the valves, a certain amount of work is required to charge the cylinder with air and exhaust the products of combustion.

4. There is considerable heat transfer between the gases in the cylinder and the cylinder walls.

5. There are irreversibilities associated with pressure and temperature gradients.

Example 10.7

The compression ratio in an air-standard Otto cycle is 10. At the beginning of the compression stoke, the pressure is 0.1 MPa and the temperature is 15 °C. The heat transfer to the air per cycle is 1800 kJ/kg air. Determine

1. The pressure and temperature at the end of each process of the cycle.
2. The thermal efficiency.
3. The mean effective pressure.

$$\begin{aligned}
\textit{Control mass}: &\quad \text{Air inside cylinder.} \\
\textit{Diagram}: &\quad \text{Fig. 10.16.} \\
\textit{State information}: &\quad P_1 = 0.1 \text{ MPa}, \qquad T_1 = 288.2 \text{ K.} \\
\textit{Process information}: &\quad \text{Four processes known (Table 10.2). Also, } r_v = 10 \text{ and} \\
&\quad q_H = 1800 \text{ kJ/kg.} \\
\textit{Model}: &\quad \text{Ideal gas, constant specific heat, value at 300 K.}
\end{aligned}$$

Analysis

The second law for compression process 1–2 is

$$\text{Entropy Eq.:} \quad s_2 = s_1$$

$$\frac{T_2}{T_1} = \left(\frac{V_1}{V_2}\right)^{k-1} \quad \text{and} \quad \frac{P_2}{P_1} = \left(\frac{V_1}{V_2}\right)^{k}$$

The energy equation for heat addition process 2–3 is

$$q_H = {}_2q_3 = u_3 - u_2 = C_v(T_3 - T_2)$$

The second law for expansion process 3–4 is

$$s_4 = s_3$$

so that

$$\frac{T_3}{T_4} = \left(\frac{V_4}{V_3}\right)^{k-1} \quad \text{and} \quad \frac{P_3}{P_4} = \left(\frac{V_4}{V_3}\right)^{k}$$

In addition,

$$\eta_{th} = 1 - \frac{1}{r_v^{k-1}}, \qquad mep = \frac{w_{net}}{v_1 - v_2}$$

Solution

Substitution yields the following:

$$v_1 = \frac{0.287 \times 288.2}{100} = 0.827 \, m^3/kg$$

$$T_2 = T_1 r_v^{k-1} = 288.2 \times 10^{0.4} = 723.9 \, K$$

$$P_2 = P_1 r_v^k = 0.1 \times 10^{1.4} = 2.512 \, MPa$$

$$v_2 = \frac{0.827}{10} = 0.0827 \, m^3/kg$$

$$_2q_3 = C_v(T_3 - T_2) = 1800 \, kJ/kg$$

$$T_3 = T_2 + {}_2q_3/C_v, \quad T_3 - T_2 = \frac{1800}{0.717} = 2510 \, K, \qquad T_3 = 3234 \, K$$

$$\frac{T_3}{T_2} = \frac{P_3}{P_2} = \frac{3234}{723.9} = 4.467, \qquad P_3 = 11.222 \, MPa$$

$$\frac{T_3}{T_4} = \left(\frac{V_4}{V_3}\right)^{k-1} = 10^{0.4} = 2.5119, \qquad T_4 = 1287.5 \, K$$

$$\frac{P_3}{P_4} = \left(\frac{V_4}{V_3}\right)^k = 10^{1.4} = 25.12, \qquad P_4 = 0.4467 \, MPa$$

$$\eta_{th} = 1 - \frac{1}{r_v^{k-1}} = 1 - \frac{1}{10^{0.4}} = 0.602 = 60.2\%$$

This can be checked by finding the heat rejected:

$$_4q_1 = C_v(T_1 - T_4) = 0.717 \times (288.2 - 1287.5) = -716.5 \, kJ/kg$$

$$\eta_{th} = 1 - \frac{716.5}{1800} = 0.602 = 60.2\%$$

$$w_{net} = 1800 - 716.5 = 1083.5 \, kJ/kg = (v_1 - v_2)mep$$

$$mep = \frac{1083.5}{(0.827 - 0.0827)} = 1456 \, kPa$$

This is a high value for mean effective pressure, largely because the two constant-volume heat-transfer processes keep the total volume change to a minimum (compared with a Brayton cycle, for example). Thus, the Otto cycle is a good model to emulate in the piston/cylinder internal-combustion engine. At the other extreme, a low mean effective pressure means a large piston displacement for a given power output, which, in turn, means high frictional losses in an actual engine.

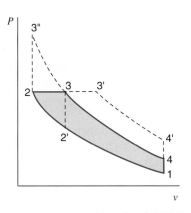
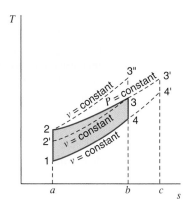

FIGURE 10.18 The air-standard diesel cycle.

10.9 THE DIESEL CYCLE

The air-standard diesel cycle is shown in Fig. 10.18. This is the ideal cycle for the diesel engine, which is also called the compression ignition engine.

In this cycle, the heat is transferred to the working fluid at constant pressure. This process corresponds to the injection, evaporation, and burning of the fuel in the actual engine. Since the gas is expanding during the heat addition in the air-standard cycle, the heat transfer must be just sufficient to maintain constant pressure. When state 3 is reached, the heat addition ceases and the gas undergoes an isentropic expansion, process 3–4, until the piston reaches BDC. As in the air-standard Otto cycle, a constant-volume rejection of heat at BDC replaces the exhaust and intake processes of the actual engine and the four processes are shown in Table 10.3.

Since work is done during the combustion process, the energy equation gives the heat transfer as

$$q_H = u_3 - u_2 + {}_2w_3 = u_3 - u_2 + P_2(v_3 - v_2) = h_3 - h_2$$

which is the only process type in which the diesel cycle is different from the Otto cycle.

The efficiency of the diesel cycle is given by the relation

$$\eta_{th} = 1 - \frac{q_L}{q_H} = 1 - \frac{C_v(T_4 - T_1)}{C_p(T_3 - T_2)} = 1 - \frac{T_1(T_4/T_1 - 1)}{kT_2(T_3/T_2 - 1)} \tag{10.13}$$

TABLE 10.3
The Diesel Cycle Processes

Process	Energy Eq.	Entropy Eq.	Process Eq.
Compression	$u_2 - u_1 = -{}_1w_2$	$s_2 - s_1 = (0/T) + 0$	$q = 0, s_1 = s_2$
Combustion	$u_3 - u_2 = q_H - {}_2w_3$	$s_3 - s_2 = \int dq_H/T + 0$	$P_3 = P_2 = C$
Expansion	$u_4 - u_3 = -{}_3w_4$	$s_4 - s_3 = (0/T) + 0$	$q = 0, s_3 = s_4$
Heat rejection	$u_1 - u_4 = -q_L$	$s_1 - s_4 = -\int dq_L/T + 0$	$v_4 = v_1 = C$

The isentropic compression ratio is greater than the isentropic expansion ratio in the diesel cycle. In addition, for a given state before compression and a given compression ratio (that is, given states 1 and 2), the cycle efficiency decreases as the maximum temperature increases. This is evident from the T–s diagram because the constant-pressure and constant-volume lines converge, and increasing the temperature from 3 to 3′ requires a large addition of heat (area 3–3′–c–b–3) and results in a relatively small increase in work (area 3–3′–4′–4–3).

A number of comparisons may be made between the Otto cycle and the diesel cycle, but here we will note only two. Consider Otto cycle 1–2–3″–4–1 and diesel cycle 1–2–3–4–1, which have the same state at the beginning of the compression stroke and the same piston displacement and compression ratio. From the T–s diagram, we see that the Otto cycle has higher efficiency. In practice, however, the diesel engine can operate on a higher compression ratio than the spark-ignition engine. The reason is that in the spark-ignition engine an air–fuel mixture is compressed, and detonation (spark knock) becomes a serious problem if too high a compression ratio is used. This problem does not exist in the diesel engine because only air is compressed during the compression stroke.

Therefore, we might compare an Otto cycle with a diesel cycle and in each case select a compression ratio that might be achieved in practice. Such a comparison can be made by considering Otto cycle 1–2′–3–4–1 and diesel cycle 1–2–3–4–1. The maximum pressure and temperature are the same for both cycles, which means that the Otto cycle has a lower compression ratio than the diesel cycle. It is evident from the T–s diagram that in this case, the diesel cycle has the higher efficiency. Thus, the conclusions drawn from a comparison of these two cycles must always be related to the basis on which the comparison has been made.

The actual compression-ignition open cycle differs from the air-standard diesel cycle in much the same way that the spark-ignition open cycle differs from the air-standard Otto cycle.

Example 10.8

An air-standard diesel cycle has a compression ratio of 20, and the heat transferred to the working fluid per cycle is 1800 kJ/kg. At the beginning of the compression process, the pressure is 0.1 MPa and the temperature is 15 °C. Determine

1. The pressure and temperature at each point in the cycle.

2. The thermal efficiency.

3. The mean effective pressure.

Control mass:	Air inside cylinder.
Diagram:	Fig. 10.18.
State information:	$P_1 = 0.1$ MPa, $T_1 = 288.2$ K.
Process information:	Four processes known (Table 10.3). Also, $r_v = 20$ and $q_H = 1800$ kJ/kg.
Model:	Ideal gas, constant specific heat, value at 300 K.

Analysis ————————————————————————————————

Entropy Eq. compression: $s_2 = s_1$

so that

$$\frac{T_2}{T_1} = \left(\frac{V_1}{V_2}\right)^{k-1} \quad \text{and} \quad \frac{P_2}{P_1} = \left(\frac{V_1}{V_2}\right)^{k}$$

The energy equation for heat addition process 2–3 is

$$q_H = {}_2q_3 = C_p(T_3 - T_2)$$

Entropy Eq. expansion: $\quad s_4 = s_3 \Rightarrow \dfrac{T_3}{T_4} = \left(\dfrac{V_4}{V_3}\right)^{k-1}$

In addition,

$$\eta_{\text{th}} = \frac{w_{\text{net}}}{q_H}, \quad \text{mep} = \frac{w_{\text{net}}}{v_1 - v_2}$$

Solution

Substitution gives

$$v_1 = \frac{0.287 \times 288.2}{100} = 0.827 \, \text{m}^3/\text{kg}$$

$$v_2 = \frac{v_1}{20} = \frac{0.827}{20} = 0.041\,35 \, \text{m}^3/\text{kg}$$

$$\frac{T_2}{T_1} = \left(\frac{V_1}{V_2}\right)^{k-1} = 20^{0.4} = 3.3145, \qquad T_2 = 955.2 \, \text{K}$$

$$\frac{P_2}{P_1} = \left(\frac{V_1}{V_2}\right)^{k} = 20^{1.4} = 66.29, \qquad P_2 = 6.629 \, \text{MPa}$$

$$q_H = {}_2q_3 = C_p(T_3 - T_2) = 1800 \, \text{kJ/kg}$$

$$T_3 - T_2 = \frac{1800}{1.004} = 1793 \, \text{K}, \qquad T_3 = 2748 \, \text{K}$$

$$\frac{V_3}{V_2} = \frac{T_3}{T_2} = \frac{2748}{955.2} = 2.8769, \qquad v_3 = 0.118\,96 \, \text{m}^3/\text{kg}$$

$$\frac{T_3}{T_4} = \left(\frac{V_4}{V_3}\right)^{k-1} = \left(\frac{0.827}{0.118\,96}\right)^{0.4} = 2.1719, \qquad T_4 = 1265 \, \text{K}$$

$$q_L = {}_4q_1 = C_v(T_1 - T_4) = 0.717(288.2 - 1265) = -700.4 \, \text{kJ/kg}$$

$$w_{\text{net}} = 1800 - 700.4 = 1099.6 \, \text{kJ/kg}$$

$$\eta_{\text{th}} = \frac{w_{\text{net}}}{q_H} = \frac{1099.6}{1800} = 61.1\,\%$$

$$\text{mep} = \frac{w_{\text{net}}}{v_1 - v_2} = \frac{1099.6}{0.827 - 0.041\,35} = 1400 \, \text{kPa}$$

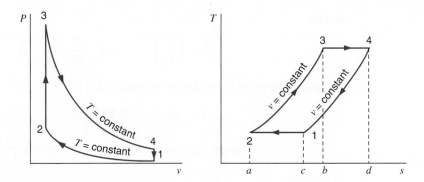

FIGURE 10.19
The air-standard
Stirling cycle.

10.10 THE STIRLING CYCLE

Another air-standard power cycle to be discussed is the Stirling cycle, which is shown on the $P-v$ and $T-s$ diagrams of Fig. 10.19. Heat is transferred to the working fluid during the constant-volume process 2–3 and also during the isothermal expansion process 3–4. Heat is rejected during the constant-volume process 4–1 and also during the isothermal compression process 1–2. Thus, this cycle is the same as the Otto cycle, with the adiabatic processes of that cycle replaced with isothermal processes. Since the Stirling cycle includes two constant-volume heat-transfer processes, keeping the total volume change during the cycle to a minimum, it is a good candidate for a piston/cylinder boundary-work application; it should have a high mean effective pressure.

Stirling-cycle engines have been developed in recent years as external combustion engines with regeneration. The significance of regeneration is noted from the ideal case shown in Fig. 10.19. Note that the heat transfer to the gas between states 2 and 3, area 2–3–b–a–2, is exactly equal to the heat transfer from the gas between states 4 and 1, area 1–4–d–c–1. Thus, in the ideal cycle, all external heat supplied Q_H takes place in the isothermal expansion process 3–4, and all external heat rejection Q_L takes place in the isothermal compression process 1–2. Since all heat is supplied and rejected isothermally, the efficiency of this cycle equals the efficiency of a Carnot cycle operating between the same temperatures. The same conclusions would be drawn in the case of an Ericsson cycle, which was discussed briefly in Section 10.4, if that cycle were to include a regenerator as well.

10.11 THE ATKINSON AND MILLER CYCLES

A cycle slightly different from the Otto cycle, the Atkinson cycle, has been proposed that has a higher expansion ratio than the compression ratio and thus can have the heat rejection process take place at constant pressure. The higher expansion ratio allows more work to be extracted, and this cycle has a higher efficiency than the Otto cycle. It is mechanically more complicated to move the piston in such a cycle, so it can be accomplished by keeping the intake valves open during part of the compression stroke, giving an actual compression

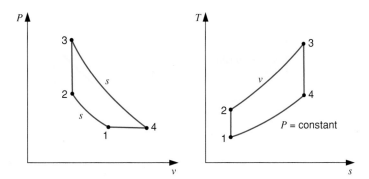

FIGURE 10.20
The Atkinson cycle.

less than the nominal one. The four processes are shown in the $P–v$ and $T–s$ diagrams in Fig. 10.20.

For the compression and expansion processes ($s = $ constant) we get

$$\frac{T_2}{T_1} = \left(\frac{v_1}{v_2}\right)^{k-1} \quad \text{and} \quad \frac{T_4}{T_3} = \left(\frac{v_3}{v_4}\right)^{k-1}$$

and the heat rejection process gives

$$P = C: \qquad T_4 = \left(\frac{v_4}{v_1}\right)T_1 \quad \text{and} \quad q_L = h_4 - h_1$$

The efficiency of the cycle becomes

$$\eta = \frac{q_H - q_L}{q_H} = 1 - \frac{q_L}{q_H} = 1 - \frac{h_4 - h_1}{u_3 - u_2}$$

$$\cong 1 - \frac{C_p}{C_v}\frac{(T_4 - T_1)}{(T_3 - T_2)} = 1 - k\frac{T_4 - T_1}{T_3 - T_2} \tag{10.14}$$

Calling the smaller compression ratio $CR_1 = (v_1/v_3)$ and the expansion ratio $CR = (v_4/v_3)$, we can express the temperatures as

$$T_2 = T_1\,CR_1^{k-1}; \qquad T_4 = \left(\frac{v_4}{v_1}\right)T_1 = \frac{CR}{CR_1}T_1 \tag{10.15}$$

and from the relation between T_3 and T_4 we can get

$$T_3 = T_4\,CR^{k-1} = \frac{CR}{CR_1}T_1\,CR^{k-1} = \frac{CR^k}{CR_1}T_1$$

Now substitute all the temperatures into Eq. 10.14 to get

$$\eta = 1 - k\frac{\dfrac{CR}{CR_1} - 1}{\dfrac{CR^k}{CR_1} - CR_1^{k-1}} = 1 - k\frac{CR - CR_1}{CR^k - CR_1^k} \tag{10.16}$$

and similarly to the other cycles, only the compression/expansion ratios are important.

As it can be difficult to ensure that $P_4 = P_1$ in the actual engine, a shorter expansion and modification using a supercharger can be approximated with a Miller cycle, which is a cycle in between the Otto cycle and the Atkinson cycle, shown in Fig. 10.21. This cycle is the approximation for the Ford Escape and the Toyota Prius hybrid car engines.

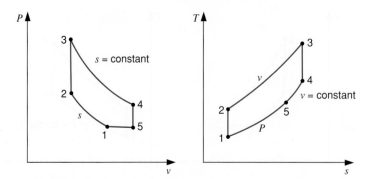

FIGURE 10.21
The Miller cycle.

Due to the extra process in the Miller cycle, the expression for the cycle efficiency is slightly more involved than the one shown for the Atkinson cycle. Both of these cycles have a higher efficiency than the Otto cycle for the same compression, but because of the longer expansion stroke, they tend to produce less power for the same-size engine. In the hybrid engine configuration, the peak power for acceleration is provided by an electric motor drawing energy from the battery.

Comment: If we determine state 1 (intake state) compression ratios CR_1 and CR, we have the Atkinson cycle completely determined. That is, only a fixed heat release will give this cycle. The heat release is a function of the air–fuel mixture, and thus the cycle is not a natural outcome of states and processes that are controlled. If the heat release is a little higher, then the cycle will be a Miller cycle; that is, the pressure will not have dropped enough when the expansion is complete. If the heat release is smaller, then the pressure is below P_1 when the expansion is done and there can be no exhaust flow against the higher pressure. From this it is clear that any practical implementation of the Atkinson cycle ends up as a Miller cycle.

We end this section with a measured cylinder pressure versus volume from a real diesel engine, as shown in Fig. 10.22. The engine is turbocharged, so the exhaust pressure is about 200 kPa in order to drive the turbine and the intake pressure is about 150 kPa delivered by a compressor. In linear coordinates, the cycle seems very similar to the Otto cycle due to the design of the injectors and the injection timing typical of modern, fast-burning diesel engines. When the cycle is plotted in log-log coordinates, we notice that the compression and expansion processes becomes nearly straight lines; they are both polytropic processes with a polytropic exponent $n \approx 1.32$ for the compression stroke and $n \approx 1.2$ for the expansion stroke. This shows that the two processes are close to being isentropic with $n \approx k$, with

FIGURE 10.22 A real diesel cycle *P–v* diagram in linear and log-log scales; measurements are from the W. E. Lay Automotive Laboratory at the University of Michigan.

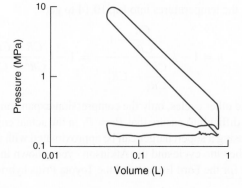

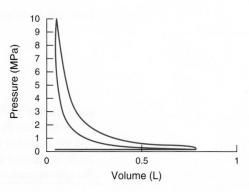

some heat transfer loss from the hot gases during the expansion versus a smaller heat loss during the compression. The higher temperature after combustion results in higher specific heats and a lower value of the ratio of specific heats k.

In-Text Concept Questions

e. How is the compression in the Otto cycle different from that in the Brayton cycle?

f. How many parameters do you need to know to completely describe the Otto cycle? How about the diesel cycle?

g. The exhaust and inlet flow processes are not included in the Otto or diesel cycles. How do these necessary processes affect the cycle performance?

10.12 COMBINED-CYCLE POWER AND REFRIGERATION SYSTEMS

There are many situations in which it is desirable to combine two cycles in series, either power systems or refrigeration systems, to take advantage of a very wide temperature range or to utilize what would otherwise be waste heat to improve efficiency. One combined power cycle, shown in Fig. 10.23 as a simple steam cycle with a liquid metal topping cycle, is often referred to as a binary cycle. The advantage of this combined system is that the liquid metal has a very low vapor pressure relative to that for water; therefore, it is possible for an isothermal boiling process in the liquid metal to take place at a high temperature, much higher than the critical temperature of water, but still at a moderate pressure. The liquid metal condenser then provides an isothermal heat source as input to the steam boiler, such that the two cycles can be closely matched by proper selection of the cycle variables, with the resulting

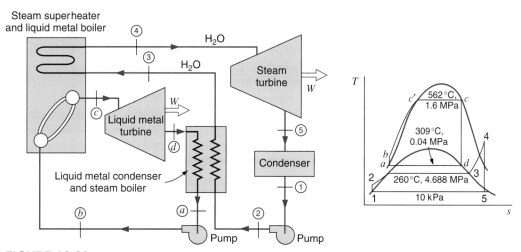

FIGURE 10.23 Liquid metal–water binary power system.

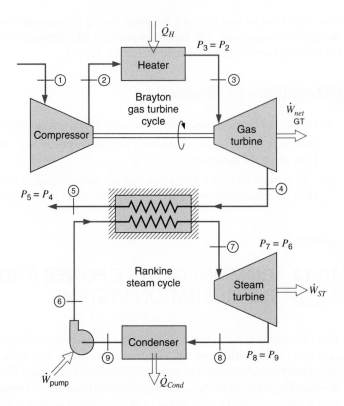

FIGURE 10.24
Combined
Brayton/Rankine-cycle
power system.

combined cycle then having a high thermal efficiency. Saturation pressures and temperatures for a typical liquid metal–water binary cycle are shown in the T–s diagram of Fig. 10.23.

A different type of combined cycle that has seen considerable attention uses the "waste heat" exhaust from a Brayton cycle gas-turbine engine (or another combustion engine such as a diesel engine) as the heat source for a steam or other vapor power cycle, in which case the vapor cycle acts as a bottoming cycle for the gas engine, in order to improve the overall thermal efficiency of the combined power system. Such a system, utilizing a gas turbine and a steam Rankine cycle, is shown in Fig. 10.24. In such a combination, there is a natural mismatch using the cooling of a noncondensing gas as the energy source to produce an isothermal boiling process plus superheating the vapor, and careful design is required to avoid a pinch point, a condition at which the gas has cooled to the vapor boiling temperature without having provided sufficient energy to complete the boiling process.

One way to take advantage of the cooling exhaust gas in the Brayton-cycle portion of the combined system is to utilize a mixture as the working fluid in the Rankine cycle. An example of this type of application is the Kalina cycle, which uses ammonia–water mixtures as the working fluid in the Rankine-type cycle. Such a cycle can be made very efficient, since the temperature differences between the two fluid streams can be controlled through careful design of the combined system.

Combined cycles are used in refrigeration systems in cases where there is a very large temperature difference between the ambient surroundings and the refrigerated space, as shown for the cascade system in Chapter 9. It can also be a coupling of a heat engine cycle providing the work to drive a refrigeration cycle, as shown in Fig. 10.25. This is what happens when a car engine produces shaft work to drive the car's air-conditioner unit or

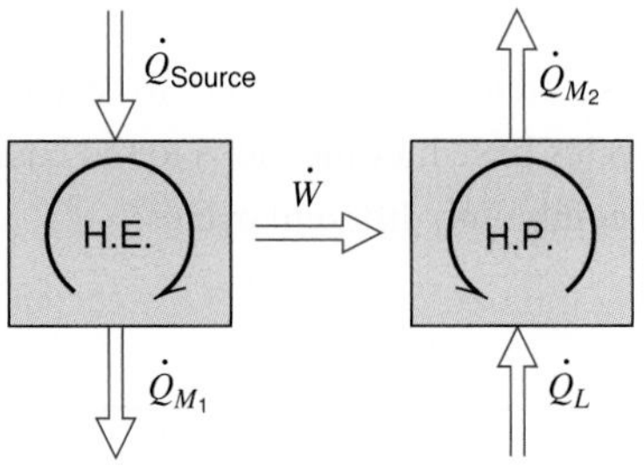

FIGURE 10.25 A heat engine–driven heat pump or refrigerator.

when electric power generated by combustion of some fuel drives a domestic refrigerator. The ammonia absorption system shown in Fig. 9.29 is such an application to greatly reduce the mechanical work input. Imagine a control volume around the left side column of devices and notice how this substitutes for the compressor in a standard refrigeration cycle. For use in remote locations, the work input can be completely eliminated, as in Fig. 10.25, with combustion of propane as the heat source to run a refrigerator without electricity.

We have described only a few combined-cycle systems here, as examples of the types of applications that can be dealt with, and the resulting improvement in overall performance that can occur. Obviously, there are many other combinations of power and refrigeration systems. Some of these are discussed in the problems at the end of the chapter.

SUMMARY

A Brayton cycle is a gas turbine producing electricity and with a modification of a jet engine producing thrust. This is a high-power, low-mass, low-volume device that is used where space and weight are at a premium cost. A high back-work ratio makes this cycle sensitive to compressor efficiency. Different variations and configurations for the Brayton cycle with regenerators and intercoolers are shown. The air-standard refrigeration cycle, the reverse of the Brayton cycle, is also covered in detail.

Piston/cylinder devices are shown for the Otto and diesel cycles modeling the gasoline and diesel engines, which can be two- or four-stroke engines. Cold air properties are used to show the influence of compression ratio on the thermal efficiency, and the mean effective pressure is used to relate the engine size to total power output. Atkinson and Miller cycles are modifications of the basic cycles that are implemented in modern hybrid engines, and these are also presented. We briefly mention the Stirling cycle as an example of an external combustion engine.

The chapter ends with a short description of combined-cycle applications. This covers stacked or cascade systems for large temperature spans and combinations of different kinds of cycles where one can be added as a topping cycle or a bottoming cycle. Often, a Rankine cycle uses exhaust energy from a Brayton cycle in larger stationary applications, and a heat engine can be used to drive a refrigerator or heat pump.

You should have learned a number of skills and acquired abilities from studying this chapter that will allow you to:

- Know the principles of gas turbines and jet engines.
- Know that real engine component processes are not reversible.
- Understand the air-standard refrigeration processes.
- Understand the basics of piston/cylinder engine configuration.

- Know the principles of the various piston/cylinder engine cycles.
- Have a sense of the most influential parameters for each type of cycle.
- Know that most real cycles have modifications to the basic cycle setup.
- Know the principle of combining different cycles.

KEY CONCEPTS AND FORMULAS

Brayton Cycle

Compression ratio	Pressure ratio $\quad r_p = P_{\text{high}}/P_{\text{low}}$
Basic cycle efficiency	$\eta = 1 - \dfrac{h_4 - h_1}{h_2 - h_3} \cong 1 - r_p^{(1-k)/k}$
Regenerator	Dual fluid heat exchanger; uses exhaust flow energy.
Cycle with regenerator	$\eta = 1 - \dfrac{h_2 - h_1}{h_3 - h_4} \cong 1 - \dfrac{T_1}{T_3} r_p^{(1-k)/k}$
Intercooler	Cooler between compressor stages; reduces work input
Jet engine	No shaft work out; kinetic energy generated in exit nozzle
Thrust	$F = \dot{m}(\mathbf{V}_e - \mathbf{V}_i) \qquad$ (momentum equation)
Propulsive power	$\dot{W} = F\,\mathbf{V}_{\text{aircraft}} = \dot{m}(\mathbf{V}_e - \mathbf{V}_i)\,\mathbf{V}_{\text{aircraft}}$

Air-Standard Refrigeration Cycle

Coefficient of performance	$\text{COP} = \beta_{\text{REF}} = \dfrac{\dot{Q}_L}{\dot{W}_{\text{net}}} = \dfrac{q_L}{w_{\text{net}}} \cong \left(r_p^{(k-1)/k} - 1\right)^{-1}$
Cooling capacity	$\dot{Q}_L$

Piston/Cylinder Power Cycles

Compression ratio	Volume ratio $\quad r_v = CR = V_{\text{max}}/V_{\text{min}}$
Displacement (one cycle)	$\Delta V = V_{\text{max}} - V_{\text{min}} = m(v_{\text{max}} - v_{\text{min}}) = SA_{\text{cyl}}$
Stroke	$S = 2R_{\text{crank}}$; piston travel in compression or expansion
Mean effective pressure	$P_{\text{meff}} = w_{\text{net}}/(v_{\text{max}} - v_{\text{min}}) = W_{\text{net}}/(V_{\text{max}} - V_{\text{min}})$
Power by one cylinder	$\dot{W} = m w_{\text{net}} \dfrac{\text{RPM}}{60} \qquad$ (times $\frac{1}{2}$ for four-stroke cycle)
Otto cycle efficiency	$\eta = 1 - \dfrac{u_4 - u_1}{u_3 - u_2} \approx 1 - r_v^{1-k}$
Diesel cycle efficiency	$\eta = 1 - \dfrac{u_4 - u_1}{h_3 - h_2} \approx 1 - \dfrac{T_1}{kT_2}\dfrac{T_4/T_1 - 1}{T_3/T_2 - 1}$
Atkinson cycle	$CR_1 = \dfrac{v_1}{v_2}$ (compression ratio); $CR = \dfrac{v_4}{v_3}$ (expansion ratio)
Atkinson cycle efficiency	$\eta = 1 - \dfrac{h_4 - h_1}{u_3 - u_2} \approx 1 - k\dfrac{CR - CR_1}{CR^k - CR_1^k}$

Combined Cycles

Topping, bottoming cycle	High- and low-temperature cycles
Cascade system	Stacked refrigeration cycles
Coupled cycles	Heat engine–driven refrigerator

CONCEPT-STUDY GUIDE PROBLEMS

10.1 Is a Brayton cycle the same as a Carnot cycle? Name the four processes.

10.2 Why is the back work ratio in the Brayton cycle much higher than that in the Rankine cycle?

10.3 For a given Brayton cycle, the cold air approximation gave a formula for the efficiency. If we use the specific heats at the average temperature for each change in enthalpy, will that give a higher or lower efficiency?

10.4 Does the efficiency of a jet engine change with altitude since the density varies?

10.5 Why are the two turbines in Fig. 10.9 not connected to the same shaft?

10.6 Why is an air refrigeration cycle not common for a household refrigerator?

10.7 Does the inlet state (P_1, T_1) have any influence on the Otto cycle efficiency? How about the power produced by a real car engine?

10.8 For a given compression ratio, does an Otto cycle have a higher or lower efficiency than a diesel cycle?

10.9 How many parameters do you need to know to completely describe the Atkinson cycle? How about the Miller cycle?

10.10 Why would one consider a combined-cycle system for a power plant? For a heat pump or refrigerator?

10.11 Can the exhaust flow from a gas turbine be useful?

10.12 Where may a heat engine–driven refrigerator be useful?

10.13 Since any heat transfer is driven by a temperature difference, how does that affect all the real cycles relative to the ideal cycles?

10.14 In an Otto cycle, the cranking mechanism dictates the volume given the crank position. Can you say something similar for the Brayton cycle?

10.15 For all the gas cycles, it is assumed that the ideal compression and expansions are isentropic. This is approximated with a polytropic process having $n = k$. The expansion after combustion will have some heat loss due to high temperature, so what does that imply for the value of n?

10.16 For all the gas cycles, it is assumed that the ideal compression and expansions are isentropic. This is approximated with a polytropic process having $n = k$. The compression in a diesel engine leads to high temperatures and thus will have some heat loss, so what does that imply for the value of n?

10.17 If we compute the efficiency of an Otto or diesel cycle, we get something like 60 % for a compression ratio of 10:1. Is a real engine efficiency close to this?

10.18 A hybrid power train couples a battery/motor with an internal combustion engine. Mention a few factors that make this combination a little more efficient.

HOMEWORK PROBLEMS

Brayton Cycles, Gas Turbines

10.19 In a Brayton cycle the inlet is at 300 K, 100 kPa, and the combustion adds 850 kJ/kg. The maximum temperature is 1500 K due to material considerations. Find the maximum permissible compression ratio and, for that ratio, the cycle efficiency using cold air properties.

10.20 A Brayton cycle has a compression ratio of 15:1 with a high temperature of 1600 K and the inlet at 290 K, 100 kPa. Use cold air properties and find the specific heat addition and specific net work output.

10.21 A large stationary Brayton cycle gas-turbine power plant delivers a power output of 25 MW to an electric generator. The minimum temperature in the cycle is 300 K, and the exhaust temperature is 750 K. The minimum pressure in the cycle is 100 kPa, and the compressor pressure ratio is 14:1. Calculate the power output of the turbine. What fraction of the turbine output is required to drive the compressor? What is the thermal efficiency of the cycle?

10.22 A Brayton cycle has air into the compressor at 95 kPa, 290 K, and has an efficiency of 50 %. The exhaust temperature is 675 K. Find the pressure ratio and the specific heat addition by the combustion for this cycle.

10.23 A Brayton cycle has inlet at 290 K, 90 kPa, and the combustion adds 1100 kJ/kg. How high can the compression ratio be so that the highest temperature is below 1700 K? Use cold air properties to determine this.

10.24 Assume a state of 1400 kPa, 2100K, into the turbine section of a Brayton cycle with an adiabatic expansion to 100 kPa and a compressor inlet temperature of 300 K. Find the missing temperatures in the cycle using Table A.7 and then give the average value of k (ratio of specific heats) for the compression and expansion processes.

10.25 Repeat Problem 10.23 using Table A.7.

10.26 A Brayton cycle produces 14 MW with an inlet state of 17 °C, 100 kPa, and a compression ratio of 16:1. The heat added in the combustion is 1160 kJ/kg. What are the highest temperature and the mass flow rate of air, assuming cold air properties?

10.27 Repeat Problem 10.22 using Table A.7; this becomes a trial-and-error process.

Regenerators, Intercoolers, and Nonideal Cycles

10.28 Would it be better to add an ideal regenerator to the Brayton cycle in Problem 10.26?

10.29 A Brayton cycle with an ideal regenerator has inlet at 290 K, 90 kPa with the highest P, T as 1170 kPa, 1700 K. Find the specific heat transfer and the cycle efficiency using cold air properties.

10.30 An ideal air-standard Brayton cycle includes an ideal regenerator. The state into the compressor is 100 kPa, 20 °C, and the pressure ratio across the compressor is 12:1. The highest cycle temperature is 1100 °C, and the air flow rate is 10 kg/s. Use cold air properties and determine the compressor work, the turbine work, and the thermal efficiency of the cycle.

10.31 Assume that the compressor in Problem 10.26 has an intercooler that cools the air to 330 K operating at 500 kPa, followed by a second stage of compression to 1600 kPa. Find the specific heat transfer in the intercooler and the total combined work required.

10.32 A two-stage air compressor has an intercooler between the two stages, as shown in Fig. P10.32. The inlet state is 100 kPa, 290 K, and the final exit pressure is 1.6 MPa. Assume that the constant-pressure intercooler cools the air to the inlet temperature, $T_3 = T_1$. It can be shown that the optimal pressure is $P_2 = (P_1 P_4)^{1/2}$ for minimum total compressor work. Find the specific compressor works and the intercooler heat transfer for the optimal P_2.

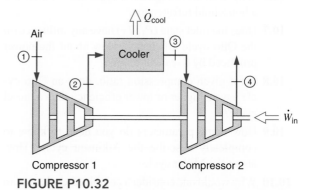

FIGURE P10.32

10.33 The gas-turbine cycle shown in Fig. P10.33 is used as an automotive engine. In the first turbine, the gas expands to pressure P_5, just low enough for this turbine to drive the compressor. The gas is then expanded through the second turbine connected to the drive wheels. The data for the engine are shown in the figure and assume that all processes are ideal. Determine the intermediate pressure P_5, the net specific work output of the engine, and the mass flow rate through the engine. Also find the air temperature entering the burner, T_3, and the thermal efficiency of the engine.

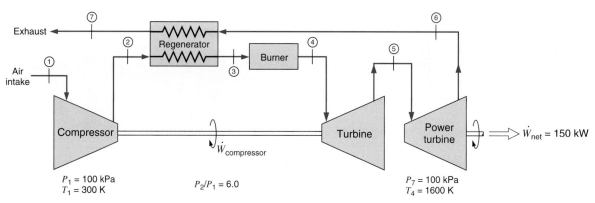

FIGURE P10.33

10.34 Repeat Problem 10.32 when the intercooler brings the air to $T_3 = 320$ K. The corrected formula for the optimal pressure is $P_2 = [P_1 P_4 (T_3/T_1)^{n/(n-1)}]^{1/2}$. See Problem 7.150, where n is the exponent in the assumed polytropic process.

10.35 Repeat Problem 10.21, but include a regenerator with 75 % efficiency in the cycle.

10.36 An air compressor has inlet at 100 kPa, 290 K, and brings it to 500 kPa, after which the air is cooled in an intercooler to 340 K by heat transfer to the ambient 290 K. Assume this first compressor stage has an isentropic efficiency of 85 % and is adiabatic. Using constant specific heat, find the compressor exit temperature and the specific entropy generation in the process.

10.37 A two-stage compressor in a gas turbine brings atmospheric air at 100 kPa, 17 °C to 500 kPa, then cools it in an intercooler to 27 °C at constant P. The second stage brings the air to 2500 kPa. Assume both stages are adiabatic and reversible. Find the combined specific work to the compressor stages. Compare that to the specific work for the case of no intercooler (i.e., one compressor from 100 to 2500 kPa).

10.38 Repeat Problem 10.21, but assume the compressor has an isentropic efficiency of 85 % and the turbine has an isentropic efficiency of 88 %.

10.39 A gas turbine with air as the working fluid has two ideal turbine sections, as shown in Fig. P10.39, the first of which drives the ideal compressor, with the second producing the power output. The compressor inlet is at 290 K, 100 kPa, and the exit is at 450 kPa. A fraction of flow, x, bypasses the burner

and the rest $(1 - x)$ goes through the burner, where 1200 kJ/kg is added by combustion. The two flows then mix before entering the first turbine and continue through the second turbine, with exhaust at 100 kPa. If the mixing should result in a temperature of 1000 K into the first turbine, find the fraction x. Find the required pressure and temperature into the second turbine and its specific power output.

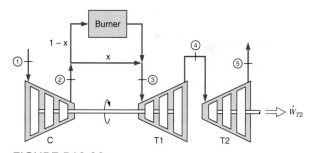

FIGURE P10.39

10.40 A gas turbine cycle has two stages of compression, with an intercooler between the stages. Air enters the first stage at 100 kPa, 300 K. The pressure ratio across each compressor stage is 4:1, and each stage has an isentropic efficiency of 82 %. Air exits the intercooler at 330 K. Calculate the temperature at the exit of each compressor stage and the total specific work required.

Ericsson Cycles

10.41 Consider an ideal air-standard Ericsson cycle that has an ideal regenerator, as shown in Fig. P10.41.

The high pressure is 1.5 MPa and the cycle efficiency is 60 %. Heat is rejected in the cycle at a temperature of 350 K, and the cycle pressure at the beginning of the isothermal compression process is 150 kPa. Determine the high temperature, the compressor work, and the turbine work per kilogram of air.

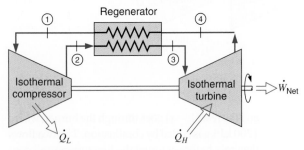

FIGURE P10.41

10.42 An air-standard Ericsson cycle has an ideal regenerator. Heat is supplied at 1000 °C and heat is rejected at 80 °C. Pressure at the beginning of the isothermal compression process is 70 kPa. The heat added is 700 kJ/kg. Find the compressor work, the turbine work, and the cycle efficiency.

Jet Engine Cycles

10.43 The Brayton cycle in Problem 10.21 is changed to be a jet engine. Find the exit velocity using cold air properties.

10.44 Consider an ideal air-standard cycle for a gas-turbine, jet propulsion unit, such as that shown in Fig. 10.11. The pressure and temperature entering the compressor are 90 kPa, 290 K. The pressure ratio across the compressor is 14:1, and the turbine inlet temperature is 1500 K. When the air leaves the turbine, it enters the nozzle and expands to 90 kPa. Determine the velocity of the air leaving the nozzle.

10.45 The turbine section in a jet engine (Fig. P10.45) receives gas (assume air) at 1200 K, 800 kPa, with an ambient atmosphere at 80 kPa. The turbine is followed by a nozzle open to the atmosphere, and all the turbine work drives a compressor. Find the turbine exit pressure so that the nozzle has an exit velocity of 800 m/s. *Hint*: take the C.V. around both the turbine and the nozzle.

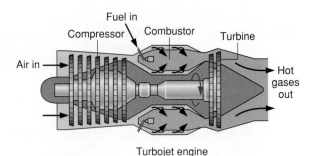

Turbojet engine

FIGURE P10.45

10.46 Given the conditions in the previous problem, what pressure could an ideal compressor generate (not the 800 kPa but higher)?

10.47 Consider a turboprop engine in which the turbine powers the compressor and a propeller. Assume the same cycle as in Problem 10.43, with a turbine exit temperature of 900 K. Find the specific work to the propeller and the exit velocity.

10.48 Consider an air-standard jet engine cycle operating in a 280 K, 100 kPa environment. The compressor requires a shaft power input of 4000 kW. Air enters the turbine state 3 at 1600 K, 2 MPa, at the rate of 9 kg/s, and the isentropic efficiency of the turbine is 85 %. Determine the pressure and temperature entering the nozzle at state 4. If the nozzle efficiency is 95 %, determine the temperature and velocity exiting the nozzle at state 5.

10.49 Solve the previous problem using the air tables.

10.50 A jet aircraft is flying at an altitude of 4900 m, where the ambient pressure is approximately 50 kPa and the ambient temperature is −20 °C. The velocity of the aircraft is 280 m/s, the pressure ratio across the compressor is 14:1, and the cycle maximum temperature is 1450 K. Assume the inlet flow goes through a diffuser to zero relative velocity at state 1. Find the temperature and pressure at state 1.

10.51 The turbine in a jet engine receives air at 1250 K, 1.5 MPa. It exhausts to a nozzle at 250 kPa, which, in turn, exhausts to the atmosphere at 100 kPa. The isentropic efficiency of the turbine is 85 % and the nozzle efficiency is 95 %. Find the nozzle inlet temperature and the nozzle exit velocity. Assume negligible kinetic energy out of the turbine.

10.52 Solve the previous problem using the air tables.

10.53 An afterburner in a jet engine similar to Fig. P10.53 adds fuel after the turbine, thus raising the pressure and temperature due to the energy of combustion. Assume a standard condition of 800 K, 250 kPa, after the turbine into the nozzle that exhausts at 95 kPa. Assume the afterburner adds 450 kJ/kg to that state, with a rise in pressure for the same specific volume, and neglect any upstream effects on the turbine. Find the nozzle exit velocity before and after the afterburner is turned on.

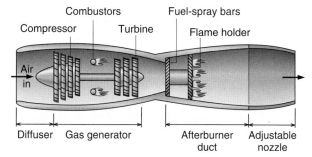

Combustors
Compressor   Turbine   Fuel-spray bars
                          Flame holder
Air in
Diffuser Gas generator   Afterburner Adjustable
                          duct   nozzle

FIGURE P10.53

Air-Standard Refrigeration Cycles

10.54 An air-standard refrigeration cycle has air into the compressor at 100 kPa, 270 K, with a compression ratio of 3:1. The temperature after heat rejection is 300 K. Find the COP and the highest cycle temperature.

10.55 A standard air refrigeration cycle has −10 °C, 100 kPa into the compressor, and the ambient cools the air down to 35 °C at 400 kPa. Find the lowest temperature in the cycle, the low-T specific heat transfer, and the specific compressor work.

10.56 The formula for the COP, assuming cold air properties, is given for the standard refrigeration cycle in Eq. 10.5. Develop the similar formula for the cycle variation with a heat exchanger, as shown in Fig. 10.14.

10.57 Assume a refrigeration cycle as shown in Fig. 10.14, with a reversible adiabatic compressor and expander. For this cycle, the low pressure is 100 kPa and the high pressure is 1.4 MPa, with constant-pressure heat exchangers (see the T–s diagram in Fig. 10.14). The temperatures are $T_4 = T_6 = -50\,°C$ and $T_1 = T_3 = 15\,°C$. Find the COP for this refrigeration cycle.

10.58 Repeat Problem 10.57, but assume an isentropic efficiency of 75 % for both the compressor and the expander.

Otto Cycles

10.59 The mean effective pressure scales with the net work and thus with the efficiency. Assume the heat transfer per unit mass is a given (it depends on the fuel–air mixture). How does the total power output then vary with the inlet conditions (P_1, T_1)?

10.60 A four-stroke gasoline engine runs at 1800 RPM with a total displacement of 3 L and a compression ratio of 10:1. The intake is at 290 K, 75 kPa, with a mean effective pressure of 600 kPa. Find the cycle efficiency and power output.

10.61 Find the missing pressures and temperatures in the cycle of Problem 10.60.

10.62 A four-stroke gasoline 4.2 L engine running at 2000 RPM has an inlet state of 85 kPa, 280 K. After combustion it is 2000 K, and the highest pressure is 5 MPa. Find the compression ratio, the cycle efficiency, and the exhaust temperature.

10.63 Find the power from the engine in Problem 10.62.

10.64 A four-stroke 2.4 L gasoline engine runs at 2500 RPM and has an efficiency of 60 %. The state before compression is 40 kPa, 280 K, and after combustion it is at 2200 K. Find the highest T and P in the cycle, the specific heat transfer added, the cycle mean effective pressure and the total power produced.

10.65 Suppose we reconsider the previous problem and, instead of the standard ideal cycle, we assume the expansion is a polytropic process with $n = 1.5$. What are the exhaust temperature and the expansion specific work?

10.66 A gasoline engine has a volumetric compression ratio of 8 and before compression has air at 280 K, 85 kPa. The combustion generates a peak pressure of 5500 kPa. Find the peak temperature, the energy added by the combustion process, and the exhaust temperature.

10.67 To approximate an actual spark-ignition engine (Fig. P10.67), consider an air-standard Otto cycle that has a heat addition of 1600 kJ/kg of air, a compression ratio of 7, and a pressure and temperature at the beginning of the compression process of 90 kPa, 10 °C. Assuming constant specific heat, with the value from Table A.5, determine the maximum pressure and temperature of the cycle, the thermal efficiency of the cycle, and the mean effective pressure.

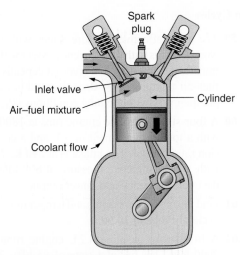

Spark plug

Inlet valve

Air–fuel mixture

Cylinder

Coolant flow

FIGURE P10.67

10.68 A 3.3 L minivan engine runs at 2000 RPM with a compression ratio of 10:1. The intake is at 50 kPa, 280 K, and after expansion it is at 750 K. Find the highest T in the cycle, the specific heat transfer added by combustion, and the mean effective pressure.

10.69 A gasoline engine takes air in at 290 K, 90 kPa and then compresses it. The combustion adds 1000 kJ/kg to the air, after which the temperature is 2050 K. Use cold air properties (i.e., constant heat capacities at 300 K) and find the compression ratio, the compression specific work, and the highest pressure in the cycle.

10.70 Answer the same three questions for the previous problem, but use variable heat capacities (use Table A.7).

10.71 A four-stroke gasoline engine (Fig. P10.71) has a compression ratio of 10:1 with four cylinders of

total displacement at 2.3 L. The inlet state is 280 K, 70 kPa and the engine is running at 2100 RPM, with the fuel adding 1400 kJ/kg in the combustion process. What is the net work in the cycle, and how much power is produced?

10.72 Assume a state of 5000 kPa, 2100 K after combustion in an Otto cycle with a compression ratio of 10:1; the intake temperature is 300 K. Find the missing temperatures in the cycle using Table A.7 and then give the average value of k (ratio of specific heats) for the compression and expansion processes.

10.73 A turbocharged engine (Fig. P10.73) runs in an Otto cycle with the lowest T at 290 K and the lowest P at 150 kPa. The highest T is 2400 K, and combustion adds 1200 kJ/kg as heat transfer. Find the compression ratio and the mean effective pressure.

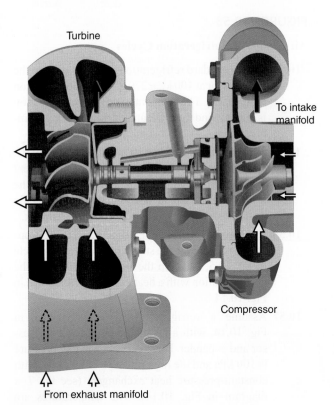

Turbine

To intake manifold

Compressor

From exhaust manifold

FIGURE P10.73

10.74 The cycle in the previous problem is used in a 2.4 L engine running at 1800 RPM. How much power does it produce?

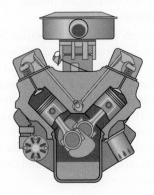

FIGURE P10.71

10.75 It is found experimentally that the power stroke expansion in an internal combustion engine can be approximated with a polytropic process, with a value of the polytropic exponent, n, somewhat larger than the specific heat ratio, k. Look at Problem 10.67, but assume that the expansion process is reversible and polytropic (instead of the isentropic expansion in the Otto cycle), with n equal to 1.50. From the average temperature during expansion, estimate the actual average k.

10.76 In the Otto cycle, all the heat transfer, q_H, occurs at constant volume. It is more realistic to assume that part of q_H occurs after the piston has started its downward motion in the expansion stroke. Therefore, consider a cycle identical to the Otto cycle, except that the first two-thirds of the total q_H occurs at constant volume and the last one-third occurs at constant pressure. Assume that the total q_H is 2100 kJ/kg, that the state at the beginning of the compression process is 90 kPa, 20 °C, and that the compression ratio is 9. Calculate the maximum pressure and temperature and the thermal efficiency of this cycle. Compare the results with those of a conventional Otto cycle having the same given variables.

10.77 A gasoline engine has a volumetric compression ratio of 9. The state before compression is 290 K, 70 kPa, and the peak cycle temperature is 1800 K. Find the pressure after expansion, the cycle net work, and the cycle efficiency using properties from Table A.7.2.

10.78 Solve Problem 10.70 using the P_r and v_r functions from Table A.7.2.

Diesel Cycles

10.79 A diesel engine has an inlet at 95 kPa, 300 K and a compression ratio of 20:1. The combustion releases 1300 kJ/kg. Find the temperature after combustion using cold air properties.

10.80 A diesel engine has a state before compression of 95 kPa, 290 K, a peak pressure of 6000 kPa, and a maximum temperature of 2400 K. Find the volumetric compression ratio and the thermal efficiency.

10.81 Find the cycle efficiency and mean effective pressure for the cycle in Problem 10.79.

10.82 The *cutoff ratio* is the ratio of v_3/v_2 (see Fig. 10.18), which is the expansion while combustion occurs at constant pressure. Determine this ratio for the cycle in Problem 10.79.

10.83 A diesel engine has a compression ratio of 20:1 with an inlet of 150 kPa, 290 K, state 1, with volume 0.5 L. The maximum cycle temperature is 1800 K. Find the maximum pressure, the net specific work, the cutoff ratio (see Problem 10.82), and the thermal efficiency.

10.84 A diesel engine (Fig. P10.84) has a bore of 0.1 m, a strike of 0.11 m, and a compression ratio of 19:1 running at 2000 RPM. Each cycle takes two revolutions and has a mean effective pressure of 1400 kPa. With a total of six cylinders, find the engine power in kilowatts and horsepower.

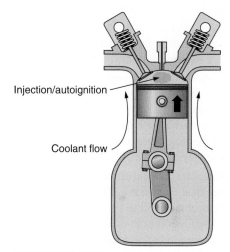

Injection/autoignition

Coolant flow

FIGURE P10.84

10.85 A supercharger is used for a two-stroke, 10 L diesel engine, so intake is 200 kPa, 320 K, and the cycle has a compression ratio of 18:1 and a mean effective pressure of 1030 kPa. The engine is 10 L running at 200 RPM. Find the power output.

10.86 A diesel engine has a state before compression of 95 kPa, 290 K, a peak pressure of 6000 kPa, and a maximum temperature of 2400 K. Use the air tables to find the volumetric compression ratio and the thermal efficiency.

10.87 At the beginning of compression in a diesel cycle, $T = 300$ K and $P = 200$ kPa. After combustion (heat addition) is complete, $T = 1500$ K and $P = 9$ MPa. Find the compression ratio, the thermal efficiency, and the mean effective pressure.

10.88 The world's largest diesel engine has displacement of 25 m³ running at 200 RPM in a two-stroke cycle producing 100 000 hp. Assume an inlet state of 200 kPa, 300 K and a compression ratio of 20:1. What is the mean effective pressure and the flow rate of air to the engine?

10.89 Solve Problem 10.80 using the P_r and V_r functions from Table A.7.2.

Stirling and Carnot Cycles

10.90 Consider an ideal Stirling-cycle engine in which the state at the beginning of the isothermal compression process is 100 kPa, 25 °C, the compression ratio is 6, and the maximum temperature in the cycle is 1100 °C. Calculate the maximum cycle pressure and the thermal efficiency of the cycle with and without regenerators.

10.91 An air-standard Stirling cycle uses helium as the working fluid. The isothermal compression brings helium from 100 kPa, 37 °C to 600 kPa. The expansion takes place at 1200 K, and there is no regenerator. Find the work and heat transfer in all of the four processes per kilogram of helium and the thermal cycle efficiency.

10.92 Consider an ideal air-standard Stirling cycle with an ideal regenerator. The minimum pressure and temperature in the cycle are 100 kPa, 25 °C, the compression ratio is 10, and the maximum temperature in the cycle is 1000 °C. Analyze each of the four processes in this cycle for work and heat transfer, and determine the overall performance of the engine.

10.93 The air-standard Carnot cycle was not shown in the text; show the T–s diagram for this cycle. In an air-standard Carnot cycle, the low temperature is 320 K and the efficiency is 60 %. If the pressure before compression and after heat rejection is 100 kPa, find the high temperature and the pressure just before heat addition.

10.94 Air in a piston/cylinder goes through a Carnot cycle in which $T_L = 26.8$ °C and the total cycle efficiency is $\eta = 2/3$. Find T_H, the specific work, and the volume ratio in the adiabatic expansion for constant C_p, C_v.

10.95 Do Problem 10.93, using values from Table A.7.1.

10.96 Do Problem 10.93, using the P_r, V_r functions in Table A.7.2.

Atkinson and Miller Cycles

10.97 An Atkinson cycle has state 1 as 150 kPa, 300 K, a compression ratio of 9, and a heat release of 1000 kJ/kg. Find the needed expansion ratio.

10.98 An Atkinson cycle has state 1 as 150 kPa, 300 K, a compression ratio of 9, and an expansion ratio of 14. Find the needed heat release in the combustion.

10.99 Repeat Problem 10.67, assuming we change the Otto cycle to an Atkinson cycle by keeping the same conditions and only increase the expansion to give a different state 4.

10.100 An Atkinson cycle has state 1 as 150 kPa, 300 K, a compression ratio of 9, and an expansion ratio of 14. Find the mean effective pressure.

10.101 A Miller cycle has state 1 as 150 kPa, 300 K, a compression ratio of 9, and an expansion ratio of 14. If P_4 is 250 kPa, find the heat release in the combustion.

10.102 A Miller cycle has state 1 as 150 kPa, 300 K, a compression ratio of 9, and a heat release of 1000 kJ/kg. Find the needed expansion ratio so that P_4 is 250 kPa.

10.103 In a Miller cycle, assume we know state 1 (intake state) compression ratios CR_1 and CR. Find an expression for the minimum allowable heat release so that $P_4 = P_5$, that is, it becomes an Atkinson cycle.

Combined Cycles

10.104 A Rankine steam power plant should operate with a high pressure of 3 MPa and a low pressure of 10 kPa, and the boiler exit temperature should be 500 °C. The available high-temperature source is the exhaust of 175 kg/s air at 600 °C from a gas turbine. If the boiler operates as a counterflowing heat exchanger where the temperature difference at the pinch point is 20 °C, find the maximum water mass flow rate possible and the air exit temperature.

10.105 A simple Rankine cycle with R-410A as the working fluid is to be used as a bottoming cycle for an electrical generating facility driven by the exhaust gas from a diesel engine as the high-temperature energy source in the R-410A boiler. Diesel inlet conditions are 100 kPa, 20 °C, the compression ratio is 20, and the maximum temperature

in the cycle is 2800 K. The R-410A leaves the bottoming cycle boiler at 80 °C, 4 MPa and the condenser pressure is 1800 kPa. The power output of the diesel engine is 1 MW. Assuming ideal cycles throughout, determine
a. The flow rate required in the diesel engine.
b. The power output of the bottoming cycle, assuming that the diesel exhaust is cooled to 200 °C in the R-410A boiler.

10.106 A small utility gasoline engine of 250 cc runs at 1500 RPM with a compression ratio of 7:1. The inlet state is 75 kPa, 17 °C and the combustion adds 1500 kJ/kg to the charge. This engine runs a heat pump using R-410A with a high pressure of 4 MPa and an evaporator operating at 0 °C. Find the rate of heating the heat pump can deliver.

10.107 Can the combined cycles in the previous problem deliver more heat than what comes from the R-410A? Find any amounts, if so, by assuming some conditions.

10.108 The power plant shown in Fig. 10.24 combines a gas-turbine cycle and a steam-turbine cycle. The following data are known for the gas-turbine cycle: air enters the compressor at 100 kPa, 25 °C, the compressor pressure ratio is 14, and the heater input rate is 60 MW; the turbine inlet temperature is 1250 °C and the exhaust pressure is 100 kPa; the cycle exhaust temperature from the heat exchanger is 200 °C. The following data are known for the steam-turbine cycle: The pump inlet state is saturated liquid at 10 kPa and the pump exit pressure is 12.5 MPa; the turbine inlet temperature is 500 °C. Determine.
a. The mass flow rate of air in the gas-turbine cycle.
b. The mass flow rate of water in the steam cycle.
c. The overall thermal efficiency of the combined cycle.

Exergy Concepts

10.109 Consider the Brayton cycle in Problem 10.26. Find all the flows and fluxes of exergy and find the overall cycle second-law efficiency. Assume the heat transfers are internally reversible processes, and neglect any external irreversibility.

10.110 A Brayton cycle has a compression ratio of 15:1 with a high temperature of 1600 K and an inlet state of 290 K, 100 kPa. Use cold air properties to find the specific net work output and the second-law efficiency if we neglect the "value" of the exhaust flow.

10.111 Reconsider the previous problem and find the second-law efficiency if we do consider the "value" of the exhaust flow.

10.112 For Problem 10.104, determine the change in exergy of the water flow and that of the air flow. Use these to determine the second-law efficiency for the boiler heat exchanger.

10.113 Determine the second-law efficiency of an ideal regenerator in the Brayton cycle.

10.114 Assume a regenerator in a Brayton cycle has an efficiency of 75%. Find an expression for the second-law efficiency.

10.115 The Brayton cycle in Problem 10.19 had a heat addition of 850 kJ/kg. What is the exergy increase in the heat addition process?

10.116 The conversion efficiency of the Brayton cycle in Eq. 10.1 was determined with cold air properties. Find a similar formula for the second-law efficiency, assuming the low-T heat rejection is assigned a zero exergy value.

10.117 Redo the previous problem for a large stationary Brayton cycle where the low-T heat rejection is used in a process application and thus has nonzero exergy.

Review Problems

10.118 Solve Problem 10.19 with variable specific heats using Table A.7.

10.119 Do Problem 10.26 using properties from Table A.7.1 instead of cold air properties.

10.120 Repeat Problem 10.33, but assume that the compressor has an efficiency of 82%, both turbines have efficiencies of 87%, and the regenerator efficiency is 70%.

10.121 Consider a gas turbine cycle with two stages of compression and two stages of expansion. The pressure ratio across each compressor stage and each turbine stage is 8:1. The pressure at the entrance of the first compressor is 100 kPa, the temperature entering each compressor is 20 °C, and the temperature entering each turbine is 1100 °C. A regenerator is also incorporated into the cycle, and it has an efficiency of 70%.

Determine the compressor work, the turbine work, and the thermal efficiency of the cycle.

10.122 A gas turbine cycle has two stages of compression, with an intercooler between the stages as shown in Fig. P10.122. Air enters the first stage at 100 kPa, 300 K. The pressure ratio across each compressor stage is 5:1, and each stage has an isentropic efficiency of 82 %. Air exits the intercooler at 330 K. The maximum cycle temperature is 1500 K, and the cycle has a single turbine stage with an isentropic efficiency of 86 %. The cycle also includes a regenerator with an efficiency of 80 %. Calculate the temperature at the exit of each compressor stage, the second-law efficiency of the turbine, and the cycle thermal efficiency.

10.123 Answer the questions in Problem 10.33 assuming that friction causes pressure drops in the burner and on both sides of the regenerator. In each case, the pressure drop is estimated to be 2 % of the inlet pressure to that component of the system, so $P_3 = 588$ kPa, $P_4 = 0.98 P_3$, and $P_6 = 102$ kPa.

10.124 A gasoline engine has a volumetric compression ratio of 9. The state before compression is 290 K, 90 kPa, and the peak cycle temperature is 1800 K. Find the pressure after expansion, the cycle net work, and the cycle efficiency using properties from Table A.7.

10.125 Consider an ideal air-standard diesel cycle in which the state before the compression process is 95 kPa, 290 K, and the compression ratio is 20. Find the maximum temperature (by iteration) in the cycle that produces a thermal efficiency of 60 %.

10.126 Find the temperature after combustion and the specific energy release by combustion in Problem 10.88 using cold air properties. This is a difficult problem, and it requires iterations.

10.127 Reevaluate the combined Brayton and Rankine cycles in Problem 10.108. For a more realistic case, assume that the air compressor, the air turbine, the steam turbine, and the pump all have an isentropic efficiency of 87 %.

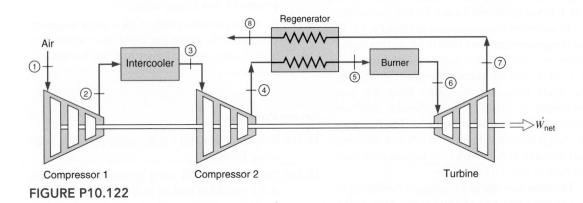

FIGURE P10.122

COMPUTER, DESIGN, AND OPEN-ENDED PROBLEMS

10.128 Write a program to solve the following problem. The effects of varying parameters on the performance of an air-standard Brayton cycle are to be determined. Consider a compressor inlet condition of 100 kPa, 20 °C, and assume constant specific heat. The thermal efficiency of the cycle and

the net specific work output should be determined for the combinations of the following variables.
a. Compressor pressure ratios of 6, 9, 12, and 15.
b. Maximum cycle temperatures of 900 °C, 1100 °C, 1300 °C, and 1500 °C.
c. Compressor and turbine isentropic efficiencies each 100 %, 90 %, 80 %, and 70 %.

10.129 The effect of adding a regenerator to the gas-turbine cycle in the previous problem is to be studied. Repeat this problem by including a regenerator with various values of the regenerator efficiency.

10.130 Write a program to simulate the Otto cycle using nitrogen as the working fluid. Use the variable specific heat given in Table A.6. The beginning of compression has a state of 100 kPa, 20 °C. Determine the net specific work output and the cycle thermal efficiency for various combinations of compression ratio and maximum cycle temperature. Compare the results with those found when constant specific heat is assumed.

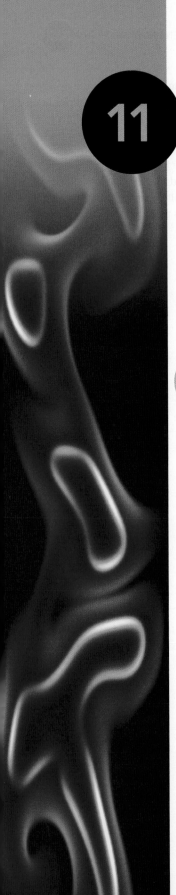

Ideal Gas Mixtures

11

Up to this point in our development of thermodynamics, we have considered primarily pure substances. A large number of thermodynamic problems involve mixtures of different pure substances. Sometimes these mixtures are referred to as *solutions*, particularly in the liquid and solid phases.

In this chapter, we shall turn our attention to various thermodynamic considerations of gas mixtures. We begin by discussing a rather simple problem: mixtures of ideal gases. This leads to a description of a simplified but very useful model of certain mixtures, such as air and water vapor, which may involve a condensed (solid or liquid) phase of one of the components.

11.1 GENERAL CONSIDERATIONS AND MIXTURES OF IDEAL GASES

Let us consider a general mixture of N components, each a pure substance, so the total mass and the total number of moles are

$$m_{\text{tot}} = m_1 + m_2 + \cdots + m_N = \sum m_i$$
$$n_{\text{tot}} = n_1 + n_2 + \cdots + n_N = \sum n_i$$

The mixture is usually described by a mass fraction (concentration)

$$c_i = \frac{m_i}{m_{\text{tot}}} \tag{11.1}$$

or a mole fraction for each component as

$$y_i = \frac{n_i}{n_{\text{tot}}} \tag{11.2}$$

which are related through the molecular mass, M_i, as $m_i = n_i M_i$. We may then convert from a mole basis to a mass basis as

$$c_i = \frac{m_i}{m_{\text{tot}}} = \frac{n_i M_i}{\sum n_j M_j} = \frac{n_i M_i / n_{\text{tot}}}{\sum n_j M_j / n_{\text{tot}}} = \frac{y_i M_i}{\sum y_j M_j} \tag{11.3}$$

and from a mass basis to a mole basis as

$$y_i = \frac{n_i}{n_{\text{tot}}} = \frac{m_i / M_i}{\sum m_j / M_j} = \frac{m_i /(M_i m_{\text{tot}})}{\sum m_j /(M_j m_{\text{tot}})} = \frac{c_i / M_i}{\sum c_j / M_j} \tag{11.4}$$

The molecular mass for the mixture becomes

$$M_{\text{mix}} = \frac{m_{\text{tot}}}{n_{\text{tot}}} = \frac{\sum n_i M_i}{n_{\text{tot}}} = \sum y_i M_i \tag{11.5}$$

which is also the denominator in Eq. 11.3.

Example 11.1

A mole-basis analysis of a gaseous mixture yields the following results:

CO_2	12.0%
O_2	4.0
N_2	82.0
CO	2.0

Determine the analysis on a mass basis and the molecular mass for the mixture.

Control mass: Gas mixture.

State: Composition known.

Solution

It is convenient to set up and solve this problem as shown in Table 11.1. The mass-basis analysis is found using Eq. 11.3, as shown in the table. It is also noted that during this calculation, the molecular mass of the mixture is found to be 30.08.

If the analysis has been given on a mass basis and the mole fractions or percentages are desired, the procedure shown in Table 11.2 is followed, using Eq. 11.4.

TABLE 11.1

Constituent	Percent by Mole	Mole Fraction	Molecular Mass	Mass kg per kmol of Mixture	Analysis on Mass Basis, Percent
CO_2	12	0.12	× 44.0	= 5.28	$\frac{5.28}{30.08}$ = 17.55
O_2	4	0.04	× 32.0	= 1.28	$\frac{1.28}{30.08}$ = 4.26
N_2	82	0.82	× 28.0	= 22.96	$\frac{22.96}{30.08}$ = 76.33
CO	2	0.02	× 28.0	= $\frac{0.56}{30.08}$	$\frac{0.56}{30.08}$ = $\frac{1.86}{100.00}$

TABLE 11.2

Constituent	Mass Fraction	Molecular Mass	kmol per kg of Mixture	Mole Fraction	Mole Percent
CO_2	0.1755	÷ 44.0	= 0.003 99	0.120	12.0
O_2	0.0426	÷ 32.0	= 0.001 33	0.040	4.0
N_2	0.7633	÷ 28.0	= 0.027 26	0.820	82.0
CO	0.0186	÷ 28.0	= $\frac{0.000\ 66}{0.033\ 24}$	$\frac{0.020}{1.000}$	$\frac{2.0}{100.0}$

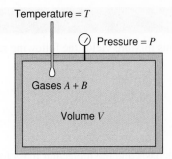

FIGURE 11.1 A mixture of two gases.

Consider a mixture of two gases (not necessarily ideal gases) such as that shown in Fig. 11.1. What properties can we experimentally measure for such a mixture? Certainly we can measure the pressure, temperature, volume, and mass of the mixture. We can also experimentally measure the composition of the mixture, and thus determine the mole and mass fractions.

Suppose that this mixture undergoes a process or a chemical reaction and we wish to perform a thermodynamic analysis of this process or reaction. What type of thermodynamic data would we use in performing such an analysis? One possibility would be to have tables of thermodynamic properties of mixtures. However, the number of different mixtures that is possible, in regard to both the substances involved and the relative amounts of each, is so great that we would need a library full of tables of thermodynamic properties to handle all possible situations. It would be much simpler if we could determine the thermodynamic properties of a mixture from the properties of the pure components. This is, in essence, the approach used in dealing with ideal gases and certain other simplified models of mixtures.

One exception to this procedure is the case where a particular mixture is encountered very frequently, the most familiar being air. Tables and charts of the thermodynamic properties of air are available. However, even in this case it is necessary to define the composition of the "air" for which the tables are given, because the composition of the atmosphere varies with altitude, with the number of pollutants, and with other variables at a given location. The composition of air on which air tables are usually based is as follows:

Component	% on Mole Basis
Nitrogen	78.10
Oxygen	20.95
Argon	0.92
CO_2 and trace elements	0.03

In this chapter, we focus on mixtures of ideal gases. We assume that each component is uninfluenced by the presence of the other components and that each component can be treated as an ideal gas. In the case of a real gaseous mixture at high pressure, this assumption would probably not be accurate because of the nature of the interaction between the molecules of the different components. In this book, we will consider only a single model in analyzing gas mixtures, namely, the Dalton model.

Dalton Model

For the Dalton model of gas mixtures, the properties of each component of the mixture are considered as though each component exists separately and independently at the temperature and volume of the mixture, as shown in Fig. 11.2. We further assume that both the gas mixture and the separated components behave according to the ideal gas model, Eqs. 2.7–2.9. In general, we would prefer to analyze gas mixture behavior on a mass basis. However, in this particular case, it is more convenient to use a mole basis, since the gas constant is then the universal gas constant for each component and also for the mixture. Thus, we may write for the mixture (Fig. 11.1)

$$PV = n\overline{R}T$$
$$n = n_A + n_B \tag{11.6}$$

and for the components (Fig. 11.2)

$$P_A V = n_A \overline{R} T$$
$$P_B V = n_B \overline{R} T \tag{11.7}$$

On substituting, we have

$$n = n_A + n_B$$
$$\frac{PV}{\overline{R}T} = \frac{P_A V}{\overline{R}T} + \frac{P_B V}{\overline{R}T} \tag{11.8}$$

or

$$P = P_A + P_B \tag{11.9}$$

where P_A and P_B are referred to as partial pressures. Thus, for a mixture of ideal gases, the pressure is the sum of the partial pressures of the individual components, where, using Eqs. 11.6 and 11.7,

$$P_A = y_A P, \qquad P_B = y_B P \tag{11.10}$$

That is, each partial pressure is the product of that component's mole fraction and the mixture pressure.

In determining the internal energy, enthalpy, and entropy of a mixture of ideal gases, the Dalton model proves useful because the assumption is made that each constituent behaves as though it occupies the entire volume by itself. Thus, the internal energy, enthalpy, and entropy can be evaluated as the sum of the respective properties of the constituent gases

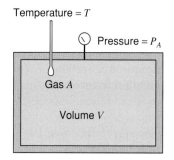
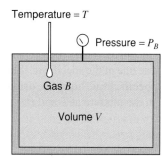

FIGURE 11.2 The Dalton model.

at the condition in which the component exists in the mixture. Since for ideal gases the internal energy and enthalpy are functions only of temperature, it follows that for a mixture of components A and B, on a mass basis,

$$
\begin{aligned}
U = mu &= m_A u_A + m_B u_B \\
&= m(c_A u_A + c_B u_B)
\end{aligned}
\tag{11.11}
$$

$$
\begin{aligned}
H = mh &= m_A h_A + m_B h_B \\
&= m(c_A h_A + c_B h_B)
\end{aligned}
\tag{11.12}
$$

In Eqs. 11.11 and 11.12, the quantities u_A, u_B, h_A, and h_B are the ideal gas properties of the components at the temperature of the mixture. For a process involving a change of temperature, the changes in these values are evaluated by one of the three models discussed in Section 3.11—involving either the ideal gas Tables A.7 or the specific heats of the components. In a similar manner to Eqs. 11.11 and 11.12, the mixture energy and enthalpy could be expressed as the sums of the component mole fractions and properties per mole.

The ideal gas mixture equation of state on a mass basis is

$$
PV = m R_{\text{mix}} T
\tag{11.13}
$$

where

$$
R_{\text{mix}} = \frac{1}{m}\left(\frac{PV}{T}\right) = \frac{1}{m}(n\overline{R}) = \overline{R}/M_{\text{mix}}
\tag{11.14}
$$

Alternatively,

$$
\begin{aligned}
R_{\text{mix}} &= \frac{1}{m}(n_A \overline{R} + n_B \overline{R}) \\[2mm]
&= \frac{1}{m}(m_A R_A + m_B R_B) \\[2mm]
&= c_A R_A + c_B R_B
\end{aligned}
\tag{11.15}
$$

The entropy of an ideal gas mixture is expressed as

$$
\begin{aligned}
S = ms &= m_A s_A + m_B s_B \\
&= m(c_A s_A + c_B s_B)
\end{aligned}
\tag{11.16}
$$

It must be emphasized that the component entropies in Eq. 11.16 must each be evaluated at the mixture temperature and the corresponding partial pressure of the component in the mixture, using Eq. 11.10 in terms of the mole fraction.

To evaluate Eq. 11.16 using the ideal gas entropy expression 6.15, it is necessary to use one of the specific heat models discussed in Section 6.7. The simplest model is constant specific heat, Eq. 6.15, using an arbitrary reference state T_0, P_0, s_{0i}, for each component i in the mixture at T and P:

$$
s_i = s_{0i} + C_{p0i} \ln\left(\frac{T}{T_0}\right) - R_i \ln\left(\frac{y_i P}{P_0}\right)
\tag{11.17}
$$

Consider a process with constant-mixture composition between state 1 and state 2, and let us calculate the entropy change for component i with Eq. 11.17.

$$(s_2 - s_1)_i = s_{0i} - s_{0i} + C_{p0i}\left[\ln\frac{T_2}{T_0} - \ln\frac{T_1}{T_0}\right] - R_i\left[\ln\frac{y_i P_2}{P_0} - \ln\frac{y_i P_1}{P_0}\right]$$

$$= 0 + C_{p0i}\ln\left[\frac{T_2}{T_0} \times \frac{T_0}{T_1}\right] - R_i\ln\left[\frac{y_i P_2}{P_0} \times \frac{P_0}{y_i P_1}\right]$$

$$= C_{p0i}\ln\frac{T_2}{T_1} - R_i\ln\frac{P_2}{P_1}$$

We observe here that this expression is very similar to Eq. 6.16 and that the reference values s_{0i}, T_0, P_0 all cancel out, as does the mole fraction.

An alternative model is to use the s_T^0 function defined in Eq. 6.18, in which case, each component entropy in Eq. 11.16 is expressed as

$$s_i = s_{Ti}^0 - R_i\ln\left(\frac{y_i P}{P_0}\right) \tag{11.18}$$

The mixture entropy could also be expressed as the sum of component properties on a mole basis.

To illustrate that mixing is an irreversible process, consider the mixing of two ideal gases, m_A and m_B, both at a given pressure and temperature, P and T, without any heat transfer or work terms. The final mixture will be at the same T (obtained from the energy equation) and at the total pressure P. The change in entropy is from Eq. 11.16 with each component entropy from Eq. 11.17, and this change equals the entropy generation by the mixing process according to the entropy equation

$$S_2 - S_1 = S_{\text{gen mix}} = m_A(s_2 - s_1)_A + m_B(s_2 - s_1)_B$$

$$= m_A\left(0 - R_A\ln\frac{P_A}{P}\right) + m_B\left(0 - R_B\ln\frac{P_B}{P}\right)$$

$$= -m_A R_A \ln y_A - m_B R_B \ln y_B$$

which can also be written in the form

$$S_{\text{gen mix}} = -n_A \overline{R} \ln y_A - n_B \overline{R} \ln y_B$$

This result can readily be generalized to include any number of components at the same temperature and pressure. The result becomes

$$S_{\text{gen mix}} = -\overline{R}\sum_k n_k \ln y_k = -n\overline{R}\sum_k y_k \ln y_k \tag{11.19}$$

The interesting thing about this equation is that the increase in entropy depends only on the number of moles of component gases and is independent of the composition of the gas. For example, when 1 mol of oxygen and 1 mol of nitrogen are mixed, the increase in entropy is the same as when 1 mol of hydrogen and 1 mol of nitrogen are mixed. But we also know that if 1 mol of nitrogen is "mixed" with another mole of nitrogen, there is no increase in entropy. The question that arises is, how dissimilar must the gases be in order to have an increase in entropy? The answer lies in our ability to distinguish between the two gases (based on their different molecular masses). The entropy increases whenever we can distinguish between the gases being mixed. When we cannot distinguish between the gases, there is no increase in entropy.

One special case that arises frequently involves an ideal gas mixture undergoing a process in which there is no change in composition. Let us also assume that the constant specific heat model is reasonable. For this case, from Eq. 11.11 on a unit mass basis, the internal energy change is

$$u_2 - u_1 = c_A C_{v0\,A}(T_2 - T_1) + c_B C_{v0\,B}(T_2 - T_1)$$
$$= C_{v0\,\text{mix}}(T_2 - T_1) \tag{11.20}$$

where

$$C_{v0\,\text{mix}} = c_A C_{v0\,A} + c_B C_{v0\,B} \tag{11.21}$$

Similarly, from Eq. 11.12, the enthalpy change is

$$h_2 - h_1 = c_A C_{p0\,A}(T_2 - T_1) + c_B C_{p0\,B}(T_2 - T_1)$$
$$= C_{p0\,\text{mix}}(T_2 - T_1) \tag{11.22}$$

where

$$C_{p0\,\text{mix}} = c_A C_{p0\,A} + c_B C_{p0\,B} \tag{11.23}$$

The entropy change for a single component was calculated from Eq. 11.17, so we substitute this result into Eq. 11.16 to evaluate the change as

$$s_2 - s_1 = c_A(s_2 - s_1)_A + c_B(s_2 - s_1)_B$$

$$= c_A C_{p0\,A} \ln \frac{T_2}{T_1} - c_A R_A \ln \frac{P_2}{P_1} + c_B C_{p0\,B} \ln \frac{T_2}{T_1} - c_B R_B \ln \frac{P_2}{P_1}$$

$$= C_{p0\,\text{mix}} \ln \frac{T_2}{T_1} - R_\text{mix} \ln \frac{P_2}{P_1} \tag{11.24}$$

The last expression used Eq. 11.15 for the mixture gas constant and Eq. 11.23 for the mixture specific heat. We see that Eqs. 11.20, 11.22, and 11.24 are the same as those for the pure substance, Eqs. 3.35, 3.44, and 6.16. So we can treat a mixture similarly to a pure substance once the mixture properties are found from the composition and the component properties in Eqs. 11.15, 11.21, and 11.23.

This also implies that all the polytropic processes in a mixture can be treated similarly to the way it is done for a pure substance (recall Sections 6.7 and 6.8). Specifically, the isentropic process where s is constant leads to the power relation between temperature and pressure from Eq. 11.24. This is similar to Eq. 6.20, provided we use the mixture specific heat and gas constant. The ratio of specific heats becomes

$$k = k_\text{mix} = \frac{C_{p\,\text{mix}}}{C_{v\,\text{mix}}} = \frac{C_{p\,\text{mix}}}{C_{p\,\text{mix}} - R_\text{mix}}$$

and the relation can then also be written as in Eq. 6.23.

Example 11.2

A mixture of 60% carbon dioxide and 40% water by mass is flowing at 400 K, 100 kPa into a reversible adiabatic compressor, where it is compressed to 1000 kPa. Find the exit temperature using (a) constant specific heats and (b) variable specific heats.

Analysis and Solution

$$\text{Energy Eq.:} \quad 0 = h_i - h_e + w_{c\,\text{in}}$$

$$\text{Entropy Eq.:} \quad 0 = s_i - s_e + 0 + 0$$

The exit states is therefore determined as

$$\text{Exit state:} \quad P_e,\ s_e = s_i$$

The mixture gas constant R_{mix} and specific heat $C_{p\,\text{mix}}$ become

$$R_{\text{mix}} = 0.6 \times 0.1889 + 0.4 \times 0.4615 = 0.297\,94 \text{ kJ/kg·K}$$

$$C_{p\,\text{mix}} = 0.6 \times 0.842 + 0.4 \times 1.872 = 1.254 \text{ kJ/kg·K}$$

$$k_{\text{mix}} = 1.254/(1.254 - 0.297\,94) = 1.3116$$

From Eq. 6.23 we get

$$T_e = T_i (P_e/P_i)^{(k-1)/k} = 400 \text{ K} \left(\frac{1000}{100}\right)^{0.3116/1.3116} = 691.25 \text{ K}$$

For variable specific heats, we must use the tabulated standard entropy functions to evaluate the change in entropy.

$$s_i - s_e = \sum c_j (s^0_{Ti} - s^0_{Te})_j - R_{\text{mix}} \ln(P_e/P_i) = 0$$

$$= 0.6(s^0_{Te\,CO_2} - 5.1196) + 0.4(s^0_{Te\,H_2O} - 11.0345) - 0.297\,94 \ln\left(\frac{1000}{100}\right)$$

Arrange the entropy equation with knowns on the right-hand side as

$$0.6\,s^0_{Te\,CO_2} + 0.4\,s^0_{Te\,H_2O} = 0.6 \times 5.1196 + 0.4 \times 11.0345 + 0.297\,94 \ln\left(\frac{1000}{100}\right)$$

$$= 8.171\,59 \text{ kJ/kg·K}$$

The left-hand-side T_e is found by trial and error as

$$\text{LHS at } 650 \text{ K} = 8.164\,66 \text{ kJ/kg·K}; \quad \text{LHS at } 700 \text{ K} = 8.2754 \text{ kJ/kg·K}$$

Now interpolate for the final result:

$$T_e = 650 \text{ K} + 50 \text{ K} \frac{8.171\,59 - 8.164\,66}{8.2754 - 8.164\,66} = 653.1 \text{ K}$$

So far, we have looked at mixtures of ideal gases as a natural extension to the description of processes involving pure substances. The treatment of mixtures for nonideal (real) gases and multiphase states is important for many technical applications—for instance, in the chemical process industry. It does require a more extensive study of the properties and general equations of state, so we will defer this subject to Chapter 12.

In-Text Concept Questions

a. Are the mass and mole fractions for a mixture ever the same?

b. For a mixture, how many component concentrations are needed?

c. Are any of the properties (P, T, v) for oxygen and nitrogen in air the same?

d. If I want to heat a flow of a four-component mixture from 300 to 310 K at constant P, how many properties and which properties do I need to know to find the heat transfer?

e. To evaluate the change in entropy between two states at different T and P values for a given mixture, do I need to find the partial pressures?

11.2 A SIMPLIFIED MODEL OF A MIXTURE INVOLVING GASES AND A VAPOR

Let us now consider a simplification, which is often a reasonable one, of the problem involving a mixture of ideal gases that is in contact with a solid or liquid phase of one of the components. The most familiar example is a mixture of air and water vapor in contact with liquid water or ice, such as is encountered in air conditioning or in drying. We are all familiar with the condensation of water from the atmosphere when it cools on a summer day.

This problem and a number of similar problems can be analyzed quite simply and with considerable accuracy if the following assumptions are made:

1. The solid or liquid phase contains no dissolved gases.

2. The gaseous phase can be treated as a mixture of ideal gases.

3. When the mixture and the condensed phase are at a given pressure and temperature, the equilibrium between the condensed phase and its vapor is not influenced by the presence of the other component. This means that when equilibrium is achieved, the partial pressure of the vapor will be equal to the saturation pressure corresponding to the temperature of the mixture.

Since this approach is used extensively and with considerable accuracy, let us give some attention to the terms that have been defined and the type of problems for which this approach is valid and relevant. In our discussion, we will refer to this as a *gas–vapor mixture*.

The dew point of a gas–vapor mixture is the temperature at which the vapor condenses or solidifies when it is cooled at constant pressure. This is shown on the T–s diagram for the vapor shown in Fig. 11.3. Suppose that the temperature of the gaseous mixture and the partial pressure of the vapor in the mixture are such that the vapor is initially superheated at state 1. If the mixture is cooled at constant pressure, the partial pressure of the vapor remains constant until point 2 is reached, and then condensation begins. The temperature at state 2 is the dew-point temperature. Line 1–3 on the diagram indicates that if the mixture is cooled at constant volume the condensation begins at point 3, which is slightly lower than the dew-point temperature.

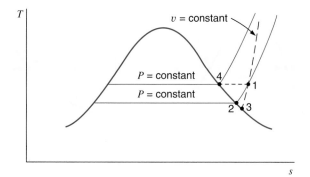

FIGURE 11.3

T–s diagram to show definition of the dew point.

If the vapor is at the saturation pressure and temperature, the mixture is referred to as a *saturated mixture*, and for an air–water vapor mixture, the term saturated air is used.

The relative humidity ϕ is defined as the ratio of the mole fraction of the vapor in the mixture to the mole fraction of vapor in a saturated mixture at the same temperature and total pressure. Since the vapor is considered an ideal gas, the definition reduces to the ratio of the partial pressure of the vapor as it exists in the mixture, P_v, to the saturation pressure of the vapor at the same temperature, P_g:

$$\phi = \frac{P_v}{P_g}$$

In terms of the numbers on the *T–s* diagram of Fig. 11.3, the relative humidity ϕ would be

$$\phi = \frac{P_1}{P_4}$$

Since we are considering the vapor to be an ideal gas, the relative humidity can also be defined in terms of specific volume or density:

$$\phi = \frac{P_v}{P_g} = \frac{\rho_v}{\rho_g} = \frac{v_g}{v_v} \qquad (11.25)$$

The humidity ratio, ω, of an air–water vapor mixture is defined as the ratio of the mass of water vapor, m_v, to the mass of dry air, m_a. The term dry air is used to emphasize that this refers only to air and not to the water vapor. The terms specific humidity or absolute humidity are used synonymously with *humidity ratio*.

$$\omega = \frac{m_v}{m_a} \qquad (11.26)$$

This definition is identical for any other gas–vapor mixture, and the subscript a refers to the gas, exclusive of the vapor. Since we consider both the vapor and the mixture to be ideal gases, a very useful expression for the humidity ratio in terms of partial pressures and molecular masses can be developed. Writing

$$m_v = \frac{P_v V}{R_v T} = \frac{P_v V M_v}{\overline{R} T}, \qquad m_a = \frac{P_a V}{R_a T} = \frac{P_a V M_a}{\overline{R} T}$$

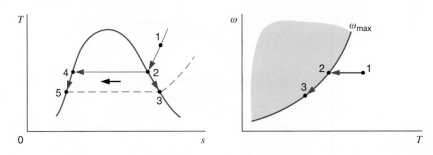

FIGURE 11.4
T–s diagram to show the cooling of a gas–vapor mixture at a constant pressure.

we have

$$\omega = \frac{P_v V / R_v T}{P_a V / R_a T} = \frac{R_a P_v}{R_v P_a} = \frac{M_v P_v}{M_a P_a} \qquad (11.27)$$

For an air–water vapor mixture, this reduces to

$$\omega = 0.622 \frac{P_v}{P_a} = 0.622 \frac{P_v}{P_{\text{tot}} - P_v} \qquad (11.28)$$

The degree of saturation is defined as the ratio of the actual humidity ratio to the humidity ratio of a saturated mixture at the same temperature and total pressure. This refers to the maximum amount of water that can be contained in moist air, which is seen from the absolute humidity in Eq. 11.28. Since the partial pressure for air $P_a = P_{\text{tot}} - P_v$ and $P_v = \phi P_g$ from Eq. 11.25, we can write

$$\omega = 0.622 \frac{\phi P_g}{P_{\text{tot}} - \phi P_g} \quad \leq \quad \omega_{\text{max}} = 0.622 \frac{P_g}{P_{\text{tot}} - P_g} \qquad (11.29)$$

The maximum humidity ratio corresponds to a relative humidity of 100 % and is a function of the total pressure (usually atmospheric) and the temperature due to P_g. This relation is also illustrated in Fig. 11.4 as a function of temperature, and the function has an asymptote at a temperature where $P_g = P_{\text{tot}}$, which is 100 °C for atmospheric pressure. The shaded regions are states not permissible, as the water vapor pressure would be larger than the saturation pressure. In a cooling process at constant total pressure, the partial pressure of the vapor remains constant until the dew point is reached at state 2; this is also on the maximum humidity ratio curve. Further cooling lowers the maximum possible humidity ratio, and some of the vapor condenses. The vapor that remains in the mixture is always saturated, and the liquid or solid is in equilibrium with it. For example, when the temperature is reduced to T_3, the vapor in the mixture is at state 3, and its partial pressure is P_g at T_3 and the liquid is at state 5 in equilibrium with the vapor.

Example 11.3

Consider 100 m³ of an air–water vapor mixture at 0.1 MPa, 35 °C, and 70 % relative humidity. Calculate the humidity ratio, dew point, mass of air, and mass of vapor.

Control mass: Mixture.

State: P, T, ϕ known; state fixed.

Analysis and Solution

From Eq. 11.25 and the steam tables, we have

$$P_v = \phi P_g = 0.70(5.628) = 3.94\,\text{kPa}$$

The dew point is the saturation temperature corresponding to this pressure, which is $28.6\,^\circ\text{C}$. The partial pressure of the air is

$$P_a = P - P_v = 100 - 3.94 = 96.06\,\text{kPa}$$

The humidity ratio can be calculated from Eq. 11.28:

$$\omega = 0.622 \times \frac{P_v}{P_a} = 0.622 \times \frac{3.94}{96.06} = 0.0255$$

The mass of air is

$$m_a = \frac{P_a V}{R_a T} = \frac{96.06 \times 100}{0.287 \times 308.2} = 108.6\,\text{kg}$$

The mass of the vapor can be calculated by using the humidity ratio or by using the ideal gas equation of state:

$$m_v = \omega m_a = 0.0255(108.6) = 2.77\,\text{kg}$$

$$m_v = \frac{P_v V}{R_v T} = \frac{3.94 \times 100}{0.4615 \times 308.2} = 2.77\,\text{kg}$$

Example 11.4

Calculate the amount of water vapor condensed if the mixture of Example 11.3 is cooled to $5\,^\circ\text{C}$ in a constant-pressure process.

Control mass: Mixture.
Initial state: Known (Example 11.3).
Final state: T known.
Process: Constant pressure.

Analysis

At the final temperature, $5\,^\circ\text{C}$, the mixture is saturated, since this is below the dew-point temperature. Therefore,

$$P_{v2} = P_{g2}, \qquad P_{a2} = P - P_{v2}$$

and

$$\omega_2 = 0.622 \frac{P_{v2}}{P_{a2}}$$

From the conservation of mass, it follows that the amount of water condensed is equal to the difference between the initial and final mass of water vapor, or

$$\text{Mass of vapor condensed} = m_a(\omega_1 - \omega_2)$$

Solution

We have

$$P_{v2} = P_{g2} = 0.8721 \text{ kPa}$$

$$P_{a2} = 100 - 0.8721 = 99.128 \text{ kPa}$$

Therefore,

$$\omega_2 = 0.622 \times \frac{0.8721}{99.128} = 0.0055$$

$$\text{Mass of vapor condensed} = m_a(\omega_1 - \omega_2) = 108.6(0.0255 - 0.0055)$$

$$= 2.172 \text{ kg}$$

11.3 THE ENERGY EQUATION APPLIED TO GAS–VAPOR MIXTURES

In applying the energy equation to gas–vapor mixtures, it is helpful to realize that because of our assumption that ideal gases are involved, the various components can be treated separately when calculating changes of internal energy and enthalpy. Therefore, in dealing with air–water vapor mixtures, the changes in enthalpy of the water vapor can be found from the steam tables and the ideal gas relations can be applied to the air. This is illustrated by the examples that follow.

In most applications involving moist air, we are dealing with conditions that are close to atmospheric temperature and pressure. The water content of the mixture is typically low on an absolute scale (recall values of ω from the previous examples), so the partial pressure of the water vapor is only a few kPa. This means that representing the water vapor as an ideal gas is an excellent approximation, and the properties (h, u) are functions of temperature only. For a flow situation, the energy equation contains terms like

$$\dot{m}h = \dot{m}_a h_a + \dot{m}_v h_v = \dot{m}_a(h_a + \omega h_v)$$

so the enthalpy per mass of dry air flowing is

$$\tilde{h} = h_a + \omega h_v = h_a + \omega h_g$$

where the enthalpy of superheated vapor equals the value for saturated vapor. For a stored energy term like mu, the same approximation is used

$$mu = m_a u_a + m_v u_v = m_a(u_a + \omega u_g)$$

The saturated vapor properties are found from Table B.1 and the air properties are from Table A.7. However, a model of constant specific heat for air is used often since the temperature range in these applications is limited.

Example 11.5

An air-conditioning unit is shown in Fig. 11.5, with pressure, temperature, and relative humidity data. Calculate the heat transfer per kilogram of dry air, assuming that changes in kinetic energy are negligible.

Control volume: Duct, excluding cooling coils.

Inlet state: Known (Fig. 11.5).

Exit state: Known (Fig. 11.5).

Process: Steady state with no kinetic or potential energy changes.

Model: Air—ideal gas, constant specific heat, value at 300 K. Water—steam tables.

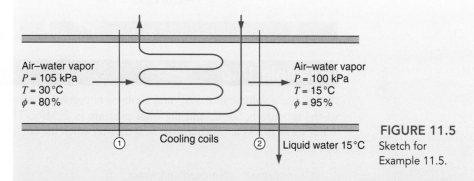

Air–water vapor
P = 105 kPa
T = 30 °C
ϕ = 80 %

Air–water vapor
P = 100 kPa
T = 15 °C
ϕ = 95 %

Cooling coils

① ② Liquid water 15 °C

FIGURE 11.5
Sketch for
Example 11.5.

Analysis

From the continuity equations for air and water, we have

$$\dot{m}_{a1} = \dot{m}_{a2}$$

$$\dot{m}_{v1} = \dot{m}_{v2} + \dot{m}_{l2}$$

The energy equation gives

$$\dot{Q}_{c.v.} + \sum \dot{m}_i h_i = \sum \dot{m}_e h_e$$

$$\dot{Q}_{c.v.} + \dot{m}_a h_{a1} + \dot{m}_{v1} h_{v1} = \dot{m}_a h_{a2} + \dot{m}_{v2} h_{v2} + \dot{m}_{l2} h_{l2}$$

If we divide this equation by $\dot{m}_a$, introduce the continuity equation for the water, and note that $\dot{m}_v = \omega \dot{m}_a$, we can write the energy equation in the form

$$\frac{\dot{Q}_{c.v.}}{\dot{m}_a} + h_{a1} + \omega_1 h_{v1} = h_{a2} + \omega_2 h_{v2} + (\omega_1 - \omega_2) h_{l2}$$

Solution

We have

$$P_{v1} = \phi_1 P_{g1} = 0.80(4.246) = 3.397 \, \text{kPa}$$

$$\omega_1 = \frac{R_a}{R_v} \frac{P_{v1}}{P_{a1}} = 0.622 \times \left(\frac{3.397}{105 - 3.4} \right) = 0.0208$$

$$P_{v2} = \phi_2 P_{g2} = 0.95(1.7051) = 1.620 \text{ kPa}$$

$$\omega_2 = \frac{R_a}{R_v} \times \frac{P_{v2}}{P_{a2}} = 0.622 \times \left(\frac{1.62}{100 - 1.62}\right) = 0.0102$$

Substituting, we obtain

$$\dot{Q}_{\text{c.v.}}/\dot{m}_a + h_{a1} + \omega_1 h_{v1} = h_{a2} + \omega_2 h_{v2} + (\omega_1 - \omega_2) h_{l2}$$

$$\dot{Q}_{\text{c.v.}}/\dot{m}_a = 1.004(15 - 30) + 0.0102(2528.9)$$

$$- 0.0208(2556.3) + (0.0208 - 0.0102)(62.99)$$

$$= -41.76 \text{ kJ/kg dry air}$$

Example 11.6

A tank has a volume of 0.5 m³ and contains nitrogen and water vapor. The temperature of the mixture is 50 °C, and the total pressure is 2 MPa. The partial pressure of the water vapor is 5 kPa. Calculate the heat transfer when the contents of the tank are cooled to 10 °C.

Control mass: Nitrogen and water.

Initial state: P_1, T_1 known; state fixed.

Final state: T_2 known.

Process: Constant volume.

Model: Ideal gas mixture; constant specific heat for nitrogen; steam tables for water.

Analysis

This is a constant-volume process. Since the work is zero, the energy equation reduces to

$$Q = U_2 - U_1 = m_{N_2} C_{v(N_2)}(T_2 - T_1) + (m_2 u_2)_v + (m_2 u_2)_l - (m_1 u_1)_v$$

This equation assumes that some of the vapor condensed. This assumption must be checked, however, as shown in the solution.

Solution

The mass of nitrogen and water vapor can be calculated using the ideal gas equation of state:

$$m_{N_2} = \frac{P_{N_2} V}{R_{N_2} T} = \frac{1995 \times 0.5}{0.2968 \times 323.2} = 10.39 \text{ kg}$$

$$m_{v1} = \frac{P_{v1} V}{R_v T} = \frac{5 \times 0.5}{0.4615 \times 323.2} = 0.016\,76 \text{ kg}$$

If condensation takes place, the final state of the vapor will be saturated vapor at 10 °C. Therefore,

$$m_{v2} = \frac{P_{v2}V}{R_v T} = \frac{1.2276 \times 0.5}{0.4615 \times 283.2} = 0.004\,70 \text{ kg}$$

Since this amount is less than the original mass of vapor, there must have been condensation. The mass of liquid that is formed, m_{l2}, is

$$m_{l2} = m_{v1} - m_{v2} = 0.016\,76 - 0.004\,70 = 0.012\,06 \text{ kg}$$

The internal energy of the water vapor is equal to the internal energy of saturated water vapor at the same temperature. Therefore,

$$u_{v_1} = 2443.5 \text{ kJ/kg}$$

$$u_{v_2} = 2389.2 \text{ kJ/kg}$$

$$u_{l2} = 42.0 \text{ kJ/kg}$$

$$\dot{Q}_{c.v.} = 10.39 \times 0.745(10 - 50) + 0.0047(2389.2)$$
$$+ 0.012\,06(42.0) - 0.016\,76(2443.5)$$
$$= -338.8 \text{ kJ}$$

11.4 THE ADIABATIC SATURATION PROCESS

An important process for an air–water vapor mixture is the adiabatic saturation process. In this process, an air–vapor mixture comes in contact with a body of water in a well-insulated duct (Fig. 11.6). If the initial humidity is less than 100 %, some of the water will evaporate and the temperature of the air–vapor mixture will decrease. If the mixture leaving the duct is saturated and if the process is adiabatic, the temperature of the mixture on leaving is known as the *adiabatic saturation temperature*. For this to take place as a steady-state process, makeup water at the adiabatic saturation temperature is added at the same rate at which water is evaporated. The pressure is assumed to be constant.

Considering the adiabatic saturation process to be a steady-state process, and neglecting changes in kinetic and potential energy, the energy equation reduces to

$$h_{a1} + \omega_1 h_{v1} + (\omega_2 - \omega_1)h_{l2} = h_{a2} + \omega_2 h_{v2}$$
$$\omega_1(h_{v1} - h_{l2}) = C_{pa}(T_2 - T_1) + \omega_2(h_{v2} - h_{l2})$$
$$\omega_1(h_{v1} - h_{l2}) = C_{pa}(T_2 - T_1) + \omega_2 h_{fg2} \qquad (11.30)$$

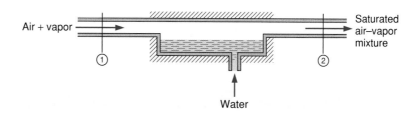

FIGURE 11.6 The adiabatic saturation process.

The most significant point to be made about the adiabatic saturation process is that the adiabatic saturation temperature, the temperature of the mixture when it leaves the duct, is a function of the pressure, temperature, and relative humidity of the entering air–vapor mixture and of the exit pressure. Thus, the relative humidity and the humidity ratio of the entering air–vapor mixture can be determined from the measurements of the pressure and temperature of the air–vapor mixture entering and leaving the adiabatic saturator. Since these measurements are relatively easy to make, this is one means of determining the humidity of an air–vapor mixture.

Example 11.7

The pressure of the mixture entering and leaving the adiabatic saturator is 0.1 MPa, the entering temperature is 30 °C, and the temperature leaving is 20 °C, which is the adiabatic saturation temperature. Calculate the humidity ratio and relative humidity of the air–water vapor mixture entering.

Control volume: Adiabatic saturator.

Inlet state: P_1, T_1 known.

Exit state: P_2, T_2 known; $\phi_2 = 100\%$; state fixed.

Process: Steady state, adiabatic saturation (Fig. 11.6).

Model: Ideal gas mixture; constant specific heat for air; steam tables for water.

Analysis

Use continuity and the energy equation, Eq. 11.30.

Solution

Since the water vapor leaving is saturated, $P_{v2} = P_{g2}$ and ω_2 can be calculated.

$$\omega_2 = 0.622 \times \left(\frac{2.339}{100 - 2.34} \right) = 0.0149$$

ω_1 can be calculated using Eq. 11.30.

$$\omega_1 = \frac{C_{pa}(T_2 - T_1) + \omega_2 h_{fg2}}{(h_{v1} - h_{l2})}$$

$$= \frac{1.004(20 - 30) + 0.0149 \times 2454.1}{2556.3 - 83.96} = 0.0107$$

$$\omega_1 = 0.0107 = 0.622 \times \left(\frac{P_{v1}}{100 - P_{v1}} \right)$$

$$P_{v1} = 1.691 \text{ kPa}$$

$$\phi_1 = \frac{P_{v1}}{P_{g1}} = \frac{1.691}{4.246} = 0.398$$

f. What happens to relative and absolute humidity when moist air is heated?

g. If I cool moist air, do I reach the dew point first in a constant-P or constant-V process?

h. What happens to relative and absolute humidity when moist air is cooled?

i. Explain in words what the absolute and relative humidity express.

j. In which direction does an adiabatic saturation process change Φ, ω, and T?

11.5 ENGINEERING APPLICATIONS—WET-BULB AND DRY-BULB TEMPERATURES AND THE PSYCHROMETRIC CHART

The humidity of air–water vapor mixtures has traditionally been measured with a device called a *psychrometer*, which uses the flow of air past wet-bulb and dry-bulb thermometers. The bulb of the wet-bulb thermometer is covered with a cotton wick saturated with water. The dry-bulb thermometer is used simply to measure the temperature of the air. The air flow can be maintained by a fan, as shown in the continuous-flow psychrometer depicted in Fig. 11.7.

The processes that take place at the wet-bulb thermometer are somewhat complicated. First, if the air–water vapor mixture is not saturated, some of the water in the wick evaporates and diffuses into the surrounding air, which cools the water in the wick. As soon as the temperature of the water drops, however, heat is transferred to the water from both the air and the thermometer, with corresponding cooling. A steady state, determined by heat and mass transfer rates, will be reached, in which the wet-bulb thermometer temperature is lower than the dry-bulb temperature.

It can be argued that this evaporative cooling process is very similar, but not identical, to the adiabatic saturation process described and analyzed in Section 11.4. In fact, the

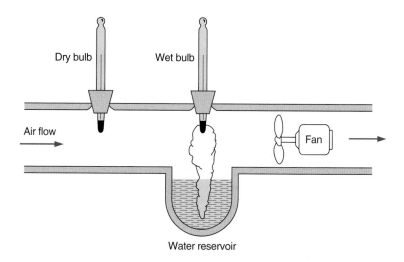

FIGURE 11.7
Steady-flow apparatus for measuring wet- and dry-bulb temperatures.

adiabatic saturation temperature is often termed the thermodynamic wet-bulb temperature. It is clear, however, that the wet-bulb temperature as measured by a psychrometer is influenced by heat and mass transfer rates, which depend, for example, on the air flow velocity and not simply on thermodynamic equilibrium properties. It does happen that the two temperatures are very close for air–water vapor mixtures at atmospheric temperature and pressure, and they will be assumed to be equivalent in this book.

In recent years, humidity measurements have been made using other phenomena and other devices, primarily electronic devices for convenience and simplicity. For example, some substances tend to change in length, in shape, in electrical capacitance, or in a number of other ways when they absorb moisture. They are therefore sensitive to the amount of moisture in the atmosphere. An instrument making use of such a substance can be calibrated to measure the humidity of air–water vapor mixtures. The instrument output can be programmed to furnish any of the desired parameters, such as relative humidity, humidity ratio, or wet-bulb temperature.

Properties of air–water vapor mixtures are given in graphical form on psychrometric charts. These are available in a number of different forms, and only the main features are considered here. It should be recalled that three independent properties—such as pressure, temperature, and mixture composition—will describe the state of this binary mixture.

A simplified version of the chart included in Appendix E, Fig. E.4, is shown in Fig. 11.8. This basic psychrometric chart is a plot of humidity ratio (ordinate) as a function of dry-bulb temperature (abscissa), with relative humidity, wet-bulb temperature, and mixture enthalpy per mass of dry air as parameters. If we fix the total pressure for which the chart is to be constructed (which in our chart is 1 bar, or 100 kPa), lines of constant relative humidity and wet-bulb temperature can be drawn on the chart, because for a given dry-bulb temperature, total pressure, and humidity ratio, the relative humidity and wet-bulb temperature are fixed. The partial pressure of the water vapor is fixed by the humidity ratio and the total pressure, and therefore a second ordinate scale that indicates the partial

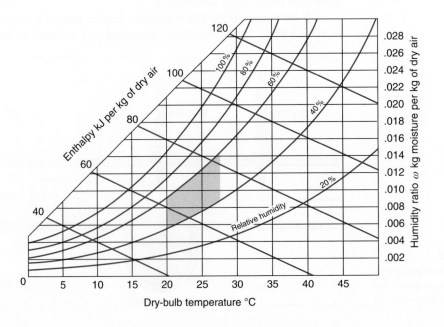

FIGURE 11.8
Psychrometric chart.

pressure of the water vapor could be constructed. It would also be possible to include the mixture-specific volume and entropy on the chart.

Most psychrometric charts give the enthalpy of an air–vapor mixture per kilogram of dry air. The values given assume that the enthalpy of the dry air is zero at $-20\,°C$, and the enthalpy of the vapor is taken from the steam tables (which are based on the assumption that the internal energy of saturated liquid is zero at $0\,°C$). The value used in the psychrometric chart is then

$$\tilde{h} \equiv h_a - h_a(-20\,°C) + \omega h_v$$

This procedure is satisfactory because we are usually concerned only with differences in enthalpy. That the lines of constant enthalpy are essentially parallel to lines of constant wet-bulb temperature is evident from the fact that the wet-bulb temperature is essentially equal to the adiabatic saturation temperature. Thus, in Fig. 11.6, if we neglect the enthalpy of the liquid entering the adiabatic saturator, the enthalpy of the air–vapor mixture leaving at a given adiabatic saturation temperature fixes the enthalpy of the mixture entering.

The chart plotted in Fig. 11.8 also indicates the human comfort zone, as the range of conditions most agreeable for human well-being. An air conditioner should then be able to maintain an environment within the comfort zone regardless of the outside atmospheric conditions to be considered adequate. Some charts are available that give corrections for variation from standard atmospheric pressures. Before using a given chart, one should fully understand the assumptions made in constructing it and should recognize that it is applicable to the particular problem at hand.

The direction in which various processes proceed for an air–water vapor mixture is shown on the psychrometric chart of Fig. 11.9. For example, a constant-pressure cooling process beginning at state 1 proceeds at constant humidity ratio to the dew point at state 2, with continued cooling below that temperature moving along the saturation line (100 % relative humidity) to point 3. Other processes could be traced out in a similar manner.

Several technical important processes involve atmospheric air that is being heated or cooled and water is added or subtracted. Special care is needed to design equipment that can withstand the condensation of water so that corrosion is avoided. In building an air conditioner, whether it is a single window unit or a central air-conditioning unit, liquid water will appear when air is being cooled below the dew point, and a proper drainage system should be arranged.

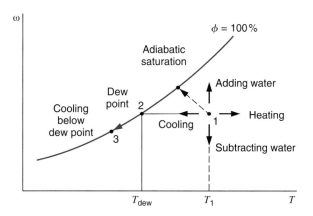

FIGURE 11.9

Processes on a psychrometric chart.

Example 11.8

Air at (1) $20\,°C$, $\phi = 40\,\%$ flows through a 400 W heater coming out at (2) $46\,°C$ and is blown over a wet surface, so it absorbs water and becomes saturated (3) without heat input. Finally, the air flows over the walls in the room, where it cools to $20\,°C$ (4) (see Fig. 11.10). We would like to know the mass flow rate of air and the rate of water condensation in the room, if any.

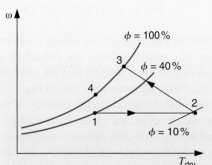

FIGURE 11.10   Process path for Example 11.8.

This is a combination of three steady flow processes:

CV. 1 to 2 Heating: $\omega_2 = \omega_1$; $q = \tilde{h}_2 - \tilde{h}_1$

CV. 2 to 3 Adiabatic saturation:

$$\omega_3 - \omega_2 = \dot{m}_{liq}/\dot{m}_a; \qquad \dot{m}_a \tilde{h}_2 + \dot{m}_{liq} h_f = \dot{m}_a \tilde{h}_3$$

CV. 3 to 4 Cooling:

$$\dot{m}_{liq} = (\omega_3 - \omega_4)\,\dot{m}_a; \qquad \dot{m}_a \tilde{h}_3 = \dot{Q}_{cool} + \dot{m}_{liq} h_f + \dot{m}_a \tilde{h}_4$$

State properties form the psychrometric chart

1: T, $\phi \Rightarrow \omega_1 = 0.0056$, $\tilde{h}_1 = 54\,kJ/kg$ dry air

2: $\omega_2 = \omega_1$, $T_2 \Rightarrow \phi_2 = 9\,\%$, $\tilde{h}_2 = 80\,kJ/kg$ dry air

3: $\phi = 100\,\%$ & adiabatic $\Rightarrow T_3 = T_{wet,2} = 21.2\,°C$, $w_3 = 0.0158$

4: $\phi = 100\,\%$, $T_4 \Rightarrow \omega_4 = 0.0148$

Solution

$$q = \tilde{h}_2 - \tilde{h}_1 = 80 - 54 = 26\,kJ/kg\ \text{dry air}$$

$$\dot{m}_a = \dot{Q}/q = 0.4\,kW/26\,kJ/kg = 0.0154\,kg/s$$

$$\dot{m}_{liq} = (\omega_3 - \omega_4)\,\dot{m}_a = (0.0158 - 0.0148) \times 0.0154 = 0.0154\,g/s$$

An example of an air-conditioning unit is shown in Fig. 11.11. It is operated in cooling mode, so the inside heat exchanger is the cold evaporator in a refrigeration cycle. The outside unit contains the compressor and the heat exchanger that functions as the condenser, rejecting

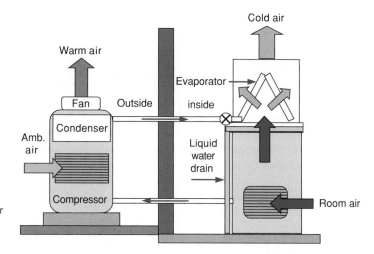

FIGURE 11.11 An air conditioner operating in cooling mode.

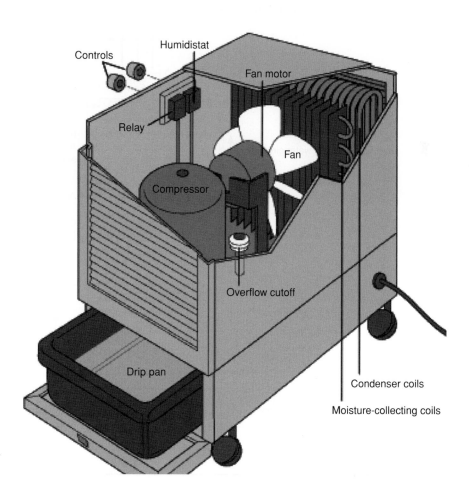

FIGURE 11.12 A household dehumidifier unit.

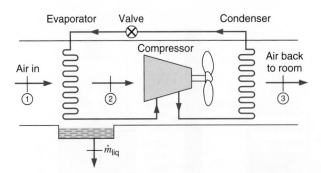

FIGURE 11.13 The dehumidifier schematic.

energy to the ambient air as the fan forces air over the warm surfaces. The same unit can function as a heat pump by reversing the two flows in a double-acting valve so that the inside heat exchanger becomes the condenser and the outside heat exchanger becomes the evaporator. In this mode, it is possible to form frost on the outside unit if the evaporator temperature is low enough.

A refrigeration cycle is also used in a smaller dehumidifier unit shown in Fig. 11.12, where a fan drives air in over the evaporator, so that it cools below the dew point and liquid water forms on the surfaces and drips into a container or drain. After some water is removed from the air, it flows over the condenser that heats the air flow, as illustrated in Fig. 11.13.

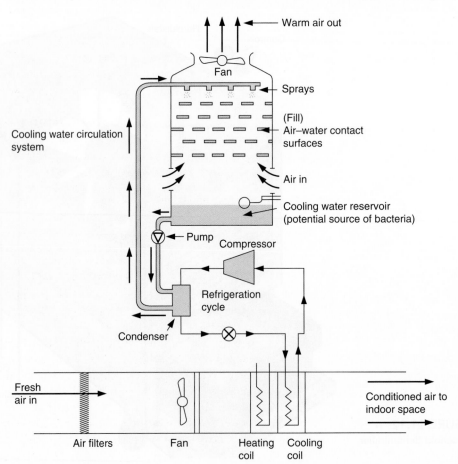

FIGURE 11.14
A cooling tower with evaporative cooling for building air-conditioning use.

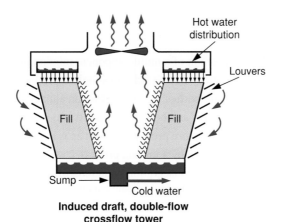

FIGURE 11.15
A cooling tower for a power plant with evaporative cooling.

This figure also shows the refrigeration cycle schematics. Looking at a control volume that includes all the components, we see that the net effect is to remove some relatively cold liquid water and add the compressor work, which heats up the air.

The cooling effect of the adiabatic saturation process is used in evaporative cooling devices to bring some water to a lower temperature than a heat exchanger alone could accomplish under a given atmospheric condition. On a larger scale, this process is used for power plants when there is no suitable large body of water to absorb the energy from the condenser. A combination with a refrigeration cycle is shown in Fig. 11.14 for building air-conditioning purposes, where the cooling tower keeps a low $T_{(high)}$ temperature for the refrigeration cycle to obtain a large COP. Much larger cooling towers are used for the power plants shown in Fig. 11.15 to make cold water to cool the condenser. As some of the water in both of these units evaporates, the water must be replenished. A large cloud is often seen rising from these towers as the water vapor condenses to form small droplets after mixing with more atmospheric air.

SUMMARY

A mixture of gases is treated from the specification of the mixture composition of the various components based on mass or on moles. This leads to the mass fractions and mole fractions, both of which can be called concentrations. The mixture has an overall average molecular mass and other mixture properties on a mass or mole basis. Further simple models includes Dalton's model of ideal mixtures of ideal gases, which leads to partial pressures as the contribution from each component to the total pressure given by the mole fraction. As entropy is sensitive to pressure, the mole fraction enters into the entropy generation by mixing. However, for processes other than mixing of different components, we can treat the mixture as we treat a pure substance by using the mixture properties.

Special treatment and nomenclature are used for moist air as a mixture of air and water vapor. The water content is quantified by the relative humidity (how close the water vapor is to a saturated state) or by the humidity ratio (also called absolute humidity). As moist air is cooled down, it eventually reaches the dew point (relative humidity is 100%), where we have saturated moist air. Vaporizing liquid water without external heat transfer gives an adiabatic saturation process also used in a process called evaporative cooling. In an actual apparatus, we can obtain wet-bulb and dry-bulb temperatures indirectly, measuring the humidity of the incoming air. These property relations are shown in a psychrometric chart.

You should have learned a number of skills and acquired abilities from studying this chapter that will allow you to:

- Handle the composition of a multicomponent mixture on a mass or mole basis.
- Convert concentrations from a mass to a mole basis and vice versa.
- Compute average properties for the mixture on a mass or mole basis.
- Know partial pressures and how to evaluate them.
- Know how to treat mixture properties (such as v, u, h, s, $C_{p\,\mathrm{mix}}$, and R_{mix}).
- Find entropy generation by a mixing process.
- Formulate the general conservation equations for mass, energy, and entropy for the case of a mixture instead of a pure substance.
- Know how to use the simplified formulation of the energy equation using the frozen heat capacities for the mixture.
- Deal with a polytropic process when the substance is a mixture of ideal gases.
- Know the special properties (ϕ, ω) describing humidity in moist air.
- Have a sense of what changes relative humidity and humidity ratio and know that you can change one and not the other in a given process.

KEY CONCEPTS AND FORMULAS

Composition

Mass concentration	$c_i = \dfrac{m_i}{m_{\mathrm{tot}}} = \dfrac{y_i M_i}{\sum y_j M_j}$
Mole concentration	$y_i = \dfrac{n_i}{n_{\mathrm{tot}}} = \dfrac{c_i/M_i}{\sum c_j/M_j}$
Molecular mass	$M_{\mathrm{mix}} = \sum y_i M_i$

Properties

Internal energy	$u_{\mathrm{mix}} = \sum c_i u_i$;	$\bar{u}_{\mathrm{mix}} = \sum y_i \bar{u}_i = u_{\mathrm{mix}} M_{\mathrm{mix}}$
Enthalpy	$h_{\mathrm{mix}} = \sum c_i h_i$;	$\bar{h}_{\mathrm{mix}} = \sum y_i \bar{h}_i = h_{\mathrm{mix}} M_{\mathrm{mix}}$
Gas constant	$R_{\mathrm{mix}} = \bar{R}/M_{\mathrm{mix}} = \sum c_i R_i$	
Specific heat frozen	$C_{v\,\mathrm{mix}} = \sum c_i C_{v\,i}$;	$\overline{C}_{v\,\mathrm{mix}} = \sum y_i \overline{C}_{v\,i}$
	$C_{v\,\mathrm{mix}} = C_{p\,\mathrm{mix}} - R_{\mathrm{mix}}$;	$\overline{C}_{v\,\mathrm{mix}} = \overline{C}_{p\,\mathrm{mix}} - \bar{R}$
	$C_{p\,\mathrm{mix}} = \sum c_i C_{p\,i}$;	$\overline{C}_{p\,\mathrm{mix}} = \sum y_i \overline{C}_{p\,i}$
Ratio of specific heats	$k_{\mathrm{mix}} = C_{p\,\mathrm{mix}}/C_{v\,\mathrm{mix}}$	
Dalton model	$P_i = y_i P_{\mathrm{tot}}$ &	$V_i = V_{\mathrm{tot}}$
Entropy	$s_{\mathrm{mix}} = \sum c_i s_i$;	$\bar{s}_{\mathrm{mix}} = \sum y_i \bar{s}_i$
Component entropy	$s_i = s_{Ti}^0 - R_i \ln[y_i P/P_0]$	$\bar{s}_i = \bar{s}_{Ti}^0 - \bar{R} \ln[y_i P/P_0]$

Air–Water Vapor Mixtures

Relative humidity $\quad \phi = \dfrac{P_v}{P_g}$

Humidity ratio $\quad \omega = \dfrac{m_v}{m_a} = 0.622\dfrac{P_v}{P_a} = 0.622\dfrac{\phi P_g}{P_{tot} - \phi P_g}$

Enthalpy per kilogram of dry air $\quad \tilde{h} = h_a + \omega h_v$

CONCEPT-STUDY GUIDE PROBLEMS

11.1 Equal masses of argon and helium are mixed. Is the molecular mass of the mixture the linear average of the two individual ones?

11.2 Constant flows of pure argon and pure helium are mixed to produce a flow of mixture mole fractions 0.25 and 0.75, respectively. Explain how to meter the inlet flows to ensure the proper ratio, assuming inlet pressures are equal to the total exit pressure and all temperatures are the same.

11.3 For a gas mixture in a tank, are the partial pressures important?

11.4 An ideal mixture at T, P is made from ideal gases at T, P by charging them into a steel tank. Assume heat is transferred, so T stays the same as the supply. How do the properties (P, v, and u) for each component increase, decrease, or remain constant?

11.5 An ideal mixture at T, P is made from ideal gases at T, P by flow into a mixing chamber with no external heat transfer and an exit at P. How do

the properties (P, v, and h) for each component increase, decrease, or remain constant?

11.6 If a certain mixture is used in a number of different processes, is it necessary to consider partial pressures?

11.7 Why is it that a set of tables for air, which is a mixture, can be used without dealing with its composition?

11.8 Develop a formula to show how the mass fraction of water vapor is connected to the humidity ratio.

11.9 For air at $110\,°C$ and 100 kPa, is there any limit on the amount of water it can hold?

11.10 Can moist air below the freezing point, say $-5\,°C$, have a dew point?

11.11 Why does a car with an air conditioner running often have water dripping out?

11.12 Moist air at $35\,°C$, $\omega = 0.0175$, and $\Phi = 50\%$ should be brought to a state of $20\,°C$, $\omega = 0.01$, and $\Phi = 70\%$. Is it necessary to add or subtract water?

HOMEWORK PROBLEMS

Mixture Composition and Properties

11.13 If oxygen is 21 % by mole of air, what is the oxygen state (P, T, v) in a room at 300 K, 100 kPa with a total volume of 60 m^3?

11.14 A 3 L liquid mixture is one-third each of water, ammonia, and ethanol by volume. Find the mass fractions and total mass of the mixture.

11.15 A flow of oxygen and one of nitrogen, both at 300 K, are mixed to produce 1 kg/s air at 300 K, 100 kPa. What are the mass and volume flow rates of each line?

11.16 A gas mixture at $20\,°C$, 125 kPa is 50 % N$_2$, 30 % H$_2$O, and 20 % O$_2$ on a mole basis. Find the mass fractions, the mixture gas constant, and the volume for 5 kg of the mixture

11.17 A slightly oxygenated air mixture is 69 % N$_2$, 1 % Ar, and 30 % O$_2$ on a mole basis. Assume a total pressure of 101 kPa and find the mass fraction of oxygen and its partial pressure.

11.18 A new refrigerant, R-407, is a mixture of 23 % R-32, 25 % R-125, and 52 % R-134a on a mass basis. Find the mole fractions, the mixture gas constant, and the mixture heat capacities for the new refrigerant.

11.19 A 100 m³ storage tank with fuel gases is at 20 °C, 100 kPa, containing a mixture of acetylene (C_2H_2), propane (C_3H_8), and butane (C_4H_{10}). A test shows that the partial pressure of the C_2H_2 is 15 kPa and that of the C_3H_8 is 65 kPa. How much mass is there of each component?

11.20 A 2 kg mixture of 25 % N_2, 50 % O_2, and 25 % CO_2 by mass is at 225 kPa and 300 K. Find the mixture gas constant and the total volume.

11.21 A diesel engine sprays fuel (assume *n*-dodecane, $C_{12}H_{26}$, $M = 170.34$ kg/kmol) into the combustion chamber so it fills with 1 mol fuel per 88 mol air. Find the fuel fraction on a mass basis and the fuel mass for a chamber that is 0.5 L at 800 K and has a total pressure of 4000 kPa.

11.22 A new refrigerant, R-410A, is a mixture of R-32 and R-125 in a 1:1 mass ratio. What are the overall molecular weight, the gas constant, and the ratio of specific heats for such a mixture?

11.23 Do Problem 11.22 for R-507a, which has a 1:1 mass ratio of R-125 and R-143a. The refrigerant R-143a has a molecular mass of 84.041 kg/kmol and $C_p = 0.929$ kJ/kg·K.

Simple Processes

11.24 A rigid container has 1 kg CO_2 at 300 K and 1 kg Ar at 400 K, both at 225 kPa. Now they are allowed to mix without any heat transfer. What is the final T, P?

11.25 The mixture in Problem 11.20 is heated to 500 K with constant volume. Find the final pressure and the total heat transfer needed using Table A.5.

11.26 The mixture in Problem 11.20 is heated to 500 K in a constant-pressure process. Find the final volume and the total heat transfer using Table A.5.

11.27 A flow of 1 kg/s argon at 300 K and another flow of 1 kg/s carbon dioxide at 1600 K, both at 150 kPa, are mixed without any heat transfer. What is the exit T, P?

11.28 A flow of 1 kg/s argon at 300 K and another flow of 1 kg/s carbon dioxide at 1600 K, both at 150 kPa, are mixed without any heat transfer. Find the exit T, P using variable specific heats.

11.29 A pipe flows 0.1 kg/s of a mixture with mass fractions of 40 % CO_2 and 60 % N_2 at 400 kPa, 300 K. Heating tape is wrapped around a section of pipe with insulation added, and 2 kW electrical power is heating the pipe flow. Find the mixture exit temperature.

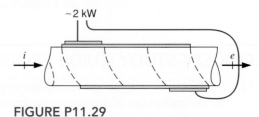

FIGURE P11.29

11.30 An insulated gas turbine receives a mixture of 10 % CO_2, 10 % H_2O, and 80 % N_2 on a mass basis at 1000 K, 500 kPa. The volume flow rate is 2 m³/s and its exhaust is at 700 K, 100 kPa. Find the power output in kW using constant specific heat from Table A.5 at 300 K.

11.31 Solve Problem 11.30 using the values of enthalpy from Table A.8.

11.32 Solve Problem 11.30 with the percentages on a mole basis and use Table A.9.

11.33 A mixture of 0.5 kg nitrogen and 0.5 kg oxygen is at 100 kPa, 300 K in a piston/cylinder maintaining constant pressure. Now 800 kJ is added by heating. Find the final temperature and the increase in entropy of the mixture using Table A.5 values.

11.34 A mixture of 0.5 kg nitrogen and 0.5 kg oxygen is at 100 kPa, 300 K in a piston/cylinder maintaining constant pressure. Now 1200 kJ is added by heating. Find the final temperature and the increase in entropy of the mixture using Table A.8 values.

11.35 A new refrigerant, R-410A, is a mixture of R-32 and R-125 in a 1:1 mass ratio. A process brings 0.5 kg R-410A from 270 K to 350 K at a constant pressure of 250 kPa in a piston/cylinder. Find the work and heat transfer.

11.36 A piston/cylinder device contains 0.1 kg of a mixture of 40 % methane and 60 % propane gases by mass at 300 K and 100 kPa. The gas is now slowly compressed in an isothermal (T = constant) process to a final pressure of 250 kPa. Show the process in a P–V diagram, and find both the work and the heat transfer in the process.

11.37 The refrigerant R-410A (see Problem 11.35) is at 100 kPa, 290 K. It is now brought to 250 kPa, 400 K in a reversible polytropic process. Find the change in specific volume, specific enthalpy, and the specific entropy for the process.

11.38 Natural gas as a mixture of 75 % methane and 25 % ethane by mass is flowing to a compressor at 17 °C, 100 kPa. The reversible adiabatic compressor brings the flow to 450 kPa. Find the exit temperature and the needed work per kilogram of flow.

11.39 A compressor brings R-410A (see Problem 11.35) from −10 °C, 125 kPa to 500 kPa in an adiabatic reversible compression. Assume ideal gas behavior and find the exit temperature and the specific work.

11.40 Two insulated tanks, A and B, are connected by a valve (Fig. P11.40). Tank A has a volume of 1 m³ and initially contains argon at 300 kPa, 10 °C. Tank B has a volume of 2 m³ and initially contains ethane at 200 kPa, 50 °C. The valve is opened and remains open until the resulting gas mixture comes to a uniform state. Determine the final pressure and temperature.

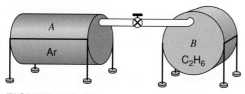

FIGURE P11.40

11.41 A steady flow of 0.1 kmol/s carbon dioxide at 1000 K in one line is mixed with 0.2 kmol/s nitrogen at 400 K in another line, both at 100 kPa. The exit mixture at 100 kPa is compressed by a reversible adiabatic compressor to 500 kPa. Use constant specific heat to find the mixing chamber's exit temperature and the needed compressor power.

11.42 A mixture of 2 kg oxygen and 2 kg argon is in an insulated piston/cylinder arrangement at 100 kPa, 300 K. The piston now compresses the mixture to 20 % of its initial volume. Find the final pressure, temperature, and the piston work.

11.43 A piston/cylinder has a 0.1 kg mixture of 25 % argon, 25 % nitrogen, and 50 % carbon dioxide by mass at a total pressure of 100 kPa and 290 K. Now the piston compresses the gases to volume seven times smaller in a polytropic process with $n = 1.3$. Find the final pressure and temperature, the work, and the heat transfer for the process.

Entropy Generation

11.44 A flow of gas A and a flow of gas B are mixed in a 1:2 mole ratio with the same T. What is the entropy generation per kilomole flow out?

11.45 A rigid container has 1 kg argon at 300 K and 1 kg argon at 400 K, both at 150 kPa. Now they are allowed to mix without any external heat transfer. What is the final T, P? Is any s generated?

11.46 What is the entropy generation in Problem 11.24?

11.47 A flow of 2 kg/s mixture of 50 % carbon dioxide and 50 % oxygen by mass is heated in a constant-pressure heat exchanger from 400 K to 1000 K by a radiation source at 1400 K. Find the rate of heat transfer and the entropy generation in the process shown in Fig. P 11.47.

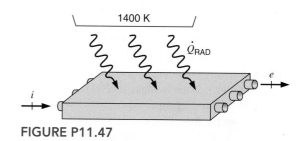

FIGURE P11.47

11.48 A flow of 1.8 kg/s steam at 400 kPa, 400 °C is mixed with 3.2 kg/s oxygen at 400 kPa, 400 K in a steady-flow mixing chamber without any heat transfer. Find the exit temperature and the rate of entropy generation.

11.49 Carbon dioxide gas at 320 K is mixed with nitrogen at 280 K in an insulated mixing chamber. Both flows are coming in at 100 kPa, and the mole ratio of carbon dioxide to nitrogen is 2:1. Find the exit temperature and the total entropy generation per kilomole of the exit mixture.

11.50 A flow of 1 kg/s carbon dioxide at 1600 K, 100 kPa is mixed with a flow of 2 kg/s water at 800 K, 100 kPa, and after the mixing it goes through a heat exchanger, where it is cooled to 500 K by a 400 K ambient (Fig. P11.50). How much heat transfer is taken out in the heat exchanger? What is the entropy generation rate for the whole process?

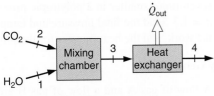

FIGURE P11.50

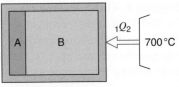

FIGURE P11.56

11.51 The only known sources of helium are the atmosphere (mole fraction approximately 5×10^{-6}) and natural gas. A large unit is being constructed to separate $100 \text{ m}^3/\text{s}$ of natural gas, assumed to be 0.001 helium mole fraction and 0.999 methane. The gas enters the unit at 150 kPa, 10 °C. Pure helium exits at 100 kPa, 20 °C, and pure methane exits at 150 kPa, 30 °C. Any heat transfer is with the surroundings at 20 °C. Is an electrical power input of 3000 kW sufficient to drive this unit?

11.52 A steady flow of 0.3 kg/s of 60 % carbon dioxide and 40 % water mixture by mass at 1200 K, 200 kPa is used in a constant-pressure heat exchanger where 300 kW is extracted from the flow. Find the exit temperature and rate of change in entropy using Table A.5.

11.53 A steady flow of 0.3 kg/s of 60 % carbon dioxide and 40 % water by mass at 1200 K, 200 kPa is used in a heat exchanger where 300 kW is extracted from the flow. Find the flow exit temperature and the rate of change of entropy using Table A.8.

11.54 A mixture of 60 % helium and 40 % nitrogen by mass enters a turbine at 1 MPa, 800 K at a rate of 2 kg/s. The adiabatic turbine has an exit pressure of 100 kPa and an isentropic efficiency of 85 %. Find the turbine work.

11.55 Three steady flows are mixed in an adiabatic chamber at 150 kPa. Flow one is 2 kg/s oxygen at 340 K, flow two is 4 kg/s nitrogen at 280 K, and flow three is 3 kg/s carbon dioxide at 310 K. All flows are at 150 kPa, the same as the total exit pressure. Find the exit temperature and the rate of entropy generation in the process.

11.56 A tank has two sides initially separated by a diaphragm (Fig. P11.56). Side A contains 1 kg water and side B contains 1.2 kg air, both at 20 °C, 100 kPa. The diaphragm is now broken and the whole tank is heated to 600 °C by a 700 °C reservoir. Find the final total pressure, heat transfer, and total entropy generation.

Air–Water Vapor Mixtures

11.57 Atmospheric air is at 100 kPa, 25 °C and relative humidity 65 %. Find the absolute humidity and the dew point of the mixture. If the mixture is heated to 35 °C, what is the new relative humidity?

11.58 A flow of 1 kg/s saturated moist air (relative humidity 100 %) at 100 kPa, 10 °C goes through a heat exchanger and comes out at 25 °C. What are the exit relative humidity and the heat transfer?

11.59 If I have air at 100 kPa and (a) −10 °C, (b) 45 °C, and (c) 110 °C, what is the maximum absolute humidity I can have?

11.60 A new high-efficiency home-heating system includes an air-to-air heat exchanger that uses energy from outgoing stale air to heat the fresh incoming air. If the outside ambient temperature is −10 °C and the relative humidity is 50 %, how much water will have to be added to the incoming air if it flows in at the rate of $1 \text{ m}^3/\text{s}$ and must eventually be conditioned to 20 °C and 45 % relative humidity?

11.61 Repeat Problem 11.59 for a total pressure of 200 kPa.

11.62 A flow of 2 kg/s completely dry air at T_1, 100 kPa is cooled down to 10 °C by spraying liquid water at 10 °C, 100 kPa into it so that it becomes saturated moist air at 10 °C. The process is steady state with no external heat transfer or work. Find the exit moist air humidity ratio and the flow rate of liquid water. Also find the dry air inlet temperature, T_1.

11.63 The products of combustion are flowing through a heat exchanger with 12 % carbon dioxide, 13 % water, and 75 % nitrogen on a volume basis at the rate of 0.1 kg/s and 100 kPa. What is the dew-point temperature? If the mixture is cooled 10 °C below the dew-point temperature, how long will it take to collect 10 kg of liquid water?

11.64 Consider a $1 \text{ m}^3/\text{s}$ flow of atmospheric air at 100 kPa, 25 °C, and 80 % relative humidity.

Assume this mixture flows into a basement room where it cools to 15 °C, 100 kPa. Find the rate of water condensing out and the exit mixture volume flow rate.

11.65 A room with air at 40 % relative humidity, 20 °C having 50 kg of dry air is made moist by boiling water to a final state of 20 °C and 80 % humidity. How much water was added to the air?

11.66 Consider a 500 L rigid tank containing an air–water vapor mixture at 100 kPa, 35 °C with 70 % relative humidity. The system is cooled until the water just begins to condense. Determine the final temperature in the tank and the heat transfer for the process.

11.67 A saturated air–water vapor mixture at 20 °C, 100 kPa is contained in a 5 m³ closed tank in equilibrium with 1 kg liquid water. The tank is heated to 80 °C. Is there any liquid water in the final state? Find the heat transfer for the process.

11.68 A flow of 0.2 kg/s liquid water at 80 °C is sprayed into a chamber together with 16 kg/s dry air at 60 °C. All the water evaporates, and the air leaves at 25 °C. What is the exit relative humidity and the heat transfer?

11.69 A water-filled reactor of 1 m³ is at 20 MPa, 360 °C, and is located inside an insulated containment room of 100 m³ that contains air at 100 kPa, 25 °C. Due to a failure, the reactor ruptures and the water fills the containment room. Find the final pressure.

11.70 In the production of ethanol from corn, the solids left after fermentation are dried in a continuous-flow oven. This process generates a flow of 15 kg/s moist air, 90 °C with 70 % relative humidity, which contains some volatile organic compounds and some particles. To remove the organic gases and the particles, the flow is sent to a thermal oxidizer, where natural gas flames bring the mixture to 800 °C (Fig. P11.70). Find the rate of heating by the natural gas burners.

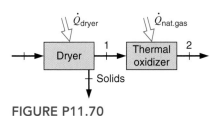

FIGURE P11.70

11.71 To reduce natural gas use in the previous problem, a suggestion is to take and cool the mixture and condense out some water before heating it again (Fig. P11.71). So, the flow is cooled from 90 °C to 50 °C and the dryer mixture is heated to 800 °C. Find the amount of water condensed out and the rate of heating by the natural gas burners for this case.

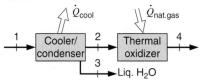

FIGURE P11.71

Tables and Formulas or Psychrometric Chart

11.72 I want to bring air at 35 °C, $\Phi = 40$ % to a state of 25 °C, $\omega = 0.012$. Do I need to add or subtract water?

11.73 A flow of moist air at 100 kPa, 40 °C, 40 % relative humidity is cooled to 15 °C in a constant-pressure device. Find the humidity ratio of the inlet and the exit flow, and the heat transfer in the device per kilogram of dry air.

11.74 Use the formulas and the steam tables to find the missing property of ϕ, ω, and T_{dry}; total pressure is 100 kPa. Repeat the answers using the psychrometric chart.
  a. $\phi = 50$ %, $\omega = 0.010$
  b. $T_{dry} = 25$ °C, $T_{wet} = 21$ °C

11.75 The discharge moist air from a clothes dryer is at 40 °C, 80 % relative humidity. The flow is guided through a pipe up through the roof and a vent to the atmosphere (Fig. P11.75). Due to heat transfer in the pipe, the flow is cooled to 24 °C by the time it reaches the vent. Find the humidity ratio in the flow out of the clothes dryer and at the vent.

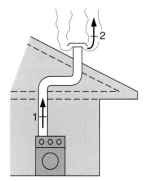

FIGURE P11.75

Find the heat transfer and any amount of liquid that may be forming per kilogram of dry air for the flow.

11.76 A steady supply of $1.0 \, \text{m}^3/\text{s}$ air at $25 \,^\circ\text{C}$, 100 kPa, 50 % relative humidity is needed to heat a building in the winter. The outdoor ambient is at $10 \,^\circ\text{C}$, 100 kPa, 50 % relative humidity. What are the required liquid water input and heat transfer rates for this purpose?

11.77 Two moist air streams with 85 % relative humidity, both flowing at a rate of 0.1 kg/s dry air, are mixed in a steady setup. One inlet flowstream is at $32.5 \,^\circ\text{C}$ and the other is at $16 \,^\circ\text{C}$. Find the exit relative humidity

11.78 A combination air cooler and dehumidification unit receives outside ambient air at $35 \,^\circ\text{C}$, 100 kPa, 90 % relative humidity. The moist air is first cooled to a low temperature, T_2, to condense the proper amount of water; assume all the liquid leaves at T_2. The moist air is then heated and leaves the unit at $20 \,^\circ\text{C}$, 100 kPa, relative humidity 30 % with volume flow rate of $0.01 \, \text{m}^3/\text{s}$. Find the temperature T_2, the mass of liquid per kilogram of dry air, and the overall heat transfer rate.

11.79 To make dry coffee powder, we spray 0.2 kg/s coffee (assume liquid water) at $80 \,^\circ\text{C}$ into a chamber, where we add 10 kg/s dry air at T. All the water should evaporate, and the air should leave at a minimum $30 \,^\circ\text{C}$; we neglect the powder. How high should T in the inlet air flow be?

11.80 An insulated tank has an air inlet, $\omega_1 = 0.0084$, and an outlet, $T_2 = 22 \,^\circ\text{C}$, $\phi_2 = 90\%$, both at 100 kPa. A third line sprays 0.25 kg/s water at $80 \,^\circ\text{C}$, 100 kPa (Fig. P11.80). For steady operation, find the outlet specific humidity, the mass flow rate of air needed, and the required air inlet temperature, T_1.

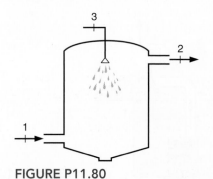

FIGURE P11.80

11.81 An air flow of 2 kg/s at $30 \,^\circ\text{C}$, relative humidity 80 %, is conditioned by taking half of the air flow, cooling it, and mixing it with the other half (Fig. P11.81). Assume the outlet flow should have a water content that is 75 % of the original flow. Find the temperature to cool to, the rate of cooling, and the final exit flow temperature.

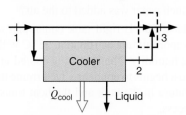

FIGURE P11.81

11.82 Moist air at $31 \,^\circ\text{C}$ and 50 % relative humidity flows over a large surface of liquid water. Find the adiabatic saturation temperature by trial and error. Hint: it is around $22.5 \,^\circ\text{C}$.

11.83 An air conditioner for an airport receives desert air at $45 \,^\circ\text{C}$, 10 % relative humidity and must deliver it to the building at $20 \,^\circ\text{C}$, 50 % relative humidity. The buildings have a cooling system with R-410A running with a high pressure of 3000 kPa and a low pressure of 1000 kPa, and their tap water is $15 \,^\circ\text{C}$. What should be done to the air? Find the needed heating/cooling per kilogram of dry air.

11.84 A flow of moist air from a domestic furnace, state 1, is at $45 \,^\circ\text{C}$, 10 % relative humidity with a flow rate of 0.05 kg/s dry air. A small electric heater adds steam at $100 \,^\circ\text{C}$, 100 kPa generated from tap water at $15 \,^\circ\text{C}$. Up in the living room the flow comes out at state 4: $30 \,^\circ\text{C}$, 60 % relative humidity. Find the power needed for the electric heater and the heat transfer to the flow from state 1 to state 4 (Fig. P11.84).

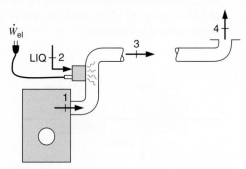

FIGURE P11.84

11.85 One means of conditioning hot summer air is evaporative cooling, which is a process similar to the adiabatic saturation process. Consider outdoor ambient air at 35 °C, 100 kPa, 30 % relative humidity. Find the lowest temperature this can generate and mention some disadvantage with this technique. Solve the problem with the energy equation and formulas and repeat it using the psychrometric chart, Fig. E.4.

11.86 A flow out of a clothes dryer of 0.05 kg/s dry air is at 40 °C and relative humidity 60 %. It flows though a heat exchanger, where it exits at 20 °C. After exiting the heat exchanger, the flow combines with another flow of 0.03 kg/s dry air at 30 °C and relative humidity 30 %. Find the dew point of state 1 (see Fig. P11.86), the heat transfer per kilogram of dry air, and the final exit state humidity ratio and relative humidity.

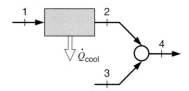

FIGURE P11.86

11.87 In a car's defrost/defog system atmospheric air, 21 °C, relative humidity 80 %, is taken in and cooled such that liquid water drips out. The now dryer air is heated to 41 °C and then blown onto the windshield, where it should have a maximum of 10 % relative humidity to remove water from the windshield. Find the dew point of the atmospheric air, specific humidity of air onto the windshield, the lowest temperature, and the specific heat transfer in the cooler.

11.88 A commercial laundry runs a dryer that has an exit flow of 0.5 kg/s moist air at 48 °C, 70 % relative humidity. To reduce the heating cost, a counterflow stack heat exchanger is used to heat a cold water line at 10 °C for the washers with the exit flow, as shown in Fig. P11.88. Assume the outgoing flow can be cooled to 25 °C. Is there a missing flow in the figure? Find the rate of energy recovered by this heat exchanger and the amount of cold water that can be heated to 30 °C.

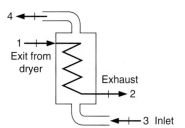

FIGURE P11.88

11.89 An indoor pool evaporates 1.512 kg/h of water, which is removed by a dehumidifier to maintain 21 °C, $\phi = 70\%$ in the room. The dehumidifier, shown in Fig. P11.89, is a refrigeration cycle in which air flowing over the evaporator cools such that liquid water drops out, and the air continues flowing over the condenser. For an air flow rate of 0.1 kg/s, the unit requires 1.4 kW input to a motor driving a fan and the compressor, and it has a cop of $\beta = Q_L/W_C = 2.0$. Find the state of the air as it returns to the room and the compressor work Q_L input.

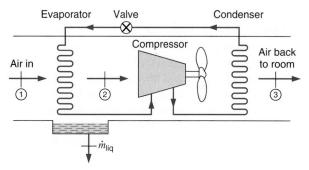

FIGURE P11.89

11.90 A moist air flow of 5 kg/min at 30 °C, $\Phi = 60\%$, 100 kPa goes through a dehumidifier in the setup shown in Problem 11.89. The air is cooled down to 15 °C and then blown over the condenser. The refrigeration cycle runs with R-134a with a low pressure of 200 kPa and a high pressure of 1000 kPa. Find the COP of the refrigeration cycle, the ratio $\dot{m}_{R\text{-}134a}/m_{air}$, and the outgoing T_3 and Φ_3.

Psychrometric Chart Only

11.91 Use the psychrometric chart to find the missing property of ϕ, ω, T_{wet}, T_{dry}.
  a. $T_{dry} = 25\,°C$, $\phi = 80\%$
  b. $\omega = 0.01$, $\phi = 100\%$

c. $T_{dry} = 20\,°C$, $\omega = 0.010$

d. $T_{dry} = 25\,°C$, $T_{wet} = 23\,°C$

11.92 Use the psychrometric chart to find the missing property of ϕ, ω, T_{wet}, T_{dry}.

a. $\phi = 50\,\%$, $\omega = 0.014$

b. $T_{wet} = 15\,°C$, $\phi = 60\,\%$

c. $\omega = 0.008$, $T_{wet} = 15\,°C$

d. $T_{dry} = 10\,°C$, $\omega = 0.006$

11.93 For each of the states in Problem 11.92, find the dew-point temperature.

11.94 Use the formulas and the steam tables to find the missing property of ϕ, ω, and T_{dry}; total pressure is 100 kPa. Repeat the answers using the psychrometric chart.

a. $\phi = 50\,\%$, $\omega = 0.006$

b. $T_{wet} = 15\,°C$, $\phi = 50\,\%$

c. $T_{dry} = 25\,°C$, $T_{wet} = 21\,°C$

11.95 An air conditioner should cool a flow of ambient moist air at $40\,°C$, $40\,\%$ relative humidity, with 0.2 kg/s flow of dry air. The exit temperature should be $20\,°C$, and the pressure is 100 kPa. Find the rate of heat transfer needed and check for the formation of liquid water.

11.96 A flow of moist air at $21\,°C$, $60\,\%$ relative humidity should be produced by mixing two different moist air flows. Flow 1 is at $10\,°C$, relative humidity $80\,\%$ and flow 2 is at $32\,°C$ and has $T_{wet} = 27\,°C$. The mixing chamber can be followed by a heater or a cooler (Fig. P11.96). No liquid water is added, and $P = 100$ kPa. Find the two controls; one is the ratio of the two mass flow rates $\dot{m}_{a1}/\dot{m}_{a2}$, and the other is the heat transfer in the heater/cooler per kilogram of dry air.

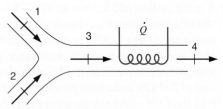

FIGURE P11.96

11.97 In a hot, dry climate, air enters an air-conditioner unit at 100 kPa, $40\,°C$, and $5\,\%$ relative humidity at a steady rate of 1.0 m^3/s. Liquid water at $20\,°C$ is sprayed into the air in the unit at the rate of 20 kg/h, and heat is rejected from the unit at the rate of 20 kW. The exit pressure is 100 kPa. What are the exit temperature and relative humidity?

11.98 Compare the weather in two places where it is cloudy and breezy. At beach A, it is $21\,°C$, 103.5 kPa, relative humidity $92\,\%$; at beach B, it is $25\,°C$, 99 kPa, relative humidity $40\,\%$. Suppose you just took a swim and came out of the water. Where would you feel more comfortable and why?

11.99 A flow of moist air at 100 kPa, $35\,°C$, $40\,\%$ relative humidity is cooled by adiabatic evaporation of liquid $20\,°C$ water to reach a saturated state. Find the amount of water added per kilogram of dry air and the exit temperature.

11.100 A flow out of a clothes dryer of 0.1 kg/s dry air is at $60\,°C$ and relative humidity $60\,\%$. It flows through a heat exchanger, where it exits at $20\,°C$. After exiting the heat exchanger, the flow combines with another flow of 0.03 kg/s dry air at $30\,°C$ and relative humidity $40\,\%$. Find the dew point of state 1 (see Fig. P11.100), the heat transfer per kilogram of dry air, and the final exit state humidity ratio and relative humidity.

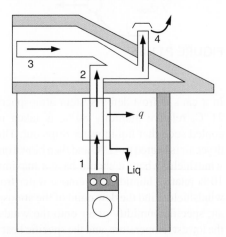

FIGURE P11.100

11.101 To refresh air in a room, a counterflow heat exchanger (see Fig. P11.101) is mounted in the wall, drawing in outside air at $0.5\,°C$, $80\,\%$ relative humidity and pushing out room air at $40\,°C$, $50\,\%$ relative humidity. Assume an exchange of 3 kg/min dry air in a steady-flow device, and also assume that the room air exits the heat exchanger at $23\,°C$ to the outside. Find the net amount of water removed from the room, any liquid flow in the heat exchanger, and T, ϕ for the fresh air entering the room.

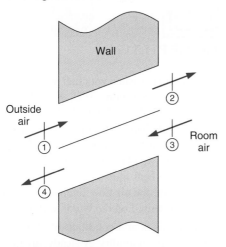

FIGURE P11.101

Exergy in Mixture

11.102 Consider several flow processes with ideal gases: (1) compression of a gas mixture from 100 kPa to 200 kPa; (2) cooling a gas mixture from 50 °C to ambient 20 °C using ambient air; (3) mixing two different gases at 100 kPa; (4) throttling a gas mixture from 125 kPa to 100 kPa. For each case, explain what happens to the exergy, whether there is any exergy destruction, and whether the composition is needed.

11.103 Find the second-law efficiency of the heat exchanger in Problem 11.47.

11.104 A mixture of 75 % carbon dioxide and 25 % water on a mole basis is flowing at 1600 K, 100 kPa into a heat exchanger, where it is used to deliver energy to a heat engine. The mixture leaves the heat exchanger at 500 K with a mass flow rate of 2 kg/min. Find the rate of energy and the rate of exergy delivered to the heat engine.

11.105 For flows with moist air where the water content is changed either by evaporation or by condensation, what happens to the exergy? Is the water vapor in air flowing over a lake in equilibrium with the liquid water?

11.106 A semipermeable membrane is used for the partial removal of oxygen from air that is blown through a grain elevator storage facility. Ambient air (79 % nitrogen, 21 % oxygen on a mole basis) is compressed to an appropriate pressure, cooled to ambient temperature 25 °C, and then fed through a bundle of hollow polymer fibers that selectively absorb oxygen, so the mixture leaving at 120 kPa, 25 °C contains only 5 % oxygen. The absorbed oxygen is bled off through the fiber walls at 40 kPa, 25 °C to a vacuum pump (Fig. P11.106). Assume the process is reversible and adiabatic, and determine the minimum inlet air pressure to the fiber bundle.

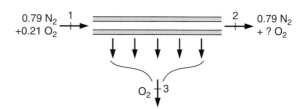

FIGURE P11.106

Review Problems

11.107 Weighing of masses gives a mixture at 60 °C, 225 kPa with 0.5 kg O_2, 1.5 kg N_2, and 0.5 kg CH_4. Find the partial pressures of each component, the mixture specific volume (mass basis), mixture molecular weight, and the total volume.

11.108 A carbureted internal combustion engine is converted to run on methane gas (natural gas). The air-fuel ratio in the cylinder is to be 20:1 on a mass basis. How many moles of oxygen per mole of methane are there in the cylinder?

11.109 A mixture of 50 % carbon dioxide and 50 % water by mass is brought from 1500 K, 1 MPa to 500 K, 200 kPa in a polytropic process through a steady-state device. Find the necessary heat transfer and work involved using values from Table A.5.

11.110 The accuracy of calculations can be improved by using a better estimate for the specific heat. Reconsider the previous problem and use $C_p = \Delta h / \Delta T$ from Table A.8 centered at 1000 K.

11.111 A large air-separation plant (Fig. P11.111) takes in ambient air (79 % N_2, 21 % O_2 by mole) at 100 kPa, 20 °C at a rate of 25 kg/s. It discharged a stream of pure O_2 gas at 200 kPa, 100 °C and a stream of pure N_2 gas at 100 kPa, 20 °C. The plant operates on an electrical power input of 2000 kW. Calculate the net rate of entropy change for the process.

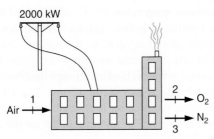

FIGURE P11.111

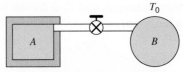

FIGURE P11.115

11.112 Take Problem 11.49 with inlet temperature of 1400 K for the carbon dioxide and 300 K for the nitrogen. Estimate the exit temperature with the specific heats from Table A.5 and use this to start iterations using Table A.9 to find the exit temperature.

11.113 A piston/cylinder has 100 kg saturated moist air at 100 kPa, 5 °C. If it is heated to 45 °C in an isobaric process, find $_1Q_2$ and the final relative humidity. If it is compressed from the initial state to 200 kPa in an isothermal process, find the mass of water condensing.

11.114 A spherical balloon has an initial diameter of 1 m and contains argon gas at 200 kPa, 40 °C. The balloon is connected by value to a 500 L rigid tank containing carbon dioxide at 100 kPa, 100 °C. The valve is opened, and eventually the balloon and tank reach a uniform state in which the pressure is 185 kPa. The balloon pressure is directly proportional to its diameter. Take the balloon and tank as a control volume, and calculate the final temperature and the heat transfer for the process.

11.115 An insulated, rigid 2 m³ tank A contains carbon dioxide gas at 200 °C, 1MPa. An uninsulated rigid 1 m³ tank B contains ethane (C_2H_6) gas at 200 kPa, room temperature 20 °C. The two tanks are connected by a one-way check valve that will allow gas to flow from A to B but not from B to A (Fig. P11.115). The valve is opened, and gas flows from tank A to B until the pressure in B reaches 500 kPa and the valve is closed. The mixture in B tank is kept at room temperature due to heat transfer. Find the total number of moles and the ethane mole fraction at the final state in tank B. Find the final temperature and pressure in tank A and the heat transfer to/from tank B.

11.116 You have just washed your hair and you now blow-dry it in a room with 23 °C, $\phi = 60\%$, (1). The dryer, 500 W, heats the air to 49 °C, (2), blows it through your hair, where the air becomes saturated (3), and then flows on to hit a window, where it cools to 15 °C (4). Find the relative humidity at state 2, the heat transfer per kilogram of dry air in the dryer, the air flow rate, and the amount of water condensed on the window, if any.

11.117 Ambient air is at a condition of 100 kPa, 35 °C, 50% relative humidity. A steady stream of air at 100 kPa, 23 °C, 70% relative humidity is to be produced by first cooling a flow of ambient air to an appropriate temperature to condense out the proper amount of water and then mix this stream adiabatically with another flow under ambient conditions. What is the ratio of the two flow rates? To what temperature must the first stream be cooled?

11.118 An air–water vapor mixture enters a steady-flow heater humidifier unit at state 1: 10 °C, 10% relative humidity, at the rate of 1 m³/s. A second air–vapor stream enters the unit at state 2: 20 °C, 20% relative humidity, at the rate of 2 m³/s. Liquid water enters at state 3: 10 °C at the rate of 400 kg/h. A single air–vapor flow exits the unit at state 4: 40 °C, as shown in Fig. P11.118. Calculate the relative humidity of the exit flow and the rate of heat transfer to the unit.

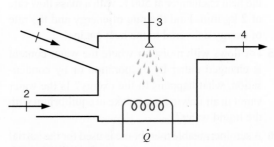

FIGURE P11.118

11.119 A dehumidifier similar to the one in Fig. P11.89 receives a flow of 0.25 kg/s dry air at 35 °C, 90%

relative humidity. It is cooled down to 20 °C as it flows over the evaporator and then is heated up again as it flows over the condenser. The standard refrigeration cycle uses R-410A with an evaporator temperature of −5 °C and a condensation pressure of 3000 kPa. Find the amount of liquid water removed and the heat transfer in the cooling process. How much compressor work is needed? What is the final air exit temperature and relative humidity?

11.120 The air conditioning by evaporative cooling in Problem 11.85 is modified by adding a dehumidification process before the water spray cooling process. This dehumidification is achieved, as shown in Fig. P11.120, by using a desiccant material, which absorbs water on one side of a rotating drum heat exchanger. The desiccant is regenerated by heating on the other side of the drum to drive the water out. The pressure is 100 kPa everywhere, and other properties are indicated on the diagram. Calculate the relative humidity of the cool air supplied to the room at state 4 and the heat transfer per unit mass of air that needs to be supplied to the heater unit.

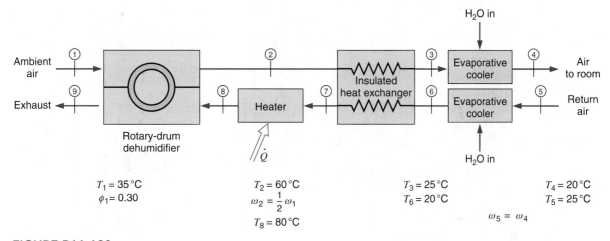

FIGURE P11.120

COMPUTER, DESIGN, AND OPEN-ENDED PROBLEMS

11.121 Write a program to solve the general case of Problem 11.40 in which the two volumes and the initial state properties of the argon and ethane are input variables. Use constant specific heat from Table A.5.

11.122 The setup in Problem 11.80 is similar to a process that can be used to produce dry powder from a slurry of water and dry material as coffee or milk. The water flow at state 3 is a mixture of 80 % liquid water and 20 % dry material on a mass basis with $C_{dry} = 0.4$ kJ/kg·K. After the water is evaporated, the dry material falls to the bottom and is removed in an additional line, $\dot{m}_{dry}$ exit at state 4. Assume a reasonable T_4 and that state 1 is heated atmospheric air. Investigate the inlet flow temperature as a function of state 1 humidity ratio.

11.123 A clothes dryer has a 60 °C, $\Phi = 90\%$ air flow out at a rate of 3 kg/min. The atmospheric conditions are 20 °C, relative humidity of 50 %. How much water is carried away and how much power is needed? To increase the efficiency, a counterflow heat exchanger is installed to preheat the incoming atmospheric air with the hot exit flow, as shown for Problem 11.88. Estimate suitable exit temperatures from the heat exchanger and investigate the design changes to the clothes dryer. (What happens to the condensed water?) How much energy can be saved this way?

11.124 Addition of steam to combustors in gas turbines and to internal-combustion engines reduces the peak temperatures and lowers emission of NO_x. Consider a modification to a gas turbine, as shown in Fig. P11.124, where the modified cycle is called the Cheng cycle. In this example, it is used for a cogenerating power plant. Assume 12 kg/s air with state 2 at 1.25 MPa, unknown temperature, is mixed with 2.5 kg/s water at 450 °C at constant pressure before the inlet to the turbine. The turbine exit temperature is $T_4 = 500$ °C, and the pressure is 125 kPa. For a reasonable turbine efficiency, estimate the required air temperature at state 2. Compare the result to the case where no steam is added to the mixing chamber and only air runs through the turbine.

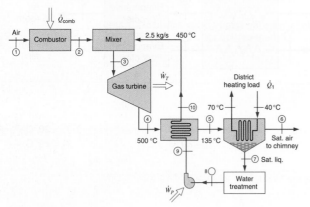

FIGURE P11.124

11.125 Consider the district water heater acting as the condenser for part of the water between states 5 and 6 in Fig. P11.124. If the temperature of the

mixture (12 kg/s air, 2.5 kg/s steam) at state 5 is 135 °C, study the district heating load, $\dot{Q}_1$, as a function of the exit temperature, T_6. Study also the sensitivity of the results with respect to the assumption that state 6 is saturated moist air.

11.126 The cogeneration gas-turbine cycle can be augmented with a heat pump to extract more energy from the turbine exhaust gas, as shown in Fig. P11.126. The heat pump upgrades the energy to be delivered at the 70 °C line for district heating. In the modified application, the first heat exchanger has exit temperature $T_{6a} = T_{7a} = 45$ °C, and the second one has $T_{6b} = T_{7b} = 36$ °C. Assume district heating line has the same exit temperature as before, so this arrangement allows for a higher flow rate. Estimate the increase in the district heating load that can be obtained and the necessary work input to the heat pump.

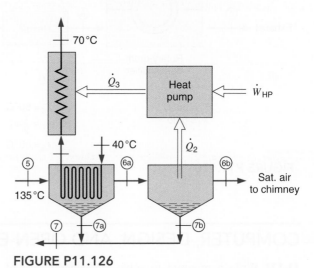

FIGURE P11.126

Thermodynamic
Property Relations

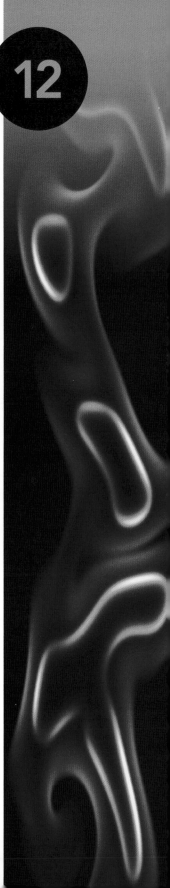

12

We have already defined and used several thermodynamic properties. Among these are pressure, specific volume, density, temperature, mass, internal energy, enthalpy, entropy, constant-pressure and constant-volume specific heats, and the Joule–Thomson coefficient. Two other properties, the Helmholtz function and the Gibbs function, will also be introduced and will be used more extensively in the following chapters. We have also had occasion to use tables of thermodynamic properties for a number of different substances.

One important question is now raised: Which thermodynamic properties can be experimentally measured? We can answer this question by considering the measurements we can make in the laboratory. Some of the properties, such as internal energy and entropy, cannot be measured directly and must be calculated from other experimental data. If we carefully consider all these thermodynamic properties, we conclude that there are only four that can be directly measured: pressure, temperature, volume, and mass.

This leads to a second question: How can values of the thermodynamic properties that cannot be measured be determined from experimental data on those properties that can be measured? In answering this question, we will develop certain general thermodynamic relations. In view of the fact that millions of such equations can be written, our study will be limited to certain basic considerations, with particular reference to the determination of thermodynamic properties from experimental data. We will also consider such related matters as generalized charts and equations of state.

 12.1 THE CLAPEYRON EQUATION

In calculating thermodynamic properties such as enthalpy or entropy in terms of other properties that can be measured, the calculations fall into two broad categories: differences in properties between two different phases and changes within a single homogeneous phase. In this section, we focus on the first of these categories, that of different phases. Let us assume that the two phases are liquid and vapor, but we will see that the results apply to other differences as well.

Consider a Carnot-cycle heat engine operating across a small temperature difference between reservoirs at T and $T - \Delta T$. The corresponding saturation pressures are P and $P - \Delta P$. The Carnot cycle operates with four steady-state devices. In the high-temperature heat-transfer process, the working fluid changes from saturated liquid at 1 to saturated vapor at 2, as shown in the two diagrams of Fig. 12.1.

From Fig. 12.1a, for reversible heat transfers,

$$q_H = Ts_{fg}; \qquad q_L = (T - \Delta T)s_{fg}$$

485

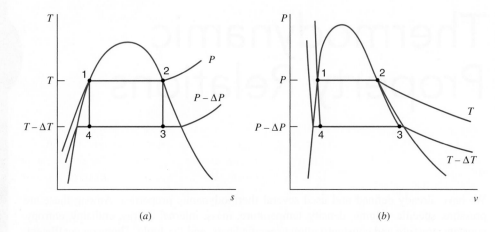

FIGURE 12.1 A Carnot cycle operating across a small temperature difference.

so that

$$w_{NET} = q_H - q_L = \Delta T s_{fg} \qquad (12.1)$$

From Fig. 12.1b, each process is steady-state and reversible, such that the work in each process is given by Eq. 7.15,

$$w = -\int v \, dP$$

Overall, for the four processes in the cycle,

$$
\begin{aligned}
w_{NET} &= 0 - \int_2^3 v \, dP + 0 - \int_4^1 v \, dP \\
&\approx -\left(\frac{v_2 + v_3}{2}\right)(P - \Delta P - P) - \left(\frac{v_1 + v_4}{2}\right)(P - P + \Delta P) \\
&\approx \Delta P\left[\left(\frac{v_2 + v_3}{2}\right) - \left(\frac{v_1 + v_4}{2}\right)\right]
\end{aligned}
\qquad (12.2)
$$

(The smaller the ΔP, the better the approximation.)

Now, comparing Eqs. 12.1 and 12.2 and rearranging,

$$\frac{\Delta P}{\Delta T} \approx \frac{s_{fg}}{\left(\dfrac{v_2 + v_3}{2}\right) - \left(\dfrac{v_1 + v_4}{2}\right)}$$

In the limit as $\Delta T \to 0$: $v_3 \to v_2 = v_g$, $v_4 \to v_1 = v_f$, which results in

$$\lim_{\Delta T \to 0} \frac{\Delta P}{\Delta T} = \frac{dP_{\text{sat}}}{dT} = \frac{s_{fg}}{v_{fg}} \qquad (12.3)$$

Since the heat addition process $1 - 2$ is at constant pressure as well as constant temperature,

$$q_H = h_{fg} = T s_{fg}$$

and the general result of Eq. 12.3 is the expression

$$\frac{dP_{\text{sat}}}{dT} = \frac{s_{fg}}{v_{fg}} = \frac{h_{fg}}{Tv_{fg}} \tag{12.4}$$

which is called the Clapeyron equation. This is a very simple relation and yet an extremely powerful one. We can experimentally determine the left-hand side of Eq. 12.4, which is the slope of the vapor pressure as a function of temperature. We can also measure the specific volumes of saturated vapor and saturated liquid at the given temperature, which means that the enthalpy change and entropy change of vaporization can both be calculated from Eq. 12.4. This establishes the means to cross from one phase to another in first- or second-law calculations, which was the goal of this development.

We could proceed along the same lines for the change of phase from solid to liquid or from solid to vapor. In each case, the result is the Clapeyron equation, in which the appropriate saturation pressure, specific volumes, entropy change, and enthalpy change are involved. For solid i to liquid f, the process occurs along the fusion line, and the result is

$$\frac{dP_{\text{fus}}}{dT} = \frac{s_{if}}{v_{if}} = \frac{h_{if}}{Tv_{if}} \tag{12.5}$$

We note that $v_{if} = v_f - v_i$ is typically a very small number, such that the slope of the fusion line is very steep. (In the case of water, v_{if} is a negative number, which is highly unusual, and the slope of the fusion line is not only steep, it is also negative.)

For sublimation, the change from solid i directly to vapor g, the Clapeyron equation has the values

$$\frac{dP_{\text{sub}}}{dT} = \frac{s_{ig}}{v_{ig}} = \frac{h_{ig}}{Tv_{ig}} \tag{12.6}$$

A special case of the Clapeyron equation involving the vapor phase occurs at low temperatures when the saturation pressure becomes very small. The specific volume v_g is then not only much larger than that of the condensed phase, liquid in Eq. 12.4 or solid in Eq. 12.6, but is also closely represented by the ideal gas equation of state (EOS). The Clapeyron equation then reduces to the form

$$\frac{dP_{\text{sat}}}{dT} = \frac{h_{fg}}{Tv_{fg}} = \frac{h_{fg}P_{\text{sat}}}{RT^2} \tag{12.7}$$

At low temperatures (not near the critical temperature), h_{fg} does not change very much with temperature. If it is assumed to be constant, then Eq. 12.7 can be rearranged and integrated over a range of temperatures to calculate a saturation pressure at a temperature at which it is not known. This point is illustrated by the following example.

Example 12.1

Determine the sublimation pressure of water vapor at $-60\,°C$ using data available in the steam tables.

Control mass: Water.

Solution

Appendix Table B.1.5 of the steam tables does not give saturation pressures for temperatures less than $-40\,°C$. However, we do notice that h_{ig} is relatively constant in this range; therefore, we proceed to use Eq. 12.7 and integrate between the limits $-40\,°C$ and $-60\,°C$.

$$\int_1^2 \frac{dP}{P} = \int_1^2 \frac{h_{ig}}{R} \frac{dT}{T^2} = \frac{h_{ig}}{R} \int_1^2 \frac{dT}{T^2}$$

$$\ln \frac{P_2}{P_1} = \frac{h_{ig}}{R}\left(\frac{T_2 - T_1}{T_1 T_2}\right)$$

Let

$$P_2 = 0.0129 \text{ kPa} \qquad T_2 = 233.2 \text{ K} \qquad T_1 = 213.2 \text{ K}$$

Then

$$\ln \frac{P_2}{P_1} = \frac{2838.9}{0.461\,52}\left(\frac{233.2 - 213.2}{233.2 \times 213.2}\right) = 2.4744$$

$$P_1 = 0.001\,09 \text{ kPa}$$

12.2 MATHEMATICAL RELATIONS FOR A HOMOGENEOUS PHASE

In the preceding section, we established the means to calculate differences in enthalpy (and therefore internal energy) and entropy between different phases in terms of properties that are readily measured. In the following sections, we will develop expressions for calculating differences in these properties within a single homogeneous phase (gas, liquid, or solid), assuming a simple compressible substance. In order to develop such expressions, it is first necessary to present a mathematical relation that will prove useful in this procedure.

Consider a variable (thermodynamic property) that is a continuous function of x and y.

$$z = f(x,\ y)$$

$$dz = \left(\frac{\partial z}{\partial x}\right)_y dx + \left(\frac{\partial z}{\partial y}\right)_x dy$$

It is convenient to write this function in the form

$$dz = M\,dx + N\,dy \qquad (12.8)$$

where

$$M = \left(\frac{\partial z}{\partial x}\right)_y$$

= partial derivative of z with respect to x (the variable y being held constant)

$$N = \left(\frac{\partial z}{\partial y}\right)_x$$

= partial derivative of z with respect to y (the variable x being held constant)

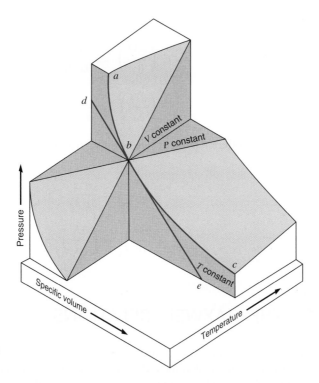

FIGURE 12.2
Schematic representation
of partial derivatives.

The physical significance of partial derivatives as they relate to the properties of a pure substance can be explained by referring to Fig. 12.2, which shows a P–v–T surface of the superheated vapor region of a pure substance. It shows constant-temperature, constant-pressure, and constant specific volume planes that intersect at point b on the surface. Thus, the partial derivative $(\partial P/\partial v)_T$ is the slope of curve abc at point b. Line de represents the tangent to curve abc at point b. A similar interpretation can be made of the partial derivatives $(\partial P/\partial T)_v$ and $(\partial v/\partial T)_p$.

If we wish to evaluate the partial derivative along a constant-temperature line, the rules for ordinary derivatives can be applied. Thus, we can write for a constant-temperature process

$$\left(\frac{\partial P}{\partial v}\right)_T = \frac{dP_T}{dv_T}$$

and the integration can be performed as usual. This point will be demonstrated later in a number of examples.

Let us return to the consideration of the relation

$$dz = M\,dx + N\,dy$$

If x, y, and z are all point functions (that is, quantities that depend only on the state and are independent of the path), the differentials are exact differentials. If this is the case, the following important relation holds:

$$\left(\frac{\partial M}{\partial y}\right)_x = \left(\frac{\partial N}{\partial x}\right)_y \tag{12.9}$$

The proof of this is

$$\left(\frac{\partial M}{\partial y}\right)_x = \frac{\partial^2 z}{\partial x \, \partial y}$$

$$\left(\frac{\partial N}{\partial x}\right)_y = \frac{\partial^2 z}{\partial y \, \partial x}$$

Since the order of differentiation makes no difference when point functions are involved, it follows that

$$\frac{\partial^2 z}{\partial x \, \partial y} = \frac{\partial^2 z}{\partial y \, \partial x}$$

$$\left(\frac{\partial M}{\partial y}\right)_x = \left(\frac{\partial N}{\partial x}\right)_y$$

12.3 THE MAXWELL RELATIONS

Consider a simple compressible control mass of fixed chemical composition. The Maxwell relations, which can be written for such a system, are four equations relating the properties P, v, T, and s. These will be found to be useful in the calculation of entropy in terms of the other measurable properties.

The Maxwell relations are most easily derived by considering the different forms of the thermodynamic property relation, which was the subject of Section 6.5. The two forms of this expression are rewritten here as

$$du = T \, ds - P \, dv \qquad (12.10)$$

and

$$dh = T \, ds + v \, dP \qquad (12.11)$$

Note that in the mathematical representation of Eq. 12.8, these expressions are of the form

$$u = u(s, v), \qquad h = h(s, P)$$

in both of which entropy is used as one of the two independent properties. This is an undesirable situation in that entropy is one of the properties that cannot be measured. We can, however, eliminate entropy as an independent property by introducing two new properties and thereby two new forms of the thermodynamic property relation. The first of these is the Helmholtz function A,

$$A = U - TS, \qquad a = u - Ts \qquad (12.12)$$

Differentiating and substituting Eq. 12.10 results in

$$da = du - T \, ds - s \, dT$$
$$= -s \, dT - P \, dv \qquad (12.13)$$

which we note is a form of the property relation utilizing T and v as the independent properties. The second new property is the Gibbs function G,

$$G = H - TS, \qquad g = h - Ts \tag{12.14}$$

Differentiating and substituting Eq. 12.11,

$$dg = dh - T\,ds - s\,dT$$
$$= -s\,dT + v\,dP \tag{12.15}$$

a fourth form of the property relation, this form using T and P as the independent properties.

Since Eqs. 12.10, 12.11, 12.13, and 12.15 are all relations involving only properties, we conclude that these are exact differentials and, therefore, are of the general form of Eq. 12.8,

$$dz = M\,dx + N\,dy$$

in which Eq. 12.9 relates the coefficients M and N,

$$\left(\frac{\partial M}{\partial y}\right)_x = \left(\frac{\partial N}{\partial x}\right)_y$$

It follows from Eq. 12.10 that

$$\left(\frac{\partial T}{\partial v}\right)_s = -\left(\frac{\partial P}{\partial s}\right)_v \tag{12.16}$$

Similarly, from Eqs. 12.11, 12.13, and 12.15 we can write

$$\left(\frac{\partial T}{\partial P}\right)_s = \left(\frac{\partial v}{\partial s}\right)_P \tag{12.17}$$

$$\left(\frac{\partial P}{\partial T}\right)_v = \left(\frac{\partial s}{\partial v}\right)_T \tag{12.18}$$

$$\left(\frac{\partial v}{\partial T}\right)_P = -\left(\frac{\partial s}{\partial P}\right)_T \tag{12.19}$$

These four equations are known as the Maxwell relations for a simple compressible mass, and the great utility of these equations will be demonstrated in later sections of this chapter. As was noted earlier, these relations will enable us to calculate entropy changes in terms of the measurable properties pressure, temperature, and specific volume.

A number of other useful relations can be derived from Eqs. 12.10, 12.11, 12.13, and 12.15. For example, from Eq. 12.10, we can write the relations

$$\left(\frac{\partial u}{\partial s}\right)_v = T, \qquad \left(\frac{\partial u}{\partial v}\right)_s = -P \qquad (12.20)$$

Similarly, from the other three equations, we have the following:

$$\left(\frac{\partial h}{\partial s}\right)_P = T, \qquad \left(\frac{\partial h}{\partial P}\right)_s = v$$

$$\left(\frac{\partial a}{\partial v}\right)_T = -P, \qquad \left(\frac{\partial a}{\partial T}\right)_v = -s$$

$$\left(\frac{\partial g}{\partial P}\right)_T = v, \qquad \left(\frac{\partial g}{\partial T}\right)_P = -s \qquad (12.21)$$

As already noted, the Maxwell relations just presented are written for a simple compressible substance. It is readily evident, however, that similar Maxwell relations can be written for substances involving other effects, such as surface or electrical effects. For example, Eq. 6.9 can be written in the form

$$dU = T\,dS - P\,dV + \mathscr{T}\,dL + \mathscr{S}\,dA + \mathscr{E}\,dZ + \cdots \qquad (12.22)$$

Thus, for a substance involving only surface effects, we can write

$$dU = T\,dS + \mathscr{S}\,dA$$

and it follows that for such a substance

$$\left(\frac{\partial T}{\partial A}\right)_S = \left(\frac{\partial \mathscr{S}}{\partial S}\right)_A$$

Other Maxwell relations could also be written for such a substance by writing the property relation in terms of different variables, and this approach could also be extended to systems having multiple effects. This matter also becomes more complex when we consider applying the property relation to a system of variable composition, a topic that will be taken up in Section 12.9.

Example 12.2

From an examination of the properties of compressed liquid water, as given in Table B.1.4 of Appendix B, we find that the entropy of compressed liquid is greater than the entropy of saturated liquid for a temperature of 0 °C and is less than that of saturated liquid for all the other temperatures listed. Explain why this follows from other thermodynamic data.

Control mass: Water.

Solution

Suppose we increase the pressure of liquid water that is initially saturated while keeping the temperature constant. The change of entropy for the water during this process can be found by integrating the following Maxwell relation, Eq. 12.19:

$$\left(\frac{\partial s}{\partial P}\right)_T = -\left(\frac{\partial v}{\partial T}\right)_P$$

Therefore, the sign of the entropy change depends on the sign of the term $(\partial v/\partial T)_P$. The physical significance of this term is that it involves the change in the specific volume of water as the temperature changes while the pressure remains constant. As water at moderate pressures and $0\,°C$ is heated in a constant-pressure process, the specific volume decreases until the point of maximum density is reached at approximately $4\,°C$, after which it increases. This is shown on a v–T diagram in Fig. 12.3. Thus, the quantity $(\partial v/\partial T)_P$ is the slope of the curve in Fig. 12.3. Since this slope is negative at $0\,°C$, the quantity $(\partial s/\partial P)_T$ is positive at $0\,°C$. At the point of maximum density the slope is zero and, therefore, the constant-pressure line shown in Fig. 6.7 crosses the saturated-liquid line at the point of maximum density.

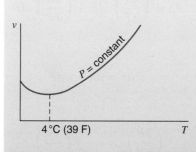

FIGURE 12.3 Sketch for Example 12.2.

12.4 THERMODYNAMIC RELATIONS INVOLVING ENTHALPY, INTERNAL ENERGY, AND ENTROPY

Let us first derive two equations, one involving C_p and the other involving C_v.

We have defined C_p as

$$C_p \equiv \left(\frac{\partial h}{\partial T}\right)_p$$

We have also noted that for a pure substance

$$T\,ds = dh - v\,dP$$

Therefore,

$$C_p = \left(\frac{\partial h}{\partial T}\right)_P = T\left(\frac{\partial s}{\partial T}\right)_P \qquad\qquad (12.23)$$

Similarly, from the definition of C_v,

$$C_v \equiv \left(\frac{\partial u}{\partial T}\right)_v$$

and the relation

$$T\,ds = du + P\,dv$$

it follows that

$$C_v = \left(\frac{\partial u}{\partial T}\right)_v = T\left(\frac{\partial s}{\partial T}\right)_v \tag{12.24}$$

We will now derive a general relation for the change of enthalpy of a pure substance. We first note that for a pure substance

$$h = h(T, P)$$

Therefore,

$$dh = \left(\frac{\partial h}{\partial T}\right)_P dT + \left(\frac{\partial h}{\partial P}\right)_T dP$$

From the relation

$$T\,ds = dh - v\,dP$$

it follows that

$$\left(\frac{\partial h}{\partial P}\right)_T = v + T\left(\frac{\partial s}{\partial P}\right)_T$$

Substituting the Maxwell relation, Eq. 12.19, we have

$$\left(\frac{\partial h}{\partial P}\right)_T = v - T\left(\frac{\partial v}{\partial T}\right)_P \tag{12.25}$$

On substituting this equation and Eq. 12.23, we have

$$dh = C_p\,dT + \left[v - T\left(\frac{\partial v}{\partial T}\right)_P\right]dP \tag{12.26}$$

Along an isobar, we have

$$dh_p = C_p\,dT_p$$

and along an isotherm

$$dh_T = \left[v - T\left(\frac{\partial v}{\partial T}\right)_P\right]dP_T \tag{12.27}$$

The significance of Eq. 12.26 is that this equation can be integrated to give the change in enthalpy associated with a change of state

$$h_2 - h_1 = \int_1^2 C_p \, dT + \int_1^2 \left[v - T\left(\frac{\partial v}{\partial T}\right)_P \right] dP \qquad (12.28)$$

The information needed to integrate the first term is a constant-pressure specific heat along one (and only one) isobar. The integration of the second integral requires that an EOS giving the relation between P, v, and T be known. Furthermore, it is advantageous to have this EOS explicit in v, for then the derivative $(\partial v/\partial T)_P$ is readily evaluated.

This matter can be further illustrated by reference to Fig. 12.4. Suppose we wish to know the change of enthalpy between states 1 and 2. We might determine this change along path 1–x–2, which consists of one isotherm, 1–x, and one isobar, x–2. Thus, we could integrate Eq. 12.28:

$$h_2 - h_1 = \int_{T_1}^{T_2} C_p \, dT + \int_{P_1}^{P_2} \left[v - T\left(\frac{\partial v}{\partial T}\right)_P \right] dP$$

Since $T_1 = T_x$ and $P_2 = P_x$, this can be written

$$h_2 - h_1 = \int_{T_x}^{T_2} C_p \, dT + \int_{P_1}^{P_x} \left[v - T\left(\frac{\partial v}{\partial T}\right)_P \right] dP$$

The second term in this equation gives the change in enthalpy along the isotherm 1–x and the first term the change in enthalpy along the isobar x–2. When these are added together, the result is the net change in enthalpy between 1 and 2. Therefore, the constant-pressure specific heat must be known along the isobar passing through 2 and x. The change in enthalpy could also be found by following path 1–y–2, in which case the constant-pressure specific heat must be known along the 1–y isobar. If the constant-pressure specific heat is known at another pressure, say, the isobar passing through m–n, the change in enthalpy can be found by following path 1–m–n–2. This involves calculating the change of enthalpy along two isotherms—1–m and n–2.

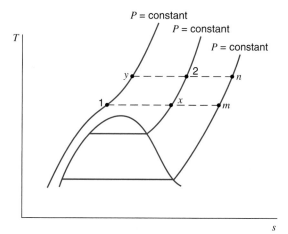

FIGURE 12.4 Sketch showing various paths by which a given change of state can take place.

Let us now derive a similar relation for the change of internal energy. All the steps in this derivation are given but without detailed comment. Note that the starting point is to write $u = u(T, v)$, whereas in the case of enthalpy the starting point was $h = h(T, P)$.

$$u = f(T, v)$$

$$du = \left(\frac{\partial u}{\partial T}\right)_v dT + \left(\frac{\partial u}{\partial v}\right)_T dv$$

$$T\,ds = du + P\,dv$$

Therefore,

$$\left(\frac{\partial u}{\partial v}\right)_T = T\left(\frac{\partial s}{\partial v}\right)_T - P \tag{12.29}$$

Substituting the Maxwell relation, Eq. 12.18, we have

$$\left(\frac{\partial u}{\partial v}\right)_T = T\left(\frac{\partial P}{\partial T}\right)_v - P$$

Therefore,

$$du = C_v\,dT + \left[T\left(\frac{\partial P}{\partial T}\right)_v - P\right]dv \tag{12.30}$$

Along an isometric, this reduces to

$$du_v = C_v\,dT_v$$

and along an isotherm, we have

$$du_T = \left[T\left(\frac{\partial P}{\partial T}\right)_v - P\right]dv_T \tag{12.31}$$

In a manner similar to that outlined earlier for changes in enthalpy, the change of internal energy for a given change of state for a pure substance can be determined from Eq. 12.30 if the constant-volume specific heat is known along one isometric and an EOS explicit in P [to obtain the derivative $(\partial P/\partial T)_v$] is available in the region involved. A diagram similar to Fig. 12.4 could be drawn, with the isobars replaced with isometrics, and the same general conclusions would be reached.

To summarize, we have derived Eqs. 12.26 and 12.30:

$$dh = C_p\,dT + \left[v - T\left(\frac{\partial v}{\partial T}\right)_P\right]dP$$

$$du = C_v\,dT + \left[T\left(\frac{\partial P}{\partial T}\right)_v - P\right]dv$$

The first of these equations concerns the change of enthalpy, the constant-pressure specific heat, and is particularly suited to an EOS explicit in v. The second equation concerns the change of internal energy and the constant-volume specific heat, and is particularly suited to an EOS explicit in P. If the first of these equations is used to determine the change of enthalpy, the internal energy is readily found by noting that

$$u_2 - u_1 = h_2 - h_1 - (P_2 v_2 - P_1 v_1)$$

If the second equation is used to find changes of internal energy, the change of enthalpy is readily found from this same relation. Which of these two equations is used to determine changes in internal energy and enthalpy will depend on the information available for specific heat and an EOS (or other P–v–T data).

Two parallel expressions can be found for the change of entropy:

$$s = s(T, P)$$

$$ds = \left(\frac{\partial s}{\partial T}\right)_P dT + \left(\frac{\partial s}{\partial P}\right)_T dP$$

Substituting Eqs. 12.19 and 12.23, we have

$$ds = C_p \frac{dT}{T} - \left(\frac{\partial v}{\partial T}\right)_P dP \qquad (12.32)$$

$$s_2 - s_1 = \int_1^2 C_p \frac{dT}{T} - \int_1^2 \left(\frac{\partial v}{\partial T}\right)_P dP \qquad (12.33)$$

Along an isobar, we have

$$(s_2 - s_1)_P = \int_1^2 C_p \frac{dT_P}{T}$$

and along an isotherm

$$(s_2 - s_1)_T = - \int_1^2 \left(\frac{\partial v}{\partial T}\right)_P dP$$

Note from Eq. 12.33 that if a constant-pressure specific heat is known along one isobar and an EOS explicit in v is available, the change of entropy can be evaluated. This is analogous to the expression for the change of enthalpy given in Eq. 12.26.

$$s = s(T, v)$$

$$ds = \left(\frac{\partial s}{\partial T}\right)_v dT + \left(\frac{\partial s}{\partial v}\right)_T dv$$

Substituting Eqs. 12.18 and 12.24 gives

$$ds = C_v \frac{dT}{T} + \left(\frac{\partial P}{\partial T}\right)_v dv \qquad (12.34)$$

$$s_2 - s_1 = \int_1^2 C_v \frac{dT}{T} + \int_1^2 \left(\frac{\partial P}{\partial T}\right)_v dv \qquad (12.35)$$

This expression for change of entropy concerns the change of entropy along an isometric where the constant-volume specific heat is known, and along an isotherm where an

EOS explicit in P is known. Thus, it is analogous to the expression for change of internal energy given in Eq. 12.30.

Example 12.3

Over a certain small range of pressures and temperatures, the EOS of a certain substance is given with reasonable accuracy by the relation

$$\frac{Pv}{RT} = 1 - C' \frac{P}{T^4}$$

or

$$v = \frac{RT}{P} - \frac{C}{T^3}$$

where C and C' are constants.

Derive an expression for the change of enthalpy and entropy of this substance in an isothermal process.

Control mass: Gas.

Solution

Since the EOS is explicit in v, Eq. 12.27 is particularly relevant to the change in enthalpy. On integrating this equation, we have

$$(h_2 - h_1)_T = \int_1^2 \left[v - T \left(\frac{\partial v}{\partial T} \right)_P \right] dP_T$$

From the EOS,

$$\left(\frac{\partial v}{\partial T} \right)_P = \frac{R}{P} + \frac{3C}{T^4}$$

Therefore,

$$(h_2 - h_1)_T = \int_1^2 \left[v - T \left(\frac{R}{P} + \frac{3C}{T^4} \right) \right] dP_T$$

$$= \int_1^2 \left[\frac{RT}{P} - \frac{C}{T^3} - \frac{RT}{P} - \frac{3C}{T^3} \right] dP_T$$

$$(h_2 - h_1)_T = \int_1^2 -\frac{4C}{T^3} \, dP_T = -\frac{4C}{T^3} (P_2 - P_1)_T$$

For the change in entropy we use Eq. 12.33, which is particularly relevant for an EOS explicit in v.

$$(s_2 - s_1)_T = -\int_1^2 \left(\frac{\partial v}{\partial T} \right)_P dP_T = -\int_1^2 \left(\frac{R}{P} + \frac{3C}{T^4} \right) dP_T$$

$$(s_2 - s_1)_T = -R \ln \left(\frac{P_2}{P_1} \right)_T - \frac{3C}{T^4} (P_2 - P_1)_T$$

In-Text Concept Questions

a. Mention two uses of the Clapeyron equation.

b. If I raise the temperature in a constant-pressure process, does g go up or down?

c. If I raise the pressure in an isentropic process, does h go up or down? Is that independent of the phase?

 12.5 **VOLUME EXPANSIVITY AND ISOTHERMAL AND ADIABATIC COMPRESSIBILITY**

The student has most likely encountered the coefficient of linear expansion in his or her studies of strength of materials. This coefficient indicates how the length of a solid body is influenced by a change in temperature while the pressure remains constant. In terms of the notation of partial derivatives, the coefficient of linear expansion, δ_T, is defined as

$$\delta_T = \frac{1}{L}\left(\frac{\delta L}{\delta T}\right)_P \tag{12.36}$$

A similar coefficient can be defined for changes in volume. Such a coefficient is applicable to liquids and gases as well as to solids. This coefficient of volume expansion, α_P, also called the *volume expansivity*, is an indication of the change in volume as temperature changes while the pressure remains constant. The definition of volume expansivity is

$$\alpha_P \equiv \frac{1}{V}\left(\frac{\partial V}{\partial T}\right)_P = \frac{1}{v}\left(\frac{\partial v}{\partial T}\right)_P = 3\,\delta_T \tag{12.37}$$

and it equals three times the coefficient of linear expansion. You should differentiate $V = L_x L_y L_z$ with temperature to prove that, which is left as a homework exercise. Notice that it is the volume expansivity that enters into the expressions for calculating changes in enthalpy, Eq. 12.26, and in entropy, Eq. 12.32.

The isothermal compressibility, β_T, is an indication of the change in volume as pressure changes while the temperature remains constant. The definition of isothermal compressibility is

$$\beta_T \equiv -\frac{1}{V}\left(\frac{\partial V}{\partial P}\right)_T = -\frac{1}{v}\left(\frac{\partial v}{\partial P}\right)_T \tag{12.38}$$

The adiabatic compressibility, β_s, is an indication of the change in volume as pressure changes while entropy remains constant; it is defined as

$$\beta_s \equiv -\frac{1}{v}\left(\frac{\partial v}{\partial P}\right)_s \tag{12.39}$$

The adiabatic bulk modulus, B_s, is the reciprocal of the adiabatic compressibility.

$$B_s \equiv -v\left(\frac{\partial P}{\partial v}\right)_s \tag{12.40}$$

The velocity of sound, c, in a medium is defined by the relation

$$c^2 = \left(\frac{\partial P}{\partial \rho}\right)_s \qquad (12.41)$$

This can also be expressed as

$$c^2 = -v^2\left(\frac{\partial P}{\partial v}\right)_s = v B_s \qquad (12.42)$$

in terms of the adiabatic bulk modulus B_s. For a compressible medium such as a gas, the speed of sound becomes modest, whereas in an incompressible state such as a liquid or a solid, it can be quite large.

The volume expansivity and isothermal and adiabatic compressibility are thermodynamic properties of a substance, and for a simple compressible substance are functions of two independent properties. Values of these properties are found in the standard handbooks of physical properties. The following examples give an indication of the use and significance of volume expansivity and isothermal compressibility.

Example 12.4

The pressure on a block of copper having a mass of 1 kg is increased in a reversible process from 0.1 to 100 MPa while the temperature is held constant at 15 °C. Determine the work done on the copper during this process, the change in entropy per kilogram of copper, the heat transfer, and the change of internal energy per kilogram.

Over the range of pressure and temperature in this problem, the following data can be used:

Volume expansivity = $\alpha_P = 5.0 \times 10^{-5}\ \text{K}^{-1}$
Isothermal compressibility = $\beta_T = 8.6 \times 10^{-12}\ \text{m}^2/\text{N}$
Specific volume = 0.000 114 m³/kg

Analysis

Control mass: Copper block.
States: Initial and final states known.
Process: Constant temperature, reversible.

The work done during the isothermal compression is

$$w = \int P\, dv_T$$

The isothermal compressibility has been defined as

$$\beta_T = -\frac{1}{v}\left(\frac{\partial v}{\partial P}\right)_T$$
$$v\beta_T\, dP_T = -dv_T$$

Therefore, for this isothermal process,

$$w = -\int_1^2 v\beta_T P\, dP_T$$

Since v and β_T remain essentially constant, this is readily integrated:

$$w = -\frac{v\beta_T}{2}(P_2^2 - P_1^2)$$

The change of entropy can be found by considering the Maxwell relation, Eq. 12.19, and the definition of volume expansivity.

$$\left(\frac{\partial s}{\partial P}\right)_T = -\left(\frac{\partial v}{\partial T}\right)_P = -\frac{v}{v}\left(\frac{\partial v}{\partial T}\right)_P = -v\alpha_P$$

$$ds_T = -v\alpha_P \, dP_T$$

This equation can be readily integrated, if we assume that v and α_P remain constant:

$$(s_2 - s_1)_T = -v\alpha_P(P_2 - P_1)_T$$

The heat transfer for this reversible isothermal process is

$$q = T(s_2 - s_1)$$

The change in internal energy follows directly from the first law.

$$(u_2 - u_1) = q - w$$

Solution

$$w = -\frac{v\beta_T}{2}(P_2^2 - P_1^2)$$

$$= -\frac{0.000\,114 \times 8.6 \times 10^{-12}}{2}(100^2 - 0.1^2) \times 10^{12}$$

$$= -4.9 \text{ J/kg}$$

$$(s_2 - s_1)_T = -v\alpha_P(P_2 - P_1)_T$$

$$= -0.000\,114 \times 5.0 \times 10^{-5}(100 - 0.1) \times 10^6$$

$$= -0.5694 \text{ J/kg·K}$$

$$q = T(s_2 - s_1) = -288.2 \times 0.5694 = -164.1 \text{ J/kg}$$

$$(u_2 - u_1) = q - w = -164.1 - (-4.9) = -159.2 \text{ J/kg}$$

12.6 REAL-GAS BEHAVIOR AND EQUATIONS OF STATE

In Sections 2.8 and 2.9 we examined the P–v–T behavior of gases, and we defined the compressibility factor in Eq. 2.12,

$$Z = \frac{Pv}{RT}$$

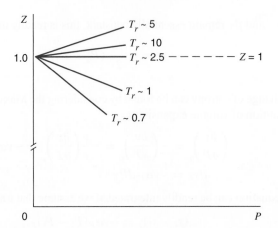

FIGURE 12.5

Low-pressure region of a compressibility chart.

We then proceeded to develop the generalized compressibility chart, presented in Appendix Fig. D.1 in terms of the reduced pressure and temperature. The generalized chart does not apply specifically to any one substance, but is instead an approximate relation that is reasonably accurate for many substances, especially those that are fairly simple in molecular structure. In this sense, the generalized compressibility chart can be viewed as one aspect of generalized behavior of substances, and also as a graphical form of EOS representing real behavior of gases and liquids over a broad range of variables.

To gain additional insight into the behavior of gases at low density, let us examine the low-pressure portion of the generalized compressibility chart in greater detail. This behavior is as shown in Fig. 12.5. The isotherms are essentially straight lines in this region, and their slope is of particular importance. Note that the slope increases as T_r increases until a maximum value is reached at a T_r of about 5, and then the slope decreases toward the $Z = 1$ line for higher temperatures. That single temperature, about 2.5 times the critical temperature, for which

$$\lim_{P \to 0} \left(\frac{\partial Z}{\partial P} \right)_T = 0 \qquad (12.43)$$

is defined as the Boyle temperature of the substance. This is the only temperature at which a gas behaves exactly as an ideal gas at low but finite pressures, since all other isotherms go to zero pressure on Fig. 12.5 with a nonzero slope. To amplify this point, let us consider the residual volume α,

$$\alpha = \frac{RT}{P} - v \qquad (12.44)$$

Multiplying this equation by P, we have

$$\alpha P = RT - Pv$$

Thus, the quantity αP is the difference between $RT-$ and Pv. Now as $P \to 0$, $Pv \to RT$. However, it does not necessarily follow that $\alpha \to 0$ as $P \to 0$. Instead, it is only required that α remain finite. The derivative in Eq. 12.43 can be written as

$$\lim_{P \to 0} \left(\frac{\partial Z}{\partial P} \right)_T = \lim_{P \to 0} \left(\frac{Z-1}{P-0} \right)$$

$$= \lim_{P \to 0} \frac{1}{RT} \left(v - \frac{RT}{P} \right) \tag{12.45}$$

$$= -\frac{1}{RT} \lim_{P \to 0} (\alpha)$$

from which we find that α tends to zero as $P \to 0$ only at the Boyle temperature, since that is the only temperature for which the isothermal slope is zero on Fig. 12.5. It is, perhaps, a somewhat surprising result that in the limit as $P \to 0$, $Pv \to RT$. In general, however, the quantity $(RT/P - v)$ does not go to zero but is instead a small difference between two large values. This does have an effect on certain other properties of the gas.

The compressibility behavior of low-density gases as noted in Fig. 12.5 is the result of intermolecular interactions and can be expressed in the form of an EOS called the virial equation, which is derived from statistical thermodynamics. The result is

$$Z = \frac{Pv}{RT} = 1 + \frac{B(T)}{v} + \frac{C(T)}{v^2} + \frac{D(T)}{v^3} + \cdots \tag{12.46}$$

where $B(T)$, $C(T)$, $D(T)$ are temperature dependent and are called virial coefficients. $B(T)$ is termed the *second virial coefficient* and is due to binary interactions on the molecular level. The general temperature dependence of the second virial coefficient is as shown for nitrogen in Fig. 12.6. If we multiply Eq. 12.46 by RT/P, the result can be rearranged to the form

$$\frac{RT}{P} - v = \alpha = -B(T)\frac{RT}{Pv} - C(T)\frac{RT}{Pv^2} \cdots \tag{12.47}$$

In the limit, as $P \to 0$,

$$\lim_{P \to 0} \alpha = -B(T) \tag{12.48}$$

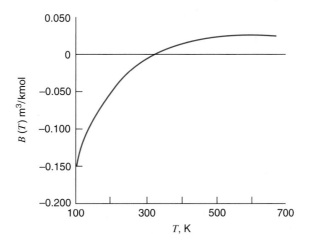

FIGURE 12.6 The second virial coefficient for nitrogen.

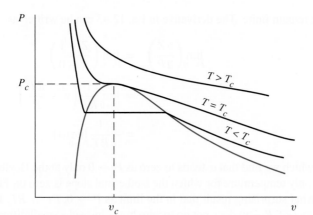

FIGURE 12.7 Plot of isotherms in the region of the critical point on pressure–volume coordinates for a typical pure substance.

and we conclude from Eqs. 12.43 and 12.45 that the single temperature at which $B(T) = 0$, Fig. 12.6, is the Boyle temperature. The second virial coefficient can be viewed as the first-order correction for nonideality of the gas, and consequently becomes of considerable importance and interest. In fact, the low-density behavior of the isotherms shown in Fig. 12.5 is directly attributable to the second virial coefficient.

Another aspect of generalized behavior of gases is the behavior of isotherms in the vicinity of the critical point. If we plot experimental data on P–v coordinates, it is found that the critical isotherm is unique in that it goes through a horizontal inflection point at the critical point, as shown in Fig. 12.7. Mathematically, this means that the first two derivatives are zero at the critical point

$$\left(\frac{\partial P}{\partial v}\right)_{T_c} = 0 \qquad \text{at C.P.} \tag{12.49}$$

$$\left(\frac{\partial^2 P}{\partial v^2}\right)_{T_c} = 0 \qquad \text{at C.P.} \tag{12.50}$$

a feature that is used to constrain many equations of state.

To this point, we have discussed the generalized compressibility chart, a graphical form of an EOS, and the virial equation, a theoretically founded EOS. We now proceed to discuss other analytical equations of state, which may be either generalized behavior in form or empirical equations, relying on specific P–v–T data of their constants. The oldest generalized equation, the van der Waals equation, is a member of the class of equations of state known as *cubic equations*, presented in Chapter 2 as Eq. 2.14. This equation was introduced in 1873 as a semitheoretical improvement over the ideal gas model. The van der Waals EOS has two constants and is written as

$$P = \frac{RT}{v - b} - \frac{a}{v^2} \tag{12.51}$$

The constant b is intended to correct for the volume occupied by the molecules, and the term a/v^2 is a correction that accounts for the intermolecular forces of attraction. As might be expected in the case of a generalized equation, the constants a and b are evaluated from the general behavior of gases. In particular, these constants are evaluated by noting that the critical isotherm passes through a point of inflection at the critical point and that the slope is zero at this point. Therefore, we take the first two derivatives with respect to v of Eq. 12.51

and set them equal to zero, according to Eqs. 12.49 and 12.50. Then this pair of equations, along with Eq. 12.51 itself, can be solved simultaneously for a, b, and v_c. The result is

$$v_c = 3b$$

$$a = \frac{27}{64}\frac{R^2 T_c^2}{P_c}$$ (12.52)

$$b = \frac{R T_c}{8 P_c}$$

The compressibility factor at the critical point for the van der Waals equation is therefore

$$Z_c = \frac{P_c v_c}{R T_c} = \frac{3}{8}$$

which is considerably higher than the actual value for any substance.

The van der Waals equation can be written in terms of the compressibility factor, $Z = Pv/RT$, and the reduced pressure and temperature in a cubic EOS:

$$Z^3 - \left(\frac{P_r}{8T_r} + 1\right) Z^2 + \left(\frac{27 P_r}{64 T_r^2}\right) Z - \frac{27 P_r^2}{512 T_r^3} = 0$$ (12.53)

It is significant that this is of the same form as the generalized compressibility chart, namely, $Z = f(P_r, T_r)$, as shown in Appendix D. The functional relation in Eq. 12.53 is quite different from the chart, which is based on the Lee–Kesler equation shown in Eq. 12.56. The concept that different substances will have the same compressibility factor at the same reduced properties (P_r, T_r) is another way of expressing the rule of corresponding states.

Another cubic EOS that is considerably more accurate than the van der Waals equation is that proposed by Redlich and Kwong in 1949:

$$P = \frac{RT}{v - b} - \frac{a}{v(v + b)T^{1/2}}$$ (12.54)

with

$$a = 0.427\,48 \; \frac{R^2 T_c^{5/2}}{P_c}$$

$$b = 0.086\,64 \; \frac{R T_c}{P_c}$$ (12.55)

The numerical values in the constants have been determined by a procedure similar to that followed in the van der Waals equation. Because of its simplicity, this equation was not sufficiently accurate to be used in the calculation of precision tables of thermodynamic properties. It has, however, been used frequently for mixture calculations and phase equilibrium correlations with reasonably good success. Several modified versions of this equation have also been utilized in recent years, two of which are given in Appendix D.

Empirical equations of state have been presented and used to represent real-substance behavior for many years. The Beattie–Bridgeman equation, containing five empirical constants, was introduced in 1928. In 1940, the Benedict–Webb–Rubin equation, commonly termed the *BWR equation*, extended that equation with three additional terms in order to better represent higher-density behavior. Several modifications of this equation have been used over the years, often to correlate gas-mixture behavior.

One particularly interesting modification of the BWR EOS is the Lee–Kesler equation, which was proposed in 1975. This equation has 12 constants and is written in terms of generalized properties as

$$Z = \frac{P_r v_r'}{T_r} = 1 + \frac{B}{v_r'} + \frac{C}{v_r'^2} + \frac{D}{v_r'^5} + \frac{c_4}{T_r^3 v_r'^2}\left(\beta + \frac{\gamma}{v_r'^2}\right)\exp\left(-\frac{\gamma}{v_r'^2}\right)$$

$$B = b_1 - \frac{b_2}{T_r} - \frac{b_3}{T_r^2} - \frac{b_4}{T_r^3}$$

$$C = c_1 - \frac{c_2}{T_r} + \frac{c_3}{T_r^3} \tag{12.56}$$

$$D = d_1 + \frac{d_2}{T_r}$$

in which the variable v_r' is not the true reduced specific volume but is, instead, defined as

$$v_r' = \frac{v}{RT_c/P_c} \tag{12.57}$$

Empirical constants for simple fluids for this equation are given in Appendix Table D.2.

When using computer software to calculate the compressibility factor Z at a given reduced temperature and reduced pressure, a third parameter, ω, the acentric factor (defined and values listed in Appendix D) can be included in order to improve the accuracy of the correlation, especially near or at saturation states. In the software, the value calculated for the simple fluid is called $Z0$, while a correction term, called the deviation $Z1$, is determined after using a different set of constants for the Lee–Kesler EOS. The overall compressibility Z is then

$$Z = Z0 + \omega Z1 \tag{12.58}$$

Finally, it should be noted that modern equations of state use a different approach to represent P–v–T behavior in calculating thermodynamic properties and tables. This subject will be discussed in detail in Section 12.11.

12.7 THE GENERALIZED CHART FOR CHANGES OF ENTHALPY AT CONSTANT TEMPERATURE

In Section 12.4, Eq. 12.27 was derived for the change of enthalpy at constant temperature.

$$(h_2 - h_1)_T = \int_1^2 \left[v - T\left(\frac{\partial v}{\partial T}\right)_P\right] dP_T$$

This equation is appropriately used when a volume-explicit EOS is known. Otherwise, it is more convenient to calculate the isothermal change in internal energy from Eq. 12.31

$$(u_2 - u_1)_T = \int_1^2 \left[T\left(\frac{\partial P}{\partial T}\right)_v - P\right] dv_T$$

and then calculate the change in enthalpy from its definition as

$$(h_2 - h_1) = (u_2 - u_1) + (P_2 v_2 - P_1 v_1)$$
$$= (u_2 - u_1) + RT(Z_2 - Z_1)$$

To determine the change in enthalpy behavior consistent with the generalized chart, Fig. D.1, we follow the second of these approaches, since the Lee–Kesler generalized EOS, Eq. 12.56, is a pressure-explicit form in terms of specific volume and temperature. Equation 12.56 is expressed in terms of the compressibility factor Z, so we write

$$P = \frac{ZRT}{v}, \qquad \left(\frac{\partial P}{\partial T}\right)_v = \frac{ZR}{v} + \frac{RT}{v}\left(\frac{\partial Z}{\partial T}\right)_v$$

Therefore, substituting into Eq. 12.31, we have

$$du = \frac{RT^2}{v}\left(\frac{\partial Z}{\partial T}\right)_v dv$$

But

$$\frac{dv}{v} = \frac{dv_r'}{v_r'}; \qquad \frac{dT}{T} = \frac{dT_r}{T_r}$$

so that, in terms of reduced variables,

$$\frac{1}{RT_c} du = \frac{T_r^2}{v_r'}\left(\frac{\partial Z}{\partial T_r}\right)_{v_r'} dv_r'$$

This expression is now integrated at constant temperature from any given state (P_r, v_r') to the ideal gas limit $(P_r^* \to 0, v_r'^* \to \infty)$(the superscript * will always denote an ideal gas state or property), causing an internal energy change or departure from the ideal gas value at the given state,

$$\frac{u^* - u}{RT_c} = \int_{v_r'}^{\infty} \frac{T_r^2}{v_r'}\left(\frac{\partial Z}{\partial T_r}\right)_{v_r'} dv_r' \qquad (12.59)$$

The integral on the right-hand side of Eq. 12.59 can be evaluated from the Lee–Kesler equation, Eq. 12.56. The corresponding enthalpy departure at the given state (P_r, v_r') is then found from integrating Eq. 12.59 to be

$$\frac{h^* - h}{RT_c} = \frac{u^* - u}{RT_c} + T_r(1 - Z) \qquad (12.60)$$

Following the same procedure as for the compressibility factor, we can evaluate Eq. 12.60 with the set of Lee–Kesler simple-fluid constants to give a simple-fluid enthalpy departure. The values for the enthalpy departure are shown graphically in Fig. D.2. Use of the enthalpy departure function is illustrated in Example 12.5.

Note that when using computer software to determine the enthalpy departure at a given reduced temperature and reduced pressure, accuracy can be improved by using the acentric factor in the same manner as was done for the compressibility factor in Eq. 12.58.

Example 12.5

Nitrogen is throttled from 20 MPa, $-70\,°C$, to 2 MPa in an adiabatic, steady-state, steady-flow process. Determine the final temperature of the nitrogen.

Control volume:	Throttling valve.
Inlet state:	P_1, T_1 known; state fixed.
Exit state:	P_2 known.
Process:	Steady-state, throttling process.
Diagram:	Figure 12.8.
Model:	Generalized charts, Fig. D.2.

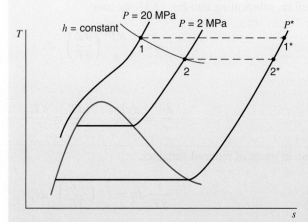

FIGURE 12.8 Sketch for Example 12.5.

Analysis

Energy equation:

$$h_1 = h_2$$

Solution

Using values from Table A.2, we have

$$P_1 = 20\text{ MPa} \qquad P_{r1} = \frac{20}{3.39} = 5.9$$

$$T_1 = 203.2\text{ K} \qquad T_{r1} = \frac{203.2}{126.2} = 1.61$$

$$P_2 = 2\text{ MPa} \qquad P_{r2} = \frac{2}{3.39} = 0.59$$

From the generalized charts, Fig. D.2, for the change in enthalpy at constant temperature, we have

$$\frac{h_1^* - h_1}{RT_c} = 2.1$$

$$h_1^* - h_1 = 2.1 \times 0.2968 \times 126.2 = 78.7\text{ kJ/kg}$$

It is now necessary to assume a final temperature and to check whether the net change in enthalpy for the process is zero. Let us assume that $T_2 = 146$ K. Then the change in enthalpy between 1* and 2* can be found from the zero-pressure, specific-heat data.

$$h_1^* - h_2^* = C_{p0}(T_1^* - T_2^*) = 1.0416(203.2 - 146) = +59.6 \, \text{kJ/kg}$$

(The variation in C_{p0} with temperature can be taken into account when necessary.)

We now find the enthalpy change between 2* and 2.

$$T_{r2} = \frac{146}{126.2} = 1.157 \qquad P_{r2} = 0.59$$

Therefore, from the enthalpy departure chart, Fig. D.2, at this state

$$\frac{h_2^* - h_2}{RT_c} = 0.5$$

$$h_2^* - h_2 = 0.5 \times 0.2968 \times 126.2 = 19.5 \, \text{kJ/kg}$$

We now check to see whether the net change in enthalpy for the process is zero.

$$h_1 - h_2 = 0 = -(h_1^* - h_1) + (h_1^* - h_2^*) + (h_2^* - h_2)$$

$$= -78.7 + 59.6 + 19.5 \approx 0$$

It essentially checks. We conclude that the final temperature is approximately 146 K. It is interesting that the thermodynamic tables for nitrogen, Table B.6, give essentially this same value for the final temperature.

12.8 THE GENERALIZED CHART FOR CHANGES OF ENTROPY AT CONSTANT TEMPERATURE

In this section, we wish to develop a generalized chart giving entropy departures from ideal gas values at a given temperature and pressure in a manner similar to that followed for enthalpy in the previous section. Once again, we have two alternatives. From Eq. 12.32, at constant temperature,

$$ds_T = -\left(\frac{\partial v}{\partial T}\right)_P dP_T$$

which is convenient for use with a volume-explicit EOS. The Lee–Kesler expression, Eq. 12.56, is, however, a pressure-explicit equation. It is therefore more appropriate to use Eq. 12.34, which is, along an isotherm,

$$ds_T = \left(\frac{\partial P}{\partial T}\right)_v dv_T$$

In the Lee–Kesler form, in terms of reduced properties, this equation becomes

$$\frac{ds}{R} = \left(\frac{\partial P_r}{\partial T_r}\right)_{v_r'} dv_r'$$

When this expression is integrated from a given state (P_r, v_r') to the ideal gas limit $(P_r^* \to 0, v_r'^* \to \infty)$, there is a problem because ideal gas entropy is a function of pressure and approaches infinity as the pressure approaches zero. We can eliminate this problem with a two-step procedure. First, the integral is taken only to a certain finite $P_r^*, v_r'^*$, which gives the entropy change

$$\frac{s_{p^*}^* - s_p}{R} = \int_{v_r'}^{v_r'^*} \left(\frac{\partial P_r}{\partial T_r}\right)_{v_r'} dv_r' \tag{12.61}$$

This integration by itself is not entirely acceptable, because it contains the entropy at some arbitrary low-reference pressure. A value for the reference pressure would have to be specified. Let us now repeat the integration over the same change of state, except this time for a hypothetical ideal gas. The entropy change for this integration is

$$\frac{s_{p^*}^* - s_p^*}{R} = +\ln \frac{P}{P^*} \tag{12.62}$$

If we now subtract Eq. 12.62 from Eq. 12.61, the result is the difference in entropy of a hypothetical ideal gas at a given state (T_r, P_r) and that of the real substance at the same state, or

$$\frac{s_p^* - s_p}{R} = -\ln \frac{P}{P^*} + \int_{v_r'}^{v_r'^* \to \infty} \left(\frac{\partial P_r}{\partial T_r}\right)_{v_r'} dv_r' \tag{12.63}$$

Here the values associated with the arbitrary reference state $P_r^*, v_r'^*$ cancel out of the right-hand side of the equation. (The first term of the integral includes the term $+\ln(P/P^*)$, which cancels the other term.) The three different states associated with the development of Eq. 12.63 are shown in Fig. 12.9.

The same procedure that was given in Section 12.7 for enthalpy departure values is followed for generalized entropy departure values. The Lee–Kesler simple-fluid constants are used in evaluating the integral of Eq. 12.63 and yield a simple-fluid entropy departure. The values for the entropy departure are shown graphically in Fig. D.3. Note that when using computer software to determine the entropy departure at a given reduced temperature

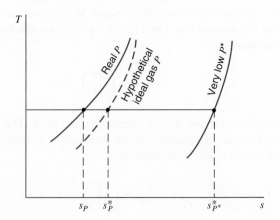

FIGURE 12.9 Real and ideal gas states and entropies.

and reduced pressure, accuracy can be improved by using the acentric factor in the same manner, as was done for the compressibility factor in Eq. 12.58 and subsequently, for the enthalpy departure in Section 12.7.

Example 12.6

Nitrogen at 8 MPa, 150 K is throttled to 0.5 MPa. After the gas passes through a short length of pipe, its temperature is measured and found to be 125 K. Determine the heat transfer and the change of entropy using the generalized charts. Compare these results with those obtained by using the nitrogen tables.

Control volume:	Throttle and pipe.
Inlet state:	P_1, T_1 known; state fixed.
Exit state:	P_2, T_2 known; state fixed.
Process:	Steady state.
Diagram:	Figure 12.10.
Model:	Generalized charts, results to be compared with those obtained with nitrogen tables.

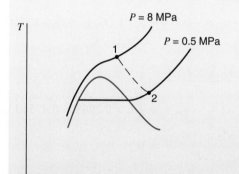

FIGURE 12.10 Sketch for Example 12.6.

Analysis

No work is done, and we neglect changes in kinetic and potential energies. Therefore, per kilogram,

Energy equation:

$$q + h_1 = h_2$$
$$q = h_2 - h_1 = -(h_2^* - h_2) + (h_2^* - h_1^*) + (h_1^* - h_1)$$

Solution

Using values from Table A.2, we have

$$P_{r1} = \frac{8}{3.39} = 2.36 \qquad T_{r1} = \frac{150}{126.2} = 1.189$$

$$P_{r2} = \frac{0.5}{3.39} = 0.147 \qquad T_{r2} = \frac{125}{126.2} = 0.99$$

From Fig. D.2,

$$\frac{h_1^* - h_1}{RT_c} = 2.5$$

$$h_1^* - h_1 = 2.5 \times 0.2968 \times 126.2 = 93.6\,\text{kJ/kg}$$

$$\frac{h_2^* - h_2}{RT_c} = 0.15$$

$$h_2^* - h_2 = 0.15 \times 0.2968 \times 126.2 = 5.6\,\text{kJ/kg}$$

Assuming a constant specific heat for the ideal gas, we have

$$h_2^* - h_1^* = C_{p0}(T_2 - T_1) = 1.0416(125 - 150) = -26.0\,\text{kJ/kg}$$

$$q = -5.6 - 26.0 + 93.6 = 62.0\,\text{kJ/kg}$$

From the nitrogen tables, Table B.6, we can find the change of enthalpy directly.

$$q = h_2 - h_1 = 123.77 - 61.92 = 61.85\,\text{kJ/kg}$$

To calculate the change of entropy using the generalized charts, we proceed as follows:

$$s_2 - s_1 = -(s_{P_2, T_2}^* - s_2) + (s_{P_2, T_2}^* - s_{P_1, T_1}^*) + (s_{P_1, T_1}^* - s_1)$$

From Fig. D.3

$$\frac{s_{P_1, T_1}^* - s_{P_1, T_1}}{R} = 1.6$$

$$s_{P_1, T_1}^* - s_{P_1, T_1} = 1.6 \times 0.2968 = 0.475\,\text{kJ/kg·K}$$

$$\frac{s_{P_2, T_2}^* - s_{P_2, T_2}}{R} = 0.1$$

$$s_{P_2, T_2}^* - s_{P_2, T_2} = 0.1 \times 0.2968 = 0.0297\,\text{kJ/kg·K}$$

Assuming a constant specific heat for the ideal gas, we have

$$s_{P_2, T_2}^* - s_{P_1, T_1}^* = C_{p0} \ln \frac{T_2}{T_1} - R \ln \frac{P_2}{P_1}$$

$$= 1.0416 \ln \frac{125}{150} - 0.2968 \ln \frac{0.5}{8}$$

$$= 0.6330\,\text{kJ/kg·K}$$

$$s_2 - s_1 = -0.0297 + 0.6330 + 0.475$$

$$= 1.078\,\text{kJ/kg·K}$$

From the nitrogen tables, Table B.6,

$$s_2 - s_1 = -5.4282 - 4.3522 = 1.0760\,\text{kJ/kg·K}$$

In-Text Concept Questions

d. If I raise the pressure in a solid at constant T, does s go up or down?

e. What does it imply if the compressibility factor is larger than 1?

f. What is the benefit of the generalized charts? Which properties must be known besides the charts themselves?

12.9 THE PROPERTY RELATION FOR MIXTURES

In Chapter 11 our consideration of mixtures was limited to ideal gases. There was no need at that point for further expansion of the subject. We now continue this subject with a view toward developing the property relations for mixtures. This subject will be particularly relevant to our consideration of chemical equilibrium in Chapter 14.

For a mixture, any extensive property X is a function of the temperature and pressure of the mixture and the number of moles of each component. Thus, for a mixture of two components,

$$X = f(T, P, n_A, n_B)$$

Therefore,

$$dX_{T,P} = \left(\frac{\partial X}{\partial n_A}\right)_{T,P,n_B} dn_A + \left(\frac{\partial X}{\partial n_B}\right)_{T,P,n_A} dn_B \tag{12.64}$$

Since at constant temperature and pressure an extensive property is directly proportional to the mass, Eq. 12.64 can be integrated to give

$$X_{T,P} = \overline{X}_A n_A + \overline{X}_B n_B \tag{12.65}$$

where

$$\overline{X}_A = \left(\frac{\partial X}{\partial n_A}\right)_{T,P,n_B}, \qquad \overline{X}_B = \left(\frac{\partial X}{\partial n_B}\right)_{T,P,n_A}$$

Here, $\overline{X}$ is defined as the partial molal property for a component in a mixture. It is particularly important to note that the partial molal property is defined under conditions of constant temperature and pressure.

The partial molal property is particularly significant when a mixture undergoes a chemical reaction. Suppose a mixture consists of components A and B, and a chemical reaction takes place so that the number of moles of A is changed by dn_A and the number of moles of B by dn_B. The temperature and the pressure remain constant. What is the change in internal energy of the mixture during this process? From Eq. 12.64, we conclude that

$$dU_{T,P} = \overline{U}_A \, dn_A + \overline{U}_B \, dn_B \tag{12.66}$$

where $\overline{U}_A$ and $\overline{U}_B$ are the partial molal internal energy of A and B, respectively. Equation 12.66 suggests that the partial molal internal energy of each component can also be defined as the internal energy of the component as it exists in the mixture.

In Section 12.3, we considered a number of property relations for systems of fixed mass such as

$$dU = T\,dS - P\,dV$$

In this equation, temperature is the intensive property or potential function associated with entropy, and pressure is the intensive property associated with volume. Suppose we have a chemical reaction such as that described in the previous paragraph. How would we modify this property relation for this situation? Intuitively, we might write the equation

$$dU = T\,dS - P\,dV + \mu_A\,dn_A + \mu_B\,dn_B \tag{12.67}$$

where μ_A is the intensive property or potential function associated with n_A, and similarly, μ_B for n_B. This potential function is called the chemical potential.

To derive an expression for this chemical potential, we examine Eq. 12.67 and conclude that it might be reasonable to write an expression for U in the form

$$U = f(S, V, n_A, n_B)$$

Therefore,

$$dU = \left(\frac{\partial U}{\partial S}\right)_{V,n_A,n_B} dS + \left(\frac{\partial U}{\partial V}\right)_{S,n_A,n_B} dV + \left(\frac{\partial U}{\partial n_A}\right)_{S,V,n_B} dn_A + \left(\frac{\partial U}{\partial n_B}\right)_{S,V,n_A} dn_B$$

Since the expressions

$$\left(\frac{\partial U}{\partial S}\right)_{V,n_A,n_B} \quad \text{and} \quad \left(\frac{\partial U}{\partial V}\right)_{S,n_A,n_B}$$

imply constant composition, it follows from Eq. 12.20 that

$$\left(\frac{\partial U}{\partial S}\right)_{V,n_A,n_B} = T \quad \text{and} \quad \left(\frac{\partial U}{\partial V}\right)_{S,n_A,n_B} = -P$$

Thus

$$dU = T\,dS - P\,dV + \left(\frac{\partial U}{\partial n_A}\right)_{S,V,n_B} dn_A + \left(\frac{\partial U}{\partial n_B}\right)_{S,V,n_A} dn_B \tag{12.68}$$

On comparing this equation with Eq. 12.67, we find that the chemical potential can be defined by the relation

$$\mu_A = \left(\frac{\partial U}{\partial n_A}\right)_{S,V,n_A}, \quad \mu_B = \left(\frac{\partial U}{\partial n_B}\right)_{S,V,n_A} \tag{12.69}$$

We can also relate the chemical potential to the partial molal Gibbs function. We proceed as follows:

$$G = U + PV - TS$$

$$dG = dU + P\,dV + V\,dP - T\,dS - S\,dT$$

Substituting Eq. 12.67 into this relation, we have

$$dG = -S\,dT + V\,dP + \mu_A\,dn_A + \mu_B\,dn_B \tag{12.70}$$

This equation suggests that we write an expression for G in the following form:

$$G = f(T, P, n_A, n_B)$$

Proceeding as we did for a similar expression for internal energy, we have

$$dG = \left(\frac{\partial G}{\partial T}\right)_{P,n_A,n_B} dT + \left(\frac{\partial G}{\partial P}\right)_{T,n_A,n_B} dP + \left(\frac{\partial G}{\partial n_A}\right)_{T,P,n_B} dn_A + \left(\frac{\partial G}{\partial n_B}\right)_{T,P,n_A} dn_B$$

$$= -S\,dT + V\,dP + \left(\frac{\partial G}{\partial n_A}\right)_{T,P,n_B} dn_A + \left(\frac{\partial G}{\partial n_B}\right)_{T,P,n_A} dn_B$$

When this equation is compared with Eq. 12.70, it follows that

$$\mu_A = \left(\frac{\partial G}{\partial n_A}\right)_{T,P,n_B}, \qquad \mu_B = \left(\frac{\partial G}{\partial n_B}\right)_{T,P,n_A}$$

Because partial molal properties are defined at constant temperature and pressure, the quantities $(\partial G/\partial n_A)_{T,P,n_B}$ and $(\partial G/\partial n_B)_{T,P,n_A}$ are the partial molal Gibbs functions for the two components. That is, the chemical potential is equal to the partial molal Gibbs function.

$$\mu_A = \overline{G}_A = \left(\frac{\partial G}{\partial n_A}\right)_{T,P,n_B}, \qquad \mu_B = \overline{G}_B = \left(\frac{\partial G}{\partial n_B}\right)_{T,P,n_A} \tag{12.71}$$

Although μ can also be defined in terms of other properties, such as in Eq. 12.69, this expression is not the partial molal internal energy, since the pressure and temperature are not constant in this partial derivative. The partial molal Gibbs function is an extremely important property in the thermodynamic analysis of chemical reactions, for at constant temperature and pressure (the conditions under which many chemical reactions occur), it is a measure of the chemical potential or the driving force that tends to make a chemical reaction take place.

As the chemical potential equals the partial molal Gibbs function, its relation to the pure substance properties involves properties of the mixture and the system as a whole. In the limit of ideal mixtures of ideal gases, such interactions are zero and the sole effect is that each component i behaves as a pure substance at the mixture temperature and the partial pressure P_i. In this limiting case the chemical potential equals the Gibbs function as

$$\mu_i = \overline{g}_i = \overline{g}(T,\,P_i) = \overline{h}_i - T\,\overline{s}_i(T,\,P_i) = \overline{g}_{i\,T,P} + \overline{R}T\ln\frac{P_i}{P}$$
$$= \overline{g}_{i\,T,P} + \overline{R}T\ln y_i \tag{12.72}$$

as the pressure sensitivity only enters in the entropy. For the general mixture, the analysis defines a fugacity (a pseudopressure) to describe the variation of the Gibbs function with the pressure, so the treatment becomes similar to the one for ideal gases, a subject not included in the current presentation.

The analysis for the gaseous mixtures can be extended to cover liquids called *solutions*, where some of the interaction effects are more pronounced than those for gases due to the smaller intermolecular distances. An example in which the interaction is related to the mixture is a liquid blend of water and ethanol that has a volume slightly smaller than the sum of the water and ethanol volumes separately and a corresponding change in the Gibbs function. An example in which an effect comes from the surface is liquid water located in a porous material like a sponge, granite, or concrete where the surface tension gives rise to a potential energy between the liquid water and the solid material. This will lower the total Gibbs function relative to the Gibbs function of liquid water outside the pores; that is, the effect is due to the interaction between the liquid and the solid substance. For an ideal solution, a result very similar to Eq. 12.72 is obtained showing that if salt (or other substances) is dissolved in water, the Gibbs function is lower ($y_i < 1$), which causes a lower freezing point.

12.10 PSEUDOPURE SUBSTANCE MODELS FOR REAL GAS MIXTURES

A basic prerequisite to the treatment of real gas mixtures in terms of pseudopure substance models is the concept and use of appropriate reference states. As an introduction to this topic, let us consider several preliminary reference state questions for a pure substance undergoing a change of state, for which it is desired to calculate the entropy change. We can express the entropy at the initial state 1 and also at the final state 2 in terms of a reference state 0, in a manner similar to that followed when dealing with the generalized-chart corrections. It follows that

$$s_1 = s_0 + (s^*_{P_0 T_0} - s_0) + (s^*_{P_1 T_1} - s^*_{P_0 T_0}) + (s_1 - s^*_{P_1 T_1}) \qquad (12.73)$$

$$s_2 = s_0 + (s^*_{P_0 T_0} - s_0) + (s^*_{P_2 T_2} - s^*_{P_0 T_0}) + (s_2 - s^*_{P_2 T_2}) \qquad (12.74)$$

These are entirely general expressions for the entropy at each state in terms of an arbitrary reference state value and a set of consistent calculations from that state to the actual desired state. One simplification of these equations would result from choosing the reference state to be a hypothetical ideal gas state at P_0 and T_0, thereby making the term

$$(s^*_{P_0 T_0} - s_0) = 0 \qquad (12.75)$$

in each equation, which results in

$$s_0 = s^*_0 \qquad (12.76)$$

It should be apparent that this choice is a reasonable one, since whatever value is chosen for the correction term, Eq. 12.75, it will cancel out of the two equations when the change $s_2 - s_1$ is calculated, and the simplest value to choose is zero. In a similar manner, the simplest value to choose for the ideal gas reference value, Eq. 12.76, is zero, and we would commonly do that if there were no restrictions on choice, such as would occur in the case of a chemical reaction.

Another point to be noted concerning reference states is related to the choice of P_0 and T_0. For this purpose, let us substitute Eqs. 12.75 and 12.76 into Eqs. 12.73 and 12.74, and also assume constant specific heat, such that those equations can be written in the form

$$s_1 = s^*_0 + C_{p0} \ln\left(\frac{T_1}{T_0}\right) - R \ln\left(\frac{P_1}{P_0}\right) + (s_1 - s^*_{P_1 T_1}) \qquad (12.77)$$

$$s_2 = s^*_0 + C_{p0} \ln\left(\frac{T_2}{T_0}\right) - R \ln\left(\frac{P_2}{P_0}\right) + (s_2 - s^*_{P_2 T_2}) \qquad (12.78)$$

Since the choice for P_0 and T_0 is arbitrary if there are no restrictions, such as would be the case with chemical reactions, it should be apparent from examining Eqs. 12.77 and 12.78 that the simplest choice would be for

$$P_0 = P_1 \quad \text{or} \quad P_2 \qquad T_0 = T_1 \quad \text{or} \quad T_2$$

FIGURE 12.11

Example of a mixing process.

It should be emphasized that inasmuch as the reference state was chosen as a hypothetical ideal gas at P_0, T_0, Eq. 12.75, it is immaterial how the real substance behaves at that pressure and temperature. As a result, there is no need to select a low value for the reference state pressure, P_0.

Let us now extend these reference state developments to include real gas mixtures. Consider the mixing process shown in Fig. 12.11, with the states and amounts of each substance as given on the diagram. Proceeding with entropy expressions as was done earlier, we have

$$\bar{s}_1 = \bar{s}_{A_0}^* + \overline{C}_{p0_A} \ln\left(\frac{T_1}{T_0}\right) - \overline{R} \ln\left(\frac{P_1}{P_0}\right) + (\bar{s}_1 - \bar{s}_{P_1 T_1}^*)A \tag{12.79}$$

$$\bar{s}_2 = \bar{s}_{B_0}^* + \overline{C}_{p0_B} \ln\left(\frac{T_2}{T_0}\right) - \overline{R} \ln\left(\frac{P_2}{P_0}\right) + (\bar{s}_2 - \bar{s}_{P_2 T_2}^*)B \tag{12.80}$$

$$\bar{s}_3 = \bar{s}_{\text{mix}_0}^* + \overline{C}_{p0_\text{mix}} \ln\left(\frac{T_3}{T_0}\right) - \overline{R} \ln\left(\frac{P_3}{P_0}\right) + (\bar{s}_3 - \bar{s}_{P_3 T_3}^*)\text{mix} \tag{12.81}$$

in which

$$\bar{s}_{\text{mix}_0}^* = y_A \bar{s}_{A_0}^* + y_B \bar{s}_{B_0}^* - \overline{R}(y_A \ln y_A + y_B \ln y_B) \tag{12.82}$$

$$\overline{C}_{p0_\text{mix}} = y_A \overline{C}_{p0_A} + y_B \overline{C}_{p0_B} \tag{12.83}$$

When Eqs. 12.79–12.81 are substituted into the equation for the entropy change,

$$n_3 \bar{s}_3 - n_1 \bar{s}_1 - n_2 \bar{s}_2$$

the arbitrary reference values, s_{A0}^*, s_{B0}^*, P_0, and T_0 all cancel out of the result, which is, of course, necessary in view of their arbitrary nature. An ideal gas entropy of mixing expression, the final term in Eq. 12.82, remains in the result, establishing, in effect, the mixture reference value relative to its components. The remarks made earlier concerning the choices for reference state and the reference state entropies apply in this situation as well.

To summarize the development to this point, we find that a calculation of real mixture properties, as, for example, using Eq. 12.81, requires the establishment of a hypothetical ideal gas reference state, a consistent ideal gas calculation to the conditions of the real mixture, and finally, a correction that accounts for the real behavior of the mixture at that state. This last term is the only place where the real behavior is introduced, and this is therefore the term that must be calculated by the pseudopure substance model to be used.

In treating a real gas mixture as a pseudopure substance, we will follow two approaches to represent the P–v–T behavior: use of the generalized charts and use of an analytical EOS. With the generalized charts, we need to have a model that provides a set of pseudocritical pressure and temperature in terms of the mixture component values. Many such models have been proposed and utilized over the years, but the simplest is that suggested by W. B. Kay in 1936, in which

$$(P_c)_{\text{mix}} = \sum_i y_i P_{ci}, \qquad (T_c)_{\text{mix}} = \sum_i y_i T_{ci} \qquad (12.84)$$

This model, known as Kay's rule, is the only pseudocritical model that we will consider in this chapter. Other models are somewhat more complicated to evaluate and use but are considerably more accurate.

The other approach to be considered involves using an analytical EOS, in which the equation for the mixture must be developed from that for the components. In other words, for an equation in which the constants are known for each component, we must develop a set of empirical combining rules that will then give a set of constants for the mixture as though it were a pseudopure substance. This problem has been studied for many equations of state, using experimental data for the real gas mixtures, and various empirical rules have been proposed. For example, for both the van der Waals equation, Eq. 12.51, and the Redlich–Kwong equation, Eq. 12.53, the two pure substance constants a and b are commonly combined according to the relations

$$a_m = \left(\sum_1 c_i a_i^{1/2} \right)^2 \qquad b_m = \sum_i c_i b_i \qquad (12.85)$$

The following example illustrates the use of these two approaches to treating real gas mixtures as pseudopure substances.

Example 12.7

A mixture of 80 % CO_2 and 20 % CH_4 (mass basis) is maintained at 310.94 K, 86.19 bar, at which condition the specific volume has been measured as 0.006 757 m^3/kg. Calculate the percent deviation if the specific volume had been calculated by (a) Kay's rule and (b) van der Waals' EOS.

Control mass: Gas mixture.

State: P, v, T known.

Model: (a) Kay's rule. (b) van der Waals' equation.

Solution

Let subscript A denote CO_2 and B denote CH_4; then from Tables A.2 and A.5

$$T_{c_A} = 304.1 \text{ K} \qquad P_{c_A} = 7.38 \text{ MPa} \qquad R_A = 0.1889 \text{ kJ/kg} \cdot \text{K}$$
$$T_{c_B} = 190.4 \text{ K} \qquad P_{c_B} = 4.60 \text{ MPa} \qquad R_B = 0.5183 \text{ kJ/kg} \cdot \text{K}$$

The gas constant from Eq. 11.15 becomes

$$R_m = \sum c_i R_i = 0.8 \times 0.1889 + 0.2 \times 0.5183 = 0.2548 \text{ kJ/kg} \cdot \text{K}$$

and the mole fractions are

$$y_A = (c_A/M_A)/ \sum (c_i/M_i) = \frac{0.8/44.01}{(0.8/44.01) + (0.2/16.043)} = 0.5932$$

$$y_B = 1 - y_A = 0.4068$$

a. For Kay's rule, Eq. 12.83,

$$T_{cm} = \sum_i y_i T_{ci} = y_A T_{cA} + y_B T_{cB}$$

$$= 0.5932(304.1) + 0.4068(190.4)$$

$$= 257.9 \, \text{k}$$

$$P_{cm} = \sum_i y_i P_{ci} = y_A P_{cA} + y_B P_{cB}$$

$$= 0.5932(7.38) + 0.4068(4.60)$$

$$= 6.249 \, \text{MPa}$$

Therefore, the pseudoreduced properties of the mixture are

$$T_{r_m} = \frac{T}{T_{cm}} = \frac{310.94}{257.9} = 1.206$$

$$P_{r_m} = \frac{P}{P_{cm}} = \frac{8.619}{6.249} = 1.379$$

From the generalized chart, Fig. D.1

$$Z_m = 0.7$$

and

$$v = \frac{Z_m R_m T}{P} = \frac{0.7 \times 0.2548 \times 310.94}{8619} = 0.006\,435 \, \text{m}^3/\text{kg}$$

The percent deviation from the experimental value is

$$\text{Percent deviation} = \left(\frac{0.006\,757 - 0.006\,435}{0.006\,757} \right) \times 100 = 4.8\,\%$$

The major factor contributing to this 5 % error is the use of the linear Kay's rule pseudocritical model, Eq. 12.84. Use of an accurate pseudocritical model and the generalized chart would reduce the error to approximately 1 %.

b. For van der Waals' equation, the pure substance constants are

$$a_A = \frac{27 R_A^2 T_{cA}^2}{64 P_{cA}} = 0.188\,64 \, \frac{\text{kPa m}^6}{\text{kg}^2}$$

$$b_A = \frac{R_A T_{cA}}{8 P_{cA}} = 0.000\,973 \, \text{m}^3/\text{kg}$$

and

$$a_B = \frac{27 R_B^2 T_{cB}^2}{64 P_{cB}} = 0.8931 \, \frac{\text{kPa m}^6}{\text{kg}^2}$$

$$b_B = \frac{R_B T_{cB}}{8 P_{cB}} = 0.002\,682 \, \text{m}^3/\text{kg}$$

Therefore, for the mixture, from Eq. 12.85,

$$a_m = (c_A\sqrt{a_A} + c_B\sqrt{a_B})^2$$

$$= (0.8\sqrt{0.188\,64} + 0.2\sqrt{0.8931})^2 = 0.2878\,\frac{\text{kPa}\cdot\text{m}^6}{\text{kg}^2}$$

$$b_m = c_A b_A + c_B b_B$$

$$= 0.8 \times 0.000\,973 + 0.2 \times 0.002\,682 = 0.001\,315\ \text{m}^3/\text{kg}$$

The EOS for the mixture of this composition is

$$P = \frac{R_m T}{v - b_m} - \frac{a_m}{v^2}$$

$$8619 = \frac{0.2548 \times 310.94}{v - 0.001\,315} - \frac{0.2878}{v^2}$$

Solving for v by trial and error,

$$v = 0.006\,326\ \text{m}^3/\text{kg}$$

$$\text{Percent derivation} = \left(\frac{0.006\,757 - 0.006\,326}{0.006\,757}\right) \times 100 = 6.4\,\%$$

As a point of interest from the ideal gas law, $v = 0.009\,19$ m³/kg, which is a deviation of 36 % from the measured value. Also, if we use the Redlich–Kwong EOS and follow the same procedure as for the van der Waals equation, the calculated specific volume of the mixture is 0.006 52 m³/kg, which is in error by 3.5 %.

We must be careful not to draw too general a conclusion from the results of this example. We have calculated percent deviation in v at only a single point for only one mixture. We do note, however, that the various methods used give quite different results. From a more general study of these models for a number of mixtures, we find that the results found here are fairly typical, at least qualitatively. Kay's rule is very useful because it is fairly accurate and yet relatively simple. The van der Waals equation is too simplified an expression to accurately represent P–v–T behavior, but it is useful to demonstrate the procedures followed in utilizing more complex analytical equations of state. The Redlich–Kwong equation is considerably better and is still relatively simple to use.

As noted in the example, the more sophisticated generalized behavior models and empirical equations of state will represent mixture P–v–T behavior to within about 1 % over a wide range of density, but they are, of course, more difficult to use than the methods considered in Example 12.7. The generalized models have the advantage of being easier to use, and they are suitable for hand computations. Calculations with the complex empirical equations of state become very involved but have the advantage of expressing the P–v–T composition relations in analytical form, which is of great value when using a computer for such calculations.

12.11 ENGINEERING APPLICATIONS— THERMODYNAMIC TABLES

For a given pure substance, tables of thermodynamic properties can be developed from experimental data in several ways. In this section, we outline the traditional procedure followed for the liquid and vapor phases of a substance and then present the more modern techniques utilized for this purpose.

Let us assume that the following data for a pure substance have been obtained in the laboratory:

1. Vapor-pressure data. That is, saturation pressures and temperatures have been measured over a wide range.
2. Pressure, specific volume, and temperature data in the vapor region. These data are usually obtained by determining the mass of the substance in a closed vessel (which means a fixed specific volume) and then measuring the pressure as the temperature is varied. This is done for a large number of specific volumes.
3. Density of the saturated liquid and the critical pressure and temperature.
4. Zero-pressure specific heat for the vapor. This might be obtained either calorimetrically or from spectroscopic data and statistical thermodynamics (see Appendix C).

From these data, a complete set of thermodynamic tables for the saturated liquid, saturated vapor, and superheated vapor can be calculated. The first step is to determine an equation for the vapor pressure curve that accurately fits the data. One form commonly used is given in terms of reduced pressure and temperature as

$$\ln P_r = [C_1 \tau_0 + C_2 \tau_0^{1.5} + C_3 \tau_0^3 + C_4 \tau_0^6]/T_r \qquad (12.86)$$

where the dimensionless temperature variable is $\tau_0 = 1 - T_r$. Once the set of constants has been determined for the given data, the saturation pressure at any temperature can be calculated from Eq. 12.86. The next step is to determine an EOS for the vapor region (including the dense fluid region above the critical point) that accurately represents the P–v–T data. It would be desirable to have an equation that is explicit in v in order to use P and T as the independent variables in calculating enthalpy and entropy changes from Eqs. 12.26 and 12.33, respectively. However, equations explicit in P, as a function of T and v, prove to be more accurate and are consequently the form used in the calculations. Therefore, at any chosen P and T (table entries), the equation is solved by iteration for v, so that the T and v can then be used as the independent variables in the subsequent calculations.

The procedure followed in determining enthalpy and entropy is best explained with the aid of Fig. 12.12. Let the enthalpy and entropy of saturated liquid at state 1 be set to zero (arbitrary reference state). The enthalpy and entropy of saturated vapor at state 2 can then be calculated from the Clapeyron equation, Eq. 12.4. The left-hand side of this equation is found by differentiating Eq. 12.86, v_g is calculated from the EOS using P_g from Eq. 12.86, and v_f is found from the experimental data for the saturated liquid phase.

From state 2, we proceed along this isotherm into the superheated vapor region. The specific volume at pressure P_3 is found by iteration from the EOS. The internal energy and entropy are calculated by integrating Eqs. 12.31 and 12.35, and the enthalpy is then calculated from its definition.

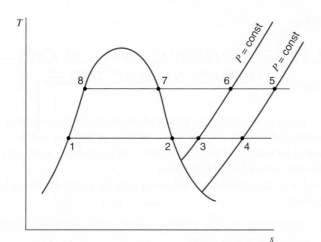

FIGURE 12.12
Sketch showing the procedure for developing a table of thermodynamic properties from experimental data.

The properties at point 4 are found in exactly the same manner. Pressure P_4 is sufficiently low that the real superheated vapor behaves essentially as an ideal gas (perhaps 1 kPa). Thus, we use this constant-pressure line to make all temperature changes for our calculations, as, for example, to point 5. Since the specific heat C_{p0} is known as a function of temperature, the enthalpy and entropy at 5 are found by integrating Eqs. 3.37 and 6.15. The properties at points 6 and 7 are found from those at point 5 in the same manner as those at points 3 and 4 were found from point 2. (The saturation pressure P_7 is calculated from the vapor-pressure equation.) Finally, the enthalpy and entropy for saturated liquid at point 8 are found from the properties at point 7 by applying the Clapeyron equation.

Thus, values for the pressure, temperature, specific volume, enthalpy, entropy, and internal energy of saturated liquid, saturated vapor, and superheated vapor can be tabulated for the entire region for which experimental data were obtained.

The modern approach to developing thermodynamic tables utilizes the Helmholtz function, defined by Eq. 12.12. Rewriting the two partial derivatives for a in Eq. 12.21 in terms of ρ instead of v, we have

$$P = \rho^2 \left(\frac{\partial a}{\partial \rho} \right)_T \tag{12.87}$$

$$s = -\left(\frac{\partial a}{\partial T} \right)_\rho \tag{12.88}$$

We now express the Helmholtz function in terms of the ideal gas contribution plus the residual (real substance) contribution,

$$a(\rho, T) = a^*(\rho, T) + a^r(\rho, T) \tag{12.89}$$

or, dividing by RT,

$$\frac{a(\rho, T)}{RT} = \alpha(\delta, \tau) = \alpha^*(\delta, \tau) + \alpha^r(\delta, \tau) \tag{12.90}$$

in terms of the reduced variables

$$\delta = \frac{\rho}{\rho_c}, \qquad \tau = \frac{T_c}{T} \tag{12.91}$$

To get an expression for the ideal gas portion α^* (or a^*/RT), we use the relations

$$a^* = h^* - RT - Ts^* \qquad (12.92)$$

in which

$$h^* = h_0^* + \int_{T_0}^{T} C_{p0}\, dT \qquad (12.93)$$

$$s^* = s_0^* + \int_{T_0}^{T} \frac{C_{p0}}{T}\, dT - R \ln\left(\frac{\rho T}{\rho_0 T_0}\right) \qquad (12.94)$$

$$\text{where} \quad \rho_0 = P_0/RT_0 \qquad (12.95)$$

and P_0, T_0, h_0^*, and s_0^* are arbitrary constants.

In these relations, the ideal gas specific heat C_{p0} must be expressed as an empirical function of temperature. This is commonly of the form of the equations in Appendix A.6, often with additional terms, some of the form of the molecular vibrational contributions as shown in Appendix C. Following selection of the expression for C_{p0}, the set of equations 12.92–12.95 gives the desired expression for α^*. This value can now be calculated at any given temperature relative to the arbitrarily selected constants.

It is then necessary to give an expression for the residual α^r. This is commonly of the form

$$\alpha^r = \sum N_k \delta^{i_k} \tau^{j_k} + \sum N_k \delta^{i_k} \tau^{j_k} \exp(-\delta^{l_k}) \qquad (12.96)$$

in which the exponents i_k and l_k are usually positive integers, while j_k is usually positive but not an integer. Depending on the substance and the accuracy of fit, each of the two summations in Eq. 12.96 may have 4 to 20 terms. The form of Eq. 12.96 is suggested by the terms in the Lee–Kesler EOS, Eq. 12.56.

We are now able to express the EOS. From Eq. 12.87,

$$Z = \frac{P}{\rho RT} = \rho\left(\frac{\partial a/RT}{\partial \rho}\right)_T = \delta\left(\frac{\partial \alpha}{\partial \delta}\right)_\tau = 1 + \delta\left(\frac{\partial \alpha^r}{\partial \delta}\right)_\tau \qquad (12.97)$$

(Note: since the ideal gas $\rho\left(\dfrac{\partial a^*}{\partial \rho}\right)_T = \dfrac{P}{\rho} = RT,\ \delta\left(\dfrac{\partial \alpha^*}{\partial \delta}\right)_\tau = 1$.)

Differentiating Eq. 12.96 and substituting into Eq. 12.97 results in the EOS as the function $Z = Z(\delta, \tau)$ in terms of the empirical coefficients and exponents of Eq. 12.96. These coefficients are now fitted to the available experimental data. Once this has been completed, the thermodynamic properties s, u, h, a, and g can be calculated directly, using the calculated value of α^* at the given T and α^r from Eq. 12.96. This gives a/RT directly from Eq. 12.90. From Eq. 12.88,

$$\frac{s}{R} = -\frac{1}{R}\left(\frac{\partial a}{\partial T}\right)_\rho = -T\left(\frac{\partial a/RT}{\partial T}\right)_\rho - \frac{a}{RT} = \tau\left(\frac{\partial \alpha}{\partial \tau}\right)_\delta - \alpha \qquad (12.98)$$

From Eqs. 12.12 and 12.98,

$$\frac{u}{RT} = \frac{s}{R} + \frac{a}{RT} = \tau\left(\frac{\partial \alpha}{\partial \tau}\right)_\delta \qquad (12.99)$$

Finally,

$$\frac{h}{RT} = \frac{u}{RT} + Z \qquad\qquad (12.100)$$

$$\frac{g}{RT} = \frac{a}{RT} + Z = \alpha + Z \qquad\qquad (12.101)$$

This last equation is particularly important, since at saturation the Gibbs functions of the liquid and vapor must be equal ($h_{fg} = Ts_{fg}$). Therefore, at the given T, the saturation pressure is the value for which the Gibbs function (from Eq. 12.101) calculated for the vapor v is equal to that calculated for the liquid v. Starting values for this iterative process are the pressure from an equation of the form 12.86, with the liquid density from given experimental data as discussed earlier in this section.

This method for using an EOS to calculate properties of both the vapor and liquid phases has the distinct advantage in accuracy of representation, in that no mathematical integrations are required in the process. Most of the substances included in CATT3 are evaluated with the Helmholtz function, as in Eq. 12.90.

SUMMARY

As an introduction to the development of property information that can be obtained experimentally, we derive the Clapeyron equation. This equation relates the slope of the two-phase boundaries in the P–T diagram to the enthalpy and specific volume change going from one phase to the other. If we measure pressure, temperature, and the specific volumes for liquid and vapor in equilibrium, we can calculate the enthalpy of evaporation. Because thermodynamic properties are functions of two variables, a number of relations can be derived from the mixed second derivatives and the Gibbs relations, which are known as Maxwell relations. Many other relations can be derived, and those that are useful let us relate thermodynamic properties to those that can be measured directly like P, v, and T and indirectly like the heat capacities.

Changes of enthalpy, internal energy, and entropy between two states are presented as integrals over properties that can be measured and thus obtained from experimental data. Some of the partial derivatives are expressed as coefficients like expansivity and compressibility, with the process as a qualifier like isothermal or isentropic (adiabatic). These coefficients, as single numbers, are useful when they are nearly constant over some range of interest, which happens for liquids and solids and thus are found in various handbooks. The speed of sound is also a property that can be measured, and it relates to a partial derivative in a nonlinear fashion.

The experimental information about a substance behavior is normally correlated in an EOS relating P–v–T to represent part of the thermodynamic surface. Starting with the general compressibility and its extension to the virial EOS, we lead up to other, more complex equations of state. We show the most versatile equations, such as the van der Waals EOS, the Redlich–Kwong EOS, and the Lee–Kesler EOS, which is shown as an extension of Benedict–Webb–Rubin (BWR), with others that are presented in Appendix D. The most accurate equations are too complex for hand calculations and are used on computers to generate tables of properties. Therefore, we do not cover those details.

As an application of the Lee–Kesler EOS for a simple fluid, we present the development of the generalized charts that can be used for substances for which we do not have a table. The charts express the deviation of the properties from an ideal gas in terms of a compressibility factor (Z) and the enthalpy and entropy departure terms. These charts are in dimensionless properties based on the properties at the critical point.

Properties for mixtures are introduced in general, and the concept of a partial molal property leads to the chemical potential derived from the Gibbs function. Real mixtures are treated on a mole basis, and we realize that a model is required to do so. We present a pseudocritical Kay's rule that predicts the critical properties for the mixture and then uses the generalized charts. Other models predict EOS parameters for the mixture and then use the EOS as for a pure substance. Typical examples here are the van der Waals and Redlich–Kwong EOSs.

Engineering applications focus on the development of tables of thermodynamic properties. The traditional procedure is covered first, followed by the more modern approach to represent properties in terms of an EOS that represents both the vapor and liquid phases.

You should have learned a number of skills and acquired abilities from studying this chapter that will allow you to:

- Apply and understand the assumptions for the Clapeyron equation.
- Use the Clapeyron equation for all three two-phase regions.
- Have a sense of what a partial derivative means.
- Understand why Maxwell relations and other relations are relevant.
- Know that the relations are used to develop expression for changes in h, u, and s.
- Know that coefficients of linear expansion and compressibility are common data useful for describing certain processes.
- Know that speed of sound is also a property.
- Be familiar with various equations of state and their use.
- Know the background for and how to use the generalized charts.
- Know that a model is needed to deal with a mixture.
- Know the pseudocritical Kay's rule and the EOS models for a mixture.
- Be familiar with the development of tables of thermodynamic properties.

KEY CONCEPTS AND FORMULAS		
Clapeyron equation	$\dfrac{dP_{\text{sat}}}{dT} = \dfrac{h'' - h'}{T(v'' - v')};$	$S\text{--}L, S\text{--}V,$ and $V\text{--}L$ regions
Maxwell relations	$dz = M\,dx + N\,dy \Rightarrow \left(\dfrac{\partial M}{\partial y}\right)_x = \left(\dfrac{\partial N}{\partial x}\right)_y$	
Change in enthalpy	$h_2 - h_1 = \displaystyle\int_1^2 C_p\,dT + \int_1^2 \left[v - T\left(\dfrac{\partial v}{\partial T}\right)_p\right]dP$	
Change in energy	$u_2 - u_1 = \displaystyle\int_1^2 C_v\,dT + \int_1^2 \left[T\left(\dfrac{\partial P}{\partial T}\right)_v - P\right]dv$	
	$= h_2 - h_1 - (P_2 v_2 - P_1 v_1)$	

Change in entropy	$s_2 - s_1 = \int_1^2 \dfrac{C_p}{T} \, dT - \int_1^2 \left(\dfrac{\partial v}{\partial T} \right)_p dP$
Virial equation	$Z = \dfrac{Pv}{RT} = 1 + \dfrac{B(T)}{v} + \dfrac{C(T)}{v^2} + \dfrac{D(T)}{v^3} + \cdots \text{(mass basis)}$
Van der Waals equation	$P = \dfrac{RT}{v-b} - \dfrac{a}{v^2} \quad (\text{mass basis})$
Redlich–Kwong equation	$P = \dfrac{RT}{v-b} - \dfrac{a}{v(v+b)T^{1/2}} \quad (\text{mass basis})$
Other equations of state	See Appendix D.
Generalized charts for h	$h_2 - h_1 = (h_2^* - h_1^*)_{I.D.G.} - RT_c(\Delta \hat{h}_2 - \Delta \hat{h}_1)$
Enthalpy departure	$\Delta \hat{h} = (h^* - h)/RT_c; \qquad h^* \text{ value for ideal gas}$
Generalized charts for s	$s_2 - s_1 = (s_2^* - s_1^*)_{I.D.G.} - R(\Delta \hat{s}_2 - \Delta \hat{s}_1)$
Entropy departure	$\Delta \hat{s} = (s^* - s)/R; \qquad s^* \text{ value for ideal gas}$
Pseudocritical pressure	$P_{c\,\text{mix}} = \sum_i y_i P_{ci}$
Pseudocritical temperature	$T_{c\,\text{mix}} = \sum_i y_i T_{ci}$
Pseudopure substance	$a_m = \left(\sum_i c_i a_i^{1/2} \right)^2 ; \quad b_m = \sum_i c_i b_i \quad (\text{mass basis})$

CONCEPT-STUDY GUIDE PROBLEMS

12.1 The slope dP/dT of the vaporization line is finite as you approach the critical point, yet h_{fg} and v_{fg} both approach zero. How can that be?

12.2 In view of Clapeyron's equation and Fig. 2.4, is there something special about ice I versus the other forms of ice?

12.3 If we take a derivative as $(\partial P/\partial T)_v$ in the two-phase region (see Figs. 2.7 and 2.8), does it matter what v is? How about T?

12.4 Sketch on a P–T diagram how a constant v line behaves in the compressed liquid region, the two-phase L–V region, and the superheated vapor region.

12.5 If the pressure is raised in an isothermal process, does h go up or down for a liquid or solid? What do you need to know if it is a gas phase?

12.6 The EOS in Example 12.3 was used as explicit in v. Is it explicit in P?

12.7 Over what ranges of states are the various coefficients in Section 12.5 most useful?

12.8 For a liquid or a solid, is v more sensitive to T or P? How about an ideal gas?

12.9 Most equations of state are developed to cover which range of states?

12.10 Is an EOS valid in the two-phase regions?

12.11 As $P \to 0$, the specific volume $v \to \infty$. For $P \to \infty$, does $v \to 0$?

12.12 Must an EOS satisfy the two conditions in Eqs. 12.49 and 12.50?

12.13 At which states are the departure terms for h and s small? What is Z there?

12.14 The departure functions for h and s as defined are always positive. What does that imply for the real-substance h and s values relative to ideal gas values?

12.15 What is the benefit of Kay's rule versus a mixture EOS?

HOMEWORK PROBLEMS

Clapeyron Equation

12.16 An approximation for the saturation pressure can be $\ln P_{sat} = A - B/T$, where A and B are constants. Which phase transition is that suitable for, and what kinds of property variations are assumed?

12.17 Verify that Clapeyron's equation is satisfied for R-410A at $10\,°C$ in Table B.4.

12.18 In a Carnot heat engine, the heat addition changes the working fluid from saturated liquid to saturated vapor at T, P. The heat rejection process occurs at lower temperature and pressure $(T - \Delta T)$, $(P - \Delta P)$. The cycle takes place in a piston/cylinder arrangement where the work is boundary work. Apply both the first and second laws with simple approximations for the integral equal to work. Then show that the relation between ΔP and ΔT results in the Clapeyron equation in the limit $\Delta T \to dT$.

12.19 Verify that Clapeyron's equation is satisfied for carbon dioxide at $6\,°C$ in Table B.3.

12.20 Use the approximation given in Problem 12.16 and Table B.1 to determine A and B for steam from properties at $25\,°C$ only. Use the equation to predict the saturation pressure at $30\,°C$ and compare this to the table value.

12.21 A certain refrigerant vapor enters a steady-flow, constant-pressure condenser at 150 kPa, $70\,°C$, at a rate of 1.5 kg/s, and it exits as saturated liquid. Calculate the rate of heat transfer from the condenser. It may be assumed that the vapor is an ideal gas and also that at saturation, $v_f \ll v_g$. The following is known:

$$\ln P_g = 8.15 - 1000/T \qquad C_{p0} = 0.7 \text{ kJ/kg·K}$$

with pressure in kPa and temperature in K. The molecular mass is 100.

12.22 Calculate the values h_{fg} and s_{fg} for nitrogen at 70 K and at 110 K from the Clapeyron equation, using the necessary pressure and specific volume values from Table B.6.1.

12.23 Find the saturation pressure for the refrigerant R-410A at $-80\,°C$, assuming it is higher than the triple-point temperature.

12.24 Ammonia at $-70\,°C$ is used in a special application at a quality of 50 %. Assume the only table available is Table B.2 that goes down to $-50\,°C$. To size a tank to hold 0.5 kg with $x = 0.5$, give your best estimate for the saturated pressure and the tank volume.

12.25 Use the approximation given in Problem 12.16 and Table B.4 to determine A and B for the refrigerant R-410A from properties at $0\,°C$ only. Use the equation to predict the saturation pressure at $5\,°C$ and compare this to the table value.

12.26 The triple point of carbon dioxide is $-56.4\,°C$. Predict the saturation pressure at that point using Table B.3.

12.27 Helium boils at 4.22 K at atmospheric pressure, 101.3 kPa, with $h_{fg} = 83.3$ kJ/kmol. By pumping a vacuum over liquid helium, the pressure can be lowered, and it may then boil at a lower temperature. Estimate the necessary pressure to produce a boiling temperature of 1 K and one of 0.5 K.

12.28 Using the properties of water at the triple point, develop an equation for the saturation pressure along the fusion line as a function of temperature.

12.29 Using thermodynamic data for water from Tables B.1.1 and B.1.5, estimate the freezing temperature of liquid water at a pressure of 30 MPa.

12.30 Ice (solid water) at $-3\,°C$, 100 kPa is compressed isothermally until it becomes liquid. Find the required pressure.

12.31 From the phase diagrams for water (Fig. 2.4) and carbon dioxide (Fig. 2.5), what can you infer for the specific volume change during melting, assuming the liquid has a higher h than the solid phase for those two substances?

12.32 A container has a double wall where the wall cavity is filled with carbon dioxide at room temperature and pressure. When the container is filled with a cryogenic liquid at 100 K, the carbon dioxide will freeze so that the wall cavity has a mixture of solid and vapor carbon dioxide at the sublimation pressure. Assume that we do not have data for carbon dioxide at 100 K, but it is known that at $-90\,°C$, $P_{sub} = 38.1$ kPa, $h_{ig} = 574.5$ kJ/kg. Estimate the pressure in the wall cavity at 100 K.

12.33 Small solid particles formed in combustion should be investigated. We would like to know the sublimation pressure as a function of temperature. The only information available is T, h_{fg} for boiling at 101.3 kPa and T, h_{if} for melting at 101.3 kPa. Develop a procedure that will allow a determination of the sublimation pressure, $P_{sub}(T)$.

Property Relations, Maxwell Relations, and Those for Enthalpy, Internal Energy, and Entropy

12.34 Use the Gibbs relation $du = T\,ds - P\,dv$ and one of Maxwell's relations to find an expression for $(\partial u/\partial P)_T$ that only has properties P, v, and T involved. What is the value of that partial derivative if you have an ideal gas?

12.35 The Joule–Thomson coefficient μ_J is a measure of the direction and magnitude of the temperature change with pressure in a throttling process. For any three properties x, y, z, use the mathematical relation

$$\left(\frac{\partial x}{\partial y}\right)_z \left(\frac{\partial y}{\partial z}\right)_x \left(\frac{\partial z}{\partial x}\right)_y = -1$$

to show the following relations for the Joule–Thomson coefficient:

$$\mu_J = \left(\frac{\partial T}{\partial P}\right)_h = \frac{T\left(\frac{\partial v}{\partial T}\right)_P - v}{C_p} = \frac{RT^2}{PC_p}\left(\frac{\partial Z}{\partial T}\right)_P$$

12.36 Find the Joule–Thomson coefficient for an ideal gas from the expression given in Problem 12.35.

12.37 Start from the Gibbs relation $dh = T\,ds + v\,dP$ and use one of the Maxwell equations to find $(\partial h/\partial v)_T$ in terms of properties P, v, and T. Then use Eq. 12.24 to also find an expression for $(\partial h/\partial T)_v$.

12.38 From Eqs. 12.23 and 12.24 and the knowledge that $C_p > C_v$, what can you conclude about the slopes of constant v and constant P curves in a T–s diagram? Notice that we are looking at functions $T(s, P, \text{or } v$ given).

12.39 Derive expressions for $(\partial T/\partial v)_u$ and for $(\partial h/\partial s)_v$ that do not contain the properties h, u, or s. Use Eq. 12.30 with $du = 0$.

12.40 Evaluate the isothermal changes in internal energy, enthalpy, and entropy for an ideal gas. Confirm the results in Chapters 3 and 6.

12.41 Develop an expression for the variation in temperature with pressure in a constant-entropy process, $(\partial T/\partial P)_s$, that only includes the properties P–v–T

and the specific heat, C_p. Follow the development of Eq. 12.32.

12.42 Use Eq. 12.34 to derive an expression for the derivative $(\partial T/\partial v)_s$. What is the general shape of a constant s process curve in a T–v diagram? For an ideal gas, can you say a little more about the shape?

12.43 Show that the P–v–T relation as $P(v-b) = RT$ satisfies the mathematical relation in Problem 12.35.

Volume Expansivity and Compressibility

12.44 What are the volume expansivity α_p, the isothermal compressibility β_T, and the adiabatic compressibility β_s for an ideal gas?

12.45 Assume that a substance has uniform properties in all directions with $V = L_x L_y L_z$. Show that volume expansivity $\alpha_p = 3\delta_T$. (*Hint*: differentiate with respect to T and divide by V.)

12.46 Determine the volume expansivity, α_p, and the isothermal compressibility, β_T, for water at $20\,°C$, 5 MPa and at $300\,°C$, 15 MPa using the steam tables.

12.47 Use the CATT3 software to solve the previous problem.

12.48 A cylinder fitted with a piston contains liquid methanol at $20\,°C$, 100 kPa, and volume 10 L. The piston is moved, compressing the methanol to 20 MPa at constant temperature. Calculate the work required for this process. The isothermal compressibility of liquid methanol at $20\,°C$ is $1.22 \times 10^{-9}\ m^2/N$.

12.49 For commercial copper at $25\,°C$ (see Table A.3), the speed of sound is about 4800 m/s. What is the adiabatic compressibility β_s?

12.50 Use Eq. 12.32 to solve for $(\partial T/\partial P)_s$ in terms of T, v, C_p, and α_p. How large a temperature change does water at $25\,°C$ ($\alpha_p = 2.1 \times 10^{-4}\ K^{-1}$) have when compressed from 100 kPa to 1000 kPa in an isentropic process?

12.51 Sound waves propagate through media as pressure waves that cause the media to go through isentropic compression and expansion processes. The speed of sound c is defined by $c^2 = (\partial P/\partial \rho)_s$ and it can be related to the adiabatic compressibility, which for liquid ethanol at $20\,°C$ is $9.4 \times 10^{-10}\ m^2/N$. Find the speed of sound at this temperature.

12.52 Use Table B.3 to find the speed of sound for carbon dioxide at 2500 kPa near 100 °C. Approximate the partial derivative numerically.

12.53 Use the CATT3 software to solve the previous problem.

12.54 Consider the speed of sound as defined in Eq. 12.42. Calculate the speed of sound for liquid water at 20 °C, 2.5 MPa, and for water vapor at 200 °C, 300 kPa, using the steam tables.

12.55 Use the CATT3 software to solve the previous problem.

12.56 Soft rubber is used as part of a motor mounting. Its adiabatic bulk modulus is $B_s = 2.82 \times 10^6$ kPa, and the volume expansivity is $\alpha_p = 4.86 \times 10^{-4}$ K^{-1}. What is the speed of sound vibrations through the rubber, and what is the relative volume change for a pressure change of 1 MPa?

12.57 Liquid methanol at 25 °C has an adiabatic compressibility of 1.05×10^{-9} m^2/N. What is the speed of sound? If it is compressed from 100 kPa to 10 MPa in an insulated piston/cylinder, what is the specific work?

12.58 Use Eq. 12.32 to solve for $(\partial T/\partial P)_s$ in terms of T, v, C_p, and α_p. How much higher does the temperature become for the compression of the methanol in Problem 12.57? Use $\alpha_p = 2.4 \times 10^{-4}$ K^{-1} for methanol at 25 °C.

12.59 Find the speed of sound for air at 20 °C, 100 kPa, using the definition in Eq. 12.42 and relations for polytropic processes in ideal gases.

Equations of State

12.60 Use Table B.3 and find the compressibility of carbon dioxide at the critical point.

12.61 Use the EOS, as shown in Example 12.3, where changes in enthalpy and entropy were found. Find the isothermal change in internal energy in a similar fashion; do not compute it from enthalpy.

12.62 Use Table B.4 to find the compressibility of R-410A at 60 °C and (a) saturated liquid, (b) saturated vapor, and (c) 3000 kPa.

12.63 Use a truncated virial EOS that includes the term with B for carbon dioxide at 20 °C, 1 MPa for which $B = -0.128$ m^3/kmol, and $T(dB/dT) = 0.266$ m^3/kmol. Find the difference between the ideal gas value and the real gas value of the internal energy.

12.64 Solve the previous problem with the values in Table B.3 and find the compressibility of the carbon dioxide at that state.

12.65 A gas is represented by the virial EOS with the first two terms, B and C. Find an expression for the work in an isothermal expansion process in a piston/cylinder.

12.66 Extend Problem 12.63 to find the difference between the ideal gas value and the real gas value of the entropy and compare it to the value in Table B.3.

12.67 Two uninsulated tanks of equal volume are connected by a valve. One tank contains a gas at a moderate pressure P_1, and the other tank is evacuated. The valve is opened and remains open for a long time. Is the final pressure P_2 greater than, equal to, or less than $P_1/2$? (*Hint*: Recall Fig. 12.5.)

12.68 Show how to find the constants in Eq. 12.52 for the van der Waals EOS.

12.69 Show that the van der Waals equation can be written as a cubic equation in the compressibility factor, as in Eq. 12.53.

12.70 Find changes in an isothermal process for u, h, and s for a gas with an EOS as $P(v - b) = RT$.

12.71 Find changes in internal energy, enthalpy, and entropy for an isothermal process in a gas obeying the van der Waals EOS.

12.72 Consider the following EOS, expressed in terms of reduced pressure and temperature: $Z = 1 + (P_r/14T_r)[1 - 6T_r^{-2}]$. What does this predict for the reduced Boyle temperature?

12.73 Use the result of Problem 12.35 to find the reduced temperature at which the Joule–Thomson coefficient is zero for a gas that follows the EOS given in Problem 12.72.

12.74 What is the Boyle temperature for this EOS with constants a and b: $P = [RT/(v - b)] - a/v^2 T$?

12.75 Determine the reduced Boyle temperature as predicted by an EOS (the experimentally observed value is about 2.5), using the van der Waals equation and the Redlich–Kwong equation. *Note*: It is helpful to use Eqs. 12.44 and 12.45 in addition to Eq. 12.43.

12.76 One early attempt to improve on the van der Waals EOS was an expression of the form

$$P = \frac{RT}{v - b} - \frac{a}{v^2 T}$$

Solve for the constants a, b, and v_c using the same procedure as for the van der Waals equation.

12.77 Develop expressions for isothermal changes in internal energy, enthalpy, and entropy for a gas obeying the Redlich–Kwong EOS.

12.78 Determine the second virial coefficient $B(T)$ using the van der Waals EOS. Also find its value at the critical temperature where the experimentally observed value is about $-0.34\, RT_c/P_c$.

12.79 Determine the second virial coefficient $B(T)$ using the Redlich–Kwong EOS. Also find its value at the critical temperature where the experimentally observed value is about $-0.34\, RT_c/P_c$.

12.80 Oxygen in a rigid tank with 1 kg is at 160 K, 4 MPa. Find the volume of the tank by iterations using the Redlich–Kwong EOS. Compare the result with the ideal gas law.

12.81 A flow of oxygen at 230 K, 5 MPa is throttled to 100 kPa in a steady-flow process. Find the exit temperature and the specific entropy generation using the Redlich–Kwong EOS and ideal gas specific heat. Notice that this becomes iterative due to the nonlinearity coupling h, P, v, and T.

Generalized Charts

12.82 How low should the pressure be so that nitrous oxide (N_2O) gas at 278.6 K can be treated as an ideal gas with 5 % accuracy or better?

12.83 Nitrous oxide (N_2O) at 278.6 K is at a pressure so that it can be in a two-phase state. Find the generalized enthalpy departure for the two saturated states of liquid and vapor.

12.84 Find the heat of evaporation, h_{fg}, for R-134a at 0 °C from the generalized charts and compare to the value in Table B.5.

12.85 A 200 L rigid tank contains propane at 9 MPa, 280 °C. The propane is allowed to cool to 50 °C as heat is transferred with the surroundings. Determine the quality at the final state and the mass of liquid in the tank, using the generalized compressibility chart, Fig. D.1.

12.86 A rigid tank contains 5 kg ethylene at 3 MPa, 30 °C. It is cooled until the ethylene reaches the saturated vapor curve. What is the final temperature?

12.87 The new refrigerant R-152a is used in a refrigerator with an evaporator at −20 °C and a condenser at 30 °C. What are the high and low pressures in this cycle?

12.88 A 4 m³ storage tank contains ethane gas at 10 MPa, 100 °C. Using the Lee–Kesler EOS, find the mass of the ethane.

12.89 The ethane gas in the storage tank from the previous problem is cooled to 0 °C. Find the new pressure.

12.90 Use the CATT3 software to solve the previous two problems when the acentric factor is used to improve the accuracy.

12.91 A geothermal power plant uses butane as saturated vapor at 80 °C into the turbine, and the condenser operates at 30 °C. Find the reversible specific turbine work.

12.92 Consider the following EOS, expressed in terms of reduced pressure and temperature: $Z = 1 + (P_r/14T_r)[1 - 6T_r^{-2}]$. What does this predict for the enthalpy departure at $P_r = 0.4$ and $T_r = 0.9$? What is it from the generalized charts?

12.93 Find the entropy departure in the previous problem.

12.94 A very-low-temperature refrigerator uses neon. From the compressor, the neon at 1.5 MPa, 80 K goes through the condenser and comes out as saturated liquid at 40 K. Find the specific heat transfer using generalized charts.

12.95 Repeat the previous problem using the CATT3 software for the neon properties.

12.96 A piston/cylinder contains 5 kg butane gas at 500 K, 5 MPa. The butane expands in a reversible polytropic process to 3 MPa, 460 K. Determine the polytropic exponent n and the work done during the process.

12.97 Calculate the heat transfer during the process described in the previous problem.

12.98 An ordinary lighter is nearly full of liquid propane with a small amount of vapor, the volume is 5 cm³, and the temperature is 23 °C. The propane is now discharged slowly such that heat transfer keeps the propane and valve flow at 23 °C. Find the initial pressure and mass of propane and the total heat transfer to empty the lighter.

12.99 A 250 L tank contains propane at 30 °C, 90 % quality. The tank is heated to 300 °C. Calculate the heat transfer during the process.

12.100 Find the heat of evaporation, h_{fg}, for isobutane ($T_c = 408.2$ K, $P_c = 3.65$ MPa, $M = 58.124$) at 12.6 °C from the generalized charts and compare to the values in the CATT3 computerized tables.

12.101 A cylinder contains ethylene, C_2H_4, at 1.536 MPa, −13 °C. It is now compressed isothermally in a reversible process to 5.12 MPa. Find the specific work and heat transfer.

12.102 Saturated vapor R-410A at 30 °C is throttled to 200 kPa in a steady-flow process. Find the exit temperature, neglecting kinetic energy, using Fig. D.2, and repeat using Table B.4.

12.103 Repeat Problem 12.91 using the CATT3 software and include the acentric factor for butane to improve the accuracy.

12.104 A cylinder contains ethylene, C_2H_4, at 1.536 MPa, −13 °C. It is now compressed in a reversible isobaric (constant-P) process to saturated liquid. Find the specific work and heat transfer.

12.105 A new refrigerant, R-123, enters a heat exchanger as saturated liquid at 40 °C and exits at 100 kPa in a steady flow. Find the specific heat transfer using Fig. D.2.

12.106 Carbon dioxide collected from a fermentation process at 5 °C, 100 kPa should be brought to 243 K, 4 MPa in a steady-flow process. Find the minimum amount of work required and the heat transfer using generalized charts. What devices are needed to accomplish this change of state?

12.107 Determine how accurate the generalized chart is for the carbon dioxide process in Problem 12.106 by using the CATT3 software for the carbon dioxide properties.

12.108 A geothermal power plant on the Raft River uses isobutane as the working fluid. The fluid enters the reversible adiabatic turbine at 160 °C, 5.475 MPa, and the condenser exit condition is saturated liquid at 33 °C. Isobutane has the properties $T_c = 408.14$ K, $P_c = 3.65$ MPa, $C_{p0} = 1.664$ kJ/kg·K, and ratio of specific heats $k = 1.094$ with a molecular mass of 58.124. Find the specific turbine work and the specific pump work.

12.109 A steady flow of oxygen at 230 K, 5 MPa is throttled to 100 kPa. Show that $T_{exit} \approx 208$ K and find the specific entropy generation.

12.110 An uninsulated piston/cylinder contains propene, C_3H_6, at ambient temperature, 19 °C, with a quality of 50 % and a volume of 10 L. The propene now expands slowly until the pressure drops to 460 kPa. Calculate the mass of the propene, the work, and the heat transfer for this process.

12.111 An alternative energy power plant has carbon dioxide at 6 MPa, 100 °C flowing into a turbine and exiting as saturated vapor at 1 MPa. Find the specific turbine work using generalized charts and repeat using Table B.3.

12.112 A distributor of bottled propane, C_3H_8, needs to bring propane from 350 K, 100 kPa to saturated liquid at 290 K in a steady-flow process. If this should be accomplished in a reversible setup given the surroundings at 300 K, find the ratio of the volume flow rates $\dot{V}_{in}/\dot{V}_{out}$, the heat specific transfer, and the work involved in the process.

12.113 An insulated piston/cylinder contains saturated vapor carbon dioxide at 0 °C and a volume of 20 L. The external force on the piston is slowly decreased, allowing the carbon dioxide to expand until the temperature reaches −30 °C. Calculate the work done by the carbon dioxide during this process using generalized charts.

12.114 A control mass of 10 kg butane gas initially at 80 °C, 500 kPa is compressed in a reversible isothermal process to one-fifth of its initial volume. What is the heat transfer in the process?

12.115 A line with a steady supply of octane, C_8H_{18}, is at 400 °C, 3 MPa. What is your best estimate for the availability in a steady-flow setup where changes in potential and kinetic energies may be neglected?

12.116 The environmentally safe refrigerant R-152a is to be evaluated as the working fluid for a heat pump system that will heat a house. It uses an evaporator temperature of −20 °C and a condensing temperature of 30 °C. Assume all processes are ideal and R-152a has a specific heat of $C_p = 0.996$ kJ/kg·K. Determine the cycle COP.

12.117 Rework the previous problem using an evaporator temperature of 0 °C.

12.118 An uninsulated compressor delivers ethylene, C_2H_4, to a pipe, $D = 10$ cm, at 10.24 MPa, 94 °C, and velocity 30 m/s. The ethylene enters the compressor at 6.4 MPa, 20.5 °C, and the work input required is 300 kJ/kg. Find the mass flow rate, the total heat transfer, and the entropy generation, assuming the surroundings are at 25 °C.

12.119 The refrigerant fluid R-123 (see Table A.2) is used in a refrigeration system that operates in the ideal refrigeration cycle, except that the compressor is neither reversible nor adiabatic. Saturated vapor at −26.5 °C enters the compressor, and superheated vapor exits at 65 °C. Heat is rejected from the compressor as 1 kW, and the R-123 flow rate is 0.1 kg/s. Saturated liquid exits the condenser at 37.5 °C. Specific heat for R-123 is $C_{p0} = 0.6$ kJ/kg·K. Find the COP.

12.120 An evacuated 100 L rigid tank is connected to a line flowing R-142b gas, chlorodifluoroethane, at 2 MPa, 100 °C. The valve is opened, allowing the gas to flow into the tank for a period of time, and then it is closed. Eventually, the tank cools to ambient temperature, 20 °C, at which point it contains 50 % liquid, 50 % vapor, by volume. Calculate the quality at the final state and the heat transfer for the process. The ideal gas specific heat of R-142b is $C_p = 0.787$ kJ/kg·K.

Mixtures

12.121 A 2 kg mixture of 50 % argon and 50 % nitrogen by mole is in a tank at 2 MPa, 180 K. How large is the volume using a model of (a) ideal gas and (b) Kay's rule with generalized compressibility charts?

12.122 A 2 kg mixture of 50 % argon and 50 % nitrogen by mole is in a tank at 2 MPa, 180 K. How large is the volume using a model of (a) ideal gas and (b) van der Waals' EOS with a, b for a mixture?

12.123 The refrigerant R-410A is a 1:1 mass ratio mixture of R-32 and R-125. Find the specific volume at 20 °C, 1200 kPa, using Kay's rule and the generalized charts, and compare it to the solution using Table B.4.

12.124 The R-410A in Problem 12.123 is flowing through a heat exchanger with an exit at 120 °C, 1200 kPa. Find the specific heat transfer using Kay's rule and the generalized charts and compare it to the solution using Table B.4.

12.125 A 2 kg mixture of 50 % argon and 50 % nitrogen by mole is in a tank at 2 MPa, 180 K. How large is the volume using a model of (a) ideal gas and (b) the Redlich–Kwong EOS with a, b for a mixture?

12.126 A modern jet engine operates so that the fuel is sprayed into air at a P, T higher than the fuel critical point. Assume we have a rich mixture of 50 % n-octane and 50 % air by moles at 600 K and 4 MPa near the nozzle exit. Do I need to treat this as a real gas mixture or is the ideal gas assumption reasonable? To answer, find Z and the enthalpy departure for the mixture, assuming Kay's rule and the generalized charts.

12.127 A mixture of 60 % ethylene and 40 % acetylene by moles is at 6 MPa, 300 K. The mixture flows through a preheater, where it is heated to 400 K at constant P. Using the Redlich–Kwong EOS with a, b for a mixture, find the inlet specific volume. Repeat using Kay's rule and the generalized charts.

12.128 For the previous problem, find the specific heat transfer using Kay's rule and the generalized charts.

12.129 A gas mixture of a known composition is required for the calibration of gas analyzers. It is desired to prepare a gas mixture of 80 % ethylene and 20 % carbon dioxide (mole basis) at 10 MPa, 25 °C, in an uninsulated, rigid 50 L tank. The tank is initially to contain carbon dioxide at 25 °C and some pressure P_1. The valve to a line flowing ethylene at 25 °C, 10 MPa, is now opened slightly and remains open until the tank reaches 10 MPa, at which point the temperature can be assumed to be 25 °C. Assume that the gas mixture so prepared can be represented by Kay's rule and the generalized charts. Given the desired final state, what is the initial pressure of the carbon dioxide, P_1?

12.130 One kilomole per second of saturated liquid methane, CH_4, at 1 MPa and 2 kmol/s of ethane, C_2H_6, at 250 °C, 1 MPa are fed to a mixing chamber, with the resultant mixture exiting at 50 °C, 1 MPa. Assume that Kay's rule applies to the mixture and determine the heat transfer in the process.

12.131 Saturated liquid ethane at $T_1 = 14$ °C is throttled into a steady-flow mixing chamber at the rate of 0.25 kmol/s. Argon gas at $T_2 = 25$ °C, 800 kPa, enters the chamber at the rate 0.75 kmol/s. Heat is transferred to the chamber from a constant-temperature source at $150\,°C$ at a rate such that a gas mixture exits the chamber at $T_3 = 120\,°C$, 800 kPa. Find the rate of heat transfer and the rate of entropy generation.

12.132 A piston/cylinder contains a gas mixture, 50% carbon dioxide and 50% ethane (C_2H_6) (mole basis), at 700 kPa, $35\,°C$, at which point the cylinder volume is 5 L. The mixture is now compressed to 5.5 MPa in a reversible isothermal process. Calculate the heat transfer and work for the process, using the following model for the gas mixture:
a. Ideal gas mixture.
b. Kay's rule and the generalized charts.

12.133 Solve the previous problem using (a) ideal gas and (b) van der Waal's EOS.

Helmholtz EOS

12.134 Verify that the ideal gas part of the Helmholtz function substituted in Eq. 12.86 does lead to the ideal gas law, as in the note after Eq. 12.97.

12.135 Gases like argon and neon have constant specific heats. Develop an expression for the ideal gas contribution to the Helmholtz function in Eq. 12.92 for these cases.

12.136 Find an expression for the change in Helmholtz function for a gas with an EOS as $P(v - b) = RT$.

12.137 Use the EOS in Example 12.3 and find an expression for isothermal changes in the Helmholtz function between two states.

12.138 Assume a Helmholtz equation as

$$a^* = C_0 + C_1 T - C_2 T \ln\left(\frac{T}{T_0}\right) + RT \ln\left(\frac{\rho}{\rho_0}\right)$$

where C_0, C_1, and C_2 are constants and T_0 and ρ_0 are reference values for temperature and density (see Eqs. 12.92–12.95). Find the properties P, u, and s from this expression. Is anything assumed for this particular form?

Review Problems

12.139 Saturated liquid ethane at 2.44 MPa enters a heat exchanger and is brought to 611 K at constant pressure, after which it enters a reversible adiabatic turbine, where it expands to 100 kPa. Find the specific heat transfer in the heat exchanger, the turbine exit temperature, and the turbine work.

12.140 A piston/cylinder initially contains propane at $T_1 = -7\,°C$, quality 50%, and volume 10 L. A valve

connecting the cylinder to a line flowing nitrogen gas at $T_i = 20\,°C$, $P_i = 1$ MPa, is opened and nitrogen flows in. When the valve is closed, the cylinder contains a gas mixture of 50% nitrogen and 50% propane on a mole basis at $T_2 = 20\,°C$, $P_2 = 500$ kPa. What is the cylinder volume at the final state, and how much heat transfer took place?

12.141 A new compound is used in an ideal Rankine cycle where saturated vapor at $200\,°C$ enters the turbine and saturated liquid at $20\,°C$ exits the condenser. The only properties known for this compound are a molecular mass of 80 kg/kmol, an ideal gas specific heat of $C_p = 0.80$ kJ/kg·K, and $T_c = 500$ K, $P_c = 5$ MPa. Find the specific work input to the pump and the cycle thermal efficiency using the generalized charts.

12.142 A 200 L rigid tank contains propane at 400 K, 3.5 MPa. A valve is opened, and propane flows out until half of the initial mass has escaped, at which point the valve is closed. During this process, the mass remaining inside the tank expands according to the relation $Pv^{1.4} = $ constant. Calculate the heat transfer to the tank during the process.

12.143 One kilogram per second water enters a solar collector at $40\,°C$ and exits at $190\,°C$, as shown in Fig. P12.143. The hot water is sprayed into a direct-contact heat exchanger (no mixing of the two fluids) used to boil the liquid butane. Pure saturated-vapor butane exits at the top at $80\,°C$ and is fed to the turbine. If the butane condenser temperature is $30\,°C$ and the turbine and pump

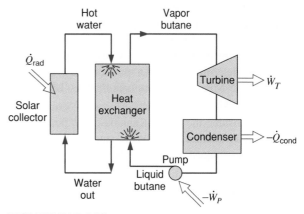

FIGURE P12.143

isentropic efficiencies are each 80 %, determine the net power output of the cycle.

12.144 A piston/cylinder contains ethane gas initially at 500 kPa, 100 L, and at ambient temperature 0 °C. The piston is moved, compressing the ethane until it is at 20 °C with a quality of 50 %. The work required is 25 % more than would have been needed for a reversible polytropic process between the same initial and final states. Calculate the heat transfer and the net entropy change for the process.

12.145 Carbon dioxide gas enters a turbine at 5 MPa, 100 °C, and exits at 1 MPa. If the isentropic efficiency of the turbine is 75 %, determine the exit temperature and the second-law efficiency, using the generalized charts.

12.146 A 10 m^3 storage tank contains methane at low temperature. The pressure inside is 700 kPa, and the tank contains 25 % liquid and 75 % vapor on a volume basis. The tank warms very slowly because heat is transferred from the ambient air.

a. What is the temperature of the methane when the pressure reaches 10 MPa?

b. Calculate the heat transferred in the process using the generalized charts.

c. Repeat parts (a) and (b) using the methane tables (Table B.7). Discuss the differences in the results.

12.147 Consider the following reference state conditions: the entropy of real saturated liquid methane at −100 °C is to be taken as 100 kJ/kmol·K, and the entropy of hypothetical ideal gas ethane at −100 °C is to be taken as 200 kJ/kmol·K. Calculate the entropy per kilomole of a real gas mixture of 50 % methane and 50 % ethane (mole basis) at 20 °C, 4 MPa, in terms of the specified reference state values, and assuming Kay's rule for the real mixture behavior.

12.148 Determine the heat transfer and the net entropy change in Problem 12.129. Use the initial pressure of the carbon dioxide to be 4.56 MPa before the ethylene is flowing into the tank.

COMPUTER, DESIGN, AND OPEN-ENDED PROBLEMS

12.149 Solve the following problem with the CATT3 software: (a) 12.81, (b) 12.85, (c) 12.86, (d) 12.108.

12.150 Write a program to obtain a plot of pressure versus specific volume at various temperatures (all on a generalized reduced basis) as predicted by the van der Waals EOS. Temperatures less than the critical temperature should be included in the results.

12.151 We wish to determine the isothermal compressibility, β_T, for a range of states of liquid water. Use the menu-driven software or write a program to determine this at a pressure of 1 MPa and at 25 MPa for temperatures of 0 °C, 100 °C, and 300 °C.

12.152 Consider the small Rankine-cycle power plant in Problem 12.141. What single change would you suggest to make the power plant more realistic?

12.153 Supercritical fluid chromatography is an experimental technique for analyzing compositions of mixtures. It utilizes a carrier fluid, often carbon

dioxide, in the dense fluid region just above the critical temperature. Write a program to express the fluid density as a function of reduced temperature and pressure in the region of $1.0 \leq T_r \leq 1.2$ in reduced temperature and $2 \leq P_r \leq 8$ in reduced pressure. The relation should be an expression curve-fitted to values consistent with the generalized compressibility charts.

12.154 It is desired to design a portable breathing system for an average-sized adult. The breather will store liquid oxygen sufficient for a 24-hour supply and will include a heater for delivering oxygen gas at ambient temperature. Determine the size of the system container and the heat exchanger.

12.155 Liquid nitrogen is used in cryogenic experiments and applications where a nonoxidizing gas is desired. Size a tank to hold 500 kg to be placed next to a building and estimate the size of an environmental (to atmospheric air) heat exchanger that can deliver nitrogen gas at a rate of 10 kg/hr at roughly ambient temperature.

12.156 The speed of sound is used in many applications. Make a list of the speed of sound at P_0, T_0 for gases, liquids, and solids. Find at least three different substances for each phase. List a number of applications where knowledge of the speed of sound can be used to estimate other quantities of interest.

12.157 Propane is used as a fuel distributed to the end consumer in a steel bottle. Make a list of design specifications for these bottles and give characteristic sizes and the amount of propane they can hold.

Combustion

13

Many thermodynamic problems involve chemical reactions. Among the most familiar of these is the combustion of hydrocarbon fuels, for this process is utilized in most of our power-generating devices. However, we can all think of a host of other processes involving chemical reactions, including those that occur in the human body.

This chapter considers a first- and second-law analysis of systems undergoing a chemical reaction. In many respects, this chapter is simply an extension of our previous consideration of the first and second laws. However, a number of new terms are introduced, and it will also be necessary to introduce the third law of thermodynamics.

In this chapter the combustion process is considered in detail. There are two reasons for this emphasis. First, the combustion process is important in many problems and devices with which the engineer is concerned. Second, the combustion process provides an excellent means of teaching the basic principles of the thermodynamics of chemical reactions. The student should keep both of these objectives in mind as the study of this chapter progresses.

Chemical equilibrium will be considered in Chapter 14; therefore, the subject of dissociation will be deferred until then.

13.1 FUELS

A thermodynamics textbook is not the place for a detailed treatment of fuels. However, some knowledge of them is a prerequisite to a consideration of combustion, and this section is therefore devoted to a brief discussion of some of the hydrocarbon fuels. Most fuels fall into one of three categories—coal, liquid hydrocarbons, or gaseous hydrocarbons.

Coal consists of the remains of vegetation deposits of past geologic ages after subjection to biochemical actions, high pressure, temperature, and submersion. The characteristics of coal vary considerably with location, and even within a given mine there is some variation in composition.

A sample of coal is analyzed on one of two bases. The proximate analysis specifies, on a mass basis, the relative amounts of moisture, volatile matter, fixed carbon, and ash; the ultimate analysis specifies, on a mass basis, the relative amounts of carbon, sulfur, hydrogen, nitrogen, oxygen, and ash. The ultimate analysis may be given on an "as-received" basis or on a dry basis. In the latter case, the ultimate analysis does not include the moisture as determined by the proximate analysis.

A number of other properties of coal are important in evaluating a coal for a given use. Some of these are the fusibility of the ash, the grindability or ease of pulverization, the weathering characteristics, and size.

Most liquid and gaseous hydrocarbon fuels are a mixture of many different hydrocarbons. For example, gasoline consists primarily of a mixture of about 40 hydrocarbons, with many others present in very small quantities. In discussing hydrocarbon fuels, therefore,

536

TABLE 13.1
Characteristics of Some of the Hydrocarbon Families

Family	Formula	Structure	Saturated
Paraffin	C_nH_{2n+2}	Chain	Yes
Olefin	C_nH_{2n}	Chain	No
Diolefin	C_nH_{2n-2}	Chain	No
Naphthene	C_nH_{2n}	Ring	Yes
Aromatic			
Benzene	C_nH_{2n-6}	Ring	No
Naphthene	C_nH_{2n-12}	Ring	No

brief consideration should be given to the most important families of hydrocarbons, which are summarized in Table 13.1.

Three concepts should be defined. The first pertains to the *structure of the molecule*. The important types are the ring and chain structures; the difference between the two is illustrated in Fig. 13.1. The same figure illustrates the definition of *saturated and unsaturated hydrocarbons*. An unsaturated hydrocarbon has two or more adjacent carbon atoms joined by a double or triple bond, whereas in a saturated hydrocarbon all the carbon atoms are joined by a single bond. The third term to be defined is an isomer. Two hydrocarbons with the same number of carbon and hydrogen atoms and different structures are called *isomers*. Thus, there are several different octanes (C_8H_{18}), each having 8 carbon atoms and 18 hydrogen atoms, but each with a different structure.

The various hydrocarbon families are identified by a common suffix. The compounds comprising the paraffin family all end in *-ane* (e.g., propane and octane). Similarly, the compounds comprising the olefin family end in *-ylene* or *-ene* (e.g., propene and octene), and the diolefin family ends in *-diene* (e.g., butadiene). The naphthene family has the same chemical formula as the olefin family but has a ring rather than a chain structure. The hydrocarbons in the naphthene family are named by adding the prefix *cyclo-* (as cyclopentane).

The aromatic family includes the benzene series (C_nH_{2n-6}) and the naphthalene series (C_nH_{2n-12}). The benzene series has a ring structure and is unsaturated.

Most liquid hydrocarbon fuels are mixtures of hydrocarbons that are derived from crude oil through distillation and cracking processes. The separation of air into its two major components, nitrogen and oxygen, using a distillation column was shown in Fig. 9.27. In a similar but much more complicated manner, a fractional distillation column is used to separate petroleum into its various constituents. This process is shown schematically in Fig. 13.2. Liquid crude oil is gasified and enters near the bottom of the distillation column. The heavier fractions have higher boiling points and condense out at the higher temperatures in the lower part of the column, while the lighter fractions condense out at the lower

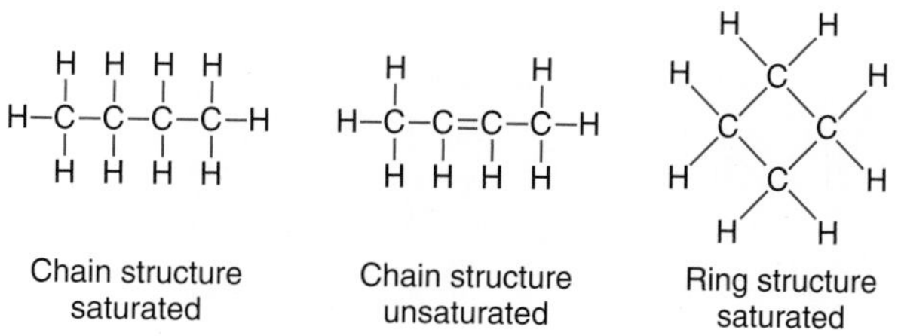

FIGURE 13.1
Molecular structure of some hydrocarbon fuels.

Chain structure saturated

Chain structure unsaturated

Ring structure saturated

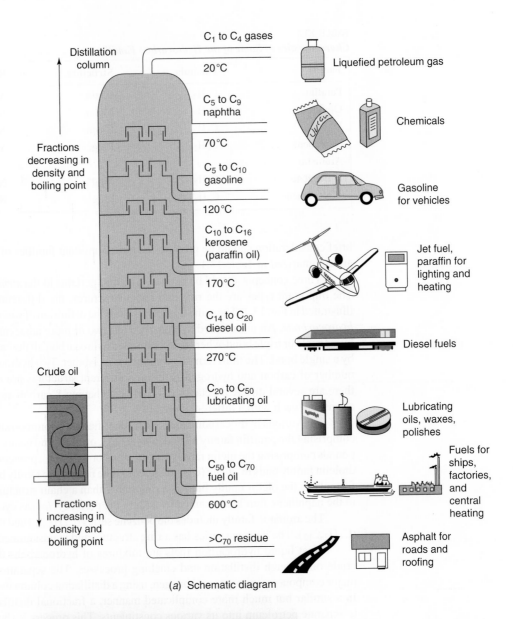

Distillation column

Fractions decreasing in density and boiling point

Crude oil

Fractions increasing in density and boiling point

C$_1$ to C$_4$ gases

20°C — Liquefied petroleum gas

C$_5$ to C$_9$ naphtha

70°C — Chemicals

C$_5$ to C$_{10}$ gasoline

120°C — Gasoline for vehicles

C$_{10}$ to C$_{16}$ kerosene (paraffin oil)

170°C — Jet fuel, paraffin for lighting and heating

C$_{14}$ to C$_{20}$ diesel oil

270°C — Diesel fuels

C$_{20}$ to C$_{50}$ lubricating oil

— Lubricating oils, waxes, polishes

C$_{50}$ to C$_{70}$ fuel oil

600°C — Fuels for ships, factories, and central heating

>C$_{70}$ residue — Asphalt for roads and roofing

(a) Schematic diagram

© Hanquan Chen/iStockphoto

(b) Photo of a distillation column in a refinery

FIGURE 13.2
Petroleum distillation column.

temperatures in the upper portion of the column. Some of the common fuels produced in this manner are gasoline, kerosene, jet engine fuel, diesel fuel, and fuel oil.

Alcohols, presently seeing increased usage as fuel in internal combustion engines, are a family of hydrocarbons in which one of the hydrogen atoms is replaced by an OH radical. Thus, methyl alcohol, or methanol, is CH_3OH, and ethanol is C_2H_5OH. Ethanol is one of the class of biofuels, produced from crops or waste matter by chemical conversion processes. There is extensive research and development in the area of biofuels at the present time, as well as in the development of processes for producing gaseous and liquid hydrocarbon fuels from coal, oil shale, and tar sands deposits. Several alternative techniques have been demonstrated to be feasible, and these resources promise to provide an increasing proportion of our fuel supplies in future years.

It should also be noted here in our discussion of fuels that there is currently a great deal of development effort to use hydrogen as a fuel for transportation usage, especially in connection with fuel cells. Liquid hydrogen has been used successfully for many years as a rocket fuel but is not suitable for vehicular use, especially because of the energy cost to produce it (at about 20 K), as well as serious transfer and storage problems. Instead, hydrogen would need to be stored as a very-high-pressure gas or in a metal hydride system. There remain many problems in using hydrogen as a fuel. It must be produced either from water or a hydrocarbon, both of which require a large energy expenditure. Hydrogen gas in air has a very broad flammability range—almost any percentage of hydrogen, small or large, is flammable. It also has a very low ignition energy; the slightest spark will ignite a mixture of hydrogen in air. Finally, hydrogen burns with a colorless flame, which can be dangerous. The incentive to use hydrogen as a fuel is that its only product of combustion or reaction is water, but it is still necessary to include the production, transfer, and storage in the overall consideration.

For the combustion of liquid fuels, it is convenient to express the composition in terms of a single hydrocarbon, even though it is a mixture of many hydrocarbons. Thus, gasoline is usually considered to be octane, C_8H_{18}, and diesel fuel is considered to be dodecane,

TABLE 13.2
Volumetric Analyses of Some Typical Gaseous Fuels

Constituent	Various Natural Gases				Producer Gas from Bituminous Coal	Carbureted Water Gas	Coke-Oven Gas
	A	B	C	D			
Methane	93.9	60.1	67.4	54.3	3.0	10.2	32.1
Ethane	3.6	14.8	16.8	16.3			
Propane	1.2	13.4	15.8	16.2			
Butanes plus[a]	1.3	4.2		7.4			
Ethene						6.1	3.5
Benzene						2.8	0.5
Hydrogen					14.0	40.5	46.5
Nitrogen		7.5		5.8	50.9	2.9	8.1
Oxygen					0.6	0.5	0.8
Carbon monoxide					27.0	34.0	6.3
Carbon dioxide					4.5	3.0	2.2

[a]This includes butane and all heavier hydrocarbons.

$C_{12}H_{26}$. The composition of a hydrocarbon fuel may also be given in terms of percentage of carbon and hydrogen.

The two primary sources of gaseous hydrocarbon fuels are natural gas wells and certain chemical manufacturing processes. Table 13.2 gives the composition of a number of gaseous fuels. The major constituent of natural gas is methane, which distinguishes it from manufactured gas.

13.2 THE COMBUSTION PROCESS

The combustion process consists of the oxidation of constituents in the fuel that are capable of being oxidized and can therefore be represented by a chemical equation. During a combustion process, the mass of each element remains the same. Thus, writing chemical equations and solving problems concerning quantities of the various constituents basically involve the conservation of mass of each element. This chapter presents a brief review of this subject, particularly as it applies to the combustion process.

Consider first the reaction of carbon with oxygen.

$$\text{Reactants} \qquad \text{Products}$$
$$C + O_2 \quad \rightarrow \quad CO_2$$

This equation states that 1 kmol of carbon reacts with 1 kmol of oxygen to form 1 kmol of carbon dioxide. This also means that 12 kg of carbon react with 32 kg of oxygen to form 44 kg of carbon dioxide. All the initial substances that undergo the combustion process are called the reactants, and the substances that result from the combustion process are called the products.

When a hydrocarbon fuel is burned, both the carbon and the hydrogen are oxidized. Consider the combustion of methane as an example.

$$CH_4 + 2O_2 \rightarrow CO_2 + 2H_2O \tag{13.1}$$

Here the products of combustion include both carbon dioxide and water. The water may be in the vapor, liquid, or solid phase, depending on the temperature and pressure of the products of combustion.

In the combustion process, many intermediate products are formed during the chemical reaction. In this book, we are concerned with the initial and final products and not with the intermediate products, but this aspect is very important in a detailed consideration of combustion.

In most combustion processes, the oxygen is supplied as air rather than as pure oxygen. The composition of air on a molal basis is approximately 21 % oxygen, 78 % nitrogen, and 1 % argon. We assume that the nitrogen and the argon do not undergo chemical reaction (except for dissociation, which will be considered in Chapter 14). They do leave at the same temperature as the other products, however, and therefore undergo a change of state if the products are at a temperature other than the original air temperature. At the high temperatures achieved in internal-combustion engines, there is actually some reaction between the nitrogen and oxygen, and this gives rise to the air-pollution problem associated with the oxides of nitrogen, commonly labeled NOx, in the engine exhaust.

In combustion calculations concerning air, the argon is usually neglected, and the air is considered to be composed of 21 % oxygen and 79 % nitrogen by volume. When this assumption is made, the nitrogen is sometimes referred to as *atmospheric nitrogen*. Atmospheric nitrogen has a molecular weight of 28.16 (which takes the argon into account) compared to 28.013 for pure nitrogen. This distinction will not be made in this text, and we will consider the 79 % nitrogen to be pure nitrogen.

The assumption that air is 21.0 % oxygen and 79.0 % nitrogen by volume leads to the conclusion that for each mole of oxygen, 79.0/21.0 = 3.76 moles of nitrogen are involved. Therefore, when the oxygen for the combustion of methane is supplied as air, the reaction can be written

$$CH_4 + 2O_2 + 2(3.76)N_2 \rightarrow CO_2 + 2H_2O + 7.52N_2 \qquad (13.2)$$

The minimum amount of air that supplies sufficient oxygen for the complete combustion of all the carbon, hydrogen, and any other elements in the fuel that may oxidize is called the theoretical air. When complete combustion is achieved with theoretical air, the products contain no oxygen. A general combustion reaction with a hydrocarbon fuel and air is thus written

$$C_xH_y + v_{O_2}(O_2 + 3.76N_2) \rightarrow v_{CO_2}CO_2 + v_{H_2O}H_2O + v_{N_2}N_2 \qquad (13.3)$$

with the coefficients to the substances called stoichiometric coefficients. The balance of atoms yields the theoretical amount of air as

$$
\begin{aligned}
\text{C:} \quad & v_{CO_2} = x \\
\text{H:} \quad & 2v_{H_2O} = y \\
\text{N}_2\text{:} \quad & v_{N_2} = 3.76 \times v_{O_2} \\
\text{O}_2\text{:} \quad & v_{O_2} = v_{CO_2} + v_{H_2O}/2 = x + y/4
\end{aligned}
$$

and the total number of moles of air for 1 mole of fuel becomes

$$n_{air} = v_{O_2} \times 4.76 = 4.76(x + y/4)$$

This amount of air is equal to 100 % theoretical air. In practice, complete combustion is not likely to be achieved unless the amount of air supplied is somewhat greater than the theoretical amount. Two important parameters often used to express the ratio of fuel and air are the air–fuel ratio (designated AF) and its reciprocal, the fuel–air ratio (designated FA). These ratios are usually expressed on a mass basis, but a mole basis is used at times.

$$AF_{mass} = \frac{m_{air}}{m_{fuel}} \qquad (13.4)$$

$$AF_{mole} = \frac{n_{air}}{n_{fuel}} \qquad (13.5)$$

They are related through the molecular masses as

$$AF_{mass} = \frac{m_{air}}{m_{fuel}} = \frac{n_{air}M_{air}}{n_{fuel}M_{fuel}} = AF_{mole}\frac{M_{air}}{M_{fuel}}$$

and a subscript *s* is used to indicate the ratio for 100 % theoretical air, also called a *stoichiometric mixture*. In an actual combustion process, an amount of air is expressed as a fraction of the theoretical amount, called *percent theoretical air*. A similar ratio named the equivalence ratio equals the actual fuel–air ratio divided by the theoretical fuel–air ratio as

$$\Phi = FA/FA_s = AF_s/AF \qquad (13.6)$$

the reciprocal of percent theoretical air. Since the percent theoretical air and the equivalence ratio are both ratios of the stoichiometric air–fuel ratio and the actual air–fuel ratio, the molecular masses cancel out and they are the same whether a mass basis or a mole basis is used.

Thus, 150 % theoretical air means that the air actually supplied is 1.5 times the theoretical air and the equivalence ratio is $^2/_3$. The complete combustion of methane with 150 % theoretical air is written

$$CH_4 + 1.5 \times 2(O_2 + 3.76\,N_2) \rightarrow CO_2 + 2\,H_2O + O_2 + 11.28\,N_2 \qquad (13.7)$$

having balanced all the stoichiometric coefficients from conservation of all the atoms.

The amount of air actually supplied may also be expressed in terms of percent excess air. The excess air is the amount of air supplied over and above the theoretical air. Thus, 150 % theoretical air is equivalent to 50 % excess air. The terms *theoretical air*, *excess air*, and *equivalence ratio* are all in current use and give equivalent information about the reactant mixture of fuel and air.

When the amount of air supplied is less than the theoretical air required, the combustion is incomplete. If there is only a slight deficiency of air, the usual result is that some of the carbon unites with the oxygen to form carbon monoxide (CO) instead of carbon dioxide (CO_2). If the air supplied is considerably less than the theoretical air, there may also be some hydrocarbons in the products of combustion.

Even when some excess air is supplied, small amounts of carbon monoxide may be present, the exact amount depending on a number of factors including the mixing and turbulence during combustion. Thus, the combustion of methane with 110 % theoretical air might be as follows:

$$CH_4 + 2(1.1)O_2 + 2(1.1)3.76\,N_2 \rightarrow$$
$$+ 0.95\,CO_2 + 0.05\,CO + 2\,H_2O + 0.225\,O_2 + 8.27\,N_2 \qquad (13.8)$$

For all the hydrocarbon fuels, the combustion generates water in the product gas mixture from which heat is extracted, resulting in a lower temperature. If the temperature drops below the dew-point temperature, water starts to condense out. Similar to the treatment of moist air, the dew point is found as the temperature for which the partial pressure of the water vapor equals the saturation pressure.

$$P_g(T_{\text{dew}}) = P_v = y_v P$$
$$y_v = \frac{v_{H_2O}}{v_{CO_2} + v_{H_2O} + v_{N_2} + v_{O_2} + \cdots} \qquad (13.9)$$

If the temperature is below the dew point, we use this to find the new lower vapor mole fraction

$$y_v' = P_g(T_{\text{new}})/P = \frac{v_{H_2O}'}{v_{CO_2} + v_{H_2O}' + v_{N_2} + v_{O_2} + \cdots} \qquad (13.10)$$

and then solve for v_{H_2O}' as the number of moles of water still vapor. The amount of liquid (or ice) formed is then the difference

$$v_{H_2O\,liq} = v_{H_2O} - v_{H_2O}'$$

The water condensed from the products of combustion usually contains some dissolved gases and therefore may be quite corrosive. For this reason, the products of combustion are often kept above the dew point until they are discharged to the atmosphere.

The material covered so far in this section is illustrated by the following examples.

Example 13.1

Calculate the theoretical air–fuel ratio for the combustion of octane, C_8H_{18}.

Solution

The combustion equation is

$$C_8H_{18} + 12.5\,O_2 + 12.5(3.76)\,N_2 \rightarrow 8\,CO_2 + 9\,H_2O + 47.0\,N_2$$

The air–fuel ratio on a mole basis is

$$AF = \frac{12.5 + 47.0}{1} = 59.5 \text{ (kilomoles of air per kilomole of fuel)}$$

The theoretical air–fuel ratio on a mass basis is found by introducing the molecular mass of the air and fuel.

$$AF = \frac{59.5(28.97)}{114.2} = 15.0 \text{ (kilograms of air per kilogram of fuel)}$$

Example 13.2

Determine the molal analysis of the products of combustion when octane, C_8H_{18}, is burned with 200 % theoretical air, and determine the dew point of the products if the pressure is 0.1 MPa.

Solution

The equation for the combustion of octane with 200 % theoretical air is

$$C_8H_{18} + 12.5(2)\,O_2 + 12.5(2)(3.76)\,N_2 \rightarrow 8\,CO_2 + 9\,H_2O + 12.5\,O_2 + 94.0\,N_2$$

Total kilomoles of product = $8 + 9 + 12.5 + 94.0 = 123.5$
Molal analysis of products:

$$
\begin{aligned}
CO_2 &= 8/123.5 = &6.47\,\% \\
H_2O &= 9/123.5 = &7.29\,\% \\
O_2 &= 12.5/123.5 = &10.12\,\% \\
N_2 &= 94/123.5 = &76.12\,\% \\
\hline
& &100.00\,\%
\end{aligned}
$$

The partial pressure of the water is $100(0.0729) = 7.29$ kPa, so the saturation temperature corresponding to this pressure is 39.7 °C, which is also the dew-point temperature.

Example 13.3

Producer gas from bituminous coal (see Table 13.2) is burned with 20 % excess air. Calculate the air–fuel ratio on a volumetric basis and on a mass basis.

Solution

To calculate the theoretical air requirement, let us write the combustion equation for the combustible substances in 1 kmol of fuel.

$$0.14\,H_2 + 0.070\,O_2 \rightarrow 0.14\,H_2O$$

$$0.27\,CO + 0.135\,O_2 \rightarrow 0.27\,CO_2$$

$$\underline{0.03\,CH_4 + 0.06\,O_2} \rightarrow 0.03\,CO_2 + 0.06\,H_2O$$

$$0.265 = \text{kilomole of oxygen required per kilomole of fuel}$$
$$\underline{-0.006} = \text{kilomole of oxygen in fuel per kilomole of fuel}$$

$$0.259 = \text{kilomole of oxygen required from air per kilomole of fuel}$$

Therefore, the complete combustion equation for 1 kmol of fuel is

$$\overbrace{0.14\,H_2 + 0.27\,CO + 0.03\,CH_4 + 0.006\,O_2 + 0.509\,N_2 + 0.045\,CO_2}^{\text{fuel}}$$

$$\overbrace{+0.259\,O_2 + 0.259(3.76)\,N_2}^{\text{air}} \rightarrow 0.20\,H_2O + 0.345\,CO_2 + 1.482\,N_2$$

$$\left(\frac{\text{kilomoles of air}}{\text{kilomoles of fuel}}\right)_{theo} = 0.259 \times \frac{1}{0.21} = 1.233$$

If the air and fuel are at the same pressure and temperature, this also represents the ratio of the volume of air to the volume of fuel.

$$\text{For 20 \% excess air,}\quad \frac{\text{kilomoles of air}}{\text{kilomoles of fuel}} = 1.233 \times 1.200 = 1.48$$

The air–fuel ratio on a mass basis is

$$AF = \frac{1.48(28.97)}{0.14(2) + 0.27(28) + 0.03(16) + 0.006(32) + 0.509(28) + 0.045(44)}$$

$$= \frac{1.48(28.97)}{24.74} = 1.73 \text{ (kilograms of air per kilogram of fuel)}$$

An analysis of the products of combustion affords a very simple method for calculating the actual amount of air supplied in a combustion process. There are various experimental methods by which such an analysis can be made. Some yield results on a "dry" basis, that is, the fractional analysis of all the components, except for water vapor. Other experimental procedures give results that include the water vapor. In this presentation we are not concerned with the experimental devices and procedures, but rather with the use of such information in a thermodynamic analysis of the chemical reaction. The following examples illustrate how an analysis of the products can be used to determine the chemical reaction and the composition of the fuel.

The basic principle in using the analysis of the products of combustion to obtain the actual fuel–air ratio is conservation of the mass of each element. Thus, in changing from reactants to products, we can make a carbon balance, hydrogen balance, oxygen balance, and nitrogen balance (plus any other elements that may be involved). Furthermore, we recognize that there is a definite ratio between the amounts of some of these elements. Thus, the ratio between the nitrogen and oxygen supplied in the air is fixed, as well as the ratio between carbon and hydrogen if the composition of a hydrocarbon fuel is known.

Example 13.4

Methane (CH_4) is burned with atmospheric air. The analysis of the products on a dry basis is as follows:

CO_2	10.00 %
O_2	2.37 %
CO	0.53 %
N_2	87.10 %
	100.00 %

Calculate the air–fuel ratio and the percent theoretical air and determine the combustion equation.

Solution

The solution consists of writing the combustion equation for 100 kmol of dry products, introducing letter coefficients for the unknown quantities, and then solving for them.

From the analysis of the products, the following equation can be written, keeping in mind that this analysis is on a dry basis.

$$a\, CH_4 + b\, O_2 + c\, N_2 \rightarrow 10.0\, CO_2 + 0.53\, CO + 2.37\, O_2 + d\, H_2O + 87.1\, N_2$$

A balance for each of the elements will enable us to solve for all the unknown coefficients:

Nitrogen balance: $c = 87.1$

Since all the nitrogen comes from the air

$$\frac{c}{b} = 3.76 \qquad b = \frac{87.1}{3.76} = 23.16$$

Carbon balance: $a = 10.00 + 0.53 = 10.53$

Hydrogen balance: $d = 2a = 21.06$

Oxygen balance: All the unknown coefficients have been solved for, and therefore the oxygen balance provides a check on the accuracy. Thus, b can also be determined by an oxygen balance

$$b = 10.00 + \frac{0.53}{2} + 2.37 + \frac{21.06}{2} = 23.16$$

Substituting these values for a, b, c, and d, we have

$$10.53\,CH_4 + 23.16\,O_2 + 87.1\,N_2 \rightarrow$$
$$10.0\,CO_2 + 0.53\,CO + 2.37\,O_2 + 21.06\,H_2O + 87.1\,N_2$$

Dividing through by 10.53 yields the combustion equation per kilomole of fuel.

$$CH_4 + 2.2\,O_2 + 8.27\,N_2 \rightarrow 0.95\,CO_2 + 0.05\,CO + 2\,H_2O + 0.225\,O_2 + 8.27\,N_2$$

The air–fuel ratio on a mole basis is

$$2.2 + 8.27 = 10.47 \text{ (kilomoles of air per kilomole of fuel)}$$

The air–fuel ratio on a mass basis is found by introducing the molecular masses.

$$AF = \frac{10.47 \times 28.97}{16.0} = 18.97 \text{ (kilograms of air per kilogram of fuel)}$$

The theoretical air–fuel ratio is found by writing the combustion equation for theoretical air.

$$CH_4 + 2\,O_2 + 2(3.76)\,N_2 \rightarrow CO_2 + 2\,H_2O + 7.52\,N_2$$

$$AF_{\text{theo}} = \frac{(2 + 7.52)28.97}{16.0} = 17.23 \text{ (kilograms of air per kilogram of fuel)}$$

The percent theoretical air is $\dfrac{18.97}{17.23} = 110\%$

Example 13.5

Coal from Jenkin, Kentucky, has the following ultimate analysis on a dry basis, percent by mass:

Component	Percent by Mass
Sulfur	0.6
Hydrogen	5.7
Carbon	79.2
Oxygen	10.0
Nitrogen	1.5
Ash	3.0

This coal is to be burned with 30% excess air. Calculate the air–fuel ratio on a mass basis.

Solution

One approach to this problem is to write the combustion equation for each of the combustible elements per 100 kg of fuel. The molal composition per 100 kg of fuel is found first.

$$\text{kilomoles of S per 100 kg fuel} = \frac{0.6}{32} = 0.02$$

$$\text{kilomoles of H}_2 \text{ per 100 kg fuel} = \frac{5.7}{2} = 2.85$$

$$\text{kilomoles of C per 100 kg fuel} = \frac{79.2}{12} = 6.60$$

$$\text{kilomoles of } O_2 \text{ per 100 kg fuel} = \frac{10}{32} = 0.31$$

$$\text{kilomoles of } N_2 \text{ per 100 kg fuel} = \frac{1.5}{28} = 0.05$$

The combustion equations for the combustible elements are now written, which enables us to find the theoretical oxygen required.

$$0.02\,S + 0.02\,O_2 \rightarrow \quad 0.02\,SO_2$$
$$2.85\,H_2 + 1.42\,O_2 \rightarrow \quad 2.85\,H_2O$$
$$6.60\,C + 6.60\,O_2 \rightarrow \quad 6.60\,CO_2$$

$$8.04 \text{ kmol } O_2 \text{ required per 100 kg fuel}$$
$$- 0.31 \text{ kmol } O_2 \text{ in fuel per 100 kg fuel}$$
$$7.73 \text{ kmol } O_2 \text{ from air per 100 kg fuel}$$

$$AF_{\text{theo}} = \frac{[7.73 + 7.73(3.76)]28.97}{100} = 10.63 \text{ (kilograms of air per kilogram of fuel)}$$

For 30 % excess air, the air–fuel ratio is

$$AF = 1.3 \times 10.63 = 13.82 \text{ (kilograms of air per kilogram of fuel)}$$

In-Text Concept Questions

a. How many kilomoles of air are needed to burn 1 kmol of carbon?

b. If I burn 1 kmol of hydrogen (H_2) with 6 kmol of air, what is the air–fuel ratio on a mole basis and what is the percent theoretical air?

c. For the 110 % theoretical air in Eq. 13.8, what is the equivalence ratio? Is that mixture rich or lean?

d. In most cases, combustion products are exhausted above the dew point. Why?

13.3 ENTHALPY OF FORMATION

In the first 12 chapters of this book, the problems always concerned a fixed chemical composition and never a change of composition through a chemical reaction. Therefore, in dealing with a thermodynamic property, we used tables of thermodynamic properties for the given substance, and in each of these tables the thermodynamic properties were given relative to some arbitrary base. In the steam tables, for example, the internal energy of saturated liquid at 0.01 °C is assumed to be zero. This procedure is quite adequate when there is no change in composition because we are concerned with the changes in the prop-

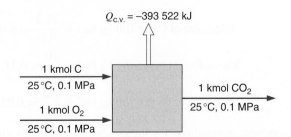

FIGURE 13.3
Example of the
combustion process.

erties of a given substance. The properties at the condition of the reference state cancel out in the calculation. When dealing with reference states in Section 12.10, we noted that for a given substance (perhaps a component of a mixture), we are free to choose a reference state condition—for example, a hypothetical ideal gas—as long as we then carry out a consistent calculation from that state and condition to the real desired state. We also noted that we are free to choose a reference state value, as long as there is no subsequent inconsistency in the calculation of the change in a property because of a chemical reaction with a resulting change in the amount of a given substance. Now that we are to include the possibility of a chemical reaction, it will become necessary to choose these reference state values on a common and consistent basis. We will use as our reference state a temperature of 25 °C, a pressure of 0.1 MPa, and a hypothetical ideal gas condition for those substances that are gases.

Consider the simple steady-state combustion process shown in Fig. 13.3. This idealized reaction involves the combustion of solid carbon with gaseous (ideal-gas) oxygen, each of which enters the control volume at the reference state, 25 °C and 0.1 MPa. The carbon dioxide (ideal gas) formed by the reaction leaves the chamber at the reference state, 25 °C and 0.1 MPa. If the heat transfer could be accurately measured, it would be found to be −393 522 kJ/kmol of carbon dioxide formed. The chemical reaction can be written

$$C + O_2 \rightarrow CO_2$$

Applying the energy equation to this process, we have

$$Q_{c.v.} + H_R = H_P$$

where the subscripts R and P refer to the reactants and products, respectively. We will find it convenient to also write the energy equation for such a process in the form

$$Q_{c.v.} + \sum_R n_i \overline{h}_i = \sum_P n_e \overline{h}_e \qquad (13.11)$$

where the summations refer, respectively, to all the reactants or all the products.

Thus, a measurement of the heat transfer would give us the difference between the enthalpy of the products and the reactants, where each is in the reference state condition. Suppose, however, that we assign the value of zero to the enthalpy of all the elements at the reference state. In this case, the enthalpy of the reactants is zero, and

$$Q_{c.v.} = H_p = -393\ 522\ \text{kJ/kmol}$$

The enthalpy of (hypothetical) ideal gas carbon dioxide at 25 °C, 0.1 MPa pressure (with reference to this arbitrary base in which the enthalpy of the elements is chosen to be zero) is called the enthalpy of formation. We designate this with the symbol $\overline{h}_f$. Thus, for carbon dioxide

$$\overline{h}_f^0 = -393\ 522 \text{ kJ/kmol}$$

The enthalpy of carbon dioxide in any other state, relative to this base in which the enthalpy of the elements is zero, would be found by adding the change of enthalpy between ideal gas at 25 °C, 0.1 MPa and the given state to the enthalpy of formation. That is, the enthalpy at any temperature and pressure, $h_{T,P}$, is

$$\overline{h}_{T,P} = (\overline{h}_f^0)_{298,0.1\,\text{MPa}} + (\Delta\overline{h})_{298,0.1\,\text{MPa}\to T,P} \quad (13.12)$$

where the term $(\Delta\overline{h})_{298,0.1\,\text{MPa}\to T,P}$ represents the difference in enthalpy between any given state and the enthalpy of ideal gas at 298.15 K, 0.1 MPa. For convenience, we usually drop the subscripts in the examples that follow.

The procedure that we have demonstrated for carbon dioxide can be applied to any compound. Table A.10 gives values of the enthalpy of formation for a number of substances in the units kJ/kmol.

The general rule for the enthalpy of formation can be stated as follows:

Enthalpy of formation of the pure elements in their ground state at the reference P_0, T_0 is zero.

Enthalpy of formation for a stable compound is negative.

Enthalpy of formation for an unstable compound is positive.

From the above rule, we may make some further observations:

1. Carbon can exist as graphite or as a crystal, as in diamonds. The graphite is the ground state, and the diamond is different but stable. Other substances can be in states that appear stable, but if they are subject to large enough disturbances they may decay into a more stable configuration. Such states are called *metastable* and they are not the ground states.

2. Substances such as ozone (O_3) or single atomic O are chemically pure, but they are not in the ground state for oxygen. The ozone would eventually decay to diatomic oxygen, and the single atomic O would combine with another to form diatomic oxygen. We can think about creating O from O_2 by forcing the two atoms apart, and that requires a work input. Thus, comparatively, the two single atoms have more energy. The work is equal to the binding energy, explaining the positive enthalpy of formation for O.

3. Stable compounds like carbon dioxide (CO_2) or water (H_2O) can be split into the pure elements if we add the chemical binding energy. Since the elements have zero formation enthalpy, it follows that these compounds will have negative formation enthalpies, and the difference is equal to the chemical binding energy.

It will be noted from Table A.10 that two values are given for the enthalpy of formation for water; one is for liquid water and the other is for gaseous (hypothetical ideal gas) water, both at the reference state of 25 °C, 0.1 MPa. It is convenient to use the hypothetical ideal gas reference in connection with the ideal gas table property changes given in Table A.9 and to use the real liquid reference in connection with real water property changes as given in the steam tables, Table B.1. The real liquid reference state properties are obtained from those at the hypothetical ideal gas reference by following the calculation procedure described in Section 12.10. The same procedure can be followed for other substances that have a saturation pressure less than 0.1 MPa at the reference temperature of 25 °C.

13.4 ENERGY ANALYSIS OF REACTING SYSTEMS

The significance of the enthalpy of formation is that it is most convenient in performing an energy analysis of a reacting system, for the enthalpies of different substances can be added or subtracted, since they are all given relative to the same base.

In such problems, we will write the energy equation for a steady-state, steady-flow process in the form

$$Q_{c.v.} + H_R = W_{c.v.} + H_P$$

or

$$Q_{c.v.} + \sum_R n_i \overline{h}_i = W_{c.v.} + \sum_P n_e \overline{h}_e$$

where R and P refer to the reactants and products, respectively. In each problem, it is necessary to choose one parameter as the basis of the scaling. Usually this is taken as 1 kmol of fuel.

Example 13.6

Consider the following reaction, which occurs in a steady-state, steady-flow process:

$$CH_4 + 2\,O_2 \rightarrow CO_2 + 2\,H_2O(l)$$

The reactants and products are each at a total pressure of 0.1 MPa and 25 °C. Determine the heat transfer per kilomole of fuel entering the combustion chamber.

Control volume:	Combustion chamber.
Inlet state:	P and T known; state fixed.
Exit state:	P and T known, state fixed.
Process:	Steady state.
Model:	Three gases ideal gases; real liquid water.

Analysis

Energy equation:

$$Q_{c.v.} + \sum_R n_i \overline{h}_i = \sum_P n_e \overline{h}_e$$

Solution

Using values from Table A.10, we have

$$\sum_R n_i \overline{h}_i = (\overline{h}_f^0)_{CH_4} = -74\,873 \text{ kJ}$$

$$\sum_P n_e \overline{h}_e = (\overline{h}_f^0)_{CO_2} + 2(\overline{h}_f^0)_{H_2O(l)}$$

$$= -393\,522 + 2(-285\,830) = -965\,182 \text{ kJ}$$

$$Q_{c.v.} = -965\,182 - (-74\,873) = -890\,309 \text{ kJ}$$

In most instances, however, the substances that comprise the reactants and products in a chemical reaction are not at a temperature of 25 °C and a pressure of 0.1 MPa (the state at which the enthalpy of formation is given). Therefore, the change of enthalpy between 25 °C and 0.1 MPa and the given state must be known. For a solid or liquid, this change of enthalpy can usually be found from a table of thermodynamic properties or from specific heat data. For gases, the change of enthalpy can usually be found by one of the following procedures.

1. Assume ideal gas behavior between 25 °C, 0.1 MPa and the given state. In this case, the enthalpy is a function of the temperature only and can be found by an equation of $\overline{C}_{p0}$ or from tabulated values of enthalpy as a function of temperature (which assumes ideal gas behavior). Table A.6 gives an equation for $\overline{C}_{p0}$ for a number of substances and Table A.9 gives values of $\overline{h}^0 - \overline{h}^0_{298}$. (That is, the $\Delta\overline{h}$ of Eq. 13.12) in kJ/kmol, ($\overline{h}^0_{298}$ refers to 25 °C or 298.15 K. For simplicity, this is designated $\overline{h}^0_{298}$.) The superscript 0 is used to designate that this is the enthalpy at 0.1 MPa pressure, based on ideal gas behavior; that is, the standard-state enthalpy.

2. If a table of thermodynamic properties is available, $\Delta\overline{h}$ can be found directly from these tables if a real substance behavior reference state is being used, such as that described above for liquid water. If a hypothetical ideal gas reference state is being used, then it is necessary to account for the real substance correction to properties at that state to gain entry to the tables.

3. If the deviation from ideal gas behavior is significant but no tables of thermodynamic properties are available, the value of $\Delta\overline{h}$ can be found from the generalized tables or charts and the values for $\overline{C}_{p0}$ or $\Delta\overline{h}$ at 0.1 MPa pressure as indicated above.

Thus, in general, for applying the energy equation to a steady-state process involving a chemical reaction and negligible changes in kinetic and potential energy, we can write

$$Q_{c.v.} + \sum_R n_i(\overline{h}^0_f + \Delta\overline{h})_i = W_{c.v.} + \sum_P n_e(\overline{h}^0_f + \Delta\overline{h})_e \qquad (13.13)$$

Example 13.7

Calculate the enthalpy of water (on a mole basis) at 3.5 MPa, 300 °C, relative to the 25 °C and 0.1 MPa base, using the following procedures.

1. Assume the steam to be an ideal gas with the value of $\overline{C}_{p0}$ given in Table A.6.
2. Assume the steam to be an ideal gas with the value for $\Delta\overline{h}$ as given in Table A.9.
3. The steam tables.
4. The specific heat behavior given in 2 above and the generalized charts.

Solution

For each of these procedures, we can write

$$\overline{h}_{T,P} = (\overline{h}^0_f + \Delta\overline{h})$$

The only difference is in the procedure by which we calculate $\Delta\overline{h}$. From Table A.10 we note that

$$(\overline{h}^0_f)_{H_2O(g)} = -241\,826 \text{ kJ/kmol}$$

1. Using the specific heat equation for $H_2O(g)$ from Table A.6,

$$C_{p0} = 1.79 + 0.107\theta + 0.586\theta^2 - 0.20\theta^3, \theta = T/1000$$

The specific heat at the average temperature

$$T_{avg} = \frac{298.15 + 573.15}{2} = 435.65 \text{ K}$$

is

$$C_{p0} = 1.79 + 0.107(0.435\,65) + 0.586(0.435\,65)^2 - 0.2(0.435\,65)^3$$
$$= 1.9313 \frac{\text{kJ}}{\text{kg·K}}$$

Therefore,

$$\Delta \overline{h} = MC_{p0} \Delta T$$
$$= 18.015 \times 1.9313(573.15 - 298.15) = 9568 \text{ kJ/kmol}$$
$$\overline{h}_{T,P} = -241\,826 + 9568 = -232\,258 \text{ kJ/kmol}$$

2. Using Table A.9 for $H_2O(g)$,

$$\Delta \overline{h} = 9539 \text{ kJ/kmol}$$
$$\overline{h}_{T,P} = -241\,826 + 9539 = -232\,287 \text{ kJ/kmol}$$

3. Using the steam tables, either the liquid reference or the gaseous reference state may be used.

For the liquid,

$$\Delta \overline{h} = 18.015(2977.5 - 104.9) = 51\,750 \text{ kJ/kmol}$$
$$\overline{h}_{T,P} = -285\,830 + 51\,750 = -234\,080 \text{ kJ/kmol}$$

For the gas,

$$\Delta \overline{h} = 18.015(2977.5 - 2547.2) = 7752 \text{ kJ/kmol}$$
$$\overline{h}_{T,P} = -241\,826 + 7752 = -234\,074 \text{ kJ/kmol}$$

The very small difference results from using the enthalpy of saturated vapor at 25 °C (which is almost but not exactly an ideal gas) in calculating the $\Delta \overline{h}$.

4. When using the generalized charts, we use the notation introduced in Chapter 12.

$$\overline{h}_{T,P} = \overline{h}_f^0 - (\overline{h}_2^* - \overline{h}_2) + (\overline{h}_2^* - \overline{h}_1^*) + (\overline{h}_1^* - \overline{h}_1)$$

where the subscript 2 refers to the state at 3.5 MPa, 300 °C, and state 1 refers to the state at 0.1 MPa, 25 °C.

From part 2, $\overline{h}_2 - \overline{h}_1^* = 9539$ kJ/kmol.

$$\overline{h}_1^* - \overline{h}_1 = 0 \qquad \text{(ideal gas reference)}$$

$$P_{r2} = \frac{3.5}{22.09} = 0.158 \qquad T_{r2} = \frac{573.2}{647.3} = 0.886$$

From the generalized enthalpy chart, Fig. D.2,

$$\frac{\overline{h}_2^* - \overline{h}_2}{RT_c} = 0.21, \qquad \overline{h}_2^* - \overline{h}_2 = 0.21 \times 8.3145 \times 647.3 = 1130 \text{ kJ/kmol}$$

$$\overline{h}_{T,P} = -241\,826 - 1130 + 9539 = -233\,417 \text{ kJ/kmol}$$

Note that if the software is used, including the acentric factor correction (value from Table D.4), as discussed in Section 12.7, the enthalpy correction is found to be 0.298 instead of 0.21 and the enthalpy is then $-233\,996$ kJ/kmol, which is considerably closer to the values found for the steam tables in procedure 3 above, the most accurate value.

The approach that is used in a given problem will depend on the data available for the given substance.

Example 13.8

A small gas turbine uses $C_8H_{18}(l)$ for fuel and 400 % theoretical air. The air and fuel enter at 25 °C, and the products of combustion leave at 900 K. The output of the engine and the fuel consumption are measured, and it is found that the specific fuel consumption is 0.25 kg/s of fuel per megawatt output. Determine the heat transfer from the engine per kilomole of fuel. Assume complete combustion.

Control volume:	Gas-turbine engine.
Inlet states:	T known for fuel and air.
Exit state:	T known for combustion products.
Process:	Steady state.
Model:	All gases ideal gases, Table A.9; liquid octane, Table A.10.

Analysis

The combustion equation is

$$C_8H_{18}(l) + 4(12.5)O_2 + 4(12.5)(3.76)N_2 \rightarrow 8CO_2 + 9H_2O + 37.5O_2 + 188.0N_2$$

Energy equation:

$$Q_{c.v.} + \sum_R n_i(\overline{h}_f^0 + \Delta\overline{h})_i = W_{c.v.} + \sum_P n_e(\overline{h}_f^0 + \Delta\overline{h})_e$$

Solution

Since the air is composed of elements and enters at 25 °C, the enthalpy of the reactants is equal to that of the fuel.

$$\sum_R n_i(\overline{h}_f^0 + \Delta\overline{h})_i = (\overline{h}_f^0)_{C_8H_{18}(l)} = -250\ 105\ \text{kJ/kmol}$$

Considering the products, we have

$$\sum_P n_e(\overline{h}_f^0 + \Delta\overline{h})_e = n_{CO_2}(\overline{h}_f^0 + \Delta\overline{h})_{CO_2} + n_{H_2O}(\overline{h}_f^0 + \Delta\overline{h})_{H_2O}$$

$$+ n_{O_2}(\Delta\overline{h})_{O_2} + n_{N_2}(\Delta\overline{h})_{N_2}$$

$$= 8(-393\ 522 + 28\ 030) + 9(-241\ 826 + 21\ 937)$$

$$+ 37.5(19\ 241) + 188(18\ 225)$$

$$= -755\ 476\ \text{kJ/kmol}$$

$$W_{c.v.} = \frac{1000\ \text{kJ/s}}{0.25\ \text{kg/s}} \times \frac{114.23\ \text{kg}}{\text{kmol}} = 456\ 920\ \text{kJ/kmol}$$

Therefore, from the energy equation,

$$Q_{c.v.} = -755\ 476 + 456\ 920 - (-250\ 105)$$

$$= -48\ 451\ \text{kJ/kmol}$$

Example 13.9

A mixture of 1 kmol of gaseous ethene and 3 kmol of oxygen at 25 °C reacts in a constant-volume bomb. Heat is transferred until the products are cooled to 600 K. Determine the amount of heat transfer from the system.

Control mass:	Constant-volume bomb.
Initial state:	T known.
Final state:	T known.
Process:	Constant volume.
Model:	Ideal gas mixtures, Tables A.9, A.10.

Analysis

The chemical reaction is

$$C_2H_4 + 3\ O_2 \rightarrow 2\ CO_2 + 2\ H_2O(g)$$

Energy equation:

$$U_P - U_R = Q$$

$$Q + \sum_R n(\overline{h}_f^0 + \Delta\overline{h} - \overline{R}T) = \sum_P n(\overline{h}_f^0 + \Delta\overline{h} - \overline{R}T)$$

Solution

Using values from Tables A.9 and A.10 gives

$$\sum_R n(\overline{h}_f^0 + \Delta\overline{h} - \overline{R}T) = (\overline{h}_f^0 - \overline{R}T)_{C_2H_4} - n_{O_2}(\overline{R}T)_{O_2} = (\overline{h}_f^0)_{C_2H_4} - 4\overline{R}T$$

$$= 52\,467 - 4 \times 8.3145 \times 298.2 = 42\,550\ \text{kJ}$$

$$\sum_P n(\overline{h}_f^0 + \Delta\overline{h} - \overline{R}T) = 2[(\overline{h}_f^0)_{CO_2} + \Delta\overline{h}_{CO_2}] + 2[(\overline{h}_f^0)_{H_2O(g)} + \Delta\overline{h}_{H_2O(g)}] - 4\overline{R}T$$

$$= 2(-393\,522 + 12\,906) + 2(-241\,826 + 10\,499)$$

$$- 4 \times 8.3145 \times 600$$

$$= -1\,243\,841\ \text{kJ}$$

Therefore,

$$Q = -1\,243\,841 - 42\,550 = -1\,286\,391\ \text{kJ}$$

For a real gas mixture, a pseudocritical method such as Kay's rule, Eq. 12.83, could be used to evaluate the nonideal gas contribution to enthalpy at the temperature and pressure of the mixture and this value added to the ideal gas mixture enthalpy at that temperature, as in the procedure developed in Section 12.10.

13.5 ENTHALPY AND INTERNAL ENERGY OF COMBUSTION; HEAT OF REACTION

The enthalpy of combustion, h_{RP}, is defined as the difference between the enthalpy of the products and the enthalpy of the reactants when complete combustion occurs at a given temperature and pressure. That is,

$$\overline{h}_{RP} = H_P - H_R$$

$$\overline{h}_{RP} = \sum_P n_e(\overline{h}_f^0 + \Delta\overline{h})_e - \sum_R n_i(\overline{h}_f^0 + \Delta\overline{h})_i \qquad (13.14)$$

The usual parameter for expressing the enthalpy of combustion is a unit mass of fuel, such as a kilogram (h_{RP}) or a kilomole ($\overline{h}_{RP}$) of fuel.

As the enthalpy of formation is fixed, we can separate the terms as

$$H = H^0 + \Delta H$$

where

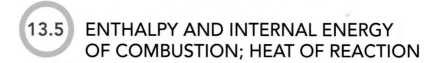

$$H_R^0 = \sum_R n_i \overline{h}_{fi}^0; \qquad \Delta H_R = \sum_R n_i \Delta\overline{h}_i$$

and

$$H_p^0 = \sum_P n_i \bar{h}^0_{fi}; \qquad \Delta H_P = \sum_P n_i \, \Delta \bar{h}_i$$

Now the difference in enthalpies is written

$$H_P - H_R = H_P^0 - H_R^0 + \Delta H_P - \Delta H_R$$
$$= \bar{h}^0_{RP} + \Delta H_P - \Delta H_R$$

explicitly showing the reference enthalpy of combustion, $\bar{h}^0_{RP}$, and the two departure terms ΔH_P and ΔH_R. The latter two terms for the products and reactants are nonzero if they exist at a state other than the reference state.

The tabulated values of the enthalpy of combustion of fuels are usually given for a temperature of 25 °C and a pressure of 0.1 MPa. The enthalpy of combustion for a number of hydrocarbon fuels at this temperature and pressure, which we designate h^0_{RP}, is given in Table 13.3.

The internal energy of combustion is defined in a similar manner.

$$\bar{u}_{RP} = U_P - U_R$$
$$= \sum_P n_e(\bar{h}^0_f + \Delta \bar{h} - P\bar{v})_e - \sum_R n_i(\bar{h}^0_f + \Delta \bar{h} - P\bar{v})_i \qquad (13.15)$$

When all the gaseous constituents can be considered as ideal gases, and the volume of the liquid and solid constituents is negligible compared to the value of the gaseous constituents, this relation for $\bar{u}_{RP}$ reduces to

$$\bar{u}_{RP} = \bar{h}_{RP} - \bar{R}T(n_{\text{gaseous products}} - n_{\text{gaseous reactants}}) \qquad (13.16)$$

Frequently, the term heating value or heat of reaction is used. This represents the heat transferred from the chamber during combustion or reaction at constant temperature. In the case of a constant-pressure or steady-flow process, we conclude from the energy equation that it is equal to the negative of the enthalpy of combustion. For this reason, this heat transfer is sometimes designated the *constant-pressure heating value* for combustion processes.

In the case of a constant-volume process, the heat transfer is equal to the negative of the internal energy of combustion. This is sometimes designated the *constant-volume heating value* in the case of combustion.

When the term *heating value* (HV) is used, the terms *higher* and *lower heating value* are used. The higher heating value (HHV) is the heat transfer with liquid water in the products, and the lower heating value (LHV) is the heat transfer with vapor water in the products.

We can now write the energy equation for a steady flow, Eq. 13.13, as

$$W_{\text{c.v.}} - Q_{\text{c.v.}} = H_R - H_P$$
$$= -\bar{h}^0_{RP} + \Delta H_R - \Delta H_P$$
$$= HV + \Delta H_R - \Delta H_P \qquad (13.17)$$

TABLE 13.3

Enthalpy of Combustion of Some Hydrocarbons at 25 °C

Units: $\frac{kJ}{kg}$		Liquid H_2O in Products		Gas H_2O in Products	
Hydrocarbon	**Formula**	**Liq. HC**	**Gas HC**	**Liq. HC**	**Gas HC**
Paraffins	C_nH_{2n+2}				
Methane	CH_4		−55 496		−50 010
Ethane	C_2H_6		−51 875		−47 484
Propane	C_3H_8	−49 973	−50 343	−45 982	−46 352
n-Butane	C_4H_{10}	−49 130	−49 500	−45 344	−45 714
n-Pentane	C_5H_{12}	−48 643	−49 011	−44 983	−45 351
n-Hexane	C_6H_{14}	−48 308	−48 676	−44 733	−45 101
n-Heptane	C_7H_{16}	−48 071	−48 436	−44 557	−44 922
n-Octane	C_8H_{18}	−47 893	−48 256	−44 425	−44 788
n-Decane	$C_{10}H_{22}$	−47 641	−48 000	−44 239	−44 598
n-Dodecane	$C_{12}H_{26}$	−47 470	−47 828	−44 109	−44 467
n-Cetane	$C_{16}H_{34}$	−47 300	−47 658	−44 000	−44 358
Olefins	C_nH_{2n}				
Ethene	C_2H_4		−50 296		−47 158
Propene	C_3H_6		−48 917		−45 780
Butene	C_4H_8		−48 453		−45 316
Pentene	C_5H_{10}		−48 134		−44 996
Hexene	C_6H_{12}		−47 937		−44 800
Heptene	C_7H_{14}		−47 800		−44 662
Octene	C_8H_{16}		−47 693		−44 556
Nonene	C_9H_{18}		−47 612		−44 475
Decene	$C_{10}H_{20}$		−47 547		−44 410
Alkylbenzenes	$C_{6+n}H_{6+2n}$				
Benzene	C_6H_6	−41 831	−42 266	−40 141	−40 576
Methylbenzene	C_7H_8	−42 437	−42 847	−40 527	−40 937
Ethylbenzene	C_8H_{10}	−42 997	−43 395	−40 924	−41 322
Propylbenzene	C_9H_{12}	−43 416	−43 800	−41 219	−41 603
Butylbenzene	$C_{10}H_{14}$	−43 748	−44 123	−41 453	−41 828
Other fuels					
Gasoline	C_7H_{17}	−48 201	−48 582	−44 506	−44 886
Diesel T-T	$C_{14.4}H_{24.9}$	−45 700	−46 074	−42 934	−43 308
JP8 jet fuel	$C_{13}H_{23.8}$	−45 707	−46 087	−42 800	−43 180
Methanol	CH_3OH	−22 657	−23 840	−19 910	−21 093
Ethanol	C_2H_5OH	−29 676	−30 596	−26 811	−27 731
Nitromethane	CH_3NO_2	−11 618	−12 247	−10 537	−11 165
Phenol	C_6H_5OH	−32 520	−33 176	−31 117	−31 774
Hydrogen	H_2		−141 781		−119 953

If the reactants and the products enter and leave under the reference conditions, the net energy out ($W_{c.v.} - Q_{c.v.}$) equals the heating value. For situations where the reactants are preheated ($\Delta H_R > 0$), the net energy out is correspondingly larger, and if the products leave at an elevated temperature, $\Delta H_P > 0$, which is typical, the net energy out is reduced. Care should be taken to evaluate all terms with the same scaling and mass or mole basis.

Example 13.10

Calculate the enthalpy of combustion of propane at 25 °C on both a kilomole and kilogram basis under the following conditions:

1. Liquid propane with liquid water in the products.
2. Liquid propane with gaseous water in the products.
3. Gaseous propane with liquid water in the products.
4. Gaseous propane with gaseous water in the products.

This example is designed to show how the enthalpy of combustion can be determined from enthalpies of formation. The enthalpy of evaporation of propane is 370 kJ/kg.

Analysis and Solution

The basic combustion equation is (nitrogen terms will cancel out in $\overline{h}_{RP}$)

$$C_3H_8 + 5\,O_2 \rightarrow 3\,CO_2 + 4\,H_2O$$

From Table A.10 $(\overline{h}_f^0)_{C_3H_8(g)} = -103\,900$ kJ/kmol. Therefore,

$$(\overline{h}_f^0)_{C_3H_8(l)} = -103\,900 - 44.097(370) = -120\,216 \text{ kJ/kmol}$$

1. Liquid propane–liquid water:

$$\overline{h}_{RP}^0 = 3(\overline{h}_f^0)_{CO_2} + 4(\overline{h}_f^0)_{H_2O(l)} - (\overline{h}_f^0)_{C_3H_8(l)}$$

$$= 3(-393\,522) + 4(-285\,830) - (-120\,216)$$

$$= -2\,203\,670 \text{ kJ/kmol} = -\frac{2\,203\,670}{44.097} = -49\,973 \text{ kJ/kg}$$

The higher heating value of liquid propane is 49 973 kJ/kg.

2. Liquid propane–gaseous water:

$$\overline{h}_{RP}^0 = 3(\overline{h}_f^0)_{CO_2} + 4(\overline{h}_f^0)_{H_2O(g)} - (\overline{h}_f^0)_{C_3H_8(l)}$$

$$= 3(-393\,522) + 4(-241\,826) - (-120\,216)$$

$$= -2\,027\,654 \text{ kJ/kmol} = -\frac{2\,027\,654}{44.097} = -45\,982 \text{ kJ/kg}$$

The lower heating value of liquid propane is 45 982 kJ/kg.

3. Gaseous propane–liquid water:

$$\overline{h}_{RP}^0 = 3(\overline{h}_f^0)_{CO_2} + 4(\overline{h}_f^0)_{H_2O(l)} - (\overline{h}_f^0)_{C_3H_8(g)}$$

$$= 3(-393\,522) + 4(-285\,830) - (-103\,900)$$

$$= -2\,219\,986 \text{ kJ/kmol} = -\frac{2\,219\,986}{44.097} = -50\,343 \text{ kJ/kg}$$

The higher heating value of gaseous propane is 50 343 kJ/kg.

4. Gaseous propane–gaseous water:

$$\overline{h}_{RP}^0 = 3(\overline{h}_f^0)_{CO_2} + 4(\overline{h}_f^0)_{H_2O(g)} - (\overline{h}_f^0)_{C_3H_8(g)}$$

$$= 3(-393\,522) + 4(-241\,826) - (-103\,900)$$

$$= -2\,043\,970 \text{ kJ/kmol} = -\frac{2\,043\,970}{44.097} = -46\,352 \text{ kJ/kg}$$

The lower heating value of gaseous propane is 46 352 kJ/kg.

Each of the four values calculated in this example corresponds to the appropriate value given in Table 13.3.

Example 13.11

Calculate the enthalpy of combustion of gaseous propane at 500 K. (At this temperature, all the water formed during combustion will be vapor.) This example will demonstrate how the enthalpy of combustion of propane varies with temperature. The average constant-pressure specific heat of propane between 25 °C and 500 K is 2.1 kJ/kg·K.

Analysis
The combustion equation is

$$C_3H_8(g) + 5\,O_2 \rightarrow 3\,CO_2 + 4\,H_2O(g)$$

The enthalpy of combustion is, from Eq. 13.13,

$$(\overline{h}_{RP})_T = \sum_P n_e(\overline{h}_f^0 + \Delta\overline{h})_e - \sum_R n_i(\overline{h}_f^0 + \Delta\overline{h})_i$$

Solution

$$\overline{h}_{R_{500}} = [\overline{h}_f^0 + \overline{C}_{p.\,av}(\Delta T)]_{C_3H_8(g)} + n_{O_2}(\Delta\overline{h})_{O_2}$$

$$= -103\,900 + 2.1 \times 44.097(500 - 298.2) + 5(6086)$$

$$= -54\,783 \text{ kJ/kmol}$$

$$\overline{h}_{P_{500}} = n_{CO_2}(\overline{h}_f^0 + \Delta\overline{h})_{CO_2} + n_{H_2O}(\overline{h}_f^0 + \Delta\overline{h})_{H_2O}$$

$$= 3(-393\,522 + 8305) + 4(-241\,826 + 6922)$$

$$= -2\,095\,267 \text{ kJ/kmol}$$

$$\overline{h}_{RP_{500}} = -2\,095\,267 - (-54\,783) = -2\,040\,484 \text{ kJ/kmol}$$

$$h_{RP_{500}} = \frac{-2\,040\,484}{44.097} = -46\,273 \text{ kJ/kg}$$

This compares with a value of $-46\,352$ at $25\,°C$.

This problem could also have been solved using the given value of the enthalpy of combustion at $25\,°C$ by noting that

$$\overline{h}_{RP_{500}} = (H_P)_{500} - (H_R)_{500}$$

$$= n_{CO_2}(\overline{h}_f^0 + \Delta h)_{CO_2} + n_{H_2O}(\overline{h}_f^0 + \Delta h)_{H_2O}$$

$$\quad - [\overline{h}_f^0 + \overline{C}_{p.\,av}(\Delta T)]_{C_3H_8(g)} - n_{O_2}(\Delta\overline{h})_{O_2}$$

$$= \overline{h}_{RP_0} + n_{CO_2}(\Delta\overline{h})_{CO_2} + n_{H_2O}(\Delta\overline{h})_{H_2O}$$

$$\quad - \overline{C}_{p.\,av}(\Delta T)_{C_3H_8(g)} - n_{O_2}(\Delta\overline{h})_{O_2}$$

$$\overline{h}_{RP_{500}} = -46\,352 \times 44.097 + 3(8305) + 4(6922)$$

$$\quad - 2.1 \times 44.097(500 - 298.2) - 5(6086)$$

$$= -2\,040\,499 \text{ kJ/kmol}$$

$$h_{RP_{500}} = \frac{-2\,040\,499}{44.097} = -46\,273 \text{ kJ/kg}$$

13.6 ADIABATIC FLAME TEMPERATURE

In many combustion applications, the peak temperature is important and the reason varies. For a heat engine, a high temperature increases the efficiency; for an incinerator or a thermal oxidizer, the peak temperature guarantees the destruction of harmful gases or particles. However, the high temperature causes formation of various forms of nitric oxide, a pollutant that is difficult to remove, and it places a strain on all materials exposed to the products. The energy in a steady flow of products is found from the energy equation, Eq. 13.17, as

$$\Delta H_P = \text{HV} + \Delta H_R + Q_{cv} - W_{cv}$$

So, for a given combustion process with no work and no heat transfer, the first two terms define the product enthalpy and the product temperature is referred to as the adiabatic flame temperature. The temperature then depends on the state of the reactants and the composition of the products, so the only well-defined adiabatic flame temperature occurs if the reactants are supplied at the reference temperature ($\Delta H_R = 0$) and the combustion is complete, so that the product composition is known.

For a given fuel and given pressure and temperature of the reactants, the maximum adiabatic flame temperature that can be achieved is with a stoichiometric mixture. The adiabatic flame temperature can be controlled by the amount of excess air that is used. This is important, for example, in gas turbines, where the maximum permissible temperature is determined by metallurgical considerations in the turbine and close control of the temperature of the products is essential.

Example 13.12 shows how the adiabatic flame temperature may be found. The dissociation that takes place in the combustion products, which has a significant effect on the adiabatic flame temperature, will be considered in the next chapter.

Example 13.12

Liquid octane at $25\,°\text{C}$ is burned with $400\,\%$ theoretical air at $25\,°\text{C}$ in a steady-state process. Determine the adiabatic flame temperature.

Control volume:	Combustion chamber.
Inlet states:	T known for fuel and air.
Process:	Steady state.
Model:	Gases ideal gases, Table A.9; liquid octane, Table A.10.

Analysis
The reaction is

$$\text{C}_8\text{H}_{18}(l) + 4(12.5)\,\text{O}_2 + 4(12.5)(3.76)\,\text{N}_2 \rightarrow 8\,\text{CO}_2 + 9\,\text{H}_2\text{O}(g) + 37.5\,\text{O}_2 + 188.0\,\text{N}_2$$

Energy equation: Since the process is adiabatic,

$$H_R = H_P$$

$$\sum_R n_i (\overline{h}_f^0 + \Delta\overline{h})_i = \sum_P n_e (\overline{h}_f^0 + \Delta\overline{h})_e$$

where $\Delta\overline{h}_e$ refers to each constituent in the products at the adiabatic flame temperature.

Solution

From Tables A.9 and A.10,

$$H_R = \sum_R n_i(\overline{h}_f^0 + \Delta\overline{h})_i = (\overline{h}_f^0)_{C_8H_{18}(l)} = -250\ 105 \text{ kJ/kmol}$$

$$H_P = \sum_P n_e(\overline{h}_f^0 + \Delta\overline{h})_e$$

$$= 8(-393\ 522 + \Delta\overline{h}_{CO_2}) + 9(-241\ 826 + \Delta\overline{h}_{H_2O}) + 37.5\ \Delta\overline{h}_{O_2} + 188.0\ \Delta\overline{h}_{N_2}$$

By trial-and-error solution, a temperature of the products is found that satisfies the energy equation. Assume that

$$T_P = 900 \text{ K}$$

$$H_P = \sum_P n_e(\overline{h}_f^0 + \Delta\overline{h})_e$$

$$= 8(-393\ 522 + 28\ 030) + 9(-241\ 826 + 21\ 892)$$

$$+ 37.5(19\ 249) + 188(18\ 222)$$

$$= -755\ 769 \text{ kJ/kmol}$$

Assume that

$$T_P = 1000 \text{ K}$$

$$H_P = \sum_P n_e(\overline{h}_f^0 + \Delta\overline{h})_e$$

$$= 8(-393\ 522 + 33\ 400) + 9(-241\ 826 + 25\ 956)$$

$$+ 37.5(22\ 710) + 188(21\ 461)$$

$$= 62\ 487 \text{ kJ/kmole}$$

Since $H_P = H_R = -250\ 105$ kJ/kmol, we find by linear interpolation that the adiabatic flame temperature is 961.8 K. Because the ideal gas enthalpy is not really a linear function of temperature, the true answer will be slightly different from this value.

In-Text Concept Questions

e. How is a fuel enthalpy of combustion connected to its enthalpy of formation?

f. What are the higher and lower heating values HHV, LHV of *n*-butane?

g. What is the value of h_{fg} for *n*-octane?

h. What happens to the adiabatic flame temperature when I burn rich and when I burn lean?

13.7 THE THIRD LAW OF THERMODYNAMICS AND ABSOLUTE ENTROPY

As we consider a second-law analysis of chemical reactions, we face the same problem we had with the first law: What base should be used for the entropy of the various substances? This problem leads directly to a consideration of the third law of thermodynamics.

The third law of thermodynamics was formulated during the early twentieth century. The initial work was done primarily by W. H. Nernst (1864–1941) and Max Planck (1858–1947). The third law deals with the entropy of substances at absolute zero temperature and, in essence, states that the entropy of a perfect crystal is zero at absolute zero. From a statistical point of view, this means that the crystal structure has the maximum degree of order. Furthermore, because the temperature is absolute zero, the thermal energy is minimum. It also follows that a substance that does not have a perfect crystalline structure at absolute zero, but instead has a degree of randomness, such as a solid solution or a glassy solid, has a finite value of entropy at absolute zero. The experimental evidence on which the third law rests is primarily data on chemical reactions at low temperatures and measurements of specific heat at temperatures approaching absolute zero. In contrast to the first and second laws, which lead, respectively, to the properties of internal energy and entropy, the third law deals only with the question of entropy at absolute zero. However, the implications of the third law are quite profound, particularly in respect to chemical equilibrium.

The relevance of the third law is that it provides an absolute base from which to measure the entropy of each substance. The entropy relative to this base is termed the *absolute entropy*. The increase in entropy between absolute zero and any given state can be found either from calorimetric data or by procedures based on statistical thermodynamics. The calorimetric method gives precise measurements of specific-heat data over the temperature range, as well as of the energy associated with phase transformations. These measurements are in agreement with the calculations based on statistical thermodynamics and observed molecular data.

Table A.10 gives the absolute entropy at 25 °C and 0.1 MPa pressure for a number of substances. Table A.9 gives the absolute entropy for a number of gases at 0.1 MPa pressure and various temperatures. For gases, the numbers in all these tables are the hypothetical ideal gas values. The pressure P^0 of 0.1 MPa is termed the standard-state pressure, and the absolute entropy as given in these tables is designated $\bar{s}^0$. The temperature is designated in kelvins with a subscript such as $\bar{s}^0_{1000}$.

If the value of the absolute entropy is known at the standard-state pressure of 0.1 MPa and a given temperature, it is a straightforward procedure to calculate the entropy change from this state (whether hypothetical ideal gas or a real substance) to another desired state following the procedure described in Section 12.10. If the substance is listed in Table A.9, then

$$\bar{s}_{T,P} = \bar{s}^0_T - \bar{R} \ln \frac{P}{P^0} + (\bar{s}_{T,P} - \bar{s}^*_{T,P}) \tag{13.18}$$

In this expression, the first term on the right side is the value from Table A.9, the second is the ideal gas term to account for a change in pressure from P^0 to P, and the third is the term that corrects for real substance behavior, as given in the generalized entropy chart in Appendix D. If the real substance behavior is to be evaluated from an equation of state or thermodynamic table of properties, the term for the change in pressure should be made to a low pressure P^*, at which ideal gas behavior is a reasonable assumption, but it is also listed in the tables. Then

$$\bar{s}_{T,P} = \bar{s}_T^0 - \bar{R} \ln \frac{P^*}{P^0} + (\bar{s}_{T,P} - \bar{s}_{T,P^*}^*) \tag{13.19}$$

If the substance is not one of those listed in Table A.9, and the absolute entropy is known only at one temperature, T_0, as given in Table A.10, for example, then it will be necessary to calculate from

$$\bar{s}_T^0 = \bar{s}_{T_0}^0 + \int_{T_0}^T \frac{\overline{C}_{p0}}{T} dT \tag{13.20}$$

and then proceed with the calculation of Eq. 13.17 or 13.19.

If Eq. 13.18 is being used to calculate the absolute entropy of a substance in a region in which the ideal gas model is a valid representation of the behavior of that substance, then the last term on the right side of Eq. 13.18 simply drops out of the calculation.

For calculation of the absolute entropy of a mixture of ideal gases at T, P, the mixture entropy is given in terms of the component partial entropies as

$$\bar{s}_{\text{mix}}^* = \sum_i y_i \overline{S}_i^* \tag{13.21}$$

where

$$\overline{S}_i^* = \bar{s}_{T_i}^0 - \bar{R} \ln \frac{P}{P^0} - \bar{R} \ln y_i = \bar{s}_{Ti}^0 - \bar{R} \ln \frac{y_i P}{P^0} \tag{13.22}$$

For a real gas mixture, a correction can be added to the ideal gas entropy calculated from Eqs. 13.21 and 13.22 by using a pseudocritical method such as was discussed in Section 12.10. The corrected expression is

$$\bar{s}_{\text{mix}} = \bar{s}_{\text{mix}}^* + (\bar{s} - \bar{s}^*)_{T,P} \tag{13.23}$$

in which the second term on the right side is the correction term from the generalized entropy chart.

13.8 SECOND-LAW ANALYSIS OF REACTING SYSTEMS

The concepts of reversible work, irreversibility, and exergy were introduced in Chapter 8. These concepts included both the first and second laws of thermodynamics. We will now develop this matter further, and we will be particularly concerned with determining the maximum work (exergy) that can be done through a combustion process and by examining the irreversibilities associated with such processes.

The reversible work for a steady-state process in which there is no heat transfer with reservoirs other than the surroundings, and also in the absence of changes in kinetic and potential energy, is, from Eq. 8.14 on a total mass basis,

$$W^{\text{rev}} = \sum m_i(h_i - T_0 s_i) - \sum m_e(h_e - T_0 s_e)$$

Applying this equation to a steady-state process that involves a chemical reaction, and introducing the symbols from this chapter, we have

$$W^{\text{rev}} = \sum_R n_i(\overline{h}_f^0 + \Delta\overline{h} - T_0\overline{s})_i - \sum_P n_e(\overline{h}_f^0 + \Delta\overline{h} - T_0\overline{s})_e \tag{13.24}$$

Similarly, the irreversibility for such a process can be written as

$$I = W^{\text{rev}} - W = \sum_P n_e T_0 \overline{s}_e - \sum_R n_i T_0 \overline{s}_i - Q_{\text{c.v.}} \tag{13.25}$$

The exergy, ψ, for a steady-flow process, in the absence of kinetic and potential energy changes, is given by Eq. 8.22 as

$$\psi = (h - T_0 s) - (h_0 - T_0 s_0)$$

We further note that if a steady-state chemical reaction takes place in such a manner that both the reactants and products are in temperature equilibrium with the surroundings, the Gibbs function ($g = h - Ts$), defined in Eq. 12.14, becomes a significant variable. For such a process, in the absence of changes in kinetic and potential energy, the reversible work is given by the relation

$$W^{\text{rev}} = \sum_R n_i \overline{g}_i - \sum_P n_e \overline{g}_e = -\Delta G \tag{13.26}$$

in which

$$\Delta G = \Delta H - T\Delta S \tag{13.27}$$

We should keep in mind that Eq. 13.26 is a special case and that the reversible work is given by Eq. 13.24 if the reactants and products are not in temperature equilibrium with the surroundings.

Let us now consider the maximum work that can be done during a chemical reaction. For example, consider 1 kmol of hydrocarbon fuel and the necessary air for complete combustion, each at 0.1 MPa pressure and 25 °C, the pressure and temperature of the surroundings. What is the maximum work that can be done as this fuel reacts with the air? From the considerations covered in Chapter 8, we conclude that the maximum work would be done if this chemical reaction took place reversibly and the products were finally in pressure and temperature equilibrium with the surroundings. We conclude that this reversible work could be calculated from the relation in Eq. 13.26,

$$W^{\text{rev}} = \sum_R n_i \overline{g}_i - \sum_P n_e \overline{g}_e = -\Delta G$$

However, since the final state is in equilibrium with the surroundings, we could consider this amount of work to be the exergy of the fuel and air.

Example 13.13

Ethene (g) at 25 °C and 0.1 MPa pressure is burned with 400 % theoretical air at 25 °C and 0.1 MPa pressure. Assume that this reaction takes place reversibly at 25 °C and that the products leave at 25 °C and 0.1 MPa pressure. To simplify this problem further, assume that the oxygen and nitrogen are separated before the reaction takes place (each at 0.1 MPa, 25 °C), that the constituents in the products are separated, and that each is at 25 °C and 0.1 MPa. Thus, the reaction takes place as shown in Fig. 13.4. This is not a realistic

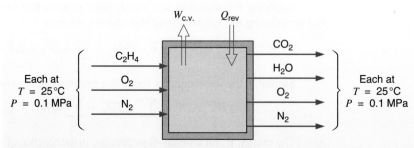

FIGURE 13.4 Sketch for Example 13.13.

situation, since the oxygen and nitrogen in the air entering are, in fact, mixed, as would also be the products of combustion exiting the chamber. This is a commonly used model, however, for the purposes of establishing a standard for comparison with other chemical reactions. For the same reason, we also assume that all the water formed is a gas (a hypothetical state at the given T and P).

Determine the reversible work for this process (that is, the work that would be done if this chemical reaction took place reversibly and isothermally).

Control volume: Combustion chamber.

Inlet states: P, T known for each gas.

Exit states: P, T known for each gas.

Model: All ideal gases, Tables A.9 and A.10.

Sketch: Figure 13.4.

Analysis

The equation for this chemical reaction is

$$C_2H_4(g) + 3(4)\,O_2 + 3(4)(3.76)\,N_2 \rightarrow 2\,CO_2 + 2\,H_2O(g) + 9\,O_2 + 45.1\,N_2$$

The reversible work for this process is equal to the decrease in Gibbs function during this reaction, Eq. 13.26. Since each component is at the standard-state pressure P^0, we write Eqs. 13.26 and 13.27 as

$$W^{\mathrm{rev}} = -\Delta G^0, \qquad \Delta G^0 = \Delta H^0 - T\Delta S^0$$

We also note that the 45.1 N_2 cancels out of both sides in these expressions, as does 9 of the 12 O_2.

Solution

Using values from Tables A.8 and A.9 at 25 °C,

$$\Delta H^0 = 2\overline{h}^0_{f\,CO_2} + 2\overline{h}^0_{f\,H_2O(g)} - \overline{h}^0_{f\,C_2H_4} - 3\overline{h}^0_{f\,O_2}$$

$$= 2(-393\,522) + 2(-241\,826) - (+52\,467) - 3(0)$$

$$= -1\,323\,163\ \text{kJ/kmol·K}$$

$$\Delta S = 2\overline{s}^0_{CO_2} + 2\overline{s}^0_{H_2O(g)} - \overline{s}^0_{C_2H_4} - 3\overline{s}^0_{O_2}$$

$$= 2(213.795) + 2(188.843) - (219.330) - 3(205.148)$$

$$= -29.516\ \text{kJ/kmol}$$

$$\Delta G^0 = -1\,323\,163 - 298.15(-29.516)$$

$$= -1\,314\,363 \text{ kJ/kmol}$$

$$W^{\text{rev}} = -\Delta G^0 = 1\,314\,363 \text{ kJ/kmol}$$

$$= \frac{1\,314\,363}{28.054} = 46\,851 \text{ kJ/kg}$$

Therefore, we might say that when 1 kg of ethene is at 25 °C and the standard-state pressure is 0.1 MPa, it has an exergy of 46 851 kJ.

Thus, it would seem logical to rate the efficiency of a device designed to do work by utilizing a combustion process, such as that of an internal-combustion engine or a steam power plant, as the ratio of the actual work to the reversible work or, in Example 13.13, the decrease in Gibbs function for the chemical reaction, instead of comparing the actual work to the heating value, as is commonly done. This is, in fact, the basic principle of the second-law efficiency, which was introduced in connection with exergy analysis in Chapter 8. As noted from Example 13.13, the difference between the decrease in Gibbs function and the heating value is small, which is typical for hydrocarbon fuels. The difference in the two types of efficiencies will, therefore, not usually be large. We must always be careful, however, when discussing efficiencies, to note the definition of the efficiency under consideration.

It is of particular interest to study the irreversibility that takes place during a combustion process. The following examples illustrate this matter. We consider the same hydrocarbon fuel that was used in Example 13.13, ethene gas at 25 °C and 0.1 MPa. We determined its exergy and found it to be 46 851 kJ/kg. Now let us burn this fuel with 400 % theoretical air in a steady-state adiabatic process. In this case, the fuel and air each enter the combustion chamber at 25 °C and the products exit at the adiabatic flame temperature, but for the purpose of illustrating the calculation procedure, let each of the three pressures be 200 kPa in this case. The result, then, is not exactly comparable to Example 13.13, but the difference is fairly minor. Since the process is adiabatic, the irreversibility for the process can be calculated directly from the increase in entropy using Eq. 13.25.

Example 13.14

Ethene gas at 25 °C and 200 kPa enters a steady-state adiabatic combustion chamber along with 400 % theoretical air at 25 °C, 200 kPa, as shown in Fig. 13.5. The product gas mixture exits at the adiabatic flame temperature and 200 kPa. Calculate the irreversibility per kilomole of ethene for this process.

Control volume:	Combustion chamber
Inlet states:	P, T known for each component gas stream
Exit state:	P, T known
Model:	All ideal gases, Tables A.9 and A.10
Sketch:	Fig. 13.5

FIGURE 13.5 Sketch for Example 13.14.

Analysis

The combustion equation is

$$C_2H_4(g) + 12\,O_2 + 12(3.76)\,N_2 \rightarrow 2\,CO_2 + 2\,H_2O(g) + 9\,O_2 + 45.1\,N_2$$

The adiabatic flame temperature is determined first.

Energy equation:

$$H_R = H_P$$

$$\sum_R n_i(\overline{h}_f^0)_i = \sum_P n_e(\overline{h}_f^0 + \Delta\overline{h})_e$$

Solution

$$52\,467 = 2(-393\,522 + \Delta\overline{h}_{CO_2}) + 2(-241\,826 + \Delta\overline{h}_{H_2O(g)}) + 9\Delta\overline{h}_{O_2} + 45.1\Delta\overline{h}_{N_2}$$

By a trial-and-error solution, we find the adiabatic flame temperature to be 1016 K. We now proceed to find the change in entropy during this adiabatic combustion process.

$$S_R = S_{C_2H_4} + S_{air}$$

From Eq. 13.17.

$$S_{C_2H_4} = 1\left(219.330 - 8.3145\ln\frac{200}{100}\right) = 213.567\ \text{kJ/K}$$

From Eqs. 13.21 and 13.22,

$$S_{air} = 12\left(205.147 - 8.3145\ln\frac{0.21 \times 200}{100}\right)$$

$$+ 45.1\left(191.610 - 8.3145\ln\frac{0.79 \times 200}{100}\right)$$

$$= 12(212.360) + 45.1(187.807) = 11\,018.416\ \text{kJ/k}$$

$$S_R = 213.567 + 11\,018.416 = 11\,231.983\ \text{kJ/K}$$

For a multicomponent product gas mixture, it is convenient to set up a table, as follows:

Comp	n_i	y_i	$\overline{R}\ln\dfrac{y_i P}{P^0}$	$\overline{s}_{Ti}^0$	$\overline{S}_i$
CO_2	2	0.0344	−22.254	270.194	292.448
H_2O	2	0.0344	−22.254	233.355	255.609
O_2	9	0.1549	−9.743	244.135	253.878
N_2	45.1	0.7763	+3.658	228.691	225.033

Then, with values from this table for n_i and $\overline{S}_i$ for each component i,

$$S_P = \sum n_i \overline{S}_i = 13\,530.004\text{ kJ/K}$$

Since this is an adiabatic process, the irreversibility is, from Eq. 13.25,

$$I = T_0(S_P - S_R) = 298.15(13\,530.004 - 11\,231.983) = 685\,155\text{ kJ/kmol}$$

$$= \frac{685\,155}{28.054} = 24\,423\text{ kJ/kg}$$

From the result of Example 13.14, we find that the irreversibility of that combustion process was 50 % of the exergy of the same fuel, as found at standard-state conditions in Example 13.13. We conclude that a typical combustion process is highly irreversible.

13.9 FUEL CELLS

The previous examples raise the question of the possibility of a reversible chemical reaction. Some reactions can be made to approach reversibility by having them take place in an electrolytic cell, as shown in Fig. 13.6 and in the text. When a potential exactly equal to the electromotive force of the cell is applied, no reaction takes place. When the applied potential

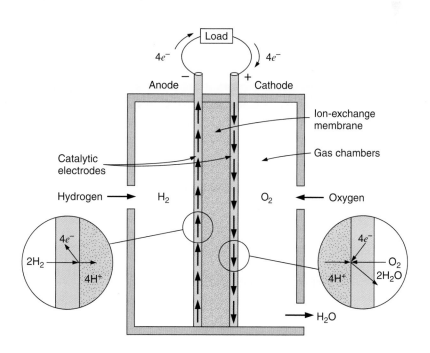

FIGURE 13.6
Schematic arrangement of an ion-exchange membrane type of fuel cell.

is increased slightly, the reaction proceeds in one direction, and if the applied potential is decreased slightly, the reaction proceeds in the opposite direction. The work done is the electrical energy supplied or delivered.

The fuel cell contains an ion-exchange membrane that separates an electrolyte from the reactant and product gases. In the basic fuel cell, hydrogen and oxygen gases are supplied separately so the hydrogen gas enters at the anode, where it dissociates into atomic hydrogen and further into hydrogen ions and electrons. The hydrogen ions travel through the membrane/electrolyte and the electrons through the external circuit, and they combine again with oxygen at the cathode to produce water. There is an electrical potential difference between the anode and cathode and the flow of electrons constitutes a current, which is how the electrical work is delivered to the external load.

Consider a reversible reaction occurring at constant temperature equal to that of its environment. The work output of the fuel cell is

$$W = -\left(\sum n_e \overline{g}_e - \sum n_i \overline{g}_i\right) = -\Delta G$$

where ΔG is the change in Gibbs function for the overall chemical reaction. We also realize that the work is given in terms of the charged electrons flowing through an electrical potential $\mathscr{E}$ as

$$W = \mathscr{E} n_e N_0 e$$

in which n_e is the number of kilomoles of electrons flowing through the external circuit and

$$N_0 e = 6.022\,136 \times 10^{26} \text{ elec/kmol} \times 1.602\,177 \times 10^{-22} \text{ kJ/elec V}$$
$$= 96\,485 \text{ kJ/kmol} \cdot \text{V}$$

Thus, for a given reaction, the maximum (reversible reaction) electrical potential $\mathscr{E}^0$ of a fuel cell at a given temperature is

$$\mathscr{E}^0 = \frac{-\Delta G}{96\,485 n_e} \tag{13.28}$$

Example 13.15

Calculate the reversible electromotive force (EMF) at 25 °C for the hydrogen–oxygen fuel cell shown in Fig. 13.6.

Solution

The anode side reaction was stated to be

$$2\,H_2 \rightarrow 4\,H^+ + 4\,e^-$$

and the cathode side reaction is

$$4\,H^+ + 4\,e^- + O_2 \rightarrow 2\,H_2O$$

Therefore, the overall reaction is, in kilomoles,

$$2\,H_2 + O_2 \rightarrow 2\,H_2O$$

for which 4 kmol of electrons flow through the external circuit. Let us assume that each component is at its standard-state pressure of 0.1 MPa and that the water formed is liquid. Then

$$\Delta H^0 = 2\overline{h}^0_{f_{H_2O_{(l)}}} - 2\overline{h}^0_{f_{H_2}} - \overline{h}^0_{f_{O_2}}$$

$$= 2(-285\,830) - 2(0) - 1(0) = -571\,660\,\text{kJ}$$

$$\Delta S^0 = 2\overline{s}^0_{H_2O_{(l)}} - 2\overline{s}^0_{H_2} - \overline{s}^0_{O_2}$$

$$= 2(69.950) - 2(130.678) - 1(205.148) = -326.604\,\text{kJ/K}$$

$$\Delta G^0 = -571\,660 - 298.15(-326.604) = -474\,283\,\text{kJ}$$

Therefore, from Eq. 13.28,

$$\mathscr{E}^0 = \frac{-(-474\,283)}{96\,485 \times 4} = 1.229\,\text{V}$$

In Example 13.15, we found the shift in the Gibbs function and the reversible EMF at 25 °C. In practice, however, many fuel cells operate at an elevated temperature where the water leaves as a gas and not as a liquid; thus, it carries away more energy. The computations can be done for a range of temperatures, leading to lower EMF as the temperature increases. This behavior is shown in Fig. 13.7.

A variety of fuel cells are being investigated for use in stationary as well as mobile power plants. The low-temperature fuel cells use hydrogen as the fuel, whereas the higher-temperature cells can use methane and carbon monoxide that are then internally reformed into hydrogen and carbon dioxide. The most important fuel cells are listed in Table 13.4 with their main characteristics.

The low-temperature fuel cells are very sensitive to being poisoned by carbon monoxide gas, so they require an external reformer and purifier to deliver hydrogen gas. The higher-temperature fuel cells can reform natural gas, mainly methane, but also ethane

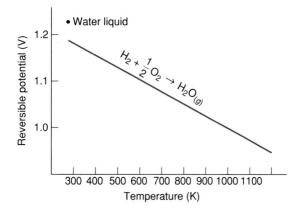

FIGURE 13.7
Hydrogen–oxygen fuel cell ideal EMF as a function of temperature.

TABLE 13.4
Fuel Cell Types

FUEL CELL	PEC	PAC	MCC	SOC
	Polymer Electrolyte	Phosphoric Acid	Molten Carbonate	Solid Oxide
T	80 °C	200 °C	650 °C	900 °C
Fuel	Hydrogen, H_2	Hydrogen, H_2	CO, hydrogen	Natural gas
Carrier	H^+	H^+	CO_3^{--}	O^{--}
Charge, n_e	$2e^-$ per H_2	$2e^-$ per H_2	$2e^-$ per H_2 $2e^-$ per CO	$8e^-$ per CH_4
Catalyst	Pt	Pt	Ni	ZrO_2
Poison	CO	CO		

and propane, as shown in Table 13.2, into hydrogen gas and carbon monoxide inside the cell. The latest research is being done with gasified coal as a fuel and operating the cell at higher pressures like 15 atm. As the fuel cell has exhaust gas with a small amount of fuel in it, additional combustion can occur and then combine the fuel cell with a gas turbine or steam power plant to utilize the exhaust gas energy. These combined-cycle power plants strive to have an efficiency of up to 60 %.

A model can be developed for the various processes that occur in a fuel cell to predict the performance. From the thermodynamic analysis, we found the theoretical voltage created by the process as the EMF from the Gibbs function. At both electrodes, there are activation losses that lower the voltage and a leak current, i_{leak}, that does not go through the cell. The electrolyte or membrane of the cell has an ohmic resistance, ASR_{ohmic}, to the ion transfer and thus also produces a loss. Finally, at high currents, there is a significant cell concentration loss that depletes one electrode for reactants and at the other electrode generates a high concentration of products, both of which increase the loss of voltage across the electrodes. The output voltage, V, generated by a fuel cell becomes

$$V = \text{EMF} - b \ln\left(\frac{i + i_{leak}}{i_0}\right) - i\,ASR_{ohmic} - c \ln\frac{i_L}{i_L - (i + i_{leak})} \qquad (13.29)$$

where i is current density [amp/cm^2], ASR_{ohmic} is the resistance [ohm cm^2] and b and c are cell constants [volts], the current densities i_0 is a reference, and i_L is the limit.

Two examples of this equation are shown in Fig. 13.8, where for the PEC (Polymer Electrolyte Cell) activation losses are high due to the low temperature and ohmic losses tend to be low. Just the opposite is the case for the high-temperature SOC (Solid Oxide Cell).

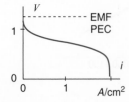
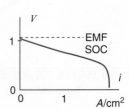

FIGURE 13.8 Simple model result from Eq. 13.29 for a low-temperature PEC and a high-temperature SOC.

As the current density increases toward the limit, the voltage drops sharply in both cases, and if the power per unit area (Vi) were shown, it would have a maximum in the middle range of current density.

This result resembles that of a heat engine with heat exchangers of a given size. As the power output is increased, the higher heat transfer requires a larger temperature difference, (recall Eqs. 5.14–5.16), which, in turn, lowers the temperature difference across the heat engine and causes it to operate with lower efficiency.

In-Text Concept Questions

i. Is the irreversibility in a combustion process significant? Explain your answer.

j. If the air–fuel ratio is larger than stoichiometric, is it more or less reversible?

k. What makes the fuel cell attractive from a power-generating point of view?

13.10 ENGINEERING APPLICATIONS

Combustion is applied in many cases where energy is needed in the form of heat or work. We use a natural gas stove, a water heater or furnace, or a propane burner for soldering, or the picnic grill, to mention a few domestic appliances with combustion that utilizes the heat. Lawn mowers, snow blowers, backup power generators, cars, and motor boats are all domestic applications where the work term is the primary output driven by a combustion process using gasoline or diesel oil as the fuel. On a larger scale, newer power plants use natural gas (methane) in gas turbines, and older plants use oil or coal as the primary fuel in the boiler-steam generator. Jet engines and rockets use combustion to generate high-speed flows for the motion of the airplane or rocket.

Most of the heat engines described in Chapter 5, and with simple models as cycles in Chapters 9 and 10, have the high-temperature heat transfer generated from a combustion process. It is thus not a heat transfer but an energy conversion process changing from the reactants to the much higher-temperature products of combustion. For the Rankine and Stirling cycles, the combustion is external to the cycle, whereas in the internal-combustion engines, as in the gasoline and diesel engines, combustion takes place in the working substance of the cycle.

In *external* combustion, the products deliver energy to the cycle by heat transfer, which cools the products, so it is never a constant-temperature source of energy. The combustion takes place in a steady flow arrangement with careful monitoring of the air–fuel mixture, including safety and pollution control aspects. In *internal* combustion, the Brayton cycle, as the model of a gas turbine, is a steady flow arrangement, and the gasoline/diesel engines are piston/cylinder engines with intermittent combustion. The latter process is somewhat difficult to control, as it involves a transient process.

A number of different parameters can be defined for evaluating the performance of an actual combustion process, depending on the nature of the process and the system considered. In the combustion chamber of a gas turbine, for example, the objective is to raise the temperature of the products to a given temperature (usually the maximum temperature the metals in the turbine can withstand). If we had a combustion process that achieved complete combustion and that was adiabatic, the temperature of the products would be the adiabatic flame temperature. Let us designate the fuel–air ratio needed to reach a given temperature under these conditions as the *ideal fuel–air ratio*. In the actual combustion chamber, the combustion will be incomplete to some extent, and there will be some heat transfer to the surroundings. Therefore, more fuel will be required to reach the given temperature, and this we designate as the *actual fuel–air ratio*. The combustion efficiency, η_{comb}, is defined here as

$$\eta_{comb} = \frac{FA_{ideal}}{FA_{actual}} \tag{13.30}$$

On the other hand, in the furnace of a steam generator (boiler), the purpose is to transfer the maximum possible amount of heat to the steam (water). In practice, the efficiency of a steam generator is defined as the ratio of the heat transferred to the steam to the higher heating value of the fuel. For a coal, this is the heating value as measured in a bomb calorimeter, which is the constant-volume heating value, and it corresponds to the internal energy of combustion. We observe a minor inconsistency, since the boiler involves a flow process, and the change in enthalpy is the significant factor. In most cases, however, the error thus introduced is less than the experimental error involved in measuring the heating value, and the efficiency of a steam generator is defined by the relation

$$\eta_{steam\ generator} = \frac{\text{heat transferred to steam per kilogram of fuel}}{\text{higher heating value of the fuel}} \tag{13.31}$$

Often the combustion of a fuel uses atmospheric air as the oxidizer, in which case the reactants also hold some water vapor. Assuming we know the humidity ratio for the moist air, ω, we would like to know the composition of air per mole of oxygen as

$$1\,O_2 + 3.76\,N_2 + x\,H_2O$$

Since the humidity ratio is $\omega = m_v/m_a$, the number of moles of water is

$$n_v = \frac{m_v}{M_v} = \frac{\omega m_a}{M_v} = \omega n_a \frac{M_a}{M_v}$$

and the number of moles of dry air per mole of oxygen is $(1 + 3.76)/1$, so we get

$$x = \frac{n_v}{n_{oxygen}} = \omega\,4.76\frac{M_a}{M_v} = 7.655\omega \tag{13.32}$$

This amount of water is found in the products together with the water produced by the oxidation of the hydrogen in the fuel.

In an internal-combustion engine, the purpose is to do work. The logical way to evaluate the performance of an internal-combustion engine would be to compare the actual work done to the maximum work that would be done by a reversible change of state from the reactants to the products. This, as we noted previously, is called the *second-law efficiency*.

In practice, however, the efficiency of an internal-combustion engine is defined as the ratio of the actual work to the negative of the enthalpy of combustion of the fuel (that is, the constant-pressure heating value). This ratio is usually called the *thermal efficiency*, η_{th}:

$$\eta_{th} = \frac{w}{-h_{RP}^0} = \frac{w}{HV} \tag{13.33}$$

When Eq. 13.33 is applied, the same scaling for the work and heating value must be used. So, if the heating value is per kilogram (kilomole) of fuel, then the work is per kilogram (kilomole) of fuel. For the work and heat transfer in the cycle analysis, we used the specific values as per kilogram of working substance, where for constant pressure combustion we have $h_P = h_R + q_H$. Since the heating value is per kilogram of fuel and q_H is per kilogram of mixture, we have

$$m_{tot} = m_{fuel} + m_{air} = m_{fuel}(1 + AF_{mass})$$

and thus

$$q_H = \frac{HV}{AF_{mass} + 1} \tag{13.34}$$

where a scaling of the HV and AF on a mass basis must be used.

The overall efficiency of a gas turbine or steam power plant is defined in the same way. It should be pointed out that in an internal-combustion engine or fuel-burning steam power plant, the fact that the combustion is itself irreversible is a significant factor in the relatively low thermal efficiency of these devices.

One other factor should be pointed out regarding efficiency. We have noted that the enthalpy of combustion of a hydrocarbon fuel varies considerably with the phase of the water in the products, which leads to the concept of higher and lower heating values. Therefore, when we consider the thermal efficiency of an engine, the heating value used to determine this efficiency must be borne in mind. Two engines made by different manufacturers may have identical performance, but if one manufacturer bases the engine's efficiency on the higher heating value and the other on the lower heating value, the latter will be able to claim a higher thermal efficiency. This claim is not significant, of course, as the performance is the same; this would be revealed by consideration of how the efficiency was defined.

The whole matter of the efficiencies of devices that undergo combustion processes is treated in detail in textbooks dealing with particular applications; our discussion is intended only as an introduction to the subject. Two examples are given, however, to illustrate these remarks.

Example 13.16

The combustion chamber of a gas turbine uses a liquid hydrocarbon fuel that has an approximate composition of C_8H_{18}. During testing, the following data are obtained:

$$T_{air} = 400 \text{ K} \qquad T_{products} = 1100 \text{ K}$$

$$\mathbf{V}_{air} = 100 \text{ m/s} \qquad \mathbf{V}_{products} = 150 \text{ m/s}$$

$$T_{fuel} = 50\,°C \qquad FA_{actual} = 0.0211 \text{ (kilograms of fuel per kilogram of air)}$$

Calculate the combustion efficiency for this process.

Control volume: Combustion chamber.
Inlet states: T known for air and fuel.
Exit state: T known.
Model: Air and products—ideal gas, Table A.9. Fuel—Table A.10.

Analysis

For the ideal chemical reaction, the heat transfer is zero. Therefore, writing the first law for a control volume that includes the combustion chamber, we have

$$H_R + KE_R = H_P + KE_P$$

$$H_R + KE_R = \sum_R n_i \left(\overline{h}_f^0 + \Delta \overline{h} + \frac{M\mathbf{V}^2}{2} \right)_i$$

$$= [\overline{h}_f^0 + \overline{C}_p(50 - 25)]_{C_8H_{18}(l)} + n_{O_2} \left(\Delta \overline{h} + \frac{M\mathbf{V}^2}{2} \right)_{O_2}$$

$$+ 3.76 n_{O_2} \left(\Delta \overline{h} + \frac{M\mathbf{V}^2}{2} \right)_{N_2}$$

$$H_P + KE_P = \sum_P n_e \left(\overline{h}_f^0 + \Delta \overline{h} + \frac{M\mathbf{V}^2}{2} \right)_e$$

$$= 8 \left(\overline{h}_f^0 + \Delta \overline{h} \frac{M\mathbf{V}^2}{2} \right)_{CO_2} + 9 \left(\overline{h}_f^0 + \Delta \overline{h} + \frac{M\mathbf{V}^2}{2} \right)_{H_2O}$$

$$+ (n_{O_2} - 12.5) \left(\Delta \overline{h} + \frac{M\mathbf{V}^2}{2} \right)_{O_2} + 3.76 n_{O_2} \left(\Delta \overline{h} + \frac{M\mathbf{V}^2}{2} \right)_{N_2}$$

Solution

$$H_R + KE_R = -250\,105 + 2.23 \times 114.23(50 - 25)$$

$$+ n_{O_2} \left[3034 + \frac{32 \times (100)^2}{2 \times 1000} \right]$$

$$+ 3.76 n_{O_2} \left[2971 + \frac{28.02 \times (100)^2}{2 \times 1000} \right]$$

$$= -243\,737 + 14\,892 n_{O_2}$$

$$H_P + KE_P = 8 \left[-393\,522 + 38\,891 + \frac{44.01 \times (150)^2}{2 \times 1000} \right]$$

$$+ 9 \left[-241\,826 + 30\,147 + \frac{18.02 \times (150)^2}{2 \times 1000} \right]$$

$$+ (n_{O_2} - 12.5) \left[26\,218 + \frac{32 \times (150)^2}{2 \times 1000} \right]$$

$$+ 3.7 n_{O_2} \left[24\,758 + \frac{28.02 \times (150)^2}{2 \times 1000} \right]$$

$$= -5\,068\,599 + 120\,853 n_{O_2}$$

Therefore,

$$-243\,737 + 14\,892 n_{O_2} = -5\,068\,599 + 120\,853 n_{O_2}$$

$$n_{O_2} = 45.53 \text{ kmol O}_2 \text{ per kilomole of fuel}$$

$$\text{kilomoles of air per kilomole of fuel} = 4.76(45.53) = 216.72$$

$$FA_{\text{ideal}} = \frac{114.23}{216.72 \times 28.97} = 0.0182 \text{ (kilogram of fuel per kilogram of air)}$$

$$\eta_{\text{comb}} = \frac{0.0182}{0.0211} \times 100 = 86.2\,\%$$

Example 13.17

In a certain steam power plant, 325 000 kg of water per hour enters the boiler at a pressure of 10 MPa and a temperature of 200 °C. Steam leaves the boiler at 8 MPa, 500 °C. The power output of the turbine is 81 000 kW. Coal is used at the rate of 26 700 kg/h and has a higher heating value of 33 250 kJ/kg. Determine the efficiency of the steam generator and the overall thermal efficiency of the plant.

In power plants, the efficiency of the boiler and the overall efficiency of the plant are based on the higher heating value of the fuel.

Solution

The efficiency of the boiler is defined by Eq. 13.31 as

$$\eta_{\text{steam generator}} = \frac{\text{heat transferred to H}_2\text{O per kilogram of fuel}}{\text{higher heating value}}$$

Therefore,

$$\eta_{\text{steam generator}} = \frac{325\,000(3398.3 - 856.0)}{26\,700 \times 33\,250} \times 100 = 93.1\,\%$$

The thermal efficiency is defined by Eq. 13.33,

$$\eta_{\text{th}} = \frac{w}{HV} = \frac{81\,000 \times 3600}{26\,700 \times 33\,250} \times 100 = 32.8\,\%$$

SUMMARY

An introduction to combustion of hydrocarbon fuels and chemical reactions in general, is given. Simple oxidation of a hydrocarbon fuel with pure oxygen or air burns the hydrogen to water and the carbon to carbon dioxide. We apply the continuity equation for each kind of atom to balance the stoichiometric coefficients of the species in the reactants and the products. The reactant mixture composition is described by the air–fuel ratio on a mass or mole basis or by the percent theoretical air or equivalence ratio according to the practice of the particular area of use. The products of a given fuel for a stoichiometric mixture and complete combustion are unique, whereas actual combustion can lead to incomplete combustion and more complex products described by measurements on a dry or wet basis. As water is part of the products, they have a dew point, so it is possible to see water condensing out from the products as they are cooled.

Due to the chemical changes from the reactants to the products, we need to measure energy from an absolute reference. Chemically pure substances (not compounds like carbon monoxide) in their ground state (graphite for carbon, not diamond form) are assigned a value of 0 for the formation enthalpy at the reference temperature and pressure (25 °C, 100 kPa). Stable compounds have a negative formation enthalpy and unstable compounds have a positive formation enthalpy. The shift in the enthalpy from the reactants to the products is the enthalpy of combustion, which is also the negative of the heating value HV. When a combustion process takes place without any heat transfer, the resulting product temperature is the adiabatic flame temperature. The enthalpy of combustion, the heating value (lower or higher), and the adiabatic flame temperature depend on the mixture (fuel and air–fuel ratio), and the reactants supply temperature. When a single unique number for these properties is used, it is understood to be for a stoichiometric mixture at the reference conditions.

Similarly to enthalpy, an absolute value of entropy is needed for the application of the second law. The absolute entropy is zero for a perfect crystal at 0 K, which is the third law of thermodynamics. The combustion process is an irreversible process; thus, a loss of exergy is associated with it. This irreversibility is increased by mixtures different from stoichiometric mixtures and by dilution of the oxygen (i.e., nitrogen in air), which lowers the adiabatic flame temperature. From the concept of flow exergy, we apply the second law to find the reversible work given by the change in Gibbs function. A process that has less irreversibility than combustion at high temperature is the chemical conversion in a fuel cell, where we approach a chemical equilibrium process (covered in detail in the following chapter). Here the energy release is directly converted into electrical power output, a system under intense study and development for future energy conversion systems.

You should have learned a number of skills and acquired abilities from studying this chapter that will allow you to:

- Write the combustion equation for the stoichiometric reaction of any fuel.
- Balance the stoichiometric coefficients for a reaction with a set of products measured on a dry basis.
- Handle the combustion of fuel mixtures as well as moist air oxidizers.
- Apply the energy equation with absolute values of enthalpy or internal energy.
- Use the proper tables for high-temperature products.
- Deal with condensation of water in low-temperature products of combustion.
- Calculate the adiabatic flame temperature for a given set of reactants.
- Know the difference between enthalpy of formation and enthalpy of combustion.
- Know the definition of the higher and lower heating values.

- Apply the second law to a combustion problem and find irreversibilities.
- Calculate the change in Gibbs function and the reversible work.
- Know how a fuel cell operates and how to find its electrical potential.
- Know some basic definition of combustion efficiencies.

KEY CONCEPTS AND FORMULAS		
	Reaction	fuel + oxidizer $\Rightarrow$ products
		hydrocarbon + air $\Rightarrow$ carbon dioxide + water + nitrogen
	Stoichiometric ratio	No excess fuel, no excess oxygen
	Stoichiometric coefficients	Factors to balance atoms between reactants and products
	Stoichiometric reaction	$C_x H_y + v_{O_2}(O_2 + 3.76 N_2)$

$$\Rightarrow v_{CO_2}CO_2 + v_{H_2O} H_2O + v_{N_2}N_2$$
$$v_{O_2} = x + y/4; \quad v_{CO_2} = x; \quad v_{H_2O} = y/2; \quad v_{N_2} = 3.76v_{O_2}$$

Air–fuel ratio
$$AF_{mass} = \frac{m_{air}}{m_{fuel}} = AF_{mole}\frac{M_{air}}{M_{fuel}}$$

Equivalence ratio
$$\Phi = \frac{FA}{FA_s} = \frac{AF_s}{AF}$$

Enthalpy of formation $\overline{h}_f^0$, zero for chemically pure substance, ground state

Enthalpy of combustion $h_{RP} = H_P - H_R$

Heating value HV $\quad HV = -h_{RP}$

Int. energy of combustion $u_{RP} = U_P - U_R = h_{RP} - RT(n_P - n_R)$ if ideal gases

Adiabatic flame temperature $H_P = H_R$ if flow; $U_P = U_R$ if constant volume

Reversible work $W^{rev} = G_R - G_P = -\Delta G = -(\Delta H - T\Delta S)$
This requires that any Q is transferred at the local T

Gibbs function $G = H - TS$

Irreversibility $i = w^{rev} - w = T_0\dot{S}_{gen}/\dot{m} = T_0 s_{gen}$
$\overline{I} = \overline{W}^{rev} - \overline{W} = T_0\dot{S}_{gen}/\dot{n} = T_0\overline{S}_{gen}$ for 1 kmol fuel

CONCEPT-STUDY GUIDE PROBLEMS

13.1 Is mass conserved in combustion? Is the number of moles constant?

13.2 Does all combustion take place with air?

13.3 Why would I sometimes need an air–fuel ratio on a mole basis? On a mass basis?

13.4 Why is there no significant difference between the number of moles of reactants and the number of products in combustion of hydrocarbon fuels with air?

13.5 Why are products measured on a dry basis?

13.6 What is the dew point of hydrogen burned with stoichiometric pure oxygen? With air?

13.7 How does the dew point change as the equivalence ratio goes from 0.9 to 1 to 1.1?

13.8 Why does combustion contribute to global warming?

13.9 What is the enthalpy of formation for oxygen as O_2? As O? For carbon dioxide?

13.10 If the nitrogen content of air can be lowered, will the adiabatic flame temperature increase or decrease?

13.11 Does the enthalpy of combustion depend on the air–fuel ratio?

13.12 Why do some fuels not have entries for liquid fuel in Table 13.3?

13.13 Is a heating value a fixed number for a fuel?

13.14 Is an adiabatic flame temperature a fixed number for a fuel?

13.15 Does it make a difference for the enthalpy of combustion whether I burn with pure oxygen or air? What about the adiabatic flame temperature?

13.16 A welder uses a bottle with acetylene and a bottle with oxygen. Why should he use the oxygen bottle instead of air?

13.17 Some gas welding is done using bottles of fuel, oxygen, and argon. Why do you think argon is used?

13.18 Is combustion a reversible process?

13.19 Is combustion with more than 100 % theoretical air more or less reversible?

HOMEWORK PROBLEMS

Fuels and the Combustion Process

13.20 In a picnic grill, gaseous propane is fed to a burner together with stoichiometric air. Find the air–fuel ratio on a mass basis and the total reactant mass for 1 kg of propane burned.

13.21 A mixture of fuels is E85, which is 85 % ethanol and 15 % gasoline (assume octane) by mass. Find the A/F ratio on a mass basis for stoichiometric combustion.

13.22 Calculate the theoretical air–fuel ratio on a mass and mole basis for the combustion of ethanol, C_2H_5OH.

13.23 Methane is burned with 125 % theoretical air. Find the composition and the dew point of the products.

13.24 Natural gas B from Table 13.2 is burned with 20 % excess air. Determine the composition of the products.

13.25 A certain fuel oil has the composition $C_{10}H_{22}$. If this fuel is burned with 150 % theoretical air, what is the composition of the products of combustion?

13.26 For complete stoichiometric combustion of gasoline, C_7H_{17}, determine the fuel molecular weight, the combustion products, and the mass of carbon dioxide produced per kilogram of fuel burned.

13.27 A sample of pine bark has the following ultimate analysis on a dry basis, percent by mass: 5.6 % H, 53.4 % C, 0.1 % S, 0.1 % N, 37.9 % O, and 2.9 % ash. This bark will be used as a fuel by burning it with 100 % theoretical air in a furnace. Determine the air–fuel ratio on a mass basis.

13.28 Liquid propane is burned with dry air. A volumetric analysis of the products of combustion yields the following volume percent composition on a dry basis: 8.6 % CO_2, 0.6 % CO, 7.2 % O_2, and 83.6 % N_2. Determine the percentage of theoretical air used in this combustion process.

13.29 The coal gasifier in an integrated gasification combined cycle (IGCC) power plant produces a gas mixture with the composition:

Product	CH_4	H_2	CO	CO_2	N_2	H_2O	H_2S	NH_3
% vol.	0.3	29.6	41.0	10.0	0.8	17.0	1.1	0.2

This gas is cooled to 40 °C, 3 MPa, and the H_2S and NH_3 are removed in water scrubbers. Assume the resulting mixture is saturated with water and sent to the combustors, determine its mixture composition and the theoretical air–fuel ratio.

13.30 In a combustion process with decane, $C_{10}H_{22}$, and air, the dry product mole fractions are 83.61 % N_2, 4.91 % O_2, 10.56 % CO_2, and 0.92 % CO. Find the equivalence ratio and the percentage of theoretical air of the reactants.

13.31 The output gas mixture of a certain air-blown coal gasifier has the composition of producer gas as listed in Table 13.2. Consider the combustion of this gas with 120 % theoretical air at 100 kPa pressure. Determine the dew point of the products and find how many kilograms of water will be condensed per kilogram of fuel if the products are cooled 10 °C below the dew-point temperature.

13.32 Methanol, CH_3OH, is burned with 125 % theoretical air in an engine, and the products are brought to 100 kPa, 30 °C. How much water is condensed per kilogram of fuel?

13.33 The hot exhaust gas from an internal-combustion engine is analyzed and found to have the following percent composition on a volumetric basis at the engine exhaust manifold: 10 % CO_2, 2 % CO, 13 % H_2O, 3 % O_2, and 72 % N_2. This gas is fed to an exhaust gas reactor and mixed with a certain amount of air to eliminate the CO, as shown in Fig. P13.33. It has been determined that a mole fraction of 10 % O_2 in the mixture at state 3 will ensure that no CO remains. What must be the ratio of flows entering the reactor?

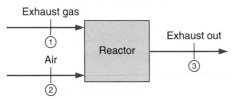

FIGURE P13.33

Energy Equation, Enthalpy of Formation

13.34 Acetylene gas, C_2H_2, is burned with stoichiometric air in a torch. The reactants are supplied at the reference conditions P_0, T_0. The products come out from the flame at 2500 K after some heat loss by radiation. Find the heat loss per kilomole of fuel.

13.35 Natural gas (methane) is burned with stoichiometric air, reactants supplied at the reference conditions P_0, T_0 in a steady-flow burner. The products come out at 800 K. If the burner should deliver 10 kW, what is the needed flow rate of natural gas in kg/s?

13.36 Butane gas and 125 % theoretical air, both at 25 °C, enter a steady-flow combustor. The products of combustion exit at 1000 K. Calculate the heat transfer from the combustor per kilomole of butane burned.

13.37 One alternative to using petroleum or natural gas as fuels is ethanol (C_2H_5OH), which is commonly produced from grain by fermentation. Consider a combustion process in which liquid ethanol is burned with 110 % theoretical air in a steady-flow process. The reactants enter the combustion chamber at 25 °C, and the products exit at 60 °C, 100 kPa. Find the heat transfer per kilomole of fuel.

13.38 Do the previous problem with the ethanol fuel delivered as a vapor.

13.39 Liquid methanol is burned with stoichiometric air, both supplied at P_0, T_0 in a constant-pressure process, and the products exit a heat exchanger at 800 K. Find the heat transfer per kilomole of fuel.

13.40 Pentene, C_5H_{10}, is burned with pure O_2 in a steady-flow process. The products at one point are brought to 700 K and used in a heat exchanger, where they are cooled to 35 °C. Find the specific heat transfer in the heat exchanger.

13.41 In a new high-efficiency furnace, natural gas, assumed to be 90 % methane and 10 % ethane (by volume), and 110 % theoretical air each enter at 25 °C, 100 kPa, and the products (assumed to be 100 % gaseous) exit the furnace at 40 °C, 100 kPa. What is the heat transfer for this process? Compare this to the performance of an older furnace where the products exit at 250 °C, 100 kPa.

13.42 Repeat the previous problem but take into account the actual phase behavior of the products exiting the furnace.

13.43 Methane, CH_4, is burned in a steady-flow adiabatic process with two different oxidizers: case A: pure O_2 and case B: a mixture of $O_2 + x$Ar. The reactants are supplied at T_0, P_0 and the products for both cases should be at 2000 K. Find the required equivalence ratio in case A and the amount of argon, x, for a stoichiometric ratio in case B.

13.44 A closed, insulated container is charged with a stoichiometric ratio of O_2 and H_2 at 25 °C and 150 kPa. After combustion, liquid water at 25 °C is sprayed in such that the final temperature is 1200 K. What is the final pressure?

13.45 In a gas turbine, natural gas (methane) and stoichiometric air flow into the combustion chamber at 1000 kPa, 500 K. Secondary air (see Fig. P13.45), also at 1000 kPa, 500 K, is added right after the combustion to result in a product mixture temperature of 1500 K. Find the air–fuel ratio mass basis for the primary reactant flow and the ratio of the secondary air to the primary air (mass flow rates ratio).

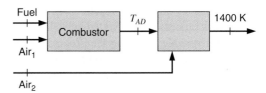

FIGURE P13.45

13.46 A rigid vessel initially contains 2 kmol of C and 2 kmol of O_2 at 25 °C, 200 kPa. Combustion occurs, and the resulting products consist of 1 kmol of CO_2, 1 kmol of CO, and excess O_2 at a temperature of 1000 K. Determine the final pressure in the vessel and the heat transfer from the vessel during the process.

Enthalpy of Combustion and Heating Value

13.47 Find the enthalpy of combustion and the heating value for pure carbon.

13.48 Phenol has an entry in Table 13.3, but it does not have a corresponding value of the enthalpy of formation in Table A.10. Can you calculate it?

13.49 Acetylene gas, C_2H_2, is burned with stoichiometric air in a torch. The reactants are supplied at the reference conditions P_0, T_0. The products come out from the flame at 2800 K after a heat loss by radiation. Find the lower heating value for the fuel, as it is not listed in Table 13.3, and the heat loss per kilomole of fuel.

13.50 Some type of wood can be characterized as $C_1H_{1.5}O_{0.7}$ with a lower heating value of 19 500 kJ/kg. Find its formation enthalpy.

13.51 Do Problem 13.37 using Table 13.3 instead of Table A.10 for the solution.

13.52 Agriculturally derived butanol, $C_4H_{10}O$, with a molecular mass of 74.12, also called *biobutanol*, has a lower heating value $LHV = 33\ 075$ kJ/kg for liquid fuel. Find its formation enthalpy.

13.53 Propylbenzene, C_9H_{12}, is listed in Table 13.3 but not in Table A.9. No molecular mass is listed in the book. Find the molecular mass, the enthalpy of formation for the liquid fuel, and the enthalpy of evaporation.

13.54 Liquid pentane is burned with dry air, and the products are measured on a dry basis as 10.1 % CO_2, 0.2 % CO, 5.9 % O_2, and remainder N_2. Find the enthalpy of formation for the fuel and the actual equivalence ratio.

13.55 Wet biomass waste from a food-processing plant is fed to a catalytic reactor, where in a steady-flow process it is converted into a low-energy fuel gas suitable for firing the processing plant boilers. The fuel gas has a composition of 50 % CH_4, 45 % CO_2, and 5 % H_2 on a volumetric basis. Determine the lower heating value of this fuel gas mixture per unit volume.

13.56 Determine the lower heating value of the gas generated from coal, as described in Problem 13.29. Do not include the components removed by the water scrubbers.

13.57 Do Problem 13.39 using Table 13.3 instead of Table A.10 for the solution.

13.58 E85 is a liquid mixture of 85 % ethanol and 15 % gasoline (assume octane) by mass. Find the lower heating value for this blend.

13.59 Assume the products of combustion in Problem 13.58 are sent out of the tailpipe and cool to ambient 20 °C. Find the fraction of the product water that will condense.

13.60 Gaseous propane and stoichiometric air are mixed and fed to a burner, both at P_0, T_0. After combustion, the products eventually cool down to T_0. How much heat was transferred for 1 kg propane?

13.61 In an experiment, a 1:1 mole ratio of propane and butane is burned in a steady flow with stoichiometric air. Both fuels and air are supplied as gases at 298 K and 100 kPa. The products are cooled to 1000 K as they give heat to some application. Find the lower heating value (per kilogram fuel mixture) and the total heat transfer for 1 kmol of fuel mixture used.

13.62 Blast furnace gas in a steel mill is available at 250 °C to be burned for the generation of steam. The composition of this gas is as follows on a volumetric basis:

Component	CH_4	H_2	CO	CO_2	N_2	H_2O
Percent by volume	0.1	2.4	23.3	14.4	56.4	3.4

Find the lower heating value (kJ/m^3) of this gas at 250 °C and ambient pressure.

13.63 Consider natural gas A in Table 13.2. Calculate the enthalpy of combustion at 25 °C, assuming that the products include vapor water. Repeat the answer for liquid water in the products.

13.64 Redo the previous problem for natural gas D in Table 13.3.

13.65 A burner receives a mixture of two fuels with mass fraction 60 % *n*-butane and 40 % methanol, both vapor. The fuel is burned with stoichiometric air. Find the product composition and the lower heating value of this fuel mixture (kJ/kg fuel mix).

13.66 Natural gas, we assume methane, is burned with 200 % theoretical air, shown in Fig. P13.66, and the reactants are supplied as gases at the reference temperature and pressure. The products are flowing through a heat exchanger, where they give off energy to some water flowing in at 20 °C, 500 kPa, and out at 700 °C, 500 kPa. The products exit at 400 K to the chimney. How much energy per kilomole fuel can the products deliver, and how many kilograms of water per kilogram of fuel can they heat?

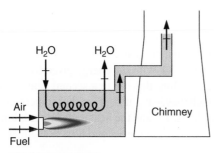

FIGURE P13.66

13.67 Liquid nitromethane is added to the air in a carburetor to make a stoichiometric mixture to which both fuel and air are added at 298 K, 100 kPa. After combustion, a constant-pressure heat exchanger brings the products to 600 K before being exhausted. Assume the nitrogen in the fuel becomes N_2 gas. Find the total heat transfer per kilomole fuel in the whole process.

13.68 Gasoline, C_7H_{17}, is burned in a steady-state burner with stoichiometric air at P_0, T_0, shown in Fig. P13.68. The gasoline is flowing as a liquid at T_0 to a carburetor, where it is mixed with air to produce a fuel air–gas mixture at T_0. The carburetor takes some heat transfer from the hot products to do the heating. After the combustion, the products go through a heat exchanger, which

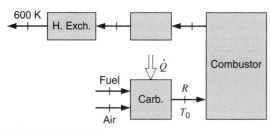

FIGURE P13.68

they leave at 600 K. The gasoline consumption is 10 kg/h. How much power is given out in the heat exchanger, and how much power does the carburetor need?

13.69 An isobaric combustion process receives gaseous benzene, C_6H_6, and air in a stoichiometric ratio at P_0, T_0. To limit the product temperature to 2000 K, liquid water is sprayed in after the combustion. Find the number of kilomole of liquid water added per kilomole of fuel and the dew point of the combined products.

13.70 A mixture of fuels is E85, which is 85 % ethanol and 15 % gasoline (assume octane) by mass. Assume we put the fuel and air, both at T_0, P_0, into a carburetor and vaporize the fuel as we mix it with stoichiometric air before it flows to an engine. Assume the engine has an efficiency as work divided by the lower heating value of 30 %, and we want it to deliver 40 kW. We use heat from the exhaust flow (500 K) for the carburetor. Find the lower heating value of this fuel (kJ/kg), the rate of fuel consumption, the heating rate needed in the carburetor, and the rate of entropy generation in the carburetor.

Adiabatic Flame Temperature

13.71 In a rocket, hydrogen is burned with air, both reactants supplied as gases at P_0, T_0. The combustion is adiabatic, and the mixture is stoichiometric (100 % theoretical air). Find the products' dew point and the adiabatic flame temperature (~2500 K).

13.72 Hydrogen gas is burned with pure O_2 in a steady-flow burner, shown in Fig. P13.72, where both reactants are supplied in a stoichiometric ratio at the reference pressure and temperature. What is the adiabatic flame temperature?

FIGURE P13.72

13.73 Some type of wood can be characterized as $C_1H_{1.5}O_{0.7}$ with a lower heating value of

19 500 kJ/kg. Find its adiabatic flame temperature when burned with stoichiometric air at 100 kPa, 298 K.

13.74 A gas turbine burns methane with 200 % theoretical air. The air and fuel come in through two separate compressors bringing them from 100 kPa, 298 K, to 1400 kPa, and after mixing they enter the combustion chamber at 600 K. Find the adiabatic flame temperature using constant specific heat for the ΔH_P terms.

13.75 Extend the solution to the previous problem by using Table A.9 for the ΔH_P terms.

13.76 Carbon is burned with air in a furnace with 200 % theoretical air, and both reactants are supplied at the reference pressure and temperature. What is the adiabatic flame temperature?

13.77 Acetylene gas at 25 °C, 100 kPa, is fed to the head of a cutting torch. Calculate the adiabatic flame temperature if the acetylene is burned with
  a. 150 % theoretical air at 25 °C.
  b. 150 % theoretical oxygen at 25 °C.

13.78 Butane gas at 25 °C is mixed with 150 % theoretical air at 600 K and is burned in an adiabatic steady-flow combustor. What is the temperature of the products exiting the combustor?

13.79 A stoichiometric mixture of benzene, C_6H_6, and air is mixed from the reactants flowing at 25 °C, 100 kPa. Find the adiabatic flame temperature. What is the error if constant-specific heat at T_0 for the products from Table A.5 is used?

13.80 What is the adiabatic flame temperature before the secondary air is added in Problem 13.45?

13.81 A special coal burner uses a stoichiometric mixture of coal and an oxygen–argon mixture (1:1 mole ratio), with the reactants supplied at the reference conditions P_0, T_0. Find the adiabatic flame temperature, assuming complete combustion.

13.82 A gas turbine burns natural gas (assume methane) where the air is supplied to the combustor at 1000 kPa, 500 K, and the fuel is at 298 K, 1000 kPa. What is the equivalence ratio and the percent theoretical air if the adiabatic flame temperature should be limited to 1800 K?

13.83 Natural gas, we assume methane, is burned with 200 % theoretical air, and the reactants are supplied as gases at the reference temperature and pressure. The products are flowing through a heat exchanger and then out the exhaust, as in Fig. P13.83. What is the adiabatic flame temperature right after combustion before the heat exchanger?

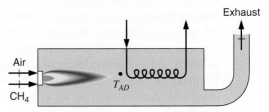

FIGURE P13.83

13.84 Solid carbon is burned with stoichiometric air in a steady-flow process. The reactants at T_0, P_0 are heated in a preheater to $T_2 = 500$ K, as shown in Fig. P13.84, with the energy given by the product gases before flowing to a second heat exchanger, which they leave at T_0. Find the temperature of the products, T_4, and the heat transfer per kilomole of fuel (4 to 5) in the second heat exchanger.

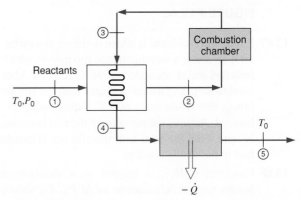

FIGURE P13.84

13.85 Gaseous ethanol, C_2H_5OH, is burned with pure oxygen in a constant-volume combustion bomb. The reactants are charged in a stoichiometric ratio at the reference condition. Assume no heat transfer and find the final temperature (>5000 K).

13.86 Liquid n-butane at T_0 is sprayed into a gas turbine, as in Fig. P13.45, with primary air flowing at 1.0 MPa, 400 K, in a stoichiometric ratio. After complete combustion, the products are at the adiabatic flame temperature, which is too high, so secondary air at 1.0 MPa, 400 K, is added, with the resulting mixture being at 1400 K. Show that $T_{AD} > 1400$ K and find the ratio of secondary to primary airflow.

13.87 The enthalpy of formation of magnesium oxide, MgO(s), is $-601\,827$ kJ/kmol at 25 °C. The melting point of magnesium oxide is approximately 3000 K, and the increase in enthalpy between 298 and 3000 K is 128 449 kJ/kmol. The enthalpy of sublimation at 3000 K is estimated at 418 000 kJ/kmol, and the specific heat of magnesium oxide vapor above 3000 K is estimated at 37.24 kJ/kmol·K.

   a. Determine the enthalpy of combustion per kilogram of magnesium.

   b. Estimate the adiabatic flame temperature when magnesium is burned with theoretical oxygen.

Second Law for the Combustion Process

13.88 Consider the combustion of hydrogen with pure O_2 in a stoichiometric ratio under steady-flow adiabatic conditions. The reactants enter separately at 298 K, 100 kPa, and the product(s) exit at a pressure of 100 kPa. What is the exit temperature, and what is the irreversibility?

13.89 Consider the combustion of methanol, CH_3OH, with 25 % excess air. The combustion products are passed through a heat exchanger and exit at 200 kPa, 400 K. Calculate the absolute entropy of the products exiting the heat exchanger, assuming all the water is vapor.

13.90 Consider the combustion of methanol, CH_3OH, with 25 % excess air. The combustion products are passed through a heat exchanger and exit at 200 kPa, 40 °C. Calculate the absolute entropy of the products exiting the heat exchanger per kilomole of methanol burned, using proper amounts of liquid and vapor water.

13.91 An inventor claims to have built a device that will take 0.001 kg/s of water from the faucet at 10 °C, 100 kPa, and produce separate streams of hydrogen and oxygen gas, each at 400 K, 175 kPa. It is stated that this device operates in a 25 °C room on 10 kW electrical power input. How do you evaluate this claim?

13.92 Propene, C_3H_6, is burned with air in a steady-flow burner with reactants at P_0, T_0. The mixture is lean, so the adiabatic flame temperature is 1800 K. Find the entropy generation per kilomole of fuel, neglecting all the partial-pressure corrections.

13.93 Hydrogen peroxide, H_2O_2, enters a gas generator at 25 °C, 500 kPa, at the rate of 0.1 kg/s and is decomposed to steam and oxygen exiting at 800 K, 500 kPa. The resulting mixture is expanded through a turbine to atmospheric pressure, 100 kPa, as shown in Fig. P13.93. Determine the power output of the turbine and the heat transfer rate in the gas generator. The enthalpy of formation of liquid H_2O_2 is $-187\,583$ kJ/kmol.

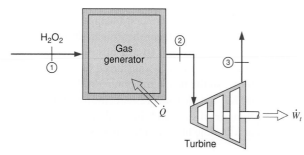

FIGURE P13.93

13.94 Graphite, C, at P_0, T_0 is burned with air coming in at P_0, 500 K, in a ratio so that the products exit at P_0, 1200 K. Find the equivalence ratio, the percent theoretical air, and the total irreversibility.

13.95 Calculate the irreversibility for the process described in Problem 13.46.

13.96 Two kilomoles of ammonia are burned in a steady-flow process with x kmol of oxygen. The products, consisting of H_2O, N_2, and the excess O_2, exit at 200 °C, 7 MPa.

   a. Calculate x if half of the H_2O in the products is condensed.

   b. Calculate the absolute entropy of the products at the exit conditions.

13.97 A flow of 0.02 kmol/s methane, CH_4, and 200 % theoretical air, both at reference conditions, are compressed separately to $P_3 = P_4 = 2$ MPa, then mixed and then burned in a steady-flow setup (like a gas turbine, see Fig. P13.97). After combustion, state 6, heat transfer goes out, so the exhaust, state 7, is at 600 K.

   a. Find T_3, T_4, and T_5.

   b. Find the total rate of irreversibility from inlet to state 5.

   c. Find the rate of heat transfer minus the work terms ($\dot{Q} - \dot{W}_1 - \dot{W}_2$).

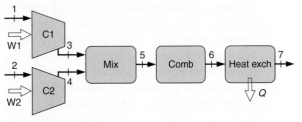

FIGURE P13.97

13.98 A closed, rigid container is charged with propene, C_3H_6, and 150 % theoretical air at 100 kPa, 298 K. The mixture is ignited and burns with complete combustion. Heat is transferred to a reservoir at 500 K, so the final temperature of the products is 700 K. Find the final pressure, the heat transfer per kilomole of fuel, and the total entropy generated per kilomole of fuel in the process.

Problems Involving Generalized Charts or Real Mixtures

13.99 Liquid butane at 25 °C is mixed with 150 % theoretical air at 600 K and is burned in an adiabatic steady-state combustor. Use the generalized charts for the liquid fuel and find the temperature of the products exiting the combustor.

13.100 A gas mixture of 50 % ethane and 50 % propane by volume enters a combustion chamber at 350 K, 10 MPa. Determine the enthalpy per kilomole of this mixture relative to the thermochemical base of enthalpy using Kay's rule.

13.101 A mixture of 80 % ethane and 20 % methane on a mole basis is throttled from 10 MPa, 65 °C, to 100 kPa and is fed to a combustion chamber, where it undergoes complete combustion with air, which enters at 100 kPa, 600 K. The amount of air is such that the products of combustion exit at 100 kPa, 1200 K. Assume that the combustion process is adiabatic and that all components behave as ideal gases except the fuel mixture, which behaves according to the generalized charts, with Kay's rule for the pseudocritical constants. Determine the percentage of theoretical air used in the process and the dew-point temperature of the products.

13.102 Saturated liquid butane enters an insulated constant-pressure combustion chamber at 25 °C, and x times theoretical oxygen gas enters at the

same P and T. The combustion products exit at 3400 K. With complete combustion, find x. What is the pressure at the chamber exit? What is the irreversibility of the process?

13.103 Liquid hexane enters a combustion chamber at 31 °C, 200 kPa, at the rate of 1 kmol/s; 200 % theoretical air enters separately at 500 K, 200 kPa. The combustion products exit at 1000 K, 200 kPa. The specific heat of ideal-gas hexane is $C_{p0} = 143$ kJ/kmol·K. Calculate the rate of irreversibility of the process.

Fuel Cells

13.104 In Example 13.15, a basic hydrogen–oxygen fuel cell reaction was analyzed at 25 °C, 100 kPa. Repeat this calculation, assuming that the fuel cell operates on air at 25 °C, 100 kPa, instead of on pure oxygen at this state.

13.105 Assume that the basic hydrogen–oxygen fuel cell operates at 600 K instead of 298 K, as in Example 13.15. Find the change in the Gibbs function and the reversible EMF it can generate.

13.106 A reversible fuel cell operating with hydrogen and pure oxygen produces water at the reference conditions P_0, T_0; this is described in Example 13.15. Find the work output and any heat transfer, both per kilomole of hydrogen. Assume an actual fuel cell operates with a second-law efficiency of 70 %, and enough heat transfer takes place to keep it at 25 °C. How much heat transfer is that per kilomole of hydrogen?

13.107 Consider a methane–oxygen fuel cell in which the reaction at the anode is

$$CH_4 + 2\,H_2O \rightarrow CO_2 + 8\,e^- + 8\,H^+$$

The electrons produced by the reaction flow through the external load, and the positive ions migrate through the electrolyte to the cathode, where the reaction is

$$8\,e^- + 8\,H^+ + 2\,O_2 \rightarrow 4\,H_2O$$

Calculate the reversible work and the reversible EMF for the fuel cell operating at 25 °C, 100 kPa.

13.108 Redo the previous problem, but assume that the fuel cell operates at 1200 K instead of at room temperature.

13.109 For a PEC operating at 350 K, the constants in Eq. 13.29 are $i_{leak} = 0.01$, $i_L = 2$, $i_0 = 0.013$

all A/cm^2, $b = 0.08$ V, $c = 0.1$ V, $ASR = 0.01$ Ω cm^2, and EMF $= 1.22$ V. Find the voltage and the power density for the current density $i = 0.25$ A/cm^2, 0.75 A/cm^2, and 1.0 A/cm^2.

13.110 Assume the PEC in the previous problem. How large an area does the fuel cell have to deliver 1 kW with a current density of 1 A/cm^2?

13.111 A solid oxide fuel cell at 900 K can be described by EMF $= 1.06$ V and the constants in Eq. 13.29 as $b = 0$ V, $c = 0.1$ V, $ASR = 0.04$ Ω cm^2, $i_{leak} = 0.01$, $i_L = 2$, $i_0 = 0.13$ all A/cm^2. Find the voltage and the power density for the current density $i = 0.25$ A/cm^2, 0.75 A/cm^2, and 1.0 A/cm^2.

13.112 Assume the SOC in the previous problem. How large an area does the fuel cell have to deliver 1 kW with a current density of 1 A/cm^2?

13.113 A PEC operating at 25 °C generates 1.0 V that also accounts for losses. For a total power of 1 kW, what is the hydrogen mass flow rate?

13.114 A basic hydrogen–oxygen fuel cell operates at 600 K instead of 298 K, as in Example 13.15. For a total power of 5 kW, find the hydrogen mass flow rate and the exergy in the exhaust flow.

13.115 Consider the fuel cell with methane in Problem 13.107. Find the work output and any heat transfer, both per kilomole of methane. Assume an actual fuel cell operates with a second-law efficiency of 75 %, and enough heat transfer takes place to keep it at 25 °C. How much heat transfer is that per kilomole of methane?

Combustion Applications and Efficiency

13.116 For the combustion of methane, 150 % theoretical air is used at 25 °C, 100 kPa, and relative humidity of 70 %. Find the composition and dew point of the products.

13.117 Pentane is burned with 120 % theoretical air in a constant-pressure process at 100 kPa. The products are cooled to ambient temperature, 20 °C. How much mass of water is condensed per kilogram of fuel? Repeat the answer, assuming that the air used in the combustion has a relative humidity of 90 %.

13.118 In an engine, a mixture of liquid octane and ethanol, mole ratio 9:1, and stoichiometric air are taken in at T_0, P_0. In the engine, the enthalpy

of combustion is used so that 30 % goes out as work, 30 % goes out as heat loss, and the rest goes out the exhaust. Find the work and heat transfer per kilogram of fuel mixture and also the exhaust temperature.

13.119 The gas-turbine cycle in Problem 10.26 has $q_H = 960$ kJ/kg air added by combustion. Assume the fuel is methane gas and q_H is from the heating value at T_0. Find the air–fuel ratio on a mass basis.

13.120 An oven heated by natural gas burners has the combustion take place inside a U-shaped steel pipe, so the heating is done from the outside surface of the pipe by radiation (Fig. P13.120). Each burner delivers 15 kW of radiation, burning 110 % theoretical air with methane. The products leave the pipe at 800 K. Find the flow (kg/s) of methane. The burner is now switched to oxygen-enriched air (30 % O_2 and 70 % N_2), so assume the same conditions as before with the same exit T. Find the new flow (kg/s) of methane needed.

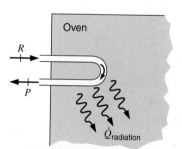

FIGURE P13.120

13.121 A slight disadvantage of the oxygen-enriched air for combustion is an increase in flame temperature, which tends to increase NO_x. Find the flame temperature for the previous problem for both cases: standard air and oxygen-enriched air.

13.122 A gas turbine burns methane with 200 % theoretical air. The air and fuel come in through two separate compressors bringing them from 100 kPa, 298 K, to 1400 kPa and enter a mixing chamber and a combustion chamber. What are the specific compressor work and q_H to be used in Brayton cycle calculation? Use constant specific heat to solve the problem.

13.123 Find the equivalent heat transfer q_H to be used in a cycle calculation for constant-pressure combustion when the fuel is (a) methane, or (b) gaseous octane. In both cases, use water vapor in the products and a stoichiometric mixture.

13.124 Consider the steady-state combustion of propane at $25\,°C$ with air at 400 K. The products exit the combustion chamber at 1200 K. Assume that the combustion efficiency is 90% and that 95% of the carbon in the propane burns to form CO_2; the remaining 5% forms CO. Determine the ideal fuel–air ratio and the heat transfer from the combustion chamber.

13.125 A gasoline engine is converted to run on propane, as shown in Fig. P13.125. Assume the propane enters the engine at $25\,°C$, at the rate of 40 kg/h. Only 90% theoretical air enters at $25\,°C$, so 90% of the C burns to form CO_2 and 10% of the C burns to form CO. The combustion products, also including H_2O, H_2, and N_2, exit the exhaust pipe at 1000 K. Heat loss from the engine (primarily to the cooling water) is 120 kW. What is the power output of the engine? What is the thermal efficiency?

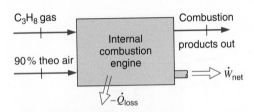

FIGURE P13.125

13.126 A gasoline engine uses liquid octane and air, both supplied at P_0, T_0, in a stoichiometric ratio. The products (complete combustion) flow out of the exhaust valve at 1100 K. Assume that the heat loss carried away by the cooling water, at $100\,°C$, is equal to the work output. Find the efficiency of the engine expressed as (work/lower heating value) and the second-law efficiency.

Review Problems

13.127 Many coals from the western United States have a high moisture content. Consider the following sample of Wyoming coal, for which the ultimate analysis on an as-received basis is, by mass:

Component	Moisture	H	C	S	N	O	Ash
% mass	28.9	3.5	48.6	0.5	0.7	12.0	5.8

This coal is burned in the steam generator of a large power plant with 150% theoretical air. Determine the air–fuel ratio on a mass basis.

13.128 A fuel, C_xH_y, is burned with dry air, and the product composition is measured on a dry mole basis to be 9.6% CO_2, 7.3% O_2, and 83.1% N_2. Find the fuel composition (x/y) and the percent theoretical air used.

13.129 Determine the higher heating value of the sample Wyoming coal, as specified in Problem 13.127.

13.130 Ethene, C_2H_4, and propane, C_3H_8, in a 1:1 mole ratio as gases are burned with 120% theoretical air in a gas turbine. Fuel is added at $25\,°C$, 1 MPa, and the air comes from the atmosphere, at $25\,°C$, 100 kPa, through a compressor to 1 MPa and is mixed with the fuel. The turbine work is such that the exit temperature is 800 K with an exit pressure of 100 kPa. Find the mixture temperature before combustion and the work, assuming an adiabatic turbine.

13.131 Consider the gas mixture fed to the combustors in the integrated gasification combined cycle power plant, as described in Problem 13.29. If the adiabatic flame temperature should be limited to 1500 K, what percent theoretical air should be used in the combustors?

13.132 Carbon monoxide, CO, is burned with 150% theoretical air, and both gases are supplied at 150 kPa and 600 K. Find the heating value and the adiabatic flame temperature.

13.133 A rigid container is charged with butene, C_4H_8, and air in a stoichiometric ratio at P_0, T_0. The charge burns in a short time, with no heat transfer to state 2. The products then cool to 1200 K, state 3. Find the final pressure, P_3, the total heat transfer, $_1Q_3$, and the temperature immediately after combustion, T_2.

13.134 Natural gas (approximate it as methane) at a rate of 0.3 kg/s is burned with 250% theoretical air in a combustor at 1 MPa, where the reactants are supplied at T_0. Steam at 1 MPa, $450\,°C$, at a rate of 2.5 kg/s is added to the products before they enter an adiabatic turbine with an exhaust pressure of 150 kPa. Determine the turbine inlet temperature and the turbine work, assuming the turbine is reversible.

13.135 The turbine in Problem 13.130 is adiabatic. Is it reversible, irreversible, or impossible?

13.136 Consider the combustion process described in Problem 13.101.

a. Calculate the absolute entropy of the fuel mixture before it is throttled into the combustion chamber.

b. Calculate the irreversibility for the overall process.

13.137 Liquid acetylene, C_2H_2, is stored in a high-pressure storage tank at ambient temperature, $25\,°C$. The liquid is fed to an insulated combustor/steam boiler at a steady rate of 1 kg/s, along with 140 % theoretical oxygen, O_2, which enters at 500 K, as shown in Fig. P13.137. The combustion products exit the unit at 500 kPa, 350 K. Liquid water enters the boiler at $10\,°C$, at the rate of 15 kg/s, and superheated steam exits at 200 kPa.

a. Calculate the absolute entropy per kilomole of liquid acetylene at the storage tank state.

b. Determine the phase(s) of the combustion products exiting the combustor boiler unit and the amount of each if more than one.

c. Determine the temperature of the steam at the boiler exit.

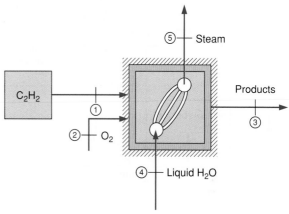

FIGURE P13.137

COMPUTER, DESIGN, AND OPEN-ENDED PROBLEMS

13.138 Write a program to study the effect of the percentage of theoretical air on the adiabatic flame temperature for a (variable) hydrocarbon fuel. Assume reactants enter the combustion chamber at $25\,°C$ and complete combustion. Use constant specific heat of the various products of combustion, and let the fuel composition and its enthalpy of formation be program inputs.

13.139 Power plants may use off-peak power to compress air into a large storage facility (see Problem 7.46). The compressed air is then used as the air supply to a gas-turbine system where it is burned with some fuel, usually natural gas. The system is then used to produce power at peak load times. Investigate such a setup and estimate the power generated with the conditions given in Problem 7.46 and combustion with 200 %–300 % theoretical air and exhaust to the atmosphere.

13.140 A car that runs on natural gas has it stored in a heavy tank with a maximum pressure of 25 MPa. Size the tank for a range of 500 km, assuming a car

engine that has a 30 % efficiency requiring about 20 kW to drive the car at 90 km/h.

13.141 The Cheng cycle, shown in Fig. P11.124, is powered by the combustion of natural gas (essentially methane) being burned with 250 %–300 % theoretical air. In the case with a single water-condensing heat exchanger, where $T_6 = 40\,°C$ and $\Phi_6 = 100\,\%$, is any makeup water needed at state 8 or is there a surplus? Does the humidity in the compressed atmospheric air at state 1 make any difference? Study the problem over a range of air–fuel ratios.

13.142 The cogenerating power plant shown in Problem 9.67 burns 170 kg/s air with natural gas, CH_4. The setup is shown in Fig. P13.142, where a fraction of the air flow out of the compressor with pressure ratio 15.8:1 is used to preheat the feedwater in the steam cycle. The fuel flow rate is 3.2 kg/s. Analyze the system, determining the total heat transfer to the steam cycle from the turbine exhaust gases, the heat transfer in the preheater, and the gas turbine inlet temperature.

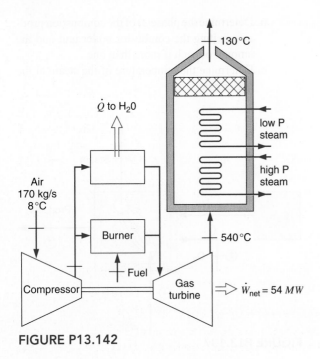

130 °C

$\dot{Q}$ to H$_2$O

low P steam

high P steam

Air
170 kg/s
8 °C

Burner

Fuel

540 °C

Compressor

Gas turbine

$\dot{W}_{net} = 54$ MW

FIGURE P13.142

13.143 Consider the combustor in the Cheng cycle (see Problems 11.124 and 13.134). Atmospheric air is compressed to 1.25 MPa, state 1. It is burned with natural gas, CH$_4$, with the products leaving at state 2. The fuel should add a total of about 15 MW to the cycle, with an air flow of 12 kg/s. For a compressor with an intercooler, estimate the temperatures T_1, T_2 and the fuel flow rate.

13.144 Study the coal gasification process that will produce methane, CH$_4$, or methanol, CH$_3$OH. What is involved in such a process? Compare the heating values of the gas products with those of the original coal. Discuss the merits of this conversion.

13.145 When a power plant burns coal or some blends of oil, the combustion process can generate pollutants as SO$_x$ and NO$_x$. Investigate the use of scrubbers to remove these products. Explain the processes that take place and the effect on the power plant operation (energy, exhaust pressures, etc.).

Phase and Chemical Equilibrium

14

Up to this point, we have assumed that we are dealing either with systems that are in equilibrium or with those in which the deviation from equilibrium is infinitesimal, as in a quasi-equilibrium or reversible process. For irreversible processes, we made no attempt to describe the state of the system during the process but dealt only with the initial and final states of the system, in the case of a control mass, or the inlet and exit states, as well, in the case of a control volume. For any case, we either considered the system to be in equilibrium throughout or at least made the assumption of local equilibrium.

In this chapter we examine the criteria for equilibrium and from them derive certain relations that will enable us, under certain conditions, to determine the properties of a system when it is in equilibrium. The specific case we will consider is that involving chemical equilibrium in a single phase (homogeneous equilibrium) as well as certain related topics.

 ## 14.1 REQUIREMENTS FOR EQUILIBRIUM

As a general requirement for equilibrium, we postulate that a system is in equilibrium when there is no possibility that it can do any work when it is isolated from its surroundings. In applying this criterion, it is helpful to divide the system into two or more subsystems and consider the possibility of doing work by any conceivable interaction between these subsystems. For example, in Fig. 14.1 a system has been divided into two systems and an engine, of any conceivable variety, placed between these subsystems. A system may be so defined as to include the immediate surroundings. In this case, we can let the immediate surroundings be a subsystem and thus consider the general case of the equilibrium between a system and its surroundings.

The first requirement for equilibrium is that the two subsystems have the same temperature; otherwise, we could operate a heat engine between the two systems and do work. Thus, we conclude that one requirement for equilibrium is that a system must be at a uniform temperature to be in equilibrium. It is also evident that there must be no unbalanced mechanical forces between the two systems, or else one could operate a turbine or piston engine between the two systems and do work.

We would like to establish general criteria for equilibrium that would apply to all simple compressible substances, including those that undergo chemical reactions. We will find that the Gibbs function is a particularly significant property in defining the criteria for equilibrium.

591

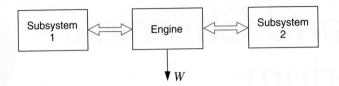

FIGURE 14.1
Two subsystems that communicate through an engine.

Let us first consider a qualitative example to illustrate this point. Consider a natural gas well that is 1 km deep, and let us assume that the temperature of the gas is constant throughout the well. Suppose we have analyzed the composition of the gas at the top of the well, and we would like to know the composition of the gas at the bottom of the well. Furthermore, let us assume that equilibrium conditions prevail in the well. If this is true, we would expect that an engine such as that shown in Fig. 14.2 (which operates on the basis of the pressure and composition change with elevation and does not involve combustion) would not be capable of doing any work.

If we consider a steady-state process for a control volume around this engine, the reversible work for the change of state from i to e is given by Eq. 8.14 on a total mass basis:

$$\dot{W}^{\text{rev}} = \dot{m}_i\left(h_i + \frac{\mathbf{V}_i^2}{2} + gZ_i - T_0 s_i\right) - \dot{m}_e\left(h_e + \frac{\mathbf{V}_e^2}{2} + gZ_e - T_0 s_e\right)$$

Furthermore, since $T_i = T_e = T_0 = $ constant, this reduces to the form of the Gibbs function $g = h - Ts$, Eq. 12.14, and the reversible work is

$$\dot{W}^{\text{rev}} = \dot{m}_i\left(g_i + \frac{\mathbf{V}_i^2}{2} + gZ_i\right) - \dot{m}_e\left(g_e + \frac{\mathbf{V}_e^2}{2} + gZ_e\right)$$

However,

$$\dot{W}^{\text{rev}} = 0, \qquad \dot{m}_i = \dot{m}_e \qquad \text{and} \qquad \frac{\mathbf{V}_i^2}{2} = \frac{\mathbf{V}_e^2}{2}$$

Then we can write

$$g_i + gZ_i = g_e + gZ_e$$

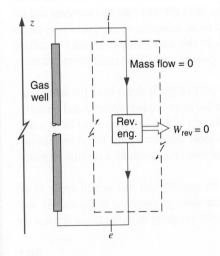

FIGURE 14.2
Illustration showing the relation between reversible work and the criteria for equilibrium.

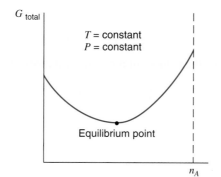

FIGURE 14.3
Illustration of the requirement for chemical equilibrium.

and the requirement for equilibrium in the well between two levels that are a distance dZ apart would be

$$dg_T + g\,dZ_T = 0$$

In contrast to a deep gas well, most of the systems that we consider are of such size that ΔZ is negligibly small, and therefore we consider the pressure to be uniform throughout.

This leads to the general statement of equilibrium that applies to simple compressible substances that may undergo a change in chemical composition, namely, that at equilibrium

$$dG_{T,P} = 0 \qquad (14.1)$$

In the case of a chemical reaction, it is helpful to think of the equilibrium state as the state in which the Gibbs function is a minimum. For example, consider a control mass consisting initially of n_A moles of substance A and n_B moles of substance B, which react in accordance with the relation

$$v_A A + v_B B \rightleftharpoons v_C C + v_D D$$

Let the reaction take place at constant pressure and temperature. If we plot G for this control mass as a function of n_A, the number of moles of A present, we would have a curve as shown in Fig. 14.3. At the minimum point on the curve, $dG_{T,P} = 0$, and this will be the equilibrium composition for this system at the given temperature and pressure. The subject of chemical equilibrium will be developed further in Section 14.4.

14.2 EQUILIBRIUM BETWEEN TWO PHASES OF A PURE SUBSTANCE

As another example of this requirement for equilibrium, let us consider the equilibrium between two phases of a pure substance. Consider a control mass consisting of two phases of a pure substance at equilibrium. We know that under these conditions, the two phases are at the same pressure and temperature. Consider the change of state associated with a transfer of dn moles from phase 1 to phase 2 while the temperature and pressure remain constant. That is,

$$dn^1 = -dn^2$$

The Gibbs function of this control mass is given by

$$G = f(T, P, n^1, n^2)$$

where n^1 and n^2 designate the number of moles in each phase. Therefore,

$$dG = \left(\frac{\partial G}{\partial T}\right)_{P,n^1,n^2} dT + \left(\frac{\partial G}{\partial P}\right)_{T,n^1,n^2} dP + \left(\frac{\partial G}{\partial n^1}\right)_{T,P,n^2} dn^1 + \left(\frac{\partial G}{\partial n^2}\right)_{T,P,n^1} dn^2$$

By definition,

$$\left(\frac{\partial G}{\partial n^1}\right)_{T,P,n^2} = \overline{g}^1 \qquad \left(\frac{\partial G}{\partial n^2}\right)_{T,P,n^1} = \overline{g}^2$$

Therefore, at constant temperature and pressure,

$$dG = \overline{g}^1 \, dn^1 + \overline{g}^2 \, dn^2 = dn^1(\overline{g}^1 - \overline{g}^2)$$

Now at equilibrium (Eq. 14.1)

$$dG_{T,P} = 0$$

Therefore, at equilibrium, we have

$$\overline{g}^1 = \overline{g}^2 \tag{14.2}$$

That is, under equilibrium conditions, the Gibbs function of each phase of a pure substance is equal. Let us check this by determining the Gibbs function of saturated liquid (water) and saturated vapor (steam) at 300 kPa. From the steam tables:

For the liquid:

$$g_f = h_f - Ts_f = 561.47 - 406.7 \times 1.6718 = -118.4 \text{ kJ/kg}$$

For the vapor:

$$g_g = h_g - Ts_g = 2725.3 - 406.7 \times 6.9919 = -118.4 \text{ kJ/kg}$$

Equation 14.2 can also be derived by applying the relation

$$T \, ds = dh - v \, dP$$

to the change of phase that takes place at constant pressure and temperature. For this process, this relation can be integrated as follows:

$$\int_f^g T \, ds = \int_f^g dh$$
$$T(s_g - s_f) = (h_g - h_f)$$
$$h_f - Ts_f = h_g - Ts_g$$
$$g_f = g_g$$

The Clapeyron equation, which was derived in Section 12.1, can be derived by an alternate method by considering the fact that the Gibbs functions of two phases in equilibrium are equal. In Chapter 12, we considered the relation (Eq. 12.15) for a simple compressible substance:

$$dg = v \, dP - s \, dT$$

Consider a control mass that consists of a saturated liquid and a saturated vapor in equilibrium, and let this system undergo a change of pressure, dP. The corresponding change in temperature, as determined from the vapor-pressure curve, is dT. Both phases will undergo the change in Gibbs function, dg, but since the phases always have the same value of the Gibbs function when they are in equilibrium, it follows that

$$dg_f = dg_g$$

But, from Eq. 12.15,

$$dg = v\,dP - s\,dT$$

it follows that

$$dg_f = v_f\,dP - s_f\,dT$$
$$dg_g = v_g\,dP - s_g\,dT$$

Since

$$dg_f = dg_g$$

it follows that

$$v_f\,dP - s_f\,dT = v_g\,dP - s_g\,dT$$
$$dP(v_g - v_f) = dT(s_g - s_f)$$
$$\frac{dP}{dT} = \frac{s_{fg}}{v_{fg}} = \frac{h_{fg}}{Tv_{fg}} \tag{14.3}$$

In summary, when different phases of a pure substance are in equilibrium, each phase has the same value of the Gibbs function per unit mass. This fact is relevant to different solid phases of a pure substance and is important in metallurgical applications of thermodynamics. Example 14.1 illustrates this principle.

Example 14.1

What pressure is required to make diamonds from graphite at a temperature of 25 °C? The following data are given for a temperature of 25 °C and a pressure of 0.1 MPa.

	Graphite	Diamond
g	0	2867.8 kJ/kmol
v	0.000 444 m³/kg	0.000 284 m³/kg
β_T	0.304×10^{-6} 1/MPa	0.016×10^{-6} 1/MPa

Analysis and Solution

The basic principle in the solution is that graphite and diamond can exist in equilibrium when they have the same value of the Gibbs function. At 0.1 MPa pressure, the Gibbs function of the diamond is greater than that of the graphite. However, the rate of increase in Gibbs function with pressure is greater for the graphite than for the diamond; therefore, at some pressure they can exist in equilibrium. Our problem is to find this pressure.

We have already considered the relation

$$dg = v\, dP - s\, dT$$

Since we are considering a process that takes place at constant temperature, this reduces to

$$dg_T = v\, dP_T \qquad\qquad (a)$$

Now at any pressure P and the given temperature, the specific volume can be found from the following relation, which utilizes isothermal compressibility factor.

$$v = v^0 + \int_{P=0.1}^{P} \left(\frac{\partial v}{\partial P}\right)_T dP = v^0 + \int_{P=0.1}^{P} \frac{v}{v}\left(\frac{\partial v}{\partial P}\right)_T dP$$

$$= v^0 - \int_{P=0.1}^{P} v\beta_T\, dP \qquad\qquad (b)$$

The superscript ^{0} will be used in this example to indicate the properties at a pressure of 0.1 MPa and a temperature of 25 °C.

The specific volume changes only slightly with pressure, so that $v \approx v^0$. Also, we assume that β_T is constant and that we are considering a very high pressure. With these assumptions, this equation can be integrated to give

$$v = v^0 - v^0 \beta_T P = v^0(1 - \beta_T P) \qquad\qquad (c)$$

We can now substitute this into Eq. (a) to give the relation

$$dg_T = [v^0(1 - \beta_T P)]\, dP_T$$

$$g - g^0 = v^0(P - P^0) - v^0\beta_T \frac{(P^2 - P^{02})}{2} \qquad\qquad (d)$$

If we assume that $P^0 \ll P$, this reduces to

$$g - g^0 = v^0\left(P - \frac{\beta_T P^2}{2}\right) \qquad\qquad (e)$$

For the graphite, $g^0 = 0$ and we can write

$$g_G = v_G^0\left[P - (\beta_T)_G \frac{P^2}{2}\right]$$

For the diamond, g^0 has a definite value and we have

$$g_D = g_D^0 + v_D^0\left[P - (\beta_T)_D \frac{P^2}{2}\right]$$

But at equilibrium, the Gibbs function of the graphite and diamond are equal:

$$g_G = g_D$$

Therefore,

$$v_G^0 \left[P - (\beta_T)_G \frac{P^2}{2} \right] = g_D^0 + v_D^0 \left[P - (\beta_T)_D \frac{P^2}{2} \right]$$

$$(v_G^0 - v_D^0)P - [v_G^0(\beta_T)_G - v_D^0(\beta_T)_D]\frac{P^2}{2} = g_D^0$$

$$(0.000\,444 - 0.000\,284)P$$

$$-(0.000\,444 \times 0.304 \times 10^{-6} - 0.000\,284 \times 0.016 \times 10^{-6})P^2/2 = \frac{2867.8}{12.011 \times 1000}$$

Solving this for P, we find

$$P = 1493 \text{ MPa}$$

That is, at 1493 MPa, 25 °C, graphite and diamond can exist in equilibrium, and the possibility exists for conversion from graphite to diamonds.

14.3 METASTABLE EQUILIBRIUM

Although the limited scope of this book precludes an extensive treatment of metastable equilibrium, a brief introduction to the subject is presented in this section. Let us first consider an example of metastable equilibrium.

Consider a slightly superheated vapor, such as steam, expanding in a convergent-divergent nozzle, as shown in Fig. 14.4. Assuming the process is reversible and adiabatic, the

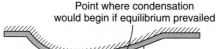

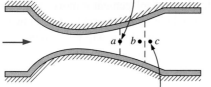

Point where condensation would begin if equilibrium prevailed

Point where condensation occurs very abruptly

FIGURE 14.4
Illustration of supersaturation in a nozzle.

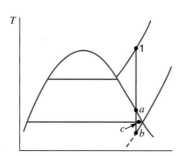
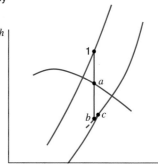

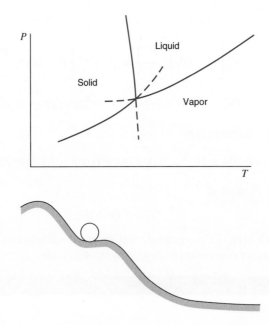

FIGURE 14.5
Metastable states for
solid–liquid–vapor
equilibrium.

FIGURE 14.6
Schematic diagram
illustrating a metastable
state.

steam will follow path 1–*a* on the *T–s* diagram, and at point *a* we would expect condensation to occur. However, if point *a* is reached in the divergent section of the nozzle, it is observed that no condensation occurs until point *b* is reached, and at this point the condensation occurs very abruptly in what is referred to as a condensation shock. Between points *a* and *b* the steam exists as a vapor, but the temperature is below the saturation temperature for the given pressure. This is known as a metastable state. The possibility of a metastable state exists with any phase transformation. The dotted lines on the equilibrium diagram shown in Fig. 14.5 represent possible metastable states for solid–liquid–vapor equilibrium.

The nature of a metastable state is often pictured schematically by the kind of diagram shown in Fig. 14.6. The ball is in a stable position (the "metastable state") for small displacements, but with a large displacement it moves to a new equilibrium position. The steam expanding in the nozzle is in a metastable state between *a* and *b*. This means that droplets smaller than a certain critical size will reevaporate, and only when droplets larger than this critical size have formed (this corresponds to moving the ball out of the depression) will the new equilibrium state appear.

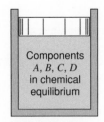

Components
A, *B*, *C*, *D*
in chemical
equilibrium

FIGURE 14.7
Schematic
diagram for
consideration of
chemical
equilibrium.

14.4 CHEMICAL EQUILIBRIUM

We now turn our attention to chemical equilibrium and consider first a chemical reaction involving only one phase. This is referred to as a homogeneous chemical reaction. It may be helpful to visualize this as a gaseous phase, but the basic considerations apply to any phase.

Consider a vessel, Fig. 14.7, that contains four compounds, *A*, *B*, *C*, and *D*, which are in chemical equilibrium at a given pressure and temperature. For example, these might consist of CO_2, H_2, CO, and H_2O in equilibrium. Let the number of moles of each component be designated n_A, n_B, n_C, and n_D. Furthermore, let the chemical reaction that takes place between these four constituents be

$$v_A A + v_B B \rightleftharpoons v_C C + v_D D \tag{14.4}$$

where the v's are the stoichiometric coefficients. It should be emphasized that there is a very definite relation between the v's (the stoichiometric coefficients), whereas the n's (the number of moles present) for any constituent can be varied simply by varying the amount of that component in the reaction vessel.

Let us now consider how the requirement for equilibrium, namely, that $dG_{T,P} = 0$ at equilibrium, applies to a homogeneous chemical reaction. Let us assume that the four components are in chemical equilibrium and then assume that from this equilibrium state, while the temperature and pressure remain constant, the reaction proceeds an infinitesimal amount toward the right as Eq. 14.4 is written. This results in a decrease in the moles of A and B and an increase in the moles of C and D. Let us designate the degree of reaction by ε and define the degree of reaction by the relations

$$dn_A = -v_A \, d\varepsilon$$
$$dn_B = -v_B \, d\varepsilon$$
$$dn_C = +v_C \, d\varepsilon$$
$$dn_D = +v_D \, d\varepsilon \tag{14.5}$$

That is, the change in the number of moles of any component during a chemical reaction is given by the product of the stoichiometric coefficients (the v's) and the degree of reaction.

Let us evaluate the change in the Gibbs function associated with this chemical reaction that proceeds to the right in the amount $d\varepsilon$. In doing so we use, as would be expected, the Gibbs function of each component in the mixture—the partial molal Gibbs function (or its equivalent, the chemical potential):

$$dG_{T,P} = \overline{G}_C \, dn_C + \overline{G}_D \, dn_D + \overline{G}_A \, dn_A + \overline{G}_B \, dn_B$$

Substituting Eq. 14.5, we have

$$dG_{T,P} = (v_C \overline{G}_C + v_D \overline{G}_D - v_A \overline{G}_A - v_B \overline{G}_B) \, d\varepsilon \tag{14.6}$$

We now need to develop expressions for the partial molal Gibbs functions in terms of properties that we are able to calculate. From the definition of the Gibbs function, Eq. 12.14,

$$G = H - TS$$

For a mixture of two components A and B, we differentiate this equation with respect to n_A at constant T, P, and n_B, which results in

$$\left(\frac{\partial G}{\partial n_A} \right)_{T,P,n_B} = \left(\frac{\partial H}{\partial n_A} \right)_{T,P,n_B} - T \left(\frac{\partial S}{\partial n_A} \right)_{T,P,n_B}$$

All three of these quantities satisfy the definition of partial molal properties according to Eq. 12.65, such that

$$\overline{G}_A = \overline{H}_A - T\overline{S}_A \tag{14.7}$$

For an ideal gas mixture, enthalpy is not a function of pressure, and

$$\overline{H}_A = \overline{h}_{ATP} = \overline{h}^{0}_{ATP^0} \tag{14.8}$$

Entropy is, however, a function of pressure, so that the partial entropy of A can be expressed by Eq. 13.22 in terms of the standard-state value,

$$\overline{S}_A = \overline{s}_{ATP_A = y_A P}$$

$$= \overline{s}^0_{ATP^0} - \overline{R} \ln\left(\frac{y_A P}{P^0}\right) \tag{14.9}$$

Now, substituting Eqs. 14.8 and 14.9 into Eq. 14.7,

$$\overline{G}_A = \overline{h}^0_{ATP^0} - T\overline{s}^0_{ATP^0} + \overline{R}T \ln\left(\frac{y_A P}{P^0}\right)$$

$$= \overline{g}^0_{ATP^0} + \overline{R}T \ln\left(\frac{y_A P}{P^0}\right) \tag{14.10}$$

Equation 14.10 is an expression for the partial Gibbs function of a component in a mixture in terms of a specific reference value, the pure-substance standard-state Gibbs function at the same temperature, and a function of the temperature, pressure, and composition of the mixture. This expression can be applied to each of the components in Eq. 14.6, resulting in

$$dG_{TP} = \left\{ v_C\left[\overline{g}^0_C + \overline{R}T \ln\left(\frac{y_C P}{P^0}\right)\right] + v_D\left[\overline{g}^0_D + \overline{R}T \ln\left(\frac{y_D P}{P^0}\right)\right] \right.$$

$$\left. - v_A\left[\overline{g}^0_A + \overline{R}T \ln\left(\frac{y_A P}{P^0}\right)\right] - v_B\left[\overline{g}^0_B + \overline{R}T \ln\left(\frac{y_B P}{P^0}\right)\right] \right\} d\varepsilon \tag{14.11}$$

Let us define ΔG^0 as follows:

$$\Delta G^0 = v_C \overline{g}^0_C + v_D \overline{g}^0_D - v_A \overline{g}^0_A - v_B \overline{g}^0_B \tag{14.12}$$

That is, ΔG^0 is the change in the Gibbs function that would occur if the chemical reaction given by Eq. 14.4 (which involves the stoichiometric amounts of each component) proceeded completely from left to right, with the reactants A and B initially separated and at temperature T and the standard-state pressure, and the products C and D finally separated and at temperature T and the standard-state pressure. Note also that ΔG^0 for a given reaction is a function of only the temperature. This will be most important to bear in mind as we proceed with our developments of homogeneous chemical equilibrium. Let us now digress from our development to consider an example involving the calculation of ΔG^0.

Example 14.2

Determine the value of ΔG^0 for the reaction $2H_2O \rightleftharpoons 2H_2 + O_2$ at 25 °C and at 2000 K, with the water in the gaseous phase.

Solution

At any given temperature, the standard-state Gibbs function change of Eq. 14.12 can be calculated from the relation

$$\Delta G^0 = \Delta H^0 - T\,\Delta S^0$$

At 25 °C,

$$\Delta H^0 = 2\overline{h}^0_{fH_2} + \overline{h}^0_{fO_2} - 2\overline{h}^0_{fH_2O(g)}$$
$$= 2(0) + 1(0) - 2(-241\,826) = 483\,652 \text{ kJ}$$
$$\Delta S^0 = 2\overline{s}^0_{H_2} + \overline{s}^0_{O_2} - 2s^0_{H_2O(g)}$$
$$= 2(130.678) + 1(205.148) - 2(188.834) = 88.836 \text{ kJ/K}$$

Therefore, at 25 °C,

$$\Delta G^0 = 483\,652 - 298.15(88.836) = 457\,166 \text{ kJ}$$

At 2000 K,

$$\Delta H^0 = 2\left(\overline{h}^0_{2000} - \overline{h}^0_{298}\right)_{H_2} + \left(\overline{h}^0_{2000} - \overline{h}^0_{298}\right)_{O_2} - 2\left(\overline{h}^0_f + \overline{h}^0_{2000} - \overline{h}^0_{298}\right)_{H_2O}$$
$$= 2(52\,942) + (59\,176) - 2(-241\,826 + 72\,788)$$
$$= 503\,136 \text{ kJ}$$
$$\Delta S^0 = 2\left(\overline{s}^0_{2000}\right)_{H_2} + \left(\overline{s}^0_{2000}\right)_{O_2} - 2\left(\overline{s}^0_{2000}\right)_{H_2O}$$
$$= 2(188.419) + (268.748) - 2(264.769)$$
$$= 116.048 \text{ kJ/K}$$

Therefore,

$$\Delta G^0 = 503\,136 - 2000 \times 116.048 = 271\,040 \text{ kJ}$$

Returning now to our development, substituting Eq. 14.12 into Eq. 14.11 and rearranging, we can write

$$dG_{T,P} = \left\{\Delta G^0 + \overline{R}T \ln\left[\frac{y_C^{v_C} y_D^{v_D}}{y_A^{v_A} y_B^{v_B}}\left(\frac{P}{P^0}\right)^{v_C+v_D-v_A-v_B}\right]\right\} d\varepsilon \qquad (14.13)$$

At equilibrium $dG_{T,P} = 0$. Therefore, since $d\varepsilon$ is arbitrary,

$$\ln\left[\frac{y_C^{v_C} y_D^{v_D}}{y_A^{v_A} y_B^{v_B}}\left(\frac{P}{P^0}\right)^{v_C+v_D-v_A-v_B}\right] = -\frac{\Delta G^0}{\overline{R}T} \qquad (14.14)$$

For convenience, we define the equilibrium constant K as

$$\ln K = -\frac{\Delta G^0}{\overline{R}T} \qquad (14.15)$$

which we note must be a function of temperature only for a given reaction, since ΔG^0 is given by Eq. 14.12 in terms of the properties of the pure substances at a given temperature and the standard-state pressure.

Combining Eqs. 14.14 and 14.15, we have

$$K = \frac{y_C^{v_C} y_D^{v_D}}{y_A^{v_A} y_B^{v_B}}\left(\frac{P}{P^0}\right)^{v_C+v_D-v_A-v_B} \qquad (14.16)$$

which is the chemical equilibrium equation corresponding to the reaction equation, Eq. 14.4.

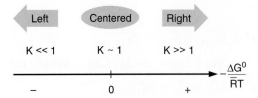

FIGURE 14.8 The shift in the reaction with the change in the Gibbs function.

From the equilibrium constant definition in Eqs. 14.15 and 14.16 we can draw a few conclusions. If the shift in the Gibbs function is large and positive, $\ln K$ is large and negative, leading to a very small value of K. At a given P in Eq. 14.16, this leads to relatively small values of the right-hand side (RHS) (component C and D) concentrations relative to the left-hand-side (LHS) component concentrations; the reaction is shifted to the left. The opposite is the case of a shift in the Gibbs function that is large and negative, giving a large value of K and the reaction is shifted to the right, as shown in Fig. 14.8. If the shift in Gibbs function is zero, then $\ln K$ is zero and K is exactly equal to 1. The reaction is in the middle, with all concentrations of the same order of magnitude, unless the stoichiometric coefficients are extreme.

The other trends we can see are the influences of the temperature and pressure. For a higher temperature but the same shift in the Gibbs function, the absolute value of $\ln K$ is smaller, which means K is closer to 1 and the reaction is more centered. For low temperatures, the reaction is shifted toward the side with the smallest Gibbs function G^0. The pressure has an influence only if the power in Eq. 14.16 is different from zero. That is so when the number of moles on the RHS ($v_C + v_D$) is different from the number of moles on the LHS ($v_A + v_B$). Assuming we have more moles on the RHS, then, we see that the power is positive. So, if the pressure is larger than the reference pressure, the whole pressure factor is larger than 1, which reduces the RHS concentrations as K is fixed for a given temperature. All other combinations can be examined in a similar fashion, and the result is that a higher pressure pushes the reaction toward the side with fewer moles, and a lower pressure pushes the reaction toward the side with more moles. The reaction tries to counteract the externally imposed pressure variation.

Example 14.3

Determine the equilibrium constant K, expressed as $\ln K$, for the reaction $2H_2O \rightleftharpoons 2H_2 + O_2$ at $25\,^\circ C$ and at 2000 K.

Solution

We have already found, in Example 14.2, ΔG^0 for this reaction at these two temperatures. Therefore, at $25\,^\circ C$,

$$(\ln K)_{298} = -\frac{\Delta G^0_{298}}{\overline{R}T} = \frac{-457\,166}{8.3145 \times 298.15} = -184.42$$

At 2000 K, we have

$$(\ln K)_{2000} = -\frac{\Delta G^0_{2000}}{\overline{R}T} = \frac{-271\,040}{8.3145 \times 2000} = -16.299$$

Table A.11 gives the values of the equilibrium constant for a number of reactions. Note again that for each reaction, the value of the equilibrium constant is determined from the properties of each of the pure constituents at the standard-state pressure and is a function of temperature only.

For other reaction equations, the chemical equilibrium constant can be calculated as in Example 14.3. Sometimes you can write a reaction scheme as a linear combination of the elementary reactions that are already tabulated, as, for example, in Table A.11. Assume we can write a reaction III as a linear combination of reaction I and reaction II, which means

$$
\begin{aligned}
\text{LHS}_\text{III} &= a\,\text{LHS}_\text{I} + b\,\text{LHS}_\text{II} \\
\text{RHS}_\text{III} &= a\,\text{RHS}_\text{I} + b\,\text{RHS}_\text{II}
\end{aligned}
\tag{14.17}
$$

From the definition of the shift in the Gibbs function, Eq. 14.12, it follows that

$$
\Delta G^0_\text{III} = G^0_\text{III RHS} - G^0_\text{III LHS} = a\,\Delta G^0_\text{I} + b\,\Delta G^0_\text{II}
$$

Then from the definition of the equilibrium constant in Eq. 14.15, we get

$$
\ln K_\text{III} = -\frac{\Delta G^0_\text{III}}{\overline{R}T} = -a\frac{\Delta G^0_\text{I}}{\overline{R}T} - b\frac{\Delta G^0_\text{II}}{\overline{R}T} = a\ln K_\text{I} + b\ln K_\text{II}
$$

or

$$
K_\text{III} = K_\text{I}^a K_\text{II}^b
\tag{14.18}
$$

Example 14.4

Show that the equilibrium constant for the reaction called the *water-gas reaction*

$$
\text{III: } H_2 + CO_2 \rightleftharpoons H_2O + CO
$$

can be calculated from values listed in Table A.11.

Solution

Using the reaction equations from Table A.11,

$$
\begin{aligned}
\text{I: } 2CO_2 &\rightleftharpoons 2CO + O_2 \\
\text{II: } 2H_2O &\rightleftharpoons 2H_2 + O_2
\end{aligned}
$$

It is seen that

$$
\text{III} = \tfrac{1}{2}\text{I} - \tfrac{1}{2}\text{II} = \tfrac{1}{2}(\text{I} - \text{II})
$$

so that

$$
K_\text{III} = \left(\frac{K_\text{I}}{K_\text{II}}\right)^{\frac{1}{2}}
$$

where K_III is calculated from the Table A.11 values

$$
\ln K_\text{III} = \tfrac{1}{2}(\ln K_\text{I} - \ln K_\text{II})
$$

We now consider a number of examples that illustrate the procedure for determining the equilibrium composition for a homogeneous reaction and the influence of certain variables on the equilibrium composition.

Example 14.5

One kilomole of carbon at 25 °C and 0.1 MPa pressure reacts with 1 kmol of oxygen at 25 °C and 0.1 MPa pressure to form an equilibrium mixture of CO_2, CO, and O_2 at 3000 K, 0.1 MPa pressure, in a steady-state process. Determine the equilibrium composition and the heat transfer for this process.

Control volume:　Combustion chamber.
Inlet states:　P, T known for carbon and for oxygen.
Exit state:　P, T known.
Process:　Steady state.
Sketch:　Figure 14.9.
Model:　Table A.10 for carbon; Tables A.9 and A.10 for ideal gases.

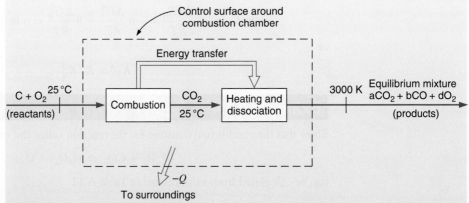

FIGURE 14.9 Sketch for Example 14.5.

Analysis and Solution

It is convenient to view the overall process as though it occurs in two separate steps, a combustion process followed by a heating and dissociation of the combustion product CO_2, as indicated in Fig. 14.9. This two-step process is represented as

$$\text{Combustion:} \quad C + O_2 \rightarrow CO_2$$
$$\text{Dissociation reaction:} \quad 2CO_2 \rightleftharpoons 2CO + O_2$$

That is, the energy released by the combustion of C and O_2 heats the CO_2 formed to high temperature, which causes dissociation of part of the CO_2 to CO and O_2. Thus, the overall reaction can be written

$$C + O_2 \rightarrow aCO_2 + bCO + dO_2$$

where the unknown coefficients a, b, and d must be found by solution of the equilibrium equation associated with the dissociation reaction. Once this is accomplished, we can write the energy equation for a control volume around the combustion chamber to calculate the heat transfer.

From the combustion equation we find that the initial composition for the dissociation reaction is 1 kmol CO_2. Therefore, letting $2z$ be the number of kilomoles of CO_2 dissociated, we find

$$2CO_2 \rightleftharpoons 2CO + O_2$$

Initial:	1	0	0
Change:	$-2z$	$+2z$	$+z$
At equilibrium:	$(1 - 2z)$	$2z$	z

Therefore, the overall reaction is

$$C + O_2 \rightarrow (1 - 2z)CO_2 + 2zCO + zO_2$$

and the total number of kilomoles at equilibrium is

$$n = (1 - 2z) + 2z + z = 1 + z$$

The equilibrium mole fractions are

$$y_{CO_2} = \frac{1 - 2z}{1 + z} \qquad y_{CO} = \frac{2z}{1 + z} \qquad y_{O_2} = \frac{z}{1 + z}$$

From Table A.11, we find that the value of the equilibrium constant at 3000 K for the dissociation reaction considered here is

$$\ln K = -2.217 \qquad K = 0.1089$$

Substituting these quantities along with $P = 0.1$ MPa into Eq. 14.16, we have the equilibrium equation,

$$K = 0.1089 = \frac{y_{CO}^2 y_{O_2}}{y_{CO_2}^2} \left(\frac{P}{P^0} \right)^{2+1-2} = \frac{\left(\dfrac{2z}{1+z} \right)^2 \left(\dfrac{z}{1+z} \right)}{\left(\dfrac{1-2z}{1+z} \right)^2} \quad (1)$$

or, in more convenient form,

$$\frac{K}{P/P^0} = \frac{0.1089}{1} = \left(\frac{2z}{1-2z} \right)^2 \left(\frac{z}{1+z} \right)$$

To obtain the physically meaningful root of this mathematical relation, we note that the number of moles of each component must be greater than zero. Thus, the root of interest to us must lie in the range

$$0 \le z \le 0.5$$

Solving the equilibrium equation by trial and error, we find

$$z = 0.2189$$

Therefore, the overall process is

$$C + O_2 \rightarrow 0.5622\,CO_2 + 0.4378\,CO + 0.2189\,O_2$$

where the equilibrium mole fractions are

$$y_{CO_2} = \frac{0.5622}{1.2189} = 0.4612$$

$$y_{CO} = \frac{0.4378}{1.2189} = 0.3592$$

$$y_{O_2} = \frac{0.2189}{1.2189} = 0.1796$$

The heat transfer from the combustion chamber to the surroundings can be calculated using the enthalpies of formation and Table A.9. For this process

$$H_R = (\overline{h}^0_f)_C + (\overline{h}^0_f)_{O_2} = 0 + 0 = 0$$

The equilibrium products leave the chamber at 3000 K. Therefore,

$$H_P = n_{CO_2}(\overline{h}^0_f + \overline{h}^0_{3000} - \overline{h}^0_{298})_{CO_2}$$
$$+ n_{CO}(\overline{h}^0_f + \overline{h}^0_{3000} - \overline{h}^0_{298})_{CO}$$
$$+ n_{O_2}(\overline{h}^0_f + \overline{h}^0_{3000} - \overline{h}^0_{298})_{O_2}$$
$$= 0.5622(-393\,522 + 152\,853)$$
$$+ 0.4378(-110\,527 + 93\,504)$$
$$+ 0.2189(98\,013)$$
$$= -121\,302 \text{ kJ}$$

Substituting into the energy equation gives

$$Q_{c.v.} = H_P - H_R$$
$$= -121\,302 \text{ kJ/kmol C burned}$$

Example 14.6

One kilomole of carbon at 25 °C reacts with 2 kmol of oxygen at 25 °C to form an equilibrium mixture of CO_2, CO, and O_2 at 3000 K, 0.1 MPa pressure. Determine the equilibrium composition.

Control volume: Combustion chamber.
Inlet states: T known for carbon and for oxygen.
Exit state: P, T known.
Process: Steady state.
Model: Ideal gas mixture at equilibrium.

Analysis and Solution

The overall process can be imagined to occur in two steps, as in the previous example. The combustion process is

$$C + 2O_2 \rightarrow CO_2 + O_2$$

and the subsequent dissociation reaction is

	$2CO_2 \rightleftharpoons$	$2CO +$	O_2
Initial:	1	0	1
Change:	$-2z$	$+2z$	$+z$
At equilibrium:	$(1 - 2z)$	$2z$	$(1 + z)$

We find that in this case, the overall process is

$$C + 2O_2 \rightarrow (1 - 2z)CO_2 + 2zCO + (1 + z)O_2$$

and the total number of kilomoles at equilibrium is

$$n = (1 - 2z) + 2z + (1 + z) = 2 + z$$

The mole fractions are

$$y_{CO_2} = \frac{1 - 2z}{2 + z} \qquad y_{CO} = \frac{2z}{2 + z} \qquad y_{O_2} = \frac{1 + z}{2 + z}$$

The equilibrium constant for the reaction $2CO_2 \rightleftharpoons 2CO + O_2$ at 3000 K was found in Example 14.5 to be 0.1089. Therefore, with these expressions, quantities, and $P = 0.1$ MPa substituted, the equilibrium equation is

$$K = 0.1089 = \frac{y_{CO}^2 y_{O_2}}{y_{CO_2}^2} \left(\frac{P}{P^0} \right)^{2+1-2} = \frac{\left(\frac{2z}{2 + z} \right)^2 \left(\frac{1 + z}{2 + z} \right)}{\left(\frac{1 - 2z}{2 + z} \right)^2} \quad (1)$$

or

$$\frac{K}{P/P^0} = \frac{0.1089}{1} = \left(\frac{2z}{1 - 2z} \right)^2 \left(\frac{1 + z}{2 + z} \right)$$

We note that in order for the number of kilomoles of each component to be greater than zero,

$$0 \leq z \leq 0.5$$

Solving the equilibrium equation for z, we find

$$z = 0.1553$$

so that the overall process is

$$C + 2O_2 \rightarrow 0.6894\,CO_2 + 0.3106\,CO + 1.1553\,O_2$$

When we compare this result with that of Example 14.5, we notice that there is more CO_2 and less CO. The presence of additional O_2 shifts the dissociation reaction more to the left side.

The mole fractions of the components in the equilibrium mixture are

$$y_{CO_2} = \frac{0.6894}{2.1553} = 0.320$$

$$y_{CO} = \frac{0.3106}{2.1553} = 0.144$$

$$y_{O_2} = \frac{1.1553}{2.1553} = 0.536$$

The heat transferred from the chamber in this process could be found by the same procedure followed in Example 14.5, considering the overall process.

In-Text Concept Questions

a. For a mixture of O_2 and O, the pressure is increased at constant T; what happens to the composition?

b. For a mixture of O_2 and O, the temperature is increased at constant P; what happens to the composition?

c. For a mixture of O_2 and O, I add some argon, keeping constant T, P; what happens to the moles of O?

14.5 SIMULTANEOUS REACTIONS

In developing the equilibrium equation and equilibrium constant expressions of Section 14.4, it was assumed that there was only a single chemical reaction equation relating the substances present in the system. To demonstrate the more general situation in which there is more than one chemical reaction, we will now analyze a case involving two simultaneous reactions by a procedure analogous to that followed in Section 14.4. These results are then readily extended to systems involving several simultaneous reactions.

Consider a mixture of substances A, B, C, D, L, M, and N as indicated in Fig. 14.10. These substances are assumed to exist at a condition of chemical equilibrium at temperature T and pressure P, and are related by the two independent reactions

$$(1) \quad v_{A1}A + v_B B \rightleftharpoons v_C C + v_D D \tag{14.19}$$

$$(2) \quad v_{A2}A + v_L L \rightleftharpoons v_M M + v_N N \tag{14.20}$$

We have considered the situation where one of the components (substance A) is involved in each of the reactions in order to demonstrate the effect of this condition on the resulting equations. As in the previous section, the changes in amounts of the components are related by the various stoichiometric coefficients (which are not the same as the number of moles of each substance present in the vessel). We also realize that the coefficients v_{A1} and v_{A2} are not necessarily the same. That is, substance A does not, in general, take part in each of the reactions to the same extent.

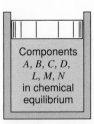

Components
A, B, C, D,
L, M, N
in chemical
equilibrium

FIGURE 14.10

Sketch demonstrating simultaneous reactions.

Development of the requirement for equilibrium is completely analogous to that of Section 14.4. We consider that each reaction proceeds an infinitesimal amount toward the right side. This results in a decrease in the number of moles of A, B, and L and an increase in the moles of C, D, M, and N. Letting the degrees of reaction be ε_1 and ε_2 for reactions 1 and 2, respectively, the changes in the number of moles are, for infinitesimal shifts from the equilibrium composition,

$$
\begin{aligned}
dn_A &= -v_{A1}\, d\varepsilon_1 - v_{A2}\, d\varepsilon_2 \\
dn_B &= -v_B\, d\varepsilon_1 \\
dn_L &= -v_L\, d\varepsilon_2 \\
dn_C &= +v_C\, d\varepsilon_1 \\
dn_D &= +v_D\, d\varepsilon_1 \\
dn_M &= +v_M\, d\varepsilon_2 \\
dn_N &= +v_N\, d\varepsilon_2
\end{aligned}
\tag{14.21}
$$

The change in Gibbs function for the mixture in the vessel at constant temperature and pressure is

$$
dG_{T,P} = \overline{G}_A\, dn_A + \overline{G}_B\, dn_B + \overline{G}_C\, dn_C + \overline{G}_D\, dn_D + \overline{G}_L\, dn_L + \overline{G}_M\, dn_M + \overline{G}_N\, dn_N
$$

Substituting the expressions of Eq. 14.21 and collecting terms,

$$
\begin{aligned}
dG_{T,P} = &(v_C\overline{G}_C + v_D\overline{G}_D - v_{A_1}\overline{G}_A - v_B\overline{G}_B)\, d\varepsilon_1 \\
&+ (v_M\overline{G}_M + v_N\overline{G}_N - v_{A_2}\overline{G}_A - v_L\overline{G}_L)\, d\varepsilon_2
\end{aligned}
\tag{14.22}
$$

It is convenient to again express each of the partial molal Gibbs functions in terms of

$$
\overline{G}_i = \overline{g}_i^0 + \overline{R}T \ln\left(\frac{y_i P}{P^0}\right)
$$

Equation 14.22 written in this form becomes

$$
\begin{aligned}
dG_{T,P} = &\left\{\Delta G_1^0 + \overline{R}T \ln\left[\frac{y_C^{v_C} y_D^{v_D}}{y_A^{v_{A1}} y_B^{v_B}}\left(\frac{P}{P^0}\right)^{v_C+v_D-v_{A1}-v_B}\right]\right\} d\varepsilon_1 \\
&+ \left\{\Delta G_2^0 + \overline{R}T \ln\left[\frac{y_M^{v_M} y_N^{v_N}}{y_A^{v_{A2}} y_L^{v_L}}\left(\frac{P}{P^0}\right)^{v_M+v_N-v_{A2}-v_L}\right]\right\} d\varepsilon_2
\end{aligned}
\tag{14.23}
$$

In this equation, the standard-state change in Gibbs function for each reaction is defined as

$$
\Delta G_1^0 = v_C\overline{g}_C^0 + v_D\overline{g}_D^0 - v_{A1}\overline{g}_A^0 - v_B\overline{g}_B^0
\tag{14.24}
$$

$$
\Delta G_2^0 = v_M\overline{g}_M^0 + v_N\overline{g}_N^0 - v_{A2}\overline{g}_A^0 - v_L\overline{g}_L^0
\tag{14.25}
$$

Equation 14.23 expresses the change in Gibbs function of the system at constant T, P, for infinitesimal degrees of reaction of both reactions 1 and 2, Eqs. 14.19 and 14.20. The requirement for equilibrium is that $dG_{T,P} = 0$. Therefore, since reactions 1 and 2 are independent, $d\varepsilon_1$ and $d\varepsilon_2$ can be independently varied. It follows that at equilibrium, each of the bracketed terms of Eq. 14.23 must be zero. Defining equilibrium constants for the two reactions by

$$\ln K_1 = -\frac{\Delta G_1^0}{\overline{R}T} \tag{14.26}$$

and

$$\ln K_2 = -\frac{\Delta G_2^0}{\overline{R}T} \tag{14.27}$$

we find that, at equilibrium

$$K_1 = \frac{y_C^{\nu_C} y_D^{\nu_D}}{y_A^{\nu_{A1}} y_B^{\nu_B}} \left(\frac{P}{P^0}\right)^{\nu_C + \nu_D - \nu_{A1} - \nu_B} \tag{14.28}$$

and

$$K_2 = \frac{y_M^{\nu_M} y_N^{\nu_N}}{y_A^{\nu_{A2}} y_L^{\nu_L}} \left(\frac{P}{P^0}\right)^{\nu_M + \nu_N - \nu_{A2} - \nu_L} \tag{14.29}$$

These equations for the equilibrium composition of the mixture must be solved simultaneously. The following example demonstrates and clarifies this procedure.

Example 14.7

One kilomole of water vapor is heated to 3000 K, 0.1 MPa pressure. Determine the equilibrium composition, assuming that H_2O, H_2, O_2, and OH are present.

Control volume: Heat exchanger.
Exit state: P, T known.
Model: Ideal gas mixture at equilibrium.

Analysis and Solution
There are two independent reactions relating the four components of the mixture at equilibrium. These can be written as

$$(1): 2\,H_2O \rightleftharpoons 2\,H_2 + O_2$$
$$(2): 2\,H_2O \rightleftharpoons H_2 + 2\,OH$$

Let $2a$ be the number of kilomoles of water dissociating according to reaction 1 during the heating, and let $2b$ be the number of kilomoles of water dissociating according to reaction 2. Since the initial composition is 1 kmol water, the changes according to the two reactions are

$$(1): 2\,H_2O \rightleftharpoons 2\,H_2 + O_2$$
Change: $-2a$ $\quad +2a + a$

$$(2): 2\,H_2O \rightleftharpoons H_2 + 2\,OH$$
Change: $-2b$ $\quad +b + 2b$

Therefore, the number of kilomoles of each component at equilibrium is its initial number plus the change, so that at equilibrium

$$
\begin{aligned}
n_{H_2O} &= 1 - 2a - 2b \\
n_{H_2} &= 2a + b \\
n_{O_2} &= a \\
\underline{n_{OH} = 2b} \\
n &= 1 + a + b
\end{aligned}
$$

The overall chemical reaction that occurs during the heating process can be written

$$
H_2O \rightarrow (1 - 2a - 2b)H_2O + (2a + b)H_2 + aO_2 + 2bOH
$$

The RHS of this expression is the equilibrium composition of the system. Since the number of kilomoles of each substance must necessarily be greater than zero, we find that the possible values of a and b are restricted to

$$
a \geq 0
$$

$$
b \geq 0
$$

$$
(a + b) \leq 0.5
$$

The two equilibrium equations are, assuming that the mixture behaves as an ideal gas,

$$
K_1 = \frac{y_{H_2}^2 y_{O_2}}{y_{H_2O}^2} \left(\frac{P}{P^0} \right)^{2+1-2}
$$

$$
K_2 = \frac{y_{H_2} y_{OH}^2}{y_{H_2O}^2} \left(\frac{P}{P^0} \right)^{1+2-2}
$$

Since the mole fraction of each component is the ratio of the number of kilomoles of the component to the total number of kilomoles of the mixture, these equations can be written in the form

$$
K_1 = \frac{\left(\dfrac{2a + b}{1 + a + b} \right)^2 \left(\dfrac{a}{1 + a + b} \right)}{\left(\dfrac{1 - 2a - 2b}{1 + a + b} \right)^2} \left(\frac{P}{P^0} \right)
$$

$$
= \left(\frac{2a + b}{1 - 2a - 2b} \right)^2 \left(\frac{a}{1 + a + b} \right) \left(\frac{P}{P^0} \right)
$$

and

$$
K_2 = \frac{\left(\dfrac{2a + b}{1 + a + b} \right) \left(\dfrac{2b}{1 + a + b} \right)^2}{\left(\dfrac{1 - 2a - 2b}{1 + a + b} \right)^2} \left(\frac{P}{P^0} \right)
$$

$$
= \left(\frac{2a + b}{1 + a + b} \right) \left(\frac{2b}{1 - 2a - 2b} \right)^2 \left(\frac{P}{P^0} \right)
$$

giving two equations in the two unknowns a and b, since $P = 0.1$ MPa and the values of K_1, K_2 are known. From Table A.11 at 3000 K, we find

$$K_1 = 0.002\,062 \qquad K_2 = 0.002\,893$$

Therefore, the equations can be solved simultaneously for a and b. The values satisfying the equations are

$$a = 0.0534 \qquad b = 0.0551$$

Substituting these values into the expressions for the number of kilomoles of each component and of the mixture, we find the equilibrium mole fractions to be

$$y_{H_2O} = 0.7063$$
$$y_{H_2} = 0.1461$$
$$y_{O_2} = 0.0482$$
$$y_{OH} = 0.0994$$

The procedure followed in this section can readily be extended to equilibrium systems having more than two independent reactions. In each case, the number of simultaneous equilibrium equations is equal to the number of independent reactions. The expression and solution of the resulting large set of nonlinear equations require a formal mathematical iterative technique and are carried out on a computer. A different approach is typically followed in situations including a large number of chemical species. This involves the direct minimization of the system Gibbs function G with respect to variations in all of the species assumed to be present at the equilibrium state (for example, in Example 14.7, these would be H_2O, H_2, O_2, and OH). In general, this is $dG = \Sigma \overline{G}_i dn_i$, in which the $\overline{G}_i$ are each given by Eq. 14.10 and the dn_i are the variations in moles. However, the number of changes in moles are not all independent, as they are subject to constraints on the total number of atoms of each element present (in Example 14.7, these would be H and O). This process then results in a set of nonlinear equations equal to the sum of the number of elements and the number of species. Again, this set of equations requires a formal iterative solution procedure, but this technique is more straightforward and simpler than that utilizing the equilibrium constants and equations in situations involving a large number of chemical species.

14.6 COAL GASIFICATION

The processes involved in the gasification of coal (or other biomass) begin with heating the solid material to around 300–400 °C such that pyrolysis results in a solid char (essentially carbon) plus volatile gases (CO_2, CO, H_2O, H_2, some light hydrocarbons) and tar. In the gasifier, the char reacts with a small amount of oxygen and steam in the reactions

$$C + 0.5\,O_2 \rightarrow CO \qquad \text{which produces heat} \qquad (14.30)$$

$$C + H_2O \rightarrow H_2 + CO \qquad (14.31)$$

The resulting gas mixture of H_2 and CO is called *syngas*.

Then using appropriate catalysts, there is the water–gas shift equilibrium reaction

$$CO + H_2O \rightleftharpoons H_2 + CO_2 \tag{14.32}$$

and the methanation equilibrium reaction

$$CO + 3\,H_2 \rightleftharpoons CH_4 + H_2O \tag{14.33}$$

Solution of the two equilibrium equations, Eqs. 14.32 and 14.33, depends on the initial amounts of O_2 and H_2O that were used to react with char in Eqs. 14.30 and 14.31 and are, of course, strongly dependent on temperature and pressure. Relatively low T and high P favor the formation of CH_4, while high T and low P favor the formation of H_2 and CO. Time is also a factor, as the mixture may not have time to come to equilibrium in the gasifier. The entire process is quite complex but is one that has been thoroughly studied for many years. Finally, it should be pointed out that there are several different processes by which syngas can be converted to liquid fuels; this is also an ongoing field of research and development.

14.7 IONIZATION

In this section, we consider the equilibrium of systems that are made up of ionized gases, or plasmas, a field that has been studied and applied increasingly in recent years. In previous sections we discussed chemical equilibrium, with a particular emphasis on molecular dissociation, as, for example, the reaction

$$N_2 \rightleftharpoons 2N$$

which occurs to an appreciable extent for most molecules only at high temperature, of the order of magnitude 3000 to 10 000 K. At still higher temperatures, such as those found in electric arcs, the gas becomes ionized. That is, some of the atoms lose an electron according to the reaction

$$N \rightleftharpoons N^+ + e^-$$

where N^+ denotes a singly ionized nitrogen atom, one that has lost one electron and consequently has a positive charge, and e^- represents the free electron. As the temperature rises still higher, many of the ionized atoms lose another electron according to the reaction

$$N^+ \rightleftharpoons N^{++} + e^-$$

and thus become doubly ionized. As the temperature continues to rise, the process continues until a temperature is reached at which all the electrons have been stripped from the nucleus.

Ionization generally is appreciable only at high temperature. However, dissociation and ionization both tend to occur to greater extents at low pressure; consequently, dissociation and ionization may be appreciable in such environments as the upper atmosphere, even at moderate temperature. Other effects, such as radiation, will also cause ionization, but these effects are not considered here.

The problems of analyzing the composition in a plasma become much more difficult than for an ordinary chemical reaction, for in an electric field, the free electrons in the mixture do not exchange energy with the positive ions and neutral atoms at the same rate that they do with the field. Consequently, in a plasma in an electric field, the electron gas is not at exactly the same temperature as the heavy particles. However, for moderate fields, assuming a condition of thermal equilibrium in the plasma is a reasonable approximation, at least for preliminary calculations. Under this condition, we can treat the ionization equilibrium in exactly the same manner as an ordinary chemical equilibrium analysis.

At these extremely high temperatures, we may assume that the plasma behaves as an ideal gas mixture of neutral atoms, positive ions, and electron gas. Thus, for the ionization of some atomic species A,

$$A \rightleftharpoons A^+ + e^- \tag{14.34}$$

we may write the ionization equilibrium equation in the form

$$K = \frac{y_{A^+} y_{e^-}}{y_A} \left(\frac{P}{P^0} \right)^{1+1-1} \tag{14.35}$$

The ionization-equilibrium constant K is defined in the ordinary manner

$$\ln K = -\frac{\Delta G^0}{\overline{R}T} \tag{14.36}$$

and is a function of temperature only. The standard-state Gibbs function change for reaction 14.34 is found from

$$\Delta G^0 = \overline{g}_{A^+}^0 + \overline{g}_{e^-}^0 - \overline{g}_A^0 \tag{14.37}$$

The standard-state Gibbs function for each component at the given plasma temperature can be calculated using the procedures of statistical thermodynamics, so that ionization–equilibrium constants can be tabulated as functions of temperature.

The ionization–equilibrium equation, Eq. 14.35, is then solved in the same manner as an ordinary chemical-reaction equilibrium equation.

Example 14.8

Calculate the equilibrium composition if argon gas is heated in an arc to 10 000 K, 1 kPa, assuming the plasma to consist of Ar, Ar$^+$, e^-. The ionization–equilibrium constant for the reaction

$$Ar \rightleftharpoons Ar^+ + e^-$$

at this temperature is 0.000 42.

Control volume:	Heating arc.
Exit state:	P, T known.
Model:	Ideal gas mixture at equilibrium.

Analysis and Solution

Consider an initial composition of 1 kmol neutral argon, and let z be the number of kilomoles ionized during the heating process. Therefore,

$$Ar \rightleftharpoons Ar^+ + e^-$$

Initial:	1	0	0
Change:	$-z$	$+z$	$+z$
Equilibrium:	$(1-z)$	z	z

and

$$n = (1 - z) + z + z = 1 + z$$

Since the number of kilomoles of each component must be positive, the variable z is restricted to the range

$$0 \le z \le 1$$

The equilibrium mole fractions are

$$y_{Ar} = \frac{n_{Ar}}{n} = \frac{1 - z}{1 + z}$$

$$y_{Ar^+} = \frac{n_{Ar^+}}{n} = \frac{z}{1 + z}$$

$$y_{e^-} = \frac{n_{e^-}}{n} = \frac{z}{1 + z}$$

The equilibrium equation is

$$K = \frac{y_{Ar^+} y_{e^-}}{y_{Ar}} \left(\frac{P}{P^0} \right)^{1+1-1} = \frac{\left(\dfrac{z}{1 + z} \right) \left(\dfrac{z}{1 + z} \right)}{\left(\dfrac{1 - z}{1 + z} \right)} \left(\frac{P}{P^0} \right)$$

so that, at 10 000 K, 1 kPa,

$$0.000\,42 = \left(\frac{z^2}{1 - z^2} \right) (0.01)$$

Solving,

$$z = 0.2008$$

and the composition is found to be

$$y_{Ar} = 0.6656$$

$$y_{Ar^+} = 0.1672$$

$$y_{e^-} = 0.1672$$

14.8 ENGINEERING APPLICATIONS

Chemical reactions and equilibrium conditions become important in many industrial processes that occur during energy conversion, like combustion. As the temperatures in the combustion products are high, a number of chemical reactions may take place that would not occur at lower temperatures. Typical examples of these are dissociations that require substantial energy to proceed and thus have a profound effect on the resulting mixture temperature. To promote chemical reactions in general, catalytic surfaces are used in many reactors, which could be platinum pellets, as in a three-way catalytic converter on a car exhaust system. We have previously shown some of the reactions that are important in coal gasification, and some of the home works have a few reactions used in the production of synthetic fuels from biomass or coal. Production of hydrogen for fuel cell applications is part of this class of processes (recall Eqs.14.31–14.33), and for this it is important to examine the effect of both the temperature and the pressure on the final equilibrium mixture.

One of the chemical reactions that is important in the formation of atmospheric pollutants is the formation of NO_x (nitrogen-oxygen combinations), which takes place in all combustion processes that utilize fuel and air. Formation of NO_x happens at higher temperatures and consists of nitric oxide (NO) and nitrogen dioxide (NO_2); usually NO is the major contributor. This forms from the nitrogen in the air through the following reactions, called the *extended Zeldovich* mechanism:

$$1: \quad O + N_2 \rightleftharpoons NO + N$$
$$2: \quad N + O_2 \rightleftharpoons NO + O \qquad (14.38)$$
$$3: \quad N + OH \rightleftharpoons NO + H$$

Adding the first two reactions equals the elementary reaction listed in Table A.11 as

$$4: \quad O_2 + N_2 \rightleftharpoons 2\,NO$$

In equilibrium, the rate of the forward reaction equals the rate of the reverse reaction. However, in nonequilibrium that is not the case, which is what happens when NO is being formed. For smaller concentrations of NO, the forward reaction rates are much larger than the reverse rates, and they are all sensitive to temperature and pressure. With a model for the reactions rates and the concentrations, the rate of formation of NO can be described as

$$\frac{dy_{NO}}{dt} = \frac{y_{NOe}}{\tau_{NO}} \qquad (14.39)$$

$$\tau_{NO} = CT(P/P_0)^{-1/2}\exp\left(\frac{58\,300\,\text{K}}{T}\right) \qquad (14.40)$$

where $C = 8 \times 10^{-16}$ sK^{-1}, y_{NOe} is the equilibrium NO concentration and τ_{NO} is the time constant in seconds. For peak T and P, as is typical in an engine, the time scale becomes short (1 ms), so the equilibrium concentration is reached very quickly. As the gases expand and T, P decrease, the time scale becomes large, typically for the reverse reaction that removes NO, and the concentration is frozen at the high level. The equilibrium concentration for NO is found from reaction 4 equilibrium constant K_4 (see Table A.11), according to Eq.14.16:

$$y_{NOe} = [K_4\, y_{O2e}\, y_{N2e}]^{1/2} \qquad (14.41)$$

To model the total process, including the reverse reaction rates, a more detailed model of the combustion product mixture, including the water–gas reaction, is required.

This simple model does illustrate the importance of the chemical reactions and the high sensitivity of NO formation to peak temperature and pressure, which are the primary focus in any attempt to design low-emission combustion processes. One way of doing this is by steam injection, shown in Problems 11.124 and 13.134. Another way is a significant bypass flow, as in Problem 13.142. In both cases, the product temperature is reduced as much as possible without making the combustion unstable.

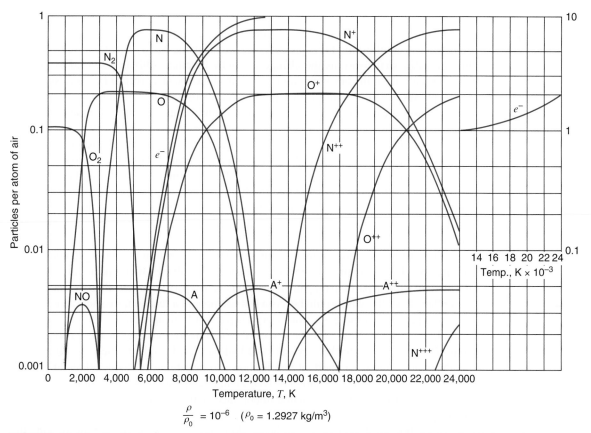

FIGURE 14.11 Equilibrium composition of air. (W. E. Moeckel and K. C. Weston, NACA TN 4265, 1958.)

A final example of an application is simultaneous reactions, including dissociations and ionization in several steps. When ionization of a gas occurs it becomes a plasma, and to a first approximation we again make the assumption of thermal equilibrium and treat it as an ideal gas. The equations for the many simultaneous reactions are solved by minimizing the Gibbs function, as explained in the end of Section 14.5. Figure 14.11 shows the equilibrium composition of air at high temperature and very low density, and indicates the overlapping regions of the various dissociations and ionization processes. Notice, for instance, that beyond 3000 K there is virtually no diatomic oxygen left, and below that temperature only O and NO are formed.

In-Text Concept Questions

d. When dissociations occur after combustion, does T go up or down?

e. For nearly all the dissociations and ionization reactions, what happens to the composition when the pressure is raised?

f. How does the time scale for NO formation change when P is lower at the same T?

g. Which atom in air ionizes first as T increases? What is the explanation?

SUMMARY

A short introduction is given to equilibrium in general, with application to phase equilibrium and chemical equilibrium. From previous analysis with the energy equation, we have found the reversible shaft work as the change in Gibbs function. This is extended to give the equilibrium state as the one with minimum Gibbs function at a given T, P. This also applies to two phases in equilibrium, so each phase has the same Gibbs function.

Chemical equilibrium is formulated for a single equilibrium reaction, assuming the components are all ideal gases. This leads to an equilibrium equation tying together the mole fractions of the components, the pressure, and the reaction constant. The reaction constant is related to the shift in the Gibbs function from the reactants (LHS) to the products (RHS) at a temperature T. As T or P changes, the equilibrium composition will shift according to its sensitivity to T and P. For very large equilibrium constants, the reaction is shifted toward the RHS, and for very small ones it is shifted toward the LHS. We show how elementary reactions can be used in linear combinations and how to find the equilibrium constant for this new reaction.

In most real systems of interest, there are multiple reactions coming to equilibrium simultaneously with a fairly large number of species involved. Often, species are present in the mixture without participating in the reactions, causing a dilution, so all mole fractions are lower than they otherwise would be. As a last example of a reaction, we show an ionization process where one or more electrons can be separated from an atom.

In the final sections, we show special reactions to consider for the gasification of coal, which also leads to the production of hydrogen and synthetic fuels. At higher temperatures, ionization is important and is shown to be similar to dissociations in the way the reactions are treated. Formation of NO_x at high temperature is an example of reactions that are rate sensitive and of particular importance in all processes that involve combustion with air.

You should have learned a number of skills and acquired abilities from studying this chapter that will allow you to:

- Apply the principle of a minimum Gibbs function to a phase equilibrium.
- Understand that the concept of equilibrium can include other effects, such as elevation, surface tension, and electrical potentials, as well as the concept of metastable states.
- Understand that the chemical equilibrium is written for ideal gas mixtures.
- Understand the meaning of the shift in Gibbs function due to the reaction.
- Know when the absolute pressure has an influence on the composition.
- Know the connection between the reaction scheme and the equilibrium constant.
- Understand that all species are present and influence the mole fractions.
- Know that a dilution with an inert gas has an effect.
- Understand the coupling between the chemical equilibrium and the energy equation.
- Intuitively know that most problems must be solved by iterations.
- Be able to treat a dissociation added to a combustion process.
- Be able to treat multiple simultaneous reactions.
- Know that syngas can be formed from an original fuel.
- Know what an ionization process is and how to treat it.
- Know that pollutants like NO_x form in a combustion process.

KEY CONCEPTS AND FORMULAS		
	Gibbs function	$g \equiv h - Ts$
	Equilibrium	Minimum g for given $T, P \Rightarrow dG_{TP} = 0$
	Phase equilibrium	$g_f = g_g$
	Equilibrium reaction	$v_A A + v_B B \Leftrightarrow v_C C + v_D D$
	Change in Gibbs function	$\Delta G^0 = v_C \bar{g}_C^0 + v_D \bar{g}_D^0 - v_A \bar{g}_A^0 - v_B \bar{g}_B^0$ evaluate at T, P^0
	Equilibrium constant	$K = e^{-\Delta G^0 / \bar{R}T}$
		$K = \dfrac{y_C^{v_C} y_D^{v_D}}{y_A^{v_A} y_B^{v_B}} \left(\dfrac{P}{P^0} \right)^{v_C + v_D - v_A - v_B}$
	Mole fractions	$y_i = n_i / n_{\text{tot}}$ (n_{tot} includes nonreacting species)
	Reaction scheme	Reaction scheme III $= a\,\text{I} + b\,\text{II} \Rightarrow K_{\text{III}} = K_{\text{I}}^a K_{\text{II}}^b$
	Dilution	reaction the same, y's are smaller
	Simultaneous reactions	$K_1, K_2, \dots$ and more y's

CONCEPT-STUDY GUIDE PROBLEMS

14.1 Is the concept of equilibrium limited to thermodynamics?

14.2 How does the Gibbs function vary with quality as you move from liquid to vapor?

14.3 How is a chemical equilibrium process different from a combustion process?

14.4 Must P and T be held fixed to obtain chemical equilibrium?

14.5 The change in the Gibbs function, ΔG^0, for a reaction is a function of which property?

14.6 In a steady-flow burner, T is not controlled; which properties are?

14.7 In a closed rigid-combustion bomb, which properties are held fixed?

14.8 Is the dissociation of water pressure sensitive?

14.9 At 298 K, $K = \exp(-184)$ for the water dissociation; what does that imply?

14.10 If a reaction is insensitive to pressure, prove that it is also insensitive to dilution effects at a given T.

14.11 For a pressure-sensitive reaction, an inert gas is added (dilution); how does the reaction shift?

14.12 In a combustion process, is the adiabatic flame temperature affected by reactions?

14.13 In equilibrium, the Gibbs function of the reactants and the products is the same; how about the energy?

14.14 Does a dissociation process require energy or does it give out energy?

14.15 If I consider the nonfrozen (composition can vary) specific heat but still assume that all components are ideal gases, does that C become a function of temperature? Of pressure?

14.16 What is K for the water–gas reaction in Example 14.4 at 1200 K?

14.17 What would happen to the concentrations of the monatomic species like O and N if the pressure is higher in Fig. 14.11?

HOMEWORK PROBLEMS

Equilibrium and Phase Equilibrium

14.18 Carbon dioxide at 15 MPa is injected into the top of a 5 km deep well in connection with an enhanced oil-recovery process. The fluid column standing in the well is at a uniform temperature of 40 °C. What is the pressure at the bottom of the well, assuming ideal gas behavior?

14.19 Consider a 2 km deep gas well containing a gas mixture of methane and ethane at a uniform temperature of 30 °C. The pressure at the top of the well is 14 MPa, and the composition on a mole basis is 90 % methane, 10 % ethane. Each component is in equilibrium (top to bottom), with $dG + g\,dZ = 0$, and assume ideal gas; so, for each component, Eq. 14.10 applies. Determine the pressure and composition at the bottom of the well.

14.20 A container has liquid water at 20 °C, 100 kPa, in equilibrium with a mixture of water vapor and dry air also at 20 °C, 100 kPa. What is the water vapor pressure and what is the saturated water vapor pressure?

14.21 Using the same assumptions as those in developing Eq. d in Example 14.1, develop an expression for pressure at the bottom of a deep column of liquid in terms of the isothermal compressibility, β_T. For liquid water at 20 °C, we know that $\beta_T = 0.0005$ [1/MPa]. Use the answer to the first question to estimate the pressure in the Pacific Ocean at a depth of 3 km.

Chemical Equilibrium, Equilibrium Constant

14.22 Which of the reactions listed in Table A.11 are pressure sensitive?

14.23 Calculate the equilibrium constant for the reaction $O_2 \rightleftharpoons 2O$ at temperatures of 298 K and 6000 K. Verify the result with Table A.11.

14.24 Calculate the equilibrium constant for the reaction $H_2 \rightleftharpoons 2H$ at a temperature of 2000 K, using properties from Table A.9. Compare the result with the value listed in Table A.11.

14.25 For the dissociation of oxygen, $O_2 \Leftrightarrow 2O$, around 2000 K, we want a mathematical expression for the equilibrium constant, $K(T)$. Assume constant specific heat, at 2000 K, for O_2 and O from Table A.9 and develop the expression from Eqs. 14.12 and 14.15.

14.26 Find K for $CO_2 \Leftrightarrow CO + \frac{1}{2}O_2$ at 2000 K using Table A.11.

14.27 Plot to scale the values of ln K versus $1/T$ for the reaction $2CO_2 \rightleftharpoons 2CO + O_2$. Write an equation for ln K as a function of temperature.

14.28 Consider the reaction $2CO_2 \Leftrightarrow 2CO + O_2$ obtained after heating 1 kmol CO_2 to 3000 K. Find the equilibrium constant from the shift in Gibbs function and verify its value with the entry in Table A.11. What is the mole fraction of CO at 3000 K, 100 kPa?

14.29 Carbon dioxide is heated at 100 kPa. What should the temperature be to see a mole fraction of CO as 0.25? For that temperature, what will the mole fraction of CO be if the pressure is 200 kPa?

14.30 Assume that a diatomic gas like O_2 or N_2 dissociates at a pressure different from P^0. Find an expression for the fraction of the original gas that has dissociated at any T, assuming equilibrium.

14.31 Hydrogen gas is heated from room temperature to 4000 K, 500 kPa, at which state the diatomic species has partially dissociated to the monatomic form. Determine the equilibrium composition at this state.

14.32 Consider the dissociation of oxygen, $O_2 \Leftrightarrow 2O$, starting with 1 kmol oxygen at 298 K and heating it at constant pressure 100 kPa. At which temperature will we reach a concentration of monatomic oxygen of 10 %?

14.33 Redo Problem 14.32 for a total pressure of 40 kPa.

14.34 Redo Problem 14.32, but start with 1 kmol oxygen and 1 kmol helium at 298 K, 100 kPa.

14.35 Calculate the equilibrium constant for the reaction $2CO_2 \rightleftharpoons 2CO + O_2$ at 3000 K using values from Table A.9, and compare the result to Table A.11.

14.36 Find the equilibrium constant for $CO + \frac{1}{2}O_2 \Leftrightarrow CO_2$ at 2200 K using Table A.11.

14.37 Pure oxygen is heated from 25 °C to 3200 K in a steady-state process at a constant pressure of 200 kPa. Find the exit composition and the heat transfer.

14.38 Nitrogen gas, N_2, is heated to 4000 K, 10 kPa. What fraction of the N_2 is dissociated to N at this state?

14.39 Assume we have air at 400 kPa, 2000 K (as 21 % O_2 and 79 % N_2) and we can neglect dissociations of O_2 and N_2. What is the equilibrium mole fraction of NO? Find the enthalpy difference in the gases due to the formation of the NO.

14.40 One kilomole Ar and one kilomole O_2 are heated at a constant pressure of 100 kPa to 3200 K, where they come to equilibrium. Find the final mole fractions for Ar, O_2, and O.

14.41 Air (assumed to be 79 % nitrogen and 21 % oxygen) is heated in a steady-state process at a constant pressure of 100 kPa, and some NO is formed (disregard dissociations of N_2 and O_2). At what temperature will the mole fraction of NO be 0.001?

14.42 Pure oxygen is heated from 25 °C, 100 kPa, to 3200 K in a constant-volume container. Find the final pressure, composition, and heat transfer.

14.43 Find the equilibrium constant for the reaction $2NO + O_2 \rightleftharpoons 2NO_2$ from the elementary reactions in Table A.11 to answer the question: which of the nitrogen oxides, NO or NO_2, is more stable at ambient conditions? What about at 2000 K?

14.44 Assume the equilibrium mole fractions of oxygen and nitrogen are close to those in air. Find the equilibrium mole fraction for NO at 3000 K, 500 kPa, disregarding dissociations.

14.45 The combustion products from burning pentane, C_5H_{12}, with pure oxygen in a stoichiometric ratio exit at 2400 K, 100 kPa. Consider the dissociation of only CO_2 and find the equilibrium mole fraction of CO.

14.46 A mixture flows with 2 kmol/s CO_2, 1 kmol/s Ar, and 1 kmol/s CO at 298 K and it is heated to 3000 K at constant 100 kPa. Assume the dissociation of CO_2 is the only equilibrium process to be considered. Find the exit equilibrium composition and the heat transfer rate.

14.47 A mixture of 1 kmol CO_2, 2 kmol CO, and 2 kmol O_2, at 25 °C, 150 kPa, is heated in a constant-pressure, steady-state process to 3000 K. Assuming that only these substances are present in the exiting chemical equilibrium mixture, determine the composition of that mixture.

14.48 Acetylene gas, C_2H_2, is burned with stoichiometric air in a torch. The reactants are supplied at the reference conditions P_0, T_0. The products come out from the flame at 2800 K after a small heat loss by radiation. Consider the dissociation of CO_2 into CO and O_2 and no others. Find the equilibrium composition of the products. Are there any other reactions that should be considered?

14.49 Consider combustion of CH_4 with O_2 forming CO_2 and H_2O as the products. Find the equilibrium constant for the reaction at 1000 K. Use an average specific heat of $C_p = 52$ kJ/kmol·K for the fuel and Table A.9 for the other components.

14.50 Water from the combustion of hydrogen and pure oxygen is at 3800 K and 50 kPa. Assume we only have H_2O, O_2, and H_2 as gases. Find the equilibrium composition.

14.51 Repeat Problem 14.47 for an initial mixture that also includes 2 kmol N_2, which does not dissociate during the process.

14.52 Catalytic gas generators are frequently used to decompose a liquid, providing a desired gas mixture (spacecraft control systems, fuel cell gas supply, and so forth). Consider feeding pure liquid hydrazine, N_2H_4, to a gas generator, from which exits a gas mixture of N_2, H_2, and NH_3 in chemical equilibrium at 100 °C, 350 kPa. Calculate the mole fractions of the species in the equilibrium mixture.

14.53 Complete combustion of hydrogen and pure oxygen in a stoichiometric ratio at P_0, T_0 to form water would result in a computed adiabatic flame temperature of 4990 K for a steady-state setup. How should the adiabatic flame temperature be found if the equilibrium reaction $2H_2 + O_2 \rightleftharpoons 2H_2O$ is considered? Disregard all other possible reactions (dissociations) and show the final equation(s) to be solved.

14.54 Consider the water–gas reaction in Example 14.4. Find the equilibrium constant at 500, 1000, 1200, and 1400 K. What can you infer from the result?

14.55 A piston/cylinder contains 0.1 kmol H_2 and 0.1 kmol Ar gas at 25 °C, 200 kPa. It is heated in a constant-pressure process, so the mole fraction of atomic hydrogen, H, is 10 %. Find the final temperature and the heat transfer needed.

14.56 The van't Hoff equation

$$d \ln K = \frac{\Delta H^0}{RT^2} dT_{p^0}$$

relates the chemical equilibrium constant K to the enthalpy of reaction ΔH^0. From the value of K in Table A.11 for the dissociation of hydrogen at 2000 K and the value of ΔH^0 calculated from Table A.9 at 2000 K, use the van't Hoff equation to predict the equilibrium constant at 2400 K.

14.57 A gas mixture of 1 kmol CO, 1 kmol N_2, and 1 kmol O_2 at 25 °C, 150 kPa, is heated in a constant-pressure, steady-state process. The exit mixture can be assumed to be in chemical equilibrium with CO_2, CO, O_2, and N_2 present. The mole fraction of CO_2 at this point is 0.176. Calculate the heat transfer for the process.

14.58 A tank contains 0.1 kmol H_2 and 0.1 kmol Ar gas at 25 °C, 200 kPa, and the tank maintains constant volume. To what T should it be heated to have a mole fraction of atomic hydrogen, H, of 10 %?

14.59 A stoichiometric combustion of butane C_4H_8 and air results in only half of the C atoms burning to CO; the other half generates CO. This means the products should contain a mixture of H_2O, CO, CO_2, O_2, and N_2, which, after some heat transfer, are at 1000 K. Write the combustion equation with no hydrogen in the products and then use the water–gas reaction to estimate the amount of hydrogen present.

14.60 A liquid fuel can be produced from a lighter fuel in a catalytic reactor according to

$$C_2H_4 + H_2O \Leftrightarrow C_2H_5OH$$

Show that the equilibrium constant is $\ln K = -6.691$ at 700 K, using $C_p = 63$ kJ/kmol·K for ethylene and $C_p = 115$ kJ/kmol·K for ethanol at 500 K.

14.61 A rigid container initially contains 2 kmol CO and 2 kmol O_2 at 25 °C, 100 kPa. The content is then heated to 3000 K, at which point an equilibrium mixture of CO_2, CO, and O_2 exists. Disregard other possible species and determine the final pressure, the equilibrium composition, and the heat transfer for the process.

14.62 Use the information in Problem 14.90 to estimate the enthalpy of reaction, ΔH^0, at 700 K using the van't Hoff equation (see Problem 14.56) with finite differences for the derivatives.

14.63 One kilomole of CO_2 and 1 kmol of H_2 at room temperature and 200 kPa are heated to 1200 K, 200 kPa. Use the water–gas reaction to determine the mole fraction of CO. Neglect dissociations of H_2 and O_2.

14.64 A step in the production of a synthetic liquid fuel from organic waste material is the following conversion process at 5 MPa: 1 kmol ethylene gas (converted from the waste) at 25 °C and 2 kmol steam at 300 °C enter a catalytic reactor. An ideal gas mixture of ethanol, ethylene, and water in equilibrium (see Problem 14.60) leaves the reactor at 700 K, 5 MPa. Determine the composition of the mixture.

14.65 A special coal burner uses a stoichiometric mixture of coal and an oxygen–argon mixture (1:1 mole ratio) with the reactants supplied at the reference conditions P_0, T_0. Consider the dissociation of CO_2 into CO and O_2 and no others. Find the equilibrium composition of the products for $T = 4800$ K. Is the final temperature, including dissociations, higher or lower than 4800 K?

14.66 Acetylene gas at 25 °C is burned with 140 % theoretical air, which enters the burner at 25 °C, 100 kPa, 80 % relative humidity. The combustion products form a mixture of CO_2, H_2O, N_2, O_2, and NO in chemical equilibrium at 2200 K, 100 kPa. This mixture is then cooled to 1000 K very rapidly, so that the composition does not change. Determine the mole fraction of NO in the products and the heat transfer for the overall process.

14.67 Coal is burned with stoichiometric air, with the reactants supplied at the reference conditions P_0, T_0. If no dissociations are considered, the adiabatic flame temperature is found to be 2461 K. What is it if the dissociation of CO_2 is included?

14.68 An important step in the manufacture of chemical fertilizer is the production of ammonia according to the reaction $N_2 + 3H_2 \Leftrightarrow 2NH_3$. Show that the equilibrium constant is $K = 6.202$ at 150 °C.

14.69 Consider the previous reaction in equilibrium at 150 °C, 5 MPa. For an initial composition of 25 % nitrogen, 75 % hydrogen, on a mole basis, calculate the equilibrium composition.

14.70 At high temperature, NO can form oxygen and nitrogen. Natural gas (methane) is burned with 150 % theoretical air at 100 kPa, and the product temperature is 2000 K. Neglect other reactions and find the equilibrium concentration of NO. Does the formation of the NO change the temperature?

14.71 Methane at 25 °C, 100 kPa, is burned with 200 % theoretical oxygen at 400 K, 100 kPa, in an adiabatic steady-state process, and the products of combustion exit at 100 kPa. Assume that the only significant dissociation reaction in the products is that of CO_2 going to CO and O_2. Determine the equilibrium composition of the products and also their temperature at the combustor exit.

14.72 Calculate the irreversibility for the adiabatic combustion process described in the previous problem.

14.73 Consider the stoichiometric combustion of pure carbon with air in a constant-pressure process at 100 kPa. Find the adiabatic flame temperature (no equilibrium reactions). Then find the temperature the mixture should be heated/cooled to so that the concentrations of CO and CO_2 are the same.

14.74 Hydrides are rare earth metals, M, that have the ability to react with hydrogen to form a different substance, MH_x, with a release of energy. The hydrogen can then be released, the reaction reversed, by heat addition to the MH_x. In this reaction, only the hydrogen is a gas, so the formula developed for the chemical equilibrium is inappropriate. Show that the proper expression to be used instead of Eq. 14.14 is

$$\ln\left(P_{H2}/P_0\right) = \Delta G^0/RT$$

when the reaction is scaled to 1 kmol of H_2.

Simultaneous Reactions

14.75 For the process in Problem 14.46, should the dissociation of oxygen also be considered? Provide a verbal answer but one supported by number(s).

14.76 Which other reactions should be considered in Problem 14.53, and which components will be present in the final mixture?

14.77 Redo Problem 14.46 and include the oxygen dissociation.

14.78 Ethane is burned with 150 % theoretical air in a gas-turbine combustor. The products exiting consist of a mixture of CO_2, H_2O, O_2, N_2, and NO in chemical equilibrium at 1800 K, 1 MPa. Determine the mole fraction of NO in the products. Is it reasonable to ignore CO in the products?

14.79 A mixture of 1 kmol H_2O and 1 kmol O_2 at 400 K is heated to 3000 K, 200 kPa, in a steady-state process. Determine the equilibrium composition at the outlet of the heat exchanger, assuming that the mixture consists of H_2O, H_2, O_2, and OH.

14.80 Assume dry air (79 % N_2 and 21 % O_2) is heated to 2000 K in a steady-flow process at 200 kPa and only the reactions listed in Table A.11 (and their linear combinations) are possible. Find the final composition (anything smaller than 1 ppm is neglected) and the heat transfer needed for 1 kmol of air in.

14.81 One kilomole of water vapor at 100 kPa, 400 K, is heated to 3000 K in a constant-pressure flow process. Determine the final composition, assuming that H_2O, H_2, H, O_2, and OH are present at equilibrium.

14.82 Water from the combustion of hydrogen and pure oxygen is at 3800 K and 50 kPa. Assume we only have H_2O, O_2, OH, and H_2 as gases with the two simple water dissociation reactions active. Find the equilibrium composition.

14.83 Methane is burned with theoretical oxygen in a steady-state process, and the products exit the combustion chamber at 3200 K, 700 kPa. Calculate the equilibrium composition at this state, assuming that only CO_2, CO, H_2O, H_2, O_2, and OH are present.

14.84 Butane is burned with 200 % theoretical air, and the products of combustion, an equilibrium mixture containing only CO_2, H_2O, O_2, N_2, NO, and NO_2, exits from the combustion chamber at 1400 K, 2 MPa. Determine the equilibrium composition at this state.

14.85 One kilomole of air (assumed to be 78 % N_2, 21 % O_2, and 1 % Ar) at room temperature is heated to 4000 K, 200 kPa. Find the equilibrium composition at this state, assuming that only N_2, O_2, NO, O, and Ar are present.

14.86 Acetylene gas and x times theoretical air ($x > 1$) at room temperature and 500 kPa are burned at constant pressure in an adiabatic flow process. The flame temperature is 2600 K, and the combustion products are assumed to consist of N_2, O_2, CO_2, H_2O, CO, and NO. Determine the value of x.

Gasification

14.87 One approach to using hydrocarbon fuels in a fuel cell is to "reform" the hydrocarbon to obtain hydrogen, which is then fed to the fuel cell. As part of the analysis of such a procedure, consider the reaction $CH_4 + H_2O \rightleftharpoons 3H_2 + CO$. Determine the equilibrium constant for this reaction at a temperature of 800 K.

14.88 A coal gasifier produces a mixture of 1 CO and 2 H_2 that is fed to a catalytic converter to produce methane. This is the methanation reaction in Eq. 14.33 with an equilibrium constant at 600 K of $K = 1.83 \times 10^6$. What is the composition of the exit flow, assuming a pressure of 600 kPa?

14.89 Gasification of char (primarily carbon) with steam following coal pyrolysis yields a gas mixture of 1 kmol CO and 1 kmol H_2. We wish to upgrade the H_2 content of this syngas fuel mixture, so it is fed to an appropriate catalytic reactor along with 1 kmol of H_2O. Exiting the reactor is a chemical equilibrium gas mixture of CO, H_2, H_2O, and CO_2 at 600 K, 500 kPa. Determine the equilibrium composition. *Note*: See Example 14.4.

14.90 The equilibrium reaction with methane as $CH_4 \rightleftharpoons C + 2H_2$ has $\ln K = -0.3362$ at 800 K, and $\ln K = -4.607$ at 600 K. Noting the relation of K to temperature, show how you would interpolate $\ln K$ in $(1/T)$ to find K at 700 K and compare that to a linear interpolation.

14.91 One approach to using hydrocarbon fuels in a fuel cell is to "reform" the hydrocarbon to obtain hydrogen, which is then fed to the fuel cell. As a part of the analysis of such a procedure, consider the reaction $CH_4 + H_2O \Leftrightarrow CO + 3H_2$. One kilomole each of methane and water are fed to a catalytic reformer. A mixture of CH_4, H_2O, H_2, and CO exits in chemical equilibrium at 800 K, 100 kPa. Determine the equilibrium composition of this mixture using an equilibrium constant of $K = 0.0237$.

14.92 Consider a gasifier that receives 4 kmol CO, 3 kmol H_2, and 3.76 kmol N_2 and brings the mixture to equilibrium at 900 K, 1 MPa, with the following reaction:

$$2\,CO + 2\,H_2 \Leftrightarrow CH_4 + CO_2$$

which is the sum of Eqs. 14.32 and 14.33. If the equilibrium constant is $K = 2.679$, find the exit flow composition.

14.93 Consider the production of a synthetic fuel (methanol) from coal. A gas mixture of 50 % CO and 50 % H_2 leaves a coal gasifier at 500 K, 1 MPa, and enters a catalytic converter. A gas mixture of methanol, CO, and H_2 in chemical equilibrium with the reaction $CO + 2H_2 \rightleftharpoons CH_3OH$ leaves the converter at the same temperature and pressure, where it is known that $\ln K = -5.119$.

a. Calculate the equilibrium composition of the mixture leaving the converter.

b. Would it be more desirable to operate the converter at ambient pressure?

Ionization

14.94 At 10 000 K, the ionization reaction for argon as $Ar \Leftrightarrow Ar^+ + e^-$ has an equilibrium constant of $K = 4.2 \times 10^{-4}$. What should the pressure be for a mole concentration of argon ions (Ar^+) of 10 %?

14.95 Repeat the previous problem, assuming the argon constitutes 1 % of a gas mixture where we neglect any reactions of other gases and find the pressure that will give a mole concentration of Ar^+ of 0.1 %.

14.96 Operation of an MHD converter requires an electrically conducting gas. A helium gas "seeded" with 1.0 mole percent cesium, as shown in Fig. P14.96, is used where the cesium is partly ionized $(Cs \rightleftharpoons Cs^+ + e^-)$ by heating the mixture to 1800 K, 1 MPa, in a nuclear reactor to provide free electrons. No helium is ionized in this process, so the mixture entering the converter consists of He, Cs, Cs^+, and e^-. Determine the mole fraction of electrons in the mixture at 1800 K, where $\ln K = 1.402$ for the cesium ionization reaction described.

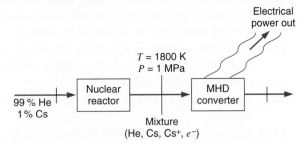

FIGURE P14.96

14.97 One kilomole of Ar gas at room temperature is heated to 20 000 K, 100 kPa. Assume that the plasma in this condition consists of an equilibrium mixture of Ar, Ar^+, Ar^{++}, and e^- according to the simultaneous reactions

(1) $Ar \rightleftharpoons Ar^+ + e^-$ (2) $Ar^+ \rightleftharpoons Ar^{++} + e^-$

The ionization equilibrium constants for these reactions at 20 000 K have been calculated from spectroscopic data as $\ln K_1 = 3.11$ and $\ln K_2 = -4.92$. Determine the equilibrium composition of the plasma.

14.98 At 10 000 K, the two ionization reactions for N and Ar as

(1) $Ar \Leftrightarrow Ar^+ + e^-$ (2) $N \Leftrightarrow N^+ + e^-$

have equilibrium constants of $K_1 = 4.2 \times 10^{-4}$ and $K_2 = 6.3 \times 10^{-4}$, respectively. If we start out with 1 kmol Ar and 0.5 kmol N_2, what is the equilibrium composition at a pressure of 10 kPa?

14.99 Plot to scale the equilibrium composition of nitrogen at 10 kPa over the temperature range 5000 K to 15 000 K, assuming that N_2, N, N^+, and e^- are present. For the ionization reaction N $\rightleftharpoons$ $N^+ + e^-$, the ionization equilibrium constant K has been calculated from spectroscopic data as

T [K]	10 000	12 000	14 000	16 000
$100K$	6.26×10^{-2}	1.51	15.1	92

Applications

14.100 Are the three reactions in the Zeldovich mechanism pressure sensitive if we look at equilibrium conditions?

14.101 Assume air is at 3000 K, 1 MPa. Find the time constant for NO formation. Repeat for 2000 K, 800 kPa.

14.102 Consider air at 2600 K, 1 MPa. Find the equilibrium concentration of NO, neglecting dissociations of oxygen and nitrogen.

14.103 Redo the previous problem but include the dissociation of oxygen and nitrogen.

14.104 Calculate the equilibrium constant for the first reaction in the Zeldovich mechanism at 2600 K, 500 kPa. Notice that this is not listed in Table A.11.

14.105 Find the equilibrium constant for the reaction $2NO + O_2 \Leftrightarrow 2NO_2$ from the elementary reaction in Table A.11 to answer these two questions: Which nitrogen oxide, NO or NO_2, is more stable at 25 °C, 100 kPa? At what T do we have an equal amount of each?

14.106 If air at 300 K is brought to 2600 K, 1 MPa, instantly, find the formation rate of NO.

14.107 Estimate the concentration of oxygen atoms in air at 3000 K, 100 kPa, and 0.0001 kPa. Compare this to the result in Fig. 14.11.

14.108 At what temperature range does air become a plasma?

Review Problems

14.109 In a test of a gas-turbine combustor, saturated-liquid methane at 115 K is burned with excess air to hold the adiabatic flame temperature to 1600 K. It is assumed that the products consist of a mixture of CO_2, H_2O, N_2, O_2, and NO in chemical equilibrium. Determine the percent excess air used in the combustion and the percentage of NO in the products.

14.110 Find the equilibrium constant for the reaction in Problem 14.92.

14.111 A space heating unit in Alaska uses propane combustion as the heat supply. Liquid propane comes from an outside tank at −44 °C, and the air supply is also taken in from the outside at −44 °C. The air flow regulator is misadjusted, so only 90 % of the theoretical air enters the combustion chamber, resulting in incomplete combustion. The products exit at 1000 K as a chemical equilibrium gas mixture, including only CO_2, CO, H_2O, H_2, and N_2. Find the composition of the products. *Hint*: Use the water gas reaction in Example 14.4.

14.112 Derive the van't Hoff equation given in Problem 14.56, using Eqs. 14.12 and 14.15. *Note*: The $d(\overline{g}/T)$ at constant P^0 for each component can be expressed using the relations in Eqs. 12.18 and 12.19.

14.113 Find the equilibrium constant for Eq. 14.33 at 600 K (see Problem 14.88).

14.114 Combustion of stoichiometric benzene, C_6H_6, and air at 80 kPa with a slight heat loss gives a flame temperature of 2400 K. Consider the dissociation of CO_2 to CO and O_2 as the only equilibrium process possible. Find the fraction of the CO_2 that is dissociated.

14.115 One kilomole of liquid oxygen, O_2, at 93 K, and x kmol of gaseous hydrogen, H_2, at 25 °C, are fed to a combustion chamber (x is greater than 2) such that there is excess hydrogen for the combustion process. There is a heat loss from the chamber of 1000 kJ per kilomole of reactants. Products exit the chamber at chemical equilibrium at 3800 K, 400 kPa, and are assumed to include only H_2O, H_2, and O.

  a. Determine the equilibrium composition of the products and x, the amount of H_2 entering the combustion chamber.

  b. Should another substance(s) have been included in part (a) as being present in the products? Justify your answer.

14.116 Dry air is heated from 25 °C to 4000 K in a 100 kPa constant-pressure process. List the possible reactions that may take place and determine the equilibrium composition. Find the required heat transfer.

14.117 Saturated liquid butane (*note*: use generalized charts) enters an insulated constant-pressure combustion chamber at 25 °C, and *x* times theoretical oxygen gas enters at the same pressure and temperature. The combustion products exit at 3400 K. Assuming that the products are a chemical equilibrium gas mixture that includes CO, what is *x*?

COMPUTER, DESIGN, AND OPEN-ENDED PROBLEMS

14.118 Write a program to solve the general case of Problem 14.64, in which the relative amount of steam input and the reactor temperature and pressure are program input variables and use constant specific heats.

14.119 Write a program to solve the following problem. One kilomole of carbon at 25 °C is burned with *b* kmol of oxygen in a constant-pressure adiabatic process. The products consist of an equilibrium mixture of CO_2, CO, and O_2. We wish to determine the flame temperature for various combinations of *b* and the pressure *P*, assuming constant specific heat for the components from Table A.5.

14.120 Study the chemical reactions that take place when CFC-type refrigerants are released into the atmosphere. The chlorine may create compounds such as HCl and $ClONO_2$ that react with the ozone, O_3.

14.121 Examine the chemical equilibrium that takes place in an engine where CO and various nitrogen–oxygen compounds summarized as NO_x may be formed. Study the processes for a range of air–fuel ratios and temperatures for typical fuels. Are there important reactions not listed in the book?

14.122 A number of products may be produced from the conversion of organic waste that can be used as fuel (see Problem 14.64). Study the subject and make a list of the major products that are formed and the conditions at which they are formed in desirable concentrations.

14.123 The hydrides, as explained in Problem 14.74, can store large amounts of hydrogen. The penalty for the storage is that energy must be supplied when the hydrogen is released. Investigate the literature for quantitative information about the quantities and energy involved in such a hydrogen storage.

14.124 Excess air or steam addition is often used to lower the peak temperature in combustion to limit formation of pollutants like NO. Study the steam addition to the combustion of natural gas, as in the Cheng cycle (see Problem 11.124), assuming the steam is added before the combustion. How does this affect the peak temperature and the NO concentration?

Compressible Flow

15

This chapter deals with the thermodynamic aspects of simple compressible flows through nozzles and passages. Several of the cycles covered in Chapters 9 and 10 have flow inside components, where it goes through nozzles or diffusers. For instance, a set of nozzles inside a steam turbine converts a high-pressure steam flow into a lower-pressure, high-velocity flow that enters the passage between the rotating blades. After several sections, the flow goes through a diffuser-like chamber and another set of nozzles. The flow in a fan-jet has several locations where a high-speed compressible gas flows; it passes first through a diffuser followed by a fan and compressor, then through passages between turbine blades, and finally exits through a nozzle. A final example of a flow that must be treated as compressible is the flow through a turbocharger in a diesel engine; the flow continues further through the intake system and valve openings to end up in a cylinder. The proper analysis of these processes is important for an accurate evaluation of the mass flow rate, the work, heat transfer, or kinetic energy involved, and feeds into the design and operating behavior of the overall system.

All of the examples mentioned here are complicated with respect to the flow geometry and the flowing media, so we will use a simplifying model. In this chapter we will treat one-dimensional flow of a pure substance that we will also assume behaves as an ideal gas for most of the developments. This allows us to focus on the important aspects of a compressible flow, which is influenced by the sonic velocity, and the Mach number appears as an important variable for this type of flow.

15.1 STAGNATION PROPERTIES

In dealing with problems involving flow, many discussions and equations can be simplified by introducing the concept of the isentropic stagnation state and the properties associated with it. The isentropic stagnation state is the state a flowing fluid would attain if it underwent a reversible adiabatic deceleration to zero velocity. This state is designated in this chapter with the subscript 0. From the energy equation for a steady-state process, we conclude that

$$h + \frac{\mathbf{V}^2}{2} = h_0 \tag{15.1}$$

The actual and isentropic stagnation states for a typical gas or vapor are shown on the *h–s* diagram of Fig. 15.1. Sometimes it is advantageous to make a distinction between the actual and isentropic stagnation states. The actual stagnation state is the state achieved after an actual deceleration to zero velocity (as at the nose of a body placed in a fluid stream), and there may be irreversibilities associated with the deceleration process. Therefore, the term stagnation property is sometimes reserved for the properties associated with the actual state, and the term total property is used for the isentropic stagnation state.

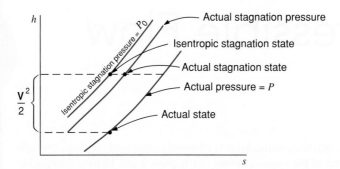

FIGURE 15.1 An h–s diagram illustrating the definition of stagnation state.

It is evident from Fig. 15.1 that the enthalpy is the same for both the actual and isentropic stagnation states (assuming that the actual process is adiabatic). Therefore, for an ideal gas, the actual stagnation temperature is the same as the isentropic stagnation temperature. However, the actual stagnation pressure may be less than the isentropic stagnation pressure. For this reason, the term total pressure (meaning isentropic stagnation pressure) has particular meaning compared to actual stagnation pressure.

Example 15.1

Air flows in a duct at a pressure of 150 kPa with a velocity of 200 m/s. The temperature of the air is 300 K. Determine the isentropic stagnation pressure and temperature.

Analysis and Solution

If we assume that the air is an ideal gas with constant specific heat as given in Table A.5, the calculation is as follows. From Eq. 15.1

$$\frac{\mathbf{V}^2}{2} = h_0 - h = C_{p0}(T_0 - T)$$

$$\frac{(200)^2}{2 \times 1000} = 1.004(T_0 - 300)$$

$$T_0 = 319.9 \, \text{K}$$

The stagnation pressure can be found from the relation

$$\frac{T_0}{T} = \left(\frac{P_0}{P}\right)^{(k-1)/k}$$

$$\frac{319.9}{300} = \left(\frac{P_0}{150}\right)^{0.286}$$

$$P_0 = 187.8 \, \text{kPa}$$

The air tables, Table A.7, which are calculated from Table A.8, could also have been used, and then the variation of specific heat with temperature would have been taken into account. Since the actual and stagnation states have the same entropy, we proceed as follows: Using Table A.7.2,

$$T = 300\,\text{K} \qquad h = 300.47\,\text{kJ/kg} \qquad P_r = 1.1146$$

$$h_0 = h + \frac{\mathbf{V}^2}{2} = 300.47 + \frac{(200)^2}{2 \times 1000} = 320.47\,\text{kJ/kg}$$

$$T_0 = 319.9\,\text{K} \qquad P_{r0} = 1.3956$$

$$P_0 = P P_{r0}/P_r = 150 \times \frac{1.3956}{1.1146} = 187.8\,\text{kPa}$$

15.2 THE MOMENTUM EQUATION FOR A CONTROL VOLUME

Before proceeding, it will be advantageous to develop the momentum equation for the control volume. Newton's second law states that the sum of the external forces acting on a body in a given direction is proportional to the rate of change of momentum in the given direction. Writing this in equation form for the x-direction, we have

$$\frac{d(m\mathbf{V}_x)}{dt} \propto \sum F_x$$

For the system of units used in this book, the proportionality can be written directly as an equality.

$$\frac{d(m\mathbf{V}_x)}{dt} = \sum F_x \tag{15.2}$$

Equation 15.2 has been written for a body of fixed mass, or in thermodynamic parlance, for a control mass. We now proceed to write the momentum equation for a control volume, and we follow a procedure similar to that used in writing the continuity equation and the first and second laws of thermodynamics for a control volume.

Consider the control volume shown in Fig. 15.2 to be fixed relative to its coordinate frame. Each flow that enters or leaves the control volume possesses an amount of momentum per unit mass, so that it adds or subtracts a rate of momentum to or from the control volume.

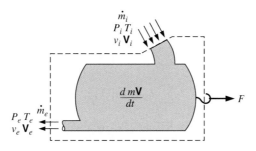

FIGURE 15.2
Development of the momentum equation for a control volume.

Writing the momentum equation in a rate form similar to the balance equations for mass, energy, and entropy, Eqs. 4.1, 4.7, and 7.2, respectively, results in an expression of the form

$$\text{Rate of change} = \sum F_x + \text{in} - \text{out} \tag{15.3}$$

Only forces acting on the mass inside the control volume (for example, gravity) or on the control volume surface (for example, friction or piston forces) and the flow of mass carrying momentum can contribute to a change of momentum. Momentum is conserved, so that it cannot be created or destroyed, as was previously stated for the other control volume developments.

The momentum equation in the x-direction from the form of Eq. 15.3 becomes

$$\frac{d(m\mathbf{V}_x)}{dt} = \sum F_x + \sum \dot{m}_i \mathbf{V}_{ix} - \sum \dot{m}_e \mathbf{V}_{ex} \tag{15.4}$$

Similarly, for the y- and z-directions,

$$\frac{d(m\mathbf{V}_y)}{dt} = \sum F_y + \sum \dot{m}_i \mathbf{V}_{iy} - \sum \dot{m}_e \mathbf{V}_{ey} \tag{15.5}$$

and

$$\frac{d(m\mathbf{V}_z)}{dt} = \sum F_z + \sum \dot{m}_i \mathbf{V}_{iz} - \sum \dot{m}_e \mathbf{V}_{ez} \tag{15.6}$$

In the case of a control volume with no mass flow rates in or out (i.e., a control mass), these equations reduce to the form of Eq. 15.2 for each direction.

In this chapter, we will be concerned primarily with steady-state processes in which there is a single flow with uniform properties into the control volume and a single flow with uniform properties out of the control volume. The steady-state assumption means that the rate of momentum change for the control volume terms in Eqs. 15.4, 15.5, and 15.6 is equal to zero. That is,

$$\frac{d(m\mathbf{V}_x)_{\text{c.v.}}}{dt} = 0 \qquad \frac{d(m\mathbf{V}_y)_{\text{c.v.}}}{dt} = 0 \qquad \frac{d(m\mathbf{V}_z)_{\text{c.v.}}}{dt} = 0 \tag{15.7}$$

Therefore, for the steady-state process the momentum equation for the control volume, assuming uniform properties at each state, reduces to the form

$$\sum F_x = \sum \dot{m}_e (\mathbf{V}_e)_x - \sum \dot{m}_i (\mathbf{V}_i)_x \tag{15.8}$$

$$\sum F_y = \sum \dot{m}_e (\mathbf{V}_e)_y - \sum \dot{m}_i (\mathbf{V}_i)_y \tag{15.9}$$

$$\sum F_z = \sum \dot{m}_e (\mathbf{V}_e)_z - \sum \dot{m}_i (\mathbf{V}_i)_z \tag{15.10}$$

Furthermore, for the special case in which there is a single flow into and out of the control volume, these equations reduce to

$$\sum F_x = \dot{m}[(\mathbf{V}_e)_x - (\mathbf{V}_i)_x] \tag{15.11}$$

$$\sum F_y = \dot{m}[(\mathbf{V}_e)_y - (\mathbf{V}_i)_y] \tag{15.12}$$

$$\sum F_z = \dot{m}[(\mathbf{V}_e)_z - (\mathbf{V}_i)_z] \tag{15.13}$$

Example 15.2

On a level floor, a man is pushing a wheelbarrow (Fig. 15.3) into which sand is falling at the rate of 1 kg/s. The man is walking at the rate of 1 m/s, and the sand has a velocity of 10 m/s as it falls into the wheelbarrow. Determine the force the man must exert on the wheelbarrow and the force the floor exerts on the wheelbarrow due to the falling sand.

Analysis and Solution

Consider a control surface around the wheelbarrow. Consider first the x-direction. From Eq. 15.4

$$\sum F_x = \frac{d(m\mathbf{V}_x)_{\text{c.v.}}}{dt} + \sum \dot{m}_e(\mathbf{V}_e)_x - \sum \dot{m}_i(\mathbf{V}_i)_x$$

Let us analyze this problem from the point of view of an observer riding on the wheelbarrow. For this observer, $\mathbf{V}_x$ of the material in the wheelbarrow is zero and therefore,

$$\frac{d(m\mathbf{V}_x)_{\text{c.v.}}}{dt} = 0$$

However, for this observer the sand crossing the control surface has an x-component velocity of -1 m/s, and $\dot{m}$, the mass flow out of the control volume, is -1 kg/s. Therefore,

$$F_x = (-1 \text{ kg/s}) \times (-1 \text{ m/s}) = 1 \text{ N}$$

If one considers this from the point of view of an observer who is stationary on the earth's surface, we conclude that $\mathbf{V}_x$ of the falling sand is zero and therefore

$$\sum \dot{m}_e(\mathbf{V}_e)_x - \sum \dot{m}_i(\mathbf{V}_i)_x = 0$$

However, for this observer there is a change of momentum within the control volume, namely,

$$\sum F_x = \frac{d(m\mathbf{V}_x)_{\text{c.v.}}}{dt} = (1 \text{ m/s}) \times (1 \text{ kg/s}) = 1 \text{ N}$$

Next, consider the vertical (y) direction.

$$\sum F_y = \frac{d(m\mathbf{V}_y)_{\text{c.v.}}}{dt} + \sum \dot{m}_e(\mathbf{V}_e)_y - \sum \dot{m}_i(\mathbf{V}_i)_y$$

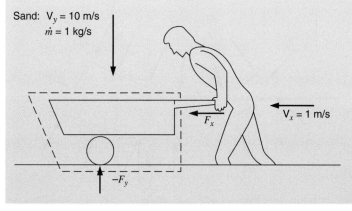

FIGURE 15.3

Sketch for Example 15.2.

For both the stationary and moving observers, the first term drops out because $\mathbf{V}_y$ of the mass within the control volume is zero. However, for the mass crossing the control surface, $\mathbf{V}_y = 10$ m/s and

$$\dot{m} = -1 \text{ kg/s}$$

Therefore

$$F_y = (10 \text{ m/s}) \times (-1 \text{ kg/s}) = -10 \text{ N}$$

The minus sign indicates that the force is in the opposite direction to $\mathbf{V}_y$.

15.3 FORCES ACTING ON A CONTROL SURFACE

In the previous section, we considered the momentum equation for the control volume. We now wish to evaluate the net force on a control surface that causes this change in momentum. Let us do this by considering the control mass shown in Fig. 15.4, which involves a pipe bend. The control surface is designated by the dotted lines and is so chosen that at the point where the fluid crosses the system boundary, the flow is perpendicular to the control surface. The shear forces at the section where the fluid crosses the boundary of the control surface are assumed to be negligible. Figure 15.4a shows the velocities, and Fig. 15.4b shows the

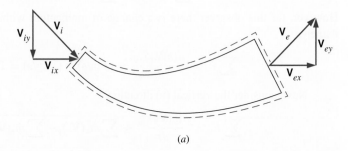

(a)

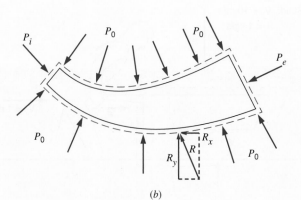

FIGURE 15.4 Forces acting on a control surface.

(b)

forces involved. The force R is the result of all external forces on the control mass, except for the pressure of all surroundings. The pressure of the surroundings, P_0, acts on the entire boundary except at A_i and A_e, where the fluid crosses the control surface; P_i and P_e represent the absolute pressures at these points.

The net forces acting on the system in the x- and y-directions, F_x and F_y, are the sum of the pressure forces and the external force R in their respective directions. The influence of the pressure of the surroundings, P_0, is most easily taken into account by noting that it acts over the entire control mass boundary except at A_i and A_e. Therefore, we can write

$$\sum F_x = (P_i A_i)_x - (P_0 A_i)_x + (P_e A_e)_x - (P_0 A_e)_x + R_x$$

$$\sum F_y = (P_i A_i)_y - (P_0 A_i)_y + (P_e A_e)_y - (P_0 A_e)_y + R_y$$

This equation may be simplified by combining the pressure terms.

$$\sum F_x = [(P_i - P_0)A_i]_x + [(P_e - P_0)A_e]_x + R_x$$

$$\sum F_y = [(P_i - P_0)A_i]_y + [(P_e - P_0)A_e]_y + R_y \qquad (15.14)$$

The proper sign for each pressure and force must, of course, be used in all calculations.

Equations 15.8, 15.9, and 15.14 may be combined to give

$$\sum F_x = \sum \dot{m}_e(\mathbf{V}_e)_x - \sum \dot{m}_i(\mathbf{V}_i)_x$$

$$= \sum [(P_i - P_0)A_i]_x + \sum [(P_e - P_0)A_e]_x + R_x$$

$$\sum F_y = \sum \dot{m}_e(\mathbf{V}_e)_y - \sum \dot{m}_i(\mathbf{V}_i)_y$$

$$= \sum [(P_i - P_0)A_i]_y + \sum [(P_e - P_0)A_e]_y + R_y \qquad (15.15)$$

If there is a single flow across the control surface, Eqs. 15.11, 15.12, and 15.14 can be combined to give

$$\sum F_x = \dot{m}(\mathbf{V}_e - \mathbf{V}_i)_x = [(P_i - P_0)A_i]_x + [(P_e - P_0)A_e]_x + R_x$$

$$\sum F_y = \dot{m}(\mathbf{V}_e - \mathbf{V}_i)_y = [(P_i - P_0)A_i]_y + [(P_e - P_0)A_e]_y + R_y \qquad (15.16)$$

A similar equation could be written for the z-direction. These equations are very useful in analyzing the forces involved in a control-volume analysis.

Example 15.3

A jet engine is being tested on a test stand (Fig. 15.5). The inlet area to the compressor is 0.2 m^2, and air enters the compressor at 95 kPa, 100 m/s. The pressure of the atmosphere is 100 kPa. The exit area of the engine is 0.1 m^2, and the products of combustion leave the exit plane at a pressure of 125 kPa and a velocity of 450 m/s. The air–fuel ratio is 50 kilograms of air to one kilogram of fuel, and the fuel enters with a low velocity. The rate of air flow entering the engine is 20 kg/s. Determine the thrust, R_x, on the engine.

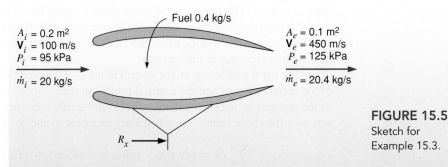

FIGURE 15.5
Sketch for
Example 15.3.

Analysis and Solution

In the solution that follows, it is assumed that forces and velocities to the right are positive.

Using Eq. 15.16

$$R_x + [(P_i - P_0)A_i]_x + [(P_e - P_0)A_e]_x = (\dot{m}_e\mathbf{V}_e - \dot{m}_i\mathbf{V}_i)_x$$

$$R_x + [(95 - 100) \times 0.2] - [(125 - 100) \times 0.1] = \frac{20.4 \times 450 - 20 \times 100}{1000}$$

$$R_x = 10.68\,\text{kN}$$

(Note that the momentum of the fuel entering has been neglected.)

15.4 ADIABATIC, ONE-DIMENSIONAL, STEADY-STATE FLOW OF AN INCOMPRESSIBLE FLUID THROUGH A NOZZLE

A nozzle is a device in which the kinetic energy of a fluid is increased in an adiabatic process. This increase involves a decrease in pressure and is accomplished by the proper change in flow area. A diffuser is a device that has the opposite function, namely, to increase the pressure by decelerating the fluid. In this section, we discuss both nozzles and diffusers, but to minimize words we shall use only the term *nozzle*.

Consider the nozzle shown in Fig. 15.6, and assume an adiabatic, one-dimensional, steady-state process of an incompressible fluid. From the continuity equation, we conclude that

$$\dot{m}_e = m_i = \rho A_i\mathbf{V}_i = \rho A_e\mathbf{V}_e$$

or

$$\frac{A_i}{A_e} = \frac{\mathbf{V}_e}{\mathbf{V}_i} \qquad (15.17)$$

The energy equation for this process is

$$h_e - h_i + \frac{\mathbf{V}_e^2 - \mathbf{V}_i^2}{2} + (Z_e - Z_i)g = 0 \qquad (15.18)$$

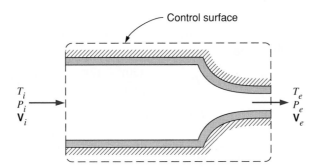

FIGURE 15.6
Schematic sketch of a nozzle.

From the entropy equation, we conclude that $s_e \geq s_i$, where the equality holds for a reversible process. Therefore, from the relation

$$T \, ds = dh - v \, dP$$

we conclude that for the reversible process

$$h_e - h_i = \int_i^e v \, dP \qquad (15.19)$$

If we assume that the fluid is incompressible, Eq. 15.19 can be integrated to give

$$h_e - h_i = v(P_e - P_i) \qquad (15.20)$$

Substituting this in Eq. 15.18, we have

$$v(P_e - P_i) + \frac{\mathbf{V}_e^2 - \mathbf{V}_i^2}{2} + (Z_e - Z_i)g = 0 \qquad (15.21)$$

This is, of course, the Bernoulli equation, which was derived in Section 7.3, Eq. 7.17. For the reversible, adiabatic, one-dimensional, steady-state flow of an incompressible fluid through a nozzle, the Bernoulli equation represents a combined statement of the energy and entropy equations.

Example 15.4

Water enters the diffuser in a pump casing with a velocity of 30 m/s, a pressure of 350 kPa, and a temperature of 25 °C. It leaves the diffuser with a velocity of 7 m/s and a pressure of 600 kPa. Determine the exit pressure for a reversible diffuser with these inlet conditions and exit velocity. Determine the increase in enthalpy, internal energy, and entropy for the actual diffuser.

Analysis and Solution

Consider first a control surface around a reversible diffuser with the given inlet conditions and exit velocity. Equation 15.21, the Bernoulli equation, is a statement of the energy and entropy equations for this process. Since there is no change in elevation, this equation reduces to

$$v[(P_e)_s - P_i] + \frac{\mathbf{V}_e^2 - \mathbf{V}_i^2}{2} = 0$$

where $(P_e)_s$ represents the exit pressure for the reversible diffuser. From the steam tables, $v = 0.001\,003$ m^3/kg.

$$P_{es} - P_i = \frac{(30)^2 - (7)^2}{0.001\,003 \times 2 \times 1000} = 424\text{ kPa}$$

$$P_{es} = 774\text{ kPa}$$

Next, consider a control surface around the actual diffuser. The change in enthalpy can be found from the energy equation for this process, Eq. 15.18.

$$h_e - h_i = \frac{\mathbf{V}_i^2 - \mathbf{V}_e^2}{2} = \frac{(30)^2 - (7)^2}{2 \times 1000} = 0.4255\text{ kJ/kg}$$

The change in internal energy can be found from the definition of enthalpy, $h_e - h_i = (u_e - u_i) + (P_e v_e - P_i v_i)$.

Thus, for an incompressible fluid

$$u_e - u_i = h_e - h_i - v(P_e - P_i)$$
$$= 0.4255 - 0.001\,003(600 - 350)$$
$$= 0.174\,75\text{ kJ/kg}$$

The change of entropy can be approximated from the familiar relation

$$T\,ds = du + P\,dv$$

by assuming that the temperature is constant (which is approximately true in this case) and noting that for an incompressible fluid, $dv = 0$. With these assumptions

$$s_e - s_i = \frac{u_e - u_i}{T} = \frac{0.174\,75}{298.2} = 0.000\,586\text{ kJ/kg·K}$$

Since this is an irreversible adiabatic process, the entropy will increase, as the above calculation indicates.

15.5 VELOCITY OF SOUND IN AN IDEAL GAS

When a pressure disturbance occurs in a compressible fluid, the disturbance travels with a velocity that depends on the state of the fluid. A sound wave is a very small pressure disturbance; the velocity of sound, also called the *sonic velocity*, is an important parameter in compressible-fluid flow. We proceed now to determine an expression for the sonic velocity of an ideal gas in terms of the properties of the gas.

Let a disturbance be set up by the movement of the piston at the end of the tube, Fig. 15.7a. A wave travels down the tube with a velocity c, which is the sonic velocity. Assume that after the wave has passed, the properties of the gas have changed an infinitesimal amount and that the gas is moving with the velocity $d\mathbf{V}$ toward the wave front.

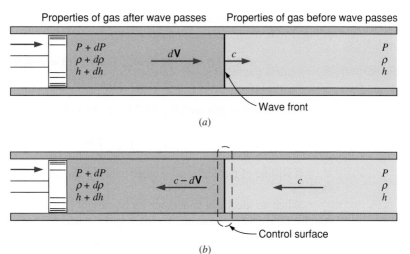

Properties of gas after wave passes   Properties of gas before wave passes

(a)

FIGURE 15.7
Diagram illustrating sonic velocity. (a) Stationary observer. (b) Observer traveling with wave front.

(b)

In Fig. 15.7b, this process is shown from the point of view of an observer who travels with the wave front. Consider the control surface shown in Fig. 15.7b. From the energy equation for this steady-state process, we can write

$$h + \frac{c^2}{2} = (h + dh) + \frac{(c - d\mathbf{V})^2}{2}$$

$$dh - c \, d\mathbf{V} = 0 \qquad (15.22)$$

From the continuity equation, we can write

$$\rho A c = (\rho + d\rho) A (c - d\mathbf{V})$$

$$c \, d\rho - \rho \, d\mathbf{V} = 0 \qquad (15.23)$$

Consider also the relation between properties

$$T \, ds = dh - \frac{dP}{\rho}$$

If the process is isentropic, $ds = 0$, and this equation can be combined with Eq. 15.22 to give the relation

$$\frac{dP}{\rho} - c \, d\mathbf{V} = 0 \qquad (15.24)$$

This can be combined with Eq. 15.23 to give the relation

$$\frac{dP}{d\rho} = c^2$$

Since we have assumed the process to be isentropic, this is better written as a partial derivative.

$$\left(\frac{\partial P}{\partial \rho} \right)_s = c^2 \qquad (15.25)$$

An alternate derivation is to introduce the momentum equation. For the control volume of Fig. 15.7b, the momentum equation is

$$PA - (P + dP)A = \dot{m}(c - d\mathbf{V} - c) = \rho Ac(c - d\mathbf{V} - c)$$

$$dP = \rho c \, d\mathbf{V} \tag{15.26}$$

On combining this with Eq. 15.23, we obtain Eq. 15.25.

$$\left(\frac{\partial P}{\partial \rho}\right)_s = c^2$$

It will be of particular advantage to solve Eq. 15.25 for the velocity of sound in an ideal gas.

When an ideal gas undergoes an isentropic change of state, we found in Chapter 6 that, for this process, assuming constant specific heat

$$\frac{dP}{P} - k\frac{d\rho}{\rho} = 0$$

or

$$\left(\frac{\partial P}{\partial \rho}\right)_s = \frac{kP}{\rho}$$

Substituting this equation in Eq. 15.25, we have an equation for the velocity of sound in an ideal gas,

$$c^2 = \frac{kP}{\rho} \tag{15.27}$$

Since for an ideal gas

$$\frac{P}{\rho} = RT$$

this equation may also be written

$$c^2 = kRT \tag{15.28}$$

Example 15.5

Determine the velocity of sound in air at 300 K and at 1000 K.

Analysis and Solution

Using Eq. 15.28

$$c = \sqrt{kRT}$$

$$= \sqrt{1.4 \times 0.287 \times 300 \times 1000} = 347.2 \text{ m/s}$$

Similarly, at 1000 K, using $k = 1.4$,

$$c = \sqrt{1.4 \times 0.287 \times 1000 \times 1000} = 633.9 \text{ m/s}$$

Note the significant increase in sonic velocity as the temperature increases.

The Mach number, M, is defined as the ratio of the actual velocity $\mathbf{V}$ to the sonic velocity c.

$$M = \frac{\mathbf{V}}{c} \qquad (15.29)$$

When $M > 1$, the flow is supersonic; when $M < 1$, the flow is subsonic; and when $M = 1$, the flow is sonic. The importance of the Mach number as a parameter in fluid-flow problems will be evident in the sections that follow.

In-Text Concept Questions

a. Is the stagnation temperature always higher than the free stream temperature? Why?

b. By looking at Eq.15.25, rank the speed of sound for a solid, a liquid, and a gas.

c. Does the speed of sound in an ideal gas depend on pressure? What about a real gas?

15.6 REVERSIBLE, ADIABATIC, ONE-DIMENSIONAL FLOW OF AN IDEAL GAS THROUGH A NOZZLE

A nozzle or diffuser with both a converging and a diverging section is shown in Fig. 15.8. The minimum cross-sectional area is called the throat.

Our first consideration concerns the conditions that determine whether a nozzle or diffuser should be converging or diverging and the conditions that prevail at the throat. For the control volume shown, the following relations can be written:

Energy equation:

$$dh + \mathbf{V}\,d\mathbf{V} = 0 \qquad (15.30)$$

Property relation:

$$T\,ds = dh - \frac{dP}{\rho} = 0 \qquad (15.31)$$

Continuity equation:

$$\rho A \mathbf{V} = \dot{m} = \text{constant}$$

$$\frac{d\rho}{\rho} + \frac{dA}{A} + \frac{d\mathbf{V}}{\mathbf{V}} = 0 \qquad (15.32)$$

FIGURE 15.8
One-dimensional, reversible, adiabatic steady flow through a nozzle.

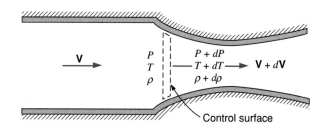

Combining Eqs. 15.30 and 15.31, we have

$$dh = \frac{dP}{\rho} = -\mathbf{V}\,d\mathbf{V}$$

$$d\mathbf{V} = -\frac{1}{\rho \mathbf{V}}\,dP$$

Substituting this in Eq. 15.32,

$$\frac{dA}{A} = \left(-\frac{d\rho}{\rho} - \frac{d\mathbf{V}}{\mathbf{V}}\right) = -\frac{d\rho}{\rho}\left(\frac{dP}{dP}\right) + \frac{1}{\rho \mathbf{V}^2}\,dP$$

$$= \frac{-dP}{\rho}\left(\frac{d\rho}{dP} - \frac{1}{\mathbf{V}^2}\right) = \frac{dP}{\rho}\left(-\frac{1}{(dP/d\rho)} + \frac{1}{\mathbf{V}^2}\right)$$

Since the flow is isentropic

$$\frac{dP}{d\rho} = c^2 = \frac{\mathbf{V}^2}{M^2}$$

and therefore

$$\frac{dA}{A} = \frac{dP}{\rho \mathbf{V}^2}\,(1 - M^2) \tag{15.33}$$

This is a very significant equation, for from it we can draw the following conclusions about the proper shape for nozzles and diffusers:

For a nozzle, $dP < 0$. Therefore,

for a subsonic nozzle, $M < 1 \Rightarrow dA < 0$, and the nozzle is converging;

for a supersonic nozzle, $M > 1 \Rightarrow dA > 0$, and the nozzle is diverging.

For a diffuser, $dP > 0$. Therefore,

for a subsonic diffuser, $M < 1 \Rightarrow dA > 0$, and the diffuser is diverging;

for a supersonic diffuser, $M > 1 \Rightarrow dA < 0$, and the diffuser is converging.

When $M = 1$, $dA = 0$, which means that sonic velocity can be achieved only at the throat of a nozzle or diffuser. These conclusions are summarized in Fig. 15.9.

We will now develop a number of relations between the actual properties, stagnation properties, and Mach number. These relations are very useful in dealing with isentropic flow of an ideal gas in a nozzle.

Equation 15.1 gives the relation between enthalpy, stagnation enthalpy, and kinetic energy.

$$h + \frac{\mathbf{V}^2}{2} = h_0$$

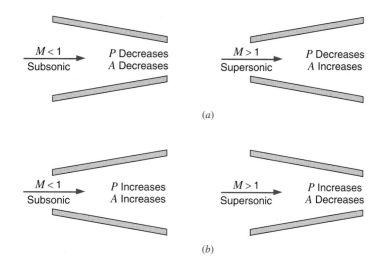

FIGURE 15.9
Required area changes
for (a) nozzles and
(b) diffusers.

For an ideal gas with constant specific heat, Eq. 15.1 can be written as

$$\mathbf{V}^2 = 2C_{p0}(T_0 - T) = 2\frac{kRT}{k-1}\left(\frac{T_0}{T} - 1\right)$$

Since

$$c^2 = kRT$$

$$\mathbf{V}^2 = \frac{2c^2}{k-1}\left(\frac{T_0}{T} - 1\right)$$

$$\frac{\mathbf{V}^2}{c^2} = M^2 = \frac{2}{k-1}\left(\frac{T_0}{T} - 1\right)$$

$$\frac{T_0}{T} = 1 + \frac{(k-1)}{2}M^2 \tag{15.34}$$

For an isentropic process,

$$\left(\frac{T_0}{T}\right)^{k/(k-1)} = \frac{P_0}{P} \qquad \left(\frac{T_0}{T}\right)^{1/(k-1)} = \frac{\rho_0}{\rho}$$

Therefore,

$$\frac{P_0}{P} = \left[1 + \frac{(k-1)}{2}M^2\right]^{k/(k-1)} \tag{15.35}$$

$$\frac{\rho_0}{\rho} = \left[1 + \frac{(k-1)}{2}M^2\right]^{1/(k-1)} \tag{15.36}$$

Values of P/P_0, ρ/ρ_0, and T/T_0 are given as a function of M in Table 15.3 on page 662 for the value $k = 1.40$.

TABLE 15.1
Critical Pressure, Density, and Temperature Ratios for Isentropic Flow of an Ideal Gas

	k = 1.1	**k = 1.2**	**k = 1.3**	**k = 1.4**	**k = 1.67**
P^*/P_0	0.5847	0.5644	0.5457	0.5283	0.4867
ρ^*/ρ_0	0.6139	0.6209	0.6276	0.6340	0.6497
T^*/T_0	0.9524	0.9091	0.8696	0.8333	0.7491

The conditions at the throat of the nozzle can be found by noting that $M = 1$ at the throat. The properties at the throat are denoted by an asterisk (*). Therefore,

$$\frac{T^*}{T_0} = \frac{2}{k+1} \tag{15.37}$$

$$\frac{P^*}{P_0} = \left(\frac{2}{k+1}\right)^{k/(k-1)} \tag{15.38}$$

$$\frac{\rho^*}{\rho_0} = \left(\frac{2}{k+1}\right)^{1/(k-1)} \tag{15.39}$$

These properties at the throat of a nozzle when $M = 1$ are frequently referred to as critical pressure, critical temperature, and critical density, and the ratios given by Eqs. 15.37, 15.38, and 15.39 are referred to as the *critical-temperature ratio*, *critical-pressure ratio*, and *critical-density ratio*. Table 15.1 gives these ratios for various values of k.

15.7 MASS FLOW RATE OF AN IDEAL GAS THROUGH AN ISENTROPIC NOZZLE

We now consider the mass rate of flow per unit area, $\dot{m}/A$, in a nozzle. From the continuity equation, we proceed as follows:

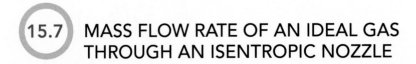

$$\frac{\dot{m}}{A} = \rho \mathbf{V} = \frac{P\mathbf{V}}{RT}\sqrt{\frac{kT_0}{kT_0}}$$

$$= \frac{P\mathbf{V}}{\sqrt{kRT}}\sqrt{\frac{k}{R}}\sqrt{\frac{T_0}{T}}\sqrt{\frac{1}{T_0}}$$

$$= \frac{PM}{\sqrt{T_0}}\sqrt{\frac{k}{R}}\sqrt{1 + \frac{k-1}{2}M^2} \tag{15.40}$$

By substituting Eq. 15.35 into Eq. 15.40, the flow per unit area can be expressed in terms of stagnation pressure, stagnation temperature, Mach number, and gas properties.

$$\frac{\dot{m}}{A} = \frac{P_0}{\sqrt{T_0}}\sqrt{\frac{k}{R}} \times \frac{M}{\left(1 + \dfrac{k-1}{2}M^2\right)^{(k+1)/2(k-1)}} \tag{15.41}$$

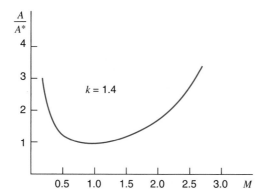

FIGURE 15.10 Area ratio as a function of the Mach number for a reversible, adiabatic nozzle.

At the throat, $M = 1$; therefore, the flow per unit area at the throat, $\dot{m}/A^*$, can be found by setting $M = 1$ in Eq. 15.41.

$$\frac{\dot{m}}{A^*} = \frac{P_0}{\sqrt{T_0}} \sqrt{\frac{k}{R}} \times \frac{1}{\left(\dfrac{k+1}{2}\right)^{(k+1)/2(k-1)}} \qquad (15.42)$$

From this result, it follows that the flow rate through a nozzle that reaches sonic conditions at the throat depends on the stagnation properties P_0, T_0, the fluid properties k, R, and the throat area A^*. For a given fluid and nozzle, only the stagnation properties are important for the flow rate.

The area ratio A/A^* can be obtained by dividing Eq. 15.42 by Eq. 15.41.

$$\frac{A}{A^*} = \frac{1}{M} \left[\left(\frac{2}{k+1} \right) \left(1 + \frac{k-1}{2} M^2 \right) \right]^{(k+1)/2(k-1)} \qquad (15.43)$$

The area ratio A/A^* is the ratio of the area at the point where the Mach number is M to the throat area, and values of A/A^* as a function of Mach number are given in Table 15.3. Figure 15.10 shows a plot of A/A^* vs. M, which is in accordance with our previous conclusion that a subsonic nozzle is converging and a supersonic nozzle is diverging.

The final point to be made regarding the isentropic flow of an ideal gas through a nozzle involves the effect of varying the back pressure (the pressure outside the nozzle exit) on the mass rate of flow.

Consider first a convergent nozzle, as shown in Fig. 15.11, which also shows the pressure ratio P/P_0 along the length of the nozzle. The conditions upstream are the stagnation

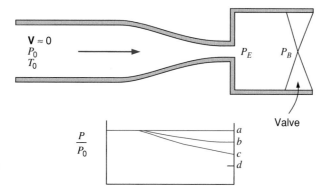

FIGURE 15.11
Pressure ratio as a function of back pressure for a convergent nozzle.

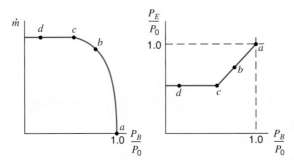

FIGURE 15.12 Mass rate of flow and exit pressure as a function of back pressure for a convergent nozzle.

conditions, which are assumed to be constant. The pressure at the exit plane of the nozzle is designated P_E and the back pressure P_B. Let us consider how the mass rate of flow $\dot{m}$ and the exit plane pressure P_E/P_0 vary as the back pressure P_B is decreased. These quantities are plotted in Fig. 15.12.

When $P_B/P_0 = 1$, there is, of course, no flow, and $P_E/P_0 = 1$ as designated by point a. Next, let the back pressure P_B be lowered to that designated by point b so that P_B/P_0 is greater than the critical-pressure ratio. The mass rate of flow has a certain value and $P_E = P_B$. The exit Mach number is less than 1. Next, let the back pressure be lowered to the critical pressure, designated by point c. The Mach number at the exit is now unity, and P_E is equal to P_B. When P_B is decreased below the critical pressure, designated by point d, there is no further increase in the mass rate of flow, P_E remains constant at a value equal to the critical pressure, and the exit Mach number is unity. The drop in pressure from P_E to P_B takes place outside the nozzle exit. Under these conditions the nozzle is said to be choked, which means that for given stagnation conditions, the nozzle is passing the maximum possible mass flow.

Consider next a convergent-divergent nozzle in a similar arrangement, Fig. 15.13. Point a designates the conditions when $P_B = P_0$ and there is no flow. When P_B is decreased to the pressure indicated by point b, so that P_B/P_0 is less than 1 but considerably greater than the critical-pressure ratio, the velocity increases in the convergent section, but $M < 1$ at the throat. Therefore, the diverging section acts as a subsonic diffuser in which the pressure increases and velocity decreases. Point c designates the back pressure at which $M = 1$ at the throat, but the diverging section acts as a subsonic diffuser (with $M = 1$ at the inlet) in which the pressure increases and velocity decreases. Point d designates one other back pressure that permits isentropic flow, and in this case, the diverging section acts as a supersonic nozzle, with a decrease in pressure and an increase in velocity. Between the back pressures designated by points c and d, an isentropic solution is not possible, and shock

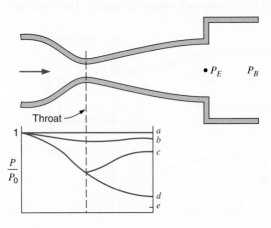

FIGURE 15.13 Nozzle pressure ratio as a function of back pressure for a reversible, convergent-divergent nozzle.

waves will be present. This matter is discussed in the section that follows. When the back pressure is decreased below that designated by point d, the exit-plane pressure P_E remains constant and the drop in pressure from P_E to P_B takes place outside the nozzle. This is designated by point e.

Example 15.6

A convergent nozzle has an exit area of 500 mm². Air enters the nozzle with a stagnation pressure of 1000 kPa and a stagnation temperature of 360 K. Determine the mass rate of flow for back pressures of 800 kPa, 528 kPa, and 300 kPa, assuming isentropic flow.

Analysis and Solution

For air, $k = 1.4$ and Table 15.3 may be used. The critical-pressure ratio, P^*/P_0, is 0.528. Therefore, for a back pressure of 528 kPa, $M = 1$ at the nozzle exit and the nozzle is choked. Decreasing the back pressure below 528 kPa will not increase the flow.

For a back pressure of 528 kPa,

$$\frac{T^*}{T_0} = 0.8333 \qquad T^* = 300 \text{ K}$$

At the exit

$$\mathbf{V} = c = \sqrt{kRT}$$
$$= \sqrt{1.4 \times 0.287 \times 300 \times 1000} = 347.2 \text{ m/s}$$
$$\rho^* = \frac{P^*}{RT^*} = \frac{528}{0.287 \times 300} = 6.1324 \text{ kg/m}^3$$
$$\dot{m} = \rho A \mathbf{V}$$

Applying this relation to the throat section

$$\dot{m} = 6.1324 \times 500 \times 10^{-6} \times 347.2 = 1.0646 \text{ kg/s}$$

For a back pressure of 800 kPa, $P_E/P_0 = 0.8$ (subscript E designates the properties in the exit plane). From Table 15.3

$$M_E = 0.573 \qquad T_E/T_0 = 0.9381$$
$$T_E = 337.7 \text{ K}$$
$$c_E = \sqrt{kRT_E} = \sqrt{1.4 \times 0.287 \times 337.7 \times 1000} = 368.4 \text{ m/s}$$
$$\mathbf{V}_E = M_E c_E = 211.1 \text{ m/s}$$
$$\rho_E = \frac{P_E}{RT_E} = \frac{800}{0.287 \times 337.7} = 8.2542 \text{ kg/m}^3$$
$$\dot{m} = \rho A \mathbf{V}$$

Applying this relation to the exit section,

$$\dot{m} = 8.2542 \times 500 \times 10^{-6} \times 211.1 = 0.8712 \text{ kg/s}$$

For a back pressure less than the critical pressure, which in this case is 528 kPa, the nozzle is choked and the mass rate of flow is the same as that for the critical pressure. Therefore, for an exhaust pressure of 300 kPa, the mass rate of flow is 1.0646 kg/s.

Example 15.7

A converging-diverging nozzle has an exit area to throat area ratio of 2. Air enters this nozzle with a stagnation pressure of 1000 kPa and a stagnation temperature of 360 K. The throat area is 500 mm². Determine the mass rate of flow, exit pressure, exit temperature, exit Mach number, and exit velocity for the following conditions:

a. Sonic velocity at the throat, diverging section acting as a nozzle. (Corresponds to point *d* in Fig. 15.13.)

b. Sonic velocity at the throat, diverging section acting as a diffuser. (Corresponds to point *c* in Fig. 15.13.)

Analysis and Solution

(a) In Table 15.3, we find that there are two Mach numbers listed for $A/A^* = 2$. One of these is greater than unity and one is less than unity. When the diverging section acts as a supersonic nozzle, we use the value for $M > 1$. The following are from Table 15.3:

$$\frac{A_E}{A^*} = 2.0 \qquad M_E = 2.197 \qquad \frac{P_E}{P_0} = 0.0939 \qquad \frac{T_E}{T_0} = 0.5089$$

Therefore,

$$P_E = 0.0939(1000) = 93.9 \text{ kPa}$$

$$T_E = 0.5089(360) = 183.2 \text{ K}$$

$$c_E = \sqrt{kRT_E} = \sqrt{1.4 \times 0.287 \times 183.2 \times 1000} = 271.3 \text{ m/s}$$

$$\mathbf{V}_E = M_E c_E = 2.197(271.3) = 596.1 \text{ m/s}$$

The mass rate of flow can be determined by considering either the throat section or the exit section. However, in general, it is preferable to determine the mass rate of flow from conditions at the throat. Since in this case, $M = 1$ at the throat, the calculation is identical to the calculation for the flow in the convergent nozzle of Example 15.6 when it is choked.

(b) The following are from Table 15.3.

$$\frac{A_E}{A^*} = 2.0 \qquad M = 0.308 \qquad \frac{P_E}{P_0} = 0.0936 \qquad \frac{T_E}{T_0} = 0.9812$$

$$P_E = 0.0936(1000) = 936 \text{ kPa}$$

$$T_E = 0.9812(360) = 353.3 \text{ K}$$

$$c_E = \sqrt{kRT_E} = \sqrt{1.4 \times 0.287 \times 353.3 \times 1000} = 376.8 \text{ m/s}$$

$$\mathbf{V}_E = M_E c_E = 0.308(376.3) = 116 \text{ m/s}$$

Since $M = 1$ at the throat, the mass rate of flow is the same as in (a), which is also equal to the flow in the convergent nozzle of Example 15.6 when it is choked.

In the example above, a solution assuming isentropic flow is not possible if the back pressure is between 936 and 93.9 kPa. If the back pressure is in this range, there will be either a normal shock in the nozzle or oblique shock waves outside the nozzle. The matter of normal shock waves is considered in the following section.

In-Text Concept Questions

d. Can a convergent adiabatic nozzle produce a supersonic flow?

e. To maximize the mass flow rate of air through a given nozzle, which properties should I try to change and in which direction, higher or lower?

f. How do the stagnation temperature and pressure change in a reversible isentropic flow?

15.8 NORMAL SHOCK IN AN IDEAL GAS FLOWING THROUGH A NOZZLE

A shock wave involves an extremely rapid and abrupt change of state. In a normal shock this change of state takes place across a plane normal to the direction of the flow. Figure 15.14 shows a control surface that includes such a normal shock. We can now determine the relations that govern the flow. Assuming steady-state, steady-flow, we can write the following relations, where subscripts x and y denote the conditions upstream and downstream of the shock, respectively. Note that no heat or work crosses the control surface.

Energy equation:

$$h_x + \frac{\mathbf{V}_x^2}{2} = h_y + \frac{\mathbf{V}_y^2}{2} = h_{0x} = h_{0y} \tag{15.44}$$

Continuity equation:

$$\frac{\dot{m}}{A} = \rho_x \mathbf{V}_x = \rho_y \mathbf{V}_y \tag{15.45}$$

Momentum equation:

$$A(P_x - P_y) = \dot{m}(\mathbf{V}_y - \mathbf{V}_x) \tag{15.46}$$

Entropy equation: Since the process is adiabatic

$$s_y - s_x = s_{\text{gen}} \geq 0 \tag{15.47}$$

The energy and continuity equations can be combined to give an equation that, when plotted on the h–s diagram, is called the Fanno line. Similarly, the momentum and continuity equations can be combined to give an equation the plot of which on the h–s diagram is known as the Rayleigh line. Both of these lines are shown on the h–s diagram of Fig. 15.15. It can be shown that the point of maximum entropy on each line, points a and b, corresponds to $M = 1$. The lower part of each line corresponds to supersonic velocities and the upper part to subsonic velocities.

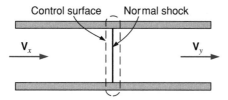

FIGURE 15.14
One-dimensional normal shock.

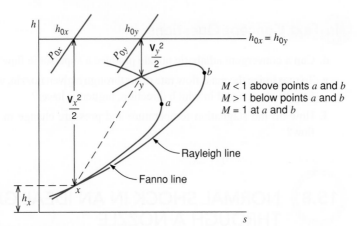

FIGURE 15.15
End states for a
one-dimensional normal
shock on an h–s diagram.

The two points where all three equations are satisfied are points x and y, x being in the supersonic region and y in the subsonic region. Since the entropy equation requires that $s_y - s_x \geq 0$ in an adiabatic process, we conclude that the normal shock can proceed only from x to y. This means that the velocity changes from supersonic ($M > 1$) before the shock to subsonic ($M < 1$) after the shock.

The equations governing normal shock waves will now be developed. If we assume ideal gas, then $h = h(T)$ and we conclude from the energy equation, Eq.15.44, that

$$T_{0x} = T_{0y} \tag{15.48}$$

That is, there is no change in stagnation temperature across a normal shock. Introducing Eq. 15.34

$$\frac{T_{0x}}{T_x} = 1 + \frac{k-1}{2}M_x^2 \qquad \frac{T_{0y}}{T_y} = 1 + \frac{k-1}{2}M_y^2$$

and substituting into Eq. 15.48, we have

$$\frac{T_y}{T_x} = \frac{1 + \dfrac{k-1}{2}M_x^2}{1 + \dfrac{k-1}{2}M_y^2} \tag{15.49}$$

The equation of state, the definition of the Mach number, and the relation $c = \sqrt{kRT}$ can be introduced into the continuity equation as follows:

$$\rho_x \mathbf{V}_x = \rho_y \mathbf{V}_y$$

But

$$\rho_x = \frac{P_x}{RT_x} \qquad \rho_y = \frac{P_y}{RT_y}$$

$$\frac{T_y}{T_x} = \frac{P_y \mathbf{V}_y}{P_x \mathbf{V}_x} = \frac{P_y M_y c_y}{P_x M_x c_x} = \frac{P_y M_y \sqrt{T_y}}{P_x M_x \sqrt{T_x}}$$

$$= \left(\frac{P_y}{P_x}\right)^2 \left(\frac{M_y}{M_x}\right)^2 \tag{15.50}$$

Combining Eqs. 15.49 and 15.50, which involves combining the energy equations and the continuity equation, gives the equation of the Fanno line.

$$\frac{P_y}{P_x} = \frac{M_x\sqrt{1 + \dfrac{k-1}{2}M_x^2}}{M_y\sqrt{1 + \dfrac{k-1}{2}M_y^2}}$$ (15.51)

The momentum and continuity equations can be combined as follows to give the equation of the Rayleigh line.

$$P_x - P_y = \frac{\dot{m}}{A}(\mathbf{V}_y - \mathbf{V}_x) = \rho_y \mathbf{V}_y^2 - \rho_x \mathbf{V}_x^2$$

$$P_x + \rho_x \mathbf{V}_x^2 = P_y + \rho_y \mathbf{V}_y^2$$

$$P_x + \rho_x M_x^2 c_x^2 = P_y + \rho_y M_y^2 c_y^2$$

$$P_x + \frac{P_x M_x^2}{RT_x}(kRT_x) = P_y + \frac{P_y M_y^2}{RT_y}(kRT_y)$$

$$P_x(1 + kM_x^2) = P_y(1 + kM_y^2)$$

$$\frac{P_y}{P_x} = \frac{1 + kM_x^2}{1 + kM_y^2}$$ (15.52)

Equations 15.51 and 15.52 can be combined to give the following equation relating M_x and M_y:

$$M_y^2 = \frac{M_x^2 + \dfrac{2}{k-1}}{\dfrac{2k}{k-1}M_x^2 - 1}$$ (15.53)

Table 15.4, on page 663, gives the normal shock function, which includes M_y as a function of M_x. This table applies to an ideal gas with a value $k = 1.4$. Note that M_x is always supersonic and M_y is always subsonic, which agrees with the previous statement that in a normal shock the velocity changes from supersonic to subsonic. Tables 15.3 and 15.4 also give the pressure, density, temperature, and stagnation pressure ratios across a normal shock as a function of M_x. These are found from Eqs. 15.49 and 15.50 and the equation of state. Note that there is always a drop in stagnation pressure across a normal shock and an increase in the static pressure.

Example 15.8

Consider the convergent-divergent nozzle of Example 15.7, in which the diverging section acts as a supersonic nozzle (Fig. 15.16). Assume that a normal shock stands in the exit plane of the nozzle. Determine the static pressure and temperature and the stagnation pressure just downstream of the normal shock.

Sketch: Figure 15.16.

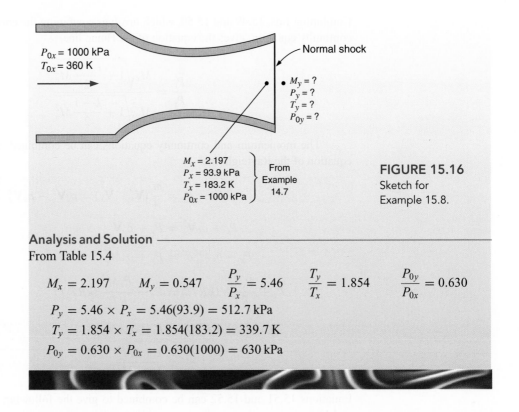

$P_{0x} = 1000$ kPa
$T_{0x} = 360$ K

Normal shock

$M_y = ?$
$P_y = ?$
$T_y = ?$
$P_{0y} = ?$

$M_x = 2.197$
$P_x = 93.9$ kPa
$T_x = 183.2$ K
$P_{0x} = 1000$ kPa

From Example 14.7

FIGURE 15.16
Sketch for Example 15.8.

Analysis and Solution
From Table 15.4

$$M_x = 2.197 \qquad M_y = 0.547 \qquad \frac{P_y}{P_x} = 5.46 \qquad \frac{T_y}{T_x} = 1.854 \qquad \frac{P_{0y}}{P_{0x}} = 0.630$$

$$P_y = 5.46 \times P_x = 5.46(93.9) = 512.7 \text{ kPa}$$
$$T_y = 1.854 \times T_x = 1.854(183.2) = 339.7 \text{ K}$$
$$P_{0y} = 0.630 \times P_{0x} = 0.630(1000) = 630 \text{ kPa}$$

In light of this example, we can conclude the discussion concerning the flow through a convergent-divergent nozzle. Figure 15.13 is repeated here as Fig. 15.17 for convenience, except that points f, g, and h have been added. Consider point d. We have already noted that with this back pressure, the exit plane pressure P_E is just equal to the back pressure P_B, and isentropic flow is maintained in the nozzle. Let the back pressure be raised to that designated by point f. The exit-plane pressure P_E is not influenced by this increase in back pressure, and the increase in pressure from P_E to P_B takes place outside the nozzle. Let the back pressure be raised to that designated by point g, which is just sufficient to cause a normal shock to

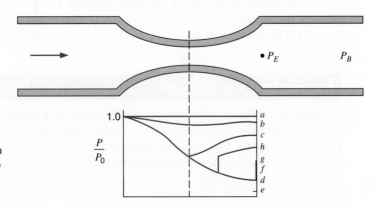

FIGURE 15.17
Nozzle pressure ratio as a function of back pressure for a convergent-divergent nozzle.

stand in the exit plane of the nozzle. The exit-plane pressure P_E (downstream of the shock) is equal to the back pressure P_B, and $M < 1$ leaving the nozzle. This is the case in Example 15.8. Now let the back pressure be raised to that corresponding to point h. As the back pressure is raised from g to h, the normal shock moves into the nozzle as indicated. Since $M < 1$ downstream of the normal shock, the diverging part of the nozzle that is downstream of the shock acts as a subsonic diffuser. As the back pressure is increased from h to c, the shock moves further upstream and disappears at the nozzle throat where the back pressure corresponds to c. This is reasonable since there are no supersonic velocities involved when the back pressure corresponds to c, and hence, no shock waves are possible.

Example 15.9

Consider the convergent-divergent nozzle of Examples 15.7 and 15.8. Assume that there is a normal shock wave standing at the point where $M = 1.5$. Determine the exit-plane pressure, temperature, and Mach number. Assume isentropic flow except for the normal shock (Fig. 15.18).

Sketch: Figure 15.18.

Analysis and Solution

The properties at point x can be determined from Table 15.3, because the flow is isentropic to point x.

$$M_x = 1.5 \qquad \frac{P_x}{P_{0x}} = 0.2724 \qquad \frac{T_x}{T_{0x}} = 0.6897 \qquad \frac{A_x}{A_x^*} = 1.1762$$

Therefore,

$$P_x = 0.2724(1000) = 272.4 \text{ kPa}$$
$$T_x = 0.6897(360) = 248.3 \text{ K}$$

The properties at point y can be determined from the normal shock functions in Table 15.4.

$$M_y = 0.7011 \qquad \frac{P_y}{P_x} = 2.4583 \qquad \frac{T_y}{T_x} = 1.320 \qquad \frac{P_{0y}}{P_{0x}} = 0.9298$$

$$P_y = 2.4583 \, P_x = 2.4583(272.4) = 669.6 \text{ kPa}$$
$$T_y = 1.320 \, T_x = 1.320(248.3) = 327.8 \text{ K}$$
$$P_{0y} = 0.9298 \, P_{0x} = 0.9298(1000) = 929.8 \text{ kPa}$$

Since there is no change in stagnation temperature across a normal shock,

$$T_{0x} = T_{0y} = 360 \text{ K}$$

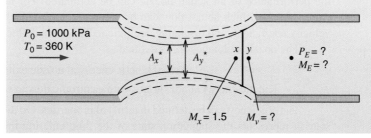

FIGURE 15.18
Sketch for
Example 15.9.

From y to E, the diverging section acts as a subsonic diffuser. In solving this problem, it is convenient to think of the flow at y as having come from an isentropic nozzle having a throat area A_y^*. Such a hypothetical nozzle is shown by the dotted line. From the table of isentropic flow functions, Table 15.3, we find the following for $M_y = 0.7011$:

$$M_y = 0.7011 \qquad \frac{A_y}{A_y^*} = 1.0938 \qquad \frac{P_y}{P_{0y}} = 0.7202 \qquad \frac{T_y}{T_{0y}} = 0.9105$$

From the statement of the problem

$$\frac{A_E}{A_x^*} = 2.0$$

Also, since the flow from y to E is isentropic,

$$\frac{A_E}{A_E^*} = \frac{A_E}{A_y^*} = \frac{A_E}{A_x^*} \times \frac{A_x^*}{A_x} \times \frac{A_x}{A_y} \times \frac{A_y}{A_y^*}$$

$$= \frac{A_E}{A_y^*} = 2.0 \times \frac{1}{1.1762} \times 1 \times 1.0938 = 1.860$$

From the table of isentropic flow functions for $A/A^* = 1.860$ and $M < 1$

$$M_E = 0.339 \qquad \frac{P_E}{P_{0E}} = 0.9222 \qquad \frac{T_E}{T_{0E}} = 0.9771$$

$$\frac{P_E}{P_{0E}} = \frac{P_E}{P_{0y}} = 0.9222$$

$$P_E = 0.9222(P_{0y}) = 0.9222(929.8) = 857.5 \text{ kPa}$$

$$T_E = 0.9771(T_{0E}) = 0.9771(360) = 351.7 \text{ K}$$

In considering the normal shock, we have ignored the effect of viscosity and thermal conductivity, which are certain to be present. The actual shock wave will occur over some finite thickness. However, the development as given here gives a very good qualitative picture of normal shocks and also provides a basis for fairly accurate quantitative results.

15.9 NOZZLE AND DIFFUSER COEFFICIENTS

Up to this point, we have considered only isentropic flow and normal shocks. As was pointed out in Chapter 7, isentropic flow through a nozzle provides a standard to which the performance of an actual nozzle can be compared. For nozzles, the three important parameters by which actual flow can be compared to the ideal flow are nozzle efficiency, velocity coefficient, and discharge coefficient. These are defined as follows:

The nozzle efficiency η_N is defined as

$$\eta_N = \frac{\text{Actual kinetic energy at nozzle exit}}{\text{Kinetic energy at nozzle exit with isentropic flow to same exit pressure}} \qquad (15.54)$$

The efficiency can be defined in terms of properties. On the h–s diagram of Fig. 15.19, state $0i$ represents the stagnation state of the fluid entering the nozzle; state e represents

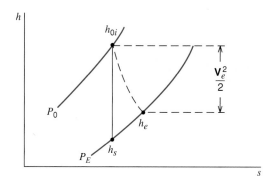

FIGURE 15.19 An h–s diagram showing the effects of irreversibility in a nozzle.

the actual state at the nozzle exit; and state s represents the state that would have been achieved at the nozzle exit if the flow had been reversible and adiabatic to the same exit pressure. Therefore, in terms of these states, the nozzle efficiency is

$$\eta_N = \frac{h_{0i} - h_e}{h_{0i} - h_s}$$

Nozzle efficiencies vary, in general, from 90 % to 99 %. Large nozzles usually have higher efficiencies than small nozzles, and nozzles with straight axes have higher efficiencies than nozzles with curved axes. The irreversibilities, which cause the departure from isentropic flow, are primarily due to frictional effects and are confined largely to the boundary layer. The rate of change of cross-sectional area along the nozzle axis (that is, the nozzle contour) is an important parameter in the design of an efficient nozzle, particularly in the divergent section. Detailed consideration of this matter is beyond the scope of this text, and the reader is referred to standard references on the subject.

The velocity coefficient, C_v, is defined as

$$C_v = \frac{\text{Actual velocity at nozzle exit}}{\text{Velocity at nozzle exit with isentropic flow to same exit pressure}} \quad (15.55)$$

It follows that the velocity coefficient is equal to the square root of the nozzle efficiency

$$C_v = \sqrt{\eta_N} \quad (15.56)$$

The coefficient of discharge, C_D, is defined by the relation

$$C_D = \frac{\text{Actual mass rate of flow}}{\text{Mass rate of flow with isentropic flow}}$$

In determining the mass rate of flow with isentropic conditions, the actual back pressure is used if the nozzle is not choked. If the nozzle is choked, the isentropic mass rate of flow is based on isentropic flow and sonic velocity at the minimum section (that is, sonic velocity at the exit of a convergent nozzle and at the throat of a convergent–divergent nozzle).

The performance of a diffuser is usually given in terms of diffuser efficiency, which is best defined with the aid of an h–s diagram. On the h–s diagram of Fig. 15.20, states 1 and 01 are the actual and stagnation states of the fluid entering the diffuser. States 2 and 02 are the actual and stagnation states of the fluid leaving the diffuser. State 3 is not attained in the diffuser, but it is the state that has the same entropy as the initial state and the pressure

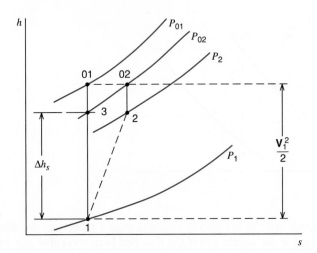

FIGURE 15.20 An h–s diagram showing the definition of diffuser efficiency.

of the isentropic stagnation state leaving the diffuser. The efficiency of the diffuser η_D is defined as

$$\eta_D = \frac{\Delta h_s}{\mathbf{V}_1^2/2} = \frac{h_3 - h_1}{h_{01} - h_1} = \frac{h_3 - h_1}{h_{02} - h_1} \tag{15.57}$$

If we assume an ideal gas with constant specific heat, this reduces to

$$\eta_D = \frac{T_3 - T_1}{T_{02} - T_1} = \frac{\dfrac{(T_3 - T_1)}{T_1} T_1}{\dfrac{\mathbf{V}_1^2}{2C_{p_0}}}$$

$$C_{p_0} = \frac{kR}{k-1} \qquad T_1 = \frac{c_1^2}{kR} \qquad \mathbf{V}_1^2 = M_1^2 c_1^2 \qquad \frac{T_3}{T_1} = \left(\frac{P_{02}}{P_1}\right)^{(k-1)/k}$$

Therefore,

$$\eta_D = \frac{\left(\dfrac{P_{02}}{P_1}\right)^{(k-1)/k} - 1}{\dfrac{k-1}{2} M_1^2}$$

$$\left(\frac{P_{02}}{P_1}\right)^{(k-1)/k} = \left(\frac{P_{01}}{P_1}\right)^{(k-1)/k} \times \left(\frac{P_{02}}{P_{01}}\right)^{(k-1)/k}$$

$$\left(\frac{P_{02}}{P_1}\right)^{(k-1)/k} = \left(1 + \frac{k-1}{2} M_1^2\right)\left(\frac{P_{02}}{P_{01}}\right)^{(k-1)/k}$$

$$\eta_D = \frac{\left(1 + \dfrac{k-1}{2} M_1^2\right)\left(\dfrac{P_{02}}{P_{01}}\right)^{(k-1)/k} - 1}{\dfrac{k-1}{2} M_1^2} \tag{15.58}$$

15.10 NOZZLES AND ORIFICES AS FLOW-MEASURING DEVICES

The mass rate of flow of a fluid flowing in a pipe is frequently determined by measuring the pressure drop across a nozzle or orifice in the line, as shown in Fig. 15.21. The ideal process for such a nozzle or orifice is assumed to be isentropic flow through a nozzle that has the measured pressure drop from inlet to exit and a minimum cross-sectional area equal to the minimum area of the nozzle or orifice. The actual flow is related to the ideal flow by the coefficient of discharge, which is defined by Eq. 15.57.

The pressure difference measured across an orifice depends on the location of the pressure taps, as indicated in Fig. 15.21. Since the ideal flow is based on the measured pressure difference, it follows that the coefficient of discharge depends on the locations of the pressure taps. Also, the coefficient of discharge for a sharp-edged orifice is considerably less than that for a well-rounded nozzle, primarily due to a contraction of the stream, known as the *vena contracta*, as it flows through a sharp-edged orifice.

There are two approaches to determining the discharge coefficient of a nozzle or orifice. One is to follow a standard design procedure, such as the ones established by the American Society of Mechanical Engineers,[1] and use the coefficient of discharge given for a particular design. A more accurate method is to calibrate a given nozzle or orifice and determine the discharge coefficient for a given installation by accurately measuring the actual mass rate of flow. The procedure to be followed will depend on the accuracy desired and other factors involved (such as time, expense, availability of calibration facilities) in a given situation.

For incompressible fluids flowing through an orifice, the ideal flow for a given pressure drop can be found by the procedure outlined in Section 15.4. Actually, it is advantageous to combine Eqs. 15.17 and 15.21 to give the following relation, which is valid for reversible flow:

$$v(P_2 - P_1) + \frac{\mathbf{V}_2^2 - \mathbf{V}_1^2}{2} = v(P_2 - P_1) + \frac{\mathbf{V}_2^2 - (A_2/A_1)^2 \mathbf{V}_2^2}{2} = 0 \qquad (15.59)$$

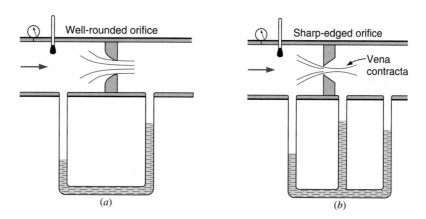

FIGURE 15.21
Nozzles and orifices as flow-measuring devices.

---

[1] *Fluid Meters, Their Theory and Application*, ASME, 1959; *Flow Measurement*, ASME, 1959.

or

$$v(P_2 - P_1) + \frac{\mathbf{V}_2^2}{2}\left[1 - \left(\frac{A_2}{A_1}\right)^2\right] = 0$$

$$\mathbf{V}_2 = \sqrt{\frac{2v(P_1 - P_2)}{[1 - (A_2/A_1)^2]}} \qquad (15.60)$$

For an ideal gas it is frequently advantageous to use the following simplified procedure when the pressure drop across an orifice or nozzle is small. Consider the nozzle shown in Fig. 15.22. From the energy equation, we conclude that

$$h_i + \frac{\mathbf{V}_i^2}{2} = h_e + \frac{\mathbf{V}_e^2}{2}$$

Assuming constant specific heat, this reduces to

$$\frac{\mathbf{V}_e^2 - \mathbf{V}_i^2}{2} = h_i - h_e = C_{p0}(T_i - T_e)$$

Let ΔP and ΔT be the decrease in pressure and temperature across the nozzle. Since we are considering reversible adiabatic flow, we note that

$$\frac{T_e}{T_i} = \left(\frac{P_e}{P_i}\right)^{(k-1)/k}$$

or

$$\frac{T_i - \Delta T}{T_i} = \left(\frac{P_i - \Delta P}{P_i}\right)^{(k-1)/k}$$

$$1 - \frac{\Delta T}{T_i} = \left(1 - \frac{\Delta P}{P_i}\right)^{(k-1)/k}$$

Using the binomial expansion on the right side of the equation, we have

$$1 - \frac{\Delta T}{T_i} = 1 - \frac{k-1}{k}\frac{\Delta P}{P_i} - \frac{k-1}{2k^2}\frac{\Delta P^2}{P_i^2}\cdots$$

If $\Delta P/P_i$ is small, this reduces to

$$\frac{\Delta T}{T_i} = \frac{k-1}{k}\frac{\Delta P}{P_i}$$

Substituting this into the first-law equation, we have

$$\frac{\mathbf{V}_e^2 - \mathbf{V}_i^2}{2} = C_{p0}\frac{k-1}{k}\Delta P\frac{T_i}{P_i}$$

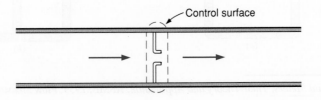

FIGURE 15.22
Analysis of a nozzle as a flow-measuring device.

But for an ideal gas

$$C_{p0} = \frac{kR}{k-1} \qquad \text{and} \qquad v_i = R\frac{T_i}{P_i}$$

Therefore,

$$\frac{\mathbf{V}_e^2 - \mathbf{V}_i^2}{2} = v_i\,\Delta P$$

which is the same as Eq. 15.59, which was developed for incompressible flow. Therefore, when the pressure drop across a nozzle or orifice is small, the flow can be calculated with high accuracy by assuming incompressible flow.

The Pitot tube, Fig. 15.23, is an important instrument for measuring the velocity of a fluid. In calculating the flow with a Pitot tube, it is assumed that the fluid is decelerated isentropically in front of the Pitot tube; therefore, the stagnation pressure of the free stream can be measured.

Applying the energy equation to this process, we have

$$h + \frac{\mathbf{V}^2}{2} = h_0$$

If we assume incompressible flow for this isentropic process, the energy equation reduces to (because $T\,ds = dh - v\,dP$)

$$\frac{\mathbf{V}^2}{2} = h_0 - h = v(P_0 - P)$$

or

$$\mathbf{V} = \sqrt{2v(P_0 - P)} \tag{15.61}$$

If we consider the compressible flow of an ideal gas with constant specific heat, the velocity can be found from the relation

$$\frac{\mathbf{V}^2}{2} = h_0 - h = C_{p0}(T_0 - T) = C_{p0}T\left(\frac{T_0}{T} - 1\right)$$

$$= C_{p0}T\left[\left(\frac{P_0}{P}\right)^{(k-1)/k} - 1\right] \tag{15.62}$$

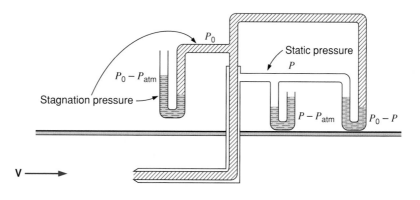

FIGURE 15.23
Schematic arrangement
of a Pitot tube.

It is of interest to know the error introduced by assuming incompressible flow when using the Pitot tube to measure the velocity of an ideal gas. To do so, we introduce Eq. 15.35 and rearrange it as follows:

$$\frac{P_0}{P} = \left(1 + \frac{k-1}{2}M^2\right)^{k/(k-1)} = \left[1 + \left(\frac{k-1}{2}\right)\left(\frac{\mathbf{V}^2}{c^2}\right)\right]^{k/(k-1)} \tag{15.63}$$

But

$$\frac{\mathbf{V}^2}{2} + C_{p0}T = C_{p0}T_0$$

$$\frac{\mathbf{V}^2}{2} + \frac{kRc^2}{(k-1)kR} = \frac{kRc_0^2}{(k-1)kR}$$

$$1 + \frac{2c^2}{(k-1)\mathbf{V}^2} = \frac{2c_0^2}{(k-1)\mathbf{V}^2} \qquad \text{where} \qquad c_0 = \sqrt{kRT_0}$$

$$\frac{c^2}{\mathbf{V}^2} = \frac{k-1}{2}\left[\left(\frac{2}{k-1}\right)\left(\frac{c_0^2}{\mathbf{V}^2}\right) - 1\right] = \frac{c_0^2}{\mathbf{V}^2} - \frac{k-1}{2}$$

or

$$\frac{c^2}{\mathbf{V}^2} = \frac{c_0^2}{\mathbf{V}^2} - \frac{k-1}{2} \tag{15.64}$$

Substituting this into Eq. 15.63 and rearranging,

$$\frac{P}{P_0} = \left[1 - \frac{k-1}{2}\left(\frac{\mathbf{V}}{c_0}\right)^2\right]^{k/(k-1)} \tag{15.65}$$

Expanding this equation by the binomial theorem, and including terms through $(\mathbf{V}/c_0)^4$, we have

$$\frac{P}{P_0} = 1 - \frac{k}{2}\left(\frac{\mathbf{V}}{c_0}\right)^2 + \frac{k}{8}\left(\frac{\mathbf{V}}{c_0}\right)^4$$

On rearranging this, we have

$$\frac{P_0 - P}{\rho_0\mathbf{V}^2/2} = 1 - \frac{1}{4}\left(\frac{\mathbf{V}}{c_0}\right)^2 \tag{15.66}$$

For incompressible flow, the corresponding equation is

$$\frac{P_0 - P}{\rho_0\mathbf{V}^2/2} = 1$$

Therefore, the second term on the right side of Eq. 15.66 represents the error involved if incompressible flow is assumed. The error in pressure for a given velocity and the error in velocity for a given pressure that would result from assuming incompressible flow are given in Table 15.2.

TABLE 15.2

Errors in pressure for a given velocity and in velocity for a given pressure resulting from assuming incompressible flow.

V/c_0	Approximate Room-Temperature Velocity, m/s	Error in Pressure for a Given Velocity, %	Error in Velocity for a Given Pressure, %
0.0	0	0	0
0.1	35	0.25	−0.13
0.2	70	1.0	−0.5
0.3	105	2.25	−1.2
0.4	140	4.0	−2.1
0.5	175	6.25	−3.3

In-Text Concept Questions

g. Which of the cases in Fig. 15.17(*a–h*) have entropy generation and which do not?

h. How do the stagnation temperature and pressure change in an adiabatic nozzle flow with an efficiency of less than 100 %?

i. Table 15.4 has a column for P_{0y}/P_{0x}; why is there not one for T_{0y}/T_{0x}?

j. How high can a gas velocity (Mach number) be and still allow us to treat it as incompressible flow within 2 % error?

SUMMARY

A short introduction is given to compressible flow in general, with particular application to flow through nozzles and diffusers. We start with the introduction of the isentropic stagnation state (recall the stagnation enthalpy from Chapter 4), which becomes important for the subsequent material. The momentum equation is formulated for a general control volume from which we can infer forces that must act on a control volume due to the presence of the flow of momentum. A special case is the thrust exerted on a jet engine due to the higher flow of momentum out.

The flow through a nozzle is introduced first as an incompressible flow, already covered in Chapter 7, leading to the Bernoulli equation. Then we cover the concept of the velocity of sound, which is the speed at which isentropic pressure waves travel. The speed of sound, c, is a thermodynamic property, which for an ideal gas can be expressed explicitly in terms of other properties. As we analyze the compressible flow through a nozzle, we discover the significantly different behavior of the flow depending on the Mach number. For a Mach number less than 1, it is subsonic flow and a converging nozzle increases the velocity, whereas for a Mach number larger than 1, it is supersonic (hypersonic) flow and a diverging nozzle is needed to increase the velocity. Similar conclusions apply to a diffuser. With a large enough pressure ratio across the nozzle, we have $M = 1$ at the throat (smallest area), at which location we have the critical properties (T^*, P^*, and ρ^*). The resulting mass flow rate through a convergent and convergent-divergent nozzle is discussed in detail as a function of the back pressure. Several different types of reversible and adiabatic—thus, isentropic—flows are possible, ranging from subsonic flow everywhere to sonic at the throat only and then subsonic followed by supersonic flow in the diverging section. The mass flow rate is maximum when the nozzle is choked, and $M = 1$ at the throat, then further decrease in the back pressure will not result in any larger mass flow rate.

For back pressures for which an isentropic solution is not possible, a shock may be present. We cover the normal shocks and the relations across the shock satisfying the continuity equation and energy equation (Fanno line), as well as the momentum equation (Rayleigh line). The flow through a shock goes from supersonic to subsonic and there is a drop in the stagnation pressure, while there is an increase in entropy across the shock. With a possible shock in the diverging section or at the exit plane or outside the nozzle, we can do the flow analysis for all possible back pressures, as shown in Figure 15.17.

In the last two sections, we consider the more practical aspects of using nozzles or diffusers. They are characterized by coefficiencies or flow coefficients, which are useful because they are constant over a range of conditions. Nozzles or orifices are used in a number of different forms for the measurement of flow rates, and it is important to know when to treat the flow as compressible.

You should have learned a number of skills and acquired abilities from studying this chapter that will allow you to:

- Find the stagnation flow properties for a given flow.
- Apply the momentum equation to a general control volume.
- Know the simplification for an incompressible flow and how to treat it.
- Know the velocity of sound and how to calculate it for an ideal gas.
- Know the importance of the Mach number and what it implies.
- Know the isentropic property relations and how properties like pressure, temperature, and density vary with the Mach number.
- Realize that the flow area and the Mach number are connected and how.
- Find the mass flow rate through a nozzle for an isentropic flow.
- Know what a choked flow is and under which conditions it happens.
- Know what a normal shock is and when to expect it.
- Be able to connect the properties before and after the shock.
- Know how to relate the properties across the shock to the upstream and downstream properties.
- Realize the importance of the stagnation properties and when they are varying.
- Treat a nozzle or diffuser flow from knowledge of the efficiency or the flow coefficient.
- Know how nozzles or orifices are used as measuring devices.

KEY CONCEPTS AND FORMULAS

Stagnation enthalpy	$h_0 = h + \dfrac{1}{2}\mathbf{V}^2$
Momentum equation, x-direction	$\dfrac{d(m\mathbf{V}_x)}{dt} = \sum F_x + \sum \dot{m}_i \mathbf{V}_{ix} - \sum \dot{m}_e \mathbf{V}_{ex}$
Bernoulli equation	$v(P_e - P_i) + \dfrac{1}{2}(\mathbf{V}_e^2 - \mathbf{V}_i^2) + (Z_e - Z_i)g = 0$
Speed of sound ideal gas	$c = \sqrt{kRT}$
Mach number	$M = \mathbf{V}/c$

Area pressure relation

$$\frac{dA}{A} = \frac{dP}{\rho \mathbf{V}^2}(1 - M^2)$$

Isentropic Relations Between Local Properties at M and Stagnation Properties

Pressure relation

$$P_0 = P\left[1 + \frac{k-1}{2}M^2\right]^{k/(k-1)}$$

Density relation

$$\rho_0 = \rho\left[1 + \frac{k-1}{2}M^2\right]^{1/(k-1)}$$

Temperature relation

$$T_0 = T\left[1 + \frac{k-1}{2}M^2\right]$$

Mass flow rate

$$\dot{m} = AP_0\sqrt{\frac{k}{RT_0}}M \bigg/ \left[1 + \frac{k-1M^2}{2}\right]^{(k+1)/2(k-1)}$$

Critical temperature

$$T^* = T_0\frac{2}{k+1}$$

Critical pressure

$$P^* = P_0\left[\frac{2}{k+1}\right]^{k/(k-1)}$$

Critical density

$$\rho* = \rho_0\left[\frac{2}{k+1}\right]^{1/(k-1)}$$

Critical mass flow rate

$$\dot{m} = A^*P_0\sqrt{\frac{k}{RT_0}}\left[\frac{2}{k+1}\right]^{(k+1)/2(k-1)}$$

Normal shock

$$M_y^2 = \left[M_x^2 + \frac{2}{k-1}\right]\bigg/\left[\frac{2k}{k-1}M_x^2 - 1\right]$$

$$\frac{P_y}{P_x} = \frac{1+kM_x^2}{1+kM_y^2}$$

$$\frac{T_y}{T_x} = \frac{1+\frac{k-1}{2}M_x^2}{1+\frac{k-1}{2}M_y^2}$$

$$P_{0y} = P_y\left[1 + \frac{k-1}{2}M_y^2\right]^{k/(k-1)}$$

$$s_y - s_x = C_p\ln\frac{T_y}{T_x} - R\ln\frac{P_y}{P_x} > 0$$

Nozzle efficiency

$$\eta_N = \frac{h_{0i} - h_e}{h_{0i} - h_s}$$

Discharge coefficient

$$C_D = \frac{\dot{m}_{\text{actual}}}{\dot{m}_s}$$

Diffuser efficiency

$$\eta_D = \frac{\Delta h_s}{\mathbf{V}_1^2/2}$$

TABLE 15.3

One-Dimensional Isentropic Compressible-Flow Functions for an Ideal Gas with Constant Specific Heat and Molecular Mass and k = 1.4

M	M*	A/A*	P/P_0	ρ/ρ_0	T/T_0
0.0	0.00000	∞	1.00000	1.00000	1.00000
0.1	0.10944	5.82183	0.99303	0.99502	0.99800
0.2	0.21822	2.96352	0.97250	0.98028	0.99206
0.3	0.32572	2.03506	0.93947	0.95638	0.98232
0.4	0.43133	1.59014	0.89561	0.92427	0.96899
0.5	0.53452	1.33984	0.84302	0.88517	0.95238
0.6	0.63481	1.18820	0.78400	0.84045	0.93284
0.7	0.73179	1.09437	0.72093	0.79158	0.91075
0.8	0.82514	1.03823	0.65602	0.73999	0.88652
0.9	0.91460	1.00886	0.59126	0.68704	0.86059
1.0	1.0000	1.00000	0.52828	0.63394	0.83333
1.1	1.0812	1.00793	0.46835	0.58170	0.80515
1.2	1.1583	1.03044	0.41238	0.53114	0.77640
1.3	1.2311	1.06630	0.36091	0.48290	0.74738
1.4	1.2999	1.11493	0.31424	0.43742	0.71839
1.5	1.3646	1.17617	0.27240	0.39498	0.68966
1.6	1.4254	1.25023	0.23527	0.35573	0.66138
1.7	1.4825	1.33761	0.20259	0.31969	0.63371
1.8	1.5360	1.43898	0.17404	0.28682	0.60680
1.9	1.5861	1.55526	0.14924	0.25699	0.58072
2.0	1.6330	1.68750	0.12780	0.23005	0.55556
2.1	1.6769	1.83694	0.10935	0.20580	0.53135
2.2	1.7179	2.00497	0.93522E-01	0.18405	0.50813
2.3	1.7563	2.19313	0.79973E-01	0.16458	0.48591
2.4	1.7922	2.40310	0.68399E-01	0.14720	0.46468
2.5	1.8257	2.63672	0.58528E-01	0.13169	0.44444
2.6	1.8571	2.89598	0.50115E-01	0.11787	0.42517
2.7	1.8865	3.18301	0.42950E-01	0.10557	0.40683
2.8	1.9140	3.50012	0.36848E-01	0.94626E-01	0.38941
2.9	1.9398	3.84977	0.31651E-01	0.84889E-01	0.37286
3.0	1.9640	4.23457	0.27224E-01	0.76226E-01	0.35714
3.5	2.0642	6.78962	0.13111E-01	0.45233E-01	0.28986
4.0	2.1381	10.7188	0.65861E-02	0.27662E-01	0.23810
4.5	2.1936	16.5622	0.34553E-02	0.17449E-01	0.19802
5.0	2.2361	25.0000	0.18900E-02	0.11340E-01	0.16667
6.0	2.2953	53.1798	0.63336E-03	0.51936E-02	0.12195
7.0	2.3333	104.143	0.24156E-03	0.26088E-02	0.09259
8.0	2.3591	190.109	0.10243E-03	0.14135E-02	0.07246
9.0	2.3772	327.189	0.47386E-04	0.81504E-03	0.05814
10.0	2.3905	535.938	0.23563E-04	0.49482E-03	0.04762
∞	2.4495	∞	0.0	0.0	0.0

TABLE 15.4

One-Dimensional Normal Shock Functions for an Ideal Gas with Constant Specific Heat and Molecular Mass and $k = 1.4$

M_x	M_y	P_y/P_x	ρ_y/ρ_x	T_y/T_x	P_{0y}/P_{0x}	P_{0y}/P_x
1.00	1.000 00	1.0000	1.0000	1.0000	1.000 00	1.8929
1.05	0.953 13	1.1196	1.0840	1.0328	0.999 85	2.0083
1.10	0.911 77	1.2450	1.1691	1.0649	0.998 93	2.1328
1.15	0.875 02	1.3763	1.2550	1.0966	0.996 69	2.2661
1.20	0.842 17	1.5133	1.3416	1.1280	0.992 80	2.4075
1.25	0.812 64	1.6563	1.4286	1.1594	0.987 06	2.5568
1.30	0.785 96	1.8050	1.5157	1.1909	0.979 37	2.7136
1.35	0.761 75	1.9596	1.6028	1.2226	0.969 74	2.8778
1.40	0.739 71	2.1200	1.6897	1.2547	0.958 19	3.0492
1.45	0.719 56	2.2863	1.7761	1.2872	0.944 84	3.2278
1.50	0.701 09	2.4583	1.8621	1.3202	0.929 79	3.4133
1.55	0.684 10	2.6362	1.9473	1.3538	0.913 19	3.6057
1.60	0.668 44	2.8200	2.0317	1.3880	0.895 20	3.8050
1.65	0.653 96	3.0096	2.1152	1.4228	0.875 99	4.0110
1.70	0.640 54	3.2050	2.1977	1.4583	0.855 72	4.2238
1.75	0.628 09	3.4063	2.2791	1.4946	0.834 57	4.4433
1.80	0.616 50	3.6133	2.3592	1.5316	0.812 68	4.6695
1.85	0.605 70	3.8263	2.4381	1.5693	0.790 23	4.9023
1.90	0.595 62	4.0450	2.5157	1.6079	0.767 36	5.1418
1.95	0.586 18	4.2696	2.5919	1.6473	0.744 20	5.3878
2.00	0.577 35	4.5000	2.6667	1.6875	0.720 87	5.6404
2.05	0.569 06	4.7362	2.7400	1.7285	0.697 51	5.8996
2.10	0.561 28	4.9783	2.8119	1.7705	0.674 20	6.1654
2.15	0.553 95	5.2263	2.8823	1.8132	0.651 05	6.4377
2.20	0.547 06	5.4800	2.9512	1.8569	0.628 14	6.7165
2.25	0.540 55	5.7396	3.0186	1.9014	0.605 53	7.0018
2.30	0.534 41	6.0050	3.0845	1.9468	0.583 29	7.2937
2.35	0.528 61	6.2762	3.1490	1.9931	0.561 48	7.5920
2.40	0.523 12	6.5533	3.2119	2.0403	0.540 14	7.8969
2.45	0.517 92	6.8363	3.2733	2.0885	0.519 31	8.2083
2.50	0.512 99	7.1250	3.3333	2.1375	0.499 01	8.5261
2.55	0.508 31	7.4196	3.3919	2.1875	0.479 28	8.8505
2.60	0.503 87	7.7200	3.4490	2.2383	0.460 12	9.1813
2.70	0.495 63	8.3383	3.5590	2.3429	0.423 59	9.8624
2.80	0.488 17	8.9800	3.6636	2.4512	0.389 46	10.569
2.90	0.481 38	9.6450	3.7629	2.5632	0.357 73	11.302
3.00	0.475 19	10.333	3.8571	2.6790	0.328 34	12.061
4.00	0.434 96	18.500	4.5714	4.0469	0.138 76	21.068
5.00	0.415 23	29.000	5.0000	5.8000	0.061 72	32.653
10.00	0.387 58	116.50	5.7143	20.387	0.003 04	129.22

CONCEPT-STUDY GUIDE PROBLEMS

15.1 Which temperature does a thermometer or thermocouple measure? Would you ever need to correct that?

15.2 A jet engine thrust is found from the overall momentum equation. Where is the actual force acting (it is not a long-range force in the flow)?

15.3 Most compressors have a small diffuser at the exit to reduce the high gas velocity near the rotating blades and increase the pressure in the exit flow. What does this do to the stagnation pressure?

15.4 A diffuser is a divergent nozzle used to reduce a flow velocity. Is there a limit for the Mach number for it to work this way?

15.5 Sketch the variation in $\mathbf{V}$, T, P, ρ, and M for a subsonic flow into a convergent nozzle with $M = 1$ at the exit plane.

15.6 Sketch the variation in $\mathbf{V}$, T, P, ρ, and M for a sonic ($M = 1$) flow into a divergent nozzle with $M = 2$ at the exit plane.

15.7 Can any low enough backup pressure generate an isentropic supersonic flow?

15.8 Is there any benefit in operating a nozzle choked?

15.9 Can a shock be located upstream from the throat?

15.10 The high-velocity exit flow in Example 15.7 is at 183 K. Can that flow be used to cool a room?

15.11 A convergent-divergent nozzle is presented for an application that requires a supersonic exit flow. What features of the nozzle do you look at first?

15.12 To increase the flow through a choked nozzle, the flow can be heated/cooled or compressed/ expanded (four processes) before or after the nozzle. Explain which of these eight possibilities will help and which will not.

15.13 Suppose a convergent-divergent nozzle is operated as case h in Fig. 15.17. What kind of nozzle could have the same exit pressure, but with a reversible flow?

HOMEWORK PROBLEMS

Stagnation Properties

15.14 A stationary thermometer measures $80\,^\circ\mathrm{C}$ in an air flow that has a velocity of 200 m/s. What is the actual flow temperature?

15.15 Steam leaves a nozzle with a pressure of 500 kPa, a temperature of $350\,^\circ\mathrm{C}$, and a velocity of 250 m/s. What are the isentropic stagnation pressure and temperature?

15.16 Steam at 1600 kPa, $300\,^\circ\mathrm{C}$, flows so that it has a stagnation (total) pressure of 1800 kPa. Find the velocity and the stagnation temperature.

15.17 An object from space enters the earth's upper atmosphere at 5 kPa, 100 K, with a relative velocity of 2500 m/s or more. Estimate the object's surface temperature.

15.18 The products of combustion of a jet engine leave the engine with a velocity relative to the plane of 500 m/s, a temperature of $525\,^\circ\mathrm{C}$, and a pressure of 75 kPa. Assuming that $k = 1.32$, $C_p = 1.15\ \mathrm{kJ/kg\cdot K}$ for the products, determine the stagnation pressure and temperature of the products relative to the airplane.

15.19 Steam is flowing to a nozzle with a pressure of 400 kPa. The stagnation pressure and temperature are measured to be 600 kPa and $350\,^\circ\mathrm{C}$. What are the flow velocity and temperature?

15.20 A meteorite melts and burns up at a temperature of 3000 K. If it hits air at 5 kPa, 50 K, how high a velocity should it have to experience such a temperature?

15.21 Air leaves a compressor in a pipe with a stagnation temperature and pressure of $150\,^\circ\mathrm{C}$, 300 kPa, and a velocity of 125 m/s. The pipe has a cross-sectional area of $0.02\ \mathrm{m}^2$. Determine the static temperature and pressure and the mass flow rate.

15.22 I drive down the highway at 110 km/h on a $25\,^\circ\mathrm{C}$, 101.3 kPa day. I put my hand, cross-sectional area $0.01\ \mathrm{m}^2$, flat out the window. What is the force on my hand and what temperature do I feel?

15.23 A stagnation pressure of 110 kPa is measured for an air flow where the pressure is 100 kPa and the temperature is 20 °C in the approach flow. What is the incoming velocity?

Momentum Equation and Forces

15.24 A 4 cm inner-diameter pipe has an inlet flow of 10 kg/s water at 20 °C, 200 kPa. After a 90 degree bend, as shown in Fig. P15.24, the exit flow is at 20 °C, 190 kPa. Neglect gravitational effects and find the anchoring forces, F_x and F_y.

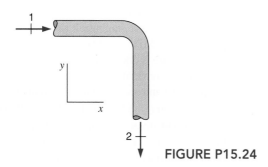

FIGURE P15.24

15.25 A jet engine receives a flow of 150 m/s air at 75 kPa, 5 °C, across an area of 0.6 m² with an exit flow at 450 m/s, 75 kPa, 800 K. Find the mass flow rate and thrust.

15.26 How large a force must be applied to a squirt gun to have 0.1 kg/s water flow out at 20 m/s? What pressure inside the chamber is needed?

15.27 A jet engine at takeoff has air at 20 °C, 100 kPa, coming at 35 m/s through the 1.5 m diameter inlet. The exit flow is at 1200 K, 100 kPa, through the exit nozzle of 0.4 m diameter. Neglect the fuel flow rate and find the net force (thrust) on the engine.

15.28 A water turbine using nozzles is located at the bottom of Hoover Dam, 175 m below the surface of Lake Mead. The water enters the nozzles at a stagnation pressure corresponding to the column of water above it minus 20 % due to losses. The temperature is 15 °C, and the water leaves at standard atmospheric pressure. If the flow through the nozzle is reversible and adiabatic, determine the velocity and kinetic energy per kilogram of water leaving the nozzle.

15.29 A water cannon sprays 1 kg/s liquid water at a velocity of 100 m/s horizontally out from a nozzle. It is driven by a pump that receives the water from a tank at 15 °C, 100 kPa. Neglect elevation differences and the kinetic energy of the water flow in the pump and hose to the nozzle. Find the nozzle exit area, the required pressure out of the pump, and the horizontal force needed to hold the cannon.

15.30 An irrigation pump takes water from a lake and discharges it through a nozzle, as shown in Fig. P15.30. At the pump exit, the pressure is 900 kPa, and the temperature is 20 °C. The nozzle is located 15 m above the pump, and the atmospheric pressure is 100 kPa. Assuming reversible flow through the system, determine the velocity of the water leaving the nozzle.

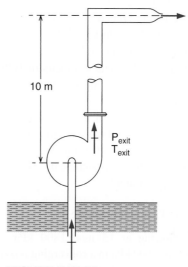

FIGURE P15.30

15.31 A water tower on a farm holds 1 m³ liquid water at 20 °C, 100 kPa, in a tank on top of a 5 m tall tower. A pipe leads to the ground level with a tap that can open a 1.5 cm diameter hole. Neglect friction and pipe losses, and estimate the time it will take to empty the tank of water.

Adiabatic 1-D Flow and Velocity of Sound

15.32 Find the speed of sound for air at 100 kPa at the two temperatures, 0 °C and 30 °C. Repeat the answer for carbon dioxide and argon gases.

15.33 Find the expression for the anchoring force, R_x, for an incompressible flow like the one in Fig. 15.6. Show that it can be written as

$$R_x = \frac{\mathbf{V}_i - \mathbf{V}_e}{\mathbf{V}_i + \mathbf{V}_e}(P_i A_i + P_e A_e)$$

15.34 Estimate the speed of sound for steam directly from Eq. 15.25 and the steam tables for a state of 6 MPa, 400 °C. Use table values at 5 and 7 MPa at the same entropy as the wanted state. Equation 15.25 is then solved by finite difference. Also find the answer for the speed of sound, assuming steam is an ideal gas.

15.35 Use the CATT3 software to solve the previous problem.

15.36 If the sound of thunder is heard 5 s after the lightning is seen and the temperature is 20 °C, how far away is the lightning?

15.37 Find the speed of sound for carbon dioxide at 2500 kPa, 60 °C, using either the tables or the CATT3 software (same procedure as in Problem 15.34) and compare that with Eq. 15.28.

15.38 A jet flies at an altitude of 12 km where the air is at −40 °C, 45 kPa, with a velocity of 1000 km/h. Find the Mach number and the stagnation temperature on the nose.

15.39 The speed of sound in liquid water at 25 °C is about 1500 m/s. Find the stagnation pressure and temperature for a $M = 0.1$ flow at 25 °C, 100 kPa. Is it possible to get a significant Mach number flow of liquid water?

Reversible Flow Through a Nozzle

15.40 Steam flowing at 15 m/s, 1800 kPa, 300 °C, expands to 1600 kPa in a converging nozzle. Find the exit velocity and area ratio, A_e/A_i.

15.41 A convergent nozzle has a minimum area of 0.1 m² and receives air at 175 kPa, 1000 K, flowing at 100 m/s. What is the back pressure that will produce the maximum flow rate? Find that flow rate.

15.42 A convergent-divergent nozzle has a throat area of 100 mm² and an exit area of 175 mm². The inlet flow is helium at a total pressure of 1 MPa and a stagnation temperature of 375 K. What is the back pressure that will produce a sonic condition at the throat but a subsonic condition everywhere else?

15.43 To what pressure should the steam in Problem 15.40 expand to reach Mach 1? Use constant specific heats to solve this problem.

15.44 A jet plane travels through the air with a speed of 1000 km/h at an altitude of 6 km, where the pressure is 40 kPa and the temperature is −12 °C.

Consider the inlet diffuser of the engine, where air leaves with a velocity of 100 m/s. Determine the pressure and temperature leaving the diffuser and the ratio of inlet to exit area of the diffuser, assuming the flow to be reversible and adiabatic.

15.45 Air flows into a convergent-divergent nozzle with an exit area of 1.59 times the throat area of 0.005 m². The inlet stagnation state is 1 MPa, 600 K. Find the back pressure that will cause subsonic flow throughout the entire nozzle with $M = 1$ at the throat. What is the mass flow rate?

15.46 A nozzle is designed assuming reversible adiabatic flow with an exit Mach number of 2.8 while flowing air with a stagnation pressure and temperature of 2 MPa and 150 °C, respectively. The mass flow rate is 5 kg/s, and k may be assumed to be 1.40 and constant. Determine the exit pressure, temperature and area, and the throat area.

15.47 An air flow at 600 kPa, 600 K, $M = 0.3$ flows into a convergent-divergent nozzle with $M = 1$ at the throat. Assume a reversible flow with an exit area twice the throat area and find the exit pressure and temperature for subsonic exit flow to exist.

15.48 Air at 150 kPa, 290 K, expands to the atmosphere at 100 kPa through a convergent nozzle with an exit area of 0.01 m². Assume an ideal nozzle. What is the percent error in mass flow rate if the flow is assumed to be incompressible?

15.49 Find the exit pressure and temperature for supersonic exit flow to exist in the nozzle flow of Problem 15.47.

15.50 Air is expanded in a nozzle from a stagnation state of 2 MPa, 600 K, to a back pressure of 1.9 MPa. If the exit cross-sectional area is 0.003 m², find the mass flow rate.

15.51 A 1 m³ insulated tank contains air at 1 MPa, 560 K. The air in the tank is now discharged through a small convergent nozzle to the atmosphere at 100 kPa. The nozzle has an exit area of 2×10^{-5} m².
a. Find the initial mass flow rate out of the tank.
b. Find the mass flow rate when half of the mass has been discharged.

15.52 A convergent-divergent nozzle has a throat diameter of 0.05 m and an exit diameter of 0.1 m. The inlet stagnation state is 500 kPa, 500 K. Find the back pressure that will lead to the maximum possible flow rate and the mass flow rate for three different gases: air, hydrogen, and carbon dioxide.

15.53 Air is expanded in a nozzle from a stagnation state of 2 MPa, 600 K, to a static pressure of 200 kPa. The mass flow rate through the nozzle is 5 kg/s. Assume the flow is reversible and adiabatic and determine the throat and exit areas for the nozzle.

15.54 Air flows into a convergent-divergent nozzle with an exit area 2.0 times the throat area of 0.005 m^2. The inlet stagnation state is 1.2 MPa, 600 K. Find the back pressure that will cause a reversible supersonic exit flow with $M = 1$ at the throat. What is the mass flow rate?

15.55 What is the exit pressure that will allow a reversible subsonic exit flow in the previous problem?

15.56 Helium flows at 500 kPa, 500 K, with 100 m/s into a convergent-divergent nozzle. Find the throat pressure and temperature for reversible flow and $M = 1$ at the throat.

15.57 Assume the same tank and conditions as in Problem 15.51. After some flow out of the nozzle, flow becomes subsonic. Find the mass in the tank and the mass flow rate out at that instant.

15.58 A given convergent nozzle operates so that it is choked with stagnation inlet flow properties of 400 kPa, 400 K. To increase the flow, a reversible adiabatic compressor is added before the nozzle to increase the stagnation flow pressure to 500 kPa. What happens to the flow rate?

15.59 A 1 m^3 uninsulated tank contains air at 1 MPa, 560 K. The air in the tank is now discharged through a small convergent nozzle to the atmosphere at 100 kPa, while heat transfer from some source keeps the air temperature in the tank at 560 K. The nozzle has an exit area of 2×10^{-5} m^2.
a. Find the initial mass flow rate out of the tank.
b. Find the mass flow rate when half of the mass has been discharged.

15.60 Assume the same tank and conditions as in Problem 15.59. After some flow out, the nozzle flow becomes subsonic. Find the mass in the tank and the mass flow rate out at that instant.

Normal Shocks

15.61 The products of combustion (use air) enter a convergent nozzle of a jet engine at a total pressure of 125 kPa and a total temperature of 650 °C. The atmospheric pressure is 45 kPa, and the flow is adiabatic, with a rate of 25 kg/s. Determine the exit area of the nozzle.

15.62 Redo the previous problem for a mixture with $k = 1.3$ and a molecular mass of 31.

15.63 At what Mach number will the normal shock occur in the nozzle of Problem 15.52 flowing with air if the back pressure is halfway between the pressures at c and d in Fig. 15.17?

15.64 Consider the nozzle of Problem 15.53 and determine what back pressure will cause a normal shock to stand in the exit plane of the nozzle. This is case g in Fig. 15.17. What is the mass flow rate under these conditions?

15.65 A normal shock in air has an upstream total pressure of 500 kPa, a stagnation temperature of 500 K, and $M_x = 1.4$. Find the downstream stagnation pressure.

15.66 How much entropy per kilogram of flow is generated in the shock in Example 15.9?

15.67 Consider the diffuser of a supersonic aircraft flying at $M = 1.4$ at such an altitude that the temperature is -20 °C and the atmospheric pressure is 50 kPa. Consider two possible ways in which the diffuser might operate, and for each case, calculate the throat area required for a flow of 50 kg/s.
a. The diffuser operates as reversible adiabatic with subsonic exit velocity.
b. A normal shock stands at the entrance to the diffuser. Except for the normal shock, the flow is reversible and adiabatic, and the exit velocity is subsonic. This is shown in Fig. P15.67. Assume a convergent-divergent diffuser with $M = 1$ at the throat.

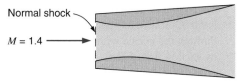

FIGURE P15.67

15.68 A flow into a normal shock in air has a total pressure of 400 kPa, a stagnation temperature of 600 K, and $M_x = 1.2$. Find the upstream temperature T_x, the specific entropy generation in the shock, and the downstream velocity.

15.69 Consider the nozzle in Problem 15.42. What should the back pressure be for a normal shock to stand at the exit plane (this is case *g* in Fig.15.17)? What is the exit velocity after the shock?

15.70 Find the specific entropy generation in the shock of the previous problem.

Nozzles, Diffusers, and Orifices

15.71 Steam at 600 kPa, 300 °C, is fed to a set of convergent nozzles in a steam turbine. The total nozzle exit area is 0.005 m², and the nozzles have a discharge coefficient of 0.94. The mass flow rate should be estimated from the pressure drop across the nozzles, which is measured to be 200 kPa. Determine the mass flow rate.

15.72 Air enters a diffuser with a velocity of 200 m/s, a static pressure of 70 kPa, and a temperature of −6 °C. The velocity leaving the diffuser is 60 m/s, and the static pressure at the diffuser exit is 80 kPa. Determine the static temperature at the diffuser exit and the diffuser efficiency. Compare the stagnation pressures at the inlet and the exit.

15.73 Repeat Problem 15.44, assuming a diffuser efficiency of 80 %.

15.74 A sharp-edged orifice is used to measure the flow of air in a pipe. The pipe diameter is 100 mm, and the diameter of the orifice is 25 mm. Upstream of the orifice, the absolute pressure is 150 kPa, and the temperature is 35 °C. The pressure drop across the orifice is 15 kPa, and the coefficient of discharge is 0.62. Determine the mass flow rate in the pipeline.

15.75 A critical nozzle is used for the accurate measurement of the flow rate of air. Exhaust from a car engine is diluted with air, so its temperature is 50 °C at a total pressure of 100 kPa. It flows through the nozzle with a throat area of 700 mm² by suction from a blower. Find the needed suction pressure that will lead to critical flow in the nozzle, the mass flow rate, and the blower work, assuming the blower exit is at atmospheric pressure, 100 kPa.

15.76 Air is expanded in a nozzle from 700 kPa, 200 °C, to 150 kPa in a nozzle having an efficiency of 90 %. The mass flow rate is 4 kg/s. Determine the exit area of the nozzle, the exit velocity, and the increase of entropy per kilogram of air.

Compare these results with those of a reversible adiabatic nozzle.

15.77 Steam at a pressure of 1 MPa and a temperature of 400 °C expands in a nozzle to a pressure of 200 kPa. The nozzle efficiency is 90 %, and the mass flow rate is 10 kg/s. Determine the nozzle exit area and the exit velocity.

15.78 Steam at 800 kPa, 350 °C, flows through a convergent-divergent nozzle that has a throat area of 350 mm². The pressure at the exit plane is 150 kPa, and the exit velocity is 800 m/s. The flow from the nozzle entrance to the throat is reversible and adiabatic. Determine the exit area of the nozzle, the overall nozzle efficiency, and the entropy generation in the process.

15.79 A convergent nozzle with an exit diameter of 2 cm has an air inlet flow of 20 °C, 101 kPa (stagnation conditions). The nozzle has an isentropic efficiency of 95 %, and the pressure drop is measured to be a 50 cm water column. Find the mass flow rate, assuming compressible adiabatic flow. Repeat this calculation for incompressible flow.

15.80 The coefficient of discharge of a sharp-edged orifice is determined at one set of conditions by the use of an accurately calibrated gasometer. The orifice has a diameter of 20 mm, and the pipe diameter is 50 mm. The absolute upstream pressure is 200 kPa, and the pressure drop across the orifice is 82 mm Hg. The temperature of the air entering the orifice is 25 °C, and the mass flow rate measured with the gasometer is 2.4 kg/min. What is the coefficient of discharge of the orifice under these conditions?

15.81 A convergent nozzle is used to measure the flow of air to an engine. The atmosphere is 100 kPa, 25 °C. The nozzle used has a minimum area of 2000 mm², and the coefficient of discharge is 0.95. A pressure difference across the nozzle is measured to be 2.5 kPa. Find the mass flow rate, assuming incompressible flow. Also find the mass flow rate, assuming compressible adiabatic flow.

Review Problems

15.82 Atmospheric air is at 20 °C, 100 kPa, with zero velocity. An adiabatic reversible compressor takes atmospheric air in through a pipe with a cross-sectional area of 0.1 m² at a rate of 1 kg/s. It is compressed up to a measured stagnation pressure of 500 kPa and leaves through a pipe with a cross-sectional area of 0.01 m². What are the required compressor work and the air velocity, static pressure, and temperature in the exit pipeline?

15.83 The nozzle in Problem 15.46 will have a throat area of 0.001 272 m² and an exit area 3.5 times as large. Suppose the back pressure is raised to 1.4 MPa and the flow remains isentropic, except for a normal shock wave. Verify that the shock Mach number (M_x) is close to 2 and find the exit Mach number, the temperature, and the mass flow rate through the nozzle.

15.84 At what Mach number will the normal shock occur in the nozzle of Problem 15.53 if the back pressure is 1.4 MPa? (Trial and error on M_x.)

COMPUTER, DESIGN, AND OPEN-ENDED PROBLEMS

15.85 Develop a program that calculates the stagnation pressure and temperature from a static pressure, temperature, and velocity. Assume the fluid is air with constant specific heats. If the inverse relation is sought, one of the three properties in the flow must be given. Include that case also.

15.86 Use the menu-driven software to solve Problem 15.78. Find from the menu-driven steam tables the ratio of specific heats at the inlet and the speed of sound from its definition in Eq. 15.28.

15.87 (Adv.) Develop a program that will track the process in time as described in Problems 15.51 and 15.53. Investigate the time it takes to bring the tank pressure to 125 kPa as a function of the size of the nozzle exit area. Plot several of the key variables as functions of time.

15.88 A pump can deliver liquid water at an exit pressure of 400 kPa using 0.5 kW of power. Assume that the inlet is water at 100 kPa, 15 °C, and that the pipe size is the same for the inlet and exit. Design a nozzle to be mounted on the exit line so that the water exit velocity is at least 20 m/s. Show the exit velocity and mass flow rate as functions of the nozzle exit area with the same power to the pump.

15.89 In all the problems in the text, the efficiency of a pump or compressor has been given as a constant. In reality, it is a function of the mass flow rate and the fluid state through the device. Examine the literature for the characteristics of a real air compressor (blower).

15.90 The throttle plate in a carburetor severely restricts the air flow where at idle it is critical flow. For normal atmospheric conditions, estimate the inlet temperature and pressure to the cylinder of the engine.

15.91 For an experiment in the laboratory, the air flow rate should be measured. The range should be 0.05 to 0.10 kg/s, and the flow should be delivered to the experiment at 110 kPa. Size one (or two in parallel) convergent nozzle(s) that sit(s) in a plate. The air is drawn through the nozzle(s) by suction of a blower that delivers the air at 110 kPa. What should be measured, and what accuracy can be expected?

15.92 An afterburner in a jet engine adds fuel that is burned after the turbine but before the exit nozzle that accelerates the gases. Examine the effect on nozzle exit velocity of having a higher inlet temperature but the same pressure as without the afterburner. Are these nozzles operating with subsonic or supersonic flow?

Contents of Appendix

671

E FIGURES

751

SI Units: Single-State Properties

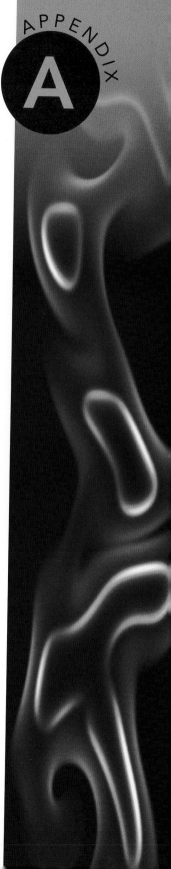

TABLE A.1
Conversion Factors

Area (*A*)

$1\ mm^2 = 1.0 \times 10^{-6}\ m^2$

$1\ cm^2\ = 1.0 \times 10^{-4}\ m^2 = 0.1550\ in.^2$ $1\ in.^2 = 6.4516\ cm^2 = 6.4516 \times 10^{-4}\ m^2$

$1\ m^2\ \ \ = 10.7639\ ft^2$ $1\ ft^2\ \ = 0.092\ 903\ m^2$

Conductivity (*k*)

$1\ W/m \cdot K = 1\ J/s \cdot m \cdot K$

$\qquad\qquad = 0.577\ 789\ Btu/h \cdot ft \cdot {}^\circ R$ $1\ Btu/h \cdot ft \cdot R = 1.730\ 735\ W/m \cdot K$

Density (*ρ*)

$1\ kg/m^3 = 0.06242797\ lbm/ft^3$ $1\ lbm/ft^3 = 16.018\ 46\ kg/m^3$

$1\ g/cm^3 = 1000\ kg/m^3$

$1\ g/cm^3 = 1\ kg/L$

Energy (*E, U*)

$1\ J\qquad = 1\ N \cdot m = 1\ kg \cdot m^2/s^2$

$1\ J\qquad = 0.737\ 562\ lbf \cdot ft$ $1\ lbf \cdot ft = 1.355\ 818\ J$

$1\ cal\ (Int.) = 4.186\ 81\ J$ $= 1.285\ 07 \times 10^{-3} Btu$

                           $1\ Btu\ (Int.) = 1.055\ 056\ kJ$

$1\ erg\qquad = 1.0 \times 10^{-7}\ J$ $= 778.1693\ lbf \cdot ft$

$1\ eV\qquad = 1.602\ 177\ 33 \times 10^{-19}\ J$

Force (*F*)

$1\ N\ = 0.224\ 809\ lbf$ $1\ lbf = 4.448\ 222\ N$

$1\ kp = 9.806\ 65\ N\ (1\ kgf)$

Gravitation

$g = 9.806\ 65\ m/s^2$

                       $g = 32.174\ 05\ ft/s^2$

Specific heat (*C_p*, *C_v*, *C*), specific entropy (*s*)

$1\ kJ/kg \cdot K = 0.238\ 846\ Btu/lbm \cdot {}^\circ R$ $1\ Btu/lbm \cdot {}^\circ R = 4.1868\ kJ/kg \cdot K$

Heat flux (per unit area)

$1\ W/m^2 = 0.316\ 998\ Btu/h \cdot ft^2$ $1\ Btu/h \cdot ft^2 = 3.154\ 59\ W/m^2$

TABLE A.1 (*continued*)
Conversion Factors

Heat-transfer coefficient (*h*)

$1 \text{ W/m}^2 \cdot \text{K} = 0.176\,11 \text{ Btu/h} \cdot \text{ft}^2 \cdot {}^\circ\text{R}$ $1 \text{ Btu/h} \cdot \text{ft}^2 \cdot {}^\circ\text{R} = 5.678\,26 \text{ W/m}^2 \cdot \text{K}$

Length (*L*)

1 mm = 0.001 m = 0.1 cm	
1 cm = 0.01 m = 10 mm = 0.3970 in.	1 in. = 2.54 cm = 0.0254 m
1 m = 3.280 84 ft = 39.370 in.	1 ft = 0.3048 m
1 km = 0.621 371 mi	1 mi = 1.609 344 km
1 mi = 1609.3 m (US statute)	1 yd = 0.9144 m

Mass (*m*)

1 kg = 2.204 623 lbm	1 lbm = 0.453 592 kg
1 tonne = 1000 kg	1 slug = 14.5939 kg
1 grain = 6.479 89 × 10⁻⁵ kg	1 ton = 2000 lbm

$1 \text{ grain} = 6.479\,89 \times 10^{-5} \text{ kg}$

Moment (torque, *T*)

$1 \text{ N} \cdot \text{m} = 0.737\,562 \text{ lbf} \cdot \text{ft}$ $1 \text{ lbf} \cdot \text{ft} = 1.355\,818 \text{ N} \cdot \text{m}$

Momentum (*mV*)

$1 \text{ kg} \cdot \text{m/s} = 7.232\,94 \text{ lbm} \cdot \text{ft/s}$ $1 \text{ lbm} \cdot \text{ft/s} = 0.138\,256 \text{ kg} \cdot \text{m/s}$
$= 0.224\,809 \text{ lbf} \cdot \text{s}$

Power ($\dot{Q}$, $\dot{W}$)

1 W = 1 J/s = 1 N·m/s	1 lbf·ft/s = 1.355 818 W
= 0.737 562 lbf·ft/s	= 4.626 24 Btu/h
1 kW = 3412.14 Btu/h	1 Btu/s = 1.055 056 kW
1 hp (metric) = 0.735 499 kW	1 hp (UK) = 0.7457 kW
	= 550 lbf·ft/s
	= 2544.43 Btu/h
1 ton of refrigeration = 3.516 85 kW	1 ton of refrigeration = 12 000 Btu/h

Pressure (*P*)

1 Pa = 1 N/m² = 1 kg/m·s²	1 lbf/in.² = 6.894 757 kPa
1 bar = 1.0 × 10⁵ Pa = 100 kPa	
1 arm = 101.325 kPa	1 atm = 14.695 94 lbf/in.²
= 1.013 25 bar	= 29.921 in. Hg [32°F]
= 760 mm Hg [0°C]	= 33.8995 ft H₂O [4°C]
= 10.332 56 m H₂O [4°C]	
1 torr = 1 mm Hg [0°C]	
1 mm Hg [0°C] = 0.133 322 kPa	
1 m H₂O [4°C] = 9.806 38 kPa	

Specific energy (*e*, *u*)

1 kJ/kg = 0.429 92 Btu/lbm	1 Btu/lbm = 2.326 kJ/kg
= 334.55 lbf·ft/lbm	1 lbf·ft/lbm = 2.989 07 × 10⁻³ kJ/kg
	= 1.285 07 × 10⁻³ Btu/lbm

TABLE A.1   (*continued*)
Conversion Factors

Specific kinetic energy ($\frac{1}{2}\mathbf{V}^2$)
$1\ m^2/s^2 = 0.001\ kJ/kg$
$1\ kJ/kg = 1000\ m^2/s^2$

$1\ ft^2/s^2 = 3.9941 \times 10^{-5}\ Btu/lbm$
$1\ Btu/lbm = 250\ 37\ ft^2/s^2$

Specific potential energy (Zg)
$1\ m \cdot g_{std} = 9.806\ 65 \times 10^{-3}\ kJ/kg$
$\qquad = 4.216\ 07 \times 10^{-3}\ Btu/lbm$

$1\ ft \cdot g_{std} = 1.0\ lbf \cdot ft/lbm$
$\qquad = 0.001\ 285\ Btu/lbm$
$\qquad = 0.002\ 989\ kJ/kg$

Specific volume (v)
$1\ cm^3/g = 0.001\ m^3/kg$
$1\ cm^3/g = 1\ L/kg$
$1\ m^3/kg = 16.018\ 46\ ft^3/lbm$

$1\ ft^3/lbm = 0.062\ 428\ m^3/kg$

Temperature (T)
$1\ K = 1°C = 1.8\ R = 1.8\ F$
$TC = TK - 273.15$
$\quad = (TF - 32)/1.8$
$TK = TR/1.8$

$1\ R = (5/9)\ K$
$TF = TR - 459.67$
$\quad = 1.8\ TC + 32$
$TR = 1.8\ TK$

Universal Gas Constant
$\bar{R} = N_0 k = 8.314\ 51\ kJ/kmol \cdot K$
$\quad = 1.985\ 89\ kcal/kmol \cdot K$
$\quad = 82.0578\ atm \cdot L/kmol \cdot K$

$\bar{R} = 1.985\ 89\ Btu/lbmol \cdot R$
$\quad = 1545.36\ lbf \cdot ft/lbmol \cdot R$
$\quad = 0.730\ 24\ atm \cdot ft^3/lbmol \cdot R$
$\quad = 10.7317\ (lbf/in.^2) \cdot ft^3/lbmol \cdot R$

Velcoity ($\mathbf{V}$)
$1\ m/s = 3.6\ km/h$
$\quad = 3.280\ 84\ ft/s$
$\quad = 2.236\ 94\ mi/h$
$1\ km/h = 0.277\ 78\ m/s$
$\quad = 0.911\ 34\ ft/s$
$\quad = 0.621\ 37\ mi/h$

$1\ ft/s = 0.681\ 818\ mi/h$
$\quad = 0.3048\ m/s$
$\quad = 1.097\ 28\ km/h$
$1\ mi/h = 1.466\ 67\ ft/s$
$\quad = 0.447\ 04\ m/s$
$\quad = 1.609\ 344\ km/h$

Volume (V)
$1\ m^3 = 35.3147\ ft^3$
$1\ L = 1\ dm^3 = 0.001\ m^3$
$1\ Gal\ (US) = 3.785\ 412\ L$
$\quad = 3.785\ 412 \times 10^{-3}\ m^3$

$1\ ft^3 = 2.831\ 685 \times 10^{-2}\ m^3$
$1\ in.^3 = 1.6387 \times 10^{-5}\ m^3$
$1\ Gal\ (UK) = 4.546\ 090\ L$

TABLE A.2
Critical Constants

Substance	Formula	Molec. Mass	Temp. (K)	Press. (MPa)	Vol. (m^3/kg)
Ammonia	NH_3	17.031	405.5	11.35	0.00426
Argon	Ar	39.948	150.8	4.87	0.00188
Bromine	Br_2	159.808	588	10.30	0.000796
Carbon dioxide	CO_2	44.01	304.1	7.38	0.00212
Carbon monoxide	CO	28.01	132.9	3.50	0.00333
Chlorine	Cl_2	70.906	416.9	7.98	0.00175
Fluorine	F_2	37.997	144.3	5.22	0.00174
Helium	He	4.003	5.19	0.227	0.0143
Hydrogen (normal)	H_2	2.016	33.2	1.30	0.0323
Krypton	Kr	83.80	209.4	5.50	0.00109
Neon	Ne	20.183	44.4	2.76	0.00206
Nitric oxide	NO	30.006	180	6.48	0.00192
Nitrogen	N_2	28.013	126.2	3.39	0.0032
Nitrogen dioxide	NO_2	46.006	431	10.1	0.00365
Nitrous oxide	N_2O	44.013	309.6	7.24	0.00221
Oxygen	O_2	31.999	154.6	5.04	0.00229
Sulfur dioxide	SO_2	64.063	430.8	7.88	0.00191
Water	H_2O	18.015	647.3	22.12	0.00317
Xenon	Xe	131.30	289.7	5.84	0.000902
Acetylene	C_2H_2	26.038	308.3	6.14	0.00433
Benzene	C_6H_6	78.114	562.2	4.89	0.00332
n-Butane	C_4H_{10}	58.124	425.2	3.80	0.00439
Chlorodifluoroethane (142b)	CH_3CClF_2	100.495	410.3	4.25	0.00230
Chlorodifluoromethane (22)	$CHClF_2$	86.469	369.3	4.97	0.00191
Dichlorofluoroethane (141)	CH_3CCl_2F	116.95	481.5	4.54	0.00215
Dichlorotrifluoroethane (123)	$CHCl_2CF_3$	152.93	456.9	3.66	0.00182
Difluoroethane (152a)	CHF_2CH_3	66.05	386.4	4.52	0.00272
Difluoromethane (32)	CF_2H_2	52.024	351.3	5.78	0.00236
Ethane	C_2H_6	30.070	305.4	4.88	0.00493
Ethyl alcohol	C_2H_5OH	46.069	513.9	6.14	0.00363
Ethylene	C_2H_4	28.054	282.4	5.04	0.00465
n-Heptane	C_7H_{16}	100.205	540.3	2.74	0.00431
n-Hexane	C_6H_{14}	86.178	507.5	3.01	0.00429
Methane	CH_4	16.043	190.4	4.60	0.00615
Methyl alcohol	CH_3OH	32.042	512.6	8.09	0.00368
n-Octane	C_8H_{18}	114.232	568.8	2.49	0.00431
Pentafluoroethane (125)	CHF_2CF_3	120.022	339.2	3.62	0.00176
n-Pentane	C_5H_{12}	72.151	469.7	3.37	0.00421
Propane	C_3H_8	44.094	369.8	4.25	0.00454
Propene	C_3H_6	42.081	364.9	4.60	0.00430
Refrigerant mixture	R-410A	72.585	344.5	4.90	0.00218
Tetrafluoroethane (134a)	CF_3CH_2F	102.03	374.2	4.06	0.00197

TABLE A.3
Properties of Selected Solids at 25 °C

Substance	ρ (kg/m³)	C_p (kJ/kg·K)
Asphalt	2120	0.92
Brick, common	1800	0.84
Carbon, diamond	3250	0.51
Carbon, graphite	2000–2500	0.61
Coal	1200–1500	1.26
Concrete	2200	0.88
Glass, plate	2500	0.80
Glass, wool	20	0.66
Granite	2750	0.89
Ice (0 °C)	917	2.04
Paper	700	1.2
Plexiglass	1180	1.44
Polystyrene	920	2.3
Polyvinyl chloride	1380	0.96
Rubber, soft	1100	1.67
Sand, dry	1500	0.8
Salt, rock	2100–2500	0.92
Silicon	2330	0.70
Snow, firm	560	2.1
Wood, hard (oak)	720	1.26
Wood, soft (pine)	510	1.38
Wool	100	1.72
Metals		
Aluminum	2700	0.90
Brass, 60–40	8400	0.38
Copper, commercial	8300	0.42
Gold	19300	0.13
Iron, cast	7272	0.42
Iron, 304 St Steel	7820	0.46
Lead	11340	0.13
Magnesium, 2 % Mn	1778	1.00
Nickel, 10 % Cr	8666	0.44
Silver, 99.9 % Ag	10524	0.24
Sodium	971	1.21
Tin	7304	0.22
Tungsten	19300	0.13
Zinc	7144	0.39

TABLE A.4
*Properties of Some Liquids at 25 °C**

Substance	ρ (kg/m³)	C_p (kJ/kg·K)
Ammonia	604	4.84
Benzene	879	1.72
Butane	556	2.47
CCl_4	1584	0.83
CO_2	680	2.9
Ethanol	783	2.46
Gasoline	750	2.08
Glycerine	1260	2.42
Kerosene	815	2.0
Methanol	787	2.55
n-Octane	692	2.23
Oil engine	885	1.9
Oil light	910	1.8
Propane	510	2.54
R-12	1310	0.97
R-22	1190	1.26
R-32	961	1.94
R-125	1191	1.41
R-134a	1206	1.43
R-410A	1059	1.69
Water	997	4.18
Liquid metals		
Bismuth, Bi	10040	0.14
Lead, Pb	10660	0.16
Mercury, Hg	13580	0.14
NaK (56/44)	887	1.13
Potassium, K	828	0.81
Sodium, Na	929	1.38
Tin, Sn	6950	0.24
Zinc, Zn	6570	0.50

*Or T_{melt} if higher.

TABLE A.5

Properties of Various Ideal Gases at 25 °C, 100 kPa* (SI Units)

Gas	Chemical Formula	Molecular Mass (kg/kmol)	R (kJ/kg·K)	ρ (kg/m³)	C_{p0} (kJ/kg·K)	C_{v0} (kJ/kg·K)	$k = \dfrac{C_p}{C_v}$
Steam	H_2O	18.015	0.4615	0.0231	1.872	1.410	1.327
Acetylene	C_2H_2	26.038	0.3193	1.05	1.699	1.380	1.231
Air	—	28.97	0.287	1.169	1.004	0.717	1.400
Ammonia	NH_3	17.031	0.4882	0.694	2.130	1.642	1.297
Argon	Ar	39.948	0.2081	1.613	0.520	0.312	1.667
Butane	C_4H_{10}	58.124	0.1430	2.407	1.716	1.573	1.091
Carbon dioxide	CO_2	44.01	0.1889	1.775	0.842	0.653	1.289
Carbon monoxide	CO	28.01	0.2968	1.13	1.041	0.744	1.399
Ethane	C_2H_6	30.07	0.2765	1.222	1.766	1.490	1.186
Ethanol	C_2H_5OH	46.069	0.1805	1.883	1.427	1.246	1.145
Ethylene	C_2H_4	28.054	0.2964	1.138	1.548	1.252	1.237
Helium	He	4.003	2.0771	0.1615	5.193	3.116	1.667
Hydrogen	H_2	2.016	4.1243	0.0813	14.209	10.085	1.409
Methane	CH_4	16.043	0.5183	0.648	2.254	1.736	1.299
Methanol	CH_3OH	32.042	0.2595	1.31	1.405	1.146	1.227
Neon	Ne	20.183	0.4120	0.814	1.03	0.618	1.667
Nitric oxide	NO	30.006	0.2771	1.21	0.993	0.716	1.387
Nitrogen	N_2	28.013	0.2968	1.13	1.042	0.745	1.400
Nitrous oxide	N_2O	44.013	0.1889	1.775	0.879	0.690	1.274
n-Octane	C_8H_{18}	114.23	0.07279	0.092	1.711	1.638	1.044
Oxygen	O_2	31.999	0.2598	1.292	0.922	0.662	1.393
Propane	C_3H_8	44.094	0.1886	1.808	1.679	1.490	1.126
R-12	CCl_2F_2	120.914	0.06876	4.98	0.616	0.547	1.126
R-22	$CHClF_2$	86.469	0.09616	3.54	0.658	0.562	1.171
R-32	CF_2H_2	52.024	0.1598	2.125	0.822	0.662	1.242
R-125	CHF_2CF_3	120.022	0.06927	4.918	0.791	0.722	1.097
R-134a	CF_3CH_2F	102.03	0.08149	4.20	0.852	0.771	1.106
R-410A	—	72.585	0.11455	2.967	0.809	0.694	1.165
Sulfur dioxide	SO_2	64.059	0.1298	2.618	0.624	0.494	1.263
Sulfur trioxide	SO_3	80.053	0.10386	3.272	0.635	0.531	1.196

*Or saturation pressure if it is less than 100 kPa.

TABLE A.6
*Constant-Pressure Specific Heats of Various Ideal Gases**

	$C_{p0} = C_0 + C_1\theta + C_2\theta^2 + C_3\theta^3$		(kJ/kg·K)	$\theta = T(\text{Kelvin})/1000$	
Gas	Formula	C_0	C_1	C_2	C_3
Steam	H_2O	1.79	0.107	0.586	−0.20
Acetylene	C_2H_2	1.03	2.91	−1.92	0.54
Air	—	1.05	−0.365	0.85	−0.39
Ammonia	NH_3	1.60	1.4	1.0	−0.7
Argon	Ar	0.52	0	0	0
Butane	C_4H_{10}	0.163	5.70	−1.906	−0.049
Carbon dioxide	CO_2	0.45	1.67	−1.27	0.39
Carbon monoxide	CO	1.10	−0.46	1.0	−0.454
Ethane	C_2H_6	0.18	5.92	−2.31	0.29
Ethanol	C_2H_5OH	0.2	4.65	−1.82	0.03
Ethylene	C_2H_4	0.136	5.58	−3.0	0.63
Helium	He	5.193	0	0	0
Hydrogen	H_2	13.46	4.6	−6.85	3.79
Methane	CH_4	1.2	3.25	0.75	−0.71
Methanol	CH_3OH	0.66	2.21	0.81	−0.89
Neon	Ne	1.03	0	0	0
Nitric oxide	NO	0.98	−0.031	0.325	−0.14
Nitrogen	N_2	1.11	−0.48	0.96	−0.42
Nitrous oxide	N_2O	0.49	1.65	−1.31	0.42
n-Octane	C_8H_{18}	−0.053	6.75	−3.67	0.775
Oxygen	O_2	0.88	−0.0001	0.54	−0.33
Propane	C_3H_8	−0.096	6.95	−3.6	0.73
R-12[†]	CCl_2F_2	0.26	1.47	−1.25	0.36
R-22[†]	$CHClF_2$	0.2	1.87	−1.35	0.35
R-32[†]	CF_2H_2	0.227	2.27	−0.93	0.041
R-125[†]	CHF_2CF_3	0.305	1.68	−0.284	0
R-134a[†]	CF_3CH_2F	0.165	2.81	−2.23	1.11
Sulfur dioxide	SO_2	0.37	1.05	−0.77	0.21
Sulfur trioxide	SO_3	0.24	1.7	−1.5	0.46

*Approximate forms valid from 250 K to 1200 K.
[†]Formula limited to maximum 500 K.

TABLE A.7.1
Ideal Gas Properties of Air, Standard Entropy at 0.1 MPa (1 Bar) Pressure

T (K)	u (kJ/kg)	h (kJ/kg)	s_T^0 (kJ/kg·K)	T (K)	u (kJ/kg)	h (kJ/kg)	s_T^0 (kJ/kg·K)
200	142.77	200.17	6.46260	1100	845.45	1161.18	8.24449
220	157.07	220.22	6.55812	1150	889.21	1219.30	8.29616
240	171.38	240.27	6.64535	1200	933.37	1277.81	8.34596
260	185.70	260.32	6.72562	1250	977.89	1336.68	8.39402
280	200.02	280.39	6.79998	1300	1022.75	1395.89	8.44046
290	207.19	290.43	6.83521	1350	1067.94	1455.43	8.48539
298.15	213.04	298.62	6.86305	1400	1113.43	1515.27	8.52891
300	214.36	300.47	6.86926	1450	1159.20	1575.40	8.57111
320	228.73	320.58	6.93413	1500	1205.25	1635.80	8.61208
340	243.11	340.70	6.99515	1550	1251.55	1696.45	8.65185
360	257.53	360.86	7.05276	1600	1298.08	1757.33	8.69051
380	271.99	381.06	7.10735	1650	1344.83	1818.44	8.72811
400	286.49	401.30	7.15926	1700	1391.80	1879.76	8.76472
420	301.04	421.59	7.20875	1750	1438.97	1941.28	8.80039
440	315.64	441.93	7.25607	1800	1486.33	2002.99	8.83516
460	330.31	462.34	7.30142	1850	1533.87	2064.88	8.86908
480	345.04	482.81	7.34499	1900	1581.59	2126.95	8.90219
500	359.84	503.36	7.38692	1950	1629.47	2189.19	8.93452
520	374.73	523.98	7.42736	2000	1677.52	2251.58	8.96611
540	389.69	544.69	7.46642	2050	1725.71	2314.13	8.99699
560	404.74	565.47	7.50422	2100	1774.06	2376.82	9.02721
580	419.87	586.35	7.54084	2150	1822.54	2439.66	9.05678
600	435.10	607.32	7.57638	2200	1871.16	2502.63	9.08573
620	450.42	628.38	7.61090	2250	1919.91	2565.73	9.11409
640	465.83	649.53	7.64448	2300	1968.79	2628.96	9.14189
660	481.34	670.78	7.67717	2350	2017.79	2692.31	9.16913
680	496.94	692.12	7.70903	2400	2066.91	2755.78	9.19586
700	512.64	713.56	7.74010	2450	2116.14	2819.37	9.22208
720	528.44	735.10	7.77044	2500	2165.48	2883.06	9.24781
740	544.33	756.73	7.80008	2550	2214.93	2946.86	9.27308
760	560.32	778.46	7.82905	2600	2264.48	3010.76	9.29790
780	576.40	800.28	7.85740	2650	2314.13	3074.77	9.32228
800	592.58	822.20	7.88514	2700	2363.88	3138.87	9.34625
850	633.42	877.40	7.95207	2750	2413.73	3203.06	9.36980
900	674.82	933.15	8.01581	2800	2463.66	3267.35	9.39297
950	716.76	989.44	8.07667	2850	2513.69	3331.73	9.41576
1000	759.19	1046.22	8.13493	2900	2563.80	3396.19	9.43818
1050	802.10	1103.48	8.19081	2950	2613.99	3460.73	9.46025
1100	845.45	1161.18	8.24449	3000	2664.27	3525.36	9.48198

TABLE A.7.2
The Isentropic Relative Pressure and Relative Volume Functions

T[K]	P_r	v_r	T[K]	P_r	v_r	T[K]	P_r	v_r
200	0.2703	493.47	700	23.160	20.155	1900	1327.5	0.95445
220	0.3770	389.15	720	25.742	18.652	1950	1485.8	0.87521
240	0.5109	313.27	740	28.542	17.289	2000	1658.6	0.80410
260	0.6757	256.58	760	31.573	16.052	2050	1847.1	0.74012
280	0.8756	213.26	780	34.851	14.925	2100	2052.1	0.68242
290	0.9899	195.36	800	38.388	13.897	2150	2274.8	0.63027
298.15	1.0907	182.29	850	48.468	11.695	2200	2516.2	0.58305
300	1.1146	179.49	900	60.520	9.9169	2250	2777.5	0.54020
320	1.3972	152.73	950	74.815	8.4677	2300	3059.9	0.50124
340	1.7281	131.20	1000	91.651	7.2760	2350	3364.6	0.46576
360	2.1123	113.65	1050	111.35	6.2885	2400	3693.0	0.43338
380	2.5548	99.188	1100	134.25	5.4641	2450	4046.2	0.40378
400	3.0612	87.137	1150	160.73	4.7714	2500	4425.8	0.37669
420	3.6373	77.003	1200	191.17	4.1859	2550	4833.0	0.35185
440	4.2892	68.409	1250	226.02	3.6880	2600	5269.5	0.32903
460	5.0233	61.066	1300	265.72	3.2626	2650	5736.7	0.30805
480	5.8466	54.748	1350	310.74	2.8971	2700	6236.2	0.28872
500	6.7663	49.278	1400	361.62	2.5817	2750	6769.7	0.27089
520	7.7900	44.514	1450	418.89	2.3083	2800	7338.7	0.25443
540	8.9257	40.344	1500	483.16	2.0703	2850	7945.1	0.23921
560	10.182	36.676	1550	554.96	1.8625	2900	8590.7	0.22511
580	11.568	33.436	1600	634.97	1.6804	2950	9277.2	0.21205
600	13.092	30.561	1650	723.86	1.52007	3000	10007	0.19992
620	14.766	28.001	1700	822.33	1.37858			
640	16.598	25.713	1750	931.14	1.25330			
660	18.600	23.662	1800	1051.05	1.14204			
680	20.784	21.818	1850	1182.9	1.04294			
700	23.160	20.155	1900	1327.5	0.95445			

The relative pressure and relative volume are temperature functions calculated with two scaling constants A_1, A_2.

$$P_r = \exp[s_T^0/R - A_1]; \qquad v_r = A_2 T/P_r$$

such that for an isentropic process ($s_1 = s_2$)

$$\frac{P_2}{P_1} = \frac{P_{r2}}{P_{r1}} = \frac{e^{s_{T_2}^0/R}}{e^{s_{T_1}^0/R}} \approx \left(\frac{T_2}{T_1}\right)^{C_p/R} \quad \text{and} \quad \frac{v_2}{v_1} = \frac{v_{r2}}{v_{r1}} \approx \left(\frac{T_1}{T_2}\right)^{C_v/R}$$

where the near equalities are for the constant specific heat approximation.

TABLE A.8

Ideal Gas Properties of Various Substances, Entropies at 0.1 MPa (1 Bar) Pressure, Mass Basis

T (K)	Nitrogen, Diatomic (N_2) $R = 0.2968$ kJ/kg·K $M = 28.013$ kg/kmol			Oxygen, Diatomic (O_2) $R = 0.2598$ kJ/kg·K $M = 31.999$ kg/kmol		
	u (kJ/kg)	h (kJ/kg)	s_T^0 (kJ/kg·K)	u (kJ/kg)	h (kJ/kg)	s_T^0 (kJ/kg·K)
200	148.39	207.75	6.4250	129.84	181.81	6.0466
250	185.50	259.70	6.6568	162.41	227.37	6.2499
300	222.63	311.67	6.8463	195.20	273.15	6.4168
350	259.80	363.68	7.0067	228.37	319.31	6.5590
400	297.09	415.81	7.1459	262.10	366.03	6.6838
450	334.57	468.13	7.2692	296.52	413.45	6.7954
500	372.35	520.75	7.3800	331.72	461.63	6.8969
550	410.52	573.76	7.4811	367.70	510.61	6.9903
600	449.16	627.24	7.5741	404.46	560.36	7.0768
650	488.34	681.26	7.6606	441.97	610.86	7.1577
700	528.09	735.86	7.7415	480.18	662.06	7.2336
750	568.45	791.05	7.8176	519.02	713.90	7.3051
800	609.41	846.85	7.8897	558.46	766.33	7.3728
850	650.98	903.26	7.9581	598.44	819.30	7.4370
900	693.13	960.25	8.0232	638.90	872.75	7.4981
950	735.85	1017.81	8.0855	679.80	926.65	7.5564
1000	779.11	1075.91	8.1451	721.11	980.95	7.6121
1100	867.14	1193.62	8.2572	804.80	1090.62	7.7166
1200	957.00	1313.16	8.3612	889.72	1201.53	7.8131
1300	1048.46	1434.31	8.4582	975.72	1313.51	7.9027
1400	1141.35	1556.87	8.5490	1062.67	1426.44	7.9864
1500	1235.50	1680.70	8.6345	1150.48	1540.23	8.0649
1600	1330.72	1805.60	8.7151	1239.10	1654.83	8.1389
1700	1426.89	1931.45	8.7914	1328.49	1770.21	8.2088
1800	1523.90	2058.15	8.8638	1418.63	1886.33	8.2752
1900	1621.66	2185.58	8.9327	1509.50	2003.19	8.3384
2000	1720.07	2313.68	8.9984	1601.10	2120.77	8.3987
2100	1819.08	2442.36	9.0612	1693.41	2239.07	8.4564
2200	1918.62	2571.58	9.1213	1786.44	2358.08	8.5117
2300	2018.63	2701.28	9.1789	1880.17	2477.79	8.5650
2400	2119.08	2831.41	9.2343	1974.60	2598.20	8.6162
2500	2219.93	2961.93	9.2876	2069.71	2719.30	8.6656
2600	2321.13	3092.81	9.3389	2165.50	2841.07	8.7134
2700	2422.66	3224.03	9.3884	2261.94	2963.49	8.7596
2800	2524.50	3355.54	9.4363	2359.01	3086.55	8.8044
2900	2626.62	3487.34	9.4825	2546.70	3210.22	8.8478
3000	2729.00	3619.41	9.5273	2554.97	3334.48	8.8899

TABLE A.8 (*continued*)
Ideal Gas Properties of Various Substances, Entropies at 0.1 MPa (1 Bar) Pressure, Mass Basis

T (K)	Carbon Dioxide (CO_2) $R = 0.1889$ kJ/kg·K $M = 44.010$ kg/kmol			Water (H_2O) $R = 0.4615$ kJ/kg·K $M = 18.015$ kg/kmol		
	u (kJ/kg)	h (kJ/kg)	s_T^0 (kJ/kg·K)	u (kJ/kg)	h (kJ/kg)	s_T^0 (kJ/kg·K)
200	97.49	135.28	4.5439	276.38	368.69	9.7412
250	126.21	173.44	4.7139	345.98	461.36	10.1547
300	157.70	214.38	4.8631	415.87	554.32	10.4936
350	191.78	257.90	4.9972	486.37	647.90	10.7821
400	228.19	303.76	5.1196	557.79	742.40	11.0345
450	266.69	351.70	5.2325	630.40	838.09	11.2600
500	307.06	401.52	5.3375	704.36	935.12	11.4644
550	349.12	453.03	5.4356	779.79	1033.63	11.6522
600	392.72	506.07	5.5279	856.75	1133.67	11.8263
650	437.71	560.51	5.6151	935.31	1235.30	11.9890
700	483.97	616.22	5.6976	1015.49	1338.56	12.1421
750	531.40	673.09	5.7761	1097.35	1443.49	12.2868
800	579.89	731.02	5.8508	1180.90	1550.13	12.4244
850	629.35	789.93	5.9223	1266.19	1658.49	12.5558
900	676.69	849.72	5.9906	1353.23	1768.60	12.6817
950	730.85	910.33	6.0561	1442.03	1880.48	12.8026
1000	782.75	971.67	6.1190	1532.61	1994.13	12.9192
1100	888.55	1096.36	6.2379	1719.05	2226.73	13.1408
1200	996.64	1223.34	6.3483	1912.42	2466.25	13.3492
1300	1106.68	1352.28	6.4515	2112.47	2712.46	13.5462
1400	1218.38	1482.87	6.5483	2318.89	2965.03	13.7334
1500	1331.50	1614.88	6.6394	2531.28	3223.57	13.9117
1600	1445.85	1748.12	6.7254	2749.24	3487.69	14.0822
1700	1561.26	1882.43	6.8068	2972.35	3756.95	14.2454
1800	1677.61	2017.67	6.8841	3200.17	4030.92	14.4020
1900	1794.78	2153.73	6.9577	3432.28	4309.18	14.5524
2000	1912.67	2290.51	7.0278	3668.24	4591.30	14.6971
2100	2031.21	2427.95	7.0949	3908.08	4877.29	14.8366
2200	2150.34	2565.97	7.1591	4151.28	5166.64	14.9712
2300	2270.00	2704.52	7.2206	4397.56	5459.08	15.1012
2400	2390.14	2843.55	7.2798	4646.71	5754.37	15.2269
2500	2510.74	2983.04	7.3368	4898.49	6052.31	15.3485
2600	2631.73	3122.93	7.3917	5152.73	6352.70	15.4663
2700	2753.10	3263.19	7.4446	5409.24	6655.36	15.5805
2800	2874.81	3403.79	7.4957	5667.86	6960.13	15.6914
2900	2996.84	3544.71	7.5452	5928.44	7266.87	15.7990
3000	3119.18	3685.95	7.5931	6190.86	7575.44	15.9036

TABLE A.9
Ideal Gas Properties of Various Substances (SI Units), Entropies at 0.1 MPa (1 Bar) Pressure, Mole Basis

T K	Nitrogen, Diatomic (N_2) $\bar{h}^0_{f,298} = 0$ kJ/kmol $M = 28.013$ kg/kmol		Nitrogen, Monatomic (N) $\bar{h}^0_{f,298} = 472\,680$ kJ/kmol $M = 14.007$ kg/kmol	
	$(\bar{h}-\bar{h}^0_{298})$ kJ/kmol	$\bar{s}^0_T$ kJ/kmol·K	$(\bar{h}-\bar{h}^0_{298})$ kJ/kmol	$\bar{s}^0_T$ kJ/kmol·K
0	−8670	0	−6197	0
100	−5768	159.812	−4119	130.593
200	−2857	179.985	−2040	145.001
298	0	191.609	0	153.300
300	54	191.789	38	153.429
400	2971	200.181	2117	159.409
500	5911	206.740	4196	164.047
600	8894	212.177	6274	167.837
700	11937	216.865	8353	171.041
800	15046	221.016	10431	173.816
900	18223	224.757	12510	176.265
1000	21463	228.171	14589	178.455
1100	24760	231.314	16667	180.436
1200	28109	234.227	18746	182.244
1300	31503	236.943	20825	183.908
1400	34936	239.487	22903	185.448
1500	38405	241.881	24982	186.883
1600	41904	244.139	27060	188.224
1700	45430	246.276	29139	189.484
1800	48979	248.304	31218	190.672
1900	52549	250.234	33296	191.796
2000	56137	252.075	35375	192.863
2200	63362	255.518	39534	194.845
2400	70640	258.684	43695	196.655
2600	77963	261.615	47860	198.322
2800	85323	264.342	52033	199.868
3000	92715	266.892	56218	201.311
3200	100134	269.286	60420	202.667
3400	107577	271.542	64646	203.948
3600	115042	273.675	68902	205.164
3800	122526	275.698	73194	206.325
4000	130027	277.622	77532	207.437
4400	145078	281.209	86367	209.542
4800	160188	284.495	95457	211.519
5200	175352	287.530	104843	213.397
5600	190572	290.349	114550	215.195
6000	205848	292.984	124590	216.926

TABLE A.9 (*continued*)
Ideal Gas Properties of Various Substances (SI Units), Entropies at 0.1 MPa (1 Bar) Pressure, Mole Basis

	Oxygen, Diatomic (O_2) $\bar{h}^0_{f,298} = 0$ kJ/kmol $M = 31.999$ kg/kmol		Oxygen, Monatomic (O) $\bar{h}^0_{f,298} = 249\,170$ kJ/kmol $M = 16.00$ kg/kmol	
T K	$(\bar{h} - \bar{h}^0_{298})$ kJ/kmol	$\bar{s}^0_T$ kJ/kmol·K	$(\bar{h} - \bar{h}^0_{298})$ kJ/kmol	$\bar{s}^0_T$ kJ/kmol·K
0	−8683	0	−6725	0
100	−5777	173.308	−4518	135.947
200	−2868	193.483	−2186	152.153
298	0	205.148	0	161.059
300	54	205.329	41	161.194
400	3027	213.873	2207	167.431
500	6086	220.693	4343	172.198
600	9245	226.450	6462	176.060
700	12499	231.465	8570	179.310
800	15836	235.920	10671	182.116
900	19241	239.931	12767	184.585
1000	22703	243.579	14860	186.790
1100	26212	246.923	16950	188.783
1200	29761	250.011	19039	190.600
1300	33345	252.878	21126	192.270
1400	36958	255.556	23212	193.816
1500	40600	258.068	25296	195.254
1600	44267	260.434	27381	196.599
1700	47959	262.673	29464	197.862
1800	51674	264.797	31547	199.053
1900	55414	266.819	33630	200.179
2000	59176	268.748	35713	201.247
2200	66770	272.366	39878	203.232
2400	74453	275.708	44045	205.045
2600	82225	278.818	48216	206.714
2800	90080	281.729	52391	208.262
3000	98013	284.466	56574	209.705
3200	106022	287.050	60767	211.058
3400	114101	289.499	64971	212.332
3600	122245	291.826	69190	213.538
3800	130447	294.043	73424	214.682
4000	138705	296.161	77675	215.773
4400	155374	300.133	86234	217.812
4800	172240	303.801	94873	219.691
5200	189312	307.217	103592	221.435
5600	206618	310.423	112391	223.066
6000	224210	313.457	121264	224.597

TABLE A.9 (*continued*)
Ideal Gas Properties of Various Substances (SI Units), Entropies at 0.1 MPa (1 Bar) Pressure, Mole Basis

	Carbon Dioxide (CO_2) $\bar{h}^0_{f,298} = -393\,522$ kJ/kmol $M = 44.01$ kg/kmol		Carbon Monoxide (CO) $\bar{h}^0_{f,298} = -110\,527$ kJ/kmol $M = 28.01$ kg/kmol	
T K	$(\bar{h} - \bar{h}^0_{298})$ kJ/kmol	$\bar{s}^0_T$ kJ/kmol·K	$(\bar{h} - \bar{h}^0_{298})$ kJ/kmol	$\bar{s}^0_T$ kJ/kmol·K
0	−9364	0	−8671	0
100	−6457	179.010	−5772	165.852
200	−3413	199.976	−2860	186.024
298	0	213.794	0	197.651
300	69	214.024	54	197.831
400	4003	225.314	2977	206.240
500	8305	234.902	5932	212.833
600	12906	243.284	8942	218.321
700	17754	250.752	12021	223.067
800	22806	257.496	15174	227.277
900	28030	263.646	18397	231.074
1000	33397	269.299	21686	234.538
1100	38885	274.528	25031	237.726
1200	44473	279.390	28427	240.679
1300	50148	283.931	31867	243.431
1400	55895	288.190	35343	246.006
1500	61705	292.199	38852	248.426
1600	67569	295.984	42388	250.707
1700	73480	299.567	45948	252.866
1800	79432	302.969	49529	254.913
1900	85420	306.207	53128	256.860
2000	91439	309.294	56743	258.716
2200	103562	315.070	64012	262.182
2400	115779	320.384	71326	265.361
2600	128074	325.307	78679	268.302
2800	140435	329.887	86070	271.044
3000	152853	334.170	93504	273.607
3200	165321	338.194	100962	276.012
3400	177836	341.988	108440	278.279
3600	190394	345.576	115938	280.422
3800	202990	348.981	123454	282.454
4000	215624	352.221	130989	284.387
4400	240992	358.266	146108	287.989
4800	266488	363.812	161285	291.290
5200	292112	368.939	176510	294.337
5600	317870	373.711	191782	297.167
6000	343782	378.180	207105	299.809

TABLE A.9 (*continued*)
Ideal Gas Properties of Various Substances (SI Units), Entropies at 0.1 MPa (1 Bar) Pressure, Mole Basis

	Water (H_2O) $\bar{h}^0_{f,298} = -241\ 826$ kJ/kmol $M = 18.015$ kg/kmol		Hydroxyl (OH) $\bar{h}^0_{f,298} = 38\ 987$ kJ/kmol $M = 17.007$ kg/kmol	
T K	$(\bar{h} - \bar{h}^0_{298})$ kJ/kmol	$\bar{s}^0_T$ kJ/kmol·K	$(\bar{h} - \bar{h}^0_{298})$ kJ/kmol	$\bar{s}^0_T$ kJ/kmol·K
0	−9904	0	−9172	0
100	−6617	152.386	−6140	149.591
200	−3282	175.488	−2975	171.592
298	0	188.835	0	183.709
300	62	189.043	55	183.894
400	3450	198.787	3034	192.466
500	6922	206.532	5991	199.066
600	10499	213.051	8943	204.448
700	14190	218.739	11902	209.008
800	18002	223.826	14881	212.984
900	21937	228.460	17889	216.526
1000	26000	232.739	20935	219.735
1100	30190	236.732	24024	222.680
1200	34506	240.485	27159	225.408
1300	38941	244.035	30340	227.955
1400	43491	247.406	33567	230.347
1500	48149	250.620	36838	232.604
1600	52907	253.690	40151	234.741
1700	57757	256.631	43502	236.772
1800	62693	259.452	46890	238.707
1900	67706	262.162	50311	240.556
2000	72788	264.769	53763	242.328
2200	83153	269.706	60751	245.659
2400	93741	274.312	67840	248.743
2600	104520	278.625	75018	251.614
2800	115463	282.680	82268	254.301
3000	126548	286.504	89585	256.825
3200	137756	290.120	96960	259.205
3400	149073	293.550	104388	261.456
3600	160484	296.812	111864	263.592
3800	171981	299.919	119382	265.625
4000	183552	302.887	126940	267.563
4400	206892	308.448	142165	271.191
4800	230456	313.573	157522	274.531
5200	254216	318.328	173002	277.629
5600	278161	322.764	188598	280.518
6000	302295	326.926	204309	283.227

TABLE A.9 (*continued*)
Ideal Gas Properties of Various Substances (SI Units), Entropies at 0.1 MPa (1 Bar) Pressure, Mole Basis

	Hydrogen (H_2) $\bar{h}^0_{f,298} = 0 \text{ kJ/kmol}$ $M = 2.016 \text{ kg/kmol}$		Hydrogen, Monatomic (H) $\bar{h}^0_{f,298} = 217\,999 \text{ kJ/kmol}$ $M = 1.008 \text{ kg/kmol}$	
T K	$(\bar{h} - \bar{h}^0_{298})$ kJ/kmol	$\bar{s}^0_T$ kJ/kmol·K	$(\bar{h} - \bar{h}^0_{298})$ kJ/kmol	$\bar{s}^0_T$ kJ/kmol·K
0	−8467	0	−6197	0
100	−5467	100.727	−4119	92.009
200	−2774	119.410	−2040	106.417
298	0	130.678	0	114.716
300	53	130.856	38	114.845
400	2961	139.219	2117	120.825
500	5883	145.738	4196	125.463
600	8799	151.078	6274	129.253
700	11730	155.609	8353	132.457
800	14681	159.554	10431	135.233
900	17657	163.060	12510	137.681
1000	20663	166.225	14589	139.871
1100	23704	169.121	16667	141.852
1200	26785	171.798	18746	143.661
1300	29907	174.294	20825	145.324
1400	33073	176.637	22903	146.865
1500	36281	178.849	24982	148.299
1600	39533	180.946	27060	149.640
1700	42826	182.941	29139	150.900
1800	46160	184.846	31218	152.089
1900	49532	186.670	33296	153.212
2000	52942	188.419	35375	154.279
2200	59865	191.719	39532	156.260
2400	66915	194.789	43689	158.069
2600	74082	197.659	47847	159.732
2800	81355	200.355	52004	161.273
3000	88725	202.898	56161	162.707
3200	96187	205.306	60318	164.048
3400	103736	207.593	64475	165.308
3600	111367	209.773	68633	166.497
3800	119077	211.856	72790	167.620
4000	126864	213.851	76947	168.687
4400	142658	217.612	85261	170.668
4800	158730	221.109	93576	172.476
5200	175057	224.379	101890	174.140
5600	191607	227.447	110205	175.681
6000	208332	230.322	118519	177.114

TABLE A.9 (*continued*)
Ideal Gas Properties of Various Substances (SI Units), Entropies at 0.1 MPa (1 Bar) Pressure,
Mole Basis

	Nitric Oxide (NO) $\bar{h}^0_{f,298} = 90\ 291\ \text{kJ/kmol}$ $M = 30.006\ \text{kg/kmol}$		Nitrogen Dioxide (NO$_2$) $\bar{h}^0_{f,298} = 33\ 100\ \text{kJ/kmol}$ $M = 46.005\ \text{kg/kmol}$	
T K	$(\bar{h} - \bar{h}^0_{298})$ kJ/kmol	$\bar{s}^0_T$ kJ/kmol·K	$(\bar{h} - \bar{h}^0_{298})$ kJ/kmol	$\bar{s}^0_T$ kJ/kmol·K
0	−9192	0	−10186	0
100	−6073	177.031	−6861	202.563
200	−2951	198.747	−3495	225.852
298	0	210.759	0	240.034
300	55	210.943	68	240.263
400	3040	219.529	3927	251.342
500	6059	226.263	8099	260.638
600	9144	231.886	12555	268.755
700	12308	236.762	17250	275.988
800	15548	241.088	22138	282.513
900	18858	244.985	27180	288.450
1000	22229	248.536	32344	293.889
1100	25653	251.799	37606	298.904
1200	29120	254.816	42946	303.551
1300	32626	257.621	48351	307.876
1400	36164	260.243	53808	311.920
1500	39729	262.703	59309	315.715
1600	43319	265.019	64846	319.289
1700	46929	267.208	70414	322.664
1800	50557	269.282	76008	325.861
1900	54201	271.252	81624	328.898
2000	57859	273.128	87259	331.788
2200	65212	276.632	98578	337.182
2400	72606	279.849	109948	342.128
2600	80034	282.822	121358	346.695
2800	87491	285.585	132800	350.934
3000	94973	288.165	144267	354.890
3200	102477	290.587	155756	358.597
3400	110000	292.867	167262	362.085
3600	117541	295.022	178783	365.378
3800	125099	297.065	190316	368.495
4000	132671	299.007	201860	371.456
4400	147857	302.626	224973	376.963
4800	163094	305.940	248114	381.997
5200	178377	308.998	271276	386.632
5600	193703	311.838	294455	390.926
6000	209070	314.488	317648	394.926

TABLE A.10

Enthalpy of Formation and Absolute Entropy of Various Substances at 25°C, 100 kPa Pressure

Substance	Formula	M kg/kmol	State	$\bar{h}_f^0$ kJ/kmol	$\bar{s}_f^0$ kJ/kmol·K
Acetylene	C_2H_2	26.038	gas	+226 731	200.958
Ammonia	NH_3	17.031	gas	−45 720	192.572
Benzene	C_6H_6	78.114	gas	+82 980	269.562
Carbon dioxide	CO_2	44.010	gas	−393 522	213.795
Carbon (graphite)	C	12.011	solid	0	5.740
Carbon monoxide	CO	28.011	gas	−110 527	197.653
Ethane	C_2H_6	30.070	gas	−84 740	229.597
Ethene	C_2H_4	28.054	gas	+52 467	219.330
Ethanol	C_2H_5OH	46.069	gas	−235 000	282.444
Ethanol	C_2H_5OH	46.069	liq	−277 380	160.554
Heptane	C_7H_{16}	100.205	gas	−187 900	427.805
Hexane	C_6H_{14}	86.178	gas	−167 300	387.979
Hydrogen peroxide	H_2O_2	34.015	gas	−136 106	232.991
Methane	CH_4	16.043	gas	−74 873	186.251
Methanol	CH_3OH	32.042	gas	−201 300	239.709
Methanol	CH_3OH	32.042	liq	−239 220	126.809
n-Butane	C_4H_{10}	58.124	gas	−126 200	306.647
Nitrogen oxide	N_2O	44.013	gas	+82 050	219.957
Nitromethane	CH_3NO_2	61.04	liq	−113 100	171.80
n-Octane	C_8H_{18}	114.232	gas	−208 600	466.514
n-Octane	C_8H_{18}	114.232	liq	−250 105	360.575
Ozone	O_3	47.998	gas	+142 674	238.932
Pentane	C_5H_{12}	72.151	gas	−146 500	348.945
Propane	C_3H_8	44.094	gas	−103 900	269.917
Propene	C_3H_6	42.081	gas	+20 430	267.066
Sulfur	S	32.06	solid	0	32.056
Sulfur dioxide	SO_2	64.059	gas	−296 842	248.212
Sulfur trioxide	SO_3	80.058	gas	−395 765	256.769
T-T-Diesel	$C_{14.4}H_{24.9}$	198.06	liq	−174 000	525.90
Water	H_2O	18.015	gas	−241 826	188.834
Water	H_2O	18.015	liq	−285 830	69.950

TABLE A.11
Logarithms to the Base e of the Equilibrium Constant K

For the reaction $v_A A + v_B B \rightleftharpoons v_C C + v_D D$, the equilibrium constant K is defined as

$$K = \frac{y_C^{v_C} y_D^{v_D}}{y_A^{v_A} y_B^{v_B}} \left(\frac{P}{P^0}\right)^{v_C + v_D - v_A - v_B}, \quad P^0 = 0.1 \text{ MPa}$$

Temp K	$H_2 \rightleftharpoons 2H$	$O_2 \rightleftharpoons 2O$	$N_2 \rightleftharpoons 2N$	$2H_2O \rightleftharpoons 2H_2 + O_2$	$2H_2O \rightleftharpoons H_2 + 2OH$	$2CO_2 \rightleftharpoons 2CO + O_2$	$N_2 + O_2 \rightleftharpoons 2NO$	$N_2 + 2O_2 \rightleftharpoons 2NO_2$
298	−164.003	−186.963	−367.528	−184.420	−212.075	−207.529	−69.868	−41.355
500	−92.830	−105.623	−213.405	−105.385	−120.331	−115.234	−40.449	−30.725
1000	−39.810	−45.146	−99.146	−46.321	−51.951	−47.052	−18.709	−23.039
1200	−30.878	−35.003	−80.025	−36.363	−40.467	−35.736	−15.082	−21.752
1400	−24.467	−27.741	−66.345	−29.222	−32.244	−27.679	−12.491	−20.826
1600	−19.638	−22.282	−56.069	−23.849	−26.067	−21.656	−10.547	−20.126
1800	−15.868	−18.028	−48.066	−19.658	−21.258	−16.987	−9.035	−19.577
2000	−12.841	−14.619	−41.655	−16.299	−17.406	−13.266	−7.825	−19.136
2200	−10.356	−11.826	−36.404	−13.546	−14.253	−10.232	−6.836	−18.773
2400	−8.280	−9.495	−32.023	−11.249	−11.625	−7.715	−6.012	−18.470
2600	−6.519	−7.520	−28.313	−9.303	−9.402	−5.594	−5.316	−18.214
2800	−5.005	−5.826	−25.129	−7.633	−7.496	−3.781	−4.720	−17.994
3000	−3.690	−4.356	−22.367	−6.184	−5.845	−2.217	−4.205	−17.805
3200	−2.538	−3.069	−19.947	−4.916	−4.401	−0.853	−3.755	−17.640
3400	−1.519	−1.932	−17.810	−3.795	−3.128	0.346	−3.359	−17.496
3600	−0.611	−0.922	−15.909	−2.799	−1.996	1.408	−3.008	−17.369
3800	0.201	−0.017	−14.205	−1.906	−0.984	2.355	−2.694	−17.257
4000	0.934	0.798	−12.671	−1.101	−0.074	3.204	−2.413	−17.157
4500	2.483	2.520	−9.423	0.602	1.847	4.985	−1.824	−16.953
5000	3.724	3.898	−6.816	1.972	3.383	6.397	−1.358	−16.797
5500	4.739	5.027	−4.672	3.098	4.639	7.542	−0.980	−16.678
6000	5.587	5.969	−2.876	4.040	5.684	8.488	−0.671	−16.588

Source: Consistent with thermodynamic data in *JANAF Thermochemical Tables*, third edition, Thermal Group, Dow Chemical U.S.A., Midland, MI, 1985.

SI Units: Thermodynamic Tables

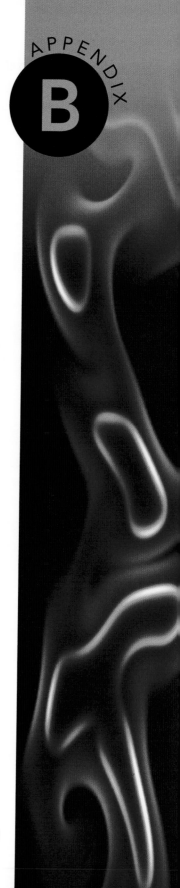

TABLE B.1
Thermodynamic Properties of Water

TABLE B.1.1
Saturated Water

Temp. (°C)	Press. (kPa)	Specific Volume, m^3/kg			Internal Energy, kJ/kg		
		Sat. Liquid v_f	Evap. v_{fg}	Sat. Vapor v_g	Sat. Liquid u_f	Evap. u_{fg}	Sat. Vapor u_g
0.01	0.6113	0.001000	206.131	206.132	0	2375.33	2375.33
5	0.8721	0.001000	147.117	147.118	20.97	2361.27	2382.24
10	1.2276	0.001000	106.376	106.377	41.99	2347.16	2389.15
15	1.705	0.001001	77.924	77.925	62.98	2333.06	2396.04
20	2.339	0.001002	57.7887	57.7897	83.94	2318.98	2402.91
25	3.169	0.001003	43.3583	43.3593	104.86	2304.90	2409.76
30	4.246	0.001004	32.8922	32.8932	125.77	2290.81	2416.58
35	5.628	0.001006	25.2148	25.2158	146.65	2276.71	2423.36
40	7.384	0.001008	19.5219	19.5229	167.53	2262.57	2430.11
45	9.593	0.001010	15.2571	15.2581	188.41	2248.40	2436.81
50	12.350	0.001012	12.0308	12.0318	209.30	2234.17	2443.47
55	15.758	0.001015	9.56734	9.56835	230.19	2219.89	2450.08
60	19.941	0.001017	7.66969	7.67071	251.09	2205.54	2456.63
65	25.03	0.001020	6.19554	6.19656	272.00	2191.12	2463.12
70	31.19	0.001023	5.04114	5.04217	292.93	2176.62	2469.55
75	38.58	0.001026	4.13021	4.13123	313.87	2162.03	2475.91
80	47.39	0.001029	3.40612	3.40715	334.84	2147.36	2482.19
85	57.83	0.001032	2.82654	2.82757	355.82	2132.58	2488.40
90	70.14	0.001036	2.35953	2.36056	376.82	2117.70	2494.52
95	84.55	0.001040	1.98082	1.98186	397.86	2102.70	2500.56
100	101.3	0.001044	1.67185	1.67290	418.91	2087.58	2506.50
105	120.8	0.001047	1.41831	1.41936	440.00	2072.34	2512.34
110	143.3	0.001052	1.20909	1.21014	461.12	2056.96	2518.09
115	169.1	0.001056	1.03552	1.03658	482.28	2041.44	2523.72
120	198.5	0.001060	0.89080	0.89186	503.48	2025.76	2529.24
125	232.1	0.001065	0.76953	0.77059	524.72	2009.91	2534.63
130	270.1	0.001070	0.66744	0.66850	546.00	1993.90	2539.90
135	313.0	0.001075	0.58110	0.58217	567.34	1977.69	2545.03
140	361.3	0.001080	0.50777	0.50885	588.72	1961.30	2550.02
145	415.4	0.001085	0.44524	0.44632	610.16	1944.69	2554.86
150	475.9	0.001090	0.39169	0.39278	631.66	1927.87	2559.54
155	543.1	0.001096	0.34566	0.34676	653.23	1910.82	2564.04
160	617.8	0.001102	0.30596	0.30706	674.85	1893.52	2568.37
165	700.5	0.001108	0.27158	0.27269	696.55	1875.97	2572.51
170	791.7	0.001114	0.24171	0.24283	718.31	1858.14	2576.46
175	892.0	0.001121	0.21568	0.21680	740.16	1840.03	2580.19
180	1002.2	0.001127	0.19292	0.19405	762.08	1821.62	2583.70
185	1122.7	0.001134	0.17295	0.17409	784.08	1802.90	2586.98
190	1254.4	0.001141	0.15539	0.15654	806.17	1783.84	2590.01

TABLE B.1.1 (*continued*)
Saturated Water

Temp. (°C)	Press. (kPa)	Enthalpy, kJ/kg			Entropy, kJ/kg·K		
		Sat. Liquid h_f	Evap. h_{fg}	Sat. Vapor h_g	Sat. Liquid s_f	Evap. s_{fg}	Sat. Vapor s_g
0.01	0.6113	0.00	2501.35	2501.35	0	9.1562	9.1562
5	0.8721	20.98	2489.57	2510.54	0.0761	8.9496	9.0257
10	1.2276	41.99	2477.75	2519.74	0.1510	8.7498	8.9007
15	1.705	62.98	2465.93	2528.91	0.2245	8.5569	8.7813
20	2.339	83.94	2454.12	2538.06	0.2966	8.3706	8.6671
25	3.169	104.87	2442.30	2547.17	0.3673	8.1905	8.5579
30	4.246	125.77	2430.48	2556.25	0.4369	8.0164	8.4533
35	5.628	146.66	2418.62	2565.28	0.5052	7.8478	8.3530
40	7.384	167.54	2406.72	2574.26	0.5724	7.6845	8.2569
45	9.593	188.42	2394.77	2583.19	0.6386	7.5261	8.1647
50	12.350	209.31	2382.75	2592.06	0.7037	7.3725	8.0762
55	15.758	230.20	2370.66	2600.86	0.7679	7.2234	7.9912
60	19.941	251.11	2358.48	2609.59	0.8311	7.0784	7.9095
65	25.03	272.03	2346.21	2618.24	0.8934	6.9375	7.8309
70	31.19	292.96	2333.85	2626.80	0.9548	6.8004	7.7552
75	38.58	313.91	2321.37	2635.28	1.0154	6.6670	7.6824
80	47.39	334.88	2308.77	2643.66	1.0752	6.5369	7.6121
85	57.83	355.88	2296.05	2651.93	1.1342	6.4102	7.5444
90	70.14	376.90	2283.19	2660.09	1.1924	6.2866	7.4790
95	84.55	397.94	2270.19	2668.13	1.2500	6.1659	7.4158
100	101.3	419.02	2257.03	2676.05	1.3068	6.0480	7.3548
105	120.8	440.13	2243.70	2683.83	1.3629	5.9328	7.2958
110	143.3	461.27	2230.20	2691.47	1.4184	5.8202	7.2386
115	169.1	482.46	2216.50	2698.96	1.4733	5.7100	7.1832
120	198.5	503.69	2202.61	2706.30	1.5275	5.6020	7.1295
125	232.1	524.96	2188.50	2713.46	1.5812	5.4962	7.0774
130	270.1	546.29	2174.16	2720.46	1.6343	5.3925	7.0269
135	313.0	567.67	2159.59	2727.26	1.6869	5.2907	6.9777
140	361.3	589.11	2144.75	2733.87	1.7390	5.1908	6.9298
145	415.4	610.61	2129.65	2740.26	1.7906	5.0926	6.8832
150	475.9	632.18	2114.26	2746.44	1.8417	4.9960	6.8378
155	543.1	653.82	2098.56	2752.39	1.8924	4.9010	6.7934
160	617.8	675.53	2082.55	2758.09	1.9426	4.8075	6.7501
165	700.5	697.32	2066.20	2763.53	1.9924	4.7153	6.7078
170	791.7	719.20	2049.50	2768.70	2.0418	4.6244	6.6663
175	892.0	741.16	2032.42	2773.58	2.0909	4.5347	6.6256
180	1002.2	763.21	2014.96	2778.16	2.1395	4.4461	6.5857
185	1122.7	785.36	1997.07	2782.43	2.1878	4.3586	6.5464
190	1254.4	807.61	1978.76	2786.37	2.2358	4.2720	6.5078

TABLE B.1.1 (*continued*)
Saturated Water

Temp. (°C)	Press. (kPa)	Specific Volume, m³/kg			Internal Energy, kJ/kg		
		Sat. Liquid v_f	Evap. v_{fg}	Sat. Vapor v_g	Sat. Liquid u_f	Evap. u_{fg}	Sat. Vapor u_g
195	1397.8	0.001149	0.13990	0.14105	828.36	1764.43	2592.79
200	1553.8	0.001156	0.12620	0.12736	850.64	1744.66	2595.29
205	1723.0	0.001164	0.11405	0.11521	873.02	1724.49	2597.52
210	1906.3	0.001173	0.10324	0.10441	895.51	1703.93	2599.44
215	2104.2	0.001181	0.09361	0.09479	918.12	1682.94	2601.06
220	2317.8	0.001190	0.08500	0.08619	940.85	1661.49	2602.35
225	2547.7	0.001199	0.07729	0.07849	963.72	1639.58	2603.30
230	2794.9	0.001209	0.07037	0.07158	986.72	1617.17	2603.89
235	3060.1	0.001219	0.06415	0.06536	1009.88	1594.24	2604.11
240	3344.2	0.001229	0.05853	0.05976	1033.19	1570.75	2603.95
245	3648.2	0.001240	0.05346	0.05470	1056.69	1546.68	2603.37
250	3973.0	0.001251	0.04887	0.05013	1080.37	1522.00	2602.37
255	4319.5	0.001263	0.04471	0.04598	1104.26	1496.66	2600.93
260	4688.6	0.001276	0.04093	0.04220	1128.37	1470.64	2599.01
265	5081.3	0.001289	0.03748	0.03877	1152.72	1443.87	2596.60
270	5498.7	0.001302	0.03434	0.03564	1177.33	1416.33	2593.66
275	5941.8	0.001317	0.03147	0.03279	1202.23	1387.94	2590.17
280	6411.7	0.001332	0.02884	0.03017	1227.43	1358.66	2586.09
285	6909.4	0.001348	0.02642	0.02777	1252.98	1328.41	2581.38
290	7436.0	0.001366	0.02420	0.02557	1278.89	1297.11	2575.99
295	7992.8	0.001384	0.02216	0.02354	1305.21	1264.67	2569.87
300	8581.0	0.001404	0.02027	0.02167	1331.97	1230.99	2562.96
305	9201.8	0.001425	0.01852	0.01995	1359.22	1195.94	2555.16
310	9856.6	0.001447	0.01690	0.01835	1387.03	1159.37	2546.40
315	10547	0.001472	0.01539	0.01687	1415.44	1121.11	2536.55
320	11274	0.001499	0.01399	0.01549	1444.55	1080.93	2525.48
325	12040	0.001528	0.01267	0.01420	1474.44	1038.57	2513.01
330	12845	0.001561	0.01144	0.01300	1505.24	993.66	2498.91
335	13694	0.001597	0.01027	0.01186	1537.11	945.77	2482.88
340	14586	0.001638	0.00916	0.01080	1570.26	894.26	2464.53
345	15525	0.001685	0.00810	0.00978	1605.01	838.29	2443.30
350	16514	0.001740	0.00707	0.00881	1641.81	776.58	2418.39
355	17554	0.001807	0.00607	0.00787	1681.41	707.11	2388.52
360	18651	0.001892	0.00505	0.00694	1725.19	626.29	2351.47
365	19807	0.002011	0.00398	0.00599	1776.13	526.54	2302.67
370	21028	0.002213	0.00271	0.00493	1843.84	384.69	2228.53
374.1	22089	0.003155	0	0.00315	2029.58	0	2029.58

TABLE B.1.1 (*continued*)
Saturated Water

Temp. (°C)	Press. (kPa)	Enthalpy, kJ/kg			Entropy, kJ/kg·K		
		Sat. Liquid h_f	Evap. h_{fg}	Sat. Vapor h_g	Sat. Liquid s_f	Evap. s_{fg}	Sat. Vapor s_g
195	1397.8	829.96	1959.99	2789.96	2.2835	4.1863	6.4697
200	1553.8	852.43	1940.75	2793.18	2.3308	4.1014	6.4322
205	1723.0	875.03	1921.00	2796.03	2.3779	4.0172	6.3951
210	1906.3	897.75	1900.73	2798.48	2.4247	3.9337	6.3584
215	2104.2	920.61	1879.91	2800.51	2.4713	3.8507	6.3221
220	2317.8	943.61	1858.51	2802.12	2.5177	3.7683	6.2860
225	2547.7	966.77	1836.50	2803.27	2.5639	3.6863	6.2502
230	2794.9	990.10	1813.85	2803.95	2.6099	3.6047	6.2146
235	3060.1	1013.61	1790.53	2804.13	2.6557	3.5233	6.1791
240	3344.2	1037.31	1766.50	2803.81	2.7015	3.4422	6.1436
245	3648.2	1061.21	1741.73	2802.95	2.7471	3.3612	6.1083
250	3973.0	1085.34	1716.18	2801.52	2.7927	3.2802	6.0729
255	4319.5	1109.72	1689.80	2799.51	2.8382	3.1992	6.0374
260	4688.6	1134.35	1662.54	2796.89	2.8837	3.1181	6.0018
265	5081.3	1159.27	1634.34	2793.61	2.9293	3.0368	5.9661
270	5498.7	1184.49	1605.16	2789.65	2.9750	2.9551	5.9301
275	5941.8	1210.05	1574.92	2784.97	3.0208	2.8730	5.8937
280	6411.7	1235.97	1543.55	2779.53	3.0667	2.7903	5.8570
285	6909.4	1262.29	1510.97	2773.27	3.1129	2.7069	5.8198
290	7436.0	1289.04	1477.08	2766.13	3.1593	2.6227	5.7821
295	7992.8	1316.27	1441.78	2758.05	3.2061	2.5375	5.7436
300	8581.0	1344.01	1404.93	2748.94	3.2533	2.4511	5.7044
305	9201.8	1372.33	1366.38	2738.72	3.3009	2.3633	5.6642
310	9856.6	1401.29	1325.97	2727.27	3.3492	2.2737	5.6229
315	10547	1430.97	1283.48	2714.44	3.3981	2.1821	5.5803
320	11274	1461.45	1238.64	2700.08	3.4479	2.0882	5.5361
325	12040	1492.84	1191.13	2683.97	3.4987	1.9913	5.4900
330	12845	1525.29	1140.56	2665.85	3.5506	1.8909	5.4416
335	13694	1558.98	1086.37	2645.35	3.6040	1.7863	5.3903
340	14586	1594.15	1027.86	2622.01	3.6593	1.6763	5.3356
345	15525	1631.17	964.02	2595.19	3.7169	1.5594	5.2763
350	16514	1670.54	893.38	2563.92	3.7776	1.4336	5.2111
355	17554	1713.13	813.59	2526.72	3.8427	1.2951	5.1378
360	18651	1760.48	720.52	2481.00	3.9146	1.1379	5.0525
365	19807	1815.96	605.44	2421.40	3.9983	0.9487	4.9470
370	21028	1890.37	441.75	2332.12	4.1104	0.6868	4.7972
374.1	22089	2099.26	0	2099.26	4.4297	0	4.4297

TABLE B.1.2
Saturated Water Pressure Entry

Press. (kPa)	Temp. (°C)	Specific Volume, m³/kg Sat. Liquid v_f	Evap. v_{fg}	Sat. Vapor v_g	Internal Energy, kJ/kg Sat. Liquid u_f	Evap. u_{fg}	Sat. Vapor u_g
0.6113	0.01	0.001000	206.131	206.132	0	2375.3	2375.3
1	6.98	0.001000	129.20702	129.20802	29.29	2355.69	2384.98
1.5	13.03	0.001001	87.97913	87.98013	54.70	2338.63	2393.32
2	17.50	0.001001	67.00285	67.00385	73.47	2326.02	2399.48
2.5	21.08	0.001002	54.25285	54.25385	88.47	2315.93	2404.40
3	24.08	0.001003	45.66402	45.66502	101.03	2307.48	2408.51
4	28.96	0.001004	34.79915	34.80015	121.44	2293.73	2415.17
5	32.88	0.001005	28.19150	28.19251	137.79	2282.70	2420.49
7.5	40.29	0.001008	19.23674	19.23775	168.76	2261.74	2430.50
10	45.81	0.001010	14.67254	14.67355	191.79	2246.10	2437.89
15	53.97	0.001014	10.02117	10.02218	225.90	2222.83	2448.73
20	60.06	0.001017	7.64835	7.64937	251.35	2205.36	2456.71
25	64.97	0.001020	6.20322	6.20424	271.88	2191.21	2463.08
30	69.10	0.001022	5.22816	5.22918	289.18	2179.22	2468.40
40	75.87	0.001026	3.99243	3.99345	317.51	2159.49	2477.00
50	81.33	0.001030	3.23931	3.24034	340.42	2143.43	2483.85
75	91.77	0.001037	2.21607	2.21711	394.29	2112.39	2496.67
100	99.62	0.001043	1.69296	1.69400	417.33	2088.72	2506.06
125	105.99	0.001048	1.37385	1.37490	444.16	2069.32	2513.48
150	111.37	0.001053	1.15828	1.15933	466.92	2052.72	2519.64
175	116.06	0.001057	1.00257	1.00363	486.78	2038.12	2524.90
200	120.23	0.001061	0.88467	0.88573	504.47	2025.02	2529.49
225	124.00	0.001064	0.79219	0.79325	520.45	2013.10	2533.56
250	127.43	0.001067	0.71765	0.71871	535.08	2002.14	2537.21
275	130.60	0.001070	0.65624	0.65731	548.57	1991.95	2540.53
300	133.55	0.001073	0.60475	0.60582	561.13	1982.43	2543.55
325	136.30	0.001076	0.56093	0.56201	572.88	1973.46	2546.34
350	138.88	0.001079	0.52317	0.52425	583.93	1964.98	2548.92
375	141.32	0.001081	0.49029	0.49137	594.38	1956.93	2551.31
400	143.63	0.001084	0.46138	0.46246	604.29	1949.26	2553.55
450	147.93	0.001088	0.41289	0.41398	622.75	1934.87	2557.62
500	151.86	0.001093	0.37380	0.37489	639.66	1921.57	2561.23
550	155.48	0.001097	0.34159	0.34268	655.30	1909.17	2564.47
600	158.85	0.001101	0.31457	0.31567	669.88	1897.52	2567.40
650	162.01	0.001104	0.29158	0.29268	683.55	1886.51	2570.06
700	164.97	0.001108	0.27176	0.27286	696.43	1876.07	2572.49
750	167.77	0.001111	0.25449	0.25560	708.62	1866.11	2574.73
800	170.43	0.001115	0.23931	0.24043	720.20	1856.58	2576.79

TABLE B.1.2 (*continued*)
Saturated Water Pressure Entry

Press. (kPa)	Temp. (°C)	Sat. Liquid h_f	Evap. h_{fg}	Sat. Vapor h_g	Sat. Liquid s_f	Evap. s_{fg}	Sat. Vapor s_g
0.6113	0.01	0.00	2501.3	2501.3	0	9.1562	9.1562
1.0	6.98	29.29	2484.89	2514.18	0.1059	8.8697	8.9756
1.5	13.03	54.70	2470.59	2525.30	0.1956	8.6322	8.8278
2.0	17.50	73.47	2460.02	2533.49	0.2607	8.4629	8.7236
2.5	21.08	88.47	2451.56	2540.03	0.3120	8.3311	8.6431
3.0	24.08	101.03	2444.47	2545.50	0.3545	8.2231	8.5775
4.0	28.96	121.44	2432.93	2554.37	0.4226	8.0520	8.4746
5.0	32.88	137.79	2423.66	2561.45	0.4763	7.9187	8.3950
7.5	40.29	168.77	2406.02	2574.79	0.5763	7.6751	8.2514
10	45.81	191.81	2392.82	2584.63	0.6492	7.5010	8.1501
15	53.97	225.91	2373.14	2599.06	0.7548	7.2536	8.0084
20	60.06	251.38	2358.33	2609.70	0.8319	7.0766	7.9085
25	64.97	271.90	2346.29	2618.19	0.8930	6.9383	7.8313
30	69.10	289.21	2336.07	2625.28	0.9439	6.8247	7.7686
40	75.87	317.55	2319.19	2636.74	1.0258	6.6441	7.6700
50	81.33	340.47	2305.40	2645.87	1.0910	6.5029	7.5939
75	91.77	384.36	2278.59	2662.96	1.2129	6.2434	7.4563
100	99.62	417.44	2258.02	2675.46	1.3025	6.0568	7.3593
125	105.99	444.30	2241.05	2685.35	1.3739	5.9104	7.2843
150	111.37	467.08	2226.46	2693.54	1.4335	5.7897	7.2232
175	116.06	486.97	2213.57	2700.53	1.4848	5.6868	7.1717
200	120.23	504.68	2201.96	2706.63	1.5300	5.5970	7.1271
225	124.00	520.69	2191.35	2712.04	1.5705	5.5173	7.0878
250	127.43	535.34	2181.55	2716.89	1.6072	5.4455	7.0526
275	130.60	548.87	2172.42	2721.29	1.6407	5.3801	7.0208
300	133.55	561.45	2163.85	2725.30	1.6717	5.3201	6.9918
325	136.30	573.23	2155.76	2728.99	1.7005	5.2646	6.9651
350	138.88	584.31	2148.10	2732.40	1.7274	5.2130	6.9404
375	141.32	594.79	2140.79	2735.58	1.7527	5.1647	6.9174
400	143.63	604.73	2133.81	2738.53	1.7766	5.1193	6.8958
450	147.93	623.24	2120.67	2743.91	1.8206	5.0359	6.8565
500	151.86	640.21	2108.47	2748.67	1.8606	4.9606	6.8212
550	155.48	655.91	2097.04	2752.94	1.8972	4.8920	6.7892
600	158.85	670.54	2086.26	2756.80	1.9311	4.8289	6.7600
650	162.01	684.26	2076.04	2760.30	1.9627	4.7704	6.7330
700	164.97	697.20	2066.30	2763.50	1.9922	4.7158	6.7080
750	167.77	709.45	2056.98	2766.43	2.0199	4.6647	6.6846
800	170.43	721.10	2048.04	2769.13	2.0461	4.6166	6.6627

TABLE B.1.2 (*continued*)
Saturated Water Pressure Entry

Press. (kPa)	Temp. (°C)	Specific Volume, m³/kg			Internal Energy, kJ/kg		
		Sat. Liquid v_f	Evap. v_{fg}	Sat. Vapor v_g	Sat. Liquid u_f	Evap. u_{fg}	Sat. Vapor u_g
850	172.96	0.001118	0.22586	0.22698	731.25	1847.45	2578.69
900	175.38	0.001121	0.21385	0.21497	741.81	1838.65	2580.46
950	177.69	0.001124	0.20306	0.20419	751.94	1830.17	2582.11
1000	179.91	0.001127	0.19332	0.19444	761.67	1821.97	2583.64
1100	184.09	0.001133	0.17639	0.17753	780.08	1806.32	2586.40
1200	187.99	0.001139	0.16220	0.16333	797.27	1791.55	2588.82
1300	191.64	0.001144	0.15011	0.15125	813.42	1777.53	2590.95
1400	195.07	0.001149	0.13969	0.14084	828.68	1764.15	2592.83
1500	198.32	0.001154	0.13062	0.13177	843.14	1751.3	2594.5
1750	205.76	0.001166	0.11232	0.11349	876.44	1721.39	2597.83
2000	212.42	0.001177	0.09845	0.09963	906.42	1693.84	2600.26
2250	218.45	0.001187	0.08756	0.08875	933.81	1668.18	2601.98
2500	223.99	0.001197	0.07878	0.07998	959.09	1644.04	2603.13
2750	229.12	0.001207	0.07154	0.07275	982.65	1621.16	2603.81
3000	233.90	0.001216	0.06546	0.06668	1004.76	1599.34	2604.10
3250	238.38	0.001226	0.06029	0.06152	1025.62	1578.43	2604.04
3500	242.60	0.001235	0.05583	0.05707	1045.41	1558.29	2603.70
4000	250.40	0.001252	0.04853	0.04978	1082.28	1519.99	2602.27
5000	263.99	0.001286	0.03815	0.03944	1147.78	1449.34	2597.12
6000	275.64	0.001319	0.03112	0.03244	1205.41	1384.27	2589.69
7000	285.88	0.001351	0.02602	0.02737	1257.51	1322.97	2580.48
8000	295.06	0.001384	0.02213	0.02352	1305.54	1264.25	2569.79
9000	303.40	0.001418	0.01907	0.02048	1350.47	1207.28	2557.75
10000	311.06	0.001452	0.01657	0.01803	1393.00	1151.40	2544.41
11000	318.15	0.001489	0.01450	0.01599	1433.68	1096.06	2529.74
12000	324.75	0.001527	0.01274	0.01426	1472.92	1040.76	2513.67
13000	330.93	0.001567	0.01121	0.01278	1511.09	984.99	2496.08
14000	336.75	0.001611	0.00987	0.01149	1548.53	928.23	2476.76
15000	342.24	0.001658	0.00868	0.01034	1585.58	869.85	2455.43
16000	347.43	0.001711	0.00760	0.00931	1622.63	809.07	2431.70
17000	352.37	0.001770	0.00659	0.00836	1660.16	744.80	2404.96
18000	357.06	0.001840	0.00565	0.00749	1698.86	675.42	2374.28
19000	361.54	0.001924	0.00473	0.00666	1739.87	598.18	2338.05
20000	365.81	0.002035	0.00380	0.00583	1785.47	507.58	2293.05
21000	369.89	0.002206	0.00275	0.00495	1841.97	388.74	2230.71
22000	373.80	0.002808	0.00072	0.00353	1973.16	108.24	2081.39
22089	374.14	0.003155	0	0.00315	2029.58	0	2029.58

TABLE B.1.2 (*continued*)
Saturated Water Pressure Entry

Press. (kPa)	Temp. (°C)	Enthalpy, kJ/kg			Entropy, kJ/kg·K		
		Sat. Liquid h_f	Evap. h_{fg}	Sat. Vapor h_g	Sat. Liquid s_f	Evap. s_{fg}	Sat. Vapor s_g
850	172.96	732.20	2039.43	2771.63	2.0709	4.5711	6.6421
900	175.38	742.82	2031.12	2773.94	2.0946	4.5280	6.6225
950	177.69	753.00	2023.08	2776.08	2.1171	4.4869	6.6040
1000	179.91	762.79	2015.29	2778.08	2.1386	4.4478	6.5864
1100	184.09	781.32	2000.36	2781.68	2.1791	4.3744	6.5535
1200	187.99	798.64	1986.19	2784.82	2.2165	4.3067	6.5233
1300	191.64	814.91	1972.67	2787.58	2.2514	4.2438	6.4953
1400	195.07	830.29	1959.72	2790.00	2.2842	4.1850	6.4692
1500	198.32	844.87	1947.28	2792.15	2.3150	4.1298	6.4448
1750	205.76	878.48	1917.95	2796.43	2.3851	4.0044	6.3895
2000	212.42	908.77	1890.74	2799.51	2.4473	3.8935	6.3408
2250	218.45	936.48	1865.19	2801.67	2.5034	3.7938	6.2971
2500	223.99	962.09	1840.98	2803.07	2.5546	3.7028	6.2574
2750	229.12	985.97	1817.89	2803.86	2.6018	3.6190	6.2208
3000	233.90	1008.41	1795.73	2804.14	2.6456	3.5412	6.1869
3250	238.38	1029.60	1774.37	2803.97	2.6866	3.4685	6.1551
3500	242.60	1049.73	1753.70	2803.43	2.7252	3.4000	6.1252
4000	250.40	1087.29	1714.09	2801.38	2.7963	3.2737	6.0700
5000	263.99	1154.21	1640.12	2794.33	2.9201	3.0532	5.9733
6000	275.64	1213.32	1571.00	2784.33	3.0266	2.8625	5.8891
7000	285.88	1266.97	1505.10	2772.07	3.1210	2.6922	5.8132
8000	295.06	1316.61	1441.33	2757.94	3.2067	2.5365	5.7431
9000	303.40	1363.23	1378.88	2742.11	3.2857	2.3915	5.6771
10000	311.06	1407.53	1317.14	2724.67	3.3595	2.2545	5.6140
11000	318.15	1450.05	1255.55	2705.60	3.4294	2.1233	5.5527
12000	324.75	1491.24	1193.59	2684.83	3.4961	1.9962	5.4923
13000	330.93	1531.46	1130.76	2662.22	3.5604	1.8718	5.4323
14000	336.75	1571.08	1066.47	2637.55	3.6231	1.7485	5.3716
15000	342.24	1610.45	1000.04	2610.49	3.6847	1.6250	5.3097
16000	347.43	1650.00	930.59	2580.59	3.7460	1.4995	5.2454
17000	352.37	1690.25	856.90	2547.15	3.8078	1.3698	5.1776
18000	357.06	1731.97	777.13	2509.09	3.8713	1.2330	5.1044
19000	361.54	1776.43	688.11	2464.54	3.9387	1.0841	5.0227
20000	365.81	1826.18	583.56	2409.74	4.0137	0.9132	4.9269
21000	369.89	1888.30	446.42	2334.72	4.1073	0.6942	4.8015
22000	373.80	2034.92	124.04	2158.97	4.3307	0.1917	4.5224
22089	374.14	2099.26	0	2099.26	4.4297	0	4.4297

TABLE B.1.3
Superheated Vapor Water

Temp. (°C)	v (m³/kg)	u (kJ/kg)	h (kJ/kg)	s (kJ/kg·K)	v (m³/kg)	u (kJ/kg)	h (kJ/kg)	s (kJ/kg·K)
	\multicolumn{4}{c}{$P = 10$ kPa (45.81 °C)}	\multicolumn{4}{c}{$P = 50$ kPa (81.33 °C)}
Sat.	14.67355	2437.89	2584.63	8.1501	3.24034	2483.85	2645.87	7.5939
50	14.86920	2443.87	2592.56	8.1749	—	—	—	—
100	17.19561	2515.50	2687.46	8.4479	3.41833	2511.61	2682.52	7.6947
150	19.51251	2587.86	2782.99	8.6881	3.88937	2585.61	2780.08	7.9400
200	21.82507	2661.27	2879.52	8.9037	4.35595	2659.85	2877.64	8.1579
250	24.13559	2735.95	2977.31	9.1002	4.82045	2734.97	2975.99	8.3555
300	26.44508	2812.06	3076.51	9.2812	5.28391	2811.33	3075.52	8.5372
400	31.06252	2968.89	3279.51	9.6076	6.20929	2968.43	3278.89	8.8641
500	35.67896	3132.26	3489.05	9.8977	7.13364	3131.94	3488.62	9.1545
600	40.29488	3302.45	3705.40	10.1608	8.05748	3302.22	3705.10	9.4177
700	44.91052	3479.63	3928.73	10.4028	8.98104	3479.45	3928.51	9.6599
800	49.52599	3663.84	4159.10	10.6281	9.90444	3663.70	4158.92	9.8852
900	54.14137	3855.03	4396.44	10.8395	10.82773	3854.91	4396.30	10.0967
1000	58.75669	4053.01	4640.58	11.0392	11.75097	4052.91	4640.46	10.2964
1100	63.37198	4257.47	4891.19	11.2287	12.67418	4257.37	4891.08	10.4858
1200	67.98724	4467.91	5147.78	11.4090	13.59737	4467.82	5147.69	10.6662
1300	72.60250	4683.68	5409.70	14.5810	14.52054	4683.58	5409.61	10.8382
	\multicolumn{4}{c}{100 kPa (99.62 °C)}	\multicolumn{4}{c}{200 kPa (120.23 °C)}
Sat.	1.69400	2506.06	2675.46	7.3593	0.88573	2529.49	2706.63	7.1271
150	1.93636	2582.75	2776.38	7.6133	0.95964	2576.87	2768.80	7.2795
200	2.17226	2658.05	2875.27	7.8342	1.08034	2654.39	2870.46	7.5066
250	2.40604	2733.73	2974.33	8.0332	1.19880	2731.22	2970.98	7.7085
300	2.63876	2810.41	3074.28	8.2157	1.31616	2808.55	3071.79	7.8926
400	3.10263	2967.85	3278.11	8.5434	1.54930	2966.69	3276.55	8.2217
500	3.56547	3131.54	3488.09	8.8341	1.78139	3130.75	3487.03	8.5132
600	4.02781	3301.94	3704.72	9.0975	2.01297	3301.36	3703.96	8.7769
700	4.48986	3479.24	3928.23	9.3398	2.24426	3478.81	3927.66	9.0194
800	4.95174	3663.53	4158.71	9.5652	2.47539	3663.19	4158.27	9.2450
900	5.41353	3854.77	4396.12	9.7767	2.70643	3854.49	4395.77	9.4565
1000	5.87526	4052.78	4640.31	9.9764	2.93740	4052.53	4640.01	9.6563
1100	6.33696	4257.25	4890.95	10.1658	3.16834	4257.01	4890.68	9.8458
1200	6.79863	4467.70	5147.56	10.3462	3.39927	4467.46	5147.32	10.0262
1300	7.26030	4683.47	5409.49	10.5182	3.63018	4683.23	5409.26	10.1982
	\multicolumn{4}{c}{300 kPa (133.55 °C)}	\multicolumn{4}{c}{400 kPa (143.63 °C)}
Sat.	0.60582	2543.55	2725.30	6.9918	0.46246	2553.55	2738.53	6.8958
150	0.63388	2570.79	2760.95	7.0778	0.47084	2564.48	2752.82	6.9299
200	0.71629	2650.65	2865.54	7.3115	0.53422	2646.83	2860.51	7.1706

TABLE B.1.3 (*continued*)
Superheated Vapor Water

Temp. (°C)	v (m³/kg)	u (kJ/kg)	h (kJ/kg)	s (kJ/kg·K)	v (m³/kg)	u (kJ/kg)	h (kJ/kg)	s (kJ/kg·K)
	300 kPa (133.55 °C)				400 kPa (143.63 °C)			
250	0.79636	2728.69	2967.59	7.5165	0.59512	2726.11	2964.16	7.3788
300	0.87529	2806.69	3069.28	7.7022	0.65484	2804.81	3066.75	7.5661
400	1.03151	2965.53	3274.98	8.0329	0.77262	2964.36	3273.41	7.8984
500	1.18669	3129.95	3485.96	8.3250	0.88934	3129.15	3484.89	8.1912
600	1.34136	3300.79	3703.20	8.5892	1.00555	3300.22	3702.44	8.4557
700	1.49573	3478.38	3927.10	8.8319	1.12147	3477.95	3926.53	8.6987
800	1.64994	3662.85	4157.83	9.0575	1.23722	3662.51	4157.40	8.9244
900	1.80406	3854.20	4395.42	9.2691	1.35288	3853.91	4395.06	9.1361
1000	1.95812	4052.27	4639.71	9.4689	1.46847	4052.02	4639.41	9.3360
1100	2.11214	4256.77	4890.41	9.6585	1.58404	4256.53	4890.15	9.5255
1200	2.26614	4467.23	5147.07	9.8389	1.69958	4466.99	5146.83	9.7059
1300	2.42013	4682.99	5409.03	10.0109	1.81511	4682.75	5408.80	9.8780
	500 kPa (15 1.86 °C)				600 kPa (158.85 °C)			
Sat.	0.37489	2561.23	2748.67	6.8212	0.31567	2567.40	2756.80	6.7600
200	0.42492	2642.91	2855.37	7.0592	0.35202	2638.91	2850.12	6.9665
250	0.47436	2723.50	2960.68	7.2708	0.39383	2720.86	2957.16	7.1816
300	0.52256	2802.91	3064.20	7.4598	0.43437	2801.00	3061.63	7.3723
350	0.57012	2882.59	3167.65	7.6328	0.47424	2881.12	3165.66	7.5463
400	0.61728	2963.19	3271.83	7.7937	0.51372	2962.02	3270.25	7.7078
500	0.71093	3128.35	3483.82	8.0872	0.59199	3127.55	3482.75	8.0020
600	0.80406	3299.64	3701.67	8.3521	0.66974	3299.07	3700.91	8.2673
700	0.89691	3477.52	3925.97	8.5952	0.74720	3477.08	3925.41	8.5107
800	0.98959	3662.17	4156.96	8.8211	0.82450	3661.83	4156.52	8.7367
900	1.08217	3853.63	4394.71	9.0329	0.90169	3853.34	4394.36	8.9485
1000	1.17469	4051.76	4639.11	9.2328	0.97883	4051.51	4638.81	9.1484
1100	1.26718	4256.29	4889.88	9.4224	1.05594	4256.05	4889.61	9.3381
1200	1.35964	4466.76	5146.58	9.6028	1.13302	4466.52	5146.34	9.5185
1300	1.45210	4682.52	5408.57	9.7749	1.21009	4682.28	5408.34	9.6906
	800 kPa (170.43 °C)				1000 kPa (179.91 °C)			
Sat.	0.24043	2576.79	2769.13	6.6627	0.19444	2583.64	2778.08	6.5864
200	0.26080	2630.61	2839.25	6.8158	0.20596	2621.90	2827.86	6.6939
250	0.29314	2715.46	2949.97	7.0384	0.23268	2709.91	2942.59	6.9246
300	0.32411	2797.14	3056.43	7.2327	0.25794	2793.21	3051.15	7.1228
350	0.35439	2878.16	3161.68	7.4088	0.28247	2875.18	3157.65	7.3010
400	0.38426	2959.66	3267.07	7.5715	0.30659	2957.29	3263.88	7.4650
500	0.44331	3125.95	3480.60	7.8672	0.35411	3124.34	3478.44	7.7621
600	0.50184	3297.91	3699.38	8.1332	0.40109	3296.76	3697.85	8.0289

TABLE B.1.3 (*continued*)
Superheated Vapor Water

Temp. (°C)	v (m³/kg)	u (kJ/kg)	h (kJ/kg)	s (kJ/kg·K)	v (m³/kg)	u (kJ/kg)	h (kJ/kg)	s (kJ/kg·K)
	\multicolumn{4}{c}{800 kPa (170.43 °C)}	\multicolumn{4}{c}{1000 kPa (179.91 °C)}
700	0.56007	3476.22	3924.27	8.3770	0.44779	3475.35	3923.14	8.2731
800	0.61813	3661.14	4155.65	8.6033	0.49432	3660.46	4154.78	8.4996
900	0.67610	3852.77	4393.65	8.8153	0.54075	3852.19	4392.94	8.7118
1000	0.73401	4051.00	4638.20	9.0153	0.58712	4050.49	4637.60	8.9119
1100	0.79188	4255.57	4889.08	9.2049	0.63345	4255.09	4888.55	9.1016
1200	0.84974	4466.05	5145.85	9.3854	0.67977	4465.58	5145.36	9.2821
1300	0.90758	4681.81	5407.87	9.5575	0.72608	4681.33	5407.41	9.4542
	\multicolumn{4}{c}{1200 kPa (187.99 °C)}	\multicolumn{4}{c}{1400 kPa (195.07 °C)}
Sat.	0.16333	2588.82	2784.82	6.5233	0.14084	2592.83	2790.00	6.4692
200	0.16930	2612.74	2815.90	6.5898	0.14302	2603.09	2803.32	6.4975
250	0.19235	2704.20	2935.01	6.8293	0.16350	2698.32	2927.22	6.7467
300	0.21382	2789.22	3045.80	7.0316	0.18228	2785.16	3040.35	6.9533
350	0.23452	2872.16	3153.59	7.2120	0.20026	2869.12	3149.49	7.1359
400	0.25480	2954.90	3260.66	7.3773	0.21780	2952.50	3257.42	7.3025
500	0.29463	3122.72	3476.28	7.6758	0.25215	3121.10	3474.11	7.6026
600	0.33393	3295.60	3696.32	7.9434	0.28596	3294.44	3694.78	7.8710
700	0.37294	3474.48	3922.01	8.1881	0.31947	3473.61	3920.87	8.1160
800	0.41177	3659.77	4153.90	8.4149	0.35281	3659.09	4153.03	8.3431
900	0.45051	3851.62	4392.23	8.6272	0.38606	3851.05	4391.53	8.5555
1000	0.48919	4049.98	4637.00	8.8274	0.41924	4049.47	4636.41	8.7558
1100	0.52783	4254.61	4888.02	9.0171	0.45239	4254.14	4887.49	8.9456
1200	0.56646	4465.12	5144.87	9.1977	0.48552	4464.65	5144.38	9.1262
1300	0.60507	4680.86	5406.95	9.3698	0.51864	4680.39	5406.49	9.2983
	\multicolumn{4}{c}{1600 kPa (201.40) °C}	\multicolumn{4}{c}{1800 kPa (207.15 °C)}
Sat.	0.12380	2595.95	2794.02	6.4217	0.11042	2598.38	2797.13	6.3793
250	0.14184	2692.26	2919.20	6.6732	0.12497	2686.02	2910.96	6.6066
300	0.15862	2781.03	3034.83	6.8844	0.14021	2776.83	3029.21	6.8226
350	0.17456	2866.05	3145.35	7.0693	0.15457	2862.95	3141.18	7.0099
400	0.19005	2950.09	3254.17	7.2373	0.16847	2947.66	3250.90	7.1793
500	0.22029	3119.47	3471.93	7.5389	0.19550	3117.84	3469.75	7.4824
600	0.24998	3293.27	3693.23	7.8080	0.22199	3292.10	3691.69	7.7523
700	0.27937	3472.74	3919.73	8.0535	0.24818	3471.87	3918.59	7.9983
800	0.30859	3658.40	4152.15	8.2808	0.27420	3657.71	4151.27	8.2258
900	0.33772	3850.47	4390.82	8.4934	0.30012	3849.90	4390.11	8.4386
1000	0.36678	4048.96	4635.81	8.6938	0.32598	4048.45	4635.21	8.6390
1100	0.39581	4253.66	4886.95	8.8837	0.35180	4253.18	4886.42	8.8290
1200	0.42482	4464.18	5143.89	9.0642	0.37761	4463.71	5143.40	9.0096
1300	0.45382	4679.92	5406.02	9.2364	0.40340	4679.44	5405.56	9.1817

TABLE B.1.3 (*continued*)
Superheated Vapor Water

Temp. (°C)	v (m³/kg)	u (kJ/kg)	h (kJ/kg)	s (kJ/kg·K)	v (m³/kg)	u (kJ/kg)	h (kJ/kg)	s (kJ/kg·K)
	2000 kPa (212.42 °C)				2500 kPa (223.99 °C)			
Sat.	0.09963	2600.26	2799.51	6.3408	0.07998	2603.13	2803.07	6.2574
250	0.11144	2679.58	2902.46	6.5452	0.08700	2662.55	2880.06	6.4084
300	0.12547	2772.56	3023.50	6.7663	0.09890	2761.56	3008.81	6.6437
350	0.13857	2859.81	3136.96	6.9562	0.10976	2851.84	3126.24	6.8402
400	0.15120	2945.21	3247.60	7.1270	0.12010	2939.03	3239.28	7.0147
450	0.16353	3030.41	3357.48	7.2844	0.13014	3025.43	3350.77	7.1745
500	0.17568	3116.20	3467.55	7.4316	0.13998	3112.08	3462.04	7.3233
600	0.19960	3290.93	3690.14	7.7023	0.15930	3287.99	3686.25	7.5960
700	0.22323	3470.99	3917.45	7.9487	0.17832	3468.80	3914.59	7.8435
800	0.24668	3657.03	4150.40	8.1766	0.19716	3655.30	4148.20	8.0720
900	0.27004	3849.33	4389.40	8.3895	0.21590	3847.89	4387.64	8.2853
1000	0.29333	4047.94	4634.61	8.5900	0.23458	4046.67	4633.12	8.4860
1100	0.31659	4252.71	4885.89	8.7800	0.25322	4251.52	4884.57	8.6761
1200	0.33984	4463.25	5142.92	8.9606	0.27185	4462.08	5141.70	8.8569
1300	0.36306	4678.97	5405.10	9.1328	0.29046	4677.80	5403.95	9.0291
	3000 kPa (233.90 °C)				4000 kPa (250.40 °C)			
Sat.	0.06668	2604.10	2804.14	6.1869	0.04978	2602.27	2801.38	6.0700
250	0.07058	2644.00	2855.75	6.2871	—	—	—	—
300	0.08114	2750.05	2993.48	6.5389	0.05884	2725.33	2960.68	6.3614
350	0.09053	2843.66	3115.25	6.7427	0.06645	2826.65	3092.43	6.5820
400	0.09936	2932.75	3230.82	6.9211	0.07341	2919.88	3213.51	6.7689
450	0.10787	3020.38	3344.00	7.0833	0.08003	3010.13	3330.23	6.9362
500	0.11619	3107.92	3456.48	7.2337	0.08643	3099.49	3445.21	7.0900
600	0.13243	3285.03	3682.34	7.5084	0.09885	3279.06	3674.44	7.3688
700	0.14838	3466.59	3911.72	7.7571	0.11095	3462.15	3905.94	7.6198
800	0.16414	3653.58	4146.00	7.9862	0.12287	3650.11	4141.59	7.8502
900	0.17980	3846.46	4385.87	8.1999	0.13469	3843.59	4382.34	8.0647
1000	0.19541	4045.40	4631.63	8.4009	0.14645	4042.87	4628.65	8.2661
1100	0.21098	4250.33	4883.26	8.5911	0.15817	4247.96	4880.63	8.4566
1200	0.22652	4460.92	5140.49	8.7719	0.16987	4458.60	5138.07	8.6376
1300	0.24206	4676.63	5402.81	8.9442	0.18156	4674.29	5400.52	8.8099

TABLE B.1.3 (*continued*)
Superheated Vapor Water

Temp. (°C)	v (m³/kg)	u (kJ/kg)	h (kJ/kg)	s (kJ/kg·K)	v (m³/kg)	u (kJ/kg)	h (kJ/kg)	s (kJ/kg·K)
	5000 kPa (263.99 °C)				6000 kPa (275.64 °C)			
Sat.	0.03944	2597.12	2794.33	5.9733	0.03244	2589.69	2784.33	5.8891
300	0.04532	2697.94	2924.53	6.2083	0.03616	2667.22	2884.19	6.0673
350	0.05194	2808.67	3068.39	6.4492	0.04223	2789.61	3042.97	6.3334
400	0.05781	2906.58	3195.64	6.6458	0.04739	2892.81	3177.17	6.5407
450	0.06330	2999.64	3316.15	6.8185	0.05214	2988.90	3301.76	6.7192
500	0.06857	3090.92	3433.76	6.9758	0.05665	3082.20	3422.12	6.8802
550	0.07368	3181.82	3550.23	7.1217	0.06101	3174.57	3540.62	7.0287
600	0.07869	3273.01	3666.47	7.2588	0.06525	3266.89	3658.40	7.1676
700	0.08849	3457.67	3900.13	7.5122	0.07352	3453.15	3894.28	7.4234
800	0.09811	3646.62	4137.17	7.7440	0.08160	3643.12	4132.74	7.6566
900	0.10762	3840.71	4378.82	7.9593	0.08958	3837.84	4375.29	7.8727
1000	0.11707	4040.35	4625.69	8.1612	0.09749	4037.83	4622.74	8.0751
1100	0.12648	4245.61	4878.02	8.3519	0.10536	4243.26	4875.42	8.2661
1200	0.13587	4456.30	5135.67	8.5330	0.11321	4454.00	5133.28	8.4473
1300	0.14526	4671.96	5398.24	8.7055	0.12106	4669.64	5395.97	8.6199
	8000 kPa (295.06 °C)				10000 kPa (311.06 °C)			
Sat.	0.02352	2569.79	2757.94	5.7431	0.01803	2544.41	2724.67	5.6140
300	0.02426	2590.93	2784.98	5.7905	—	—	—	—
350	0.02995	2747.67	2987.30	6.1300	0.02242	2699.16	2923.39	5.9442
400	0.03432	2863.75	3138.28	6.3633	0.02641	2832.38	3096.46	6.2119
450	0.03817	2966.66	3271.99	6.5550	0.02975	2943.32	3240.83	6.4189
500	0.04175	3064.30	3398.27	6.7239	0.03279	3045.77	3373.63	6.5965
550	0.04516	3159.76	3521.01	6.8778	0.03564	3144.54	3500.92	6.7561
600	0.04845	3254.43	3642.03	7.0205	0.03837	3241.68	3625.34	6.9028
700	0.05481	3444.00	3882.47	7.2812	0.04358	3434.72	3870.52	7.1687
800	0.06097	3636.08	4123.84	7.5173	0.04859	3628.97	4114.91	7.4077
900	0.06702	3832.08	4368.26	7.7350	0.05349	3826.32	4361.24	7.6272
1000	0.07301	4032.81	4616.87	7.9384	0.05832	4027.81	4611.04	7.8315
1100	0.07896	4238.60	4870.25	8.1299	0.06312	4233.97	4865.14	8.0236
1200	0.08489	4449.45	5128.54	8.3115	0.06789	4444.93	5123.84	8.2054
1300	0.09080	4665.02	5391.46	8.4842	0.07265	4660.44	5386.99	8.3783

TABLE B.1.3 (*continued*)
Superheated Vapor Water

Temp. (°C)	v (m³/kg)	u (kJ/kg)	h (kJ/kg)	s (kJ/kg·K)	v (m³/kg)	u (kJ/kg)	h (kJ/kg)	s (kJ/kg·K)
	15000 kPa (342.24 °C)				20000 kPa (365.81 °C)			
Sat.	0.01034	2455.43	2610.49	5.3097	0.00583	2293.05	2409.74	4.9269
350	0.01147	2520.36	2692.41	5.4420	—	—	—	—
400	0.01565	2740.70	2975.44	5.8810	0.00994	2619.22	2818.07	5.5539
450	0.01845	2879.47	3156.15	6.1403	0.01270	2806.16	3060.06	5.9016
500	0.02080	2996.52	3308.53	6.3442	0.01477	2942.82	3238.18	6.1400
550	0.02293	3104.71	3448.61	6.5198	0.01656	3062.34	3393.45	6.3347
600	0.02491	3208.64	3582.30	6.6775	0.01818	3174.00	3537.57	6.5048
650	0.02680	3310.37	3712.32	6.8223	0.01969	3281.46	3675.32	6.6582
700	0.02861	3410.94	3840.12	6.9572	0.02113	3386.46	3809.09	6.7993
800	0.03210	3610.99	4092.43	7.2040	0.02385	3592.73	4069.80	7.0544
900	0.03546	3811.89	4343.75	7.4279	0.02645	3797.44	4326.37	7.2830
1000	0.03875	4015.41	4596.63	7.6347	0.02897	4003.12	4582.45	7.4925
1100	0.04200	4222.55	4852.56	7.8282	0.03145	4211.30	4840.24	7.6874
1200	0.04523	4433.78	5112.27	8.0108	0.03391	4422.81	5100.96	7.8706
1300	0.04845	4649.12	5375.94	8.1839	0.03636	4637.95	5365.10	8.0441
	30000 kPa				40000 kPa			
375	0.001789	1737.75	1791.43	3.9303	0.001641	1677.09	1742.71	3.8289
400	0.002790	2067.34	2151.04	4.4728	0.001908	1854.52	1930.83	4.1134
425	0.005304	2455.06	2614.17	5.1503	0.002532	2096.83	2198.11	4.5028
450	0.006735	2619.30	2821.35	5.4423	0.003693	2365.07	2512.79	4.9459
500	0.008679	2820.67	3081.03	5.7904	0.005623	2678.36	2903.26	5.4699
550	0.010168	2970.31	3275.36	6.0342	0.006984	2869.69	3149.05	5.7784
600	0.011446	3100.53	3443.91	6.2330	0.008094	3022.61	3346.38	6.0113
650	0.012596	3221.04	3598.93	6.4057	0.009064	3158.04	3520.58	6.2054
700	0.013661	3335.84	3745.67	6.5606	0.009942	3283.63	3681.29	6.3750
800	0.015623	3555.60	4024.31	6.8332	0.011523	3517.89	3978.80	6.6662
900	0.017448	3768.48	4291.93	7.0717	0.012963	3739.42	4257.93	6.9150
1000	0.019196	3978.79	4554.68	7.2867	0.014324	3954.64	4527.59	7.1356
1100	0.020903	4189.18	4816.28	7.4845	0.015643	4167.38	4793.08	7.3364
1200	0.022589	4401.29	5078.97	7.6691	0.016940	4380.11	5057.72	7.5224
1300	0.024266	4615.96	5343.95	7.8432	0.018229	4594.28	5323.45	7.6969

TABLE B.1.4
Compressed Liquid Water

Temp. (°C)	v (m³/kg)	u (kJ/kg)	h (kJ/kg)	s (kJ/kg·K)	v (m³/kg)	u (kJ/kg)	h (kJ/kg)	s (kJ/kg·K)
	500 kPa (151.86 °C)				2000 kPa (212.42 °C)			
Sat.	0.001093	639.66	640.21	1.8606	0.001177	906.42	908.77	2.4473
0.01	0.000999	0.01	0.51	0.0000	0.000999	0.03	2.03	0.0001
20	0.001002	83.91	84.41	0.2965	0.001001	83.82	85.82	.2962
40	0.001008	167.47	167.98	0.5722	0.001007	167.29	169.30	.5716
60	0.001017	251.00	251.51	0.8308	0.001016	250.73	252.77	.8300
80	0.001029	334.73	335.24	1.0749	0.001028	334.38	336.44	1.0739
100	0.001043	418.80	419.32	1.3065	0.001043	418.36	420.45	1.3053
120	0.001060	503.37	503.90	1.5273	0.001059	502.84	504.96	1.5259
140	0.001080	588.66	589.20	1.7389	0.001079	588.02	590.18	1.7373
160	—	—	—	—	0.001101	674.14	676.34	1.9410
180	—	—	—	—	0.001127	761.46	763.71	2.1382
200	—	—	—	—	0.001156	850.30	852.61	2.3301
	5000 kPa (263.99 °C)				10000 kPa (311.06 °C)			
Sat	0.001286	1147.78	1154.21	2.9201	0.001452	1393.00	1407.53	3.3595
0	0.000998	0.03	5.02	0.0001	0.000995	0.10	10.05	0.0003
20	0.001000	83.64	88.64	0.2955	0.000997	83.35	93.32	0.2945
40	0.001006	166.93	171.95	0.5705	0.001003	166.33	176.36	0.5685
60	0.001015	250.21	255.28	0.8284	0.001013	249.34	259.47	0.8258
80	0.001027	333.69	338.83	1.0719	0.001025	332.56	342.81	1.0687
100	0.001041	417.50	422.71	1.3030	0.001039	416.09	426.48	1.2992
120	0.001058	501.79	507.07	1.5232	0.001055	500.07	510.61	1.5188
140	0.001077	586.74	592.13	1.7342	0.001074	584.67	595.40	1.7291
160	0.001099	672.61	678.10	1.9374	0.001195	670.11	681.07	1.9316
180	0.001124	759.62	765.24	2.1341	0.001120	756.63	767.83	2.1274
200	0.001153	848.08	853.85	2.3254	0.001148	844.49	855.97	2.3178
220	0.001187	938.43	944.36	2.5128	0.001181	934.07	945.88	2.5038
240	0.001226	1031.34	1037.47	2.6978	0.001219	1025.94	1038.13	2.6872
260	0.001275	1127.92	1134.30	2.8829	0.001265	1121.03	1133.68	2.8698
280					0.001322	1220.90	1234.11	3.0547
300					0.001397	1328.34	1342.31	3.2468

TABLE B.1.4 (*continued*)
Compressed Liquid Water

Temp. (°C)	v (m³/kg)	u (kJ/kg)	h (kJ/kg)	s (kJ/kg·K)	v (m³/kg)	u (kJ/kg)	h (kJ/kg)	s (kJ/kg·K)
	15000 kPa (342.24 °C)				20000 kPa (365.81 °C)			
Sat.	0.001658	1585.58	1610.45	3.6847	0.002035	1785.47	1826.18	4.0137
0	0.000993	0.15	15.04	0.0004	0.000990	0.20	20.00	0.0004
20	0.000995	83.05	97.97	0.2934	0.000993	82.75	102.61	0.2922
40	0.001001	165.73	180.75	0.5665	0.000999	165.15	185.14	0.5646
60	0.001011	248.49	263.65	0.8231	0.001008	247.66	267.82	0.8205
80	0.001022	331.46	346.79	1.0655	0.001020	330.38	350.78	1.0623
100	0.001036	414.72	430.26	1.2954	0.001034	413.37	434.04	1.2917
120	0.001052	498.39	514.17	1.5144	0.001050	496.75	517.74	1.5101
140	0.001071	582.64	598.70	1.7241	0.001068	580.67	602.03	1.7192
160	0.001092	667.69	684.07	1.9259	0.001089	665.34	687.11	1.9203
180	0.001116	753.74	770.48	2.1209	0.001112	750.94	773.18	2.1146
200	0.001143	841.04	858.18	2.3103	0.001139	837.70	860.47	2.3031
220	0.001175	929.89	947.52	2.4952	0.001169	925.89	949.27	2.4869
240	0.001211	1020.82	1038.99	2.6770	0.001205	1015.94	1040.04	2.6673
260	0.001255	1114.59	1133.41	2.8575	0.001246	1108.53	1133.45	2.8459
280	0.001308	1212.47	1232.09	3.0392	0.001297	1204.69	1230.62	3.0248
300	0.001377	1316.58	1337.23	3.2259	0.001360	1306.10	1333.29	3.2071
320	0.001472	1431.05	1453.13	3.4246	0.001444	1415.66	1444.53	3.3978
340	0.001631	1567.42	1591.88	3.6545	0.001568	1539.64	1571.01	3.6074
360					0.001823	1702.78	1739.23	3.8770
	30000 kPa				50000 kPa			
0	0.000986	0.25	29.82	0.0001	0.000977	0.20	49.03	−0.0014
20	0.000989	82.16	111.82	0.2898	0.000980	80.98	130.00	0.2847
40	0.000995	164.01	193.87	0.5606	0.000987	161.84	211.20	0.5526
60	0.001004	246.03	276.16	0.8153	0.000996	242.96	292.77	0.8051
80	0.001016	328.28	358.75	1.0561	0.001007	324.32	374.68	1.0439
100	0.001029	410.76	441.63	1.2844	0.001020	405.86	456.87	1.2703
120	0.001044	493.58	524.91	1.5017	0.001035	487.63	539.37	1.4857
140	0.001062	576.86	608.73	1.7097	0.001052	569.76	622.33	1.6915
160	0.001082	660.81	693.27	1.9095	0.001070	652.39	705.91	1.8890
180	0.001105	745.57	778.71	2.1024	0.001091	735.68	790.24	2.0793
200	0.001130	831.34	865.24	2.2892	0.001115	819.73	875.46	2.2634
220	0.001159	918.32	953.09	2.4710	0.001141	904.67	961.71	2.4419
240	0.001192	1006.84	1042.60	2.6489	0.001170	990.69	1049.20	2.6158
260	0.001230	1097.38	1134.29	2.8242	0.001203	1078.06	1138.23	2.7860
280	0.001275	1190.69	1228.96	2.9985	0.001242	1167.19	1229.26	2.9536
300	0.001330	1287.89	1327.80	3.1740	0.001286	1258.66	1322.95	3.1200
320	0.001400	1390.64	1432.63	3.3538	0.001339	1353.23	1420.17	3.2867
340	0.001492	1501.71	1546.47	3.5425	0.001403	1451.91	1522.07	3.4556
360	0.001627	1626.57	1675.36	3.7492	0.001484	1555.97	1630.16	3.6290
380	0.001869	1781.35	1837.43	4.0010	0.001588	1667.13	1746.54	3.8100

TABLE B.1.5
Saturated Solid–Saturated Vapor, Water

Temp. ($°C$)	Press. (kPa)	Specific Volume, m^3/kg			Internal Energy, kJ/kg		
		Sat. Solid v_i	Evap. v_{ig}	Sat. Vapor v_g	Sat. Solid u_i	Evap. u_{ig}	Sat. Vapor u_g
0.01	0.6113	0.0010908	206.152	206.153	−333.40	2708.7	2375.3
0	0.6108	0.0010908	206.314	206.315	−333.42	2708.7	2375.3
−2	0.5177	0.0010905	241.662	241.663	−337.61	2710.2	2372.5
−4	0.4376	0.0010901	283.798	283.799	−341.78	2711.5	2369.8
−6	0.3689	0.0010898	334.138	334.139	−345.91	2712.9	2367.0
−8	0.3102	0.0010894	394.413	394.414	−350.02	2714.2	2364.2
−10	0.2601	0.0010891	466.756	466.757	−354.09	2715.5	2361.4
−12	0.2176	0.0010888	553.802	553.803	−358.14	2716.8	2358.7
−14	0.1815	0.0010884	658.824	658.824	−362.16	2718.0	2355.9
−16	0.1510	0.0010881	785.906	785.907	−366.14	2719.2	2353.1
−18	0.1252	0.0010878	940.182	940.183	−370.10	2720.4	2350.3
−20	0.10355	0.0010874	1128.112	1128.113	−374.03	2721.6	2347.5
−22	0.08535	0.0010871	1357.863	1357.864	−377.93	2722.7	2344.7
−24	0.07012	0.0010868	1639.752	1639.753	−381.80	2723.7	2342.0
−26	0.05741	0.0010864	1986.775	1986.776	−385.64	2724.8	2339.2
−28	0.04684	0.0010861	2415.200	2415.201	−389.45	2725.8	2336.4
−30	0.03810	0.0010858	2945.227	2945.228	−393.23	2726.8	2333.6
−32	0.03090	0.0010854	3601.822	3601.823	−396.98	2727.8	2330.8
−34	0.02499	0.0010851	4416.252	4416.253	−400.71	2728.7	2328.0
−36	0.02016	0.0010848	5430.115	5430.116	−404.40	2729.6	2325.2
−38	0.01618	0.0010844	6707.021	6707.022	−408.06	2730.5	2322.4
−40	0.01286	0.0010841	8366.395	8366.396	−411.70	2731.3	2319.6

TABLE B.1.5 (*continued*)
Saturated Solid-Saturated Vapor, Water

Temp. (°C)	Press. (kPa)	Enthalpy, kJ/kg			Entropy, kJ/kg·K		
		Sat. Solid h_i	Evap. h_{ig}	Sat. Vapor h_g	Sat. Solid s_i	Evap. s_{ig}	Sat. Vapor s_g
0.01	0.6113	−333.40	2834.7	2501.3	−1.2210	10.3772	9.1562
0	0.6108	−333.42	2834.8	2501.3	−1.2211	10.3776	9.1565
−2	0.5177	−337.61	2835.3	2497.6	−1.2369	10.4562	9.2193
−4	0.4376	−341.78	2835.7	2494.0	−1.2526	10.5358	9.2832
−6	0.3689	−345.91	2836.2	2490.3	−1.2683	10.6165	9.3482
−8	0.3102	−350.02	2836.6	2486.6	−1.2839	10.6982	9.4143
−10	0.2601	−354.09	2837.0	2482.9	−1.2995	10.7809	9.4815
−12	0.2176	−358.14	2837.3	2479.2	−1.3150	10.8648	9.5498
−14	0.1815	−362.16	2837.6	2475.5	−1.3306	10.9498	9.6192
−16	0.1510	−366.14	2837.9	2471.8	−1.3461	11.0359	9.6898
−18	0.1252	−370.10	2838.2	2468.1	−1.3617	11.1233	9.7616
−20	0.10355	−374.03	2838.4	2464.3	−1.3772	11.2120	9.8348
−22	0.08535	−377.93	2838.6	2460.6	−1.3928	11.3020	9.9093
−24	0.07012	−381.80	2838.7	2456.9	−1.4083	11.3935	9.9852
−26	0.05741	−385.64	2838.9	2453.2	−1.4239	11.4864	10.0625
−28	0.04684	−389.45	2839.0	2449.5	−1.4394	11.5808	10.1413
−30	0.03810	−393.23	2839.0	2445.8	−1.4550	11.6765	10.2215
−32	0.03090	−396.98	2839.1	2442.1	−1.4705	11.7733	10.3028
−34	0.02499	−400.71	2839.1	2438.4	−1.4860	11.8713	10.3853
−36	0.02016	−404.40	2839.1	2434.7	−1.5014	11.9704	10.4690
−38	0.01618	−408.06	2839.0	2431.0	−1.5168	12.0714	10.5546
−40	0.01286	−411.70	2838.9	2427.2	−1.5321	12.1768	10.6447

TABLE B.2
Thermodynamic Properties of Ammonia

TABLE B.2.1
Saturated Ammonia

Temp. (°C)	Press. (kPa)	Specific Volume, m³/kg			Internal Energy, kJ/kg		
		Sat. Liquid v_f	Evap. v_{fg}	Sat. Vapor v_g	Sat. Liquid u_f	Evap. u_{fg}	Sat. Vapor u_g
−50	40.9	0.001424	2.62557	2.62700	−43.82	1309.1	1265.2
−45	54.5	0.001437	2.00489	2.00632	−22.01	1293.5	1271.4
−40	71.7	0.001450	1.55111	1.55256	−0.10	1277.6	1277.4
−35	93.2	0.001463	1.21466	1.21613	21.93	1261.3	1283.3
−30	119.5	0.001476	0.96192	0.96339	44.08	1244.8	1288.9
−25	151.6	0.001490	0.76970	0.77119	66.36	1227.9	1294.3
−20	190.2	0.001504	0.62184	0.62334	88.76	1210.7	1299.5
−15	236.3	0.001519	0.50686	0.50838	111.30	1193.2	1304.5
−10	290.9	0.001534	0.41655	0.41808	133.96	1175.2	1309.2
−5	354.9	0.001550	0.34493	0.34648	156.76	1157.0	1313.7
0	429.6	0.001566	0.28763	0.28920	179.69	1138.3	1318.0
5	515.9	0.001583	0.24140	0.24299	202.77	1119.2	1322.0
10	615.2	0.001600	0.20381	0.20541	225.99	1099.7	1325.7
15	728.6	0.001619	0.17300	0.17462	249.36	1079.7	1329.1
20	857.5	0.001638	0.14758	0.14922	272.89	1059.3	1332.2
25	1003.2	0.001658	0.12647	0.12813	296.59	1038.4	1335.0
30	1167.0	0.001680	0.10881	0.11049	320.46	1016.9	1337.4
35	1350.4	0.001702	0.09397	0.09567	344.50	994.9	1339.4
40	1554.9	0.001725	0.08141	0.08313	368.74	972.2	1341.0
45	1782.0	0.001750	0.07073	0.07248	393.19	948.9	1342.1
50	2033.1	0.001777	0.06159	0.06337	417.87	924.8	1342.7
55	2310.1	0.001804	0.05375	0.05555	442.79	899.9	1342.7
60	2614.4	0.001834	0.04697	0.04880	467.99	874.2	1342.1
65	2947.8	0.001866	0.04109	0.04296	493.51	847.4	1340.9
70	3312.0	0.001900	0.03597	0.03787	519.39	819.5	1338.9
75	3709.0	0.001937	0.03148	0.03341	545.70	790.4	1336.1
80	4140.5	0.001978	0.02753	0.02951	572.50	759.9	1332.4
85	4608.6	0.002022	0.02404	0.02606	599.90	727.8	1327.7
90	5115.3	0.002071	0.02093	0.02300	627.99	693.7	1321.7
95	5662.9	0.002126	0.01815	0.02028	656.95	657.4	1314.4
100	6253.7	0.002188	0.01565	0.01784	686.96	618.4	1305.3
105	6890.4	0.002261	0.01337	0.01564	718.30	575.9	1294.2
110	7575.7	0.002347	0.01128	0.01363	751.37	529.1	1280.5
115	8313.3	0.002452	0.00933	0.01178	786.82	476.2	1263.1
120	9107.2	0.002589	0.00744	0.01003	825.77	414.5	1240.3
125	9963.5	0.002783	0.00554	0.00833	870.69	337.7	1208.4
130	10891.6	0.003122	0.00337	0.00649	929.29	226.9	1156.2
132.3	11333.2	0.004255	0	0.00426	1037.62	0	1037.6

TABLE B.2.1 (*continued*)
Saturated Ammonia

Temp. (°C)	Press. (kPa)	Enthalpy, kJ/kg			Entropy, kJ/kg·K		
		Sat. Liquid h_f	Evap. h_{fg}	Sat. Vapor h_g	Sat. Liquid s_f	Evap. s_{fg}	Sat. Vapor s_g
−50	40.9	−43.76	1416.3	1372.6	−0.1916	6.3470	6.1554
−45	54.5	−21.94	1402.8	1380.8	−0.0950	6.1484	6.0534
−40	71.7	0	1388.8	1388.8	0	5.9567	5.9567
−35	93.2	22.06	1374.5	1396.5	0.0935	5.7715	5.8650
−30	119.5	44.26	1359.8	1404.0	0.1856	5.5922	5.7778
−25	151.6	66.58	1344.6	1411.2	0.2763	5.4185	5.6947
−20	190.2	89.05	1329.0	1418.0	0.3657	5.2498	5.6155
−15	236.3	111.66	1312.9	1424.6	0.4538	5.0859	5.5397
−10	290.9	134.41	1296.4	1430.8	0.5408	4.9265	5.4673
−5	354.9	157.31	1279.4	1436.7	0.6266	4.7711	5.3977
0	429.6	180.36	1261.8	1442.2	0.7114	4.6195	5.3309
5	515.9	203.58	1243.7	1447.3	0.7951	4.4715	5.2666
10	615.2	226.97	1225.1	1452.0	0.8779	4.3266	5.2045
15	728.6	250.54	1205.8	1456.3	0.9598	4.1846	5.1444
20	857.5	274.30	1185.9	1460.2	1.0408	4.0452	5.0860
25	1003.2	298.25	1165.2	1463.5	1.1210	3.9083	5.0293
30	1167.0	322.42	1143.9	1466.3	1.2005	3.7734	4.9738
35	1350.4	346.80	1121.8	1468.6	1.2792	3.6403	4.9196
40	1554.9	371.43	1098.8	1470.2	1.3574	3.5088	4.8662
45	1782.0	396.31	1074.9	1471.2	1.4350	3.3786	4.8136
50	2033.1	421.48	1050.0	1471.5	1.5121	3.2493	4.7614
55	2310.1	446.96	1024.1	1471.0	1.5888	3.1208	4.7095
60	2614.4	472.79	997.0	1469.7	1.6652	2.9925	4.6577
65	2947.8	499.01	968.5	1467.5	1.7415	2.8642	4.6057
70	3312.0	525.69	938.7	1464.4	1.8178	2.7354	4.3533
75	3709.0	552.88	907.2	1460.1	1.8943	2.6058	4.5001
80	4140.5	580.69	873.9	1454.6	1.9712	2.4746	4.4458
85	4608.6	609.21	838.6	1447.8	2.0488	2.3413	4.3901
90	5115.3	638.59	800.8	1439.4	2.1273	2.2051	4.3325
95	5662.9	668.99	760.2	1429.2	2.2073	2.0650	4.2723
100	6253.7	700.64	716.2	1416.9	2.2893	1.9195	4.2088
105	6890.4	733.87	668.1	1402.0	2.3740	1.7667	4.1407
110	7575.7	769.15	614.6	1383.7	2.4625	1.6040	4.0665
115	8313.3	807.21	553.8	1361.0	2.5566	1.4267	3.9833
120	9107.2	849.36	482.3	1331.7	2.6593	1.2268	3.8861
125	9963.5	898.42	393.0	1291.4	2.7775	0.9870	3.7645
130	10892	963.29	263.7	1227.0	2.9326	0.6540	3.5866
132.3	11333	1085.85	0	1085.9	3.2316	0	3.2316

TABLE B.2.2
Superheated Ammonia

Temp. (°C)	v (m³/kg)	u (kJ/kg)	h (kJ/kg)	s (kJ/kg·K)	v (m³/kg)	u (kJ/kg)	h (kJ/kg)	s (kJ/kg·K)
	50 kPa (−46.53 °C)				100 kPa (−33.60 °C)			
Sat.	2.1752	1269.6	1378.3	6.0839	1.1381	1284.9	1398.7	5.8401
−30	2.3448	1296.2	1413.4	6.2333	1.1573	1291.0	1406.7	5.8734
−20	2.4463	1312.3	1434.6	6.3187	1.2101	1307.8	1428.8	5.9626
−10	2.5471	1328.4	1455.7	6.4006	1.2621	1324.6	1450.8	6.0477
0	2.6474	1344.5	1476.9	6.4795	1.3136	1341.3	1472.6	6.1291
10	2.7472	1360.7	1498.1	6.5556	1.3647	1357.9	1494.4	6.2073
20	2.8466	1377.0	1519.3	6.6293	1.4153	1374.5	1516.1	6.2826
30	2.9458	1393.3	1540.6	6.7008	1.4657	1391.2	1537.7	6.3553
40	3.0447	1409.8	1562.0	6.7703	1.5158	1407.9	1559.5	6.4258
50	3.1435	1426.3	1583.5	6.8379	1.5658	1424.7	1581.2	6.4943
60	3.2421	1443.0	1605.1	6.9038	1.6156	1441.5	1603.1	6.5609
70	3.3406	1459.9	1626.9	6.9682	1.6653	1458.5	1625.1	6.6258
80	3.4390	1476.9	1648.8	7.0312	1.7148	1475.6	1647.1	6.6892
100	3.6355	1511.4	1693.2	7.1533	1.8137	1510.3	1691.7	6.8120
120	3.8318	1546.6	1738.2	7.2708	1.9124	1545.7	1736.9	6.9300
140	4.0280	1582.5	1783.9	7.3842	2.0109	1581.7	1782.8	7.0439
160	4.2240	1619.2	1830.4	7.4941	2.1093	1618.5	1829.4	7.1540
180	4.4199	1656.7	1877.7	7.6008	2.2075	1656.0	1876.8	7.2609
200	4.6157	1694.9	1925.7	7.7045	2.3057	1694.3	1924.9	7.3648
	150 kPa (−25.22 °C)				200 kPa (−18.86 °C)			
Sat.	0.7787	1294.1	1410.9	5.6983	0.5946	1300.6	1419.6	5.5979
−20	0.7977	1303.3	1422.9	5.7465	—	—	—	—
−10	0.8336	1320.7	1445.7	5.8349	0.6193	1316.7	1440.6	5.6791
0	0.8689	1337.9	1468.3	5.9189	0.6465	1334.5	1463.8	5.7659
10	0.9037	1355.0	1490.6	5.9992	0.6732	1352.1	1486.8	5.8484
20	0.9382	1372.0	1512.8	6.0761	0.6995	1369.5	1509.4	5.9270
30	0.9723	1389.0	1534.9	6.1502	0.7255	1386.8	1531.9	6.0025
40	1.0062	1406.0	1556.9	6.2217	0.7513	1404.0	1554.3	6.0751
50	1.0398	1423.0	1578.9	6.2910	0.7769	1421.3	1576.6	6.1453
60	1.0734	1440.0	1601.0	6.3583	0.8023	1438.5	1598.9	6.2133
70	1.1068	1457.2	1623.2	6.4238	0.8275	1455.8	1621.3	6.2794
80	1.1401	1474.4	1645.4	6.4877	0.8527	1473.1	1643.7	6.3437
100	1.2065	1509.3	1690.2	6.6112	0.9028	1508.2	1688.8	6.4679
120	1.2726	1544.8	1735.6	6.7297	0.9527	1543.8	1734.4	6.5869
140	1.3386	1580.9	1781.7	6.8439	1.0024	1580.1	1780.6	6.7015
160	1.4044	1617.8	1828.4	6.9544	1.0519	1617.0	1827.4	6.8123
180	1.4701	1655.4	1875.9	7.0615	1.1014	1654.7	1875.0	6.9196
200	1.5357	1693.7	1924.1	7.1656	1.1507	1693.2	1923.3	7.0239
220	1.6013	1732.9	1973.1	7.2670	1.2000	1732.4	1972.4	7.1255

TABLE B.2.2 (*continued*)
Superheated Ammonia

Temp. (°C)	v (m³/kg)	u (kJ/kg)	h (kJ/kg)	s (kJ/kg·K)	v (m³/kg)	u (kJ/kg)	h (kJ/kg)	s (kJ/kg·K)
	300 kPa (−9.24 °C)				400 kPa (−1.89 °C)			
Sat.	0.40607	1309.9	1431.7	5.4565	0.30942	1316.4	1440.2	5.3559
0	0.42382	1327.5	1454.7	5.5420	0.31227	1320.2	1445.1	5.3741
10	0.44251	1346.1	1478.9	5.6290	0.32701	1339.9	1470.7	5.4663
20	0.46077	1364.4	1502.6	5.7113	0.34129	1359.1	1495.6	5.5525
30	0.47870	1382.3	1526.0	5.7896	0.35520	1377.7	1519.8	5.6338
40	0.49636	1400.1	1549.0	5.8645	0.36884	1396.1	1543.6	5.7111
50	0.51382	1417.8	1571.9	5.9365	0.38226	1414.2	1567.1	5.7850
60	0.53111	1435.4	1594.7	6.0060	0.39550	1432.2	1590.4	5.8560
70	0.54827	1453.0	1617.5	6.0732	0.40860	1450.1	1613.6	5.9244
80	0.56532	1470.6	1640.2	6.1385	0.42160	1468.0	1636.7	5.9907
100	0.59916	1506.1	1685.8	6.2642	0.44732	1503.9	1682.8	6.1179
120	0.63276	1542.0	1731.8	6.3842	0.47279	1540.1	1729.2	6.2390
140	0.66618	1578.5	1778.3	6.4996	0.49808	1576.8	1776.0	6.3552
160	0.69946	1615.6	1825.4	6.6109	0.52323	1614.1	1823.4	6.4671
180	0.73263	1653.4	1873.2	6.7188	0.54827	1652.1	1871.4	6.5755
200	0.76572	1692.0	1921.7	6.8235	0.57321	1690.8	1920.1	6.6806
220	0.79872	1731.3	1970.9	6.9254	0.59809	1730.3	1969.5	6.7828
240	0.83167	1771.4	2020.9	7.0247	0.62289	1770.5	2019.6	6.8825
260	0.86455	1812.2	2071.6	7.1217	0.64764	1811.4	2070.5	6.9797
	500 kPa (4.13 °C)				600 kPa (9.28 °C)			
Sat.	0.25035	1321.3	1446.5	5.2776	0.21038	1325.2	1451.4	5.2133
10	0.25757	1333.5	1462.3	5.3340	0.21115	1326.7	1453.4	5.2205
20	0.26949	1353.6	1488.3	5.4244	0.22154	1347.9	1480.8	5.3156
30	0.28103	1373.0	1513.5	5.5090	0.23152	1368.2	1507.1	5.4037
40	0.29227	1392.0	1538.1	5.5889	0.24118	1387.8	1532.5	5.4862
50	0.30328	1410.6	1562.2	5.6647	0.25059	1406.9	1557.3	5.5641
60	0.31410	1429.0	1586.1	5.7373	0.25981	1425.7	1581.6	5.6383
70	0.32478	1447.3	1609.6	5.8070	0.26888	1444.3	1605.7	5.7094
80	0.33535	1465.4	1633.1	5.8744	0.27783	1462.8	1629.5	5.7778
100	0.35621	1501.7	1679.8	6.0031	0.29545	1499.5	1676.8	5.9081
120	0.37681	1538.2	1726.6	6.1253	0.31281	1536.3	1724.0	6.0314
140	0.39722	1575.2	1773.8	6.2422	0.32997	1573.5	1771.5	6.1491
160	0.41748	1612.7	1821.4	6.3548	0.34699	1611.2	1819.4	6.2623
180	0.43764	1650.8	1869.6	6.4636	0.36389	1649.5	1867.8	6.3717
200	0.45771	1689.6	1918.5	6.5691	0.38071	1688.5	1916.9	6.4776
220	0.47770	1729.2	1968.1	6.6717	0.39745	1728.2	1966.6	6.5806
240	0.49763	1769.5	2018.3	6.7717	0.41412	1768.6	2017.0	6.6808
260	0.51749	1810.6	2069.3	6.8692	0.43073	1809.8	2068.2	6.7786

TABLE B.2.2 (*continued*)
Superheated Ammonia

Temp. (°C)	v (m³/kg)	u (kJ/kg)	h (kJ/kg)	s (kJ/kg·K)	v (m³/kg)	u (kJ/kg)	h (kJ/kg)	s (kJ/kg·K)
	800 kPa (17.85 °C)				1000 kPa (24.90 °C)			
Sat.	0.15958	1330.9	1458.6	5.1110	0.12852	1334.9	1463.4	5.0304
20	0.16138	1335.8	1464.9	5.1328	—	—	—	—
30	0.16947	1358.0	1493.5	5.2287	0.13206	1347.1	1479.1	5.0826
40	0.17720	1379.0	1520.8	5.3171	0.13868	1369.8	1508.5	5.1778
50	0.18465	1399.3	1547.0	5.3996	0.14499	1391.3	1536.3	5.2654
60	0.19189	1419.0	1572.5	5.4774	0.15106	1412.1	1563.1	5.3471
70	0.19896	1438.3	1597.5	5.5513	0.15695	1432.2	1589.1	5.4240
80	0.20590	1457.4	1622.1	5.6219	0.16270	1451.9	1614.6	5.4971
100	0.21949	1495.0	1670.6	5.7555	0.17389	1490.5	1664.3	5.6342
120	0.23280	1532.5	1718.7	5.8811	0.18477	1528.6	1713.4	5.7622
140	0.24590	1570.1	1766.9	6.0006	0.19545	1566.8	1762.2	5.8834
160	0.25886	1608.2	1815.3	6.1150	0.20597	1605.2	1811.2	5.9992
180	0.27170	1646.8	1864.2	6.2254	0.21638	1644.2	1860.5	6.1105
200	0.28445	1686.1	1913.6	6.3322	0.22669	1683.7	1910.4	6.2182
220	0.29712	1726.0	1963.7	6.4358	0.23693	1723.9	1960.8	6.3226
240	0.30973	1766.7	2014.5	6.5367	0.24710	1764.8	2011.9	6.4241
260	0.32228	1808.1	2065.9	6.6350	0.25720	1806.4	2063.6	6.5229
280	0.33477	1850.2	2118.0	6.7310	0.26726	1848.8	2116.0	6.6194
300	0.34722	1893.1	2170.9	6.8248	0.27726	1891.8	2169.1	6.7137
	1200 kPa (30.94 °C)				1400 kPa (36.26 °C)			
Sat.	0.10751	1337.8	1466.8	4.9635	0.09231	1339.8	1469.0	4.9060
40	0.11287	1360.0	1495.4	5.0564	0.09432	1349.5	1481.6	4.9463
50	0.11846	1383.0	1525.1	5.1497	0.09942	1374.2	1513.4	5.0462
60	0.12378	1404.8	1553.3	5.2357	0.10423	1397.2	1543.1	5.1370
70	0.12890	1425.8	1580.5	5.3159	0.10882	1419.2	1571.5	5.2209
80	0.13387	1446.2	1606.8	5.3916	0.11324	1440.3	1598.8	5.2994
100	0.14347	1485.8	1658.0	5.5325	0.12172	1481.0	1651.4	5.4443
120	0.15275	1524.7	1708.0	5.6631	0.12986	1520.7	1702.5	5.5775
140	0.16181	1563.3	1757.5	5.7860	0.13777	1559.9	1752.8	5.7023
160	0.17071	1602.2	1807.1	5.9031	0.14552	1599.2	1802.9	5.8208
180	0.17950	1641.5	1856.9	6.0156	0.15315	1638.8	1853.2	5.9343
200	0.18819	1681.3	1907.1	6.1241	0.16068	1678.9	1903.8	6.0437
220	0.19680	1721.8	1957.9	6.2292	0.16813	1719.6	1955.0	6.1495
240	0.20534	1762.9	2009.3	6.3313	0.17551	1761.0	2006.7	6.2523
260	0.21382	1804.7	2061.3	6.4308	0.18283	1803.0	2059.0	6.3523
280	0.22225	1847.3	2114.0	6.5278	0.19010	1845.8	2111.9	6.4498
300	0.23063	1890.6	2167.3	6.6225	0.19732	1889.3	2165.5	6.5450
320	0.23897	1934.6	2221.3	6.7151	0.20450	1933.5	2219.8	6.6380

TABLE B.2.2 (*continued*)
Superheated Ammonia

Temp. (°C)	v (m³/kg)	u (kJ/kg)	h (kJ/kg)	s (kJ/kg·K)	v (m³/kg)	u (kJ/kg)	h (kJ/kg)	s (kJ/kg·K)
	1600 kPa (41.03 °C)				2000 kPa (49.37 °C)			
Sat.	0.08079	1341.2	1470.5	4.8553	0.06444	1342.6	1471.5	4.7680
50	0.08506	1364.9	1501.0	4.9510	0.06471	1344.5	1473.9	4.7754
60	0.08951	1389.3	1532.5	5.0472	0.06875	1372.3	1509.8	4.8848
70	0.09372	1412.3	1562.3	5.1351	0.07246	1397.8	1542.7	4.9821
80	0.09774	1434.3	1590.6	5.2167	0.07595	1421.6	1573.5	5.0707
100	0.10539	1476.2	1644.8	5.3659	0.08248	1466.1	1631.1	5.2294
120	0.11268	1516.6	1696.9	5.5018	0.08861	1508.3	1685.5	5.3714
140	0.11974	1556.4	1748.0	5.6286	0.09447	1549.3	1738.2	5.5022
160	0.12662	1596.1	1798.7	5.7485	0.10016	1589.9	1790.2	5.6251
180	0.13339	1636.1	1849.5	5.8631	0.10571	1630.6	1842.0	5.7420
200	0.14005	1676.5	1900.5	5.9734	0.11116	1671.6	1893.9	5.8540
220	0.14663	1717.4	1952.0	6.0800	0.11652	1713.1	1946.1	5.9621
240	0.15314	1759.0	2004.1	6.1834	0.12182	1755.2	1998.8	6.0668
260	0.15959	1801.3	2056.7	6.2839	0.12705	1797.9	2052.0	6.1685
280	0.16599	1844.3	2109.9	6.3819	0.13224	1841.3	2105.8	6.2675
300	0.17234	1888.0	2163.7	6.4775	0.13737	1885.4	2160.1	6.3641
320	0.17865	1932.4	2218.2	6.5710	0.14246	1930.2	2215.1	6.4583
340	0.18492	1977.5	2273.4	6.6624	0.14751	1975.6	2270.7	6.5505
360	0.19115	2023.3	2329.1	6.7519	0.15253	2021.8	2326.8	6.6406
	5000 kPa (88.90 °C)				10000 kPa (125.20 °C)			
Sat.	0.02365	1323.2	1441.4	4.3454	0.00826	1206.8	1289.4	3.7587
100	0.02636	1369.7	1501.5	4.5091	—	—	—	—
120	0.03024	1435.1	1586.3	4.7306	—	—	—	—
140	0.03350	1489.8	1657.3	4.9068	0.01195	1341.8	1461.3	4.1839
160	0.03643	1539.5	1721.7	5.0591	0.01461	1432.2	1578.3	4.4610
180	0.03916	1586.9	1782.7	5.1968	0.01666	1500.6	1667.2	4.6617
200	0.04174	1633.1	1841.8	5.3245	0.01842	1560.3	1744.5	4.8287
220	0.04422	1678.9	1900.0	5.4450	0.02001	1615.8	1816.0	4.9767
240	0.04662	1724.8	1957.9	5.5600	0.02150	1669.2	1884.2	5.1123
260	0.04895	1770.9	2015.6	5.6704	0.02290	1721.6	1950.6	5.2392
280	0.05123	1817.4	2073.6	5.7771	0.02424	1773.6	2015.9	5.3596
300	0.05346	1864.5	2131.8	5.8805	0.02552	1825.5	2080.7	5.4746
320	0.05565	1912.1	2190.3	5.9809	0.02676	1877.6	2145.2	5.5852
340	0.05779	1960.3	2249.2	6.0786	0.02796	1930.0	2209.6	5.6921
360	0.05990	2009.1	2308.6	6.1738	0.02913	1982.8	2274.1	5.7955
380	0.06198	2058.5	2368.4	6.2668	0.03026	2036.1	2338.7	5.8960
400	0.06403	2108.4	2428.6	6.3576	0.03137	2089.8	2403.5	5.9937
420	0.06606	2159.0	2489.3	6.4464	0.03245	2143.9	2468.5	6.0888
440	0.06806	2210.1	2550.4	6.5334	0.03351	2198.5	2533.7	6.1815

TABLE B.3
Thermodynamic Properties of Carbon Dioxide

TABLE B.3.1
Saturated Carbon Dioxide

Temp. (°C)	Press. (kPa)	Specific Volume, m³/kg			Internal Energy, kJ/kg		
		Sat. Liquid v_f	Evap. v_{fg}	Sat. Vapor v_g	Sat. Liquid u_f	Evap. u_{fg}	Sat. Vapor u_g
−50.0	682.3	0.000866	0.05492	0.05579	−20.55	302.26	281.71
−48	739.5	0.000872	0.05075	0.05162	−16.64	298.86	282.21
−46	800.2	0.000878	0.04694	0.04782	−12.72	295.42	282.69
−44	864.4	0.000883	0.04347	0.04435	−8.80	291.94	283.15
−42	932.5	0.000889	0.04029	0.04118	−4.85	288.42	283.57
−40	1004.5	0.000896	0.03739	0.03828	−0.90	284.86	283.96
−38	1080.5	0.000902	0.03472	0.03562	3.07	281.26	284.33
−36	1160.7	0.000909	0.03227	0.03318	7.05	277.60	284.66
−34	1245.2	0.000915	0.03002	0.03093	11.05	273.90	284.95
−32	1334.2	0.000922	0.02794	0.02886	15.07	270.14	285.21
−30	1427.8	0.000930	0.02603	0.02696	19.11	266.32	285.43
−28	1526.1	0.000937	0.02425	0.02519	23.17	262.45	285.61
−26	1629.3	0.000945	0.02261	0.02356	27.25	258.51	285.75
−24	1737.5	0.000953	0.02110	0.02205	31.35	254.50	285.85
−22	1850.9	0.000961	0.01968	0.02065	35.48	250.41	285.89
−20	1969.6	0.000969	0.01837	0.01934	39.64	246.25	285.89
−18	2093.8	0.000978	0.01715	0.01813	43.82	242.01	285.84
−16	2223.7	0.000987	0.01601	0.01700	48.04	237.68	285.73
−14	2359.3	0.000997	0.01495	0.01595	52.30	233.26	285.56
−12	2501.0	0.001007	0.01396	0.01497	56.59	228.73	285.32
−10	2648.7	0.001017	0.01303	0.01405	60.92	224.10	285.02
−8	2802.7	0.001028	0.01216	0.01319	65.30	219.35	284.65
−6	2963.2	0.001040	0.01134	0.01238	69.73	214.47	284.20
−4	3130.3	0.001052	0.01057	0.01162	74.20	209.46	283.66
−2	3304.2	0.001065	0.00985	0.01091	78.74	204.29	283.03
0	3485.1	0.001078	0.00916	0.01024	83.34	198.96	282.30
2	3673.3	0.001093	0.00852	0.00961	88.01	193.44	281.46
4	3868.8	0.001108	0.00790	0.00901	92.76	187.73	280.49
6	4072.0	0.001124	0.00732	0.00845	97.60	181.78	279.38
8	4283.1	0.001142	0.00677	0.00791	102.54	175.57	278.11
10	4502.2	0.001161	0.00624	0.00740	107.60	169.07	276.67
12	4729.7	0.001182	0.00573	0.00691	112.79	162.23	275.02
14	4965.8	0.001205	0.00524	0.00645	118.14	154.99	273.13
16	5210.8	0.001231	0.00477	0.00600	123.69	147.26	270.95
18	5465.1	0.001260	0.00431	0.00557	129.48	138.95	268.43
20	5729.1	0.001293	0.00386	0.00515	135.56	129.90	265.46
22	6003.1	0.001332	0.00341	0.00474	142.03	119.89	261.92
24	6287.7	0.001379	0.00295	0.00433	149.04	108.55	257.59
26	6583.7	0.001440	0.00247	0.00391	156.88	95.20	252.07
28	6891.8	0.001526	0.00193	0.00346	166.20	78.26	244.46
30	7213.7	0.001685	0.00121	0.00290	179.49	51.83	231.32
31.0	7377.3	0.002139	0.0	0.00214	203.56	0.0	203.56

TABLE B.3.1 (*continued*)
Saturated Carbon Dioxide

Temp. (°C)	Press. (kPa)	Enthalpy, kJ/kg			Entropy, kJ/kg·K		
		Sat. Liquid h_f	Evap. h_{fg}	Sat. Vapor h_g	Sat. Liquid s_f	Evap. s_{fg}	Sat. Vapor s_g
−50.0	682.3	−19.96	339.73	319.77	−0.0863	1.5224	1.4362
−48	739.5	−16.00	336.38	320.38	−0.0688	1.4940	1.4252
−46	800.2	−12.02	332.98	320.96	−0.0515	1.4659	1.4144
−44	864.4	−8.03	329.52	321.49	−0.0342	1.4380	1.4038
−42	932.5	−4.02	326.00	321.97	−0.0171	1.4103	1.3933
−40	1004.5	0	322.42	322.42	0	1.3829	1.3829
−38	1080.5	4.04	318.78	322.82	0.0170	1.3556	1.3726
−36	1160.7	8.11	315.06	323.17	0.0339	1.3285	1.3624
−34	1245.2	12.19	311.28	323.47	0.0507	1.3016	1.3523
−32	1334.2	16.30	307.42	323.72	0.0675	1.2748	1.3423
−30	1427.8	20.43	303.48	323.92	0.0842	1.2481	1.3323
−28	1526.1	24.60	299.46	324.06	0.1009	1.2215	1.3224
−26	1629.3	28.78	295.35	324.14	0.1175	1.1950	1.3125
−24	1737.5	33.00	291.15	324.15	0.1341	1.1686	1.3026
−22	1850.9	37.26	286.85	324.11	0.1506	1.1421	1.2928
−20	1969.6	41.55	282.44	323.99	0.1672	1.1157	1.2829
−18	2093.8	45.87	277.93	323.80	0.1837	1.0893	1.2730
−16	2223.7	50.24	273.30	323.53	0.2003	1.0628	1.2631
−14	2359.3	54.65	268.54	323.19	0.2169	1.0362	1.2531
−12	2501.0	59.11	263.65	322.76	0.2334	1.0096	1.2430
−10	2648.7	63.62	258.61	322.23	0.2501	0.9828	1.2328
−8	2802.7	68.18	253.43	321.61	0.2668	0.9558	1.2226
−6	2963.2	72.81	248.08	320.89	0.2835	0.9286	1.2121
−4	3130.3	77.50	242.55	320.05	0.3003	0.9012	1.2015
−2	3304.2	82.26	236.83	319.09	0.3173	0.8734	1.1907
0	3485.1	87.10	230.89	317.99	0.3344	0.8453	1.1797
2	3673.3	92.02	224.73	316.75	0.3516	0.8167	1.1683
4	3868.8	97.05	218.30	315.35	0.3690	0.7877	1.1567
6	4072.0	102.18	211.59	313.77	0.3866	0.7580	1.1446
8	4283.1	107.43	204.56	311.99	0.4045	0.7276	1.1321
10	4502.2	112.83	197.15	309.98	0.4228	0.6963	1.1190
12	4729.7	118.38	189.33	307.72	0.4414	0.6640	1.1053
14	4965.8	124.13	181.02	305.15	0.4605	0.6304	1.0909
16	5210.8	130.11	172.12	302.22	0.4802	0.5952	1.0754
18	5465.1	136.36	162.50	298.86	0.5006	0.5581	1.0588
20	5729.1	142.97	152.00	294.96	0.5221	0.5185	1.0406
22	6003.1	150.02	140.34	290.36	0.5449	0.4755	1.0203
24	6287.7	157.71	127.09	284.80	0.5695	0.4277	0.9972
26	6583.7	166.36	111.45	277.80	0.5971	0.3726	0.9697
28	6891.8	176.72	91.58	268.30	0.6301	0.3041	0.9342
30	7213.7	191.65	60.58	252.23	0.6778	0.1998	0.8776
31.0	7377.3	219.34	0.0	219.34	0.7680	0.0	0.7680

TABLE B.3.2
Superheated Carbon Dioxide

Temp. (°C)	v (m³/kg)	u (kJ/kg)	h (kJ/kg)	s (kJ/kg·K)	v (m³/kg)	u (kJ/kg)	h (kJ/kg)	s (kJ/kg·K)
	400 kPa (NA)				800 kPa (−46.00 °C)			
Sat.	—	—	—	—	0.04783	282.69	320.95	1.4145
−40	0.10499	292.46	334.46	1.5947	0.04966	287.05	326.78	1.4398
−20	0.11538	305.30	351.46	1.6646	0.05546	301.13	345.49	1.5168
0	0.12552	318.31	368.51	1.7295	0.06094	314.92	363.67	1.5859
20	0.13551	331.57	385.77	1.7904	0.06623	328.73	381.72	1.6497
40	0.14538	345.14	403.29	1.8482	0.07140	342.70	399.82	1.7094
60	0.15518	359.03	421.10	1.9033	0.07648	356.90	418.09	1.7660
80	0.16491	373.25	439.21	1.9561	0.08150	371.37	436.57	1.8199
100	0.17460	387.80	457.64	2.0069	0.08647	386.11	455.29	1.8714
120	0.18425	402.67	476.37	2.0558	0.09141	401.15	474.27	1.9210
140	0.19388	417.86	495.41	2.1030	0.09631	416.47	493.52	1.9687
160	0.20348	433.35	514.74	2.1487	0.10119	432.07	513.03	2.0148
180	0.21307	449.13	534.36	2.1930	0.10606	447.95	532.80	2.0594
200	0.22264	465.20	554.26	2.2359	0.11090	464.11	552.83	2.1027
220	0.23219	481.55	574.42	2.2777	0.11573	480.52	573.11	2.1447
240	0.24173	498.16	594.85	2.3183	0.12056	497.20	593.64	2.1855
260	0.25127	515.02	615.53	2.3578	0.12537	514.12	614.41	2.2252
	1000 kPa (−40.12 °C)				1400 kPa (−30.58 °C)			
Sat.	0.03845	283.94	322.39	1.3835	0.02750	285.37	323.87	1.3352
−20	0.04342	298.89	342.31	1.4655	0.02957	294.04	335.44	1.3819
0	0.04799	313.15	361.14	1.5371	0.03315	309.42	355.83	1.4595
20	0.05236	327.27	379.63	1.6025	0.03648	324.23	375.30	1.5283
40	0.05660	341.46	398.05	1.6633	0.03966	338.90	394.42	1.5914
60	0.06074	355.82	416.56	1.7206	0.04274	353.62	413.45	1.6503
80	0.06482	370.42	435.23	1.7750	0.04575	368.48	432.52	1.7059
100	0.06885	385.26	454.11	1.8270	0.04870	383.54	451.72	1.7588
120	0.07284	400.38	473.22	1.8768	0.05161	398.83	471.09	1.8093
140	0.07680	415.77	492.57	1.9249	0.05450	414.36	490.66	1.8579
160	0.08074	431.43	512.17	1.9712	0.05736	430.14	510.44	1.9046
180	0.08465	447.36	532.02	2.0160	0.06020	446.17	530.45	1.9498
200	0.08856	463.56	552.11	2.0594	0.06302	462.45	550.68	1.9935
220	0.09244	480.01	572.46	2.1015	0.06583	478.98	571.14	2.0358
240	0.09632	496.72	593.04	2.1424	0.06863	495.76	591.83	2.0770
260	0.10019	513.67	613.86	2.1822	0.07141	512.77	612.74	2.1169
280	0.10405	530.86	634.90	2.2209	0.07419	530.01	633.88	2.1558

TABLE B.3.2 (*continued*)
Superheated Carbon Dioxide

Temp. (°C)	v (m³/kg)	u (kJ/kg)	h (kJ/kg)	s (kJ/kg·K)	v (m³/kg)	u (kJ/kg)	h (kJ/kg)	s (kJ/kg·K)
	2000 kPa (−19.50 °C)				3000 kPa (−5.55 °C)			
Sat.	0.01903	285.88	323.95	1.2804	0.01221	284.09	320.71	1.2098
0	0.02193	303.24	347.09	1.3684	0.01293	290.52	329.32	1.2416
20	0.02453	319.37	368.42	1.4438	0.01512	310.21	355.56	1.3344
40	0.02693	334.88	388.75	1.5109	0.01698	327.61	378.55	1.4104
60	0.02922	350.19	408.64	1.5725	0.01868	344.14	400.19	1.4773
80	0.03143	365.49	428.36	1.6300	0.02029	360.30	421.16	1.5385
100	0.03359	380.90	448.07	1.6843	0.02182	376.35	441.82	1.5954
120	0.03570	396.46	467.85	1.7359	0.02331	392.42	462.35	1.6490
140	0.03777	412.22	487.76	1.7853	0.02477	408.57	482.87	1.6999
160	0.03982	428.18	507.83	1.8327	0.02619	424.87	503.44	1.7485
180	0.04186	444.37	528.08	1.8784	0.02759	441.34	524.12	1.7952
200	0.04387	460.79	548.53	1.9226	0.02898	457.99	544.92	1.8401
220	0.04587	477.43	569.17	1.9653	0.03035	474.83	565.88	1.8835
240	0.04786	494.31	590.02	2.0068	0.03171	491.88	587.01	1.9255
260	0.04983	511.41	611.08	2.0470	0.03306	509.13	608.30	1.9662
280	0.05180	528.73	632.34	2.0862	0.03440	526.59	629.78	2.0057
300	0.05377	546.26	653.80	2.1243	0.03573	544.25	651.43	2.0442
	6000 kPa (21.98 °C)				10 000 kPa			
Sat.	0.00474	261.97	290.42	1.0206	—	—	—	—
20	—	—	—	—	0.00117	118.12	129.80	0.4594
40	0.00670	298.62	338.82	1.1806	0.00159	184.23	200.14	0.6906
60	0.00801	322.51	370.54	1.2789	0.00345	277.63	312.11	1.0389
80	0.00908	342.74	397.21	1.3567	0.00451	312.82	357.95	1.1728
100	0.01004	361.47	421.69	1.4241	0.00530	338.20	391.24	1.2646
120	0.01092	379.47	445.02	1.4850	0.00598	360.19	419.96	1.3396
140	0.01176	397.10	467.68	1.5413	0.00658	380.54	446.38	1.4051
160	0.01257	414.56	489.97	1.5939	0.00715	399.99	471.46	1.4644
180	0.01335	431.97	512.06	1.6438	0.00768	418.94	495.73	1.5192
200	0.01411	449.40	534.04	1.6913	0.00819	437.61	519.49	1.5705
220	0.01485	466.91	556.01	1.7367	0.00868	456.12	542.91	1.6190
240	0.01558	484.52	578.00	1.7804	0.00916	474.58	566.14	1.6652
260	0.01630	502.27	600.05	1.8226	0.00962	493.03	589.26	1.7094
280	0.01701	520.15	622.19	1.8634	0.01008	511.53	612.32	1.7518
300	0.01771	538.18	644.44	1.9029	0.01053	530.11	635.37	1.7928
320	0.01840	556.37	666.80	1.9412	0.01097	548.77	658.46	1.8324

TABLE B.4
Thermodynamic Properties of R-410A

TABLE B.4.1
Saturated R-410A

Temp. (°C)	Press. (kPa)	Specific Volume, m³/kg			Internal Energy, kJ/kg		
		Sat. Liquid v_f	Evap. v_{fg}	Sat. Vapor v_g	Sat. Liquid u_f	Evap. u_{fg}	Sat. Vapor u_g
−60	64.1	0.000727	0.36772	0.36845	−27.50	256.41	228.91
−55	84.0	0.000735	0.28484	0.28558	−20.70	251.89	231.19
−51.4	101.3	0.000741	0.23875	0.23949	−15.78	248.59	232.81
−50	108.7	0.000743	0.22344	0.22418	−13.88	247.31	233.43
−45	138.8	0.000752	0.17729	0.17804	−7.02	242.67	235.64
−40	175.0	0.000762	0.14215	0.14291	−0.13	237.95	237.81
−35	218.4	0.000771	0.11505	0.11582	6.80	233.14	239.94
−30	269.6	0.000781	0.09392	0.09470	13.78	228.23	242.01
−25	329.7	0.000792	0.07726	0.07805	20.82	223.21	244.03
−20	399.6	0.000803	0.06400	0.06480	27.92	218.07	245.99
−15	480.4	0.000815	0.05334	0.05416	35.08	212.79	247.88
−10	573.1	0.000827	0.04470	0.04553	42.32	207.36	249.69
−5	678.9	0.000841	0.03764	0.03848	49.65	201.75	251.41
0	798.7	0.000855	0.03182	0.03267	57.07	195.95	253.02
5	933.9	0.000870	0.02699	0.02786	64.60	189.93	254.53
10	1085.7	0.000886	0.02295	0.02383	72.24	183.66	255.90
15	1255.4	0.000904	0.01955	0.02045	80.02	177.10	257.12
20	1444.2	0.000923	0.01666	0.01758	87.94	170.21	258.16
25	1653.6	0.000944	0.01420	0.01514	96.03	162.95	258.98
30	1885.1	0.000968	0.01208	0.01305	104.32	155.24	259.56
35	2140.2	0.000995	0.01025	0.01124	112.83	147.00	259.83
40	2420.7	0.001025	0.00865	0.00967	121.61	138.11	259.72
45	2728.3	0.001060	0.00723	0.00829	130.72	128.41	259.13
50	3065.2	0.001103	0.00597	0.00707	140.27	117.63	257.90
55	3433.7	0.001156	0.00482	0.00598	150.44	105.34	255.78
60	3836.9	0.001227	0.00374	0.00497	161.57	90.70	252.27
65	4278.3	0.001338	0.00265	0.00399	174.59	71.59	246.19
70	4763.1	0.001619	0.00124	0.00286	194.53	37.47	232.01
71.3	4901.2	0.00218	0	0.00218	215.78	0	215.78

TABLE B.4.1 (*continued*)
Saturated R-410A

Temp. (°C)	Press. (kPa)	Enthalpy, kJ/kg			Entropy, kJ/kg·K		
		Sat. Liquid h_f	Evap. h_{fg}	Sat. Vapor h_g	Sat. Liquid s_f	Evap. s_{fg}	Sat. Vapor s_g
−60	64.1	−27.45	279.96	252.51	−0.1227	1.3135	1.1907
−55	84.0	−20.64	275.83	255.19	−0.0912	1.2644	1.1732
−51.4	101.3	−15.70	272.78	257.08	−0.0688	1.2301	1.1613
−50	108.7	−13.80	271.60	257.80	−0.0603	1.2171	1.1568
−45	138.8	−6.92	267.27	260.35	−0.0299	1.1715	1.1416
−40	175.0	0	262.83	262.83	0	1.1273	1.1273
−35	218.4	6.97	258.26	265.23	0.0294	1.0844	1.1139
−30	269.6	13.99	253.55	267.54	0.0585	1.0428	1.1012
−25	329.7	21.08	248.69	269.77	0.0871	1.0022	1.0893
−20	399.6	28.24	243.65	271.89	0.1154	0.9625	1.0779
−15	480.4	35.47	238.42	273.90	0.1435	0.9236	1.0671
−10	573.1	42.80	232.98	275.78	0.1713	0.8854	1.0567
−5	678.9	50.22	227.31	277.53	0.1989	0.8477	1.0466
0	798.7	57.76	221.37	279.12	0.2264	0.8104	1.0368
5	933.9	65.41	215.13	280.55	0.2537	0.7734	1.0272
10	1085.7	73.21	208.57	281.78	0.2810	0.7366	1.0176
15	1255.4	81.15	201.64	282.79	0.3083	0.6998	1.0081
20	1444.2	89.27	194.28	283.55	0.3357	0.6627	0.9984
25	1653.6	97.59	186.43	284.02	0.3631	0.6253	0.9884
30	1885.1	106.14	178.02	284.16	0.3908	0.5872	0.9781
35	2140.2	114.95	168.94	283.89	0.4189	0.5482	0.9671
40	2420.7	124.09	159.04	283.13	0.4473	0.5079	0.9552
45	2728.3	133.61	148.14	281.76	0.4765	0.4656	0.9421
50	3065.2	143.65	135.93	279.58	0.5067	0.4206	0.9273
55	3433.7	154.41	121.89	276.30	0.5384	0.3715	0.9099
60	3836.9	166.28	105.04	271.33	0.5729	0.3153	0.8882
65	4278.3	180.32	82.95	263.26	0.6130	0.2453	0.8583
70	4763.1	202.24	43.40	245.64	0.6752	0.1265	0.8017
71.3	4901.2	226.46	0	226.46	0.7449	0	0.7449

TABLE B.4.2
Superheated R-410A

Temp. (°C)	v (m³/kg)	u (kJ/kg)	h (kJ/kg)	s (kJ/kg·K)	v (m³/kg)	u (kJ/kg)	h (kJ/kg)	s (kJ/kg·K)
	50 kPa (−64.34 °C)				100 kPa (−51.65 °C)			
Sat.	0.46484	226.90	250.15	1.2070	0.24247	232.70	256.94	1.1621
−60	0.47585	229.60	253.40	1.2225	—	—	—	—
−40	0.52508	241.94	268.20	1.2888	0.25778	240.40	266.18	1.2027
−20	0.57295	254.51	283.16	1.3504	0.28289	253.44	281.73	1.2667
0	0.62016	267.52	298.53	1.4088	0.30723	266.72	297.44	1.3265
20	0.66698	281.05	314.40	1.4649	0.33116	280.42	313.54	1.3833
40	0.71355	295.15	330.83	1.5191	0.35483	294.64	330.12	1.4380
60	0.75995	309.84	347.83	1.5717	0.37833	309.40	347.24	1.4910
80	0.80623	325.11	365.43	1.6230	0.40171	324.75	364.92	1.5425
100	0.85243	340.99	383.61	1.6731	0.42500	340.67	383.17	1.5928
120	0.89857	357.46	402.38	1.7221	0.44822	357.17	401.99	1.6419
140	0.94465	374.50	421.74	1.7701	0.47140	374.25	421.39	1.6901
160	0.99070	392.12	441.65	1.8171	0.49453	391.89	441.34	1.7372
180	1.03671	410.28	462.12	1.8633	0.51764	410.07	461.84	1.7835
200	1.08270	428.98	483.11	1.9087	0.54072	428.79	482.86	1.8289
220	1.12867	448.19	504.63	1.9532	0.56378	448.02	504.40	1.8734
240	1.17462	467.90	526.63	1.9969	0.58682	467.74	526.42	1.9172
	150 kPa (−43.35 °C)				200 kPa (−37.01 °C)			
Sat.	0.16540	236.36	261.17	1.1368	0.12591	239.09	264.27	1.1192
−40	0.16851	238.72	263.99	1.1489	—	—	—	—
−20	0.18613	252.34	280.26	1.2159	0.13771	251.18	278.72	1.1783
0	0.20289	265.90	296.33	1.2770	0.15070	265.06	295.20	1.2410
20	0.21921	279.78	312.66	1.3347	0.16322	279.13	311.78	1.2995
40	0.23525	294.12	329.40	1.3899	0.17545	293.59	328.68	1.3553
60	0.25112	308.97	346.64	1.4433	0.18750	308.53	346.03	1.4090
80	0.26686	324.37	364.40	1.4950	0.19943	324.00	363.89	1.4610
100	0.28251	340.35	382.72	1.5455	0.21127	340.02	382.28	1.5117
120	0.29810	356.89	401.60	1.5948	0.22305	356.60	401.21	1.5611
140	0.31364	374.00	421.04	1.6430	0.23477	373.74	420.70	1.6094
160	0.32915	391.66	441.03	1.6902	0.24645	391.43	440.72	1.6568
180	0.34462	409.87	461.56	1.7366	0.25810	409.66	461.28	1.7032
200	0.36006	428.60	482.61	1.7820	0.26973	428.41	482.35	1.7487
220	0.37548	447.84	504.16	1.8266	0.28134	447.67	503.93	1.7933
240	0.39089	467.58	526.21	1.8705	0.29293	467.41	526.00	1.8372
260	0.40628	487.78	548.73	1.9135	0.30450	487.63	548.53	1.8803

TABLE B.4.2 (*continued*)
Superheated R-410A

Temp. (°C)	v (m³/kg)	u (kJ/kg)	h (kJ/kg)	s (kJ/kg·K)	v (m³/kg)	u (kJ/kg)	h (kJ/kg)	s (kJ/kg·K)
	300 kPa (−27.37 °C)				400 kPa (−19.98 °C)			
Sat.	0.08548	243.08	268.72	1.0949	0.06475	246.00	271.90	1.0779
−20	0.08916	248.71	275.46	1.1219	—	—	—	—
0	0.09845	263.33	292.87	1.1881	0.07227	261.51	290.42	1.1483
20	0.10720	277.81	309.96	1.2485	0.07916	276.44	308.10	1.2108
40	0.11564	292.53	327.22	1.3054	0.08571	291.44	325.72	1.2689
60	0.12388	307.65	344.81	1.3599	0.09207	306.75	343.58	1.3242
80	0.13200	323.25	362.85	1.4125	0.09828	322.49	361.80	1.3773
100	0.14003	339.37	381.38	1.4635	0.10440	338.72	380.48	1.4288
120	0.14798	356.03	400.43	1.5132	0.11045	355.45	399.64	1.4788
140	0.15589	373.23	420.00	1.5617	0.11645	372.72	419.30	1.5276
160	0.16376	390.97	440.10	1.6093	0.12241	390.51	439.47	1.5752
180	0.17159	409.24	460.72	1.6558	0.12834	408.82	460.16	1.6219
200	0.17940	428.03	481.85	1.7014	0.13424	427.64	481.34	1.6676
220	0.18719	447.31	503.47	1.7462	0.14012	446.96	503.01	1.7125
240	0.19496	467.09	525.58	1.7901	0.14598	466.76	525.15	1.7565
260	0.20272	487.33	548.15	1.8332	0.15182	487.03	547.76	1.7997
280	0.21046	508.02	571.16	1.8756	0.15766	507.74	570.81	1.8422
	500 kPa (−13.89 °C)				600 kPa (−8.67 °C)			
Sat.	0.05208	248.29	274.33	1.0647	0.04351	250.15	276.26	1.0540
0	0.05651	259.59	287.84	1.1155	0.04595	257.54	285.12	1.0869
20	0.06231	275.02	306.18	1.1803	0.05106	273.56	304.20	1.1543
40	0.06775	290.32	324.20	1.2398	0.05576	289.19	322.64	1.2152
60	0.07297	305.84	342.32	1.2959	0.06023	304.91	341.05	1.2722
80	0.07804	321.72	360.74	1.3496	0.06455	320.94	359.67	1.3265
100	0.08302	338.05	379.56	1.4014	0.06877	337.38	378.65	1.3787
120	0.08793	354.87	398.84	1.4517	0.07292	354.29	398.04	1.4294
140	0.09279	372.20	418.60	1.5007	0.07701	371.68	417.89	1.4786
160	0.09760	390.05	438.85	1.5486	0.08106	389.58	438.22	1.5266
180	0.10238	408.40	459.59	1.5954	0.08508	407.98	459.03	1.5736
200	0.10714	427.26	480.83	1.6413	0.08907	426.88	480.32	1.6196
220	0.11187	446.61	502.55	1.6862	0.09304	446.26	502.08	1.6646
240	0.11659	466.44	524.73	1.7303	0.09700	466.11	524.31	1.7088
260	0.12129	486.73	547.37	1.7736	0.10093	486.42	546.98	1.7521
280	0.12598	507.46	570.45	1.8161	0.10486	507.18	570.09	1.7947
300	0.13066	528.62	593.95	1.8578	0.10877	528.36	593.62	1.8365

TABLE B.4.2 (*continued*)
Superheated R-410A

Temp. (°C)	v (m³/kg)	u (kJ/kg)	h (kJ/kg)	s (kJ/kg·K)	v (m³/kg)	u (kJ/kg)	h (kJ/kg)	s (kJ/kg·K)
	800 kPa (0.05 °C)				1000 kPa (7.25 °C)			
Sat.	0.03262	253.04	279.14	1.0367	0.02596	255.16	281.12	1.0229
20	0.03693	270.47	300.02	1.1105	0.02838	267.11	295.49	1.0730
40	0.04074	286.83	319.42	1.1746	0.03170	284.35	316.05	1.1409
60	0.04429	303.01	338.44	1.2334	0.03470	301.04	335.75	1.2019
80	0.04767	319.36	357.49	1.2890	0.03753	317.73	355.27	1.2588
100	0.05095	336.03	376.79	1.3421	0.04025	334.65	374.89	1.3128
120	0.05415	353.11	396.42	1.3934	0.04288	351.91	394.79	1.3648
140	0.05729	370.64	416.47	1.4431	0.04545	369.58	415.04	1.4150
160	0.06039	388.65	436.96	1.4915	0.04798	387.70	435.68	1.4638
180	0.06345	407.13	457.90	1.5388	0.05048	406.28	456.76	1.5113
200	0.06649	426.10	479.30	1.5850	0.05294	425.33	478.27	1.5578
220	0.06951	445.55	501.15	1.6302	0.05539	444.84	500.23	1.6032
240	0.07251	465.46	523.46	1.6746	0.05781	464.80	522.62	1.6477
260	0.07549	485.82	546.21	1.7181	0.06023	485.21	545.43	1.6914
280	0.07846	506.61	569.38	1.7607	0.06262	506.05	568.67	1.7341
300	0.08142	527.83	592.97	1.8026	0.06501	527.30	592.31	1.7761
	1200 kPa (13.43 °C)				1400 kPa (18.88 °C)			
Sat.	0.02145	256.75	282.50	1.0111	0.01819	257.94	283.40	1.0006
20	0.02260	263.39	290.51	1.0388	0.01838	259.18	284.90	1.0057
40	0.02563	281.72	312.48	1.1113	0.02127	278.93	308.71	1.0843
60	0.02830	299.00	332.96	1.1747	0.02371	296.88	330.07	1.1505
80	0.03077	316.06	352.98	1.2331	0.02593	314.35	350.64	1.2105
100	0.03311	333.24	372.97	1.2881	0.02801	331.80	371.01	1.2666
120	0.03537	350.69	393.13	1.3408	0.03000	349.46	391.46	1.3199
140	0.03756	368.51	413.59	1.3915	0.03192	367.43	412.13	1.3712
160	0.03971	386.75	434.40	1.4407	0.03380	385.79	433.12	1.4208
180	0.04183	405.43	455.62	1.4886	0.03565	404.56	454.47	1.4690
200	0.04391	424.55	477.24	1.5353	0.03746	423.77	476.21	1.5160
220	0.04597	444.12	499.29	1.5809	0.03925	443.41	498.36	1.5618
240	0.04802	464.14	521.77	1.6256	0.04102	463.49	520.92	1.6066
260	0.05005	484.60	544.66	1.6693	0.04278	483.99	543.88	1.6505
280	0.05207	505.48	567.96	1.7122	0.04452	504.91	567.25	1.6936
300	0.05407	526.77	591.66	1.7543	0.04626	526.25	591.01	1.7358
320	0.05607	548.47	615.75	1.7956	0.04798	547.97	615.14	1.7772

TABLE B.4.2 (*continued*)
Superheated R-410A

Temp. (°C)	v (m³/kg)	u (kJ/kg)	h (kJ/kg)	s (kJ/kg·K)	v (m³/kg)	u (kJ/kg)	h (kJ/kg)	s (kJ/kg·K)
	1800 kPa (28.22 °C)				2000 kPa (32.31 °C)			
Sat.	0.01376	259.38	284.15	0.9818	0.01218	259.72	284.09	0.9731
40	0.01534	272.67	300.29	1.0344	0.01321	269.07	295.49	1.0099
60	0.01754	292.34	323.92	1.1076	0.01536	289.90	320.62	1.0878
80	0.01945	310.76	345.77	1.1713	0.01717	308.88	343.22	1.1537
100	0.02119	328.84	366.98	1.2297	0.01880	327.30	364.91	1.2134
120	0.02283	346.93	388.03	1.2847	0.02032	345.64	386.29	1.2693
140	0.02441	365.24	409.17	1.3371	0.02177	364.12	407.66	1.3223
160	0.02593	383.85	430.51	1.3875	0.02317	382.86	429.20	1.3732
180	0.02741	402.82	452.16	1.4364	0.02452	401.94	450.99	1.4224
200	0.02886	422.19	474.14	1.4839	0.02585	421.40	473.10	1.4701
220	0.03029	441.97	496.49	1.5301	0.02715	441.25	495.55	1.5166
240	0.03170	462.16	519.22	1.5753	0.02844	461.50	518.37	1.5619
260	0.03309	482.77	542.34	1.6195	0.02970	482.16	541.56	1.6063
280	0.03447	503.78	565.83	1.6627	0.03095	503.21	565.12	1.6497
300	0.03584	525.19	589.70	1.7051	0.03220	524.66	589.05	1.6922
320	0.03720	546.98	613.94	1.7467	0.03343	546.49	613.35	1.7338
340	0.03855	569.15	638.54	1.7875	0.03465	568.69	637.99	1.7747
	3000 kPa (49.07 °C)				4000 kPa (61.90 °C)			
Sat.	0.00729	258.19	280.06	0.9303	0.00460	250.37	268.76	0.8782
60	0.00858	274.96	300.70	0.9933	—	—	—	—
80	0.01025	298.38	329.12	1.0762	0.00661	285.02	311.48	1.0028
100	0.01159	319.07	353.84	1.1443	0.00792	309.62	341.29	1.0850
120	0.01277	338.84	377.16	1.2052	0.00897	331.39	367.29	1.1529
140	0.01387	358.32	399.92	1.2617	0.00990	352.14	391.75	1.2136
160	0.01489	377.80	422.49	1.3150	0.01076	372.51	415.53	1.2698
180	0.01588	397.46	445.09	1.3661	0.01156	392.82	439.05	1.3229
200	0.01683	417.37	467.85	1.4152	0.01232	413.25	462.52	1.3736
220	0.01775	437.60	490.84	1.4628	0.01305	433.88	486.10	1.4224
240	0.01865	458.16	514.11	1.5091	0.01377	454.79	509.85	1.4696
260	0.01954	479.08	537.69	1.5541	0.01446	475.99	533.83	1.5155
280	0.02041	500.37	561.59	1.5981	0.01514	497.51	558.08	1.5601
300	0.02127	522.01	585.81	1.6411	0.01581	519.37	582.60	1.6037
320	0.02212	544.02	610.37	1.6833	0.01647	541.55	607.42	1.6462
340	0.02296	566.37	635.25	1.7245	0.01712	564.06	632.54	1.6879
360	0.02379	589.07	660.45	1.7650	0.01776	586.90	657.95	1.7286

TABLE B.5
Thermodynamic Properties of R-134a

TABLE B.5.1
Saturated R-134a

Temp. (°C)	Press. (kPa)	Specific Volume, m³/kg			Internal Energy, kJ/kg		
		Sat. Liquid v_f	Evap. v_{fg}	Sat. Vapor v_g	Sat. Liquid u_f	Evap. u_{fg}	Sat. Vapor u_g
−70	8.3	0.000675	1.97207	1.97274	119.46	218.74	338.20
−65	11.7	0.000679	1.42915	1.42983	123.18	217.76	340.94
−60	16.3	0.000684	1.05199	1.05268	127.52	216.19	343.71
−55	22.2	0.000689	0.78609	0.78678	132.36	214.14	346.50
−50	29.9	0.000695	0.59587	0.59657	137.60	211.71	349.31
−45	39.6	0.000701	0.45783	0.45853	143.15	208.99	352.15
−40	51.8	0.000708	0.35625	0.35696	148.95	206.05	355.00
−35	66.8	0.000715	0.28051	0.28122	154.93	202.93	357.86
−30	85.1	0.000722	0.22330	0.22402	161.06	199.67	360.73
−26.3	101.3	0.000728	0.18947	0.19020	165.73	197.16	362.89
−25	107.2	0.000730	0.17957	0.18030	167.30	196.31	363.61
−20	133.7	0.000738	0.14576	0.14649	173.65	192.85	366.50
−15	165.0	0.000746	0.11932	0.12007	180.07	189.32	369.39
−10	201.7	0.000755	0.09845	0.09921	186.57	185.70	372.27
−5	244.5	0.000764	0.08181	0.08257	193.14	182.01	375.15
0	294.0	0.000773	0.06842	0.06919	199.77	178.24	378.01
5	350.9	0.000783	0.05755	0.05833	206.48	174.38	380.85
10	415.8	0.000794	0.04866	0.04945	213.25	170.42	383.67
15	489.5	0.000805	0.04133	0.04213	220.10	166.35	386.45
20	572.8	0.000817	0.03524	0.03606	227.03	162.16	389.19
25	666.3	0.000829	0.03015	0.03098	234.04	157.83	391.87
30	771.0	0.000843	0.02587	0.02671	241.14	153.34	394.48
35	887.6	0.000857	0.02224	0.02310	248.34	148.68	397.02
40	1017.0	0.000873	0.01915	0.02002	255.65	143.81	399.46
45	1160.2	0.000890	0.01650	0.01739	263.08	138.71	401.79
50	1318.1	0.000908	0.01422	0.01512	270.63	133.35	403.98
55	1491.6	0.000928	0.01224	0.01316	278.33	127.68	406.01
60	1681.8	0.000951	0.01051	0.01146	286.19	121.66	407.85
65	1889.9	0.000976	0.00899	0.00997	294.24	115.22	409.46
70	2117.0	0.001005	0.00765	0.00866	302.51	108.27	410.78
75	2364.4	0.001038	0.00645	0.00749	311.06	100.68	411.74
80	2633.6	0.001078	0.00537	0.00645	319.96	92.26	412.22
85	2926.2	0.001128	0.00437	0.00550	329.35	82.67	412.01
90	3244.5	0.001195	0.00341	0.00461	339.51	71.24	410.75
95	3591.5	0.001297	0.00243	0.00373	351.17	56.25	407.42
100	3973.2	0.001557	0.00108	0.00264	368.55	28.19	396.74
101.2	4064.0	0.001969	0	0.00197	382.97	0	382.97

TABLE B.5.1 (*continued*)
Saturated R-134a

Temp. (°C)	Press. (kPa)	Enthalpy, kJ/kg			Entropy, kJ/k-K		
		Sat. Liquid h_f	Evap. h_{fg}	Sat. Vapor h_g	Sat. Liquid s_f	Evap. s_{fg}	Sat. Vapor s_g
−70	8.3	119.47	235.15	354.62	0.6645	1.1575	1.8220
−65	11.7	123.18	234.55	357.73	0.6825	1.1268	1.8094
−60	16.3	127.53	233.33	360.86	0.7031	1.0947	1.7978
−55	22.2	132.37	231.63	364.00	0.7256	1.0618	1.7874
−50	29.9	137.62	229.54	367.16	0.7493	1.0286	1.7780
−45	39.6	143.18	227.14	370.32	0.7740	0.9956	1.7695
−40	51.8	148.98	224.50	373.48	0.7991	0.9629	1.7620
−35	66.8	154.98	221.67	376.64	0.8245	0.9308	1.7553
−30	85.1	161.12	218.68	379.80	0.8499	0.8994	1.7493
−26.3	101.3	165.80	216.36	382.16	0.8690	0.8763	1.7453
−25	107.2	167.38	215.57	382.95	0.8754	0.8687	1.7441
−20	133.7	173.74	212.34	386.08	0.9007	0.8388	1.7395
−15	165.0	180.19	209.00	389.20	0.9258	0.8096	1.7354
−10	201.7	186.72	205.56	392.28	0.9507	0.7812	1.7319
−5	244.5	193.32	202.02	395.34	0.9755	0.7534	1.7288
0	294.0	200.00	198.36	398.36	1.0000	0.7262	1.7262
5	350.9	206.75	194.57	401.32	1.0243	0.6995	1.7239
10	415.8	213.58	190.65	404.23	1.0485	0.6733	1.7218
15	489.5	220.49	186.58	407.07	1.0725	0.6475	1.7200
20	572.8	227.49	182.35	409.84	1.0963	0.6220	1.7183
25	666.3	234.59	177.92	412.51	1.1201	0.5967	1.7168
30	771.0	241.79	173.29	415.08	1.1437	0.5716	1.7153
35	887.6	249.10	168.42	417.52	1.1673	0.5465	1.7139
40	1017.0	256.54	163.28	419.82	1.1909	0.5214	1.7123
45	1160.2	264.11	157.85	421.96	1.2145	0.4962	1.7106
50	1318.1	271.83	152.08	423.91	1.2381	0.4706	1.7088
55	1491.6	279.72	145.93	425.65	1.2619	0.4447	1.7066
60	1681.8	287.79	139.33	427.13	1.2857	0.4182	1.7040
65	1889.9	296.09	132.21	428.30	1.3099	0.3910	1.7008
70	2117.0	304.64	124.47	429.11	1.3343	0.3627	1.6970
75	2364.4	313.51	115.94	429.45	1.3592	0.3330	1.6923
80	2633.6	322.79	106.40	429.19	1.3849	0.3013	1.6862
85	2926.2	332.65	95.45	428.10	1.4117	0.2665	1.6782
90	3244.5	343.38	82.31	425.70	1.4404	0.2267	1.6671
95	3591.5	355.83	64.98	420.81	1.4733	0.1765	1.6498
100	3973.2	374.74	32.47	407.21	1.5228	0.0870	1.6098
101.2	4064.0	390.98	0	390.98	1.5658	0	1.5658

TABLE B.5.2
Superheated R-134a

Temp. (°C)	v (m³/kg)	u (kJ/kg)	h (kJ/kg)	s (kJ/kg·K)	v (m³/kg)	u (kJ/kg)	h (kJ/kg)	s (kJ/kg·K)
	50 kPa (−40.67 °C)				100 kPa (−26.54 °C)			
Sat.	0.36889	354.61	373.06	1.7629	0.19257	362.73	381.98	1.7456
−20	0.40507	368.57	388.82	1.8279	0.19860	367.36	387.22	1.7665
−10	0.42222	375.53	396.64	1.8582	0.20765	374.51	395.27	1.7978
0	0.43921	382.63	404.59	1.8878	0.21652	381.76	403.41	1.8281
10	0.45608	389.90	412.70	1.9170	0.22527	389.14	411.67	1.8578
20	0.47287	397.32	420.96	1.9456	0.23392	396.66	420.05	1.8869
30	0.48958	404.90	429.38	1.9739	0.24250	404.31	428.56	1.9155
40	0.50623	412.64	437.96	2.0017	0.25101	412.12	437.22	1.9436
50	0.52284	420.55	446.70	2.0292	0.25948	420.08	446.03	1.9712
60	0.53941	428.63	455.60	2.0563	0.26791	428.20	454.99	1.9985
70	0.55595	436.86	464.66	2.0831	0.27631	436.47	464.10	2.0255
80	0.57247	445.26	473.88	2.1096	0.28468	444.89	473.36	2.0521
90	0.58896	453.82	483.26	2.1358	0.29302	453.47	482.78	2.0784
100	0.60544	462.53	492.81	2.1617	0.30135	462.21	492.35	2.1044
110	0.62190	471.41	502.50	2.1874	0.30967	471.11	502.07	2.1301
120	0.63835	480.44	512.36	2.2128	0.31797	480.16	511.95	2.1555
130	0.65479	489.63	522.37	2.2379	0.32626	489.36	521.98	2.1807
	150 kPa (−17.29 °C)				200 kPa (−10.22 °C)			
Sat.	0.13139	368.06	387.77	1.7372	0.10002	372.15	392.15	1.7320
−10	0.13602	373.44	393.84	1.7606	0.10013	372.31	392.34	1.7328
0	0.14222	380.85	402.19	1.7917	0.10501	379.91	400.91	1.7647
10	0.14828	388.36	410.60	1.8220	0.10974	387.55	409.50	1.7956
20	0.15424	395.98	419.11	1.8515	0.11436	395.27	418.15	1.8256
30	0.16011	403.71	427.73	1.8804	0.11889	403.10	426.87	1.8549
40	0.16592	411.59	436.47	1.9088	0.12335	411.04	435.71	1.8836
50	0.17168	419.60	445.35	1.9367	0.12776	419.11	444.66	1.9117
60	0.17740	427.76	454.37	1.9642	0.13213	427.31	453.74	1.9394
70	0.18308	436.06	463.53	1.9913	0.13646	435.65	462.95	1.9666
80	0.18874	444.52	472.83	2.0180	0.14076	444.14	472.30	1.9935
90	0.19437	453.13	482.28	2.0444	0.14504	452.78	481.79	2.0200
100	0.19999	461.89	491.89	2.0705	0.14930	461.56	491.42	2.0461
110	0.20559	470.80	501.64	2.0963	0.15355	470.50	501.21	2.0720
120	0.21117	479.87	511.54	2.1218	0.15777	479.58	511.13	2.0976
130	0.21675	489.08	521.60	2.1470	0.16199	488.81	521.21	2.1229
140	0.22231	498.45	531.80	2.1720	0.16620	498.19	531.43	2.1479

TABLE B.5.2 (*continued*)
Superheated R-134a

Temp. (°C)	v (m³/kg)	u (kJ/kg)	h (kJ/kg)	s (kJ/kg·K)	v (m³/kg)	u (kJ/kg)	h (kJ/kg)	s (kJ/kg·K)
	300 kPa (0.56 °C)				400 kPa (8.84 °C)			
Sat.	0.06787	378.33	398.69	1.7259	0.05136	383.02	403.56	1.7223
10	0.07111	385.84	407.17	1.7564	0.05168	383.98	404.65	1.7261
20	0.07441	393.80	416.12	1.7874	0.05436	392.22	413.97	1.7584
30	0.07762	401.81	425.10	1.8175	0.05693	400.45	423.22	1.7895
40	0.08075	409.90	434.12	1.8468	0.05940	408.70	432.46	1.8195
50	0.08382	418.09	443.23	1.8755	0.06181	417.03	441.75	1.8487
60	0.08684	426.39	452.44	1.9035	0.06417	425.44	451.10	1.8772
70	0.08982	434.82	461.76	1.9311	0.06648	433.95	460.55	1.9051
80	0.09277	443.37	471.21	1.9582	0.06877	442.58	470.09	1.9325
90	0.09570	452.07	480.78	1.9850	0.07102	451.34	479.75	1.9595
100	0.09861	460.90	490.48	2.0113	0.07325	460.22	489.52	1.9860
110	0.10150	469.87	500.32	2.0373	0.07547	469.24	499.43	2.0122
120	0.10437	478.99	510.30	2.0631	0.07767	478.40	509.46	2.0381
130	0.10723	488.26	520.43	2.0885	0.07985	487.69	519.63	2.0636
140	0.11008	497.66	530.69	2.1136	0.08202	497.13	529.94	2.0889
150	0.11292	507.22	541.09	2.1385	0.08418	506.71	540.38	2.1139
160	0.11575	516.91	551.64	2.1631	0.08634	516.43	550.97	2.1386
	500 kPa (15.66 °C)				600 kPa (21.52 °C)			
Sat.	0.04126	386.82	407.45	1.7198	0.03442	390.01	410.66	1.7179
20	0.04226	390.52	411.65	1.7342	—	—	—	—
30	0.04446	398.99	421.22	1.7663	0.03609	397.44	419.09	1.7461
40	0.04656	407.44	430.72	1.7971	0.03796	406.11	428.88	1.7779
50	0.04858	415.91	440.20	1.8270	0.03974	414.75	438.59	1.8084
60	0.05055	424.44	449.72	1.8560	0.04145	423.41	448.28	1.8379
70	0.05247	433.06	459.29	1.8843	0.04311	432.13	457.99	1.8666
80	0.05435	441.77	468.94	1.9120	0.04473	440.93	467.76	1.8947
90	0.05620	450.59	478.69	1.9392	0.04632	449.82	477.61	1.9222
100	0.05804	459.53	488.55	1.9660	0.04788	458.82	487.55	1.9492
110	0.05985	468.60	498.52	1.9924	0.04943	467.94	497.59	1.9758
120	0.06164	477.79	508.61	2.0184	0.05095	477.18	507.75	2.0019
130	0.06342	487.13	518.83	2.0440	0.05246	486.55	518.03	2.0277
140	0.06518	496.59	529.19	2.0694	0.05396	496.05	528.43	2.0532
150	0.06694	506.20	539.67	2.0945	0.05544	505.69	538.95	2.0784
160	0.06869	515.95	550.29	2.1193	0.05692	515.46	549.61	2.1033
170	0.07043	525.83	561.04	2.1438	0.05839	525.36	560.40	2.1279

TABLE B.5.2 (*continued*)
Superheated R-134a

Temp. (°C)	v (m³/kg)	u (kJ/kg)	h (kJ/kg)	s (kJ/kg·K)	v (m³/kg)	u (kJ/kg)	h (kJ/kg)	s (kJ/kg·K)
	800 kPa (31.30 °C)				1000 kPa (39.37 °C)			
Sat.	0.02571	395.15	415.72	1.7150	0.02038	399.16	419.54	1.7125
40	0.02711	403.17	424.86	1.7446	0.02047	399.78	420.25	1.7148
50	0.02861	412.23	435.11	1.7768	0.02185	409.39	431.24	1.7494
60	0.03002	421.20	445.22	1.8076	0.02311	418.78	441.89	1.7818
70	0.03137	430.17	455.27	1.8373	0.02429	428.05	452.34	1.8127
80	0.03268	439.17	465.31	1.8662	0.02542	437.29	462.70	1.8425
90	0.03394	448.22	475.38	1.8943	0.02650	446.53	473.03	1.8713
100	0.03518	457.35	485.50	1.9218	0.02754	455.82	483.36	1.8994
110	0.03639	466.58	495.70	1.9487	0.02856	465.18	493.74	1.9268
120	0.03758	475.92	505.99	1.9753	0.02956	474.62	504.17	1.9537
130	0.03876	485.37	516.38	2.0014	0.03053	484.16	514.69	1.9801
140	0.03992	494.94	526.88	2.0271	0.03150	493.81	525.30	2.0061
150	0.04107	504.64	537.50	2.0525	0.03244	503.57	536.02	2.0318
160	0.04221	514.46	548.23	2.0775	0.03338	513.46	546.84	2.0570
170	0.04334	524.42	559.09	2.1023	0.03431	523.46	557.77	2.0820
180	0.04446	534.51	570.08	2.1268	0.03523	533.60	568.83	2.1067
	1200 kPa (46.31 °C)				1400 kPa (52.42 °C)			
Sat.	0.01676	402.37	422.49	1.7102	0.01414	404.98	424.78	1.7077
50	0.01724	406.15	426.84	1.7237	—	—	—	—
60	0.01844	416.08	438.21	1.7584	0.01503	413.03	434.08	1.7360
70	0.01953	425.74	449.18	1.7908	0.01608	423.20	445.72	1.7704
80	0.02055	435.27	459.92	1.8217	0.01704	433.09	456.94	1.8026
90	0.02151	444.74	470.55	1.8514	0.01793	442.83	467.93	1.8333
100	0.02244	454.20	481.13	1.8801	0.01878	452.50	478.79	1.8628
110	0.02333	463.71	491.70	1.9081	0.01958	462.17	489.59	1.8914
120	0.02420	473.27	502.31	1.9354	0.02036	471.87	500.38	1.9192
130	0.02504	482.91	512.97	1.9621	0.02112	481.63	511.19	1.9463
140	0.02587	492.65	523.70	1.9884	0.02186	491.46	522.05	1.9730
150	0.02669	502.48	534.51	2.0143	0.02258	501.37	532.98	1.9991
160	0.02750	512.43	545.43	2.0398	0.02329	511.39	543.99	2.0248
170	0.02829	522.50	556.44	2.0649	0.02399	521.51	555.10	2.0502
180	0.02907	532.68	567.57	2.0898	0.02468	531.75	566.30	2.0752

TABLE B.5.2 (*continued*)
Superheated R-134a

Temp. (°C)	v (m³/kg)	u (kJ/kg)	h (kJ/kg)	s (kJ/kg·K)	v (m³/kg)	u (kJ/kg)	h (kJ/kg)	s (kJ/kg·K)
	1600 kPa (57.90 °C)				2000 kPa (67.48 °C)			
Sat.	0.01215	407.11	426.54	1.7051	0.00930	410.15	428.75	1.6991
60	0.01239	409.49	429.32	1.7135	—	—	—	—
70	0.01345	420.37	441.89	1.7507	0.00958	413.37	432.53	1.7101
80	0.01438	430.72	453.72	1.7847	0.01055	425.20	446.30	1.7497
90	0.01522	440.79	465.15	1.8166	0.01137	436.20	458.95	1.7850
100	0.01601	450.71	476.33	1.8469	0.01211	446.78	471.00	1.8177
110	0.01676	460.57	487.39	1.8762	0.01279	457.12	482.69	1.8487
120	0.01748	470.42	498.39	1.9045	0.01342	467.34	494.19	1.8783
130	0.01817	480.30	509.37	1.9321	0.01403	477.51	505.57	1.9069
140	0.01884	490.23	520.38	1.9591	0.01461	487.68	516.90	1.9346
150	0.01949	500.24	531.43	1.9855	0.01517	497.89	528.22	1.9617
160	0.02013	510.33	542.54	2.0115	0.01571	508.15	539.57	1.9882
170	0.02076	520.52	553.73	2.0370	0.01624	518.48	550.96	2.0142
180	0.02138	530.81	565.02	2.0622	0.01676	528.89	562.42	2.0398
	3000 kPa (86.20 °C)				4000 kPa (100.33 °C)			
Sat.	0.00528	411.83	427.67	1.6759	0.00252	394.86	404.94	1.6036
90	0.00575	418.93	436.19	1.6995	—	—	—	—
100	0.00665	433.77	453.73	1.7472	—	—	—	—
110	0.00734	446.48	468.50	1.7862	0.00428	429.74	446.84	1.7148
120	0.00792	458.27	482.04	1.8211	0.00500	445.97	465.99	1.7642
130	0.00845	469.58	494.91	1.8535	0.00556	459.63	481.87	1.8040
140	0.00893	480.61	507.39	1.8840	0.00603	472.19	496.29	1.8394
150	0.00937	491.49	519.62	1.9133	0.00644	484.15	509.92	1.8720
160	0.00980	502.30	531.70	1.9415	0.00683	495.77	523.07	1.9027
170	0.01021	513.09	543.71	1.9689	0.00718	507.19	535.92	1.9320
180	0.01060	523.89	555.69	1.9956	0.00752	518.51	548.57	1.9603
	6000 kPa				10000 kPa			
90	0.001059	328.34	334.70	1.4081	0.000991	320.72	330.62	1.3856
100	0.001150	346.71	353.61	1.4595	0.001040	336.45	346.85	1.4297
110	0.001307	368.06	375.90	1.5184	0.001100	352.74	363.73	1.4744
120	0.001698	396.59	406.78	1.5979	0.001175	369.69	381.44	1.5200
130	0.002396	426.81	441.18	1.6843	0.001272	387.44	400.16	1.5670
140	0.002985	448.34	466.25	1.7458	0.001400	405.97	419.98	1.6155
150	0.003439	465.19	485.82	1.7926	0.001564	424.99	440.63	1.6649
160	0.003814	479.89	502.77	1.8322	0.001758	443.77	461.34	1.7133
170	0.004141	493.45	518.30	1.8676	0.001965	461.65	481.30	1.7589
180	0.004435	506.35	532.96	1.9004	0.002172	478.40	500.12	1.8009

TABLE B.6
Thermodynamic Properties of Nitrogen

TABLE B.6.1
Saturated Nitrogen

Temp. (K)	Press. (kPa)	Specific Volume, m³/kg			Internal Energy, kJ/kg		
		Sat. Liquid v_f	Evap. v_{fg}	Sat. Vapor v_g	Sat. Liquid u_f	Evap. u_{fg}	Sat. Vapor u_g
63.1	12.5	0.001150	1.48074	1.48189	−150.92	196.86	45.94
65	17.4	0.001160	1.09231	1.09347	−147.19	194.37	47.17
70	38.6	0.001191	0.52513	0.52632	−137.13	187.54	50.40
75	76.1	0.001223	0.28052	0.28174	−127.04	180.47	53.43
77.3	101.3	0.001240	0.21515	0.21639	−122.27	177.04	54.76
80	137.0	0.001259	0.16249	0.16375	−116.86	173.06	56.20
85	229.1	0.001299	0.10018	0.10148	−106.55	165.20	58.65
90	360.8	0.001343	0.06477	0.06611	−96.06	156.76	60.70
95	541.1	0.001393	0.04337	0.04476	−85.35	147.60	62.25
100	779.2	0.001452	0.02975	0.03120	−74.33	137.50	63.17
105	1084.6	0.001522	0.02066	0.02218	−62.89	126.18	63.29
110	1467.6	0.001610	0.01434	0.01595	−50.81	113.11	62.31
115	1939.3	0.001729	0.00971	0.01144	−37.66	97.36	59.70
120	2513.0	0.001915	0.00608	0.00799	−22.42	76.63	54.21
125	3208.0	0.002355	0.00254	0.00490	−0.83	40.73	39.90
126.2	3397.8	0.003194	0	0.00319	18.94	0	18.94

Temp. (K)	Press. (kPa)	Enthalpy, kJ/kg			Entropy, kJ/kg·K		
		Sat. Liquid h_f	Evap. h_{fg}	Sat. Vapor h_g	Sat. Liquid s_f	Evap. s_{fg}	Sat. Vapor s_g
63.1	12.5	−150.91	215.39	64.48	2.4234	3.4109	5.8343
65	17.4	−147.17	213.38	66.21	2.4816	3.2828	5.7645
70	38.6	−137.09	207.79	70.70	2.6307	2.9684	5.5991
75	76.1	−126.95	201.82	74.87	2.7700	2.6909	5.4609
77.3	101.3	−122.15	198.84	76.69	2.8326	2.5707	5.4033
80	137.0	−116.69	195.32	78.63	2.9014	2.4415	5.3429
85	229.1	−106.25	188.15	81.90	3.0266	2.2135	5.2401
90	360.8	−95.58	180.13	84.55	3.1466	2.0015	5.1480
95	541.1	−84.59	171.07	86.47	3.2627	1.8007	5.0634
100	779.2	−73.20	160.68	87.48	3.3761	1.6068	4.9829
105	1084.6	−61.24	148.59	87.35	3.4883	1.4151	4.9034
110	1467.6	−48.45	134.15	85.71	3.6017	1.2196	4.8213
115	1939.3	−34.31	116.19	81.88	3.7204	1.0104	4.7307
120	2513.0	−17.61	91.91	74.30	3.8536	0.7659	4.6195
125	3208.0	6.73	48.88	55.60	4.0399	0.3910	4.4309
126.2	3397.8	29.79	0	29.79	4.2193	0	4.2193

TABLE B.6.2
Superheated Nitrogen

Temp. (K)	v (m³/kg)	u (kJ/kg)	h (kJ/kg)	s (kJ/kg·K)	v (m³/kg)	u (kJ/kg)	h (kJ/kg)	s (kJ/kg·K)
	100 kPa (77.24 K)				200 kPa (83.62 K)			
Sat.	0.21903	54.70	76.61	5.4059	0.11520	58.01	81.05	5.2673
100	0.29103	72.84	101.94	5.6944	0.14252	71.73	100.24	5.4775
120	0.35208	87.94	123.15	5.8878	0.17397	87.14	121.93	5.6753
140	0.41253	102.95	144.20	6.0501	0.20476	102.33	143.28	5.8399
160	0.47263	117.91	165.17	6.1901	0.23519	117.40	164.44	5.9812
180	0.53254	132.83	186.09	6.3132	0.26542	132.41	185.49	6.1052
200	0.59231	147.74	206.97	6.4232	0.29551	147.37	206.48	6.2157
220	0.65199	162.63	227.83	6.5227	0.32552	162.31	227.41	6.3155
240	0.71161	177.51	248.67	6.6133	0.35546	177.23	248.32	6.4064
260	0.77118	192.39	269.51	6.6967	0.38535	192.14	269.21	6.4900
280	0.83072	207.26	290.33	6.7739	0.41520	207.04	290.08	6.5674
300	0.89023	222.14	311.16	6.8457	0.44503	221.93	310.94	6.6393
350	1.03891	259.35	363.24	7.0063	0.51952	259.18	363.09	6.8001
400	1.18752	296.66	415.41	7.1456	0.59392	296.52	415.31	6.9396
450	1.33607	334.16	467.77	7.2690	0.66827	334.04	467.70	7.0630
500	1.48458	371.95	520.41	7.3799	0.74258	371.85	520.37	7.1740
600	1.78154	448.79	626.94	7.5741	0.89114	448.71	626.94	7.3682
700	2.07845	527.74	735.58	7.7415	1.03965	527.68	735.61	7.5357
800	2.37532	609.07	846.60	7.8897	1.18812	609.02	846.64	7.6839
900	2.67217	692.79	960.01	8.0232	1.33657	692.75	960.07	7.8175
1000	2.96900	778.78	1075.68	8.1451	1.48501	778.74	1075.75	7.9393

TABLE B.6.2 *(continued)*
Superheated Nitrogen

Temp. (K)	v (m³/kg)	u (kJ/kg)	h (kJ/kg)	s (kJ/kg·K)	v (m³/kg)	u (kJ/kg)	h (kJ/kg)	s (kJ/kg·K)
	400 kPa (91.22 K)				600 kPa (96.37 K)			
Sat.	0.05992	61.13	85.10	5.1268	0.04046	62.57	86.85	5.0411
100	0.06806	69.30	96.52	5.2466	0.04299	66.41	92.20	5.0957
120	0.08486	85.48	119.42	5.4556	0.05510	83.73	116.79	5.3204
140	0.10085	101.06	141.40	5.6250	0.06620	99.75	139.47	5.4953
160	0.11647	116.38	162.96	5.7690	0.07689	115.34	161.47	5.6422
180	0.13186	131.55	184.30	5.8947	0.08734	130.69	183.10	5.7696
200	0.14712	146.64	205.49	6.0063	0.09766	145.91	204.50	5.8823
220	0.16228	161.68	226.59	6.1069	0.10788	161.04	225.76	5.9837
240	0.17738	176.67	247.62	6.1984	0.11803	176.11	246.92	6.0757
260	0.19243	191.64	268.61	6.2824	0.12813	191.13	268.01	6.1601
280	0.20745	206.58	289.56	6.3600	0.13820	206.13	289.05	6.2381
300	0.22244	221.52	310.50	6.4322	0.14824	221.11	310.06	6.3105
350	0.25982	258.85	362.78	6.5934	0.17326	258.52	362.48	6.4722
400	0.29712	296.25	415.10	6.7331	0.19819	295.97	414.89	6.6121
450	0.33437	333.81	467.56	6.8567	0.22308	333.57	467.42	6.7359
500	0.37159	371.65	520.28	6.9678	0.24792	371.45	520.20	6.8471
600	0.44595	448.55	626.93	7.1622	0.29755	448.40	626.93	7.0416
700	0.52025	527.55	735.65	7.3298	0.34712	527.43	735.70	7.2093
800	0.59453	608.92	846.73	7.4781	0.39666	608.82	846.82	7.3576
900	0.66878	692.67	960.19	7.6117	0.44618	692.59	960.30	7.4912
1000	0.74302	778.68	1075.89	7.7335	0.49568	778.61	1076.02	7.6131
	800 kPa (100.38 K)				1000 kPa (103.73 K)			
Sat.	0.03038	63.21	87.52	4.9768	0.02416	63.35	87.51	4.9237
120	0.04017	81.88	114.02	5.2191	0.03117	79.91	111.08	5.1357
140	0.04886	98.41	137.50	5.4002	0.03845	97.02	135.47	5.3239
160	0.05710	114.28	159.95	5.5501	0.04522	113.20	158.42	5.4772
180	0.06509	129.82	181.89	5.6793	0.05173	128.94	180.67	5.6082
200	0.07293	145.17	203.51	5.7933	0.05809	144.43	202.52	5.7234
220	0.08067	160.40	224.94	5.8954	0.06436	159.76	224.11	5.8263
240	0.08835	175.54	246.23	5.9880	0.07055	174.98	245.53	5.9194
260	0.09599	190.63	267.42	6.0728	0.07670	190.13	266.83	6.0047
280	0.10358	205.68	288.54	6.1511	0.08281	205.23	288.04	6.0833
300	0.11115	220.70	309.62	6.2238	0.08889	220.29	309.18	6.1562
350	0.12998	258.19	362.17	6.3858	0.10401	257.86	361.87	6.3187
400	0.14873	295.69	414.68	6.5260	0.11905	295.42	414.47	6.4591
500	0.18609	371.25	520.12	6.7613	0.14899	371.04	520.04	6.6947
600	0.22335	448.24	626.93	6.9560	0.17883	448.09	626.92	6.8895
700	0.26056	527.31	735.76	7.1237	0.20862	527.19	735.81	7.0573
800	0.29773	608.73	846.91	7.2721	0.23837	608.63	847.00	7.2057
900	0.33488	692.52	960.42	7.4058	0.26810	692.44	960.54	7.3394
1000	0.37202	778.55	1076.16	7.5277	0.29782	778.49	1076.30	7.4614

TABLE B.6.2 (*continued*)
Superheated Nitrogen

Temp. (K)	v (m³/kg)	u (kJ/kg)	h (kJ/kg)	s (kJ/kg·K)	v (m³/kg)	u (kJ/kg)	h (kJ/kg)	s (kJ/kg·K)
	1500 kPa (110.38 K)				2000 kPa (115.58 K)			
Sat.	0.01555	62.17	85.51	4.8148	0.01100	59.25	81.25	4.7193
120	0.01899	74.26	102.75	4.9650	0.01260	66.90	92.10	4.8116
140	0.02452	93.36	130.15	5.1767	0.01752	89.37	124.40	5.0618
160	0.02937	110.44	154.50	5.3394	0.02144	107.55	150.43	5.2358
180	0.03393	126.71	177.60	5.4755	0.02503	124.42	174.48	5.3775
200	0.03832	142.56	200.03	5.5937	0.02844	140.66	197.53	5.4989
220	0.04260	158.14	222.05	5.6987	0.03174	156.52	219.99	5.6060
240	0.04682	173.57	243.80	5.7933	0.03496	172.15	242.08	5.7021
260	0.05099	188.87	265.36	5.8796	0.03814	187.62	263.90	5.7894
280	0.05512	204.10	286.78	5.9590	0.04128	202.97	285.53	5.8696
300	0.05922	219.27	308.10	6.0325	0.04440	218.24	307.03	5.9438
350	0.06940	257.03	361.13	6.1960	0.05209	256.21	360.39	6.1083
400	0.07949	294.73	413.96	6.3371	0.05971	294.05	413.47	6.2500
450	0.08953	332.53	466.82	6.4616	0.06727	331.95	466.49	6.3750
500	0.09953	370.54	519.84	6.5733	0.07480	370.05	519.65	6.4870
600	0.11948	447.71	626.92	6.7685	0.08980	447.33	626.93	6.6825
700	0.13937	526.89	735.94	6.9365	0.10474	526.59	736.07	6.8507
800	0.15923	608.39	847.22	7.0851	0.11965	608.14	847.45	6.9994
900	0.17906	692.24	960.83	7.2189	0.13454	692.04	961.13	7.1333
1000	0.19889	778.32	1076.65	7.3409	0.14942	778.16	1077.01	7.2553
	3000 kPa (123.61 K)				10000 kPa			
Sat.	0.00582	46.03	63.47	4.5032	—	—	—	—
140	0.01038	79.98	111.13	4.8706	0.00200	0.84	20.87	4.0373
160	0.01350	101.35	141.85	5.0763	0.00291	47.44	76.52	4.4088
180	0.01614	119.68	168.09	5.2310	0.00402	82.44	122.65	4.6813
200	0.01857	136.78	192.49	5.3596	0.00501	108.21	158.35	4.8697
220	0.02088	153.24	215.88	5.4711	0.00590	129.86	188.88	5.0153
240	0.02312	169.30	238.66	5.5702	0.00672	149.42	216.64	5.1362
260	0.02531	185.10	261.02	5.6597	0.00749	167.77	242.72	5.2406
280	0.02746	200.72	283.09	5.7414	0.00824	185.34	267.69	5.3331
300	0.02958	216.21	304.94	5.8168	0.00895	202.38	291.90	5.4167
350	0.03480	254.57	358.96	5.9834	0.01067	243.57	350.26	5.5967
400	0.03993	292.70	412.50	6.1264	0.01232	283.59	406.79	5.7477
500	0.05008	369.06	519.29	6.3647	0.01551	362.42	517.48	5.9948
600	0.06013	446.57	626.95	6.5609	0.01861	441.47	627.58	6.1955
700	0.07012	525.99	736.35	6.7295	0.02167	521.96	738.65	6.3667
800	0.08008	607.67	847.92	6.8785	0.02470	604.42	851.43	6.5172
900	0.09003	691.65	961.73	7.0125	0.02771	689.02	966.15	6.6523
1000	0.09996	777.85	1077.72	7.1347	0.03072	775.68	1082.84	6.7753

TABLE B.7
Thermodynamic Properties of Methane

TABLE B.7.1
Saturated Methane

Temp. (K)	P (kPa)	Specific Volume, m³/kg			Internal Energy, kJ/kg		
		v_f	v_{fg}	v_g	u_f	u_{fg}	u_g
90.7	11.7	0.002215	3.97941	3.98163	−358.10	496.59	138.49
95	19.8	0.002243	2.44845	2.45069	−343.79	488.62	144.83
100	34.4	0.002278	1.47657	1.47885	−326.90	478.96	152.06
105	56.4	0.002315	0.93780	0.94012	−309.79	468.89	159.11
110	88.2	0.002353	0.62208	0.62443	−292.50	458.41	165.91
111.7	101.3	0.002367	0.54760	0.54997	−286.74	454.85	168.10
115	132.3	0.002395	0.42800	0.43040	−275.05	447.48	172.42
120	191.6	0.002439	0.30367	0.30610	−257.45	436.02	178.57
125	269.0	0.002486	0.22108	0.22357	−239.66	423.97	184.32
130	367.6	0.002537	0.16448	0.16701	−221.65	411.25	189.60
135	490.7	0.002592	0.12458	0.12717	−203.40	397.77	194.37
140	641.6	0.002653	0.09575	0.09841	−184.86	383.42	198.56
145	823.7	0.002719	0.07445	0.07717	−165.97	368.06	202.09
150	1040.5	0.002794	0.05839	0.06118	−146.65	351.53	204.88
155	1295.6	0.002877	0.04605	0.04892	−126.82	333.61	206.79
160	1592.8	0.002974	0.03638	0.03936	−106.35	314.01	207.66
165	1935.9	0.003086	0.02868	0.03177	−85.06	292.30	207.24
170	2329.3	0.003222	0.02241	0.02563	−62.67	267.81	205.14
175	2777.6	0.003393	0.01718	0.02058	−38.75	239.47	200.72
180	3286.4	0.003623	0.01266	0.01629	−12.43	205.16	192.73
185	3863.2	0.003977	0.00846	0.01243	18.47	159.49	177.96
190	4520.5	0.004968	0.00300	0.00797	69.10	67.01	136.11
190.6	4599.2	0.006148	0	0.00615	101.46	0	101.46

TABLE B.7.1 (*continued*)
Saturated Methane

Temp. (K)	P (kPa)	Enthalpy, kJ/kg			Entropy, kJ/kg·K		
		h_f	h_{fg}	h_g	s_f	s_{fg}	s_g
90.7	11.7	−358.07	543.12	185.05	4.2264	5.9891	10.2155
95	19.8	−343.75	537.18	193.43	4.3805	5.6545	10.0350
100	34.4	−326.83	529.77	202.94	4.5538	5.2977	9.8514
105	56.4	−309.66	521.82	212.16	4.7208	4.9697	9.6905
110	88.2	−292.29	513.29	221.00	4.8817	4.6663	9.5480
111.7	101.3	−286.50	510.33	223.83	4.9336	4.5706	9.5042
115	132.3	−274.74	504.12	229.38	5.0368	4.3836	9.4205
120	191.6	−256.98	494.20	237.23	5.1867	4.1184	9.3051
125	269.0	−238.99	483.44	244.45	5.3321	3.8675	9.1996
130	367.6	−220.72	471.72	251.00	5.4734	3.6286	9.1020
135	490.7	−202.13	458.90	256.77	5.6113	3.3993	9.0106
140	641.6	−183.16	444.85	261.69	5.7464	3.1775	8.9239
145	823.7	−163.73	429.38	265.66	5.8794	2.9613	8.8406
150	1040.5	−143.74	412.29	268.54	6.0108	2.7486	8.7594
155	1295.6	−123.09	393.27	270.18	6.1415	2.5372	8.6787
160	1592.8	−101.61	371.96	270.35	6.2724	2.3248	8.5971
165	1935.9	−79.08	347.82	268.74	6.4046	2.1080	8.5126
170	2329.3	−55.17	320.02	264.85	6.5399	1.8824	8.4224
175	2777.6	−29.33	287.20	257.87	6.6811	1.6411	8.3223
180	3286.4	−0.53	246.77	246.25	6.8333	1.3710	8.2043
185	3863.2	33.83	192.16	226.00	7.0095	1.0387	8.0483
190	4520.5	91.56	80.58	172.14	7.3015	0.4241	7.7256
190.6	4599.2	129.74	0	129.74	7.4999	0	7.4999

TABLE B.7.2
Superheated Methane

Temp. (K)	v (m³/kg)	u (kJ/kg)	h (kJ/kg)	s (kJ/kg·K)	v (m³/kg)	u (kJ/kg)	h (kJ/kg)	s (kJ/kg·K)
	100 kPa (111.50K)				200 kPa (120.61 K)			
Sat.	0.55665	167.90	223.56	9.5084	0.29422	179.30	238.14	9.2918
125	0.63126	190.21	253.33	9.7606	0.30695	186.80	248.19	9.3736
150	0.76586	230.18	306.77	10.1504	0.37700	227.91	303.31	9.7759
175	0.89840	269.72	359.56	10.4759	0.44486	268.05	357.02	10.1071
200	1.02994	309.20	412.19	10.7570	0.51165	307.88	410.21	10.3912
225	1.16092	348.90	464.99	11.0058	0.57786	347.81	463.38	10.6417
250	1.29154	389.12	518.27	11.2303	0.64370	388.19	516.93	10.8674
275	1.42193	430.17	572.36	11.4365	0.70931	429.36	571.22	11.0743
300	1.55215	472.36	627.58	11.6286	0.77475	471.65	626.60	11.2670
325	1.68225	516.00	684.23	11.8100	0.84008	515.37	683.38	11.4488
350	1.81226	561.34	742.57	11.9829	0.90530	560.77	741.83	11.6220
375	1.94220	608.58	802.80	12.1491	0.97046	608.07	802.16	11.7885
400	2.07209	657.89	865.10	12.3099	1.03557	657.41	864.53	11.9495
425	2.20193	709.36	929.55	12.4661	1.10062	708.92	929.05	12.1059

TABLE B.7.2 (*continued*)
Superheated Methane

Temp. (K)	v (m³/kg)	u (kJ/kg)	h (kJ/kg)	s (kJ/kg·K)	v (m³/kg)	u (kJ/kg)	h (kJ/kg)	s (kJ/kg·K)
	400 kPa (131.42 K)				600 kPa (138.72 K)			
Sat.	0.15427	191.01	252.72	9.0754	0.10496	197.54	260.51	8.9458
150	0.18233	223.16	296.09	9.3843	0.11717	218.08	288.38	9.1390
175	0.21799	264.61	351.81	9.7280	0.14227	261.03	346.39	9.4970
200	0.25246	305.19	406.18	10.0185	0.16603	302.44	402.06	9.7944
225	0.28631	345.61	460.13	10.2726	0.18911	343.37	456.84	10.0525
250	0.31978	386.32	514.23	10.5007	0.21180	384.44	511.52	10.2830
275	0.35301	427.74	568.94	10.7092	0.23424	426.11	566.66	10.4931
300	0.38606	470.23	624.65	10.9031	0.25650	468.80	622.69	10.6882
325	0.41899	514.10	681.69	11.0857	0.27863	512.82	680.00	10.8716
350	0.45183	559.63	740.36	11.2595	0.30067	558.48	738.88	11.0461
375	0.48460	607.03	800.87	11.4265	0.32264	605.99	799.57	11.2136
400	0.51731	656.47	863.39	11.5879	0.34456	655.52	862.25	11.3754
425	0.54997	708.05	928.04	11.7446	0.36643	707.18	927.04	11.5324
450	0.58260	761.85	994.89	11.8974	0.38826	761.05	994.00	11.6855
475	0.61520	817.89	1063.97	12.0468	0.41006	817.15	1063.18	11.8351
500	0.64778	876.18	1135.29	12.1931	0.43184	875.48	1134.59	11.9816
525	0.68033	936.67	1208.81	12.3366	0.45360	936.03	1208.18	12.1252
	800 kPa (144.40 K)				1000 kPa (149.13 K)			
Sat.	0.07941	201.70	265.23	8.8505	0.06367	204.45	268.12	8.7735
150	0.08434	212.53	280.00	8.9509	0.06434	206.28	270.62	8.7902
175	0.10433	257.30	340.76	9.3260	0.08149	253.38	334.87	9.1871
200	0.12278	299.62	397.85	9.6310	0.09681	296.73	393.53	9.5006
225	0.14050	341.10	453.50	9.8932	0.11132	338.79	450.11	9.7672
250	0.15781	382.53	508.78	10.1262	0.12541	380.61	506.01	10.0028
275	0.17485	424.47	564.35	10.3381	0.13922	422.82	562.04	10.2164
300	0.19172	467.36	620.73	10.5343	0.15285	465.91	618.76	10.4138
325	0.20845	511.55	678.31	10.7186	0.16635	510.26	676.61	10.5990
350	0.22510	557.33	737.41	10.8938	0.17976	556.18	735.94	10.7748
375	0.24167	604.95	798.28	11.0617	0.19309	603.91	797.00	10.9433
400	0.25818	654.57	861.12	11.2239	0.20636	653.62	859.98	11.1059
425	0.27465	706.31	926.03	11.3813	0.21959	705.44	925.03	11.2636
450	0.29109	760.24	993.11	11.5346	0.23279	759.44	992.23	11.4172
475	0.30749	816.40	1062.40	11.6845	0.24595	815.66	1061.61	11.5672
500	0.32387	874.79	1133.89	11.8311	0.25909	874.10	1133.19	11.7141
525	0.34023	935.38	1207.56	11.9749	0.27221	934.73	1206.95	11.8580
550	0.35657	998.14	1283.45	12.1161	0.28531	997.53	1282.84	11.9992

TABLE B.7.2 (*continued*)
Superheated Methane

Temp. (K)	v (m³/kg)	u (kJ/kg)	h (kJ/kg)	s (kJ/kg·K)	v (m³/kg)	u (kJ/kg)	h (kJ/kg)	s (kJ/kg·K)
		1500 kPa (158.52 K)				2000 kPa (165.86 K)		
Sat.	0.04196	207.53	270.47	8.6215	0.03062	207.01	268.25	8.4975
175	0.05078	242.64	318.81	8.9121	0.03504	229.90	299.97	8.6839
200	0.06209	289.13	382.26	9.2514	0.04463	280.91	370.17	9.0596
225	0.07239	332.85	441.44	9.5303	0.05289	326.64	432.43	9.3532
250	0.08220	375.70	499.00	9.7730	0.06059	370.67	491.84	9.6036
275	0.09171	418.65	556.21	9.9911	0.06796	414.40	550.31	9.8266
300	0.10103	462.27	613.82	10.1916	0.07513	458.59	608.85	10.0303
325	0.11022	507.04	672.37	10.3790	0.08216	503.80	668.12	10.2200
350	0.11931	553.30	732.26	10.5565	0.08909	550.40	728.58	10.3992
375	0.12832	601.30	793.78	10.7263	0.09594	598.69	790.57	10.5703
400	0.13728	651.24	857.16	10.8899	0.10274	648.87	854.34	10.7349
425	0.14619	703.26	922.54	11.0484	0.10949	701.08	920.06	10.8942
450	0.15506	757.43	990.02	11.2027	0.11620	755.43	987.84	11.0491
475	0.16391	813.80	1059.66	11.3532	0.12289	811.94	1057.72	11.2003
500	0.17273	872.37	1131.46	11.5005	0.12955	870.64	1129.74	11.3480
525	0.18152	933.12	1205.41	11.6448	0.13619	931.51	1203.88	11.4927
550	0.19031	996.02	1281.48	11.7864	0.14281	994.51	1280.13	11.6346
		4000 kPa (186.10 K)				8000 kPa		
Sat.	0.01160	172.96	219.34	8.0035	—	—	—	—
200	0.01763	237.70	308.23	8.4675	0.00412	55.58	88.54	7.2069
225	0.02347	298.52	392.39	8.8653	0.00846	217.30	284.98	8.1344
250	0.02814	349.08	461.63	9.1574	0.01198	298.05	393.92	8.5954
275	0.03235	396.67	526.07	9.4031	0.01469	357.88	475.39	8.9064
300	0.03631	443.48	588.73	9.6212	0.01705	411.71	548.15	9.1598
325	0.04011	490.62	651.07	9.8208	0.01924	463.52	617.40	9.3815
350	0.04381	538.70	713.93	10.0071	0.02130	515.02	685.39	9.5831
375	0.04742	588.18	777.86	10.1835	0.02328	567.12	753.34	9.7706
400	0.05097	639.34	843.24	10.3523	0.02520	620.38	821.95	9.9477
425	0.05448	692.38	910.31	10.5149	0.02707	675.14	891.71	10.1169
450	0.05795	747.43	979.23	10.6725	0.02891	731.63	962.92	10.2796
475	0.06139	804.55	1050.12	10.8258	0.03072	789.99	1035.75	10.4372
500	0.06481	863.78	1123.01	10.9753	0.03251	850.28	1110.34	10.5902
525	0.06820	925.11	1197.93	11.1215	0.03428	912.54	1186.74	10.7393
550	0.07158	988.53	1274.86	11.2646	0.03603	976.77	1264.99	10.8849
575	0.07495	1053.98	1353.77	11.4049	0.03776	1042.96	1345.07	11.0272

Ideal Gas Specific Heat

Three types of energy storage or possession were identified in Section 1.8, of which two, translation and intramolecular energy, are associated with the individual molecules. These comprise the ideal gas model, with the third type, the system intermolecular potential energy, then accounting for the behavior of real (nonideal gas) substances. This appendix deals with the ideal gas contributions. Since these contribute to the energy, and therefore also the enthalpy, they also contribute to the specific heat of each gas. The different possibilities can be grouped according to the intramolecular energy contributions as follows:

C.1 MONATOMIC GASES (INERT GASES AR, HE, NE, XE, KR; ALSO N, O, H, CL, F, ...)

$$h = h_{\text{translation}} + h_{\text{electronic}} = h_t + h_e$$

$$\frac{dh}{dT} = \frac{dh_t}{dT} + \frac{dh_e}{dT}, \qquad C_{p0} = C_{p0t} + C_{p0e} = \frac{5}{2}R + f_e(T)$$

where the electronic contribution, $f_e(T)$, is usually small, except at very high T (common exceptions are O, Cl, F).

C.2 DIATOMIC AND LINEAR POLYATOMIC GASES (N₂, O₂, CO, OH, ..., CO₂, N₂O, ...)

In addition to translational and electronic contributions to specific heat, these also have molecular rotation (about the center of mass of the molecule) and also $(3a - 5)$ independent modes of molecular vibration of the a atoms in the molecule relative to one another, such that

$$C_{p0} = C_{p0t} + C_{p0r} + C_{p0v} + C_{p0e} = \frac{5}{2}R + R + f_v(T) + f_e(T)$$

where the vibrational contribution is

$$f_v(T) = R \sum_{i=1}^{3a-5} [x_i^2 e^{x_i}/(e^{x_i} - 1)^2], \qquad x_i = \frac{\theta_i}{T}$$

and the electronic contribution, $f_e(T)$, is usually small, except at very high T (common exceptions are O₂, NO, OH).

Example C.1

N_2, $3a - 5 = 1$ vibrational mode, with $\theta_i = 3392$ K.

At $T = 300$ K, $C_{p0} = 0.742 + 0.2968 + 0.0005 + \approx 0 = 1.0393$ kJ/kg·K.

At $T = 1000$ K, $C_{p0} = 0.742 + 0.2968 + 0.123 + \approx 0 = 1.1618$ kJ/kg·K.

(an increase of 11.8% from 300 K).

Example C.2

CO_2, $3a - 5 = 4$ vibrational modes, with $\theta_i = 960$ K, 960 K, 1993 K, 3380 K

At $T = 300$ K, $C_{p0} = 0.4723 + 0.1889 + 0.1826 + \approx 0 = 0.8438$ kJ/kg·K.

At $T = 1000$ K, $C_{p0} = 0.4723 + 0.1889 + 0.5659 + \approx 0 = 1.2271$ kJ/kg·K.

(an increase of 45.4% from 300 K).

C.3 NONLINEAR POLYATOMIC MOLECULES (H_2O, NH_3, CH_4, C_2H_6, ...)

Contributions to specific heat are similar to those for linear molecules, except that the rotational contribution is larger and there are $(3a - 6)$ independent vibrational modes, such that

$$C_{p0} = C_{p0t} + C_{p0r} + C_{p0v} + C_{p0e} = \frac{5}{2} R + \frac{3}{2} R + f_v(T) + f_e(T)$$

where the vibrational contribution is

$$f_v(T) = R \sum_{i=1}^{3a-6} [x_i^2 e^{x_i}/(e^{x_i} - 1)^2], \qquad x_i = \frac{\theta_i}{T}$$

and $f_e(T)$ is usually small, except at very high temperatures.

Example C.3

CH_4, $3a - 6 = 9$ vibrational modes, with $\theta_i = 4196$ K, 2207 K (two modes), 1879 K (three), 4343 K (three)

At $T = 300$ K, $C_{p0} = 1.2958 + 0.7774 + 0.1527 + \approx 0 = 2.2259$ kJ/kg·K.

At $T = 1000$ K, $C_{p0} = 1.2958 + 0.7774 + 2.4022 + \approx 0 = 4.4754$ kJ/kg·K.

(an increase of 101.1% from 300 K).

Equations of State

APPENDIX

D

Some of the most used pressure-explicit equations of state can be shown in a form with two parameters. This form is known as a cubic equation of state and contains as a special case the ideal-gas law:

$$P = \frac{RT}{v-b} - \frac{a}{v^2 + cbv + db^2}$$

where (a, b) are parameters and (c, d) define the model as shown in the following table with the acentric factor (ω) and

$$b = b_0 R T_c / P_c \qquad \text{and} \qquad a = a_0 R^2 T_c^2 / P_c$$

The acentric factor is defined by the saturation pressure at a reduced temperature $T_r = 0.7$

$$\omega = -\frac{ln\ P_r^{\text{sat}}\ \text{at}\ T_r = 0.7}{ln\ 10} - 1$$

TABLE D.1
Equations of State

Model	c	d	b_0	a_0
Ideal gas	0	0	0	0
van der Waals	0	0	1/8	27/64
Redlich–Kwong	1	0	0.08664	$0.42748\ T_r^{-1/2}$
Soave	1	0	0.08664	$0.42748[1 + f(1 - T_r^{1/2})]^2$
Peng–Robinson	2	−1	0.0778	$0.45724[1 + f(1 - T_r^{1/2})]^2$

$$f = 0.48 + 1.574\omega - 0.176\omega^2 \qquad \text{for Soave}$$
$$f = 0.37464 + 1.54226\ \omega - 0.26992\omega^2 \qquad \text{for Peng–Robinson}$$

TABLE D.2
The Lee–Kesler Equation of State

The Lee–Kesler generalized equation of state is

$$Z = \frac{P_r v_r'}{T_r} = 1 + \frac{B}{v_r'} + \frac{C}{v_r'^2} + \frac{D}{v_r'^5} + \frac{c_4}{T_r^3 v_r'^2}\left(\beta + \frac{\gamma}{v_r'^2}\right)\exp\left(-\frac{\gamma}{v_r'^2}\right)$$

$$B = b_1 - \frac{b_2}{T_r} - \frac{b_3}{T_r^2} - \frac{b_4}{T_r^3}$$

$$C = c_1 - \frac{c_2}{T_r} + \frac{c_3}{T_r^3}$$

$$D = d_1 + \frac{d_2}{T_r}$$

in which

$$T_r = \frac{T}{T_c}, \quad P_r = \frac{P}{P_c}, \quad v_r' = \frac{v}{RT_c/P_c}$$

The set of constants is as follows:

Constant	Simple Fluids	Constant	Simple Fluids
b_1	0.118 119 3	c_3	0.0
b_2	0.265 728	c_4	0.042 724
b_3	0.154 790	$d_1 \times 10^4$	0.155 488
b_4	0.030 323	$d_2 \times 10^4$	0.623 689
c_1	0.023 674 4	β	0.653 92
c_2	0.018 698 4	γ	0.060 167

TABLE D.3
Saturated Liquid–Vapor Compressibilities, Lee–Kesler Simple Fluid

T_r	0.40	0.50	0.60	0.70	0.80	0.85	0.90	0.95	1
P_r sat	2.7E-4	4.6E-3	0.028	0.099	0.252	0.373	0.532	0.737	1
Z_f	6.5E-5	9.5E-4	0.0052	0.017	0.042	0.062	0.090	0.132	0.29
Z_g	0.999	0.988	0.957	0.897	0.807	0.747	0.673	0.569	0.29

TABLE D.4
Acentric Factor for Some Substances

Substance		ω	Substance		ω
Ammonia	NH_3	0.25	Water	H_2O	0.344
Argon	Ar	0.001	*n*-Butane	C_4H_{10}	0.199
Bromine	Br_2	0.108	Ethane	C_2H_6	0.099
Helium	He	−0.365	Methane	CH_4	0.011
Neon	Ne	−0.029	R-32	CF_2H_2	0.277
Nitrogen	N_2	0.039	R-125	CHF_2CF_3	0.305

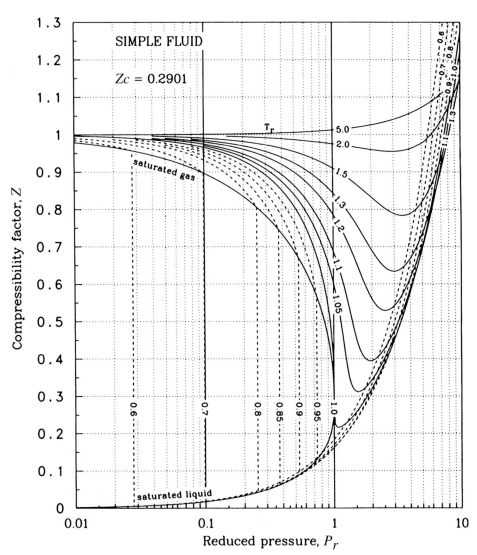

FIGURE D.1 Lee–Kesler simple fluid compressibility factor.

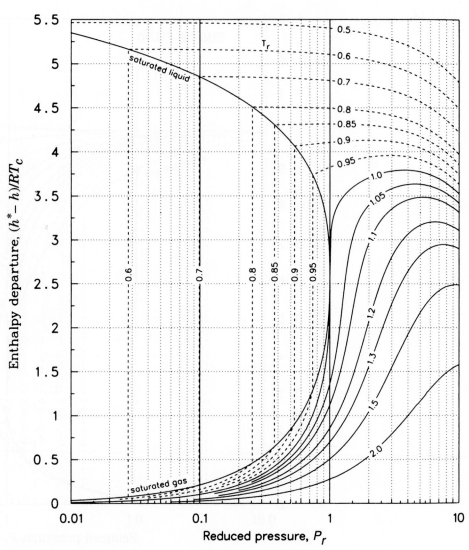

FIGURE D.2 Lee–Kesler simple fluid enthalpy departure.

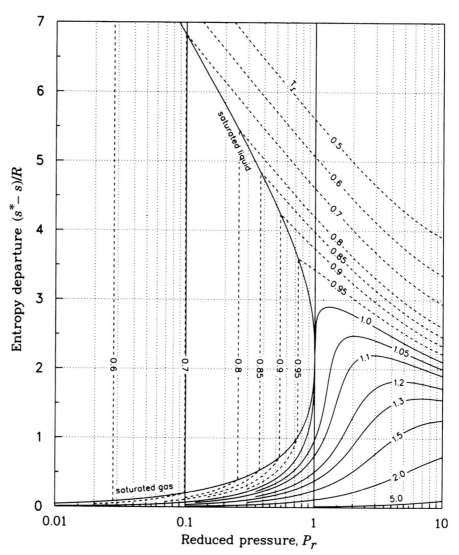

FIGURE D.3 Lee–Kesler simple fluid entropy departure.

Figures

APPENDIX

E

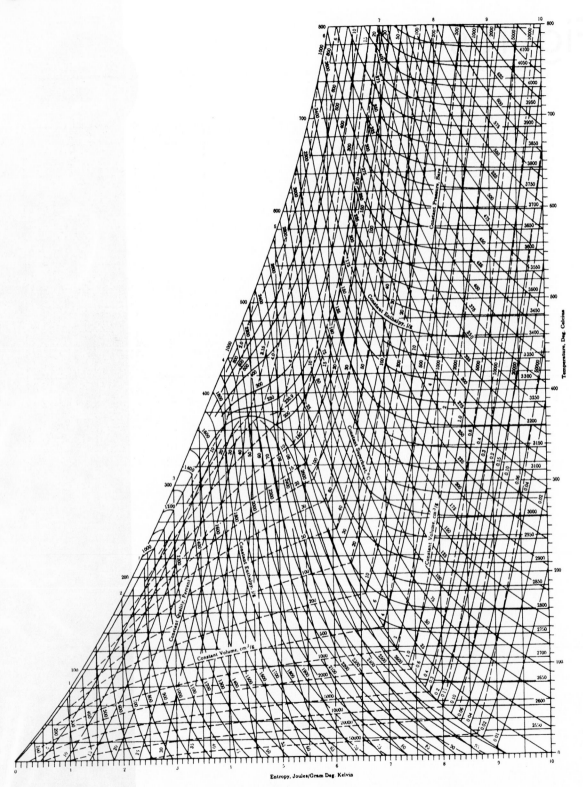

FIGURE E.1 Temperature–entropy diagram for water.
Keenan, Keyes, Hill, & Moore. STEAM TABLES (International Edition–Metric Units). Copyright © 1969, John Wiley & Sons, Inc.

ENTHALPY   (Btu/lb)

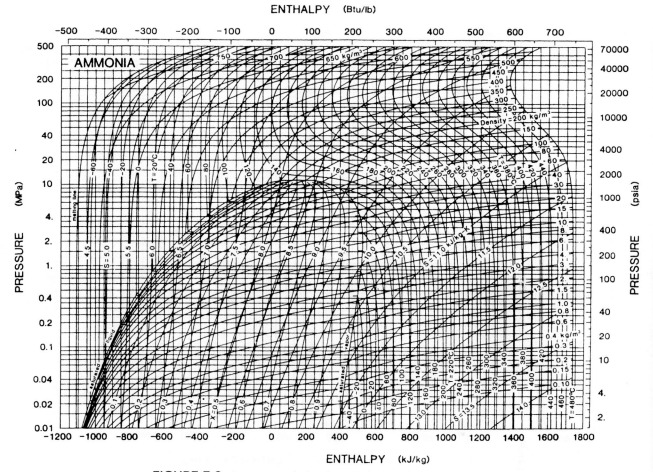

FIGURE E.2   Pressure–enthalpy diagram for ammonia.

ENTHALPY (Btu/lb)

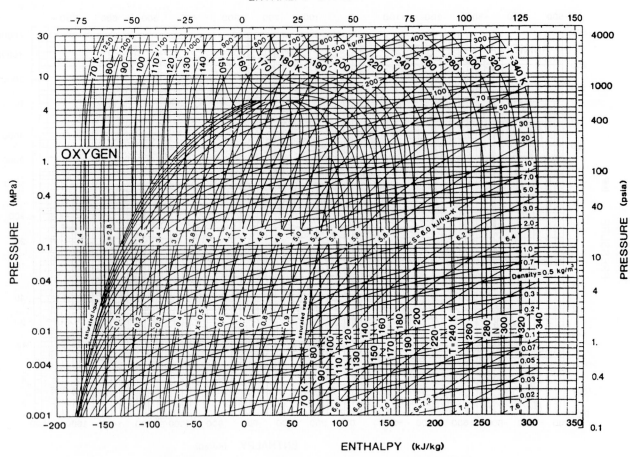

FIGURE E.3 Pressure–enthalpy diagram for oxygen.

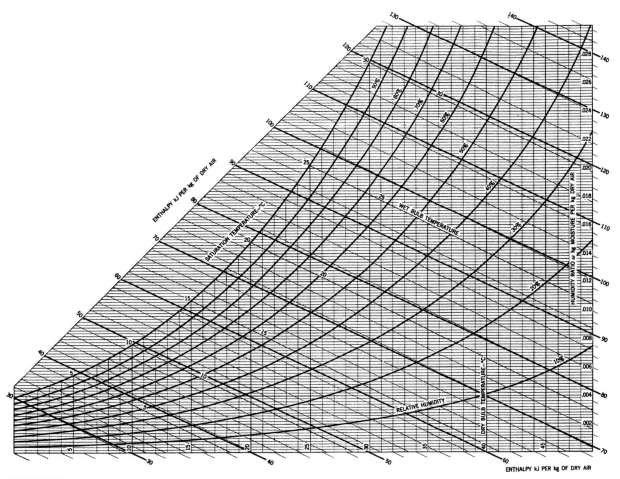

FIGURE E.4 Psychrometric chart.

Answers to Selected Problems

1.23	0.406 kmol	2.64	96.4 kPa
1.28	6000 N, 3.8 s	2.66	506.9 kPa
1.31	272 N	2.68	6.8, 20, 53%
1.33	11×10^6 kg	2.70	a. 8.711 kg, 38.715 kg
1.36	16 m³/kmol		b. 3.33 MPa
1.38	700 N	2.77	1.04 MPa
1.41	1752 kg	2.80	0.022, 0.0189
1.49	$P = P_0 + (H - h)\rho g$	2.82	0.304 m³/kg
1.52	26.3 kPa	2.85	10 357 kPa, 10 000 kPa
1.54	1346 kPa	2.88	0.003 33 m³/kg
1.56	0.12 kPa	2.91	10 820 kPa, 8040 kPa, 8000 kPa
1.57	295 m	2.97	0.283
1.63	106.4 kPa	2.98	a. 0.0156 m³
1.66	8.33 kg		b. 0.000 726 m³
1.75	0.005 m	2.101	641 °C
1.77	23.94 kPa	2.104	10%, 1.1%
1.80	6.0 MPa	2.105	256.7 kPa, -31.3 °C
		2.107	84.5 kPa
2.18	9123 kPa, -1 °C	2.109	x = undef, 1200 kPa, 2033 kPa,
2.23	All super heated vapor		0.03257 m³/kg
2.27	0.9893		
2.36	0.000 969 m³/kg, 0.0296 m³/kg	3.28	31 kJ
2.38	35.7 kg	3.30	5311 m
2.41	0.05 m, 120.2 °C	3.32	31 kJ, 0.0386 m³
2.43	(190, 1782) kPa,	3.35	5.1 m
	(0.622, 0.0707) m³/kg	3.37	500 N, 300 N, 25 J
2.46	0.5746 m³/kg	3.40	50 N, 2.5 J
2.48	rise, fall	3.43	233.9 °C, 161.7 kJ
2.50	1.893 kg	3.46	-128.7 kJ
2.52	5 hours, 824 kPa	3.50	7.424 kg, 401.2 kJ
2.55	212 °C, more	3.57	1000 K or 725 °C
2.59	argon $= 0.2081$ kJ/kg·K; hydrogen	3.66	a. -0.03, 0.47 kJ/kg
	$= 4.124\ 256$ kJ/kg·K		b. -0.12, 1.88 kJ/kg
2.61	Y, Y, N, N, Y		

3.73   −18.5 kJ, −213.9 kJ
3.78   995 kJ
3.84   569 kPa
3.98   80 °C
3.102   392, 612, 604 all kJ/kg
3.105   1238, 1764 both kJ/kg
3.108   36, 45 both kJ/kg
3.110   51 °C, 81 kPa
3.115   2.323 kg, 3.484 kg, 736 K, 613 kPa
3.118   70.64 kJ, −17.6 kJ
3.123   40.5 kJ
3.126 1.17×10^{-5} kg, 10 cm^3, 2.3 J, 0.91 J, 13.6 m/s
3.129   −7907 kJ
3.140   9.81 s
3.143   15 h
3.146   0.53 °C/min
3.149   −0.17 K/s
3.152   −8.92 μJ
3.156   12.6 J
3.159   1000 kPa, 218 kJ, 744 kJ
3.162   360 kPa, −92.6 kJ
3.166   143.6 °C, 0.4625 m^3, 145 kJ
3.180   14.7 kg, 120 kJ, 2988 kJ

4.12   0.84 m/s, 0.0126 m^3/s
4.17   6.37 m/s
4.19   360 K, 306 K, 330 K
4.22   890 K
4.26   22.9 °C, 144 kPa
4.30   Vapor = 5.86 %
       Liquid = 94.1 %
4.40   −9.9 kW
4.41   −46.3 kJ/kg
4.44   −317 kJ/kg, 307 kJ/kg
4.50   15 kW
4.53   3235 kJ/kg
4.55   1.35 kW, 1.62 kW
4.61   0.0079 kg/s
4.64   31 m/s, 20.12 °C
4.65   1.57 kg/s, 196 kW
4.76   1.8 kg/s
4.77   367 K

4.78   0.258 kg/s, 4.2 m^3/s
4.82   120 °C, 3 m^3/s
4.83   2.069 kg/s
4.86   193 kW
4.88   1357 K
4.94   13.75 MW, 67 MW
4.97   0.35 kW, 11.7 kW, 7.3 kW
4.100   T$_2$ > 20 °C, No
4.103   −379,636 kJ
4.106   13.3 kJ
4.108   41 MJ
4.110   8.9 kg, 25.5 MJ
4.114   2.66 m^3/s, 39.5 MW
4.119   126 °C, −2.62 MW
4.121   8405 kJ, 225 MJ
4.122   238 MJ, 203 MJ

5.15   5.25 kJ, 2
5.17   0.225
5.19   2.91
5.23   2.33
5.25   1.53 g/s, 42.9 kW
5.32   26.81 kJ, 35.75 s
5.36   1st: Y, Y, Y; 2nd: Y, N, N
5.39   45%
5.45   7.8
5.46   100 MJ
5.50   2.56 kW
5.54   98 W
5.56   62 kJ, 9.85 kJ
5.62   impossible
5.68   5.1%, 3.8%
5.74   6 kW, 0.31 kg/s
5.75   38.8 °C
5.77   0.00267 K/s
5.79   3.43
5.81   3.33, 49.7 kJ/kg
5.83   335 kJ, 60 kJ
5.87   10.9 kW
5.92   153 kJ

6.18   a) n.a.   b) OK   c) $\dot{w}$ = 2.53 kW
6.22   4.8116 kJ/kg·K
       4.2422 kJ/kg·K
       0.7619 kJ/kg·K

6.28	3.9391 kJ/kg·K
6.30	Δu, Δs = (23.2, 0.776) (26, 1.1) (28.3, 1.85)
6.32	61 °C, −48.9 kJ/kg
6.36	77.2 °C, −46.84 kJ/kg
6.43	2000 kPa, 471.2 kJ
6.47	0.385 m³
6.51	137.5 °C, 0.1794 kJ/K
6.56	334.6 kJ/kg, 1 kJ/kg·K, same
6.61	81.95 MJ
6.66	772 K, −267 kJ/kg 400 K, −264 kJ/kg
6.69	2.78, 2.725, 2.335 kJ/kg·K
6.75	450 K, −112.5 kJ/kg 460 K, −110.7 kJ/kg
6.78	143 K, −624 kJ/kg
6.80	360 K, 0.5571 kJ/K
6.83	Negative
6.91	1.8 kJ, −0.96 kJ
6.96	312 °C, 0.225 kJ/K
6.97	191.7 MJ, 654 kJ/K
6.105	Yes, it is possible
6.111	0.202 kJ/K
6.112	−58 kJ, −519 kJ, 0.022 kJ/K
6.119	133 kPa, 300 K, 0.034 kJ/K
6.125	200 kPa, 428 K, 0.0068 m³, 0.173 J/K
6.128	300 kPa, 400 K, 0.52 kJ/K
6.129	0.365 kJ/K
6.130	1.303, 0.0218 m³, −21.3 kJ, −5.1 kJ, 0.0036 kJ/K
6.135	0.1 kW/K, 0.1 kW/K
6.138	0.555, 0.309, 0.994 W/K
6.141	0.0782 kJ/K
6.143	26.3 kJ/K
6.147	442 °C, 1.72 kJ/K
6.151	3.33 kJ, 30.43 kJ, 9 kJ
7.15	629.4 m/s
7.18	358 kPa, 1.78 × 10⁻⁴ m²
7.25	−2.74 kW (i.e., out)
7.28	706 K, 558 kJ/kg, 662 K, 540 kJ/kg
7.29	69.3 kW, 69.3 kW

7.35	1071.8 K
7.44	0.4249 kg, 29.57 kJ
7.52	0.2 m
7.58	42.4 m/s
7.63	18.44 MPa, −849 kJ/kg, −104 kJ/kg
7.66	1612 kPa, 1977 K, 200 MPa, 1977 K
7.67	No
7.71	0.017 kJ/kg·K
7.75	764 kW, 0.624 kW/K
7.77	47.3 kg/min, 8.9 kJ/min·K
7.78	0, 187.1 kJ/kg, 0.163 kJ/kg·K
7.81	327 K, 0.036 kW/K
7.86	120.2 °C, 1.54 kW/K
7.88	No
7.93	443 K, 0.023 kW/K
7.98	0.0371 kg, 4.864 kJ, 0.000 484 kJ/K
7.103	331 K, 0.0107 kJ/K
7.106	495 °C, 0
7.109	533 m/s
7.110	50 kW
7.114	69.53 kJ/kg
7.120	587 kPa
7.123	411 kPa, 758 K
7.126	17.3 m/s, 0.8 kg/s
7.128	129 kPa, 313 K
7.131	281 °C, 0.724 kW/K
7.132	141.5 kJ/kg in, 431 K, 532 m/s
7.134	Yes
7.135	108 kW, 103 kW
7.138	2.675 kg, 450 kJ, 1276 kJ, 106 kPa
8.18	5.25 kW, 10.16 kW
8.23	1484 kJ/kg, 1637 kJ/kg
8.24	−38.9 kJ/kg
8.30	8.56 kg, 1592 kJ
8.33	550 W
8.35	197 kJ/kg, 197 kJ/kg
8.39	621 K, −333 kJ/kg
8.42	302.2 K, 18,191 kJ
8.46	46.3 °C, 19.8 kJ/kg
8.47	5.02 kg, 747 kJ

8.52	59.6 kJ/kg, 59.6 kJ/kg		9.76	4386 kW
8.54	−216 kJ/kg		9.80	5.06, 5.43
8.56	2.46 kJ/kg		9.83	2.24, 223 W
8.59	153.3 kJ/kg, 48 kJ/kg		9.85	11.3 kW, 0.0094 kW/K
8.62	1788, 219, 1.5, 21.6 all kJ/kg		9.87	1.83, 1.44
8.66	300.6 K, −44 kJ		9.90	It is the same
8.68	64.6 kJ, 1286 kJ		9.93	0.9
8.70	1500 W		9.110	11.39, 0.529
8.73	0.55 kW		9.114	Overall cycle OK, turbine impossible
8.74	62 W			
8.78	Destr.: 43.3 kW (inside), 14.1 kW (wall), 20.8 kW (radiator)		10.20	975 kJ/kg, 525 kJ/kg
8.85	0.31		10.22	11.314, 1350 K
8.87	0.659, 0.663		10.26	1597 K, 26.7 kg/s
8.90	0.835, 0.884		10.30	0.565
8.92	0.315, 0.672		10.31	130 kJ/kg, 318 kJ/kg
8.95	0.9		10.35	166 MW, 0.4, 0.582
8.96	0.51		10.38	214 MW, 0.533, 0.386
8.98	0.61		10.43	1012 m/s
8.102	263 kJ, 112 kJ, 164.6 kJ		10.45	340.7 kPa
8.109	4.67 m/s		10.48	1157 K, 504 kPa, 750 K, 904 m/s
8.111	14.9 W, 32.8 W, 50 W			
			10.51	824 K, 602 m/s
9.16	0.326, 0.54		10.54	2.71, 219 K
9.20	0.102		10.57	0.57
9.31	529 °C, 6.49 MW, 16.48 MW		10.60	0.6, 21.6 kW
9.34	0.1427		10.66	2677 K, 1458 kJ/kg, 1165 K
9.35	0.0434		10.69	7.67, −262 kJ/kg, 4883 kPa
9.39	0.1046, 34 kW		10.73	9.93, 819 kPa
9.40	3 kg/s, 1836 kg/s		10.74	52 kW
9.45	0.1913, 5.04 kJ/kg, 4.5 kJ/kg		10.79	2289 K
			10.80	19.32, 0.619
9.46	0.191, 4903 kW		10.84	121 kW, 162 hp
9.52	0.271, 0.256		10.86	20.2, 0.553
9.53	3.8, 2609, 719, 1893 all kJ/kg, 0.274		10.91	−1154, 2773, 4466, −2773 all kJ/kg, 0.458
9.54	15.2 kW		10.94	900 K, 430 kJ/kg, 15.6
9.58	3.02 kJ/kg·K		10.97	19.4
9.64	40.3 °C, 29.2 MW, 11.6 MW		10.99	3127 K, 6958 kPa, 0.654, 428 kPa
9.68	9102 kW			
9.70	a. 771 kPa, 201.7 kPa b. 1885 kPa, 573 kPa		10.102	13.5
9.73	45.9 °C, 22 °C, 6.2		10.105	0.79 kg/s, 51 kW

10.108   58.3 kg/s, 6.259 kg/s, 0.634

10.111   1

10.122   514 K, 565 K, 0.93, 0.405

10.126   1540.5 K, 548 kJ/kg

11.15 $0.18 \text{ m}^3/\text{s}$, $0.68 \text{ m}^3/\text{s}$

11.16   0.543, 0.209, 0.248, 0.322 kJ/kg·K, 5.065 m^3

11.17   0.2835 kJ/kg·K, 30.3 kPa

11.27   1103.7 K, 150 kPa

11.29   320.8 K

11.30   1096 kW

11.32   1247 kW

11.34   1400.6 K, 1.6367 kJ/kg·K

11.37 $-0.149 \text{ m}^3/\text{kg}$, 88.7 kJ/kg, 0.154 kJ/kg·K

11.41   573 K, 90 kW

11.44   5.292 kJ/kmol·K

11.51   Yes

11.53   616 K, -0.339 kW/K

11.56   698 kPa, 3748 kJ, 5.3 kJ/K

11.58   39%, 15.2 kW

11.64   0.0061 kg/s

11.66   28 °C, -2.77 kJ

11.73   0.0189, 0.0108, 46 kJ/kg

11.77   94%

11.80   0.015, 36.2 kg/s, 36.5 °C

11.85   21.4 °C

11.87   17.3 °C, 0.0044, -39 kJ/kg

11.90   4.07, 0.206, 49.3 °C, 15%

11.93   (16.8, 12, 10.9, 6.5) °C

11.96   3.77, 6.43 kJ/kg

11.101   0.06 kg/min, 0.0162 kg/min, 32.5 °C, 12%

11.104   55 kW, 38 kW

11.106   141 kPa

11.109 -880, 476 kJ/kg

11.112   1089 K, 1164 K

11.114   361 K, -2.4 kJ

12.21   151 kW out

12.24   11 kPa, 2.2 m^3

12.27 2.2×10^{-3} Pa

12.30   40.5 MPa

12.36   0

12.48   2.44 kJ

12.51   1166 m/s

12.54   1415 m/s, 506 m/s

12.57   1100 m/s, -66.7 J/kg

12.60   0.27

12.63 $u\text{-}u^* = -6.4$ kJ/kg

12.66   0.022 vs 0.0148 kJ/kg·K

12.72   2.45

12.75 $3.375 \, T_c$, $2.9 \, T_c$

12.78 $0.125 \, (1 - 27 \, T_c/8 \, T) \, RT_c/P_c$, $-0.297 \, RT_c/P_c$

12.81   208 K, 0.987 kJ/kg·K

12.88   173 kg

12.92 $0.606 \, RT_c$

12.93 -0.47 R

12.96   0.998, 125 kJ

12.98   1.06 MPa, 0.0024 kg, 0.753 kJ

12.99   3391 kJ

12.108   66.8 kJ/kg, 11 kJ/kg

12.115   296.5 kJ/kg

12.117   8.58

12.121 0.044 m^3, 0.0407 m^3

12.126   0.87, 28.51 kJ/kg

12.128   286 kJ/kg

12.130 -8309 kW

12.133   a. -7.71 kJ, -7.71 kJ
         b. -9.93 kJ, -7.81 kJ

12.139   935 kJ/kg, 368 K, 418 kJ/kg

12.143   62.6 kW

12.146   254 K, 470 MJ, 259 K, 452 MJ

13.25 $11 \, H_2O + 10 \, CO_2 + 87.42 \, N_2 + 7.75 \, O_2$

13.26   101.2, 3.044 kg/kg

13.30   0.8, 125%

13.33   0.718 (kilomole of air per kilomole of gas)

13.37 −1 215 860 kJ/kmol

13.42 −915 MJ/kmol,
       −778 MJ/kmol

13.46 838 kPa, −453 MJ

13.48 −158 065 kJ/kmol,
       −96 232 kJ/kmol

13.54 −172 998 kJ/kmol, 0.74

13.64 −1 196 121 and
       −1 310 223 kJ/kmol

13.66 + 740 519 kJ/kmol, 12 kg/kg

13.71 72.6 °C, 2525 K

13.74 1931 K

13.78 2048 K

13.82 0.59, 169 %

13.84 2461 K, −393 522 kJ/kmol

13.86 1.43

13.91 Impossible

13.93 38.7 kW, −83.3 kW

13.96 5.76, 1414 kJ/K

13.99 2039 K

13.102 2.594, 380 kPa, 676 MJ

13.105 427 995 kJ/4 kmol e⁻, 1.109 V

13.107 817 903 kJ, 1.06 V

13.112 1053 cm²

13.116 2.324 H_2O + 1 CO_2 + 11.28 N_2
        + 1 O_2, 53.8 °C

13.118 13 101 kJ/kg, 13 101 kJ/kg,
        1216 K

13.123 2760 kJ/kg, 2799 kJ/kg

13.126 −4.081 kW, 0.139

13.127 9.444 kg/kg

13.129 20 986 kJ/kg

13.131 238 % theo. air

13.134 1139 K, 8710 kW

13.137 140.7 kJ/kmol·K, 433 °C

14.18 34.4 MPa

14.21 29.68 MPa

14.24 exp(−12.8407)

14.27 linear in 1/T

14.32 2980 K

14.36 exp(5.116)

14.41 1444 K

14.42 1108 kPa, 93.7 % O_2, 6.3 % O,
       97.7 MJ/kmol

14.49 exp(154.665)

14.52 21.8 % N_2, 9.1 % H_2, 69.1 % NH_3

14.56 exp(−8.293)

14.58 3617 K

14.64 1.4 % C_2H_5OH, 32.4 % C_2H_4,
       66.2 % H_2O

14.66 0.006 55, −836 MJ

14.71 8.7 % CO_2, 10.3 % CO_2, 37.9 %
       H_2O, 43.1 % O_2

14.78 0.0024, Yes

14.81 66.1 % H_2O, 12.9 % H_2, 5.4 % O_2,
       9.9 % OH, 5.7 % H

14.84 6.2 % CO_2, 7.8 % H_2O, 75.9 % N_2,
       10.1 % O_2, 0.06 % NO, 0.001 % NO_2

14.87 exp(−3.7411) = 0.0237

14.90 exp(−2.1665) vs exp(−2.4716)

14.93 5.8 % CH_3OH, 50 % CO, 44.2 %
       H_2, no

14.96 0.0097

14.102 2.7 %

14.105 NO_2, 703 K

14.108 10–12 000 K

14.111 11.1 % CO_2, 1.5 % CO, 70.7 %
        N_2, 14 % H_2O, 2.7 % H_2

14.113 0.4

14.117 1.96

15.15 556 kPa, 365 °C

15.18 127 kPa, 907 K

15.21 142.2 °C, 281 kPa, 5.9 kg/s

15.24 −205 N, −193 N

15.27 61920 N

15.30 36 m/s

15.36 1716 m

15.39 11 350 kPa, 27.7 °C, no

15.42 906 kPa

15.45 896 kPa, 8.251 kg/s

15.48 25 %

15.51 0.0342 kg/s, 0.0149 kg/s

15.54 112.8 kPa, 9.9 kg/s

15.57 1.895 kg, 0.0082 kg/s

15.60   1.178 kg, 0.012 24 kg/s

15.63   2.41

15.66   0.0206 kJ/kg·K

15.69   627.6 m/s

15.72   279.3 K, 0.608

15.75   52.83 kPa, 0.157 kg/s

15.78 6.115×10^{-4} m², 0.167 kJ/kg·K

15.81   0.1454 kg/s, 0.1433 kg/s

15.84   1.756

Index

Fundamental Physical Constants

Avogadro	$N_0 = 6.022\,1415 \times 10^{23}\ \text{mol}^{-1}$
Boltzmann	$k = 1.380\,6505 \times 10^{-23}\ \text{J K}^{-1}$
Planck	$h = 6.626\,0693 \times 10^{-34}\ \text{Js}$
Gas Constant	$\overline{R} = N_0 k = 8.314\,472\ \text{J mol}^{-1}\,\text{K}^{-1}$
Atomic Mass Unit	$m_0 = 1.660\,538\,86 \times 10^{-27}\ \text{kg}$
Velocity of light	$c = 2.997\,924\,58 \times 10^8\ \text{ms}^{-1}$
Electron Charge	$e = 1.602\,176\,53 \times 10^{-19}\ \text{C}$
Electron Mass	$m_e = 9.109\,3826 \times 10^{-31}\ \text{kg}$
Proton Mass	$m_p = 1.672\,621\,71 \times 10^{-27}\ \text{kg}$
Gravitation (Std.)	$g = 9.806\,65\ \text{ms}^{-2}$
Stefan Boltzmann	$\sigma = 5.670\,400 \times 10^{-8}\ \text{W m}^{-2}\,\text{K}^{-4}$

Mol here is gram mol.

Prefixes

10^{-1}	deci	d
10^{-2}	centi	c
10^{-3}	milli	m
10^{-6}	micro	μ
10^{-9}	nano	n
10^{-12}	pico	p
10^{-15}	femto	f
10^1	deka	da
10^2	hecto	h
10^3	kilo	k
10^6	mega	M
10^9	giga	G
10^{12}	tera	T
10^{15}	peta	P

Concentration

10^{-6} parts per million ppm